Paleoanthropology of the Balkans and Anatolia

Vertebrate Paleobiology and Paleoanthropology Series

Edited by

Eric Delson
Vertebrate Paleontology, American Museum of Natural History
New York, NY, USA
delson@amnh.org

Eric J. Sargis
Anthropology, Yale University
New Haven, CT 06520, USA
eric.sargis@yale.edu

Focal topics for volumes in the series will include systematic paleontology of all vertebrates (from agnathans to humans), phylogeny reconstruction, functional morphology, Paleolithic archaeology, taphonomy, geochronology, historical biogeography, and biostratigraphy. Other fields (e.g., paleoclimatology, paleoecology, ancient DNA, total organismal community structure) may be considered if the volume theme emphasizes paleobiology (or archaeology). Fields such as modeling of physical processes, genetic methodology, nonvertebrates or neontology are out of our scope.

Volumes in the series may either be monographic treatments (including unpublished but fully revised dissertations) or edited collections, especially those focusing on problem-oriented issues, with multidisciplinary coverage where possible.

More information about this series at http://www.springer.com/series/6978

Paleoanthropology of the Balkans and Anatolia

Human Evolution and its Context

Edited by

Katerina Harvati

Paleoanthropology, Department of Geosciences, Eberhard Karls Universität Tübingen, Tübingen, Germany

Senckenberg Center for Human Evolution and Paleoenvironment, Eberhard Karls Universität Tübingen, Tübingen, Germany

PhD Program in Anthropology, City University of New York, Graduate Center and New York Consortium in Evolutionary Primatology, New York, NY, USA

Mirjana Roksandic

Department of Anthropology, University of Winnipeg, Winnipeg, MB, Canada

Springer

Editors
Katerina Harvati
Paleoanthropology, Department of Geosciences
Eberhard Karls Universität Tübingen
Tübingen, Germany

Senckenberg Center for Human Evolution
and Paleoenvironment
Eberhard Karls Universität Tübingen
Tübingen, Germany

PhD Program in Anthropology
City University of New York
Graduate Center and New York Consortium
in Evolutionary Primatology
New York, NY, USA

Mirjana Roksandic
Department of Anthropology
University of Winnipeg
Winnipeg, MB, Canada

ISSN 1877-9077 ISSN 1877-9085 (electronic)
Vertebrate Paleobiology and Paleoanthropology
ISBN 978-94-024-0873-7 ISBN 978-94-024-0874-4 (eBook)
DOI 10.1007/978-94-024-0874-4

Library of Congress Control Number: 2016955953

Cover illustration: The cover shows the Petralona cranium on the left (Photo K. Harvati, Copyright E. Delson) and the Kokkinopilos handaxe, shown in three views (Photos and Copyright V. Tourloukis). In the background is a view of the Megalopolis basin, central Peloponnese (Photo and Copyright A. Vlachopoulos).

Printed on acid-free paper

This Springer imprint is published by Springer Nature
The registered company is Springer Science+Business Media B.V.
The registered company address is: Van Godewijckstraat 30, 3311 GX Dordrecht, The Netherlands

Preface

The timing, route, and origin of the earliest human dispersal into Europe, the number of Eurasian hominin taxa during the Pleistocene, the evolution and possible late survival of the Neanderthal lineage, and the Late Pleistocene arrival of modern humans in Europe continue to be central themes of discussion and research in paleoanthropology. However, in these discussions there is a glaring lack of primary data from one of the most relevant geographic regions of Europe: the Balkans. This area, together with neighboring Anatolia, is at the geographic center of the hypothesized dispersals and is often considered the most likely migration route into the continent. Furthermore, together with the Italian and Iberian peninsulae, it is one of the main refugia where fauna and flora, as well as, presumably, human populations, would have been able to survive during glacial times. This region, therefore, has been crucial in shaping the course of human evolution in Europe. Nevertheless, despite its geographic significance, it does not enjoy a strong paleoanthropological tradition, and, with a few exceptions, paleolithic research was neglected there until recent years.

This lack of past research and promise for future findings are recurring themes throughout this volume, whose goals are to present a comprehensive review of the paleoanthropological records in the Balkans and Anatolia, report recent results, provide information on the paleoenvironmental and geological background, and, where possible, attempt a regional synthesis. The volume is based on the lectures presented during the conference "*Human Evolution in the Southern Balkans*," organized by Katerina Harvati and Vangelis Tourloukis in Tübingen on December 6–8, 2012, as part of the ERC Starting Grant project "Paleoanthropology at the Gates of Europe: Human Evolution in the Southern Balkans" (PaGE). PaGE, directed by K. Harvati, is a 5-year research program aiming to increase, through systematic fieldwork, the number of paleoanthropological findings from Greece and to help reassess the human fossil record from the region. The ultimate goal of PaGE is to help shed light on open questions in European paleoanthropology by providing new primary data and to develop a research network among scholars working in these fields in South Eastern Europe. First and foremost, this network comprises the close collaborating partners of PaGE: Drs. E. Panagopoulou and A. Darlas from the Ephoreia of Paleoanthropology and Speleology (Greek Ministry of Culture), Profs. C. Doukas and G. Koutessi-Philippaki from the National and Kapodistrian University of Athens, Profs. G. Koufos and D. Kostopoulos from the Aristotle University of Thessaloniki, and Dr. P. Karkanas from the Wiener Laboratory, American School of Classical Studies at Athens. The PaGE 2012 conference, organized at the end of the first year of the project, brought together several research teams from across the region to present the state of the art of paleoanthropological research in their countries, showcase their most recent work, and discuss their future plans. Scholars representing various institutions from Greece, Turkey, Bulgaria, Serbia, Croatia, and Romania and their collaborating partners from Canada, the USA, UK, France, and Germany all gathered in snowy Tübingen at the imposing medieval setting of the Fürstenzimmer, Castle Hohentübingen, for 2 days of talks and lively discussion. Most of the articles presented during the conference, as well as some additions to the original program, are collected here as chapters of this volume.

The volume is organized into three parts. The first part (The Human Fossil Record: Chaps. 1–6) deals with this record from Greece, the Central Balkans, Croatia, Romania, Bulgaria, and Turkey. The second part (The Archaeological Record: Chaps. 7–14) presents the paleolithic record from the same countries, following the same order. Two chapters are devoted to new paleolithic research in Greece, while one presents a synthesis of the record of the region. Part 3 (Paleoenvironments, Biogeography, Chronology: Chaps. 15–18) provides the paleoenvironmental, geological, and biogeographic background to the regional Paleolithic.

In the first part, Chap. 1 (Harvati 2016) presents an overview of the Greek human fossil record, incorporating some recent work on material from Kalamakia and Megalopolis and placing it within the broad framework of the European record. Although Greek human paleontology is better known than that of many of the other Balkan countries, most of it samples different phases of the Neanderthal lineage. Earlier hominins, as well as Upper Paleolithic humans, are not known, with a few possible exceptions. Chapter 2 (Roksandic 2016) presents the fossil record from the Central Balkans, highlighting the recent fossil human find from Mala Balanica. Roksandic puts forth the possibility for an alternative course for human evolution in this part of Europe, different from the one proposed by the accretion hypothesis for the Western part of the continent. Chapter 3 (Janković et al. 2016) presents the Croatian hominin record. Croatia is the only country in the region with a strong paleoanthropological tradition, and Janković et al. present the material from Krapina and Vindija and outline the contributions of Croatian paleoanthropology to the development of the discipline, including the significance of the Vindija remains to the Neanderthal genome project. Chapter 4 (Harvati and Roksandic 2016) presents an overview of the fossil human record from Romania, as well as a new comparative geometric morphometric analysis of the Upper Paleolithic Romanian mandibular remains (Oase 1 and Muierii 1), in light of the new findings of recent Neanderthal ancestry for the former specimen. The results highlight the difficulties in assessing admixture from skeletal morphology. Chapter 5 (Strait et al. 2016) reviews the scant fossil human record from Bulgaria, most of which appears to have been lost. Strait et al. develop testable hypotheses for human dispersals into Eurasia, to be assessed against future discoveries. Chapter 6 (Aytek and Harvati 2016) is a review of the human fossil record from Turkey, including a preliminary comparative 3D geometric morphometric analysis of the Kocabaş *Homo erectus* specimen. Results show affinities with Eurasian *H. erectus* and *H. heidelbergensis*, but no particular similarities with early African *H. erectus*.

Part 2 starts with two chapters on the Greek paleolithic record. In Chap. 7, Darlas and Psathi (2016) present their new work at Upper Paleolithic cave sites in Mani, Southern Greece, where excavations are currently under way. These new sites are all the more important because of an extreme scarcity of evidence dating from this period in Greece. The authors present a summary of new results, including radiometric dates for two of the caves. Chapter 8 (Galanidou et al. 2016) is a report on the newly discovered Lower Paleolithic site Rodafnidia on Lesvos. Galanidou et al. present the results of their first field seasons at Rodafnidia, including a short description of the Acheulian material discovered at the site and preliminary dating results. Acheulian lithics are extremely rare in Greece and elsewhere in the region, and the authors find parallels for the Rodafnidia material in the Near East and Africa. Chapter 9 (Mihailović and Bogićević 2016) describes the paleolithic record of the Central Balkans, concentrating on the Lower to Middle Paleolithic transition in the region. The authors propose that the first appearance of the Charentian in Europe in the Middle Pleistocene could be linked to demographic factors, migrations, and cultural transmission with the Near East. Chapter 10 (Karavanić et al. 2016) discusses the evidence for the Middle to Upper Paleolithic transition from Croatia. Karavanić et al. present the evidence from Vindija in particular detail, discussing alternative hypotheses about the transition in this site. Chapter 11 (Doboş and Iovita 2016) critically addresses the evidence for Lower Paleolithic sites in Romania, most of which is deemed to be unreliable. The authors further report on the recent results of their Lower Danube Survey for Paleolithic Sites, and particularly on the Dealul Guran site, dated to OIS11. Chapter 12 (Ivanova 2016) presents evidence for the Lower Paleolithic in Bulgaria by summarizing the Lower Paleolithic assemblages from Kozarnika cave and critically evaluating their dating. Furthermore, the chapter draws attention to possible Lower

Paleolithic assemblages from open-air sites in the Rhodope Mountains. Chapter 13 (Dinçer 2016) summarizes the evidence for the Lower Paleolithic in Turkey. Dinçer insists on reconceptualizing Anatolia as a challenging environment that required substantial behavioral adaptations from the migrating hominins, and not just as a transit route, and suggests that the early human presence in Anatolia was sporadic and ephemeral, leading to continuous occupation only in later phases of the Middle Pleistocene. Chapter 14 (Sitlivy 2016) synthesizes the current debate on the Middle to Early Upper Paleolithic transition in the Balkans and the surrounding areas on the basis of technological variability, innovations, and changes in lithic technologies. These issues are examined from the point of view of understanding the reduction sequence as a key insight into technological changes that underpin this important transition.

In the third part, the authors provide a synthesis of current paleoenvironmental evidence for the Balkans. In Chap. 15, Koufos and Kostopoulos (2016) present their research on large mammal evolution in Greece. They posit a shift in environmental conditions leading to open grasslands during the late Early Pleistocene and suggest that humans may have entered Europe at this time as part of an Asian, rather than African, faunal dispersal event. Chapter 16 (Spassov 2016) continues in the same vein, also examining the evidence for the timing and the route of possible early human dispersals into Europe, focusing on recently published faunal data from Bulgaria and the Balkans. Chapter 17 (Doukas and Papayianni 2016) provides an overview of micromammalian fauna in Greece and its potential for providing relevant environmental and chronological information for hominin-bearing sites. The authors call for establishing a Balkan-specific biochronology of micro-mammals. Chapter 18 (Tourloukis 2016), the final paper in the volume, examines the spatiotemporal distribution of Lower Paleolithic sites in the Mediterranean as a function of landscape dynamics which influence both the distribution of desirable site locations and their potential for preservation and visibility in the archaeological record, in an effort to assess whether the extremely small number of known Lower Paleolithic sites in Greece might be due not only to past research priorities but also to geological factors. The geological perspective put forth by Tourloukis offers a new tool in efforts to locate such sites in the Balkans.

We are grateful to all the participants of the "*Human Evolution in the Southern Balkans*" conference and all the contributors to this volume for their outstanding presentations, critical discussions, and excellent chapters, as well as the many colleagues who carefully reviewed each chapter. We also thank Vangelis Tourloukis for co-organizing the conference and co-chairing sessions; Nicholas Conard for giving the Keynote lecture of the first evening of the conference; Monika Doll for her superb organizational skills, which made the conference possible; Thomas Rein, who volunteered his time to put together the program, abstract book, and conference poster; Laura McCarty for her help during the conference and with the copyediting of this volume; Joshua Linder for copyediting help; Sibylle Wolf for giving the tour of the Castle Museum for the conference guests; and all the University of Tübingen students and fellows who were instrumental for the smooth running of the conference: Cathi Bauer, Judith Beier, Michael Francken, Lisa Kellner, Panos Kritikakis, Marlijn Noback, Heike Scherf, and Bernd Trautmann. We thank the Editors of the Vertebrate Paleobiology and Paleoanthropology Series, Eric Delson and Eric Sargis, for agreeing to publish this volume and for their help with various issues of editorial nature, and all the colleagues who kindly gave their time and effort to provide reviews of the manuscripts. We are deeply grateful to the University of Tübingen President, Professor Dr. Bernd Engler, and Vice President for Research, Professor Dr. Peter Grathwohl, for their continuing support. Funding for the conference was provided by the European Research Council (ERC StG 283503 "PaGE"). Finally, for their unwavering patience and support throughout the years, we owe our deepest gratitude to our families and our spouses, Elias and Ivan.

Katerina Harvati
Baden-Württemberg, Germany
Mirjana Roksandic
Winnipeg, MB, Canada
December 2015

References

Aytek, I. A., & Harvati, K. (2016). The human fossil record from Turkey. In K. Harvati & M. Roksandic (Eds.), *Paleoanthropology of the Balkans and Anatolia: Human evolution and its context* (pp. 79–91). Dordrecht: Springer.

Darlas, A., & Psathi, E. (2016). The Middle and Upper Paleolithic on the Western coast of the Mani Peninsula (Southern Greece). In K. Harvati & M. Roksandic (Eds.), *Paleoanthropology of the Balkans and Anatolia: Human evolution and its context* (pp. 95–117). Dordrecht: Springer.

Dinçer, B. (2016). The Lower Paleolithic in Turkey: Anatolia and hominin dispersals out of Africa. In K. Harvati & M. Roksandic (Eds.), *Paleoanthropology of the Balkans and Anatolia: Human evolution and its context* (pp. 213–228). Dordrecht: Springer.

Doboş, A., & Iovita, R. (2016). The Lower Paleolithic of Romania revisited: New evidence from the site of Dealul Guran. In K. Harvati & M. Roksandic (Eds.), *Paleoanthropology of the Balkans and Anatolia: Human evolution and its context* (pp. 171–186). Dordrecht: Springer.

Doukas, C. S., & Papayianni, K. (2016). Small mammals in the Plio/Pleistocene sediments of Greece. In K. Harvati & M. Roksandic (Eds.), *Paleoanthropology of the Balkans and Anatolia: Human evolution and its context* (pp. 291–302). Dordrecht: Springer.

Galanidou, N., Athanassas, C., Cole, J., Iliopoulos, G., Katerinopoulos, A., Magganas, A., et al. (2016). The Acheulean site at Rodafnidia, Lisvori on Lesbos, Greece: Part I. In K. Harvati & M. Roksandic (Eds.), *Paleoanthropology of the Balkans and Anatolia: Human evolution and its context* (pp. 119–138). Dordrecht: Springer.

Harvati, K. (2016). Paleoanthropology in Greece: Recent findings and interpretations. In K. Harvati & M. Roksandic (Eds.), *Paleoanthropology of the Balkans and Anatolia: Human evolution and its context* (pp. 3–14). Dordrecht: Springer.

Harvati, K., & Roksandic, M. (2016). The human fossil record from Romania: Early Upper Paleolithic Europeans and possible evidence for Neanderthal admixture. In K. Harvati & M. Roksandic (Eds.), *Paleoanthropology of the Balkans and Anatolia: Human evolution and its context* (pp. 51–68). Dordrecht: Springer.

Ivanova, S. (2016). A route through the Balkans and implications for the earliest settlement of Europe. In K. Harvati & M. Roksandic (Eds.), *Paleoanthropology of the Balkans and Anatolia: Human evolution and its context* (pp. 187–211). Dordrecht: Springer.

Janković, I., Ahern, J. C. M., Karavanić· I., & Smith, F. H. (2016). The importance of Croatian Pleistocene hominin finds in the study of human evolution. In K. Harvati & M. Roksandic (Eds.), *Paleoanthropology of the Balkans and Anatolia: Human evolution and its context* (pp. 35–50). Dordrecht: Springer.

Karavanić, I., Šošić, R., Ahern, J. C. M., Čondić, N., Janković, I., Zubčić, K., et al. (2016). Recent research on the Middle/Upper Paleolithic interface in Croatia in the context of Central and SE Europe. In K. Harvati & M. Roksandic (Eds.), *Paleoanthropology of the Balkans and Anatolia: Human evolution and its context* (pp. 153–169). Dordrecht: Springer.

Koufos, G. D., & Kostopoulos, D. S. (2016). The Plio-Pleistocene large mammal record of Greece: Implications for early human dispersals into Europe. In K. Harvati & M. Roksandic (Eds.), *Paleoanthropology of the Balkans and Anatolia: Human evolution and its context* (pp. 269–280). Dordrecht: Springer.

Mihailović, D., & Bogićević, K. (2016). Technological changes and population movements in the late Lower and early Middle Palaeolithic of the Central Balkans. In K. Havarti & M. Roksandic (Eds.), *Paleoanthropology of the Balkans and Anatolia: Human evolution and its context* (pp. 139–151). Dordrecht: Springer.

Roksandic, M. (2016). The role of the Central Balkans in the peopling of Europe: Paleoanthropological evidence. In K. Harvati & M. Roksandic (Eds.), *Paleoanthropology of the Balkans and Anatolia: Human evolution and its context* (pp. 15–33). Dordrecht: Springer.

Sitlivy, V. (2016). Technological variability of the Middle-to-Upper Paleolithic transition: Examples from the Balkans and neighboring regions. In K. Harvati & M. Roksandic (Eds.), *Paleoanthropology of the Balkans and Anatolia: Human evolution and its context* (pp. 229–265). Dordrecht: Springer.

Spassov, N. (2016). Southeastern Europe as a route for the earliest dispersal of *Homo* towards Europe: Ecological conditions and the timing of the first human occupation of Europe. In K. Harvati & M. Roksandic (Eds.), *Paleoanthropology of the Balkans and Anatolia: Human evolution and its context* (pp. 281–290). Dordrecht: Springer.

Strait, D. S., Orr, C. M., Hodgkins, J., Spassov, N., Gurova, M., Miller, C., et al. (2016). The human fossil record of Bulgaria, and the formulation of biogeographic hypotheses. In K. Harvati & M. Roksandic (Eds.), *Paleoanthropology of the Balkans and Anatolia: Human evolution and its context* (pp. 69–78). Dordrecht: Springer.

Tourloukis, V. (2016). On the spatio-temporal distribution of Mediterranean Lower Paleolithic sites: A geoarchaeological perspective. In K. Harvati & M. Roksandic (Eds.), *Paleoanthropology of the Balkans and Anatolia: Human evolution and its context* (pp. 303–323). Dordrecht: Springer.

Contents

Contributors

James C.M. Ahern Department of Anthropology, University of Wyoming, Laramie, WY, USA

Constantin Athanassas Centre de Recherche and d'Enseignement de Géosciences de l'Environnement (C.E.R.E.G.E.), Europôle Méditerranéen de l'Arbois, Aix-en-Provence, France

Ahmet İ. Aytek Department of Anthropology, Faculty of Arts and Science, Mehmet Akif Ersoy University, Istiklal Campus, Burdur, Turkey

Paleoanthropology, Department of Geosciences, Eberhard Karls, Universität Tübingen, Tübingen, Germany

Senckenberg Center for Human Evolution and Paleoenvironment, Eberhard Karls Universität Tübingen, Tübingen, Germany

Katarina Bogićević Department of Paleontology, Faculty of Mining and Geology, University of Belgrade, Belgrade, Serbia

Natalija Čondić Archaeological Museum Zadar, Zadar, Croatia

James Cole School of Environment and Technology, University of Brighton, Brighton, UK

Andreas Darlas Ephoreia of Paleoanthropology and Speleology, Athens, Greece

Berkay Dinçer Prehistory Department, Faculty of Letters, Istanbul University, Istanbul, Turkey

Adrian Doboş Department of Palaeolithic Archaeology, Institute of Archaeology "Vasile Parvan", Bucharest, Romania

Constantin S. Doukas Department of Historical Geology and Paleontology, University of Athens, Athens, Greece

Nena Galanidou Department of History and Archaeology, University of Crete, Rethymno, Greece

Maria Gurova National Institute of Archaeology and Museum, Bulgarian Academy of Sciences, Sofia, Bulgaria

Katerina Harvati Paleoanthropology, Department of Geosciences, Eberhard Karls Universität Tübingen, Tübingen, Germany

Senckenberg Center for Human Evolution and Paleoenvironment, Eberhard Karls Universität Tübingen, Tübingen, Germany

PhD Program in Anthropology, City University of New York Graduate Center and New York Consortium in Evolutionary Primatology, New York, NY, USA

Jamie Hodgkins Department of Anthropology, University of Colorado, Denver, CO, USA

Giorgos Iliopoulos Department of Geology, University of Patras, Rio Patras, Greece

Radu Iovita MONREPOS Archaeological Research Centre and Museum for Human Behavioural Evolution, Römisch-Germanisches Zentralmuseum, Leibniz-Forschungsinstitut für Archäologie, Neuwied, Germany

Stefanka Ivanova National Institute of Archaeology and Museum, Bulgarian Academy of Sciences, Sofia, Bulgaria

Ivor Janković Department of Anthropology, University of Wyoming, Laramie, WY, USA

Institute for Anthropological Research, Zagreb, Croatia

Ivor Karavanić Department of Archaeology, University of Zagreb, Zagreb, Croatia

Athanasios Katerinopoulos Faculty of Geology and GeoEnvironment, University of Athens, Athens, Greece

Dimitris S. Kostopoulos Department of Geology, Laboratory of Geology and Palaeontology, Aristotle University of Thessaloniki, Thessaloniki, Greece

George D. Koufos Department of Geology, Laboratory of Geology and Palaeontology, Aristotle University of Thessaloniki, Thessaloniki, Greece

Andreas Magganas Faculty of Geology and GeoEnvironment, University of Athens, Athens, Greece

John McNabb Department of Archaeology, University of Southampton, Southampton, UK

Dušan Mihailović Department of Archaeology, Faculty of Philosophy, University of Belgrade, Belgrade, Serbia

Christopher Miller Institut für Naturwissenschaftliche Archäologie, Eberhard Karls University of Tübingen, Tübingen, Germany

Caley M. Orr Department of Cell and Developmental Biology, University of Colorado School of Medicine, Aurora, CO, USA

K. Papayianni Wiener Laboratory, ASCSA, Muséum National d'Histoire Naturelle, Paris, France

Eleni Psathi Ephoreia of Antiquities of Arcadia, Tripolis, Greece

Mirjana Roksandic Department of Anthropology, University of Winnipeg, Winnipeg, MB, Canada

Valéry Sitlivy Liège, Belgium

Fred H. Smith Department of Sociology and Anthropology, Illinois State University, Bloomington-Normal, IL, USA

Rajna Šošić-Klindžić Department of Archaeology, University of Zagreb, Zagreb, Croatia

Nikolai Spassov National Museum of Natural History, Bulgarian Academy of Sciences, Sofia, Bulgaria

David S. Strait Department of Anthropology, Washington University, St. Louis, MO, USA

V. Tourloukis Paleoanthropology, Department of Geosciences, Eberhard Karls Universität Tübingen, Tübingen, Germany

Senckenberg Center for Human Evolution and Paleoenvironment, Eberhard Karls Universität Tübingen, Tübingen, Germany

Tsanko Tzankov Department of Geography, Ecology and the Preservation of the Environment, South-Western University "Neofit Rilski", Blagoevgrad, Bulgaria

Krunoslav Zubčić Croatian Conservation Institute, Zagreb, Croatia

Part I
The Human Fossil Record

Chapter 1
Paleoanthropology in Greece: Recent Findings and Interpretations

Katerina Harvati

Abstract Greece lies at the crossroads between Europe, Asia, and Africa, and represents a logical gateway through which early human populations might have repeatedly passed on the way to and from Europe. It also represents one of the three European Mediterranean peninsulas which acted as a refugium for fauna, flora and, very likely, human populations during glacial times. Evidence from this region is therefore essential in order to test hypotheses about the course of human evolution in Europe. Despite the importance of the region, paleoanthropological research has until recently been relatively neglected. In recent years, however, renewed research efforts have produced new human fossils from Greece, recovered from excavated contexts. This chapter reviews the Greek human fossil evidence in the context of broader questions in European paleoanthropology.

Keywords Neanderthals • Upper Paleolithic • Modern humans • *Homo heidelbergensis*

Introduction

The European human fossil record continues to produce unexpected discoveries even after more than a century of study. These finds are reshaping our knowledge of human evolution on the continent. Over the last 20 years, views on the short *vs.* long chronology of human presence in Europe have shifted radically toward the latter, although the identity of the earliest colonizers and their evolutionary fate remain elusive (e.g., Carbonell et al. 2008; Bermúdez de Castro et al. 2011; Toro-Moyano et al. 2013). Our understanding of the Neanderthal lineage is now clearer than ever, although paleoanthropologists still struggle with the classification and relationships of the earlier, Middle Pleistocene, European hominins (e.g., Harvati et al. 2010; Freidline et al. 2012; Arsuaga et al. 2014). Finally, the advent of modern humans, *Homo sapiens*, in Europe around 40–45 ka (Benazzi et al. 2011; Higham et al. 2011a, b, 2013), and the replacement of local populations of *H. neanderthalensis* that may have survived later in Southern European refugia, raises questions about possible interactions between the two species, about the level of cultural and/or biological exchanges that might have occurred, and about the causes for the Neanderthal extinction.

Within this research landscape, crucial primary evidence from the geographic region representing both a major dispersal corridor to and from Europe and a Mediterranean refugium for both fauna and flora (the Southern Balkans in general and Greece in particular) is missing. This unfortunate situation is likely due to the lack of a strong tradition in basic Paleolithic research in the region. Nowhere is this data gap more evident than in the human fossil record (similarly to the situation in the Central Balkans, Bulgaria, and Anatolia, see Aytek and Harvati 2016; Roksandic 2016; Strait et al. 2016). This chapter reviews the existing human fossil evidence from Greece in the framework of the research questions outlined above.

Neanderthals and Early Modern Humans

Greece lies directly on one of the proposed dispersal routes of modern humans coming into Europe from the Near East and Africa. However, Upper Paleolithic sites are rare in

K. Harvati (✉)
Paleoanthropology, Department of Geosciences, Eberhard Karls Universität Tübingen, Tübingen, Germany

Senckenberg Center for Human Evolution and Paleoenvironment, Eberhard Karls Universität Tübingen, Tübingen, Germany

PhD Program in Anthropology, City University of New York Graduate Center and New York Consortium in Evolutionary Primatology, New York, NY, USA
e-mail: katerina.harvati@ifu.uni-tuebingen.de

Katerina Harvati and Mirjana Roksandic (eds.), *Paleoanthropology of the Balkans and Anatolia*, Vertebrate Paleobiology and Paleoanthropology, DOI 10.1007/978-94-024-0874-4_1

Greece, and to date, no definitive early Upper Paleolithic human remains are known. A skeleton from the site of Apidima (Cave Γ) in the Mani region (Fig. 1.1; see below) has been proposed to represent an early Upper Paleolithic burial (Pitsios 1985, 1995) and was reportedly found associated with lithics tentatively assigned to the Aurignacian (Darlas 1995). However, no convincing chronological assessment exists for Cave Γ, and neither the skeleton nor the lithic and faunal material from this cave have been described in detail. Therefore, the identity of the specimen and its chronology and cultural affiliation must remain uncertain until further research (Harvati et al. 2009).

Fig. 1.1 Map of Greece showing the approximate geographic location of the sites mentioned in the text. Adapted from Harvati et al. (2009)

The Klisoura site in the Northern Peloponnese has recently yielded a lithic industry closely resembling the Uluzzian (Fig. 1.1; Koumouzelis et al. 2001). This technocomplex was recently dated to *ca.* 42–45 cal kBP at the Grotta del Cavallo site in Italy, where it was associated with modern human remains (Benazzi et al. 2011). If we accept that the Uluzzian was produced by modern humans, its presence in Greece might indicate the arrival of modern humans in this region. Since the Klisoura Uluzzian layer is capped by the Campanian Ignimbrite (CI; Stiner et al. 2010; Lowe et al. 2012), it may testify to a modern human arrival predating 40 ka. Douka et al. (2014) recently obtained the radiocarbon date of 39.9–38.5 cal kBP (OxA-21068) from a shell bead from the Uluzzian layer at Klisoura, a date consistent with the presence of the CI above it. On the basis of new dates for multiple sites and their Bayesian statistical modeling, these authors concluded that the Uluzzian appeared *ca.* 45 ka in both Italy and Greece, where it persisted until *ca.* 39 ka. Unfortunately, no human remains have been found in the Uluzzian layer at Klisoura. Therefore, its attribution to early modern humans cannot be confirmed at this site (see also Sitlivy 2016). Similarly, the first Aurignacian layer at Franchthi in the Argolid, Northern Peloponnese, postdates the CI likely only by *ca.* two millennia (Fig. 1.1; Farrand 2000; Stiner and Munro 2011), again suggesting an early modern human presence, although no human remains have been recovered from this layer.

In contrast to the sparse early Upper Paleolithic fossil record, Neanderthal remains have recently been identified at Lakonis and Kalamakia, two sites in the Mani peninsula, Southern Peloponnese (Table 1.1; see also Darlas and Psathi 2016). Lakonis consists of a series of caves and collapsed

Table 1.1 Summary of the human fossil record from Greece up to the early Upper Paleolithic. Adapted from Harvati et al. 2009, 2013

Site	Hominins	Taxon	Age (ka)	Method	Assoc. Lithics
Megalopolis (Peloponnese)	Isolated LUM3	*Homo* sp.	Possibly Early/ Middle Pleistocene	Faunal, Paleomagnetism	–
Petralona Cave (Macedonia)	Petralona cranium	*H. heidelbergensis*	>240 ka	ESR/ U/Th, Faunal	–
Apidima Cave A (Mani)	LAO 1/S1 and LAO 1/S2 partial crania	*H. heidelbergensis*— *H. neanderthalensis*	Late Middle- Early Late Pleistocene	Geomorphology	–
Lakonis Site 1 (Mani)	LKH1, isolated LLM3	*H. neanderthalensis*	42–48 ka (cal)	AMS ^{14}C on charcoal	Initial Upper Paleolithic
Kalamakia Cave (Mani)	KAL1-KAL14 Isolated teeth: LUP3, LUM3, L?UP4, LLP4, RUM2, RLP4, RUI2, LUI1, LUdi2, L?Udi1; occipital fragment; right fibula shaft fragment; subadult lumbar vertebra; left navicular bone	*H. neanderthalensis*	>40–100 ka	AMS ^{14}C on charcoal, U/Th on marine shell	Mousterian
Apidima Cave Γ (Mani)	Partial skeleton LAO 1/S3	*H. sapiens*	Late Pleistocene	–	Possibly Aurignacian

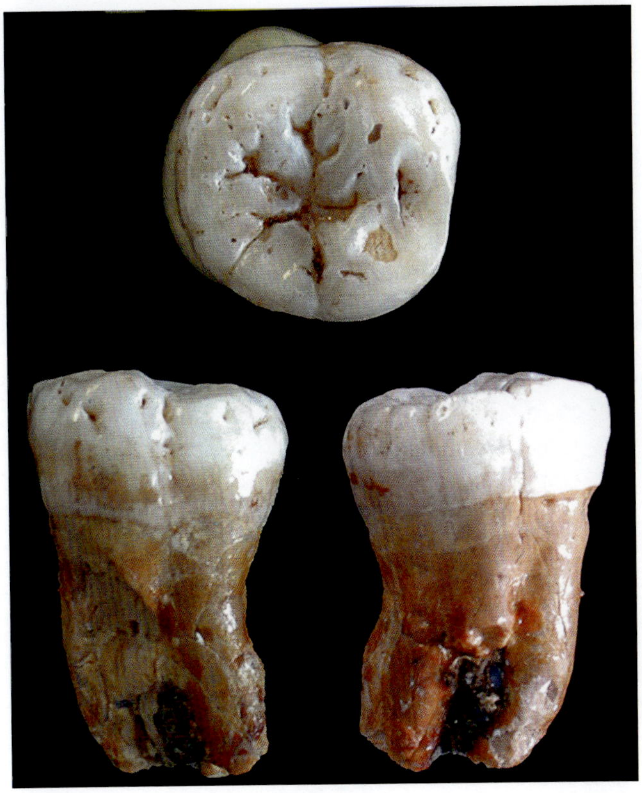

Fig. 1.2 The LKH1 Neanderthal lower third molar. *Top*: Occlusal view; *Bottom Left*: Buccal view; *Bottom right*: Lingual view. Photographs copyright K. Harvati

caves along the shoreline on the Eastern coast of Mani, near the town of Gytheion (Fig. 1.1). It was excavated between 1997 and 2011 by a team from the Ephoreia of Paleoanthropology and Speleology, Greek Ministry of Culture. The site preserves a rich, although highly fragmented, fauna, and very rich Middle Paleolithic lithic assemblages throughout most of the stratigraphic sequence. Additionally, the topmost layer yielded an assemblage described as Initial Upper Paleolithic (hereafter IUP), dated to *ca.* 48–42 cal kBP by AMS ^{14}C on charcoal (Panagopoulou et al. 2002–2004; Elefanti et al. 2008). A single human specimen from this layer, LKH1, was recovered at Lakonis during excavation in 2002 (Panagopoulou et al. 2002–2004; Harvati et al. 2003).

LKH1 (Fig. 1.2) is a lower left third molar. Although an isolated specimen, it preserves morphology that strongly supports its taxonomic assignment as a Neanderthal. This includes a large anterior fovea, complex root morphology, a relatively enlarged pulp cavity, and a midtrigonid crest, a feature found at very high frequencies on Neanderthal, but almost never on modern human lower third molars (Fig. 1.2; Harvati et al. 2003). Similar to other Neanderthal samples, it exhibits relatively high enamel secretion rates and relatively thin enamel (Smith et al. 2009). LKH1 has also yielded important information about Neanderthal paleobiology. Strontium isotope analysis suggested that during the formation of the tooth crown this individual lived at least 20 km away from Lakonis, the site where it was found (Richards et al. 2008). Although the distance is limited, the evidence from this analysis is the first direct indication of Neanderthal mobility, and a first assessment of a minimum range on the seasonal or lifetime movements of Neanderthal groups. For the case of the Mani peninsula, it is an indication that the Neanderthal population living at Lakonis would likely have communicated with the one present at Kalamakia, on the other side of the peninsula and approximately 30 km away (see below). LKH1 also raises questions about the authorship of the IUP industry that it was associated with, and of possible contact with early modern humans in the region. The identification of the Uluzzian at Klisoura—and therefore the possibility of the presence of modern humans—some 200 km to the north, and dated to before 40 ka BP (Koumouzelis et al. 2001; Stiner et al. 2010; Douka et al. 2014), further highlights this possibility.

Kalamakia is the second site in the Mani peninsula to have yielded Neanderthal fossils. It is a karstic cave formed in the limestone cliff side and situated on the Western coast of the Mani peninsula (Fig. 1.1; see also Darlas and Psathi 2016). It was excavated from 1993 to 2006 by a joint team from the Ephoreia of Paleoanthropology and Speleology, Greek Ministry of Culture, and the Muséum national d'Histoire naturelle, Paris (de Lumley et al. 1994; Darlas and de Lumley 2004). The site has yielded Mousterian lithics with Levallois elements throughout the stratigraphic sequence, as well as a rich fauna comprising fallow deer, ibex, wild boar, and red deer; several species of carnivores; and numerous small vertebrates (Darlas and de Lumley 2004; Harvati et al. 2013). The deposits in the cave date to between 100,000 ka (U/Th radiometric dating of a marine shell at the Institut de Paléontologie Humaine in Paris, IPH Kal 9304: 109,000 + 14,000/−13,000; de Lumley et al. 1994) and >39,000 ^{14}C BP (^{14}C AMS dating on charcoal at Gif-sur-Yvette in France, GifA 94592; de Lumley et al. 1994).

Thirteen fragments of human remains were excavated from several layers of Unit IV, and one from the uppermost layer of Unit III. In total, ten isolated teeth, one cranial fragment, and three postcranial elements were recovered, representing at least eight individuals, two of them juveniles (Harvati et al. 2013). Although not all specimens are taxonomically informative, several elements preserve diagnostic Neanderthal morphology. Among the dental remains, the two lower premolars and the two upper incisors display combinations of crown features that are observed at very high frequencies among Neanderthals but not in Upper Paleolithic modern humans (Harvati et al. 2013). The two upper incisors (KAL10 and KAL11; Fig. 1.3) show a combination of shoveling, lingual tubercles, and labial convexity, considered unique to Neanderthals and pre-Neanderthals (e.g., Bailey 2007; Martinón-Torres et al. 2012). The two lower premolars

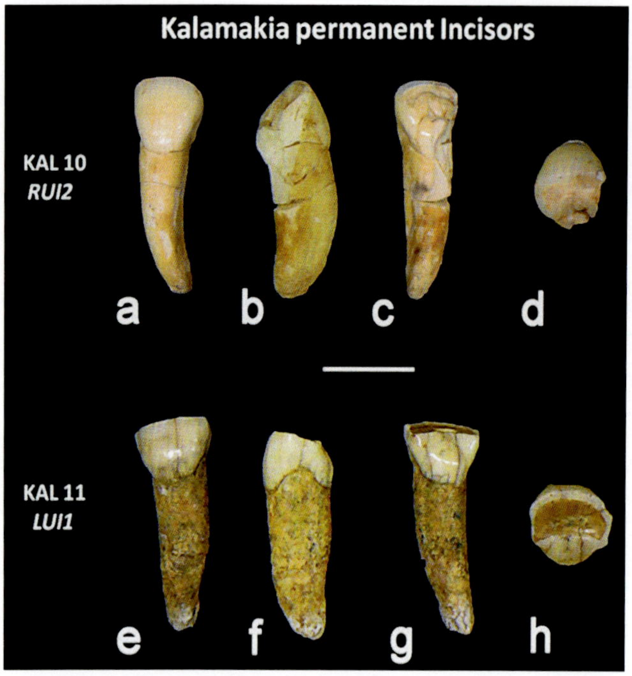

Fig. 1.3 The Kalamakia permanent incisors KAL10 (*top*) and KAL11 (*bottom*). Labial (**a**, **e**), mesial (**b**, **f**), lingual (**c**, **g**) and occlusal (**d**, **h**) views. Photographs copyright K. Harvati. Adapted from Harvati et al. (2013)

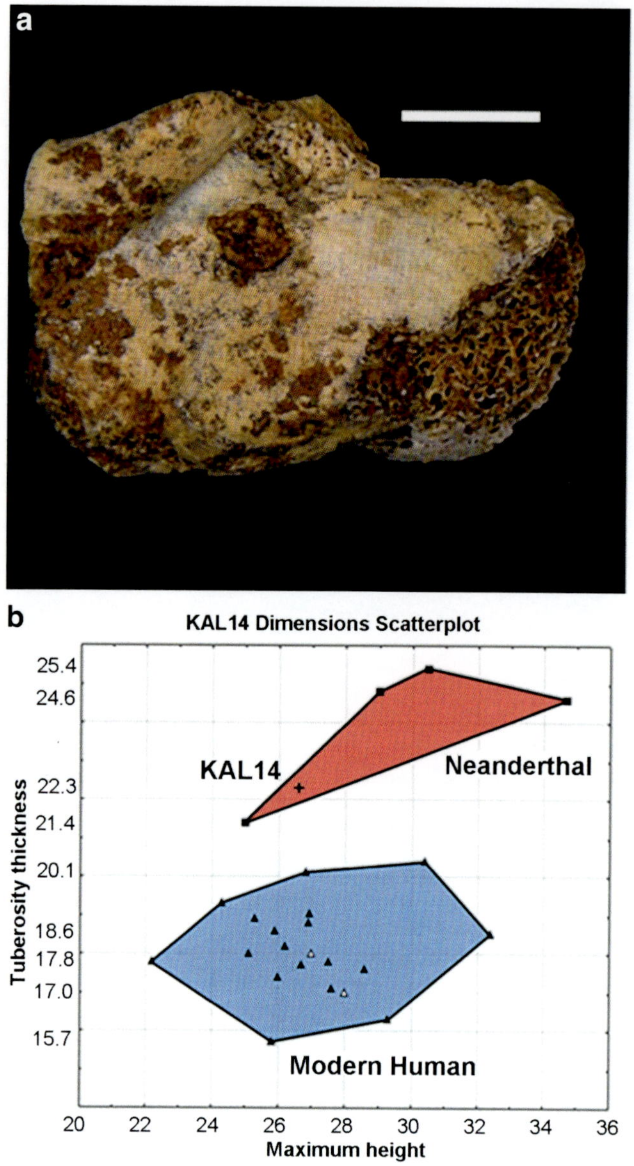

Fig. 1.4 (**a**) The Kalamakia KAL14 navicular bone. Photograph copyright K. Harvati. (**b**) Maximum thickness of navicular tuberosity plotted against minimum tuberosity thickness in Neanderthals and modern humans. Adapted from Harvati et al. (2013)

(KAL6 and KAL9) show multiple lingual cusps, a transverse crest and an asymmetric crown, again a combination thought to be unique to Neanderthals and some of their ancestors (e.g., Bailey 2007; Martinón-Torres et al. 2012). Furthermore, two additional isolated teeth, an upper third premolar (KAL2) and an upper third molar (KAL3), have crown diameters that place them closer to Neanderthals than to early modern or Upper Paleolithic modern humans (Harvati et al. 2013). Beyond the dental remains, one of the postcranial elements, the navicular bone KAL14, shows dimensions that fall within the Neanderthal, rather than the modern human range of variation (Fig. 1.4; Harvati et al. 2013; McCarty et al. 2014). This specimen also shows carnivore puncture marks, indicating that some of the human remains found at the site were scavenged, and confirming that Kalamakia was intermittently used by humans and carnivores (Harvati et al. 2013). Therefore, although the Kalamakia fossil human assemblage comprises isolated remains, it includes several dental and postcranial diagnostic elements that point to Neanderthal affinities, while none of the diagnostic elements show modern human-derived features. Their association with Mousterian lithic assemblages throughout the Kalamakia stratigraphic sequence further indicates that the Kalamakia human remains belong to a Neanderthal population.

The Kalamakia and Lakonis Neanderthals might represent the same population. Although their chronology is not completely resolved, the two sites likely overlap temporally. In addition to the strontium isotope analysis of the Lakonis Neanderthal molar (see above; Richards et al. 2008), human mobility between the two sites is suggested by the use of green andesite as a lithic raw material at Kalamakia (see Harvati et al. 2013). The source of this material is near the village of Krokees, close to Lakonis, on the eastern coast of the peninsula and is the most abundant raw material used for the production of the Lakonis lithics (Panagopoulou et al. 2002–2004). Unfortunately, the Kalamakia human sample does not preserve a lower third molar; therefore, a direct comparison of the human remains from the two sites is not possible.

Although it indicates a strong Neanderthal presence, the current evidence does not point to a late survival of this taxon in the Mani region, as might be predicted for a refugium area. If modern humans were already in the Northern Peloponnese before 40 ka, as suggested by the evidence from Klisoura (Douka et al. 2014), their presence might account for an early Neanderthal demise in the region. The current state of research, however, does not allow for any conclusions on this topic. Ongoing work in the Mani and elsewhere (see Darlas and Psathi 2008, 2016) will help test this hypothesis.

Middle-Late Pleistocene and the Origins of the Neanderthals

The Greek human fossil record presents two possible cases of early Neanderthal or pre-Neanderthal hominins that can add to the discussion of Neanderthal evolution (Table 1.1). One of these also comes from the Mani region, from the Apidima cave site.

Apidima is another karstic cave complex on the Western coast of the Mani peninsula (Fig. 1.1) in the vicinity of Kalamakia. It was investigated between 1978 and 1985 by a team from the University of Athens Medical School (Pitsios 1995; Harvati and Delson 1999; Harvati et al. 2009). As mentioned above, Cave Γ has yielded a modern human skeleton of uncertain chronology and cultural affiliation, which may represent an Upper Paleolithic burial (Pitsios 1985). One of the other caves in the complex, Cave A, has produced two human fossil crania, found encased in a block of matrix attached to the cave walls and close to the ceiling (Pitsios 1985, 1995, 1999; Harvati and Delson 1999; Harvati et al. 2009, 2011). Apidima 1 (LAO 1/S1) was partially eroded before discovery and preserves the posterior part of the neurocranium and cranial base. Apidima 2 (LAO 1/S2) is in better condition and is a relatively complete cranium (Fig. 1.5; Harvati et al. 2009). Excavation of Cave A yielded lithic artifacts likely of Middle Paleolithic character (see Harvati and Delson 1999) and a mixed fauna (Tsoukala 1999). However, these finds are not associated with the human specimens, and neither are the ESR dates produced on travertine samples from the entrances of Cave B and of the cave complex (Liritzis and Maniatis 1989; see Harvati et al. 2009, 2011). The chronological framework of these specimens is therefore uncertain.

In terms of its morphology, Apidima 2 shows a low neurocranium, a strong supra-orbital torus, a wide interorbital breadth, no canine fossa, large orbits and nasal aperture, and a prognathic face, suggesting Neanderthal or pre-Neanderthal affinities (Harvati and Delson 1999; Pitsios 2002; Harvati et al. 2009, 2011). Because of its relative gracility it has been considered to be a female, perhaps representing a female

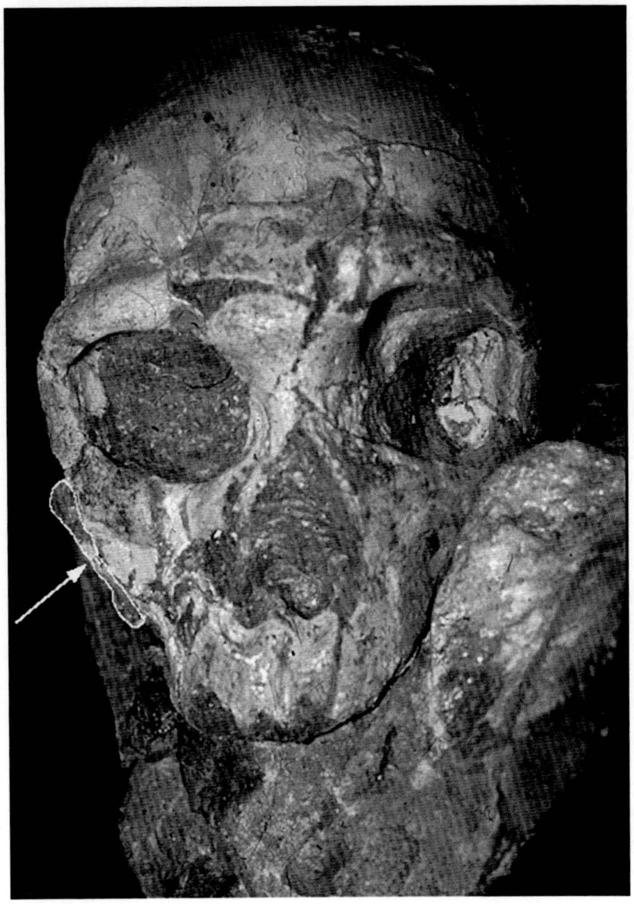

Fig. 1.5 The Apidima 2 cranium. Photo courtesy and copyright E. Delson

H. heidelbergensis comparable to Petralona in its chronology (Harvati and Delson 1999; Pitsios 2002; Harvati et al. 2009, 2011). However, the lack of detailed description or extensive metric data for either of the Apidima specimens precluded their more precise identification.

A recent study sought to elucidate the affiliations and temporal placement of these specimens by reanalyzing published original measurements for Apidima 2 and by reexamining the site's geological context (Harvati et al. 2011). This analysis pointed to strong Neanderthal affinities for Apidima 2 and found little resemblance to Petralona or *Homo heidelbergensis* in general (Fig. 1.6). It suggested a late Middle–early Late Pleistocene timeframe as most consistent with the geological setting of the site as well as with the specimen's morphology (Harvati et al. 2011). If these results are correct, the Apidima specimens do not belong to a Petralona-like population of *Homo heidelbergensis*. Instead they are early Neanderthals and likely represent the ancestors of the populations found at Kalamakia and Lakonis, pointing to a long history of Neanderthal presence in the Mani. Nonetheless, it must be kept in mind that

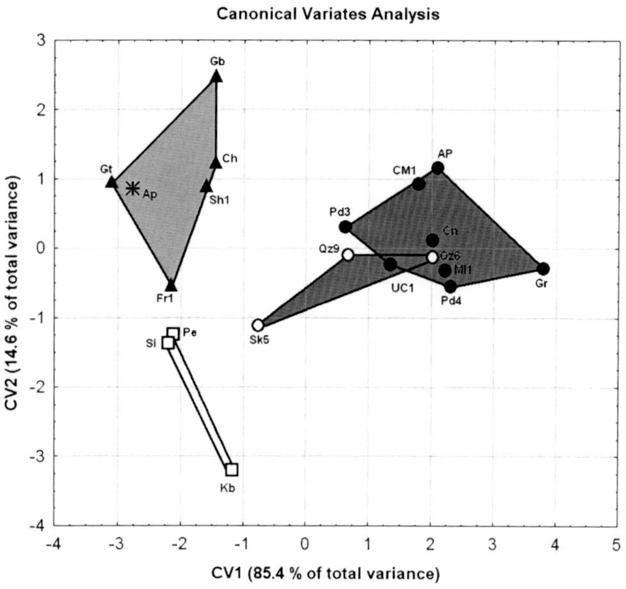

Fig. 1.6 Canonical variates analysis. *Black triangles*: Neanderthals; *open squares*: Middle Pleistocene hominins; *open circles*: Skuhl-Qafzeh; *Black circles*: Upper Paleolithic Europeans; star: Apidima 2. Adapted from Harvati et al. (2011)

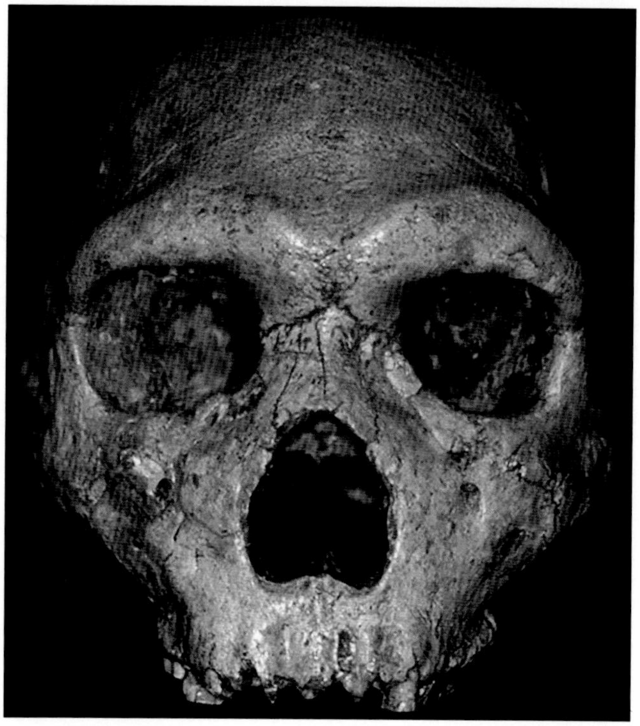

Fig. 1.7 The Petralona cranium. Photo courtesy and copyright E. Delson

these findings were obtained on the basis of a handful of published measurements. A thorough description and metric comparative study of the Apidima crania is still required to confirm our preliminary results and to assess the relationships between Apidima and Neanderthal populations from Europe and the Near East.

Undoubtedly the best known individual human fossil from Greece is the Petralona cranium (Fig. 1.7). It was discovered in 1960 in the Petralona Cave, Northern Greece (Fig. 1.1) by a group of villagers. Petralona cave was, until recently, the only excavated Middle Pleistocene site in Greece. It was joined in 2013–2015 by the Marathousa 1 site in Megalopolis, a site identified and excavated by a team from the Ephoreia of Paleoanthropology and Speleology (Greek Ministry of Culture) and the University of Tübingen, within the framework of the PaGE research (Panagopoulou et al. 2015; see also below); and by Rodafnidia (see Galanidou et al. 2016). The Petralona cranium is in excellent state of preservation and perhaps one of the most complete cranial specimens in the fossil human record of Europe. It is commonly thought to be of Middle Pleistocene age with a proposed date of *ca.* 250 ka, although there is a high degree of uncertainty about its chronological placement (see Harvati et al. 2009). Current consensus assigns Petralona to *H. heidelbergensis*, generally believed to be ancestral to Neanderthals in Europe. Along with other early European *H. heidelbergensis*, it has been described as showing incipient Neanderthal facial characteristics (e.g., Dean et al. 1998). However, it also shows strong overall similarities with African representatives of this taxon, and particularly with the Kabwe cranium (e.g., Stringer 1974; Stringer et al. 1979). Recent reappraisal of the Petralona facial morphology using landmark- and semilandmark-based geometric morphometrics to quantify the subtle morphology of such 'incipient' features could not confirm a more Neanderthal-like morphology in Petralona than in African *H. heidelbergensis* in all instances (Harvati 2009; Harvati et al. 2010; but see Freidline et al. 2012). These analyses, however, confirmed the strong similarity of this specimen with its African counterparts. For the Greek fossil record, these findings suggest contact with Africa at the time of Petralona in the Middle Pleistocene.

Early Colonization of Europe

Recent paleoanthropological discoveries have pushed back the date of the human settlement of Europe to more than a million years before present. The most important securely dated such discoveries come from two Iberian sites, Sima del Elefante and Gran Dolina, both in the Sierra de Atapuerca, Spain, and both yielding excavated human remains and artifacts dated to as early as *ca.* 1.2 Ma (Carbonell et al. 2008) and *ca.* 800 ka respectively (Bermúdez de Castro et al. 1997).

To these early sites can be added lithic assemblages and human footprints dating to *ca.* 700–900 ka, recovered in the United Kingdom at Pakefield and Happisburg (Parfitt et al. 2005, 2010; Ashton et al. 2014). Additional sites, including Pirro Nord in Italy, and Lézignan-la-Cèbe and Pont-de-Lavaud in France, are purported to have produced lithic remains of an even earlier chronology (Arzarello et al. 2009; Crochet et al. 2009; Despriée et al. 2010; Spassov 2016). A recently published isolated human deciduous tooth from Barranco-León associated with lithic assemblages and faunal remains could be as old as 1.2–1.4 Ma (Toro-Moyano et al. 2013; but see Muttoni et al. 2013).

On the basis of this evidence it has been hypothesized that Southern Europe may have been colonized earlier than the Northern or Central parts of the continent (Roebroeks 2001). Once more, South-East Europe is a logical dispersal corridor for early hominins spreading into Europe from either the Near East or the Caucasus region. With the exception of Kozarnika cave, which has yielded Lower Paleolithic artifacts suggested to be as old as 1.5 Ma (Sirakov et al. 2010; but see Kahlke et al. 2011; Spassov 2016; Ivanova 2016), evidence for early hominin presence in the region is conspicuously absent.

The existing human fossil record from Greece offers a glimpse of human presence possibly as early as the late Lower–early Middle Pleistocene at Megalopolis (Table 1.1). The Megalopolis Basin is an intramontane lacustrine basin located in central Peloponnese, in the vicinity of the town of Megalopolis (Fig. 1.1). The area has been known since the late nineteenth century for the fossiliferous deposits of its Marathousa beds, and several paleontological localities were excavated in the 1960s by the University of Athens. Stratified lithic artifacts were observed in the region by Darlas (2003), and more recently by a team from the Ephoreia of Paleoanthropology and Speleology (Greek Ministry of Culture) and the University of Tübingen working in the region within the framework of the PaGE project, most importantly at the Middle Pleistocene site of Marathousa 1 (Panagopoulou et al. 2015). A geological survey of the Megalopolis lignite beds, conducted in 1962–1965 by the Geological Society of Hannover, also collected a large fossil mammal assemblage from the Marathousa beds, which comprised a human upper third molar. Sickenberg (1975) assigned this assemblage to the early Middle Pleistocene. Due to the significant financial interest of the Megalopolis lignite beds, the geology of the region has been well studied, with a solid faunal and paleomagnetic framework (van Vugt 2000). The Sickenberg fauna can therefore be assigned to one of the fossil bearing horizons, ranging in time from *ca.* 870 ka and 850 ka (CHO 1 and 2, respectively), to 730 ka (CHO 3) and 600 ka (CHO 4; van Vugt 2000). The exact provenance of the Megalopolis tooth, however, is currently not known, and concerns have been raised that the specimen might be an intrusive *H. sapiens* (see Sickenberg 1975).

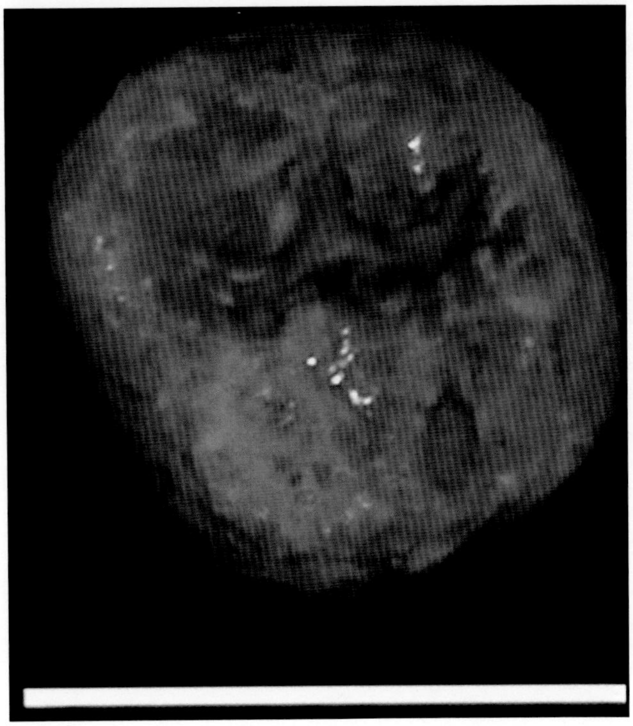

Fig. 1.8 The Megalopolis upper third molar. Photograph copyright K. Harvati

The Megalopolis human specimen is an isolated upper third molar (Fig. 1.8). With the exception of a short article on its enamel prism morphology (Xirotiris et al. 1979), there has been no study or publication of this specimen to date. Its crown appears eroded, perhaps due to acid etching, with the result that the details of the crown morphology cannot be observed, making its analysis difficult. Nonetheless, the Megalopolis M^3 is notable for its very small size. Bucco-lingual and mesio-distal crown dimensions were recorded and compared with a large comparative sample of hominin upper M^3s from the literature (Table 1.2; Fig. 1.9). Megalopolis is among the smallest ones, overlapping in its crown dimensions with modern humans, but also with the low end of the Neanderthal, *H. heidelbergensis* and *H. erectus* ranges of variation. However, when crown shape is assessed using the Crown Shape Index (BL/MD*100), Megalopolis is more similar to earlier taxa, particularly African early *Homo*, than to the later *Homo* specimens included in our comparative samples (Fig. 1.10). This analysis is intriguing and tentatively supports an early geological age for the Megalopolis tooth. However, it is very preliminary, and further work is planned to help elucidate the specimen's affinities, including the potentially taxonomically informative analysis of the crown outline and of the enamel thickness.

Table 1.2 Comparative samples used in the Megalopolis dental analysis: Measurements reported in mm

Specimen	Taxon	M-D	B-L	CBA	CSI (BL/MD*100)	Source
Megalopolis	**Megalopolis**	**9.10**	**10.30**	**93.73**	**113.19**	This study
KNM-ER 1813 right	*H. habilis*	11.70	13.50	157.95	115.38	Walker and Leakey (1993)
KNM-ER 1813 left	*H. habilis*	11.50	13.00	149.50	113.04	
Olduvai 13	*H. habilis*	12.00	13.00	156.00	108.33	Day (1986)
Dmanisi 2700 left	*H. erectus* - Dmanisi	10.00	11.90	119.00	119.00	Rightmire et al. (2006)
Dmanisi 2711 right	*H. erectus* - Dmanisi	9.80	11.90	114.60	121.43	Martinón-Torres et al. (2008)
Konso	*H. ergaster*	12.00	12.80	153.60	106.67	Suwa et al. (2007)
KNM 807	*H. ergaster*	11.00	13.00	143.00	118.18	Walker and Leakey (1993)
Sangiran S7-3d	*H. erectus*	9.50	12.00	114.00	126.32	Grine and Franzen (1994)
Sangiran S7-6	*H. erectus*	9.90	11.80	116.82	119.19	
Sangiran S7-17	*H. erectus*	9.70	11.50	111.55	118.56	
Sangiran S7-73	*H. erectus*	12.40	15.30	189.72	123.39	
Zkd46 R	*H. erectus*	9.10	10.90	99.19	119.78	Walker and Leakey (1993)
Zkd 47 L	*H. erectus*	9.40	11.30	106.22	120.21	
Zkd 48	*H. erectus*	9.90	12.00	118.80	121.21	
Zkd49 R	*H. erectus*	8.70	10.40	90.48	119.54	
Zkd112 R	*H. erectus*	10.10	12.50	126.25	123.76	
Zkd113 L	*H. erectus*	10.40	12.10	125.84	116.35	
Zkd146′	*H. erectus*	9.80	12.50	122.50	127.55	
Z.M	*H. erectus*	9.80	12.00	117.60	122.45	
ATA (SM) 171 (L)	*H. heidelbergensis*	8.80	11.80	103.80	134.09	Bermúdez de Castro (1986, 1993)
ATA (SM) 194 (R)	*H. heidelbergensis*	8.70	12.10	105.30	139.08	
ATA (SM) 274 (L)	*H. heidelbergensis*	8.00	10.10	80.80	126.25	
ATA (SM) 140 (L)	*H. heidelbergensis*	9.30	13.00	120.90	139.78	
ATA (SM) 10 (R)	*H. heidelbergensis*	8.60	11.50	98.90	133.72	
Arago XXI (R)	*H. heidelbergensis*	9.50	12.60	119.70	132.63	Condemi (1992)
Petralona (R)	*H. heidelbergensis*	10.00	12.50	125.00	125.00	Koufos (pers. comm.)
Petralona (L)	*H. heidelbergensis*	10.10	12.20	123.22	120.79	Koufos (pers. comm.)
Kabwe (L)	*H. heidelbergensis*	9.00	12.00	108.00	133.33	Day (1986)
Amud 1	*H. neanderthalensis*	8.20	11.10	91.02	135.37	Coppa et al. (2005)
Kebara 2	*H. neanderthalensis*	9.30	13.00	120.90	139.78	
Tabun C1	*H. neanderthalensis*	8.50	10.40			
Shanidar 1R	*H. neanderthalensis*	9.70	11.60	112.52	119.59	Condemi (1992)
Shanidar 2R	*H. neanderthalensis*	10.00	12.90	129.00	129.00	
Shanidar 3R	*H. neanderthalensis*	9.60	12.80	122.88	133.33	
Shanidar 5 L	*H. neanderthalensis*	9.60	13.00	124.80	135.42	
Shanidar 6R	*H. neanderthalensis*	10.60	12.20	129.32	115.09	
Tabun BC7	*H. neanderthalensis*	9.10	11.50	104.65	126.37	Coppa et al. (2005)
Hortus XI	*H. neanderthalensis*	8.7	11.60	100.92	133.33	de Lumley (1973)
Saccopastore 2 (R)	*H. neanderthalensis*	9.00	11.50	103.50	127.78	Condemi (1992)
Spy1	*H. neanderthalensis*	9.50	11.60	110.20	122.11	de Lumley (1973)
Spy 2	*H. neanderthalensis*	10.10	12.80	129.28	126.73	
La Quina 5	*H. neanderthalensis*	11.50	13.00	149.50	113.04	
Krapina	*H. neanderthalensis*	12.20	12.00	146.40	98.36	
Krapina	*H. neanderthalensis*	10.00	12.50	125.00	125.00	
Krapina	*H. neanderthalensis*	10.20	12.50	127.50	122.55	
Le Moustier	*H. neanderthalensis*	11.00	12.00	132.00	109.09	
Klasies River (L)	Early *H. sapiens*	7.60	10.50	79.80	138.16	Rightmire and Deacon (2001)
Skhul VII	Early *H. sapiens*	8.60	11.00	94.60	127.91	de Lumley (1973)
Skhul IV	Early *H. sapiens*	9.40	11.10	104.34	118.09	
Skhul 5	Early *H. sapiens*	9.10	11.80	107.38	129.67	

(continued)

Table 1.2 (continued)

Specimen	Taxon	M-D	B-L	CBA	CSI (BL/MD*100)	Source
Qafzeh 7 (R)	Early *H. sapiens*	9.40	12.10	113.74	128.72	Vandermeersch (1981)
Qafzeh 9 (R)	Early *H. sapiens*	10.90	12.10	131.89	111.01	
Qafzeh 9 (L)	Early *H. sapiens*	9.50	13.50	128.25	142.11	
Early UP France (*n*=4)	*H. sapiens*	8.80	10.80	95.04	122.73	Coppa et al. (2005)
Early UP Italy (*n*=5, BL 6)	*H. sapiens*	8.60	11.00	95.00	127.91	
Early UP Central Europe (*n*=10)	*H. sapiens*	8.90	11.70	104.13	131.46	
Late UP France (*n*=10, BL 11)	*H. sapiens*	8.80	11.60	102.90	131.82	
Late UP Italy (*n*=9, BL 12)	*H. sapiens*	8.60	11.70	101.10	136.05	
Late UP Central Europe (*n*=1)	*H. sapiens*	7.90	10.60	83.74	134.18	
European Mesolithic (*n*=9)	*H. sapiens*	8.50	10.70	90.95	125.88	

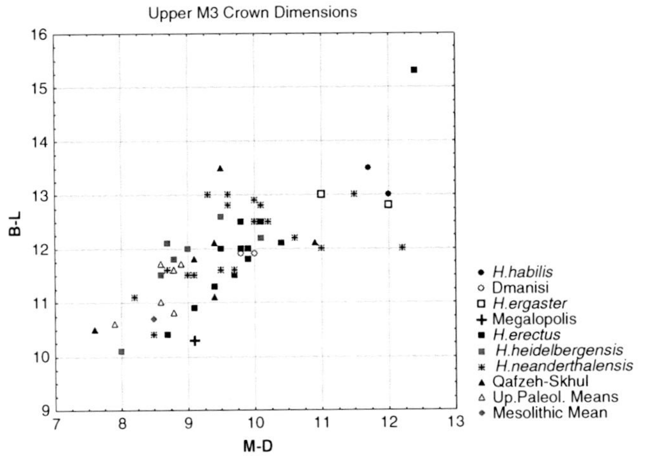

Fig. 1.9 Scatterplot of crown dimensions of fossil and recent human upper third molars. Comparative samples listed in Table 1.2. Crown dimensions from: de Lumley 1973; Vandermeersch 1981; Bermúdez de Castro 1986, 1993; Day 1986; Condemi 1992; Walker and Leakey 1993; Grine and Franzen 1994; Rightmire and Deacon 2001; Coppa et al. 2005; Rightmire et al. 2006; Suwa et al. 2007; Martinón-Torres et al. 2008

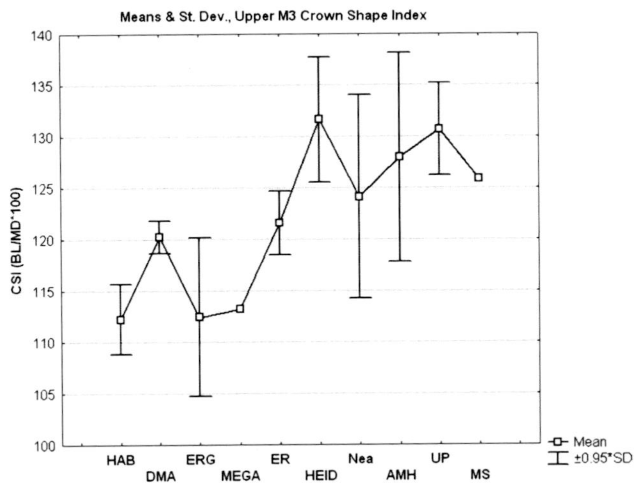

Fig. 1.10 Box plot of the Crown Shape Index (mean and standard deviation) of upper third molars among hominin samples (Table 1.2)

Conclusions

Despite many decades of relatively little research, the fossil human record of Greece is relatively rich, albeit sporadic in both time and space. Neanderthals are best represented in this record, with remains of both early and relatively late Neanderthals recovered from the Mani peninsula in Southern Greece. Most of these are isolated skeletal elements, although well-preserved cranial remains have also been recovered (Apidima). This part of the record closely mirrors that of Croatia, where both early and late Neanderthals are known from different sites (e.g., Janković et al. 2016). Comparisons with the Croatian and Near Eastern record in the context of Western European Neanderthal variation would therefore be of great interest for future study of this material.

The evidence is sparser in both earlier and younger periods of the Paleolithic. The Petralona cranium is commonly accepted as an early member of the Neanderthal lineage, belonging to *Homo heidelbergensis*. This specimen, however, is not well dated and shows strong affinities with the African Middle Pleistocene record. The affinities of the Megalopolis tooth, as well as its chronology, are also not well understood. On the opposite end of the temporal spectrum, very few Upper Paleolithic remains are currently known in Greece. This situation contrasts with other regions of the Balkans, which have yielded numerous Upper Paleolithic human fossils, such as Romania (e.g., Harvati and Roksandic 2016), highlighting the critical role of the Danube river in the dispersal of modern humans into Europe. An earlier modern human dispersal along the Mediterranean coast has been suggested on the basis of the Uluzzian sites in Italy and Greece (Mellars 2011), since the Uluzzian technocomplex was recently found to be associated with modern human remains (Benazzi et al. 2011). Nevertheless, until the discovery of taxonomically identifiable human remains from Greek Uluzzian sites, this scenario must remain hypothetical.

Perhaps the greatest shortcoming of the human fossil record from Greece is that the most complete and most important specimens were not recovered from excavations (Petralona, Apidima, Megalopolis). With the systematic excavation of Paleolithic sites in the last decades this situation has begun to change, and human fossil remains are now known also from excavated contexts. While the potential for paleoanthropological research in the country is far from having been fulfilled, we are now in a position to formulate hypotheses about the course of human evolution in the region and to target areas for future research. High research priorities include the Middle–Upper Paleolithic transition, as well as the Lower Paleolithic. Both periods are currently not well understood but are of great potential interest. An important goal of PaGE ('Paleoanthropology in the Southern Balkans')—a 5-year research program led by the author in collaboration with the Ephoreia of Paleoanthropology and Speleology of Southern Greece (Ministry of Culture), the University of Athens, and the University of Thessaloniki and funded by the European Research Council—is to help fill this research gap by conducting systematic fieldwork in areas selected for their strong research potential. These areas include, among others, the North-Western Mani peninsula, where new Paleolithic cave sites have been identified (Tourloukis et al. 2016); the Megalopolis basin, where stratified lithic artifacts have now been documented in Middle Pleistocene deposits (Panagopoulou et al. 2015); and the Mygdonia basin, Northern Greece, where new Early Pleistocene paleontological localities were identified and excavated (Konidaris et al. 2015). PaGE is one of several Paleolithic and Paleoanthropological projects currently active in the country (see e.g., Darlas and Psathi 2016; Galanidou et al. 2016). Together, their results promise to build an increasingly detailed and informative picture of human evolution in this crucial, but currently little known, region.

Acknowledgements I am deeply grateful to all my collaborators who have enabled the research summarized here: Eleni Panagopoulou, Andreas Darlas, Constantin Doukas, George Koufos, Dimitris Kostopoulos, Panagiotis Karkanas, Georgia Kourtessi-Philippaki and many others too numerous to list. I also thank the members of Tübingen Paleoanthropology and of the PaGE team, and particularly Monika Doll and Vagelis Tourloukis, for their contribution to the organization of the PaGE conference 'Human Evolution in the Southern Balkans,' where this paper was presented. K. Harvati is the PI of the European Research Council Starting Grant 'Paleoanthropology at the Gates of Europe' (PaGE; ERC-2011-StG-283503). This manuscript was greatly improved by the comments and suggestions of three anonymous reviewers and M. Roksandic.

References

Arsuaga, J. L., Martínez, I., Arnold, L. J., Aranburu, A., Gracia-Téllez, A., Sharp, W. D., et al. (2014). Neandertal roots: Cranial and chronological evidence from Sima de los Huesos. *Science, 344*, 1358–1363.

Arzarello, M., Marcolini, F., Pavia, G., Pavia, M., Petronio, C., Petrucci, M., et al. (2009). L'industrie lithique du site Pléistocène inférieur de Pirro Nord (Apricena, Italie du sud): une occupation humaine entre 1.3 et 1.7 Ma. *L'Anthropologie, 113*, 47–58.

Ashton, N., Lewis, S. G., De Groote, I., Duffy, S. M., Bates, M., Bates, R., et al. (2014). Hominin footprints from early Pleistocene deposits at Happisburgh, UK. *PLoS One, 9*, e88329.

Aytek, I. A., & Harvati, K. (2016). The human fossil record from Turkey. In K. Harvati & M. Roksandic (Eds.), *Paleoanthropology of the Balkans and Anatolia: Human evolution and its context*. Dordrecht: Springer.

Bailey, S. (2007). The evolution of non-metric dental variation in Europe. *Mitteilungen der Gesellschaft für Urgeschichte, 15*, 9–30.

Benazzi, S., Douka, K., Fornai, C., Bauer, C. C., Kullmer, O., Svoboda, J., et al. (2011). Early dispersal of modern humans in Europe and implications for Neanderthal behavior. *Nature, 479*, 525–528.

Bermúdez de Castro, J. M. (1986). Dental remains from Atapuerca (Spain) I. Metrics. *Journal of Human Evolution, 15*, 265–287.

Bermúdez de Castro, J. M. (1993). The Atapuerca dental remains. New evidence (1987–1991 excavations) and interpretations. *Journal of Human Evolution, 24*, 339–371.

Bermúdez de Castro, J. M., Arsuaga, J. L., Carbonell, E., Rosas, A., Martinez, I., & Mosquera, M. (1997). A hominid from the lower Pleistocene of Atapuerca, Spain: Possible ancestors to Neandertals and modern humans. *Science, 276*, 1392–1395.

Bermúdez de Castro, J. M., Martinón-Torres, M., Gómez-Róblez, A., Prado-Simón, L., Martín-Francés, L., Lapresa, M., et al. (2011). Early Pleistocene human mandible from Sima del Elefante (TE) cave site in Sierra de Atapuerca (Spain): A comparative morphological study. *Journal of Human Evolution, 61*, 12–25.

Carbonell, E., Bermúdez de Castro, J. M., Parés, J. M., Pérez-González, A., Cuenca-Bescós, G., Ollé, A., et al. (2008). The first hominin of Europe. *Nature, 452*, 465–469.

Condemi, S. (1992). Les Hommes Fossiles de Saccopastore et Leurs Relations Phylogénétiques. In *Cahiers de Paléoanthropologie*. Paris: CNRS Éditions.

Coppa, A., Grün, R., Stringer, C., Eggins, S., & Vargiu, R. (2005). Newly recognized Pleistocene human teeth from Tabun Cave, Israel. *Journal of Human Evolution, 49*, 301–315.

Crochet, J.-Y., Welcomme, J.-L., Ivorra, J., Ruffet, G., Boulbes, N., Capdevila, R., et al. (2009). Une nouvelle faune de vertébrés continentaux, associée à des artefacts dans le Pléistocène inférieur del'Hérault (Sud de la France), vers 1.57 Ma. *Comptes Rendus Palevol, 8*, 725–736.

Darlas, A. (1995). Τα λίθινα εργαλεία του σκελετού ΛΑΟ 1/Σ 3 (Απήδημα—Μάνη). *Acta Anthropologica, 1*, 59–62.

Darlas, A. (2003). Palaeolithic finds from the Megalopolis basin: Their relation to the fossils of the same area. In A. Vlachopoulos & K. Birtaha (Eds.), *Αργοναύτης: Τιμιτικός Τόμος για τον Καθηγητή Χρήστο Γ. Ντούμα από τους Μαθητές του στο Πανεπιστήμιο Αθηνών (1980–2000)* (pp. 27–37). Athens: Kathimerini.

Darlas, A., & de Lumley, H. (2004). La grotte de Kalamakia (Aréopolis, Grèce). Sa contribution à la connaissance du Paléolithique moyen de Grèce. *BAR, 1239*, 225–233.

Darlas, A., & Psathi, E. (2008). Le Paléolithique supérieur dans la péninsule du Mani (Péloponnèse, Grèce). In A. Darlas & D. Mihailovic (Eds.), *The Paleolithic of the Balkans* (pp. 51–59). Oxford: Archaeopress.

Darlas, A., & Psathi, E. (2016). The Middle and Upper Paleolithic on the Western coast of the Mani Peninsula (Southern Greece). In K. Harvati & M. Roksandic (Eds.), *Paleoanthropology of the Balkans and Anatolia: Human evolution and its context*. Dordrecht: Springer.

Day, M. H. (1986). *Guide to fossil man*. Chicago: University of Chicago Press.

de Lumley, M. A. (1973). *Anténéandertaliens et Néandertaliens du bassin méditerraneen occidental européen. Études Quaternaires 2*. Marseille: Université de Provence.

de Lumley, H., Darlas, A., Anglada, R., Cataliotti-Valdina, J., Desclaux, E., Dubar, M., et al. (1994). Grotte de Kalamakia (Aréopolis,

Péloponnèse). *Bulletin de Correspondance Hellenique, 118*, 535–559.

Dean, D., Hublin, J.-J., Holloway, R., & Ziegler, R. (1998). On the phylogenetic position of the pre-Neanderthal specimen from Reilingen, Germany. *Journal of Human Evolution, 34*, 485–508.

Despriée, J., Voinchet, P., Tissoux, H., Moncel, M.-H., Arzarello, M., Robin, S., et al. (2010). Lower and middle Pleistocene human settlements in the Middle Loire River Basin, Centre Region, France. *Quaternary International, 223*, 345–359.

Douka, K., Higham, T., Wood, R., Boscato, P., Gambassini, P., Karkanas, P., et al. (2014). On the chronology of the Uluzzian. *Journal of Human Evolution, 68*, 1–13.

Elefanti, P., Panagopoulou, E., & Karkanas, P. (2008). The transition from the Middle to the Upper Palaeolithic in the Southern Balkans: The evidence from the Lakonis I cave, Greece. *Eurasian Prehistory, 5*, 85–96.

Farrand, W. R. (2000). Depositional history of Franchthi Cave: Stratigraphy, sedimentology, and chronology. In *Excavations at Franchthi Cave, Greece, Fascicle 12*. Indiana: Indiana University Press.

Freidline, S. E., Gunz, P., Harvati, K., & Hublin, J.-J. (2012). Middle Pleistocene human facial morphology in an evolutionary and developmental context. *Journal of Human Evolution, 63*, 723–740.

Galanidou, N., Athanassas, C., Cole, J., Iliopoulos, G., Katerinopoulos, A., Magganas, A., & McNabb, J. (2016). The Acheulean site at Rodafnidia, Lisvori on Lesbos, Greece: Part I. In K. Harvati & M. Roksandic (Eds.), *Paleoanthropology of the Balkans and Anatolia: Human evolution and its context*. Dordrecht: Springer.

Grine, F. E., & Franzen, J. L. (1994). Fossil hominid teeth from the Sangiran dome (Java, Indonesia). *Courier Forschungsinstitut Senckenberg, 171*, 5–103.

Harvati, K. (2009). Petralona: Link between Africa and Europe? In L. Schepartz, C. Bourbou, & S. Fox (Eds.), *New directions in the skeletal biology of Greece* (pp. 31–49). Athens: Wiener Laboratory Series.

Harvati, K., Darlas, A., Bailey, S. E., Rein, T. R., El Zaatari, S., Fiorenza, L., et al. (2013). New Neanderthal remains from Mani peninsula, Southern Greece: The Kalamakia Middle Paleolithic cave site. *Journal of Human Evolution, 64*, 486–499.

Harvati, K., & Delson, E. (1999). Conference report: Paleoanthropology of the Mani Peninsula (Greece). *Journal of Human Evolution, 36*, 343–348.

Harvati, K., Hublin, J.-J., & Gunz, P. (2010). Evolution of Middle-Late Pleistocene human cranio-facial form: A 3-D approach. *Journal of Human Evolution, 59*, 445–464.

Harvati, K., Panagopoulou, E., & Karkanas, P. (2003). First Neanderthal remains from Greece: The evidence from Lakonis. *Journal of Human Evolution, 45*, 465–473.

Harvati, K., Panagopoulou, E., & Runnels, C. (2009). The paleoanthropology of Greece. *Evolutionary Anthropology, 18*, 131–143.

Harvati, K., & Roksandic, M. (2016). The human fossil record from Romania: Early Upper Paleolithic Europeans and possible evidence for Neanderthal admixture. In K. Harvati & M. Roksandic (Eds.), *Paleoanthropology of the Balkans and Anatolia: Human evolution and its context*. Dordrecht: Springer.

Harvati, K., Stringer, C., & Karkanas, P. (2011). Multivariate analysis and classification of the Apidima 2 cranium from Mani, Southern Greece. *Journal of Human Evolution, 60*, 246–250.

Higham, T., Basell, L., Jacobi, R., Wood, R., Bronk Ramsey, C., & Conard, N. J. (2011a). Testing models for the beginnings of the Aurignacian and the advent of figurative art and music: The radiocarbon chronology of Geißenklösterle. *Journal of Human Evolution, 62*, 664–676.

Higham, T., Compton, T., Stringer, C., Jacobi, R., Shapiro, B., Trinkaus, E., et al. (2011b). The earliest evidence foranatomically modern humans in northwestern Europe. *Nature, 479*, 521–524.

Higham, T., Wood, R., Moreau, L., Conard, N. J., & Bronk Ramsey, C. (2013). Comments on 'Human-climate interaction during the early Upper Palaeolithic: Testing the hypothesis of an adaptive shift between the Proto-Aurignacian and the Early Aurignaican' by Banks et al. *Journal of Human Evolution, 65*, 806–809.

Ivanova, S. (2016). A route through the Balkans and implications for the earliest settlement of Europe. In K. Harvati & M. Roksandic (Eds.), *Paleoanthropology of the Balkans and Anatolia: Human evolution and its context*. Dordrecht: Springer.

Janković, I., Ahern, J. C. M., Karavanić, I., & Smith, F. H. (2016). The importance of Croatian Pleistocene hominin finds in the study of human evolution. In K. Harvati & M. Roksandic (Eds.), *Paleoanthropology of the Balkans and Anatolia: Human evolution and its context*. Dordrecht: Springer.

Kahlke, R.-D., García, N., Kostopoulos, D., Lacombat, F., Lister, A., Mazza, P., et al. (2011). Western Palaearctic palaeoenvironmental conditions during the Early and early Middle Pleistocene inferred from large mammal communities, and implications for hominin dispersal in Europe. In J. Carrión, J. Rose & C. Stringer (Eds.), *Ecological scenarios for human evolution during the Early and early Middle Pleistocene in the western Palaearctic. Quaternary Science Reviews* (Vol. 30, pp. 1368–1395).

Konidaris, G., Tourloukis, V., Kostopoulos, D. S., Thompson, N., Giusti, D., Mihailidis, D., Koufos, G. D., & Harvati, K. (2015). Two new vertebrate localities from the Early Pleistocene of Mygdonia Basin (Macedonia, Greece): Preliminary results. *Comptes Rendus Palevol, 14*, 353–362.

Koumouzelis, M., Boleslawm, G., Kozlowski, J. K., Pawlikowski, M., Bar-Yosef, O., Albert, R. M., et al. (2001). The early Upper Paleolithic in Greece: The excavations at Klisoura cave. *Journal of Archaeological Science, 28*, 515–539.

Liritzis, Y., & Maniatis, Y. (1989). ESR experiments on Quaternary calcites and bones for dating purposes. *Journal of Radioanalytical and Nuclear Chemistry, 129*, 3–21.

Lowe, J., Barton, N., Blockley, S., Ramsey, C. B., Cullen, V. L., Davies, W., et al. (2012). Volcanic ash layers illuminate the resilience of Neanderthals and early modern humans to natural hazards. *Proceedings of the National Academy of Sciences of the United States of America, 109*, 13532–13537.

Martinón-Torres, M., Bermúdez de Castro, J. M., Gómez-Robles, A., Margvelashvili, A., Prado, L., Lordkipanidze, D., et al. (2008). Dental remains from Dmanisi (Republic of Georgia): Morphological analysis and comparative study. *Journal of Human Evolution, 55*, 249–273.

Martinón-Torres, M., Bermúdez de Castro, J. M., Gómez-Robles, A., Prado-Símon, L., & Arsuaga, J. L. (2012). Morphological description and comparison of the dental remains from Atapuerca-Sima de los Huesos site (Spain). *Journal of Human Evolution, 62*, 7–58.

McCarty, L., Rein, T. R., Scherf, H., Darlas, A., & Harvati, K. (2014). Geometric morphometric and trabecular bone analysis of the Kalamakia Neanderthal navicular bone: A comparative study. *American Journal of Physical Anthropology, 153*(S58), 180.

Mellars, P. (2011). The earliest modern humans in Europe. *Nature, 479*, 483–485.

Muttoni, G., Scardia, G., & Kent, D. V. (2013). A critique of evidence for human occupation of Europe older than the Jaramillo subchron (~1 Ma): Comment on 'The oldest human fossil in Europe from Orce (Spain)' by Toyo-Moyano et al. (2013). *Journal of Human Evolution, 65*, 746–749.

Panagopoulou, E., Karkanas, P., Tsartsidou, G., Kotjabopoulou, E., Harvati, K., & Ntinou, M. (2002–2004). Late Pleistocene archaeological and fossil human evidence from Laconis Cave, Southern Greece. *Journal of Field Archaeology, 29*, 323–349.

Panagopoulou, E., Tourloukis, V., Thompson, N., Athanassiou, A., Tsartsidou, G., Konidaris, G. E., Giusti, D., Karkanas, P., & Harvati, K. (2015). Marathousa 1: A new Middle Pleistocene archaeological site from Greece. *Antiquity Project Gallery, 89*(343).

Parfitt, S. A., Ashton, N. M., Lewis, S. G., Abel, R. L., Coope, G. R., Field, M. H., et al. (2010). Early Pleistocene human occupation at the edge of the boreal zone in northwest Europe. *Nature, 466*, 229–233.

Parfitt, S. A., Barendregt, R. W., Breda, M., Candy, I., Collins, M., Coope, G. R., et al. (2005). The earliest record of human activity in northern Europe. *Nature, 438*, 1008–1012.

Pitsios, T. K. (1985). Παλαιοανθρωπολογικές έρευνες στη θέση «Απήδημα» της Μέσα Μάνης. *Αρχαιολογία, 15*, 26–33.

Pitsios, T. K. (1995). Paleoanthropological Research at the cave site of Apidima, Laconia, Greece. *Acta Anthropologica, 1*, 1–180.

Pitsios, T. (1999). Paleoanthropological research at the cave site of Apidima and the surrounding region (South Peloponnese, Greece). *Anthropologischer Anzeiger, 57*, 1–11.

Pitsios, T. (2002). The fossil hominid findings from the Cave site of Apidima, South Peloponnese Greece. *Collegium Anthropologicum, 26*(Suppl.), 158.

Richards, M., Harvati, K., Grimes, V., Smith, C., Smith, T., Hublin, J.-J., et al. (2008). Isotope evidence of Neanderthal mobility. *Journal of Archaeological Science, 35*, 1251–1256.

Rightmire, P. G., & Deacon, H. J. (2001). New human teeth from Middle Stone Age deposits at Klasies River, South Africa. *Journal of Human Evolution, 41*, 535–544.

Rightmire, P. G., Lordkipanidze, D., & Vekua, A. (2006). Anatomical descriptions, comparative studies and evolutionary significance of the hominin skulls from Dmanisi, Republik of Georgia. *Journal of Human Evolution, 50*, 115–141.

Roebroeks, W. (2001). Hominid behaviour and the earliest occupation of Europe: An exploration. *Journal of Human Evolution, 41*, 437–461.

Roksandic, M. (2016). The role of the Central Balkans in the peopling of Europe: Paleoanthropological evidence. In K. Harvati, & M. Roksandic (Eds.), *Paleoanthropology of the Balkans and Anatolia: Human evolution and its context* (pp. 15–33). Dordrecht: Springer.

Sickenberg, O. (1975). Eine Säugertierfauna des tieferen Bihariums aus den Becken von Megalopolis (Peloponnes, Griechenland). *Annales Géologiques des Pays Helléniques, 27*, 26–71.

Sirakov, N., Guadelli, J.-L., Ivanova, S., Sirakova, S., Boudadi-Maligne, M., Dimitrova, I., et al. (2010). An ancient continuous human presence in the Balkans and the beginnings of human settlement in western Eurasia: A Lower Pleistocene example of the Lower Palaeolithic levels in Kozarnika cave (North-western Bulgaria). *Quaternary International, 223–224*, 94–106.

Sitlivy, V. (2016). Technological variability of the Middle-to-Upper Paleolithic transition: Examples from the Balkans and neighboring regions. In K. Harvati & M. Roksandic (Eds.), *Paleoanthropology of the Balkans and Anatolia: Human evolution and its context*. Dordrecht: Springer.

Smith, T. M., Harvati, K., Olejniczak, A. J., Reid, D. J., Hublin, J. J., & Panagopoulou, E. (2009). Dental development and enamel thickness in the Lakonis Neanderthal molar. *American Journal of Physical Anthropology, 138*, 112–118.

Spassov, N. (2016). Southeastern Europe as a route for the earliest dispersal of *homo* towards Europe: Ecological conditions and the timing of the first human occupation of Europe. In K. Harvati & M. Roksandic (Eds.), *Paleoanthropology of the Balkans and Anatolia: Human evolution and its context*. Dordrecht: Springer.

Stiner, M. C., Kozlowski, J. K., Kuhn, S. L., Karkanas, P., & Koumouzelis, M. (2010). Klissoura Cave 1 and the Upper Paleolithic of Southern Greece in cultural and ecological context. *Eurasian Prehistory, 7*, 309–321.

Stiner, M. C., & Munro, N. D. (2011). On the evolution of diet and landscape during the Upper Paleolithic through Mesolithic at Franchthi Cave (Peloponnese, Greece). *Journal of Human Evolution, 60*, 618–636.

Strait, D. S., Orr, C. M., Hodgkins, J., Spassov, N., Gurova, M., Miller, C., & Tzankov, T. (2016). The human fossil record of Bulgaria, and the formulation of biogeographic hypotheses. In K. Harvati & M. Roksandic (Eds.), *Paleoanthropology of the Balkans and Anatolia: Human evolution and its context*. Dordrecht: Springer.

Stringer, C. B. (1974). A multivariate study of the Petralona skull. *Journal of Human Evolution, 3*, 397–404.

Stringer, C. B., Howell, F. C., & Melentis, J. (1979). The significance of the fossil hominid skull from Petralona, Greece. *Journal of Archaeological Science, 6*, 235–253.

Suwa, G., Berhane, A., Haile-Selassie, Y., White, T., Katoh, S., Wolde Gabriel, G., et al. (2007). Early Pleistocene Homo erectus fossils from Konso, southern Ethiopia. *Anthropological Science, 115*, 113–151.

Toro-Moyano, I., Bienvenido Martínez-Navarro, B., Agustí, J., Souday, C., Bermúdez de Castro, J. M., Martinón-Torres, M., et al. (2013). The oldest human fossil in Europe dated to ca. 1.4 Ma at Orce (Spain). *Journal of Human Evolution, 65*, 1–9.

Tourloukis, V., Thompson, N., Garefalakis, C., Konidaris, G., Karkanas, P., Panagopoulou, E., & Harvati, K. (2016). New Middle Paleolithic sites from the Mani peninsula, Southern Greece. *Journal of Field Archaeology, 41*(1), 68–83.

Tsoukala, E. (1999). Quaternary large mammals from the Apidima caves (Lakonia, S Peloponnese, Greece). *Beiträge zur Paläontologie, 24*, 207–229.

van Vugt, N. (2000). Orbital forcing in late Neogene lacustrine basins from the Mediterranean; A magnetostratigraphic and cyclostratigraphic study (Ph.D. Dissertation, University of Utrecht, 2000).

Vandermeersch, B. (1981). Les Hommes Fossiles de Qafzeh (Israel). In *Cahiers de Paléontologie*. Paris: CNRS Éditions.

Walker, A., & Leakey, R. (1993). *The Nariokotome Homo erectus Skeleton*. Cambridge: Harvard University Press.

Xirotiris, N., Henke, W., & Symeonidis, N. (1979). Der M^3 von Megalopolis — ein Betrag zu seiner morphologischen Kennzeichnung. *Zeitschrift für Morphologie und Anthropologie, 70*, 117–122.

Chapter 2
The Role of the Central Balkans in the Peopling of Europe: Paleoanthropological Evidence

Mirjana Roksandic

Abstract The paucity of fossil human remains from the Central Balkans represents a very serious lacuna in our understanding of human evolution in the Pleistocene of Europe, which is—as a result—strongly influenced by the material from the better researched parts of the continent further to the west of the Balkans. The scant fossil record from the Central Balkans suffers from a lack of archaeological/geological context, and with the exception of the Balanica hominin (BH-1) has no associated chronological data. In this chapter, I present all of the *purported* Pleistocene specimens currently known from the area and discuss their possible affinities.

Keywords Human evolution • *Homo* • Pleistocene • Balkan Peninsula

Introduction

The last three decades have brought about important insights into human evolution in Europe. Dominated over the past 160 years by relatively abundant Upper Pleistocene fossil remains from more westerly parts of Europe and the explanatory models they engendered, the field is rapidly changing with the opening of new geographic areas to intensive research. The discovery of Dmanisi (Gabunia and Vekua 1995) demonstrated a human population outside of Africa by 1.8 Ma, and a recent publication on the Dmanisi cranium D4500 (Lordkipanidze et al. 2013) indicated greater variation among early hominins from a single locality than previously suspected. At the other end of the continent, well-dated Early Pleistocene sites and contexts emerged in Spain with the oldest hominin find in Europe dated to *ca.* 1.4 Ma at Orce (Toro-Moyano et al. 2013; but see Muttoni et al. 2013; also Spassov 2016 and references therein). Well-documented Early Pleistocene archaeological sites are also known from Italy, although no human remains have been recovered there so far (Manzi et al. 2011). Further to the east, a proposed, though contentious, date of 1.4 Ma at Kozarnika cave in Bulgaria (Ivanova 2016; Spassov 2016) would be contemporaneous with Ubeidiya in Israel (Belmaker et al. 2002). The opening of these new geographic foci to systematic survey and excavation resulted in possibly the greatest advances in human evolutionary studies in Europe over the last two decades. However, we are still far from fully understanding who the first inhabitants of the continent were; what was their relationship to fossil hominins in Asia, Africa, and later European fossil populations; how many migrations into and out of Europe occurred in the Pleistocene; where the migrants came from; and what route they took. The paleoanthropological record of the Central Balkans—currently consisting for the most part of fortuitous finds, or finds gathered from excavations that leave much to be desired—could represent a crucial piece in this puzzle.

The Central Balkans area is at the crossroads of the south-to-north and east-to-west migratory routes that run through the Balkan Peninsula (see also, e.g., Aytek and Harvati 2016; Doboş and Iovita 2016; Harvati 2016; Spassov 2016; Strait et al. 2016). At the gates to the continent, the Balkan Peninsula is the most logical route of migration from the Levant into Europe—already identified as the confirmed route of animal migrations during the colder phases of the Early Pleistocene (Belmaker et al. 2002). The Central Balkans, defined by the Morava and Vardar rivers and their tributaries, covers most of what is today Serbia (without Vojvodina, which belongs to the Pannonian basin and therefore Central Europe), Eastern Bosnia and Northern Macedonia. More than just a migratory route, this region was also an integral part of the Balkan refugium (Hewitt 2011; Griffiths et al. 2004) for temperate decid-

M. Roksandic (✉)
Department of Anthropology, University of Winnipeg, 515 Portage Avenue, Winnipeg, MB, Canada R3B 2E9
e-mail: m.roksandic@uwinnipeg.ca

Katerina Harvati and Mirjana Roksandic (eds.), *Paleoanthropology of the Balkans and Anatolia*, Vertebrate Paleobiology and Paleoanthropology, DOI 10.1007/978-94-024-0874-4_2

uous forests and associated biota (Eastwood 2004; Tzedakis 2004). The potential benefits of a more vigorous research program into the Pleistocene of the Central Balkan Peninsula cannot be overstated: the area could have played an important role in the initial peopling of the continent, in the repopulation of more northerly areas during interglacials, as well as in the demise of the Neanderthals and the advancement of modern humans. Whatever speculative role we can ascribe to the Central Balkans, the region is conspicuous by its absence in most discussions of migration(s) into and out of Europe (see, for example, a recent review by Bar-Yosef and Belfer-Cohen 2013).

Despite its likely importance and the strong tradition of archaeological research in the region, the Central Balkans Paleolithic record is scant (similar to the situation in many neighboring countries; see e.g., Aytek and Harvati 2016; Harvati 2016; Strait et al. 2016). A strong initial interest in Pleistocene-fauna and tool-bearing caves in the late 1800s–early 1900s (Cvijić 1903, 1918; Žujović 1893; Jovanović 1892) coincided with the discovery of Krapina in adjoining Croatia (Gorjanović-Kramberrger 1906; Janković et al. 2016). However, with the exception of some sporadic forays in the 1950s (Gavela 1951), this particular area of archaeology was all but forgotten until the very end of the twentieth century (Mihailović 2008; Mihailović and Bogićević 2016). Against this background, it is not surprising that the hominin fossil record is limited. Most of the purported Pleistocene specimens were uncovered in the late nineteenth and early twentieth century and subsequently lost during the First and the Second World Wars. In a recent AMS ^{14}C dating of six purported Pleistocene specimens from the Natural History Museum in Belgrade and the National Museum in Kraljevo, all were demonstrated to be of Holocene age (Roksandic et al. 2014), stressing the need for great caution in interpreting finds from old excavations.

The total tally of putative fossil hominins currently known from the Central Balkans (Fig. 2.1) includes: (1) a calotte from Bajloni's building discovered and described in 1892 (Jovanović 1892); (2) a mandible from the "loess in the vicinity of Belgrade" found in 1920 and published in 2001 (Roksandic and Dimitrijević 2001); (3) a tooth from Jerinina cave found in 1951, not described (Gavela 1951); (4) a skull fragment from the Kolubara gravel pit found in 1952, not described (Roksandic and Dimitrijević 2001); (5) a mandible found in Mala Balanica cave in 2006 (Roksandic et al. 2011). I will include in this review two additional cranial fragments: (6) a calotte from Bački Petrovac found in 1952 and published in 1966; and (7) a frontal from Žitište found in 1960 and described in 1966 (Živanović 1966; Radović et al. 2014). Both of these were found just north of the Central Balkans in the Pannonian plain of Central Europe. Popular lore mentions several more finds of which there is no mention in the published record. In addition to the specimen from "Bajloni's building" (Jovanović 1892) discussed later, there is mention of an "antediluvian man" uncovered from unspecified excavations in Cetinjska street. Since "Bajloni's building" refers to the brewery between Skadarska and Cetinjska streets in downtown Belgrade, this "antediluvian man" could potentially refer to the same specimen as the one from the Bajloni's building. A "Neanderthal" from Banovo brdo could be the one described as a "brachycephalic skull" (Žujović 1893, p. 21) uncovered from a loess deposit while excavating pylons for the bridge over the Sava river in Belgrade. Another "Neanderthal skull" from "Palata Albanija" was presumably found together with mammoth bones in 1938. The latter two specimens were recently located in the Natural History Museum in Belgrade. With the generous help of Sanja Paunović and Dr. Zoran Marković, I obtained permission to examine them and take samples for dating. Both skulls are clearly brachycephalic and therefore of post-Pleistocene age and will not be discussed in this chapter.

With the exception of the mandible from Mala Balanica, none of these specimens is associated with an archaeological context. Although unspecified stone tools were reportedly found with the Bački Petrovac specimen (Živanović 1966), given the accidental nature of the discovery, as well as the fact that the tools were neither described nor preserved, such an association cannot be confirmed. A very vague geological context reported as "*with bones of Elephas antiquus*" (Jovanović 1892, p. 30) in "*quaternary layers*" (Jovanović 1892, p. 31) has been reported for "Bajloni's building"; the Belgrade mandible was designated on its museum label as "*from the upper loess*" by its discoverer Professor Laskarev (Roksandic and Dimitrijević 2001, p. 28). The "*brachycephalic skull*" uncovered during the excavations for the Sava bridge—even according to the author—is not of Pleistocene age, although it was found in the loess deposit (Žujović 1893, p. 21): "*Under the third pylon, closer to the Austrian bank, plain river shells were unearthed as low as 12 m below the river bottom, while at the 14th meter, there was a human skull of a brachycephalous man.*" Noting other non-Pleistocene fauna in the river deposits in the area, Žujović (1893, p. 21) quite convincingly describes the taphonomic process that he considered responsible for the mixing: "*The river Sava still, within our memory, raises the plane; it still brings us deposits in which, mixed with river shells and snails, one finds fragments of horse, cattle, pig and sometimes mammoth skeletons that it unearthed from its original layers.*"

In this chapter, I will review what we know about each of the finds recorded in the scientific literature, and what we can learn about them by reexamining the very scant published measurements and descriptions. I will then offer some preliminary suggestions about the place of the Central Balkans in human evolution based on this rather limited evidence.

Fig. 2.1 Map of sites discussed in the chapter: Beograd (Belgrade) stands for both Bajloni's building calotte (BAJ in further text) and the "mandible from the loess in the vicinity of Belgrade" (RGF94/1) specimens. Inset shows the Balkan Peninsula and its relationship with the Black sea and adjoining regions; location of Belgrade and Balanica anchors the larger map in relation to well-known sites of Krapina (in Croatia) and Dmanisi (Georgia)

Materials and Methods

Before proceeding to describe the specimens in question, a note on the choice of measurements and morphological traits, as well as specimens and taxonomic groups included in the comparative sample, should be made. All the measurements were gathered from the reported original descriptions (for the more recently published material) and from large sets of data on originals by Rightmire (2008) for earlier discoveries (see Table 2.1 for the list of sources). Morphological traits of the mandible were taken from Mounier et al.'s (2009) comprehensive scoring of mandibular specimens. The choice of measurements and morphological traits was guided by the preserved morphology that could be measured or scored, or by the information available in the literature. This has of course resulted in limited comparative samples, which comprise only specimens that preserve the same measurements. In order to maximize the comparative sample, in some cases it was necessary to reduce the number of measurements used (notably for Bački Petrovac), as the alternative—i.e., to compute missing values—could introduce unknown biases.

When discussing hominin populations in the Pleistocene, the notion of "Paleo-deme" or "p-deme" (Howell 1996, 1999), which allows us to distinguish between geographically and chronologically restricted populations and discuss their possible phyletic relationships without implying or rejecting species status is the most appropriate. *Homo heidelbergensis* is a case in point, as it is differently interpreted to include European Middle Pleistocene specimens (*Homo heidelbergensis sensu stricto*), or European and African Middle Pleistocene specimens, (*Homo heidelbergensis sensu lato*), or even to extend to Asian samples (Rightmire 1998; Mounier et al. 2009; Harvati et al. 2010; Stringer 2012; Manzi 2012), or dismissed altogether (Mounier and Caparros 2015). The term Middle Pleistocene European *Homo* (MPEH) will be used here to denote European Middle Pleistocene humans with affinities to Neanderthals. Whenever possible, the comparative sample is grouped into the following categories: (1) *Homo habilis/rudolfensis*, (2) African *Homo erectus /ergaster*, (3) Early Pleistocene Eurasian *Homo*, (4) Asian *Homo erectus*, (5) Middle Pleistocene Asian *Homo*, (6) Middle Pleistocene African *Homo* (MPAfH), (7) Middle Pleistocene

Table 2.1 Linear measurements and angles used in the analysis[a]

Group/Specimen	Abbrev.	M1	M8	M29	M26	M32(5)	M10	M9	References
		Measurements used (Martin's number)[b]							
Early Pleistocene Euroasian *Homo*									
Dmanisi 2700	Dm2700	155	126	89	95	150	85	67	Lordkipanidze et al. (2006)
Dmanisi 2280	Dm2280	177	136	101	108	149	105	65	Lordkipanidze et al. (2006)
Dmanisi 3444	Dm3444	163	132	80	90	148	91	67.5	Lordkipanidze et al. (2006)
African *Homo erectus/ergaster*									
Daka	Dk	180	133	101	116	141	105	89	Asfaw et al. (2008)
KNM-ER3733	ER3733	182	142	104	119	139	110	83	Lordkipanidze et al. (2006) and Rightmire (1990)
KNM-ER3883	ER3883	182	140	101	118	140	105	80	Lordkipanidze et al. (2006) and Rightmire (1990)
Asian *Homo erectus*									
Sangiran 17	San17	207	161	118	–	–	–	–	Lordkipanidze et al. (2006)
Bukuran	Bk	194	149	110	–	–	–	–	Grimaud-Herve et al. (2012)
Sinanthropus III	Sin3	188	144	102	–	–	–	–	Weidenreich (1943)
Sinanthropus X	Sin10	190	150	115	–	–	–	–	Weidenreich (1943)
Sinanthropus XI	Sin11	192	145	106	–	–	–	–	Weidenreich (1943)
Sinanthropus XII	Sin12	195.5	147	113	–	–	–	–	Weidenreich (1943)
Ngandong 1	Ng1	198	153	114	128	141	120	106	Kaifu et al. (2008) and Rightmire (1990)
Ngandong 7	Ng2	192	147	116	125	140	116	103	Kaifu et al. (2008) and Rightmire (1990)
Ngandong 11	Ng11	203	160	120	130	138	122	112	Kaifu et al. (2008) and Rightmire (1990)
Ngandong 12	Ng12	201	151	113	121	146	114	103	Kaifu et al. (2008) and Rightmire (1990)
Middle Pleistocene African *Homo*									
Kabwe	Kb	209	149	120	139	140	118	98	Rightmire (2008) and Murrill (1981)
Elandsfontein	El	202	138	116	–	–	–	–	Rightmire (2008)
Bodo	Bd	–	–	125	144	139	119	105	Rightmire (1996, 2008)
Middle Pleistocene Asian *Homo*									
Dali	Dl	206.5	149.5	114	135	128	119	104	Wu and Athreya (2013)
Jinniushan	Jn	199	140	113	–	–	–	–	Coppens et al. (2008)
Middle Pleistocene European *Homo*									
Sima de los Huesos 4	SH4	201	164	115	126	140	126	117	Rightmire (2008)
Sima de los Huesos 5	SH5	185	146	106	114	145	118	105.7	Rightmire (2008)
Petralona	Pt	208	165	109	128	140	120	110	Rightmire (2008)
Ceprano	Cep	198	151	106	118	138	118	106	Ascenzi et al. (2000)
Upper Pleistocene *Homo sapiens*									
Skhul IV	Sk4	206	148	118	132	129.7	121	106	Vandermeersch (1981), Murrill (1981) and Cartmill and Smith (2009)
Skhul V	Sk5	193	146	106	118	130.7	114	99	Murrill (1981), Howells (1989) and Cartmill and Smith (2009)
Skhul IX	Sk9	213	145	114	130	131.6	120	96	Cartmill and Smith (2009)
Djebel Qafzeh 6	Q6	195	144	114	133	126.6	125	109.5	Vandermeersch (1981) and Howells (1989)
Djebel Qafzeh 9	Q9	–	–	115	130	133.8	117	103	Vandermeersch (1981) and Simmons et al. (1991)
Jebel Irhoud 1	JIr1	198	152	108	–	–	–	–	Howells (1989)
Upper Paleolithic *Homo sapiens*									
Predmosti 3	Pr3	202	143.4	120	137	135	128	104	Lubsen and Corruccini (2011) and Howells (1989)
Predmosti 4	Pr4	192	144	114	133	130	122	98	Lubsen and Corruccini (2011) and Howells 1989
Chancelade	Chan	–	–	111	130	128	127	101	Vandermeersch (1981) and Howells (1989)
Cro-Magnon 1	CrM1	206	153	125	147	125	126	102.5	Howells (1989) and Lubsen and Corruccini 2011
Mladeč 5	Ml5	205.6	156	116	–	–	–	–	Frayer et al. (2006)
Mladeč 6	Ml6	200.5	166.5	120.5	–	–	–	–	Frayer et al. (2006)
Mladeč 1	Ml1	198.5	141.5	114	133	123	126.5	103.5	Wolpoff et al. (2006)
Obercassel 1	Ob1	195	144	118.9	–	–	–	–	Vandermeersch (1981)
Obercassel 2	Ob2	183	134	106.4	–	–	–	–	Vandermeersch (1981)
Khvalynsk	Khv	–	–	115.9	130	136.1	115	94.2	Stansfield and Gunz (2011)
Podkumok	Pod	–	–	108.6	125.4	129.8	115	94.1	Stansfield and Gunz (2011)
Satanay	Sat	–	–	111.4	123	141.9	105	91.5	Stansfield and Gunz (2011)
Skhodnya	Skho	–	–	122.5	140.7	134.9	114	98.9	Stansfield and Gunz (2011)

(continued)

Table 2.1 (continued)

Group/Specimen	Abbrev.	M1	M8	M29	M26	M32(5)	M10	M9	References
		Measurements used (Martin's number)[b]							
Neanderthals									
La Chapelle	LCh	209	157	107	121	137	122	109	Murrill (1981) and Howells (1989)
La Ferrassie I	LF1	208	159	116	135	145	121	109	Murrill (1981) and Howells (1989)
Šal'a	Sal	–	–	110	121	138	127	105	Sládek et al. (2002)
La Quina 5	LQ5	201	139	109	–	–	–	–	Weidenreich (1943) and Cartmill and Smith (2009)
Neanderthal 1	Neand	201	147	116	–	–	–	–	Murrill (1981) and Cartmill and Smith (2009)
Shanidar 1	Sh1	207	154	111.3	119	144	128	110	Trinkaus (1983) and Howells (1989)
Shanidar 5	Sh5	–	–	118	129	147	128	103.5	Trinkaus (1983) and Simmons et al. (1991)
Tabun C1	TbC1	183	141	96	107	130.7	121.5	98	Simmons et al. (1988), Weidenreich (1943) and Cartmill and Smith (2009)
Amud	Am	215	154	120	135	138.5	124	115	Vandermeersch (1981) and Cartmill and Smith (2009)
Specimens from the Central Balkans									
Bajloni's building	BAJ	188	138	104	–	–	–	–	Jovanović (1892)
Bački Petrovac	BP	–	–	118	137	139	117	95	Živanović, (1966)

[a]All measurements are in given millimeters, except M 32 (5), which is given in degrees
[b]M numbers follow Martin and Saller (1957): Maximum cranial length (M1); Maximum cranial breadth (M8); Minimum frontal breadth (M9); Maximum frontal breadth (M10); Frontal sagittal arc (M26); Frontal sagittal chord (M29); Frontal angle (M32(5))

European *Homo* (MPEH), (8) Upper Pleistocene *Homo sapiens* from Africa/Near East, (9) Neanderthals, (10) Upper Paleolithic *Homo sapiens*.

Descriptions

"Bajloni's Building" Calotte

This specimen (hereafter BAJ) was found during the excavations of the foundations for the Bajloni's brewery building in the Old Town district of Belgrade in the late nineteenth century. The brewery opened in 1880 and the calotte must have been excavated shortly before that. It was subsequently lost in one of the many bombings of Belgrade in the early twentieth century. Professor Djordje Jovanović (1892) states that it was found two and a half meters below the current street level, on the low ledge that runs from Vidin gate to the Danube River, which he concludes was likely a Pleistocene river terrace. If we accept his claim that the specimen was found in the proximity of several teeth of *Elephas antiquus* (Falconer and Cautley 1847), a species found in Europe between 736 ka (in Italy) and 37 ka (in Netherlands) (Mol et al. 2007), the calotte could be of Pleistocene age.

According to Jovanović's (1892) description "*the skull is not complete. One can see the frontal, parietals, occipital and one temporal bone. Even fragmentary as it is, this skull is quite characteristic. On the frontal which is 104 mm long, one can observe well developed supraorbital arches (or tori). The right arch is more developed than the left. Above the right frontal arch there is a rough depression 2 cm by 3 cm. Frontal bossae are almost invisible and in the middle there is a rather well developed sagittal ridge. The forehead is so small and receding that one of our sculptors remarked—on having seen it for the first time—that the skull almost doesn't have any forehead*" (Jovanović 1892, p. 33). Further on, he notes that the "*parietal bones are asymmetrical. The right one is more convex than the left. Obelion is very large. On the temporal bone one can see the origin of a strong and well developed temporal muscle and well developed mastoid process. The circumference of the skull was 50.4 cm. The length 18.8 cm and breadth 13.8 cm and accordingly, the cranial index is 72 and the skull is dolichocephalic*" (Jovanović 1892, p. 34). Jovanović promised a more detailed analysis should there be more finds—which he did not doubt—and concluded that "*with its receding forehead, well developed supraorbital arches and well developed temporal bone the skull belonged to a far more primitive man than any so far found in Belgrade*" (Jovanović 1892, p. 34). Unfortunately, no drawings or photographs accompanied this report.

The three measurements are far from sufficient to give us a reasonable picture of the taxonomic position of the specimen. Given the lack of standardization of measurements in the late nineteenth century, to evaluate whether or not the measurements are reliable, row-standardized values were compared with averages for the specified groups (following Harvati et al. 2011). Although limited in scope, the measurements seem to be reliable (Table 2.2). Given the paucity of measurements, a principal components analysis (PCA) run on both raw data and size-adjusted data was not informative. BAJ plotted in the middle of the graph (not shown) between the

Table 2.2 Row-standardized measurements with the means for all groups and BAJ

Group	M1	M8	M29
Early Pleistocene Euroasian *Homo*	2.22	2.12	1.95
African *Homo erectus/ergaster*	2.26	2.14	2.01
Asian *Homo erectus*	2.29	2.18	2.05
Middle Pleistocene African *Homo*	2.31	2.16	2.07
Middle Pleistocene Asian *Homo*	2.31	2.16	2.05
Middle Pleistocene European *Homo*	2.30	2.19	2.04
Early *Homo sapiens* Africa/Near East	2.30	2.17	2.05
Upper Paleolithic *Homo sapiens*	2.30	2.17	2.07
Neanderthals	2.31	2.18	2.04
BAJ	2.27	2.14	2.02

Early Pleistocene and the Middle and Upper Pleistocene material, but close to Tabun C1 (a Neanderthal) and Oberkassel 2 (a modern human), both of which are very small females (Bar-Yosef and Callander 1999; Bruzek 2006, respectively).

Frontal bone morphology can be a good indicator of a specimen's general affinities (Athreya 2012). However, only one measurement, the frontal chord, is available for BAJ. Based on values in Table 2.1, at 104 mm, the frontal chord value is just below the range of values for modern humans (106–125), MPEH (106–115), MPAfH (120–125), and MPAsH (113–114) and in the lower range of values for Neanderthals (96–120) and the Asian *Homo erectus* (102–120). While it cannot be taken at face value, this observation gives some support to the description provided by Jovanović (1892) that the forehead is very low, and strengthens the suggestion that it could have been of Pleistocene age. Although descriptions are not detailed enough, frontal keeling and a well-developed mastoid process would be inconsistent with Neanderthals and could point to *Homo erectus s.l.* or robust modern humans. Given its low forehead, existence of sagittal keeling, strong attachment for the temporal muscle, and a pronounced mastoid process, we could very tentatively attribute this specimen to the plesiomorphic end of the spectrum of Middle and Upper Pleistocene variation, consistent with erectus-like and modern-human-like morphology and not consistent with Neanderthal morphology. However, the recorded measurements and the description provided are not sufficient to exclude the possibility that it is a modern human of Pleistocene or even post-Pleistocene age.

Bački Petrovac and Žitište

The other two partial calottes come from the area north of Belgrade in the Pannonian plain: Bački Petrovac and Žitište. The current whereabouts of these two specimens are not known and I could not examine them directly. According to Živanovic (1966), only one fragment of a skull was found in Žitište (Fig. 2.2) comprising the squama and a small part of the horizontal portion of the frontal bone. "*Supraorbital tori are broken; however, based on what remains of them, and given the size of the frontal sinuses, they were well-developed. Frontal eminences were not clearly marked.... The maximum width of the bone is 8 mm and the minimum 1 mm. The bone is fossilized, although it is more compact and less fragile than the other one (Bački Petrovac). Prof Škerlj maintains that this fragment belongs to the skull of a recent human*" (Živanović 1966, p. 190). Not much can be learned from this very short description. The photographs of the specimen (Fig. 2.2) do not show any indication that the frontal fragment deviates from modern human morphology, particularly as there is a clear supraorbital notch. Other than the assertion that it is fossilized (although this cannot be taken for granted given the assessment by Dr. Škerlj reported above), there is no indication that it is not a recent, post-Pleistocene human.

The calotte from Bački Petrovac (Fig. 2.3) was uncovered during the excavation of a brickyard pit in the vicinity of the village of the same name in the 1950s. The fossilized calotte came into the possession of a local schoolteacher and an amateur collector who handed it to Serbian archaeologist Miodrag Grbić. According to Grbić (as reported by Živanović 1966), it was associated with Paleolithic stone tools, which were not described or specified. The calotte consisted of an almost complete frontal, fragmentary parietals (the right one was better preserved), and a small fragment of the ethmoid bone. Živanović presented the specimen in 1960 at an unspecified meeting of Yugoslav anthropologists and published measurements and a description of the fossil in 1966 in Starinar, the main archaeological journal in the country—the same one in which the Bajloni's calotte was published in 1892. Subsequently, Živanović published another report likening this specimen to his Proto-Dinarid group of the Padina type (Živanović 1975; Radović et al. 2014). The author notes "*more pronounced superciliary arches than modern ones and a very low forehead. The skull is very long and the volume is low. Morphologically notable are much larger dimensions of the frontal bone than of parietal bones. Regardless of the very pronounced frontal dimensions, the orbits are small*" (Živanović 1966, p. 190).

It is difficult to evaluate Živanović's description on the basis of the published figures alone. Notably, a larger frontal and short parietals are inconsistent with the description of the skull as very long, with low volume. The impression that the skull is low and long could be partially due to the lack of elements that would allow for proper orientation of the skull in *norma lateralis*, demonstrated by the difference between the left and the right profile in Živanović's (1966) original figures. In addition to describing the morphology, Živanović (1966, p. 189) provided a number of measurements, most of them on the frontal bone. As previously noted, the frontal bone has been found to be a good indicator of species status

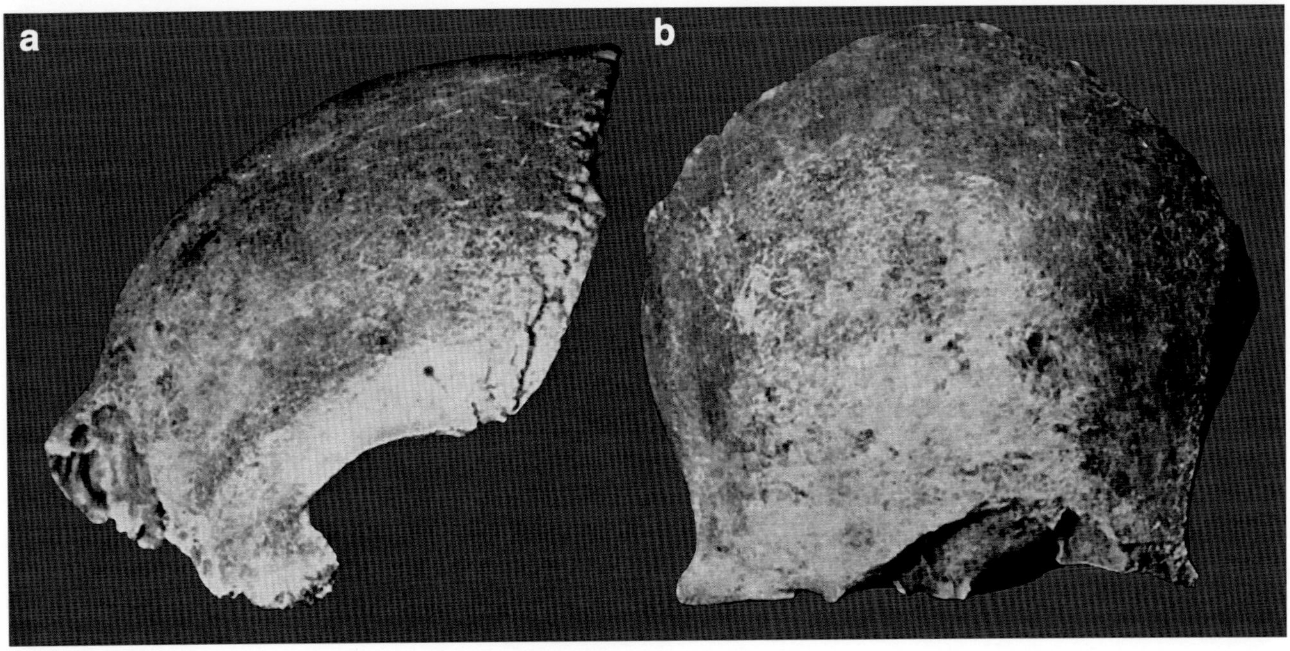

Fig. 2.2 Frontal from Žitište in (**a**) *norma frontalis* and (**b**) *norma lateralis*. Adapted from Živanović (1966)

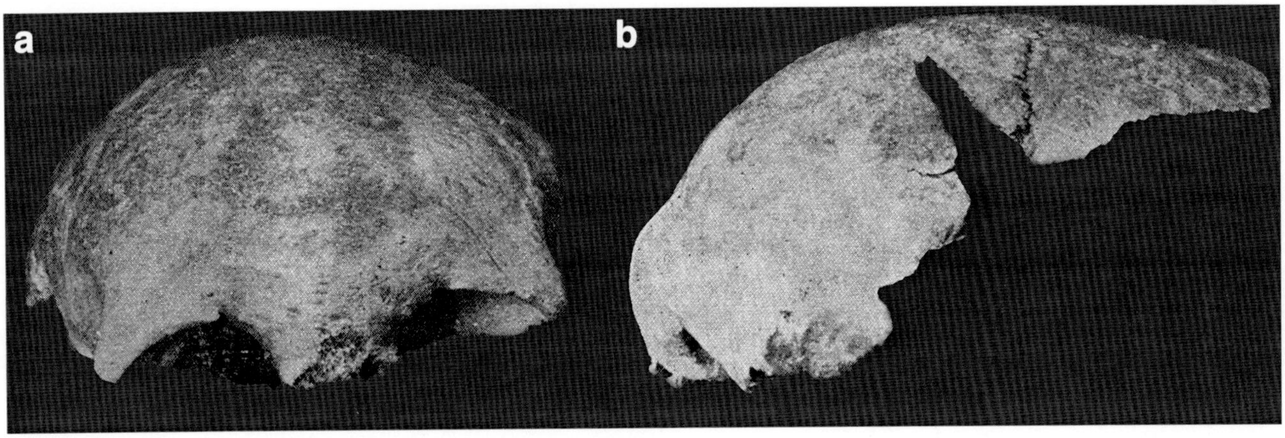

Fig. 2.3 Bački Petrovac calotte in (**a**) *norma frontalis* and (**b**) *norma lateralis*. Adapted from Živanović (1966)

in human evolution (Athreya 2012). A detailed reanalysis of these measurements is provided in a recent paper (Radović et al. 2014) and briefly summarized here.

A PCA (Fig. 2.4, Table 2.3) was performed on a variance–covariance matrix of five of the 17 measurements provided by Živanović (1966) for Bački Petrovac. Size-adjusted values were obtained by subtracting the log geometric mean of each variable for each individual from each log-transformed measurement (following Harvati et al. 2011). In order to maximize the comparative sample and strike a balance between the number of measurements and the number of specimens, measurements that are most commonly reported in the literature were selected (see Table 2.1). The optimal point at which most specimens have the greatest number of measurements was reached at five measurements, present in 33 specimens of the Middle and Upper Pleistocene ages.

The first principal component suggests that 48.4 % of total variance is due to size differences even when using size-standardized values. All variables were loading positively, with the exception of the frontal angle (Table 2.3): the low values of the eigenvector for frontal angle indicate that this variable does not have a strong influence on PC1; it is also negative as it is inversely proportional to size, since reducing the angle increases the curvature and therefore the size of the bone. Given the observed overlap between groups, size is not relevant for between-group differentiation. PC 2

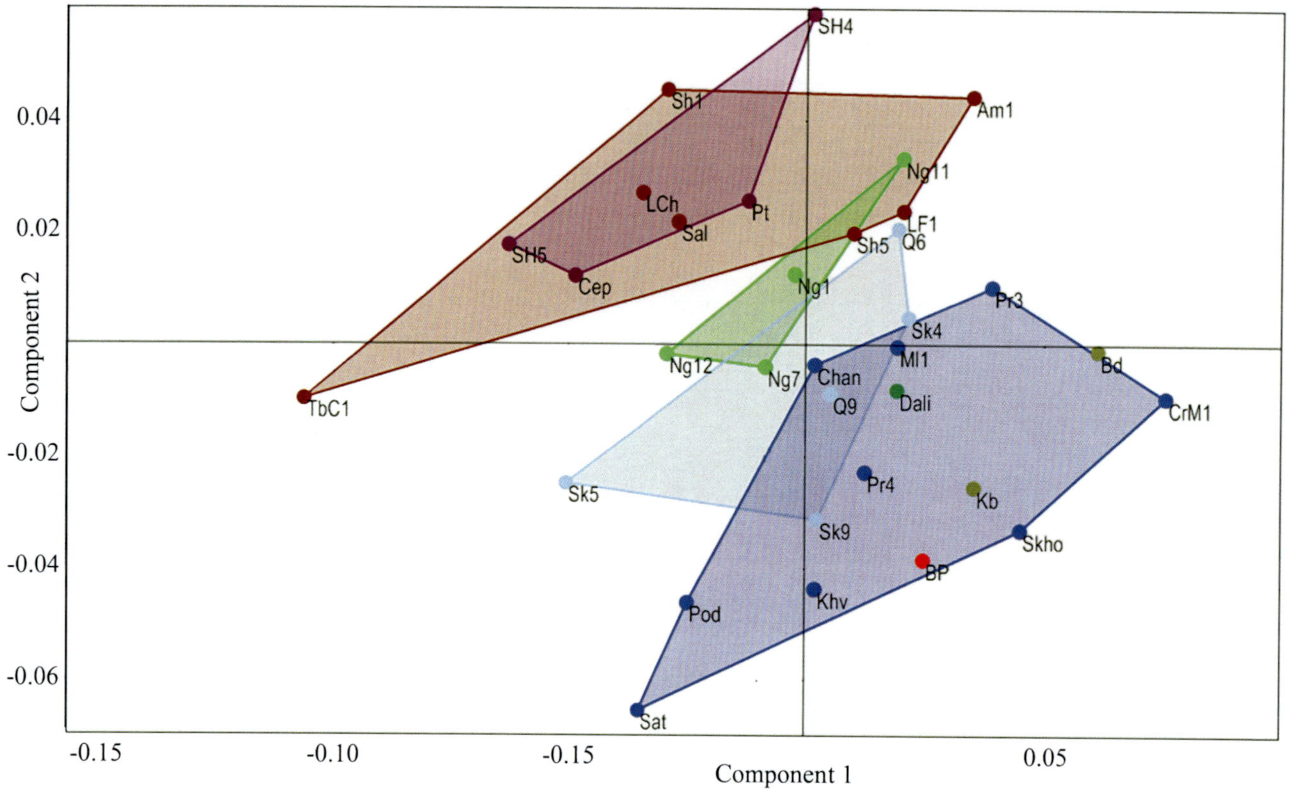

Fig. 2.4 Principal components analysis (PCA) of size-adjusted values for five frontal measurements of Bački Petrovac (BP) and a comparative sample. *Blue*: Upper Paleolithic *H. sapiens*; *Light blue*: Upper Pleistocene *H. sapiens*; *Green*: MPAsH and MPAfH; *Tan*: Neanderthals; *Purple*: MPEH

Table 2.3 Eigenvalues for size-adjusted data and loadings of variables on each axis

PC	Eigenvalue	% variance	M29_frontal chord	M26_frontal arc	Frontal angle (M-32(5))	M10_MFB	M9_min frontal
1	0.00139355	48.421	0.5975	0.779	−0.1103	0.1279	0.08746
2	0.000834579	28.999	−0.002527	−0.1397	0.2089	0.4607	0.8512
3	0.000473695	16.459	0.2754	−0.02104	0.8559	−0.4366	0.02363
4	0.000138295	4.8053	0.05955	−0.05874	0.3824	0.7621	−0.5157
5	3.79E−05	1.3155	−0.7507	0.6081	0.2558	0.0005135	0.0345

(29.0 % of the total variance) shows a contrast between breadth and length variables: the strongest positive influence is exerted by both the minimum (M9) and maximum (M10) frontal breadth and the strongest negative influence by the frontal arc. Neanderthals group together with MPEH with wider and shorter frontals and smaller difference between minimum and maximum frontal breadth, while Upper Paleolithic *H. sapiens* and African Middle Pleistocene specimens (especially Kabwe) group together on the opposite end with a larger difference between the two breadths. *H. erectus* and early modern humans are in the middle. PC3 (16.5 % of variation; not shown) represents a contrast between the frontal angle and remaining variables, with Bački Petrovac falling within the range of variation of Upper Paleolithic *H. sapiens*, close to Bodo and Kabwe, with a wider frontal angle and longer frontal chord. Since post-Pleistocene modern human variation completely overlaps with Pleistocene modern humans, until the actual remains are located and dated directly, it is not possible to say anything more definitive about the specimen, or ascertain Pleistocene affinities. A new project that aims to recover more materials from this location and the surrounding area is underway and we are still looking for the actual calotte in hope of obtaining a direct date.

Belgrade Mandible RGF94/1

A mandible unearthed in the 1920s from loess deposits in the vicinity of Belgrade is currently housed at the Faculty of Mining and Geology at the University of Belgrade (RGF 94/1). It was rediscovered in the storage drawers of the Geological collection and a description of the specimen was published by Roksandic and Dimitrijević (2001). While (glaciogenic) loess deposits in Serbia are unequivocally associated with the Pleistocene (Marković et al. 2008), new research shows that aridity in the Pannonian basin during the Holocene could produce significant eolian nonglaciogenic loess-like deposits (Sherwood et al. 2013). Given the geographic position of Belgrade on the Southern edge of the Pannonian plane, this is important to keep in mind. The evidence of fossilization has been obscured by the impregnation of the mandible with paraffin, which was performed for conservation purposes. Recently, a ^{14}C date indicating Holocene age has been obtained (Dimitrijević, pers. comm. 28/05/2013). However, at this point, it is not clear to what extent the carbon from the paraffin could have influenced the obtained date. The post-Pleistocene date would be consistent with the attribution of the specimen to an anatomically modern human (Roksandic and Dimitrijević 2001).

Even though this right semimandible is broken off at the symphysis—generally considered to be one of the most unambiguous anatomical area that separates modern human mandibles from more plesiomorphic forms (Schwartz and Tattersall 2000)—it is still possible to see the beginning of a slight exomandibular curvature at the breakage point that could indicate the existence of a bony chin (Fig. 2.5, upper right panel). There are other indicators that the mandible belongs to an anatomically modern human: there is no evidence of a retromolar space, the mental foramen is situated under the P_3/P_4 and is equidistant from the alveolar and basal margins. In addition, the P_3 is bicuspid, and tall and narrow in buccal view. It shows remarkable symmetry in the occlusal view, with a prominent lingual cusp, well-developed marginal ridges, and a clear mesiolingual groove. The central developmental groove is not present, a relatively common variant in modern humans. The mandibular P_3 has been noted for exhibiting the highest variability after the M_3 in modern humans (Cleghorn et al. 2007), but its overall symmetry is often associated with the modern human condition, while pronounced asymmetry is a plesiomorphic trait observed in 40–50 % of *H. erectus*, Neanderthals, and Middle Pleistocene *H. sapiens* (Bailey 2002). The P_4 is tricuspid with the buccal cusp the most prominent; it exhibits a

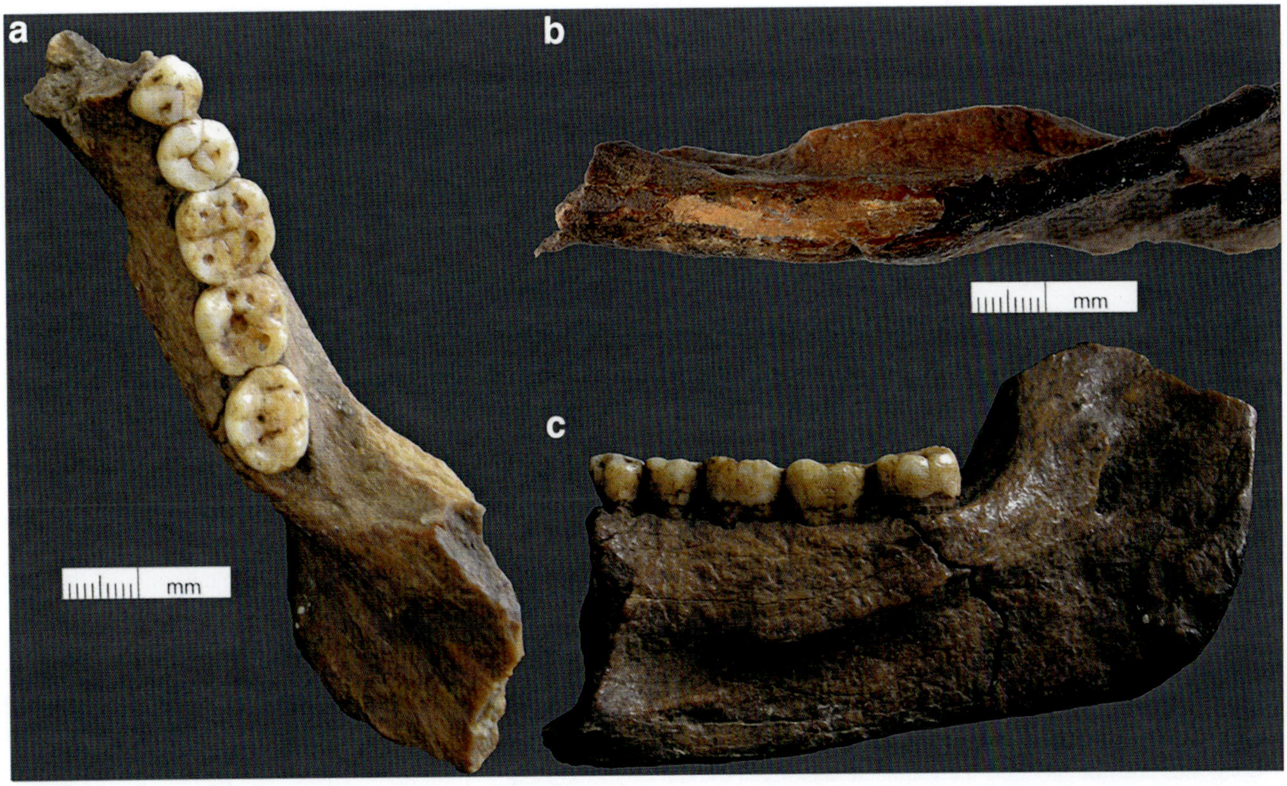

Fig. 2.5 Belgrade mandible RGF94/1. (**a**) Occlusal view, (**b**) basal view, and (**c**) endomanibular view of the specimen from the vicinity of Belgrade

pronounced mesiolingual cusp, without a mesial crest. The tooth shows no marked asymmetry. Asymmetry is predominant in Neanderthals (90 %), very rare in modern humans (6 %) but occurs in both *H. erectus* and archaic *H. sapiens* at 36 % and 33 %, respectively (as reported by Bailey 2002, although note small sample sizes). Together with the asymmetry, a mesially placed metaconid and a mesial crest are deemed distinctively Neanderthal features (Bailey and Lynch 2005). This specimen has no mesial crest, and a mesially placed metaconid on its own can be found in modern humans, albeit at somewhat lower and more variable frequencies than in Neanderthals (Bailey 2002: Table 5.6). The M_1 has four cusps, a square outline, an anterior marginal ridge without midtrigonid crest, and a "+4" pattern. The M_2 has a square outline, an anterior fovea and no midtrigonid crest, a "Y4" pattern, and a mesial and central occlusal pit. The M_3 has a six-cusp pattern with an irregular outline and a shallow anterior fovea (Hillson 1996). The teeth are tightly packed and intermolar wear facets are present. One notable feature of this mandible is the extreme development of the mylohyoid line. As can be seen in Fig. 2.5, the mylohyoid line is very strong and begins below the $M_{1,}$ forming an abrupt angle in continuation of the sublingual fossa, which is deep and oval in aspect. While not uncommon in modern humans (or Neanderthals), an exaggerated mylohyoid line is rarely mentioned in the literature and needs to be more systematically examined. Kennedy (2000) notes it for the Upper Paleolithic mandible from Bhimbetka, and Mirazón Lahr and Haydenblit (1995) for a Natufian mandible from the cave of Et-Tin. The sublingual fossa is considered as a very variable feature in modern human populations (Uchida et al. 2012).

Table 2.4 shows character states for the mandibular specimens included in the Principal Coordinate Analysis (PCO). These nonmetric traits are taken from Mounier et al. (2009) as relevant for differentiating between MPEH, Neanderthals, and modern humans in Pleistocene Europe. Mounier et al. (2009) used a larger battery of traits and therefore obtained more robust results and a better separation than observed here. This is because RGF 94/1 lacks all of the diagnostic traits of the symphyseal region and the vertical ramus (see also results for the Balanica mandible, below). Nevertheless, the PCO (Fig. 2.6) shows a separation between Neanderthals / MPEH on one hand and modern humans and *H. erectus* on the other hand. RGF94/1 falls in the modern human range of the graph overlapping with *H. erectus* and far from Neanderthal or MPEH morphology.

The Balanica Mandible

Among these fortuitous finds, the Balanica mandible (BH-1) stands out as the only specimen unearthed during controlled archaeological excavations (Roksandic et al. 2011). The mandible has recently been dated by electron spin resonance (ESR) combined with uranium series isotopic analysis (U-series), and infrared/postinfrared luminescence (IRSL) dating, to older than 392–525 ka (Rink et al. 2013). As such, it represents the oldest radiometrically dated human fossil from Eastern Europe and the Balkans. The mandible was excavated from Mala Balanica cave (N43°20.211′, E22°05.115′), part of a two-cave system located in the Sićevo gorge. The cave is situated some 332 m above sea level and currently about 100 m above the Nišava River, with the opening facing SSW across the valley, 7 m away from the entrance to the larger Velika Balanica cave. The gorge is cut through by the Nišava River, which provides an important communication route between two adjoining river valleys. BH-1 originates from layer 3b, three arbitrary 5 cm spits below the base of a pit dug in by "gold diggers" in this area between the field campaigns of the 2005 and 2006 seasons. Below the clandestine pit there are 2 m of compact, water-borne silts and clays. These fine-grained sediments are *in situ*, in their primary position relating to water pooling in this area of the cave (Morley, pers. comm. 4/15/2013). The lowest recorded artifacts were found in layers 1.5 m above the mandible. The animal teeth used for dating originate from the layer directly above the mandible and were recorded *in situ*. The concordance of all three dating techniques—ESR, U-series, and IRSL (Rink et al. 2013)—indicates that the obtained minimum date is reliable; the fact that the mandible was recovered from a layer below the obtained date suggests that the mandible could be slightly older, although probably not substantially.

The BH-1 specimen is a left hemi-mandible (Fig. 2.7), preserved from the posterior margin of the canine alveolus to the mesial aspect of the ascending ramus, with all three molars present in their sockets. The mesial portion of the mandible shows an old breakage filled with sediment, whereas all of the breaks on the distal end are fresh: the lower half of the mesolingual root of the M_3 is missing and the remaining roots are exposed due to the destruction of the adjacent endomandibular lamina. The alveoli of the P_3 and P_4 are complete and are for the most part filled with sediment. The posterior portion of the mandible seems to have been subject to water infiltration resulting in substantial fragility. Complete eruption and closure of the root apex of the M_3 indicates an adult individual, while minimal wear on the M_3 and slight to moderate wear on the M_1 and M_2 each suggest a relatively young adult. Sex could not be determined.

The highly relevant symphyseal region is missing and so is the basal margin mesially from below the mental foramen. The anterior marginal tubercle could not be observed in this specimen as the relevant area is missing. In lateral view, the basal and alveolar margins are almost parallel: the corpus measures 34.2 mm in height at the mental foramen and recedes slightly toward the M_3, where it measures 31.2 mm. The exomandibular relief is faint: a poorly defined superior

Table 2.4 Character states used in PCO analysis

Group/specimen	Abbreviations	I[a]	J	K	L	N	O	P	Q	R	S	T	U	OO	PP	UU
Early Pleistocene Euroasian *Homo*																
Dmanisi 211	Dm211	3	2	1	1	2	1	2	2	1	2	3	2	1	2	1
Dmanisi 2600	Dm2600	1	2	1	1	3	2	2	2	2	2	2	2	2	1	1
ATD6-96	ATD6-96	3	2	1	2	2	1	1	2	1	2	2	2	2	1	1
African *Homo erectus/ergaster*																
KNM-ER992	ER992	3	1	1	3	2	2	1	3	1	2	2	2	1	1	1
Asian *Homo erectus*																
Sangiran1b	San1b	3	2	1	2	2	2	1	2	1	2	2	2	2	2	1
Sinanthropus H1	SinH1	3	2	2	2	3	2	2	2	2	3	2	1	2	2	1
Middle Pleistocene African *Homo*																
Tighenif1	Tig1	3	2	1	1	2	1	2	3	1	2	1	2	1	1	1
Tighenif2	Tig2	3	1	1	2	2	1	2	2	2	2	2	2	2	1	1
Tighenif3	Tig3	3	2	2	2	3	2	2	3	2	2	1	3	1	1	1
Middle Pleistocene European *Homo*																
Mauer	Ma	2	2	2	2	3	2	2	3	3	2	1	3	1	1	1
AT-888	AT-888	3	1	3	1	2	1	2	2	3	3	1	2	3	1	1
AT-950	AT-950	3	2	2	1	3	2	2	3	3	3	1	2	1	1	2
Arago II	Ar2	3	1	3	1	2	2	1	3	3	2	1	3	1	2	2
Arago XIII	Ar13	3	2	2	1	2	1	2	3	2	2	1	2	1	1	2
Montmaurin	Mont	3	2	2	2	2	2	1	2	2	2	1	2	2	1	2
Ehringsdorf F	EhF	3	2	3	2	2	1	2	1	3	3	1	1	1	2	2
Neanderthals																
Krapina J	KrJ	3	2	3	1	3	2	2	2	3	3	1	1	3	1	1
Krapina G	KrG	3	2	3	2	3	2	2	1	3	3	1	1	3	3	2
Spy 1	Spy1	2	1	3	2	1	1	1	2	3	3	1	2	3	2	2
Regourdou	Reg	3	2	3	1	2	1	2	1	3	3	1	2	3	1	1
Bañolas	Ban	3	1	2	3	3	2	2	3	2	3	2	1	2	1	2
La Ferrassie 1	LF1	3	2	3	2	2	2	1	2	3	3	1	1	3	3	2
La Quina H5	LQH5	3	1	2	1	2	1	2	1	3	3	1	2	3	3	2
Shanidar I	Sh1	2	1	2	1	3	2	2	2	3	3	1	2	3	1	2
Amud1	Am1	3	1	3	1	2	1	2	1	3	3	1	2	3	1	1
Zafarraya	Zaf	3	2	3	1	3	2	2	1	2	3	1	1	3	2	1
Early *Homo sapiens* Upper Pleistocene																
Qafzeh 9	Q9	1	1	1	1	2	1	2	3	2	2	2	2	1	3	1
Skhul V	Sk5	2	1	1	2	1	1	1	2	2	3	1	2	2	1	1
Upper Paleolithic *Homo sapiens*																
Cro-Magnon 1	CrM1	2	1	1	2	2	1	2	2	1	2	3	2	1	2	2
Ohalo II	Oh2	3	1	1	2	2	1	2	2	1	2	2	2	2	2	2
Abri Pataud 1	AP1	3	1	1	1	1	1	1	2	1	1	2	1	2	3	1
Specimens from the Central Balkans																
Balanica 1	BH-1	3	1	1	2	2	1	1	2	1	2	2	3	2	2	1
RGF94/1	RGF94/1	3	1	1	2	1	2	1	2	2	1	2	2	2	2	2

[a]After Mounier et al. (2009): except for Balanica 1 and the RGF94/1 which were scored by the author. **(I)** Alveolar margin orientation toward inferior margin: (1) Steep (2) Slowly inclined (3) Parallel; **(J)** *Foramen mentale* number: (1) Single (2) Multiple; **(K)** *Foramen mentale* position toward the tooth row: (1) P3-P4, P4 (2) P4-M1 (3) M1; **(L)** Foramen mentale superoinferior position on the corpus: (1) Inferior (2) Midline (3) Superior; **(N)** *Sulcus intertoralis* definition of the hollowed area posterior to the foramen mentale surrounded by the marginal tori: (1) Flat surface (2) Weak: mainly defined by one torus (3) Well: defined by the two tori; **(O)** *Torus marginalis superius* relief: (1) Weak/absent (2) Swelling clearly visible; **(P)** *Torus marginalis inferius* relief: (1) Weak/absent (2) Swelling clearly visible; **(Q)** *Prominentia lateralis* relief: (1) Flat surface (2) Weak swelling (<7 mm) (3) Strong swelling (>7 mm); **(R)** *Prominentia lateralis* position along the tooth row: (1) M1 and M2 (2) M2–M3 (3) M3; **(S)** Retromolar space relationship between the anterior ramus rim and M3 in norma lateralis (1) Covered (2) Partially covered (3) Uncovered; **(T)** Retromolar area inclination (1) Horizontal (2) Inclined (3) Vertical; **(U)** *Extramolar sulcus*: Width of the gutter (1) Absence (2) Narrow gutter (3) Large gutter; **(OO)** Mylohyoid line orientation: (1) Parallel (2) Inclined (3) Diagonal; **(PP)** Mylohyoid line position at the M3 level (1) Low (2) Intermediate (3) High; **(UU)** *Submandibular fossa* depth beneath the alveolar region:(1) Shallow (2) Deep

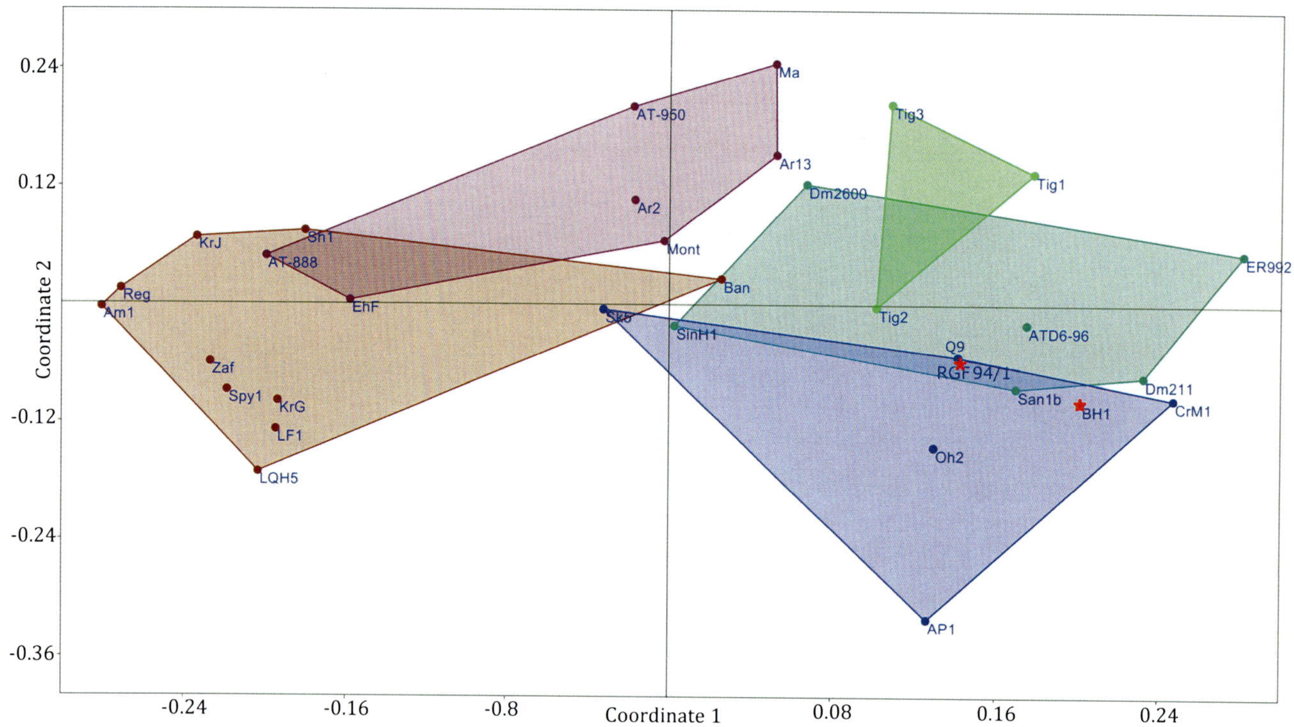

Fig. 2.6 Principal Coordinate analysis (PCO) of character states for all of the traits preserved in BH-1, RGF94/1, and comparative specimens. *Blue*: Pleistocene *H. sapiens*; *Light green*: MPAfH; *Dark green*: Early Pleistocene Eurasian Homo; *Tan*: Neanderthals; *Purple*: MPEH

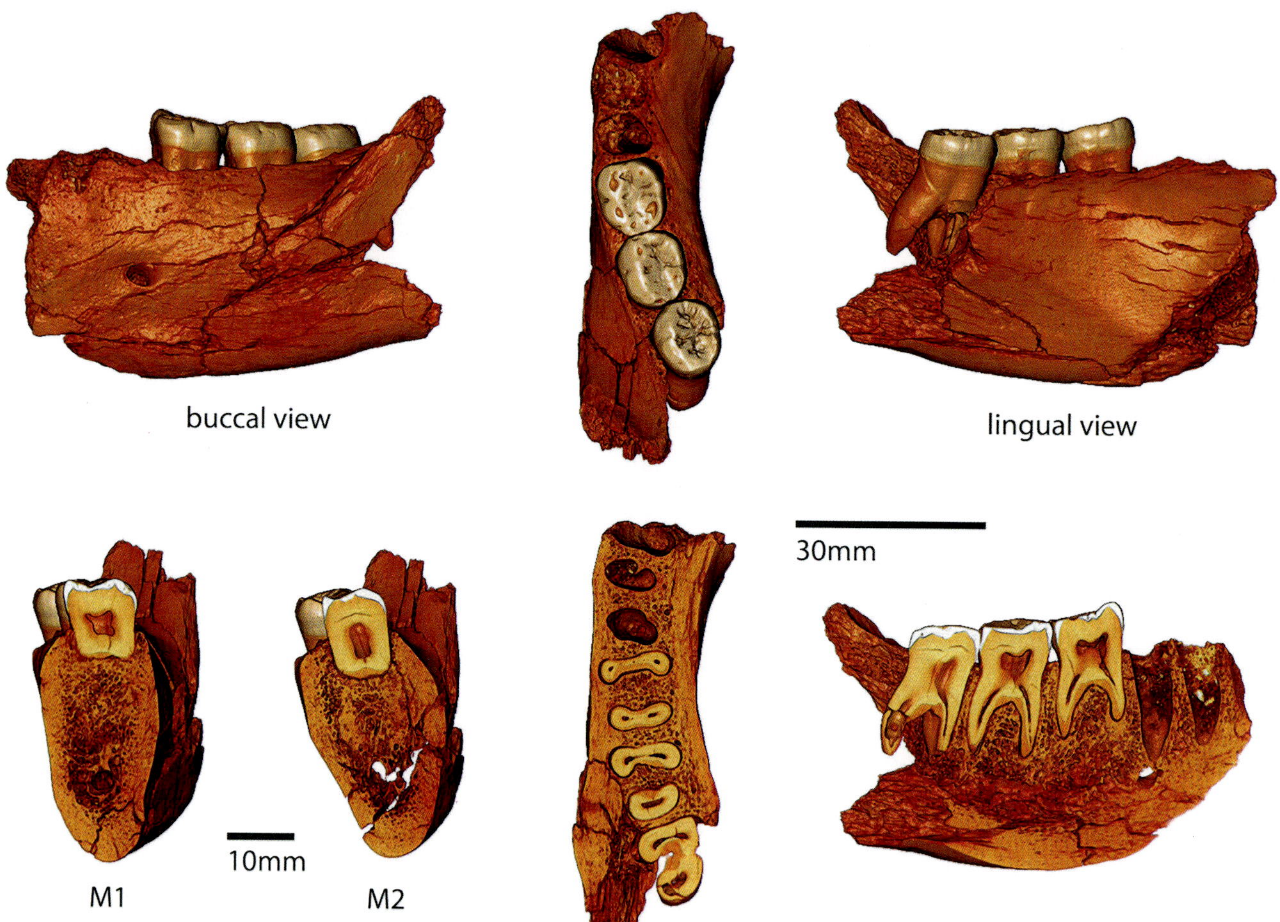

Fig. 2.7 The BH-1 specimen visualized as a volume rendering using the microtomographic images: external morphology of the mandible and internal structures visualized using sections. Reproduced with permission from Skinner et al. (2016): Fig. 1 top two rows

marginal torus represented by a slight change in the orientation of the lamina to the axes of the horizontal branch above and below the mental foramen, transitions smoothly into the lateral prominence. The latter is located at the level of M_1/M_2, equidistant from the alveolar and basal margins. The ascent of the oblique line begins just above the posterior marginal tubercle, 18.5 mm below the alveolar border at the level of M_1/M_2 vertically, and the mental foramen horizontally. The lateral prominence is more anterior than in Neanderthal and MEPH samples, where it is commonly located under the M_3 (Rosas 2001). The fragment of the exomandibular lamina of the vertical branch shows very slight relief at the masseteric fossa, with no pronounced rugosities. The reconstructed root of the vertical branch does not indicate the presence of a retro-molar space. The mental foramen is oval in shape, situated below the P_4 alveolus, almost equidistant from alveolar and basal margins. While the bone is robust, the relief of the internal surface is not marked. The alveolar border shows thickening on the lingual side from P_3 to M_2 (and possibly beyond), forming a mandibular torus just below the alveolar process, with the width decreasing mesially. The width of the subalveolar plane increases toward the middle portion of the mandible, forming a shelf-like area (oblique rather than subvertical *planum alveolare*) that extends from below the P_3 toward the canines and the symphysis. The subalveolar plane (sublingual fossa) is flat rather than concave. The submandibular fossa is moderately concave, and the expression of the mylohyoid line is moderate, presenting a change in orientation between the subalveolar plane and the submandibular fossa rather than a sharply delineated line. The level at which it begins cannot be ascertained, as the lower portion of the endomandibular face is destroyed in that area. However, it seems to extend toward the P_3. Its ascent is not steep and it is still present at the level of the mesial alveolar margin of the M_3 beyond which it can no longer be observed due to the breakage (Roksandic et al. 2011).

The mandible is thick in the bucco-lingual dimension. The width of the mandible varies from 19.1 mm at the canine alveolus, becoming more restricted toward the mental foramen (17.8 mm) and M_1 (17.5 mm) and increasing toward the M_2 (18.4 mm) and the M_3 (23. 8 mm). The occlusal view shows that the mandibular torus decreases in width from the M_3 to P_3, while the shelf-like thickening of the alveolar plane increases in width from the M_1 toward the symphysis. The extramolar sulcus is very wide, accentuated by a low and nonsteep oblique line. The substantial width of the extramolar sulcus is further accentuated by a pronounced curvature of the distal portion of dental arcade toward the sagittal plane (Roksandic et al. 2011).

Only the three left molars are present in the BH-1 specimen. Their occlusal outline is subrectangular and elongated mesiodistally. The molars have all five main cusps (protoconid, metaconid, hypoconid, entoconid, and hypoconulid), but the occlusal surface is not complex, and there are no extra fissures or crests. The hypoconulid is large and buccally aligned on all three teeth. There is an easily observed, wedge-shaped "cusp 7" (*tuberculum intermedium*) (Scott and Turner 1997) in all three molars. The mesial marginal ridge exhibits as a proper ridge in M_1 with no anterior fovea. This feature is continuous and depressed (very low) in M_2 and accompanied by an anterior fovea that is relatively poorly defined; it is represented by a wide depression rather than a deep triangular depression, as described by Scott and Turner (1997). The mesial marginal ridge shows a tubercle on the M_3 and a possible but unclear anterior fovea (Hillson 1996). The M_2 and M_3 present a distal trigonid crest that can be assessed by a short transverse fissure, slightly oblique to the buccolingual fissure. None of the teeth show a continuous midtrigonid crest—considered to be an indicator of Neanderthal affinity as it occurs in 96 % of Neanderthals (Bailey 2002). While the M_1 and M_2 have the same buccolingual width (10.9 mm) and mesiodistal length (11.5 mm), the M_3 is longer mesiodistally (12.1 mm) and narrower buccolingually (10.5 mm) (Roksandic et al. 2011).

A well-developed anterior fovea is common in Neanderthals (87 % according to Bailey 2002) and variable in modern humans (with an 83 % frequency in a sample of modern Croatians reported by Gauta et al. 2010). The presence of "cusp 7" is nondiagnostic, although it is much more common in *H. erectus* (40 %) than in Neanderthals (18.8 %), and variable in modern human populations (3–61 %) (Bailey 2002), with the highest frequencies recorded in Africa (Scott and Turner 1997). The expression of the distal trigonid crest is highly variable (Scott and Turner 1997) and according to Martinón-Torres and colleagues (2008, 2102, 2014) often underscored. It is, however, expressed in higher frequencies in the Dmanisi and Sangiran populations (Martinón-Torres et al. 2008; Martinez de Pinillos et al. 2014). The mental foramen is located under the M_1 in up to 80 % of Neanderthal specimens and 54 % of the Middle Pleistocene samples from Sima de los Huesos (Rosas 2001). This position is often interpreted to be a reflection of the development of marked midfacial prognathism (Quam and Smith 1998). The more anterior position of the mental foramen, its equidistant position in relation to the alveolar and basal margins, and the absence of a retromolar space—all plesiomorphic traits observable in *H. erectus*—reinforce the dental evidence and indicate that the mandible lacks autapomorphies of Neanderthals and their Middle Pleistocene precursors.

The results of the PCO (Table 2.5, Fig. 2.6) reveal that BH-1 plots close to Dmanisi 211, Sangiran 1B, and Upper Paleolithic modern humans. This should not be surprising, given its plesiomorphic character states and complete lack of Neanderthal morphology. Figure 2.6 shows a separation between Neanderthal / MPEH morphology on one hand and modern humans and *H. ergaster/erectus* on the other hand.

In this context, it is interesting to note the position of the Bañolas mandible, whose ambiguous morphology is well illustrated by its position on this graph close to the modern human /*H. erectus* overlap. The Bañolas mandible has been variably treated as a pre-Neanderthal, *H. heidelbergensis*, or Neanderthal, and recently as showing more modern traits (Alcázar de Velasco et al. 2011). The Atapuerca specimen ATD6-96 is placed on the *H. erectus*/modern human part of the graph, while the Sima de los Huesos specimens fall close to the Neanderthals and other MPEH specimens. The Tighenif mandibles overlap with Early Pleistocene Eurasian specimens close to the *H. erectus*/modern human convex hull, while MPEH show substantial overlap with Neanderthals. On the basis of preserved morphology, BH-1 differs significantly from the MPEH specimens generally grouped under *H. heidelbergensis* (Roksandic et al. 2011). It exhibits plesiomorphic features such as a prominent *planum alveolare*, thick mandibular corpus, wide exomolar sulcus, flat rather than concave sublingual fossa, and poorly defined relief of the submandibular fossa. There is a complete lack of derived Neanderthal features: the mental foramen is below the P_4 alveolus, equidistant from the alveolar and the basal margins, and there is no retromolar space. Dental traits are equally plesiomorphic: mesotaurodontic roots, two mesial and two distal diverticles on the M_1, "Y" fissure pattern, five main cusps, and a well-developed "cusp 7." Given the size of the mandibular body, the dentition is relatively small, and its size fits well with that of Middle Pleistocene specimens.

Table 2.5 Principal coordinates analysis matrix (using chord distance)

Axis	Eigenvalue	Percent
1	72.233	34.111
2	32.475	15.336
3	19.302	9.1151
4	16.591	7.8349
5	15.041	7.1027
6	13.098	6.1854
7	9.4069	4.4423
8	7.6781	3.6259
9	6.725	3.1758
10	5.1328	2.4239
11	3.4586	1.6333
12	2.6406	1.247
13	2.4238	1.1446[a]

[a]Other values explain less than 1 % of variation

A recent examination of the internal structure of the mandibular molars using microcomputed tomography (Fig. 2.8; Skinner et al. 2016) confirmed that the absence of Neanderthal traits in the mandibular morphology of BH-1 should not be regarded as a result of its partial preservation. Skinner et al. (2016) quantitatively assessed the enamel–dentine junction (EDJ) morphology using geometric morphometrics, molar enamel thickness, and the expression of discrete dental traits in comparison to *Homo erectus sensu lato*, *Homo neanderthalensis*, Pleistocene *Homo sapiens*, and recent *Homo sapiens*. The results of the study indicate a primitive dental morphology for BH-1 molars and confirm a lack of Neanderthal affinity.

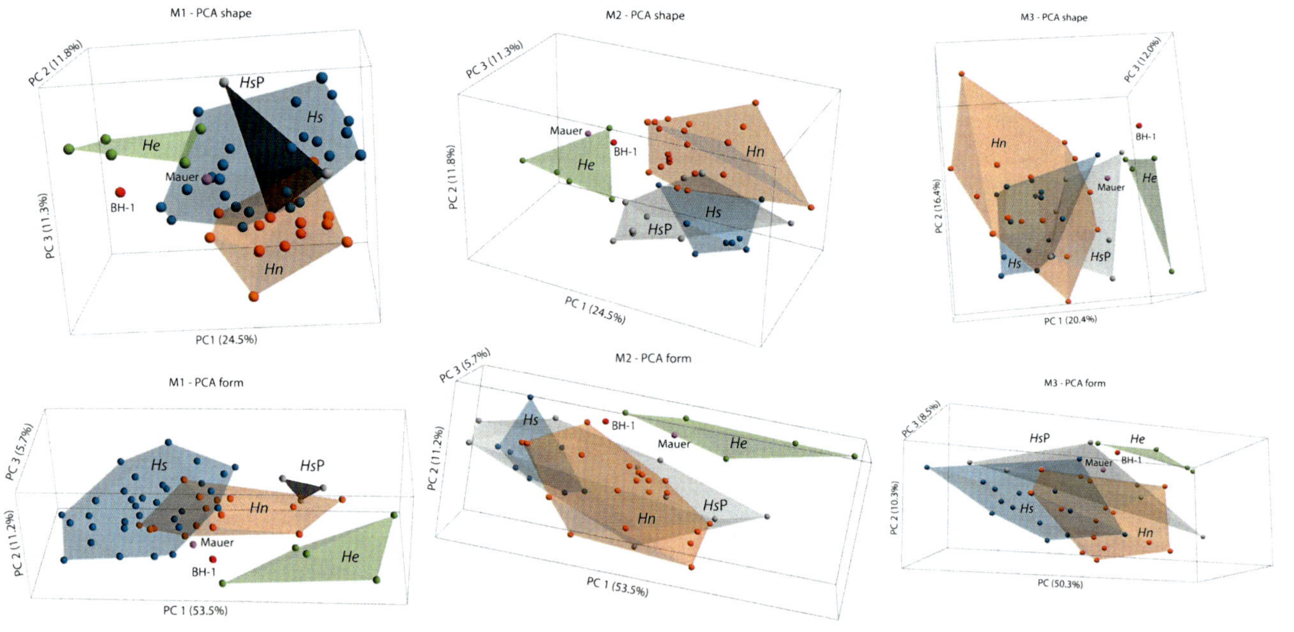

Fig. 2.8 Principal components analyses (PCA) of enamel–dentine junction (EDJ) morphology of the first, second and third molar sample in shape (*top*) and form (*bottom*) space. *Red sphere*—Balanica, He—*Homo erectus sensu lato*, Hn—*Homo neanderthalensis*, HsP—Pleistocene *Homo sapiens*, Hs—recent *Homo sapiens*. Adapted from Skinner et al. (2016)

Discussion

We have already suggested that the BH-1 mandible could play an important role in our understanding of the evolution of Middle Pleistocene hominins in Europe (Rink et al. 2013). The absence of Neanderthal traits in the BH-1 mandibular morphology could be interpreted as a result of individual variation, as mandibles are generally highly variable. Moreover, the specimen is fragmentary. However, the mandibular morphology, the dental and EDJ morphology, enamel volume, and root morphology all lack Neanderthal features, suggesting that this is not due to the fragmentary nature of the specimen. At the age range earlier than 397–525 ka, the primitive character of the mandible is not entirely unexpected. In the context of an accretion model of Neanderthal evolution (Dean et al. 1998; Hublin 2013), the traits would appear in a mosaic fashion allowing the individual within a population to exhibit Neanderthal morphology in one area of the skull while retaining plesiomorphies in other areas. A recent reevaluation of the Sima de los Huesos cranial remains (Arsuaga et al. 2014)—including seven previously unpublished skulls—confirmed the existence of Neanderthal-derived morphology in these specimens in both mandibular and cranial morphology, as well as on the EDJ. The Sima de los Huesos material is now dated to circa 430 ka by a combination of different methods (Arnold et al. 2014; Arsuaga et al. 2014). Acording to Arsuaga et al. (2014), changes in the facial skeleton preceded the changes in the braincase and conform to the expectations of the accretion model (Dean et al. 1998). The authors noted the difference between the Sima material and Ceprano and Arago which do not exhibit the same suite of Neanderthal features in the cranial skeleton and postulated several paleodemes within the EMPH.

The Balanica (BH-1) individual could be interpreted as belonging to one of these paleodemes, as we already suggested (Rink et al. 2013). Alternatively, given that the age of BH-1 hominin is only a minimum age, this individual could have belonged to an undifferentiated population, ancestral to both Neanderthal and non-Neanderthal lineages. The lack of Neanderthal traits in both the dentition and the mandible of the Mauer specimen dated to 609±40 ka (Wagner et al. 2010) is consistent with this interpretation, even as it plots closer to Western European specimens in Fig. 2.6.

If an ancestral Neanderthal population continued to develop in relative isolation over the cold periods in the west (as the evidence seems to indicate), the plesiomorphic character of the Visogliano mandible dated to 350–500 ka (Falguères et al. 2008) and the ambiguous morphology of the Ceprano calvaria dated to 353±4 ka (Nomade et al. 2011), might be explained by their geographic distance from such western populations.

When the Middle Pleistocene variability in Europe is examined in the context of geographically and chronologically defined p-demes (Howell 1996), and if we accept several successive migrations into Europe on the basis of lithic (Lycett 2009; Bar-Yosef and Belfer-Cohen 2013) and paloecological evidence (Carrión et al. 2011), one could postulate a core demographic area (Dennell et al. 2011) from which human populations were reseeded after glaciations. In this population model, which is based on demographic "sources" and "sinks," a small number of core "sources" in the south of the continent would have repopulated more northern parts during interglacials, with northern groups representing demographic "sinks." With western source populations as bearers of derived Neanderthal morphology as early as 430 ka in Sima material (Arnold et al. 2014), the observed attenuation of Neanderthal traits in the more easterly or later populations (Visogliano, Ceprano, maybe even Petralona) could be explained by admixture with a group from outside of the isolated glacial refugium, i.e., a population from Southwest Asia.

The Balkan Peninsula (and consecutively the Central Balkans)—which remained in contact with Southwest Asia during glacial times—could be perceived as belonging to this core demographic area. Alternating routes of migration within Eastern Mediterranean were open throughout the Pleistocene: the one, over the coastal areas of the Black Sea, was available during warmer phases; while the other, over the Bosporus, the Aegean and Ionian shelf was open during glaciations (see Tourloukis 2010, Fig. 6.18). Koufos et al. (2005, p. 181) consider the Eastern Mediterranean—comprised of the Balkan Peninsula, the Aegean Sea, Asia Minor, and the Middle East "as an important domain for mammal exchanges between Asia, Europe and Africa during the Neogene/Quaternary," where "migration pathways between the three continents crossed" (see also Koufos and Kostopoulos, 2016). While their analysis, similar to that of Spassov (2016), is concerned with early human migrations, there is strong evidence of contact between Eastern European and Asian micromammal fauna in the Middle Pleistocene and beyond (Van Kolfschoten and Markova 2005).

Considering these areas as a single geographic entity places emphasis on the current fossil record of Southeast Europe, which, while comparatively scant, becomes critical for understanding continent-wide processes. While isolation represented the major mechanism of evolutionary change in the west of the continent (Rightmire 1998), causing a bottleneck and fixation of derived traits, the Balkan Peninsula need not have experienced the effects of this isolation. Accordingly, the population that inhabited it and maintained contact with Southwest Asia throughout glaciations would be expected to retain a number of plesiomorphic (i.e., non-Neanderthal) traits, without precluding morphological changes associated with encephalization and tooth reduction observed in Middle Pleistocene populations on all three continents.

Conclusion

The unambiguous presence of Neanderthals in neighboring Croatia and Greece (see overviews in Janković et al. 2016; Harvati et al. 2009, 2011, 2013; Harvati 2016) leaves little doubt that Neanderthals were also present in the Central Balkans. However, we need to be alert to the possibility that the picture is more complex, and that future Balkan finds might redefine the current understanding of human evolution in Europe, still largely based on the evidence from the west of the continent. Considering the Balkans as part of the larger area open to communication throughout the Pleistocene is not only warranted, but necessary. It will, however, require a shift in our communal perception of the geography of the region. We might need to do away with the perception of the Aegean and the Black seas as barriers for movement of populations and view them as a geographic center of the Eastern Mediterranean Area (Roksandic 2015) which would encompass Southeast Europe and Southwest Asia, and which could have maintained population contact and gene exchange throughout human evolution. This hypothesis needs to be tested within a wider systematic examination and correlation of changes in micro and macro-fauna of the whole Eastern Mediterranean area throughout the Pleistocene.

With more vigorous surveys and small-scale excavations over the course of the last decade, we are slowly starting to understand the relationship of Central Balkan Paleolithic assemblages to the ones in the east and the west (Mihailović and Bogićević 2016). Whether the same chronological sequence of changes can be extended to human groups is up for discussion, and will not be possible to ascertain without further well-contextualized finds and a better understanding of the environment, faunal assemblages, and the chronology in the region. While the specimens—other than the mandible from Mala Balanica—cannot be ascertained as Pleistocene without direct dating, they demonstrate the potential of this area for discoveries from a range of time periods. Obtaining a more substantive body of evidence on human presence and the environment in the Central Balkans will be relevant for fleshing out the process of human evolution in the region and will contribute to our understanding of continent wide processes.

Acknowledgements I am deeply indebted to Dušan Mihailović who asked me to join his team in Balanica, to Predrag Radović for editing tables and graphs; to Dejana Nikitović, Dušan Mihailović and Josh Lindal who commented on the first draft; to Katerina Harvati, participants at the PaGE conference and anonymous reviewers for their constructive comments. Funding was provided by an NSERC grant (371077-2010).

References

Alcázar de Velasco, A., Arsuaga Ferreras, J. L., Martínez, I., & Bonmatí, A. (2011). Revisión de la mandíbula humana de Bañolas, Gerona, España. *Boletín de la Real Sociedad Española de Historia Natural. Sección geológica, 105*(1–4), 99–108.

Arnold, L. J., Demuro, M., Parés, J. M., Arsuaga, J. L., Aranburu, A., Bermúdez de Castro, J. M., et al. (2014). Luminescence dating and palaeomagnetic age constraint on hominins from Sima de los Huesos, Atapuerca, Spain. *Journal of Human Evolution, 67*, 85–107.

Arsuaga, J. L., Martínez, I., Arnold, L. J., Aranburu, A., Gracia-Téllez, A., Sharp, W. D., et al. (2014). Neandertal roots: Cranial and chronological evidence from Sima de los Huesos. *Science, 344*, 1358–1363.

Ascenzi, A., Mallegni, F., Manzi, G., Segre, A. G., & Naldini, E. S. (2000). A re-appraisal of Ceprano calvaria affinities with *Homo erectus*, after the new reconstruction. *Journal of Human Evolution, 39*(4), 443–450.

Asfaw, B., Gilbert, W. H., & Richards, G. D. (2008). *Homo erectus* cranial anatomy. In W. H. Gilbert & B. Asfaw (Eds.), *Homo erectus: Pleistocene evidence from the Middle Awash, Ethiopia* (pp. 265–328). Berkeley, CA: University of California Press.

Athreya, S. (2012). The frontal bone in the genus *Homo*: A survey of functional and phylogenetic sources of variation. *Journal of Anthropological Sciences, 90*, 59–80.

Aytek, I. A., & Harvati, K. (2016). The human fossil record from Turkey. In K. Harvati & M. Roksandic (Eds.), *Paleoanthropology of the Balkans and Anatolia: Human evolution and its context* (pp. 79–91). Dordrecht: Springer.

Bailey, S. E. (2002). *Neandertal dental morphology: Implications for modern human origins*. Unpublished PhD, Arizona State University, Phoenix, AZ.

Bailey, S. E., & Lynch, J. M. (2005). Diagnostic differences in mandibular P4 shape between Neandertals and anatomically modern humans. *American Journal of Physical Anthropology, 126*(3), 268–277.

Bar-Yosef, O., & Belfer-Cohen, A. (2013). Following Pleistocene road signs of human dispersals across Europe. *Quaternary International, 285*, 30–40.

Bar-Yosef, O., & Callander, J. (1999). The woman from Tabun: Garrod's doubts in historical perspective. *Journal of Human Evolution, 37*(6), 879–885.

Belmaker, M., Tchernov, E., Condemi, S., & Bar-Yosef, O. (2002). New evidence for hominid presence in the Lower Pleistocene of the Southern Levant. *Journal of Human Evolution, 43*(1), 43–56.

Bruzek, J. (2006). The assessment of sex. In E. Trinkaus & J. R. Svoboda (Eds.), *Early modern human evolution in Central Europe: The people of Dolní Vestonce and Pavlov*. New York: Oxford Univeristy Press.

Carrión, J. S., Rose, J., & Stringer, C. (2011). Early human evolution in the Western Palaearctic: Ecological scenarios. *Quaternary Science Reviews, 30*(11-12), 1281–1295.

Cartmill, M., & Smith, F. H. (2009). *The human lineage*. Hoboken, NJ: Wiley-Blackwell.

Cleghorn, B. M., Christie, W. H., & Dong, C. C. (2007). The root and root canal morphology of the human mandibular second premolar: A literature review. *Journal of Endodontics, 33*(9), 1031–1037.

Coppens, Y., Tseveendorj, D., Demeter, F., Turbat, T., & Giscard, P. H. (2008). Discovery of an archaic *Homo sapiens* skullcap in Northeast Mongolia. *Comptes Rendus Palevol, 7*, 51–60.

Cvijić, J. (1903). *Novi rezultati o glacijalnoj eposi Balkanskoga poluostrva*. Beograd: Srpska kraljevska akademija.

Cvijić, J. (1918). *La Péninsule Balkanique: Géographie Humaine*. Paris: Colin.

Dean, D., Hublin, J.-J., Holloway, R., & Ziegler, R. (1998). On the phylogenetic position of the pre-Neandertal specimen from Reilingen, Germany. *Journal of Human Evolution, 34*(5), 485–508.

Dennell, R. W., Martinón-Torres, M., & Bermúdez de Castro, J. M. (2011). Hominin variability, climatic instability and population demography in Middle Pleistocene Europe. *Quaternary Science Reviews, 30*(11–12), 1511–1524.

Doboş, A., & Iovita, R. (2016). The Lower Paleolithic of Romania revisited: New evidence from the site of Dealul Guran. In K. Harvati & M. Roksandic (Eds.), *Paleoanthropology of the Balkans and Anatolia: Human evolution and its context* (pp. 171–186). Dordrecht: Springer.

Eastwood, W. J. (2004). East Mediterranean vegetation and climate change. In H. I. Griffiths, B. Krystufek, & J. M. Reed (Eds.), *Balkan biodiversity: Pattern and process in the European hotspot* (pp. 25–48). Dordrecht: Kluwer Academic.

Falconer, H., & Cautley, P. T. S. (1847). *Fauna Antiqua Sivalensis: Proboscidea*. London: Smith Elder.

Falguères, C., Bahain, J.-J., Tozzi, C., Boschian, G., Dolo, J.-M., Mercier, N., et al. (2008). ESR/U-series chronology of the Lower Palaeolithic palaeoanthropological site of Visogliano, Trieste, Italy. *Quaternary Geochronology, 3*(4), 390–398.

Frayer, D. W., Jelínek, J., Oliva, M., & Wolpoff, M. H. (2006). Aurignacian male crania, jaws and teeth from the Mladeč Caves, Moravia, Czech Republic. In M. Teschler-Nicola (Ed.), *Early modern humans at the Moravian Gate: The Mladeč Caves and their remains* (pp. 185–272). Vienna: Springer.

Gabunia, L., & Vekua, A. (1995). A Plio-Pleistocene hominid from Dmanisi, East Georgia, Caucasus. *Nature, 373*, 509–512.

Gauta, I., Vazdar, M. A., & Vodanović, M. (2010). Human molar crown traits in Croatian medieval and contemporary populations. *Acta Stomatologica Croatica, 44*, 3e16.

Gavela, B. (1951). Iskopavanja na paleolitskom lokalitetu Pećina pod Jerininim brdom u Gracu kod Kragujevca. *Glasnik SANU, III*(2).

Gorjanović-Kramberrger, D. (1906). *Der Diluviale Menschvon Krapina in Kroatien*. Wiesbaden: C. W. Kreidel's Verlag.

Griffiths, H. I., Krystufek, B., & Reed, J. M. (Eds.). (2004). *Balkan biodiversity: Pattern and process in the European hotspot*. Dordrecht: Kluwer.

Grimaud-Herve, D., Widianto, H., Detroit, F., & Semah, F. (2012). Comparative morphological and morphometric description of the hominin calvaria from Bukuran (Sangiran, Central Java, Indonesia). *Journal of Human Evolution, 63*(5), 637–652.

Harvati, K. (2016). Paleoanthropology in Greece: Recent finds and implications. In K. Havarti & M. Roksandic (Eds.), *Paleoanthropology of the Balkans and Anatolia: Human evolution and its context* (pp. 3–14). Dordrecht: Springer.

Harvati, K., Darlas, A., Bailey, S. E., Rein, T. R., El Zaatari, S., Fiorenza, L., et al. (2013). New Neanderthal remains from Mani peninsula, Southern Greece: The Kalamakia Middle Paleolithic cave site. *Journal of Human Evolution, 64*(6), 486–499.

Harvati, K., Stringer, C., & Karkanas, P. (2011). Multivariate analysis and classification of the Apidima 2 cranium from Mani, Southern Greece. *Journal of Human Evolution, 60*(2), 246–250.

Harvati, K., Hublin, J.-J., & Gunz, P. (2010). Evolution of middle-late Pleistocene human cranio-facial form: A 3-D approach. *Journal of Human Evolution, 59*(5), 445–464.

Harvati, K., Panagopoulou, E., & Runnels, C. (2009). The paleoanthropology of Greece. *Evolutionary Anthropology: Issues, News, and Reviews, 18*(4), 131–143.

Hewitt, G. M. (2011). Mediterranean peninsulas: The evolution of hotspots. In F. E. Zachos & J. C. Habel (Eds.), *Biodiversity hotspots distribution and protection of conservation priority areas* (pp. 123–147). Berlin: Springer.

Hillson, S. (1996). *Dental anthropology*. Cambridge: Cambridge University Press.

Howell, F. (1996). Thoughts on the study and interpretation of the human fossil record. In W. Meikle, F. Howell, & N. Jablonski (Eds.), *Contemporary issues in human evolution* (Vol. Mem 21, pp. 1–46). San Francisco: California Academy of Sciences.

Howell, F. (1999). Paleo-demes, species clades, and extinctions in the Pleistocene hominin record. *Journal of Anthropological Research, 55*, 191–243.

Howells, W. W. (1989). *Skull shapes and the map. Craniometric analyses in the dispersion of modern homo* (Vol. 79). Cambridge, MA: Peabody Museum.

Hublin, J.-J. (2013). The Middle Pleistocene record. On the origin of Neandertals, modern humans and others. In D. Begun (Ed.), *A companion to paleoanthropology* (pp. 517–537). Hoboken, NJ: Wiley-Blackwell.

Ivanova, S. (2016). A route through the Balkans and implications for the earliest settlement of Europe. In K. Harvati & M. Roksandic (Eds.), *Paleoanthropology of the Balkans and Anatolia: Human evolution and its context* (pp. 187–211). Dordrecht: Springer.

Janković, I., Ahern, J. C. M., Karavanić, I., & Smith, F. H. (2016). The importance of Croatian Pleistocene hominin finds in the study of human evolution. In K. Havarti & M. Roksandic (Eds.), *Paleoanthropology of the Balkans and Anatolia: Human evolution and its context* (pp. 35–50). Dordrecht: Springer.

Jovanović, D. P. (1892). Prilozi za Paleontropologiju srpskih zemalja. *Starinar, IX*, 24–33.

Kennedy, K. A. R. (2000). *God-apes and fossil men: Paleoanthropology of South Asia*. Ann Arbor: University of Michigen Press.

Koufos, G. D. & D. S. Kostopoulos (2016). The Plio-Pleistocene Large Mammal Record of Greece: Implications for Early Human Dispersals into Europe. In K. Harvati & M. Roksandic (Eds.), *Paleoanthropology of the Balkans and Anatolia: Human evolution and its context*. Dordrecht: Springer.

Koufos, G. D., Kostopoulos, D. S., & Vlachou, T. D. (2005). Neogene/Quaternary mammalian migrations in Eastern Mediterranean. *Belgian Journal of Zoology, 135*(2), 181–190.

Lordkipanidze, D., Ponce de León, M. S., Margvelashvili, A., Rak, Y., Rightmire, G. P., Vekua, A., et al. (2013). A complete skull from Dmanisi, Georgia, and the evolutionary biology of early *Homo*. *Science, 342*(6156), 326–331.

Lordkipanidze, D., Vekua, A., Ferring, R., Rightmire, R. G., Zollikofer, C. P. E., Ponce de León, M. S., et al. (2006). A Fourth hominin skull from Dmanisi, Georgia. *The Anatomical Record, 288*, 1146–1157.

Lubsen, K. D., & Corruccini, R. S. (2011). Morphometric analysis of the Herto cranium (BOU-VP-16/1): Where does it fit? *Journal of Contemporary Anthropology, 2*, 1–16.

Lycett, S. J. (2009). Understanding ancient hominin dispersals using artefactual data: A phylogeographic analysis of Acheulean handaxes. *PLoS One, 4*(10), e7404.

Manzi, G. (2012). On the trail of the genus *Homo* between archaic and derived morphologies. *Journal of Anthropological Science, 90*, 99–116.

Manzi, G., Magri, D., & Palombo, M. R. (2011). Early-Middle Pleistocene environmental changes and human evolution in the Italian peninsula. *Quaternary Science Reviews, 30*, 1420–1438.

Marković, S. B., Bokhorst, M. P., Vandenberghe, J., McCoy, W. D., Oches, E. A., Hambach, U., et al. (2008). Late Pleistocene loess-palaeosol sequences in the Vojvodina region, north Serbia. *Journal of Quaternary Science, 23*(1), 73–84.

Martin, R., & Saller, K. (1957). *Lehrbuch der Anthropologie* (Vol. 1). Stuttgart: Gustav Fischer.

Martinez de Pinillos, M., Martinón-Torres, M., Skinner, M. M., Arsuaga, J. L., Gracia-Tellez, A., Martínez, I., et al. (2014). Trigonid crest expression in Atapuerca-Sima de los Huesos lower molars: Internal and external morphological expression and evolutionary inferences. *Comptes Rendus Palevol, 13*(3), 205–221.

Martinón-Torres, M., Bermúdez de Castro, J. M., Gomez-Robles, A., Prado-Simon, L., & Arsuaga, J. L. (2012). Morphological description and comparison of the dental remains from Atapuerca-Sima de los Huesos site (Spain). *Journal of Human Evolution, 62*(1), 7–58.

Martinón-Torres, M., de Castro, J. M. B., Gomez-Robles, A., Margvelashvili, A., Prado, L., Lordkipanidze, D., et al. (2008).

Dental remains from Dmanisi (Republic of Georgia): Morphological analysis and comparative study. *Journal of Human Evolution, 55*(2), 249–273.

Martinón-Torres, M., Martine de Pinillos, M., Skinner, M. M., Martin-Frances, L., Gracia-Tellez, A., Martínez, I., et al. (2014). Talonid crest expression at the enamel-dentine junction of hominin lower permanent and deciduous molars. *Comptes Rendus Palevol, 13*(3), 223–234.

Mihailović, D. (2008). New data about the Middle Palaeolithic of Serbia. In A. Darlas & D. Mihailović (Eds.), *The Palaeolithic of the Balkans* (British Archaeological Reports, International Series 1819) (pp. 93–100). Oxford: Archaeopress.

Mihailović, D., & Bogićević, K. (2016). Technological changes and population movements in the late Lower and early Middle Palaeolithic of the Central Balkans. In K. Havarti & M. Roksandic (Eds.), *Paleoanthropology of the Balkans and Anatolia: Human evolution and its context* (pp. 139–151). Dordrecht: Springer.

Mirazón Lahr, M., & Haydenblit, R. (1995). The human remains from the site of Et-Tin, Israel. *Paléorient, 21*, 97–111.

Mol, D., de Vos, J., & van der Plicht, J. (2007). The presence and extinction of *Elephas antiquus* Falconer and Cautley, 1847, in Europe. *Quaternary International, 169–170*, 149–153.

Mounier, A., & Caparros, M. (2015). The phylogenetic status of *Homo heidelbergensis*—A cladistic study of Middle Pleistocene hominins. *Bulletins et Mémoires de la Société d anthropologie de Paris*. Published on-line August 2015. doi:10.1007/s13219-015-0127-4.

Mounier, A., Marchal, F., & Condemi, S. (2009). Is *Homo heidelbergensis* a distinct species? New insight on the Mauer mandible. *Journal of Human Evolution, 56*(3), 219–246.

Murrill, R. I. (1981). *Petralona man: A descriptive and comparative study, with new important information on Rhodesian man*. Springfield, IL: Charles C. Thomas.

Muttoni, G., Scardia, G., & Kent, D. V. (2013). A critique of evidence for human occupation of Europe older than the Jaramillo subchron (~1 Ma): Comment on Toro-Moyano et al. (2013). *Journal of Human Evolution, 65*, 746–749.

Nomade, S., Muttoni, G., Guillou, H., Robin, E., & Scardia, G. (2011). First 40Ar/39Ar age of the Ceprano man (central Italy). *Quaternary Geochronology, 6*(5), 453–457.

Quam, R., & Smith, F. H. (1998). A reassessment of the Tabun C2 mandible. In T. Akazawa, K. Aoki, & O. Bar-Yosef (Eds.), *Neandertals and modern humans in Western Asia* (pp. 405–421). New York: Kluwer Academic.

Radović, P., Lindal, J., & Roksandic, M. (2014). A re-examination of the human fossil specimen from Bački Petrovac (Serbia). *HOMO Journal of Comparative Human Biology, 65*(4), 281–295.

Rightmire, G. P. (1990). *The evolution of Homo erectus. Comparative anatomical studies of an extinct human species*. Cambridge: Cambridge University Press.

Rightmire, G. P. (1996). The human cranium from Bodo, Ethiopia: Evidence for speciation in the Middle Pleistocene. *Journal of Human Evolution, 31*(1), 21–39.

Rightmire, G. P. (1998). Human evolution in the Middle Pleistocene: The role of *Homo heidelbergensis*. *Evolutionary Anthropology, 6*(6), 218–227.

Rightmire, G. P. (2008). *Homo* in the Middle Pleistocene: Hypodigms, variation, and species recognition. *Evolutionary Anthropology, 17*, 8–21.

Rink, W. J., Mercier, N., Mihailović, D., Morley, M. W., Thompson, J. W., & Roksandic, M. (2013). New radiometric ages for the BH-1 hominin from Balanica (Serbia): Implications for understanding the role of the Balkans in Middle Pleistocene human evolution. *PLoS One, 8*(2), e54608.

Roksandic, M. (2015). The role of Balkans in the peopling of Europe: New evidence from Serbia. In A. R. Sankhyan, (Ed.), *Proceedings of the BH13 Panel: "Exploring Human Origins" 17th IUAES Congress Manchester 2013* (pp. 63–68). Oxford: Archaeopress.

Roksandic, M., & Dimitrijević, V. (2001). A human mandible from the loess in the vicinity of Belgrade (Yugoslavia). *Human Evolution, 16*(1), 27–36.

Roksandic, M., Mihailović, D., Mercier, N., Dimitrijević, V., Morley, M. W., Rakočević, Z., et al. (2011). A human mandible (BH-1) from the Pleistocene deposits of Mala Balanica cave (Sicevo Gorge, Nis, Serbia). *Journal of Human Evolution, 61*(2), 186–196.

Roksandic, M., Radović, P., Alex, B. A., Pavić, S., Paunović, M., & Marković, Z. (2014). Looking for hominins in Museum drawers—possible Upper Pleistocene specimens from Serbia: morphological descriptions and radiocarbon dating. In D. Mihailovic (Ed.), *Palaeolithic and mesolithic research in the Central Balkans* (pp. 7–20). Belgrade: Serbian Archaeological Society.

Rosas, A. (2001). Occurrence of Neanderthal features in mandibles from the Atapuerca-SH site. *American Journal of Physical Anthropology, 114*(1), 74–91.

Schwartz, J. H., & Tattersall, I. (2000). The human chin revisited: What is it and who has it? *Journal of Human Evolution, 38*(3), 367–409.

Scott, G. R., & Turner, C. G. I. (1997). *The anthropology of modern human teeth: Dental morphology and its variation in recent human populations*. Cambridge: Cambridge University Press.

Sherwood, S. C., Windingstad, J. D., Barker, A. W., O'Shea, J. M., & Sherwood, W. C. (2013). Evidence for Holocene aeolian activity at the close of the Middle Bronze Age in the Eastern Carpathian Basin: Geoarchaeological results from the Mureş River Valley, Romania. *Geoarchaeology, 28*(2), 131–146.

Simmons, T., Falsetti, A. B., & Smith, F. H. (1991). Frontal bone morphometrics of southwest Asian Pleistocene hominids. *Journal of Human Evolution, 20*(3), 249–269.

Skinner, M. M., De Vries, D., Gunz, P., Hublin, J.-J., Kupczik, K., Klassen, R. P., & Roksandic, M. (2016). Morphological affinity of the Balanica mandible (BH-1): Molar enamel-dentine junction morphology, discrete dental traits, and enamel thickness. *Journal of Human Evolution, 93*, 63–81.

Sládek, V., Trinkaus, E., Sefcáková, A., Halouzka, R. (2002). Morphological affinities of the Sal'a 1 frontal bone. *Journal of Human Evolution, 43*(6):787–815.

Spassov, N. (2016). Southeastern Europe as a route for the earliest dispersal of *Homo* towards Europe: Ecological conditions and the timing of the first human occupation of Europe. In K. Harvati & M. Roksandic (Eds.), *Paleoanthropology of the Balkans and Anatolia: Human evolution and its context* (pp. 281–290). Dordrecht: Springer.

Stansfield, E., & Gunz, P. (2011). Skhodnya, Khvalynsk, Satanay, and Podkumokcalvaria: Possible Upper Palaeolithic hominins from European Russia. *Journal of Human Evolution, 60*(2), 129–144.

Strait, D. S., Orr, C. M., Hodgkins, J., Spassov, N., Gurova, M., Miller, C., et al. (2016). The human fossil record of Bulgaria, and the formulation of biogeographic hypotheses. In K. Harvati & M. Roksandic (Eds.), *Paleoanthropology of the Balkans and Anatolia: Human evolution and its context*. Dordrecht: Springer.

Stringer, C. (2012). The status of *Homo heidelbergensis* (Schoetensack 1908). *Evolutionary Anthropology, 21*(3), 101–107.

Toro-Moyano, I., Martínez-Navarro, B., Agustí, J., Souday, C., Bermúdez de Castro, J. M., Martinón-Torres, M., et al. (2013). The oldest human fossil in Europe dated to ca. 1.4Ma at Orce (Spain). *Journal of Human Evolution, 65*(1), 1–9.

Tourloukis, V. (2010). *The Early and Middle Pleistocene archaeological record of Greece: Current status and future prospects*. Leiden: Leiden University Press.

Trinkaus, E. (1983). *The Shanidar Neanderthals*. New York: Academic.

Tzedakis, P. C. (2004). The Balkan as prime glacial refugial territory of European temperate trees. In H. I. Griffiths, B. Krystufek, & J. M. Reed (Eds.), *Balkan biodiversity: Pattern and process in the European hotspot* (pp. 49–69). Dordrecht: Kluwer Academic.

Uchida, Y., Goto, M., Danjo, A., Yamashita, Y., & Kuraoka, A. (2012). Anatomic measurement of the depth and location of the sublingual fossa. *International Journal of Oral and Maxillofacial Surgery, 12*, 1571–1576.

Van Kolfschoten, T., & Markova, A. K. (2005). Response of the European mammalian fauna to the mid Pleistocene transition. In M. J. Head & P. L. Gibbard (Eds.), *Early-Middle Pleistocene transitions: The land-ocean evidence* (pp. 251–262). London: Geological Society, Special Publications, 247.

Vandermeersch, B. (1981). *L'Homme fossile de Qafzeh*. Paris: C.N.R.S.

Wagner, G. A., Krbetschek, M., Degering, D., Bahain, J. J., Shao, Q., Falguères, C., et al. (2010). Radiometric dating of the type-site for *Homo heidelbergensis* at Mauer, Germany. *Proceeding of the National Academy of Sciences USA, 107*(46), 19726–19730.

Weidenreich, F. (1943). *The skull of Sinanthropus pekinensis: A comparative study on a primitive hominid skull* (Vol. 10). Pehpei, Chungking: Geological Survey of China.

Wu, X., & Athreya, S. (2013). A description of the geological context, discrete traits, and linear morphometrics of the Middle Pleistocene hominin from Dali, Shaanxi Province, China. *American Journal of Physical Anthropology, 150*, 141–157.

Živanović, S. (1966). Prethodno saopštenje o nalazu dva ostatka lobanja praistorijskog čoveka iz Vojvodine. *Starinar, XV-XVI*, 187–190.

Živanović, S. (1975). Lobanja iz Bačkog Petrovca. *Starinar, XXIV-XXV*, 155–156.

Žujović, J. (1893). *Geologija Srbije* (Vol. 1). Beograd.

Chapter 3
The Importance of Croatian Pleistocene Hominin Finds in the Study of Human Evolution

Ivor Janković, James C.M. Ahern, Ivor Karavanić, and Fred H. Smith

Abstract In this chapter, we discuss Croatian sites that have yielded human skeletal remains from the Pleistocene. These include the well-known Neandertal localities Hušnjakovo (at Krapina) and Vindija cave, as well as the Late Upper Paleolithic hominin fossil site Šandalja II cave in Istria. The Krapina site played an important role in the historical development of paleoanthropology and is still the Neandertal site with the largest known minimum number of skeletal individuals to date. Finds from Vindija cave belong to one of the latest Neandertal groups in Europe and provide data for the study of both their behavioral, as well as biological characteristics (including genomics studies). The Šandalja II cave in Istria is the only site in Croatia with direct association of human skeletal finds and the late Paleolithic, an Epigravettian industry, providing us with data on the anatomy and behavior of the Late Paleolithic inhabitants of this region.

Keywords Neandertals • Paleoanthropology • Paleolithic • Mousterian

I. Janković (✉)
Institute for Anthropological Research, Gajeva 32, Zagreb 10000, Croatia
e-mail: ivor.jankovic@inantro.hr

J.C.M. Ahern
Department of Anthropology, University of Wyoming, Laramie, WY 82071, USA
e-mail: jahern@uwyo.edu

I. Karavanić
Department of Archaeology, University of Zagreb, I. Lučića 3, Zagreb 10000, Croatia
e-mail: ikaravan@ffzg.hr

F.H. Smith
Department of Sociology and Anthropology, Illinois State University, Normal, IL 61790-4660, USA
e-mail: fsmith@ilstu.edu

Introduction

Although Croatia has a rather small number of sites preserving Pleistocene human skeletal material, it has yielded some very crucial ones. The two Croatian sites with Neandertal skeletal remains (the Hušnjakovo site in Krapina and Vindija cave) are essential in any attempts to understand Neandertal variation and behavior, as well as to shed light on the complex patterns of their demise. Hušnjakovo (Krapina) has also played an important role in the historical development of paleoanthropology and has influenced our views about the role of Neandertals in human evolution. It is still the largest known Neandertal site in terms of the minimum number of skeletal individuals found; and, as the stratigraphic sequence is relatively short, it provides a rare glimpse into idiosyncratic, ontogenetic, and sex-related variation of a relatively early Neandertal "population." The Vindija Neandertal assemblage, on the other hand, has provided a sample from the final phase of the European Neandertal reign. Finally, in the Late Upper Paleolithic layers of the Šandalja II cave in Istria, human skeletal remains have been found in direct association with the Epigravettian. In this chapter, we will summarize the most important biological, and to some degree, behavioral aspects of these finds, including—where appropriate—a brief historical overview.

Only four sites in Croatia have been identified as yielding Lower Paleolithic material (Malez 1979). These include the cave Šandalja I near Pula and the three open air sites of Donje Pazarište, Punikve, and Golubovec. At the site of Šandalja I, a single chopper and an additional pebble that might have been used in the production of this tool have been found within breccia containing Villafranchian fauna (Malez 1974, 1975). Malez (1975, 1980) identified as human an incisor found in the same breccia; the specimen was later attributed to fauna by Wolpoff (1999). The chopper morphologically resembles those from the site of Le Vallonnet in France (see Karavanić and Janković 2007). However, until a

Katerina Harvati and Mirjana Roksandic (eds.), *Paleoanthropology of the Balkans and Anatolia*, Vertebrate Paleobiology and Paleoanthropology, DOI 10.1007/978-94-024-0874-4_3

revision of the faunal sequence is performed, it is impossible to put a more precise date on the find itself than attributing it to the Lower Paleolithic.

The other stone tools that have been attributed to the Lower Paleolithic on the basis of their typological properties are open air surface finds (Vuković 1962/1963; Malez 1979), making it impossible to have any insight into their original context and chronology. Several artifacts, including two handaxes, have been found at Punikve, near Ivanec, in northwestern Croatia. At Donje Pazarište in the Lika region, a single handaxe has been found alongside several ecofacts resembling tools. The attribution of the finds from Golubovec to the Lower Paleolithic is dubious at best (Karavanić and Janković 2007). The fact that the Lower Paleolithic finds from Croatia are rare probably does not reflect a lack of human habitation during the Early/Middle Pleistocene, but more likely ecological–geological–climatic fluctuation and changes in the sea level during this time, preservation of sediments from this period, and the relative lack of research in the past. Croatia is very rich in archaeological heritage, especially in the coastal region, where the abundance of monuments and sites from antiquity and the Middle Ages has resulted in most research focusing on these younger periods, similarly to other areas discussed in this book (see Harvati 2016; Roksandic 2016; Dinçer 2016). This has changed over the years, and it is to be expected that new research will result in a more detailed knowledge of the earliest phases of the Paleolithic habitation of Croatia in the future.

The Middle Paleolithic of Croatia is much better known, as there are several open air finds, as well as numerous cave sites from this period (see Fig. 3.1). These include the well-known finds from Krapina and Vindija (the only two sites that have yielded Neandertal skeletal remains), Velika pećina and Veternica. The first three sites are located in the Zagorje region (Hrvatsko Zagorje) and the fourth lies just north of Zagreb. Other important Middle Paleolithic localities are Mujina pećina; the open-air sites located between Ljubački bay and Posedarje in Dalmatia (where dozens of find sites have been identified; see, Batović 1965; Chapman et al. 1996; Karavanić and Janković 2007), and Veli rat on Dugi otok; and the Kličevica Cave near Benkovac that is currently under systematic excavation (Batović 1988; Karavanić and Čondić 2006; Karavanić and Janković 2007; Karavanić et al. 2016). Mujina pećina near Kaštela is the only Dalmatian Middle Paleolithic site with *in situ* finds that has been systematically excavated and radiometrically dated in recent years (see Karavanić 2000; Karavanić and Janković 2007; Karavanić et al. 2008).

Although these sites have yielded important data on human habitation and lifeways during the Middle Paleolithic, with the noted exception of Krapina and Vindija, no human bones have been found. A human frontal bone from Velika pećina in the Zagorje region received considerable interest in the past, as it was attributed to a (late) Neandertal (Malez 1963, 1965, 1980; Mann and Trinkaus 1974) or an early anatomically modern human with some archaic features (Smith 1976b, 1982). However, later AMS direct dating of the specimen to 5045±40 ^{14}C kBP (Smith et al. 1999) removed it (alongside numerous other specimens in recent years, cf. Street et al. 2006; Ahern et al. 2013) from the debate on modern human origins in Europe.

There are several sites in Croatia that yielded archaeological material attributable to the Upper Paleolithic (e.g., Vindija, Velika pećina, Romualdova pećina, cave Bukovac, Šandalja II, Zala cave etc.). However, human skeletal remains from this period are very scarce. According to Malez (1980) fragmentary human remains have been found in the Upper Paleolithic layers at several other sites (e.g., Romualdo cave and Vergotin cave in Istria, the Upper Cerovac cave in Lika). However, these lack secure proveniences and direct dates.

We will therefore turn our attention to the three sites that have yielded securely dated human skeletal remains from the Paleolithic: Krapina (Hušnjakovo), Vindija, and Šandalja II.

Krapina (Hušnjakov brijeg)

Neandertal discoveries at Krapina played an important role in the history of paleoanthropology. Krapina was recognized as an archaeological and paleontological site by the Croatian paleontologist Dragutin Gorjanović-Kramberger in 1899 and excavated under his direction between 1899 and 1905. The site is located on the Hušnjak hill (Hušnjakovo, or Hušnjakov brijeg) overlooking the Krapinica river, in the present-day town of Krapina in NW Croatia (Fig. 3.1). The site itself was used for sand quarrying by locals for many years prior to the discovery of Neandertal bones. In fact the reason for Gorjanović's initial visit was the discovery of remains of extinct mammals (rhinoceros and buffalo) that were sent to him by the local schoolteacher, Josip Rehorić (for a detailed insight on Gorjanović and his discovery of the Krapina site see Barić 1978; Radovčić 1988). Gorjanović never expected that his visit to the site on the 23rd of August 1899 would result in the discovery of one of Europe's most important human fossil localities. His excitement is best reflected in his own words: "*In the quaint little market town of Krapina, on the slope of Hušnjak Hill above the Kneip bathing pool, and 25 m above the Krapinica stream is an open cave that used to be filled with sand. The local inhabitants used this sand for building purposes, and many bones that used to lie in that sand are lost forever due to ignorance. I received the first animal remains, those of a rhinoceros and a buffalo, in 1895 from Mr. Rehorić, the schoolteacher, who had collected the objects with Mr. K. Semenić and sent them, without a clear description of where they were found, to the Geology*

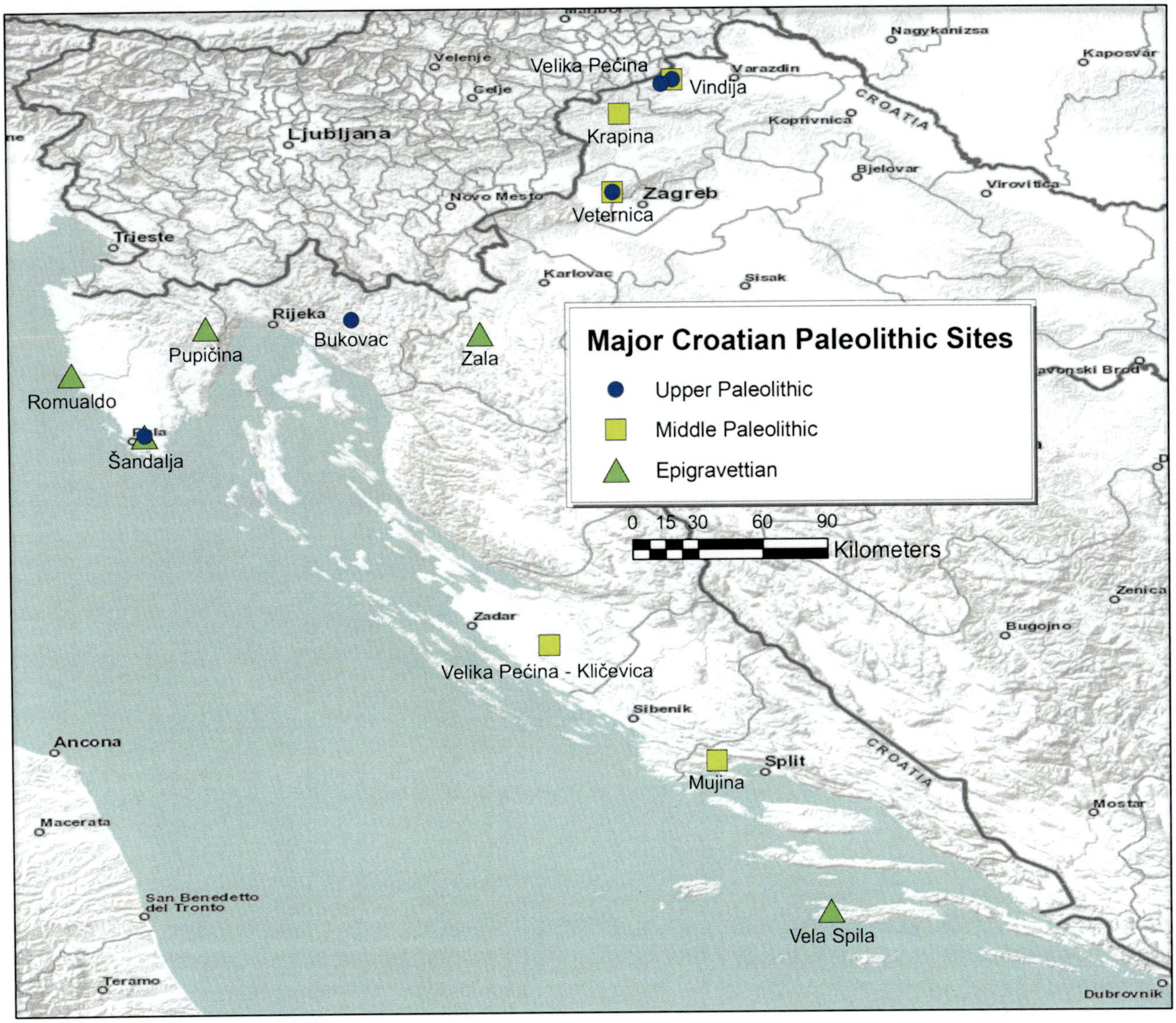

Fig. 3.1 Map of Croatia with most important Paleolithic sites

Museum. In 1899 I visited Krapina to acquaint myself with the spot where remains of animals had been unearthed. At quite some distance from the open cave it was possible to discern several dark bands running more or less parallel in the light yellow exposed sandstone cliff. Upon reaching the cliff I was struck, noting the composition of those bands as containing ash, chattered sand and charcoal, that I was looking at a whole sequence of hearths, repeated time and time again in that 8–9 m tall sandstone cliff. At once it was utterly clear that beings had resided therein who had lit fires, but nearby such a hearth site I also came across a fragment of flint-like stone that had been shaped for use. And moreover I observed bits of animal bones, and extracted—this was then the first time—a single human molar. The honorable reader can readily imagine how this discovery thrilled me beyond belief! Why, I was standing on the treshold of a primeval human settlement unlike anything previously discovered in our land" (Gorjanović-Kramberger 1918, p. 164, cited from translation in Radovčić 1988, p. 21).

Interestingly, although this does not diminish the importance of Gorjanović's discovery and recognition of the Hušnjakovo Neandertal site in any way, there is an even earlier mention of human bones from Krapina. During his travels through the region in 1792 and in 1809, a professor of chemistry and botany at Pécs University in Hungary, P. Kitaibel, discussed what he called "*petrified human bones of remarkable size*" in a letter to local pharmacist Ivan Gaj dated March 25th, 1811 (see Horvat and Ravlić 1956; Barić 1978). In the same letter he acknowledged that the bones were collected from a quarry in Krapina by count Julije Keglević. Although highly likely, it is unclear whether the bones come from the same site of later Neandertal discoveries

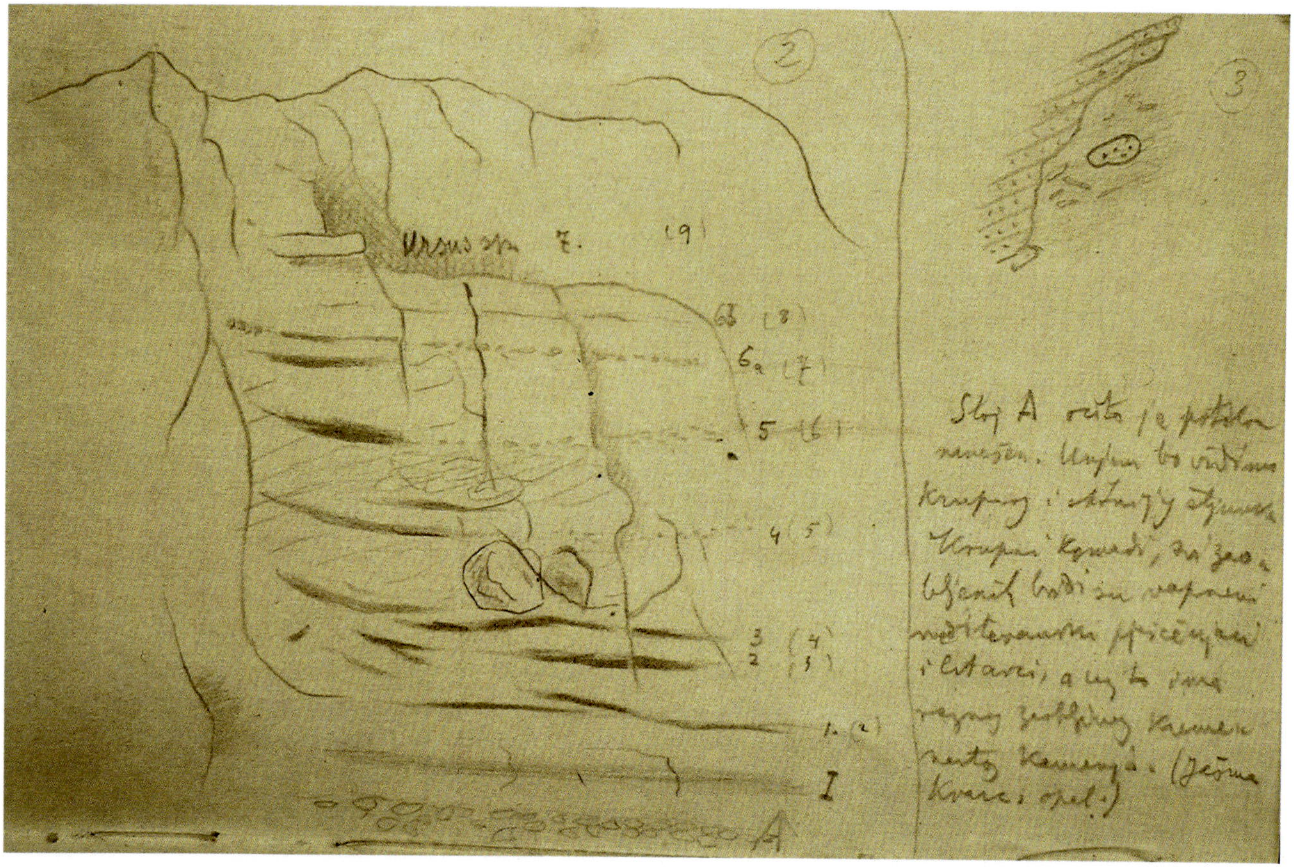

Fig. 3.2 Fascimile from Gorjanović-Kramberger's notebook showing the stratigrapy at Hušnjakovo (from Radovčić 1988, p. 22)

at Hušnjakovo hill. The whereabouts of these bones is unknown, but this event suggests that many other specimens were likely lost or destroyed prior to Gorjanović's 1899 visit.

Krapina is still the largest known Neandertal locality in terms of the number of individuals found at a single site. More than nine meters high sequence of sediments (Fig. 3.2) yielded more than 1100 human skeletal fragments (most from layers 2–4, which Gorjanović initially called the "*Homo sapiens* zone"; Fig. 3.3), over a 1000 lithic finds, as well as numerous faunal remains. Gorjanović-Kramberger (1906) estimated that around 20 individuals were represented in the Krapina sample. Later analyses variably assessed the minimum number of individuals from 24 (Gardner and Smith 2006) to as high as 82, the latter estimate based on the analysis of dental remains by M. Wolpoff (1978). The fact that the deposition at Hušnjakovo was relatively rapid, and that both sexes and individuals of various ages are present, provides a rare glimpse into idiosyncratic, ontogenetic, and sex-related variation of a relatively early Neandertal "population." We use the term population because there are distinct indications that the Krapina people were closely interrelated biologically. The Krapina sequence was once claimed to extend from the last interglacial to the middle of the last glaciation (Malez 1970). However, more recent ESR dates cluster around 130 ka regardless of their position in the deposits (Rink et al. 1995). Other hominin fossil assemblages in western Eurasia were apparently deposited within short time spans as well, including Dmanisi, Sima de los Huesos, Skhūl and el Sidrón. However, the presence of a unique morphology of the upper nasal bones (Smith and Smith 1986) and a high proportion of fourth mandibular premolar anomalies (Wolpoff 1999) at Krapina represent the type of discrete features that indicate close biological relationships. Thus, the variation exhibited at Krapina, demonstrated by any number of anatomical studies (for discussion see Smith 1976a; Wolpoff 1999; Cartmill and Smith 2009), is a reasonable approximation of what would be expected in a biological population of Neandertals. In this sense, with the exception perhaps of the Sima de los Huesos, Krapina is quite unique.

In discussing the historical importance of the Krapina fossils, one should bear in mind that not many Neandertal fossils were known prior to the discovery at Hušnjakovo. At the time, the total Neandertal sample known to science consisted of the remains from the eponymous site in the Neander valley (Kleine Feldhofer Grotte, discovered in 1856; Schaaffhausen 1857) in Germany, the two skeletons from the Belgian cave of Spy (discovered in 1886; Fraipont and Lohest 1886), and the mandibular remains from La Naulette in Belgium (discovered in 1866; Dupont 1866) and Šipka in

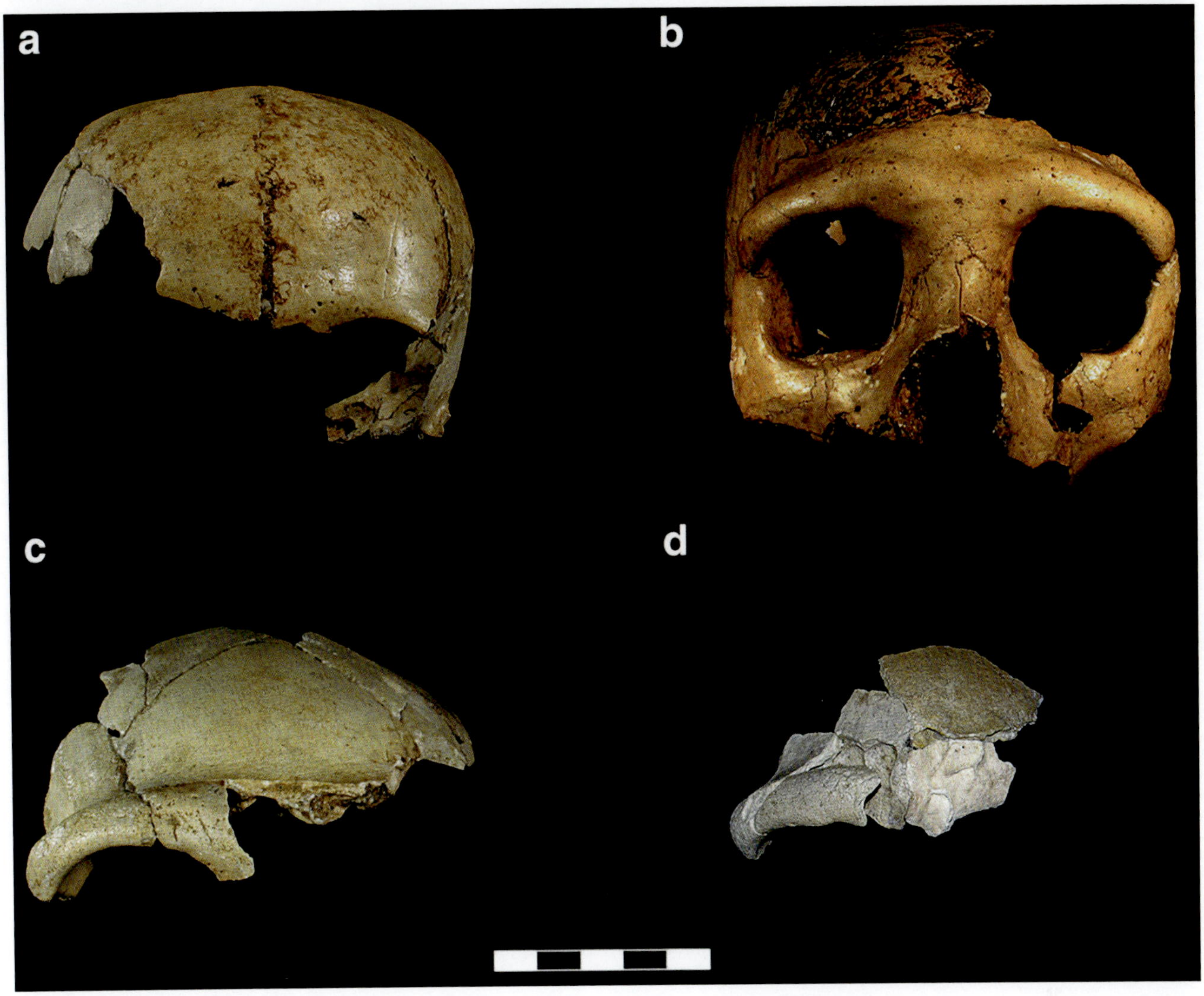

Fig. 3.3 Selected Krapina fossils. (**a**) Kr 1, (**b**) Kr 3, (**c**) Kr 4, and (**d**) Kr 27–28 (photo by J.C.M. Ahern)

Czech Republic (discovered in 1880; Schaaffhausen 1880). These were actually predated by the discoveries of what we now know represent Neandertals from Engis (discovered in 1829/1830; Huxley 1862) and Forbes Quarry on Gibraltar (discovered in 1848; Schwalbe 1906; Sollas 1907), but whose significance was not recognized at the time. Thus, we came to know these interesting Pleistocene people as Neandertals, not Engisians or Gibraltarians.

The 1856 discovery caused stir in scientific and public circles alike. This is not surprising, since the idea of evolution was in the air at the time (Bowler 1986), and the publication of Darwin's book *On the Origin of Species* ensued 3 years later. Therefore, the peculiar human bones from the Feldhofer cave needed to be explained in light of the debate—either as arguments for or against the existence of "fossil man." Following the initial description of the Feldhofer remains by Hermann Schaaffhausen (1858), one group of scientists, perhaps led most strongly by an eminent German anatomist of the time, Rudolf Virchow (1872), believed that the morphology of the remains was best explained by pathological changes, and that they derived from relatively recent times, certainly not the "diluvial age." The other group suggested that the anatomy reflected the ancient age of the bones. Quite simply, the bones were different from contemporary Europeans because they belonged to people that lived long before us. This view was held, among others, by Schaaffhausen and "Darwin's bulldog" T. H. Huxley (Schaaffhausen 1858, 1888; Huxley 1863). Despite the obvious primitive anatomy of the original Neandertal specimen, Huxley included Neandertals within our own species *Homo sapiens*, because estimates of brain size were in the range of modern humans. It is not until 1864 that a separate taxon, *Homo neanderthalensis* was suggested by William King (King 1864). It is not our aim to present a detailed overview of these early debates on the find (for discussions see Trinkaus and Shipman 1993; Cartmill and Smith 2009), but, rather, to emphasize the mind-set at the time of the discoveries at Krapina. As noted by Klaatsch (cited in Gorjanović-Kramberger 1918), it is unclear when the debate on the age or biological significance of Neandertals would have been solved without the discovery at Krapina. The fact that the sheer number of individuals present at Hušnjakovo shared the same basic anatomical features (or "peculiarities") with specimens from the Neander valley and other known sites made the suggestions of Virchow and his followers quite unlikely. Why would all the ancient humans from Europe share the same pathological features? Also, like the remains from Spy, the Hušnjakovo bones were also discovered in association with stone tools and extinct fauna. Gorjanović, a rare breed of scientist, also used the relatively new Fluorine dating technique to prove the contemporaneity of the extinct fauna and human bones (Radovčić 1988). As a reminder, the same method was used to discredit the Piltdown "fossil" as a hoax several decades later (Oakley and Hoskins 1950; Oakley and Weiner 1953). Gorjanović also correctly recognized the 1191 lithic finds from his excavations as Mousterian, a characterization confirmed by later studies (Gorjanović-Kramberger 1906, 1913; Malez 1970, 1979; Simek 1991; Simek and Smith 1997).

In addition to his careful excavation of the site for his time (see discussions in Smith 1976a; Radovčić 1988), Gorjanović used all available technological inventions in his approach to the analysis of bones. For example, he used X-rays, discovered by Roentgen only a few years earlier, to study the internal structure of the remains. This was the first use of this invention on human fossil material (Gorjanović-Kramberger 1902). Furthermore, his detailed descriptive monograph (published in 1906, only a year after the end of the excavations) was the first monograph published on Neandertals. In this monograph, as well as in his earlier papers, he hypothesized on the role of Neandertals in human evolution, giving us one of the first systematic models of modern human origins.

Alongside its historical importance, the Krapina remains have a prominent role in contemporary studies of Neandertals. They constitute the most intensively studied and most thoroughly published Neandertal collection in the world (for a comprehensive list of publications on Krapina up to 2006 see Frayer 2006; and papers in *Periodicum Biologorum* Vol. 108, Nos. 3 and 4). Because of the size and composition of the sample, the Krapina (Hušnjakovo) remains must form an integral part of any comparative study of Neandertal morphology.

Overall, the Hušnjakovo Neandertal sample exhibits "typical" Neandertal morphological features (Smith 1976a). This invalidates the suggestion that the Krapina people were "progressive" Neandertals, lacking full development of the so-called classic Neandertal morphology seen later in the Pleistocene (Howell 1957). These morphological features include shoveling of maxillary incisors, large incisors and canines, distinctive premolar and maxillary molar occlusal anatomy, the lack of a *mentum osseum* on mandibles, the presence of a retromolar space, the elongation of the skull with low forehead, a robust supraorbital region with double-arched supraorbital torus, occipital bunning, the presence of *fossa suprainiaca*, mid-facial prognathism, thick femoral cortices, elongated and thinned superior pubic rami, and other typical features of the postcranial anatomy common in most Neandertal finds (Gorjanović-Kramberger 1906; Kallay 1970a, b, 1978; Smith 1976a, 1982, 1984; Wolpoff 1978, 1979; Radovčić 1988; Bailey 2006). Another contentious issue at Krapina surrounds the purported presence of more modern humans at the site. This was first raised by Hauser in 1910 and soundly rejected by Gorjanović-Kramberger (see discussion in Smith 1976a). Somewhat later, it was suggested that the child's cranium (Krapina A, or Krapina 1) from the upper layer 8 belonged to a morphologically more modern group (Škerlj 1958) or that it might represent a transitional form (Wolpoff 1999). However, later analyses have shown that this find cannot be excluded from the variation seen in Neandertals (Minugh-Purvis et al. 2000).

After more than a century of study of the Hušnjakovo sample, there are still unresolved issues. For example, why is the sample so large, when we know that the deposition rate was relatively rapid? What is the reason that the bones are so fragmented? Gorjanović was the first to propose cannibalistic practices as possible reason for this, which later resulted in unflattering depictions of Neandertals in popular press. What is the meaning of the cut marks found on some of the bones? Was this due to defleshing in pursuit of dietary satisfaction (Gorjanović-Kramberger 1901, 1906; Tomić-Karović 1970; White and Toth 1991; Chiarelli 2004) or is it a reflection of cultural practices (e.g., secondary burial) as suggested by others (Trinkaus 1985; Russell 1987a, b; Ullrich 1989, 2006). For example, Frayer and colleagues (2006) recently suggested that distinctive cut marks on the frontal bone of the most complete cranial specimen, the Krapina C, or Krapina 3, skull reflected the presence of cultural ritual among these Neandertals. Certainly, there is additional significant work to be done on the Krapina people in the future and it will certainly increase our knowledge about Neandertals and their behavior.

Vindija

Vindija Cave is located in the Hrvatsko Zagorje region, not far from Krapina. The cave is approximately 50 m deep, 28 m wide, and over 20 m high (Fig. 3.4). The first, small-scale excavations of the site were conducted by S. Vuković in 1928 (Vuković 1935, 1949, 1950) during which he established that human habitation in various historic and prehistoric periods. Systematic excavations were conducted by M. Malez between 1974 and 1986 (Malez 1979, 1983; Malez et al. 1980; Wolpoff

Fig. 3.4 The Vindija cave (photo by J.C.M. Ahern)

et al. 1981; Smith et al. 1985), during which most of the Paleolithic material, including all Neandertal bones, was discovered. The stratigraphy of Vindija consists of over 12 m of sediment divided into 13 basic stratigraphic units (A–M), with the F, G, and K complex further subdivided into Fg, Fs, Fd/s, Fd, Fd/d, G1–G5, K1–K3 (Malez and Rukavina 1979; Paunović et al. 2001; Ahern et al. 2004). Layers A–D date to the Holocene, while the older layers E–M date to the Pleistocene. All of the Neandertal remains come from complex G, with the possible exception of the Vi 11.52 mandibular ramus described by Ahern and colleagues (2004), and the long bone shaft Vi 33.25 from which DNA was successfully extracted (Green et al. 2010).

Most of the Vindija Neandertal sequence can be bracketed between around 38 and 45.6 ^{14}C kBP (for the layer G_3 remains) and c. 33 and 35 ^{14}C kBP (for the layer G_1 remains, see Krings et al. 2000; Wild et al. 2001; Higham et al. 2006; Green et al. 2010). The 32,400 ± 800 ^{14}C BP and 32,400 + 1800 ^{14}C BP ultrafiltered AMS dates on two level G_1 specimens represent the latest direct Neandertal dates from anywhere in Europe. The G sequence corresponds to the time when the earliest groups of anatomically modern people were also present in parts of Europe. For example, dates on two presumably modern human teeth from the Grotta del Cavallo indicate the presence of modern humans in neighboring Italy 43,000–45,000 cal BP (Benazzi et al. 2011), while the earliest date for more complete early modern specimens is ca. 37,230 cal BP at Peştera cu Oase, Romania (Trinkaus et al. 2003).

The Vindija Neandertal sample, although clearly a part of the Neandertal morph as a whole (Malez et al. 1980; Wolpoff et al. 1981; Smith 1982, 1984, 1994; Smith et al. 1985; Ahern et al. 2004; Janković et al. 2006, 2011), shows interesting morphological features noted by many authors. Certain aspects of their anatomy are different from the earlier (Krapina), as well as from the so-called classic Neandertals. The Vindija remains show patterns that are either intermediate between Neandertals and anatomically modern humans, or closer to anatomically modern humans. This is most clearly seen in the supraorbital and mandibular sample (Smith and Ranyard 1980; Smith 1984, 1994; Ahern 1998;

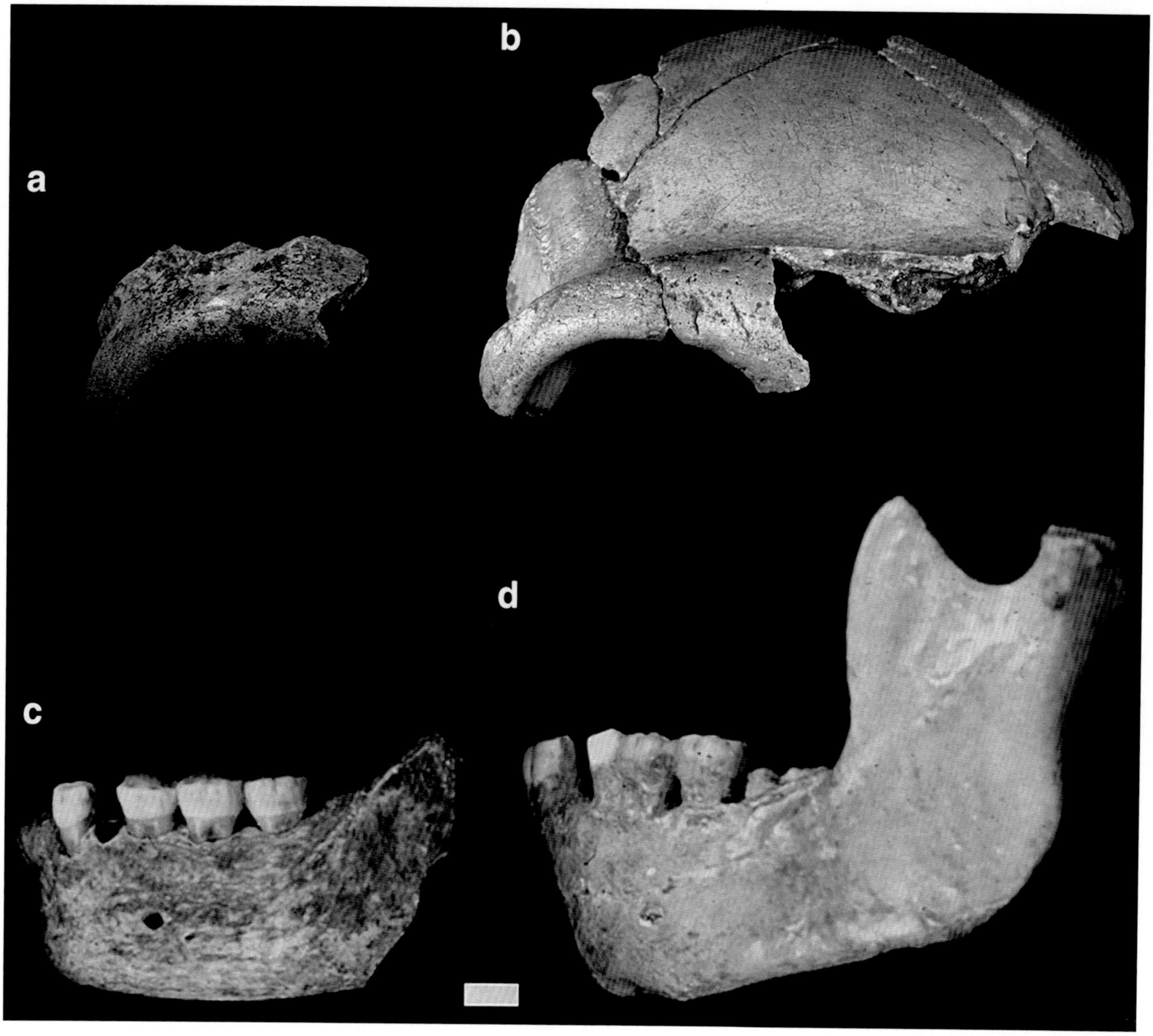

Fig. 3.5 (**a**) Vi 202 from level G_3 at Vindija compared with (**b**) Kr 4. (**c**) Vi 231 compared with (**d**) Kr 59. Scale is 1 cm (photo by J.C.M. Ahern)

Ahern et al. 2002, 2004) (see Figs. 3.5 and 3.6), and facial dimensions and projection (Wolpoff et al. 1981; Smith and Ranyard 1980; Smith 1982, 1984; Ahern 1998; Ahern et al. 2004). The Vindija supraorbital tori are reduced overall in size and exhibit a relatively greater degree of midorbital thinning and projection compared to most other Neandertals (Smith and Ranyard 1980; Smith 1984, 1994). The maxillae show narrower nasal openings and shorter alveolar processes than any other Neandertal (Smith 1992), while the mandibles show more vertical symphyses and incipient anterior basal projections (Wolpoff et al. 1981; Smith 1982, 1994; Ahern and Smith 1993; Kesterke and Ahern 2007). As documented in the references (see also Cartmill and Smith 2009), these features are consistent throughout the Vindija sample and indicate significant facial reduction compared to the average Neandertal morphology.

Some scientists have suggested that the morphological features and gracility at Vindija may be a reflection of either small size of the Vindija people, or a result of sex bias and/or overrepresentation of younger individuals (cf. Howell 1984; Bräuer 1989; Stringer and Bräuer 1994). Later studies have shown that this is not the case. Trinkaus and Smith (1995) showed, on the basis of postcranial remains, that the Vindija people were of average size for Neandertals and not unusually small. Furthermore, several studies (Ahern and Smith 1993; Smith 1994; Kesterke and Ahern 2007) have demonstrated that the Vindija pattern is not due to age or sex bias in the sample, suggesting that morphological differences seen at Vindija reflect a distinct, biologically significant pattern of change in this late Neandertal sample.

If we look at the large collection of archaeological data from Vindija (lithics and bone tools) the story becomes even

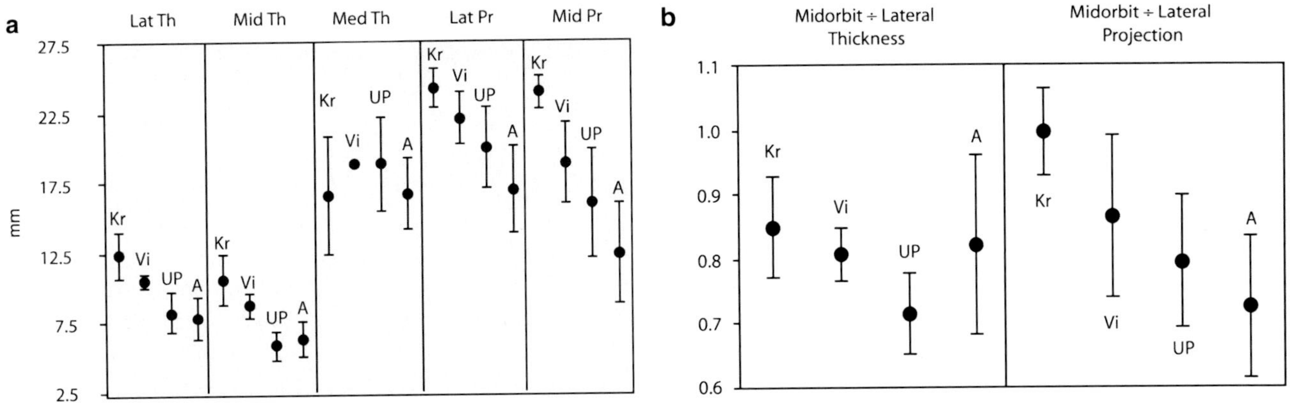

Fig. 3.6 Vindija, Krapina, and modern human supraorbital metrics and indices. (**a**) Variables: *LaTh* lateral supraorbital thickness, *MidTh* midorbit supraorbital thickness, *MedTh* medial supraorbital thickness, *LatPr* lateral supraorbital projection, and *MidPr* midorbit supraorbital projection. *Kr* Krapina, *Vi* Vindija, *UP* Upper Paleolithic, and *A* Altendorf Neolithic

more interesting. Levels G_1 and G_3 exhibit a cultural assemblage that has aspects of both Mousterian and Upper Paleolithic (including Aurignacian) elements (see Janković et al. 2011; Karavanić et al. 2016 for recent discussions). However, it is not clear whether this represents cultural reality or artificial mixture of elements from separate cultural entities. Although we recognize that problems with excavation techniques, the presence of cryoturbation and bioturbation, and other processes have disturbed some parts of the cave and caused mixing of the material from certain parts/layers of the site, these have been overemphasized and exaggerated in the past (e.g., see Bruner 2009; Zilhão 2009). Over the last two decades much work and time has been invested to clarify the stratigraphy and resolve problems regarding the Vindija sequence (see Ahern et al. 2004; Janković et al. 2011, 2012; Karavanić 1995; Karavanić and Smith 1998, 2011; Karavanić et al. 2016). There are issues with the Vindija archaeological sequence that are difficult to explain through mixing of the material alone. For example, if we look at the typological features of the stone tools from the older layers of the site (e.g., layer K), through the G_3 and to the G_1 layers, we can see the rise in the percentage of certain Upper Paleolithic tool types. Although this might partially be explained by the mixing of layers, mixing is an inadequate explanation for certain observations. Layer K contains typical Mousterian tools, and the use of Levallois and flake technology. There are virtually no Upper Paleolithic types, and the dominant raw material is local quartz. Layer G_3 consists of a mixture of Mousterian and some Upper Paleolithic tool types (e.g., end-scrapers), bifacial and blade technology is noted in the production of tools, and the frequency of chert use for the production of some of the tools is higher (Karavanić and Smith 2011). If we look at the raw material distribution at the site, starting from the lower Mousterian levels (level K) through the Upper Paleolithic levels (unit F and level E), there is a distinct change in raw material use/selection from quartz and similar low-quality material (dominant in lower levels) to the use of better quality materials (such as chert and tuff) in the upper levels of the site (Blaser et al. 2002; Ahern et al. 2004). This is seen not only for the Upper Paleolithic tool types that appear in G_3 layer, but also in the whole Mousterian assemblage from that level. The industry of layer G_1 shows an even more pronounced shift in the use of higher quality raw material (chert), while the use of quartz is less common than in earlier levels. The Upper Paleolithic elements are more abundant, and several bone points (one split-based and three full base, so called Mladeč points) were found in this layer. While caution is needed, as parts of these layers were disturbed by various processes (mainly bioturbation), the direct dating of the Vindija Neandertal mandible Vi 207 found *in situ*, just a few centimeters from the split base bone point Vi 3437, makes the argument that all this is a result of mixing of the layers even more unlikely.

The Vindija Neandertal remains have yielded another insight into biological properties of the (late) Neandertals: their genetic data. The first genetic studies on skeletal remains from Vindija yielded a partial mitochondrial sequence and were published in 2000 (Krings et al. 2000; Serre et al. 2004). This sequence showed an overall pattern similar to mtDNA sequences from other Neandertal specimens. The modern era of what is now referred to as "Neandertal genomics" started with the first successful extraction of the nuclear DNA from one tibial fragment from Vindija (Vi 33.16 from layer G_3) that was soon followed by several other successful extractions. Today, most of what we know about the Neandertal genome comes from three Vindija bones (Vi 33.16, Vi 33.25, and Vi 33.26). This is truly remarkable work, important not only for paleoanthropologists, but with much wider implications for various fields: it allows for various comparisons to be made in the future, as genetic research reveals genes or genetic complexes linked to functional traits, and allow comparisons of Neandertal and contemporary genomes. However, the real surprise, for at least part of the paleoanthropological community, was that genetic analysis of the Vindija bones suggests that living people in Europe and Asia have between 1 and 4 % of Neandertal ancestry (Green et al. 2010; but see Prüfer et al. 2014 for a slightly lower, and Lohse and Frantz 2014 for a higher estimate). Further studies also suggested some Neandertal gene flow into North Africa (see Sánchez-Quinto et al. 2012) and Siberia (Krause et al. 2007; Reich et al. 2010; Prüfer et al. 2014).

All three sets of data from the Vindija cave (Neandertal skeletal remains, cultural evidence, and genomic data) provide evidence of potential Neandertal–early modern human interaction. These late Neandertals of the Vindija cave were more "modern" in certain details of their anatomy than earlier Neandertals. Likewise, their behavior appears to have changed, becoming more "modern." All this was happening at a time when evidence suggests they overlaped temporally and geographically with anatomically modern newcomers into Europe. The genetic (genomic) data and morphological changes that are documented at Vindija may be best explained as a result of biological and cultural contact with these new human groups as they spread from their African homeland to other regions of the World. It was initially suggested that interbreeding occurred somewhere in the Near East, but interbreeding in Europe is also a possibility as, according to Sankararaman et al. (2012), interbreeding might have happened as late as 37 ka. This was recently shown by Fu et al. (2015) in their analysis of the Oase 1 genomic data (also see Harvati and Roksandic 2016). However, as new genomic studies involving ancient DNA (including Neandertal) are published, there are certain to be substantial disagreements on where and when the interbreeding happened, to what extent, what specific parts of the genome later human populations inherited from Neandertals, and what these genes regulate (see Herrera et al. 2009; Hammer et al. 2011; Abi-Rached et al. 2011; Yotova et al. 2011; Mendez et al. 2012, 2013; Sankararaman et al. 2012, 2014; Sánchez-Quinto et al. 2012; Hawks and Throckmorton 2013; Wall et al. 2013; Prüfer et al. 2014 and references therein).

When anatomically modern humans entered Europe, they came into contact with the Vindija Neandertals (and likely other, although not necessarily all Neandertal groups), which might explain the cultural change seen in the Vindija G sequence, but also at other late Neandertal sites, such as St. Césaire and Grotte du Renne at Arcy sur Cure (Leroi-Gourhan 1958; Lévêque and Vandermeersch 1980; Hedges et al. 1994; Hublin et al. 1996). Interestingly, level mixing has now been offered as an explanation for the modern-like aspects of the Châtelperronian artifacts at the Grotte du Renne as well (Higham et al. 2010; but see Hublin et al. 2012).

It is important to note that our assertions that Vindija reflects modern human gene flow into late Neandertals are based on morphology and not based on the current genetic evidence. The Neandertal genome study found evidence of a Neandertal genetic contribution to early modern Eurasians but no evidence of gene flow from early moderns to Neandertals. Unidirectional gene flow, however, seems highly unlikely. We contend that it is the morphology of the Vindija Neandertals that reflects early modern biological impact on late Neandertals. As stated earlier, there is no direct genetic evidence of this, but it is reasonable to assume that the morphological particularities exhibited by the Vindija Neandertals have their source in contact with early anatomically modern humans. As the results of modern scientific research continue to provide more and better information about the complex patterns of change and contact in the so-called Middle to Upper Paleolithic transition in Europe, the Vindija remains provide some of the crucial data to evaluate this interaction.

Šandalja II

The third site in Croatia that yielded human skeletal remains securely associated with a Paleolithic assemblage is the cave of Šandalja II near Pula, in Istria. The site is a part of a larger cave system that was discovered during quarrying in 1961 (Malez and Vogel 1969). The cave of Šandalja I, where the Villafranchian fauna containing one stone artifact (a chopper) was found, is part of the same, larger cave (Malez 1979). Therefore, we refer to the Upper Paleolithic sequence of the site as Šandalja II. Šandalja II is of major importance as an archaeological, as well as a paleontological, site. Numerous stone and bone tools, as well as faunal remains, were found during excavations by M. Malez between 1962 and 1989.

The stratigrapy of the site consists of over 8 m of sediments divided into units A–H, with subdivision of complex C and B into layers C/d, C/s, C/g, and B/d, B/s, B/g, respectively. All the collected material from strata B to H can be attributed to the Upper Paleolithic, while Bronze Age material was found in the uppermost layer A (Karavanić 1999). Malez (1979) attributed the Paleolithic sequence of the site to two distinct industries, the Aurignacian (layers G-D) and the Gravettian (layers C and D), but later revisions by Karavanić (Karavanić 1999, 2003; Karavanić and Janković 2010; see also Montet-White 1996) attributed the material from C and D layers to the Epigravettian rather than Gravettian. This was confirmed both by typological analyses (most common lythic tool types are short end-scrapers, microgravettes, backed bladelets, circular segments, and Azilian points), as well as with radiocarbon dates from layer B (layer B/s was dated to 12,539 ± 369 cal BP, see Malez and Vogel 1969; Janković et al. 2012). Revisions of the faunal sequence were performed by Miracle (1995) and Brajković (1998).

The human skeletal remains were found in layer B/s and partially described by Malez (1972, 1987) and Smith (1976b). A more detailed description was published recently by Janković and colleagues (2011, 2012). The same is true for the lithics and bone artifacts from the site, first published by Malez (1979) and later analyzed in detail in numerous

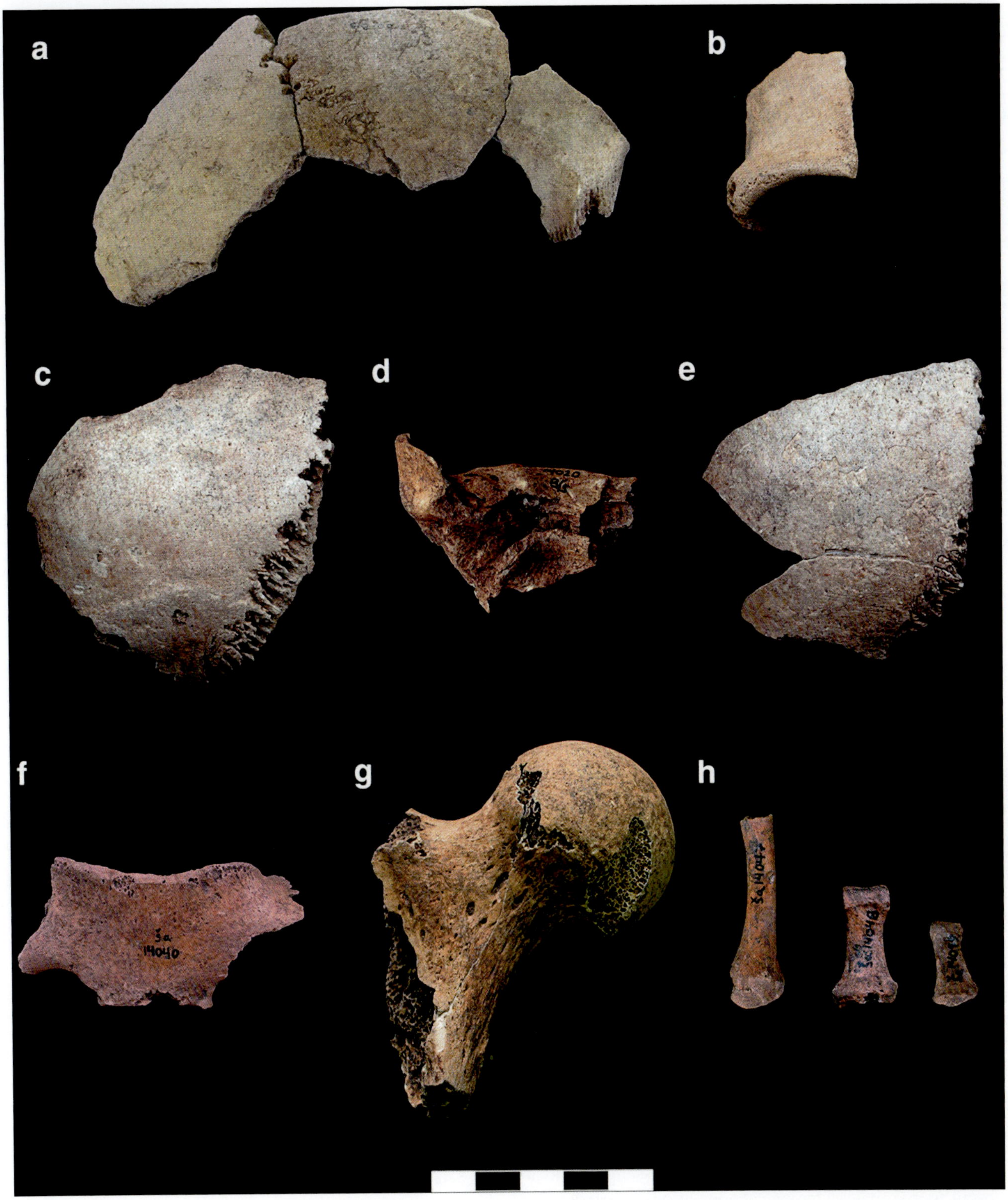

Fig. 3.7 Select Šandalja specimens. (**a**) Ša 14013 partial calotte, (**b**) anterior view of Ša 14013, (**c**) Ša 14021 parietal, (**d**) Ša 14020 temporal, (**e**) Ša 14015 parietal, (**f**) Ša 14040 sacral vertebra, (**g**) Ša 14050 proximal femur, (**h**) hand bones

studies (e.g., Karavanić 1999, 2003; Karavanić and Janković 2010; Janković et al. 2012; Karavanić et al. in press).

The human skeletal remains from the site are very fragmentary and unlikely to derive from burials (Fig. 3.7). Although Malez (1972) suggested that cannibalistic practices of the Upper Pleistocene inhabitants of the cave might explain the fragmentary state of the material, this was not confirmed by later taphonomic studies (Miracle 1995). The skeletal remains belong to a minimum of two adults (likely three; two males and one female) and one subadult (see Janković et al. 2012 for a detailed discussion of the sample). Although, based on available dental, cranial, and postcranial metrics, the Šandalja II people were rather small in size, they overlap with reported measurements for described late Upper Paleolithic samples of neighboring regions. The analyses and comparisons of the Late Epigravettian sequence (layer B/s) from the site show similarities with contemporary sites in both western and eastern parts of the Adriatic region (Janković et al. 2012). Bearing in mind that during the formation of these layers at Šandalja II the sea level was between 50 and 90 m lower than today (Miracle 1995), the exposed Adriatic plain would make a natural connection between Italy and the mountainous ridges of the plain on the eastern Adriatic shore. This may, at least in part, explain the similarities in basic tool types between regions. Some of the noted differences (e.g., percentages of certain tool types, see Janković et al. 2012) are more likely a result of variation in site function rather than of production processes and raw material selection.

Concluding Remarks

The three sites described in this chapter all provide important data for understanding various aspects of hominin behavior and evolution during the late Pleistocene; however, there are different aspects to their importance for paleoanthropological research. The site of Krapina, or Hušnjakov brijeg, played an important role in the early views on Neandertals. It was discovered at a time when paleoanthropology was still an emerging field in which Gorjanović played an important role, and the Hušnjakovo remains were crucial in establishing Neandertals as an ancient group of people that were in many aspects of their anatomy (and behavior) different from later inhabitants of Europe. Likewise, Gorjanović used the Krapina remains (and the rest of the available fossil record of the time) to present a very specific argument as to the role of Neandertals in human evolution, seeing them as a chronological step in the evolution of anatomically modern Europeans. Today, the Krapina collection provides a basis for discussion of morphology and variation, and gives us a rare glimpse into ontogeny, sex, and idiosyncratic variational aspects of a sample which closely approximates a biological population of Neandertals. Furthermore, Krapina has yielded behavioral data, as numerous tools and animal bones were found at the site, and certain aspects of the sample (i.e., cut marks on some of the Neandertal remains) provide a basis for discussion of symbolic practices.

The site of Vindija has yielded what is to date the youngest Neandertal skeletal sample in Europe, and provides insight into the morphological and genetic makeup of late European Neandertals. The Vindija Neandertals are dated to a time when anatomically modern human groups were already present in Central Europe and Italy. The fact that certain aspects of the Vindija morphology show a change toward more "modern" patterns provides potential further evidence of biological interaction between Neandertals and early modern people, but in the opposite direction to what is revealed in Vindija nuclear DNA. Furthermore, there seems to be a similar pattern of change seen in cultural data from the site, making the story more complicated (and more interesting). However, it is important to remember that the biological data are not impacted by how the cultural patterns are interpreted. In other words, the problems impacting interpretation of the cultural remains have no direct relevance to the discussion of the biological aspects of the Neandertal sample at Vindija. In fact, the Vindija genetic and morphological data, along with the recent indication that the Peştera cu Oase early modern humans reflect as much as a 6–9 % Neandertal contribution (Fu et al. 2015)—considerably more than other early modern specimens from Russia (Seguin-Orlando et al. 2014; Fu et al. 2014)—suggests to us that East Central Europe (in the sense of Ahern et al. 2013) was a major region of biological interaction between late Neandertals and early modern humans.

The site of Šandalja II is the first Late Upper Paleolithic site from the eastern Adriatic that yielded a secure association of cultural sequence and human remains. Therefore, it provides us with data for comparisons of biological/anatomical properties of these people with available contemporaneous sites in the nearby regions. The cultural sequence of the site is no less important, as both Aurignacian and Late Epigravettian layers are present (in addition to animal bones), thus providing a basis for comparisons of behavior within the Upper Paleolithic. The site of Šandalja II thus can be put in the context of behavioral and biological aspects, as well as of the population movement and contact patterns during the Late Pleistocene.

Research continues at various locations around Croatia in an attempt to augment this sample of cultural and fossil human evidence pertinent to understanding Neandertals and their interactions with early modern people. Recent evidence suggests this pattern exhibits the kind of biological and cultural complexity that we have advocated for some time. It is our belief that both the data summarized here, as well as new information from excavations carried out by our team, as well as others, will maintain the central importance of Croatian material for understanding the late Pleistocene evolution of humans.

Acknowledgements Authors would like to thank the organizers of the International symposium "Human evolution in the Southern Balkans" held in Tübingen, Germany, in December of 2012, for inviting us to participate. The organizational skills of Katerina Harvati and Vangelis Tourloukis, along with the many student volunteers in their team made our stay a most pleasant experience. We also thank Katerina Harvati and Mirjana Roksandic for editing this volume. Our results are based on research that was partially supported by the Ministry of Science, Education and Sports of the Republic of Croatia, the U.S. Fulbright Foundation, the University of Wyoming, and Illinois State University. We also thank our colleague David Strait and the anonymous reviewers for their comments. All mistakes are, of course, our responsibility.

References

Abi-Rached, L., Jobin, M. J., Kulkarni, S., McWhinnie, A., Dalva, K., Gragert, L., et al. (2011). The shaping of modern human immune systems by multiregional admixture with archaic humans. *Science, 334*(6052), 89–94.

Ahern, J. C. M. (1998). *Late Pleistocene frontals of the Hrvatsko Zagorje: An analysis of intrapopulational variation in South Central European Neandertals*. Ann Arbor: University of Michigan.

Ahern, J. C. M., Janković, I., Voisin, J.-L., & Smith, F. H. (2013). Modern human origins in Central Europe. In F. H. Smith & J. C. M. Ahern (Eds.), *The origins of modern humans: Biology reconsidered* (pp. 151–221). Hoboken: Wiley-Blackwell.

Ahern, J. C. M., Karavanić, I., Paunović, I., Janković, I., & Smith, F. H. (2004). New discoveries and interpretations of hominid fossils and artifacts from Vindija Cave, Croatia. *Journal of Human Evolution, 46*, 25–67.

Ahern, J. C. M., Lee, S.-H., & Hawks, J. (2002). The late Neandertal supraorbital fossils from Vindija Cave, Croatia: A biased sample? *Journal of Human Evolution, 43*, 419–432.

Ahern, J. C. M., & Smith, F. H. (1993). The transitional nature of the late Neanderthal mandibles from Vindija Cave, Croatia. *American Journal of Physical Anthropology, 16*, 47.

Bailey, S. A. (2006). Beyond shovel-shaped incisors: Neandertal dental morphology in a comparative context. *Periodicum Biologorum, 108*, 253–267.

Barić, L. (1978). Dragutin Gorjanović-Kramberger i otkriće krapinskog pračovjeka. In M. Malez (Ed.), *Krapinski pračovjek i evolucija hominida* (pp. 23–51). Zagreb: Jugoslavenska Akademija Znanosti i Umjetnosti.

Batović, Š. (1965). Prvi paleolitski nalazi u srednjoj Dalmaciji. *Diadora, 3*, 205–209.

Batović, Š. (1988). Paleolitički i mezolitički ostaci s Dugog otoka. *Poročilo o raziskovanju paleolita, neolita in eneolita v Sloveniji, 16*, 7–54.

Benazzi, S., Douka, K., Fornai, C., Bauer, C., Kullmer, O., Svoboda, J., et al. (2011). The early dispersal of modern humans in Europe and implications for Neanderthal behaviour. *Nature, 479*, 525–528.

Blaser, F., Kurtanjek, D., & Paunović, M. (2002). L'industrie du site néandertalien de la grotte de Vindija (Croatie): une révision des matières premières lithiques. *L'Anthropologie, 106*, 387–398.

Bowler, P. (1986). *Theories of human evolution: A century of debate 1844–1944*. Baltimore: Johns Hopkins University Press.

Brajković, D. (1998). Šandalja. *Histria Archaeologica, 29*, 5–25.

Bräuer, G. (1989). The evolution of modern humans: A comparison of the African and non-African evidence. In C. B. Stringer & P. Mellars (Eds.), *The human revolution* (pp. 121–154). Edinburgh: Edinburgh University Press.

Bruner, K. M. (2009). *Testing stratigraphic integrity of Upper and Middle Paleolithic deposits in Vindija Cave (Croatia): A chipped stone refitting analysis*. Unpublished Master Theses, University of Kansas, Lawrence.

Cartmill, M., & Smith, F. H. (2009). *The human lineage*. Hoboken: Wiley-Blackwell.

Chapman, J., Shiel, R., & Batović, Š. (1996). *The changing face of Dalmatia: Archaeological and ecological studies in a Mediterranean landscape*. London: Leicester University press, The Socety of Antiquaries of London.

Chiarelli, B. (2004). Spongiform encelopathy, cannibalism and Neanderthal extinction. *Human Evolution, 19*, 81–92.

Dinçer, B. (2016). The Lower Paleolithic in Turkey: Anatolia and hominin dispersals out of Africa. In K. Harvati & M. Roksandic (Eds.), *Paleoanthropology of the Balkans and Anatolia: Human evolution and its context* (pp. 213–228). Dordrecht: Springer.

Dupont, E. (1866). Etude sur les fouilles scientifiques executees pendant l'hiver de 1865–1866 dans les cavernes des bords de la Lesse. *Bulletin de l'Academie Royal de Belgique, 22*, 44–45.

Fraipont, J., & Lohest, M. (1886). La race humaine de Néandertal ou de Cannstadt, en Belgique. *Bulletins de l'Académie royale des sciences, des lettres et des beaux-arts de Belgique, 12*, 741–784.

Frayer, D. W. (2006). *The Krapina Neandertals. A comprehensive, centennial, illustrated bibliography*. Zagreb: Croatian Natural History Museum.

Frayer, D. W., Orschiedt, J., Cook, J., Russell, M. D., & Radovčić, J. (2006). Krapina 3: Cut marks and ritual behavior? *Periodicum Biologorum, 108*, 519–524.

Fu, Q., Hajdinjak, M., Moldovan, O. T., Constantin, S., Mallick, S., Skoglund, P., et al. (2015). An early modern human from Romania with a recent Neanderthal ancestor. *Nature, 524*(7564), 216–219.

Fu, Q., Li, H., Moorjani, P., Jay, F., Slepchenko, S., Bondarev, A., et al. (2014). Genetic sequence of a 45,000-year old modern human from western Siberia. *Nature, 514*, 445–450.

Gardner, J. C., & Smith, F. H. (2006). The paleopathology of the Krapina Neandertals. *Periodicum Biologorum, 108*, 471–484.

Gorjanović-Kramberger, D. (1901). Der paläolitische Mensch und seine Zeitgenossen aus der Diluvium von Krapina in Kroatien. *Mitteilungen der Anthropologischen Gesellschaft in Wien, 31*, 164–197.

Gorjanović-Kramberger, D. (1902). Der paläeolitische Mensch und seine Zeitgnossen aus dem Diluvium von Krapina in Kroatien II. *Mitteilungen der Anthropologischen Gesellschaft in Wien, 32*, 189–216.

Gorjanović-Kramberger, D. (1906). *Der Diluvijale Mensch von Krapina in Kroatien. Ein Bertrag zur Paläoanthropologie*. Kreidel: Wiesbaden.

Gorjanović-Kramberger, D. (1913). *Život i kultura diluvijalnoga čovjeka iz Krapine u Hrvatskoj*. Zagreb: JAZU.

Gorjanović-Kramberger, D. (1918). Pračovjek iz Krapine. *Priroda, 7*, 155–162.

Green, R. E., Krause, J., Briggs, A. W., Maričić, T., Stenzel, U., Kircher, M., et al. (2010). A draft sequence of the Neandertal genome. *Science, 328*, 710–722.

Hammer, M. F., Woerner, A. E., Mendez, F. L., Watkins, J. C., & Wall, J. D. (2011). Genetic evidence for archaic admixture in Africa. *Proceedings of the National Academy of Sciences, 108*(37), 15123–15128.

Harvati, K. (2016). Paleoanthropology in Greece: Recent findings and interpretations. In K. Harvati & M. Roksandic (Eds.), *Paleoanthropology of the Balkans and Anatolia: Human evolution and its context* (pp. 3–14). Dordrecht: Springer.

Harvati, K., & Roksandic, M. (2016). The human fossil record from Romania: Early Upper Paleolithic Europeans and possible evidence for Neanderthal admixture. In K. Harvati & M. Roksandic (Eds.), *Paleoanthropology of the Balkans and Anatolia: Human evolution and its context*. Dordrecht: Springer.

Hawks, J., & Throckmorton, Z. (2013). The relevance of archaic genomes to modern humans origins. In F. H. Smith & C. M. Ahern (Eds.), *The origins of modern humans: Biology reconsidered* (pp. 339–354). New York: Wiley.

Hedges, R. E. M., Housley, R. A., Bronk-Ramsey, C., & Van Klinken, G. J. (1994). Radiocarbon dates from the Oxford AMS system. *Archaeometry, 36*, 337–374.

Herrera, J. K., Somarelli, J. A., Lowery, R. K., & Herrera, R. J. (2009). To what extent did Neanderthals and modern humans interact? *Biological Reviews, 84*, 245–257.

Higham, T., Jacobi, R., Julien, M., David, F., Basell, L., Wood, R., et al. (2010). Chronology of the Grotte du Renne (France) and implications for the context of ornaments and human remains within the Châtelperronian. *Proceedings of the National Academy of Sciences USA, 107*, 20234–20239.

Higham, T., Ramsey, C. B., Karavanić, I., Smith, F. H., & Trinkaus, E. (2006). Revised direct radiocarbon dating of the Vindija G_1 Upper Paleolithic Neandertals. *Proceedings of the National Academy of Sciences USA, 103*, 553–557.

Horvat, J., & Ravlić, J. (1956). *Pisma Ljudevitu Gaju. Građa za povijest književnosti Hrvatske, knjiga 26*. Zagreb: JAZU.

Howell, F. C. (1957). The evolutionary significance of variation and varieties of "Neanderthal" man. *Quarterly Review of Biology, 32*, 330–347.

Howell, F. (1984). Introduction. In F. Smith & F. Spencer (Eds.), *Origins of modern humans: A world survey of the fossil evidence* (pp. xiii–xxii). New York: Wiley.

Hublin, J.-J., Spoor, F., Braun, M., Zonneveld, F., & Condemi, S. (1996). A late Neanderthal associated with Upper Palaeolithic artefacts. *Nature, 381*, 224–226.

Hublin, J.-J., Talamo, S., Julien, M., David, F., Connet, N., Bodu, N. P., et al. (2012). Radiocarbon dates from the Grotte du Renne and Saint-Césaire support a Neandertal origin for the Châtelperronian. *Proceedings of the National Academy of Sciences USA, 109*, 18743–18748.

Huxley, T. H. (1862). On fossil remains of man. *Proceedings of the Royal Society of Great Britain, 3*, 420–422.

Huxley, T. H. (1863). *Evidence as to man's place in nature*. London: Williams and Norgate.

Janković, I., Ahern, J. C. M., Karavanić, I., Stockton, T., & Smith, F. H. (2012). Epigravettian human remains and artifacts from Šandalja II, Istria, Croatia. *PaleoAnthropology, 2012*, 87–122.

Janković, I., Karavanić, I., Ahern, J. C. M., Brajković, D., Mauch Lenardić, J., & Smith, F. H. (2006). Vindija Cave and the modern human peopling of Europe. *Collegium Antropologicum, 30*, 457–466.

Janković, I., Karavanić, I., Ahern, J. C. M., Brajković, D., Mauch Lenardić, J., & Smith, F. H. (2011). Archaeological, paleontological and genomic perspectives on late European Neandertals at Vindija Cave, Croatia. In S. Condemi & G.-C. Weniger (Eds.), *Continuity and discontinuity in the peopling of Europe. One hundred fifty years of Neanderthal study* (pp. 299–313). Dordrecht, Netherlands: Springer.

Kallay, J. (1970a). Komparativne napomene o čeljustima krapinskih praljudi s obzirom na položaj među hominidima. In M. Malez (Ed.), *Krapina 1899–1969* (pp. 153–161). Zagreb: JAZU.

Kallay, J. (1970b). Osobitosti zubi krapinskih neandertalaca. In M. Malez (Ed.), *Krapina 1899–1969* (pp. 165–174). Zagreb: JAZU.

Kallay, J. (1978). Neke značajnosti zubi i njihova upotreba u krapinskog neandertalca. In M. Malez (Ed.), *Krapinski pračovjek i evolucija hominida* (pp. 145–154). Zagreb: JAZU.

Karavanić, I. (1995). Upper Paleolithic occupation levels and late-occuring Neandertal at Vindija Cave (Croatia) in the context of Central Europe and the Balkans. *Journal of Anthropological Research, 51*, 9–35.

Karavanić, I. (1999). *Gornji paleolitik Šandalje II u okviru jadranske regije*. Ph.D. Dissertation, University of Zagreb.

Karavanić, I. (2000). Research on the Middle Paleolithic in Dalmatia, Croatia. *Antiquity, 74*, 777–778.

Karavanić, I. (2003). L'industrie aurignacienne de la grotte de Šandalja II (Istrie, Croatie) dans le contexte de la région de l'est de l'Adriatique. *L'Anthropologie, 107*, 577–602.

Karavanić, I., & Čondić, N. (2006). Probno sondiranje Velike pećine u Kličevici kod Benkovca. *Obavijesti HAD-a, 38*(2), 45–50.

Karavanić, I., & Janković, I. (2007). Srednji i rani gornji paleolitik u Hrvatskoj/The Middle and Early Upper Paleolithic in Croatia. *Opuscula Archaeologica, 30*, 21–54.

Karavanić, I., & Janković, I. (2010). Šandalja II i problem orinjasijenske industrije u istočnom Jadranskom području/Šandalja II and the Problem of Aurignacian Industry in the Eastern Adriatic. In D. Komšo (Ed.), *100 Godina Arheološkog Muzeja Istre u Puli. Nova Istraživanja u Hrvatskoj* (pp. 35–59). Zagreb/Pula: Hrvatsko Arheološko Društvo/Arheološki Muzej Istre u Puli.

Karavanić, I., Miracle, P. T., Culiberg, M., Kurtanjek, D., Zupanič, J., Golubić, V., et al. (2008). The Middle Paleolithic from Mujina pećina, Dalmatia, Croatia. *Journal of Field Archaeology, 33*, 259–277.

Karavanić, I., & Smith, F. H. (1998). The Middle/Upper Paleolithic interface and the relationship of Neanderthals and early modern humans in the Hrvatsko Zagorje, Croatia. *Journal of Human Evolution, 34*, 223–248.

Karavanić, I., & Smith, F. H. (2011). Middle/Upper Paleolithic interface at Vindija Cave (Croatia) in the context of central Europe and the Adriatic. In A. P. Derevianko & M. V. Shunkov (Eds.), *Characteristic features of the Middle to Upper Paleolithic transition in Eurasia* (pp. 124–134). Novosibirsk: Institute of Archaeology and Etnography SB RAS.

Karavanić, I., Šošić, R., Ahern, J. C. M., Čondić, N., Janković, I., Zubčić, K., et al. (2016). Recent research on the Middle/Upper Paleolithic interface in Croatia in the context of Central and SE Europe. In K. Harvati & M. Roksandic (Eds.), *Paleoanthropology of the Balkans and Anatolia: Human evolution and its context* (pp. 153–169). Dordrecht: Springer.

Karavanić, I., Vukosaljević, N., Šošić, R., Kurtanjek, D., & Zupanič, J. (2013). Litička i koštana industrija epigravetijenskih slojeva Šandalje II kod Pule. *Vjesnik za arheologiju i povijest dalmatinsku*, 106, 7–73.

Kesterke, M. J., & Ahern, J. C. M. (2007). Is the late mandibular sample from Vindija Cave (Croatia) biased? *Collegium Antropologicum, 31*, 365–373.

King, W. (1864). The reputed fossil Man of Neanderthal. *Quaterly Journal of Science, 1*, 88–97.

Krause, J., Orlando, L., Serre, D., Viola, B., Prüfer, K., Richards, M., et al. (2007). Neandertals in Central Asia and Siberia. *Nature, 449*, 902–904.

Krings, M., Capelli, C., Tschentscher, F., Geisert, H., Meyer, S., von Haeseler, A., et al. (2000). A view of Neandertal genetic diversity. *Nature Genetics, 26*, 144–146.

Leroi-Gourhan, A. (1958). Etude sur des restes humains provenant des grottes d'Arcy-sur-Cure. *Annales de Paleontologie, 44*, 87–147.

Lévêque, F., & Vandermeersch, B. (1980). Découverte de restes humains dans un niveau castelperronien à Saint-Cesaire (Charente Maritime). *Comptes Rendus de l'Académie des Sciences Paris, D291*, 187–189.

Lohse, K., & Frantz, A. F. (2014). Neandertal admixture in Eurasia confirmed by Maximum likelyhood analysis of three genomes. *Genetics, 96*(4), 1241–1245.

Malez, M. (1963). Istraživanje pleistocenske stratigrafije i faune u 1962. godini. *Ljetopis jugoslavenske Akademije Znanosti i Umjetnosti, 690*, 305–313.

Malez, M. (1965). Nalazišta fosilnih hominida u Hrvatskoj. *Geoloski Vjesnik, 18*, 309–324.

Malez, M. (1970). Paleolitska kultura Krapine u svjetlu novijih istraživanja. In M. Malez (Ed.), *Krapina 1899–1969* (pp. 57–119). Zagreb: JAZU.

Malez, M. (1972). Ostaci fosilnog čovjeka iz gornjeg pleistocena Šandalje kod Pule (Istra). *Paleontologia Jugoslavica, 12*, 5–39.

Malez, M. (1974). Über die bedeutung der entdeckung von geröllgeräten in den villafranchiumschichten der Šandalja I in Istrien (Kroatien, Jugoslawien). *Bulletin scientifique Cons. Academie Yougoslavie, 19*, 79–80.

Malez, M. (1975). O značenju otkrića ostataka roda Homo u naslagama vilafranka Šandalje I kod Pule. *Rad jugoslavenske Akademije Znanosti i Umjetnosti, 371*, 181–201.

Malez, M. (1979). Nalazišta paleolitskog i mezolitskog doba u Hrvatskoj. In A. Benac (Ed.), *Praistorija Jugoslavenskih Zemalja I* (pp. 195–295). Sarajevo: Svjetlost.

Malez, M. (1980). Sites of fossil men in Yugoslavia. *Collegium Antropologicum, 4*, 13–29.

Malez, M. (1983). Razvoj kvartara, fosilnog čovjeka i njegovih materijalnih kultura na tlu Sjeverne Hrvatske. In *Posebni otisak iz knjige «Varaždinski Zbornik»*. Varaždin: JAZU.

Malez, M. (1987). Pregled paleolitičkih i mezolitičkih kultura na području Istre. In *Arheološka istraživanja u Istri i Hrvatskom primorju*. Pula: Izdanja Hrvatskog Arheološkog Društva.

Malez, M., & Rukavina, D. (1979). Položaj naslaga spilje Vindije u sustavu članjenja kvartara šireg područja Alpa. *Rad JAZU, 383*, 187–218.

Malez, M., Smith, F. H., Radovčić, J., & Rukavina, D. (1980). Upper Pleistocene hominids from Vindija Cave, Croatia, Yugoslavia. *Current Anthropology, 21*, 365–367.

Malez, M., & Vogel, J. C. (1969). Rezultati određivanja apsolutne starosti pleistocenskih naslaga Šandalje II kod Pule u Istri. *Geoloski Vjesnik, 22*, 121–133.

Mann, A., & Trinkaus, E. (1974). Neandertal and Neandertal-like fossils from the Upper Pleistocene. *Yearbook of Physical Anthropology, 17*, 169–193.

Mendez, F. L., Watkins, J. C., & Hammer, M. F. (2012). A haplotype at STAT2 introgressed from Neanderthals and serves as a candidate of positive selection in Papua New Guinea. *The American Journal of Human Genetics, 91*, 265–274.

Mendez, L. F., Watkins, J. C., & Hammer, M. F. (2013). Neandertal origin of genetic variation at the cluster of OAS immunity genes. *Molecular Biolology and Evolution, 30*(4), 798–801.

Minugh-Purvis, N., Radovčić, J., & Smith, F. H. (2000). Krapina 1: A juvenile Neandertal from the Early Later Pleistocene of Croatia. *American Journal of Physical Anthropology, 111*, 393–424.

Miracle, P. T. (1995). *Broad-spectrum adaptations re-examined: Hunter-gatherer responses to late glacial environmental changes in the eastern Adriatic*. Ph.D. Dissertation, University of Michigan.

Montet-White, A. (1996). *Le Paléolithique en ancienne Yougoslavie*. Grenoble: Jérôme Million.

Oakley, K. P., & Hoskins, C. R. (1950). New evidence on the antiquity of Piltdown man. *Nature, 165*, 379–382.

Oakley, K. P., & Weiner, J. S. (1953). Chemical examination of the Piltdown implements. *Nature, 172*, 1110.

Paunović, M., Jambrešić, G., Brajković, D., Malez, V., & Mauch Lenardić, J. (2001). Last glacial settlement of Croatia: Catalogue of fossil sites dated to the OIS 2 and 3. *Acta Geologica, 26*, 27–70.

Prüfer, K., Racimo, F., Patterson, N., Jay, F., Sankararaman, S., Sawyer, S., et al. (2014). The complete genome sequence of a Neanderthal from the Altai Mountains. *Nature, 505*, 43–49.

Radovčić, J. (1988). *Gorjanović-Kramberger i krapinski pračovjek*. Zagreb: Hrvatski prirodoslovni muzej i Školska Knjiga.

Reich, D., Green, R., Kircher, M., Krause, J., Patterson, N., Durand, E., et al. (2010). Genetic history of an archaic hominin group from Denisova Cave in Siberia. *Nature, 468*, 1053–1060.

Rink, J. W., Schwarcz, H., Smith, F. H., & Radovčić, J. (1995). ESR ages for Krapina Hominids. *Nature, 378*, 24.

Roksandic, M. (2016). The role of the Central Balkans in the peopling of Europe: Paleoanthropological evidence. In K. Harvati & M. Roksandic (Eds.), *Paleoanthropology of the Balkans and Anatolia: Human evolution and its context* (pp. 15–33). Dordrecht: Springer.

Russell, M. D. (1987a). Bone breakage in the Krapina hominid collection. *American Journal of Physical Anthropology, 72*, 373–379.

Russell, M. D. (1987b). Mortuary practices at the Krapina Neandertal site. *American Journal of Physical Anthropology, 72*, 381–397.

Sánchez-Quinto, F., Botigué, L. R., Civit, S., Arenas, C., Ávila-Arcos, M. C., Bustamante, C. D., et al. (2012). North African populations carry the signature of admixture with Neandertals. *PLoS One, 7*(10), e47765.

Sankararaman, S., Mallick, S., Dannemann, M., Prüfer, K., Kelso, J., Pääbo, S., et al. (2014). The genomic landscape of Neanderthal ancestry in present-day humans. *Nature, 507*(7492), 354–357.

Sankararaman, S., Patterson, N., Li, H., Pääbo, S., & Reich, D. (2012). The date of interbreeding between Neandertals and Modern Humans. *PLoS Genetics, 8*(10), e1002947.

Schaaffhausen, H. (1857). Vortrag vom 2. Juni zu den Menschenknochen aus dem Neandertal. Verhandl. *Naturhistorischer Verein der Rheinlande und Westfalens, 14*, 50–52.

Schaaffhausen, H. (1858). Zur Kenntniß der altensten Rassenschadel. *Mullers Archiv, 5*, 453–478.

Schaaffhausen, H. (1880). Funde in der Schipkahohle in Mahren. *Verhandlungen des naturhistorisches Vereins der preussiichen Rhenilande und Westfalens, 73*, 260–264.

Schaaffhausen, H. (1888). *Der Neanderthaler Fund*. Bonn: Marcus.

Schwalbe, G. (1906). *Studien zur Vorgeschichte des Menschen*. Stuttgart: Schweizerbart'sche.

Seguin-Orlando, A., Korneliussen, T., Sikora, M., Malaspinas, A.-S., Manica, A., Moltke, I., et al. (2014). Genomic structure in Europeans dating back at least 36,200 years. *Science, 346*, 1113–1118.

Serre, D., Langaney, A., Chech, M., Teschler-Nicola, M., Paunović, M., Hofreiter, M., et al. (2004). No evidence of Neandertal mtDNA contribution to early Modern Humans. *PLoS Biology, 2*, 313–317.

Simek, J. F. (1991). Stone tool assemblages from Krapina (Croatia, Yugoslavia). In A. Montet-White & S. Holen (Eds.), *Raw material economies among prehistoric hunter-gatherers* (pp. 59–71). Lawrence: University of Kansas.

Simek, J. F., & Smith, F. H. (1997). Chronological changes in stone tool assemblages from Krapina (Croatia). *Journal of Human Evolution, 32*, 561–575.

Škerlj, B. (1958). Were Neanderthalers the only inhabitants of Krapina? *Bulletin Scientifique Yougoslavie, 4*, 44.

Smith, F. H. (1976a). A fossil hominid frontal from Velika pećina (Croatia) and a consideration of Upper Pleistocene Hominids from Yugoslavia. *American Journal of Physical Anthropology, 44*, 127–134.

Smith, F. H. (1976a). *The Neandertal remains from Krapina: A descriptive and comparative study* (Reports of Investigation No. 15). Knoxville: University of Tennessee.

Smith, F. H. (1982). Upper Pleistocene hominid evolution in South-Central Europe: A review of the evidence and analysis of trends. *Current Anthropology, 23*, 667–703.

Smith, F. H. (1984). Fossil hominids from the Upper Pleistocene of Central Europe and the origin of modern Europeans. In F. H. Smith & F. Spencer (Eds.), *The origins of modern humans: A world survey of the fossil evidence* (pp. 137–209). New York: A. R. Liss.

Smith, F. H. & Smith, M. O. (1986), On the significance of anomalous nasal bones in the Neanderthal from Krapina. Anthropos, 23, 217–226.

Smith, F. H. (1992). The role of continuity in human evolution. In G. Bräuer & F. H. Smith (Eds.), *Continuity or replacement? Controversies in Homo sapiens evolution* (pp. 145–156). Rotterdam: Balkema.

Smith, F. H. (1994). Samples, species and speculations in the study of modern human origins. In M. H. Nitecki & D. V. Nitecki (Eds.), *Origins of anatomically modern humans* (pp. 227–249). New York: Plenum Press.

Smith, F. H., Boyd, D. C., & Malez, M. (1985). Additional Upper Pleistocene human remains from Vindija Cave, Croatia, Yugoslavia. *American Journal of Physical Anthropology, 68*, 375–383.

Smith, F. H., & Ranyard, G. C. (1980). Evolution of the supraorbital region in Upper Pleistocene fossil hominids from South-Central Europe. *American Journal of Physical Anthropology, 53*, 589–610.

Smith, F. H., Trinkaus, E., Pettitt, P. B., Karavanić, I., & Paunović, M. (1999). Direct radiocarbon dates for Vindija G1 and Velika pećina Late Pleistocene hominid remains. *Proceedings of the National Academy of Sciences USA, 96*, 12281–12286.

Sollas, W. (1907). On the cranial and facial characters of the Neanderthal race. *Philosophical Transactions of the Royal Society of London, Series B, 199*, 321–337.

Street, M., Terberger, T., & Orschiedt, J. (2006). A critical review of the German Paleolithic hominin record. *Journal of Human Evolution, 51*(6), 551–579.

Stringer, C., & Bräuer, G. (1994). Methods, misreadings, and bias. *American Anthropologist, 96*(2), 416–424.

Tomić-Karović, K. (1970). Krapinski neandertalac i kanibalizam. In M. Malez (Ed.), *Krapina 1899–1969* (pp. 181–186). Zagreb: JAZU.

Trinkaus, E. (1985). Cannibalism and burial at Krapina. *Journal of Human Evolution, 14*, 203–216.

Trinkaus, E., Moldovan, O., Milota, S., Bîlgăr, A., Sarcina, L., Athreya, S., et al. (2003). An early modern human from the Peştera cu Oase, Romania. *Proceedings of the National Academy of Sciences, 100*, 11231–11236.

Trinkaus, E., & Shipman, P. (1993). *The Neandertals: Changing the image of mankind*. New York: A.A. Knopf.

Trinkaus, E., & Smith, F. H. (1995). Body size of the Vindija Neandertals. *Journal of Human Evolution, 28*, 201–208.

Ullrich, H. (1989). Neandertal remains from Krapina and Vindija—Mortuary practices, burials or cannibalism? *Humanbiol Budapest, 19*, 15–19.

Ullrich, H. (2006). Krapina—A mortuary practice site with cannibalistic rites. *Periodicum Biologorum, 108*, 503–517.

Virchow, V. (1872). Unterschung des Neanderthal-schädels. *Zeitschrift für Ethnologie, 4*, 157–165.

Vuković, S. (1935). *Istraživanje prethistorijskog nalazišta u spilji Vindiji kod Voće*. Varaždin: Spomenica varaždinskog muzeja.

Vuković, S. (1949). Prethistorijsko nalazište spilje Vindije. *Historijski Zbornik, 2*, 243–249.

Vuković, S. (1950). Paleolitska kamena industrija spilje Vindije. *Historijski Zbornik, 3*, 241–256.

Vuković, S. (1962/1963). Paleolitska kamena industrija nalazišta Punikve kod Ivanca. *Godišnjak Gradskog muzeja Varaždin, 2-3*, 23–30.

Wall, J. D., Yang, M. A., Jay, F., Kim, S. K., Durand, E. Y., Stevison, L. S., et al. (2013). Higher levels of Neanderthal ancestry in East Asians than in Europeans. *Genetics, 194*, 199–209.

White, T. D., & Toth, N. (1991). The question of ritual cannibalism at Grotta Guattar. *Current Anthropology, 32*, 118–124.

Wild, E. M., Paunović, M., Rabeder, G., Steffan, G., & Steier, P. (2001). Age determination of fossil bones from the Vindija Neanderthal site in Croatia. *Radiocarbon, 43*, 1021–1028.

Wolpoff, M. H. (1978). The dental remains from Krapina. In M. Malez (Ed.), *Krapinski pračovjek i evolucija hominida* (pp. 119–144). Zagreb: JAZU.

Wolpoff, M. H. (1979). The Krapina dental remains. *American Journal of Physical Anthropology, 50*, 64–114.

Wolpoff, M. H. (1999). *Paleoanthropology* (2nd ed.). New York: McGraw-Hill.

Wolpoff, M. H., Smith, F. H., Malez, M., Radovčić, J., & Rukavina, D. (1981). Upper Pleistocene human remains from Vindija cave, Croatia, Yugoslavia. *American Journal of Physical Anthropology, 54*, 499–545.

Yotova, V., Lefebvre, J.-F., Moreau, C., Gbeha, E., Hovhannesyan, K., Bourgeois, S., et al. (2011). An X-linked haplotype of Neandertal origin is present among all non-African populations. *Molecular Biolology and Evolution, 28*(7), 1957–1962.

Zilhão, J. (2009). Szeletian, not Aurignacian: A review of the chronology and cultural associations of the Vindija G1 Neandertals. In M. Camps & P. Chauhan (Eds.), *Sourcebook of Paleolithic transitions* (pp. 407–426). New York: Springer.

Chapter 4
The Human Fossil Record from Romania: Early Upper Paleolithic European Mandibles and Neanderthal Admixture

Katerina Harvati and Mirjana Roksandic

Abstract The hominin fossil record of Romania comprises some of the earliest and best preserved early modern humans in Europe. As such, these fossils play an important role in our understanding of the timing of the modern human arrival in Europe, their local evolution, and their potential interactions with Neanderthal populations. In this chapter, we briefly review the Romanian human fossil record and present new 3D geometric morphometric analyses of the mandibular remains from Oase and Muierii. Our findings are discussed in the context of admixture between Neanderthals and modern humans and in light of paleogenetic results (Fu et al. 2015) indicating recent Neanderthal ancestry for Oase 1.

Keywords Modern human origins • Oase • Muierii • Interbreeding • Hybrid

Introduction

Romania forms the northeastern part of the Balkan Peninsula, on the frontier with Central Europe. Its southern border is flanked by the Danube, considered an important potential corridor for hominin dispersals into Europe from the northern shores of the Black Sea. In the northwest, the Carpathian Mountains rise from the low-lying areas to the east and the south and represent a geographic boundary between the Balkans and Central Europe. The Iron Gates Gorge cuts through the Carpathian Basin, and while the gorge itself is not navigable and the slopes are rough, the hinterland represents a mild, easily penetrable landscape with low-lying undulating banks. A number of relic Tertiary floras are preserved here (Mišić et al. 1969) and this area could have played the role of a limited, but important refugium during Quaternary glaciations.

As is the case in other areas of the Balkans, the Romanian Paleolithic is characterized by a paucity of well-documented sites. This is particularly evident for the Lower Paleolithic, where only one site, Dealul Guran, can be confidently assigned to this period (Doboş and Iovita 2016). Middle Paleolithic sites are more abundant, with 120 sites attributed to this period with some confidence, 24 of which have been systematically excavated (Pop 2013). Of these, only 13 are radiometrically dated (Mertens 1996; Cârciumaru et al. 2007; Pop 2013)—including the site of Vârtop, which preserves human footprints but no associated artifacts (Onac et al. 2005). Very few of the Middle Paleolithic sites have yielded faunal remains associated with artifacts, with the exception of Ohaba Ponor (also known as Bordul Mare) and Ripiceni (also known as Izvor) (for a review see Pop 2013).

Ohaba Ponor (see Fig. 4.1), a cave in the Streiu valley in the Southern Carpathian mountains excavated by Roska in 1923 and 1924, was reported by Istvan Gaál (1928) to have yielded a human phalanx (see also Rainer and Simionescu 1942; Necrasov and Cristescu 1965; Necrasov 1971). The Ohaba Ponor lithic assemblage spans the Middle Paleolithic (represented by Eastern Charentian, or "Cave Mousterian") and the early Upper Paleolithic (Mertens 1996; Paunescu 2000). It has been dated to 43,600 + 2800 − 2100 and >41,000 ^{14}C BP for Level IIIb and 39,200 + 4500 − 2900 ^{14}C BP for Level IIIa on the basis of uncalibrated conventional ^{14}C dates on wood charcoal and unburnt bone (Honea 1984; Mertens 1996). The second right pedal phalanx has been

K. Harvati (✉)
Paleoanthropology, Department of Geosciences, Eberhard Karls Universität Tübingen, Tübingen, Germany

Senckenberg Center for Human Evolution and Paleoenvironment, Eberhard Karls Universität Tübingen, Tübingen, Germany

PhD Program in Anthropology, City University of New York Graduate Center and New York Consortium in Evolutionary Primatology, New York, NY, USA
e-mail: katerina.harvati@ifu.uni-tuebingen.de

M. Roksandic
Department of Anthropology, University of Winnipeg, 515 Portage Avenue, Winnipeg, MB, Canada R3B 2E9
e-mail: m.roksandic@uwinnipeg.ca

Katerina Harvati and Mirjana Roksandic (eds.), *Paleoanthropology of the Balkans and Anatolia*, Vertebrate Paleobiology and Paleoanthropology, DOI 10.1007/978-94-024-0874-4_4

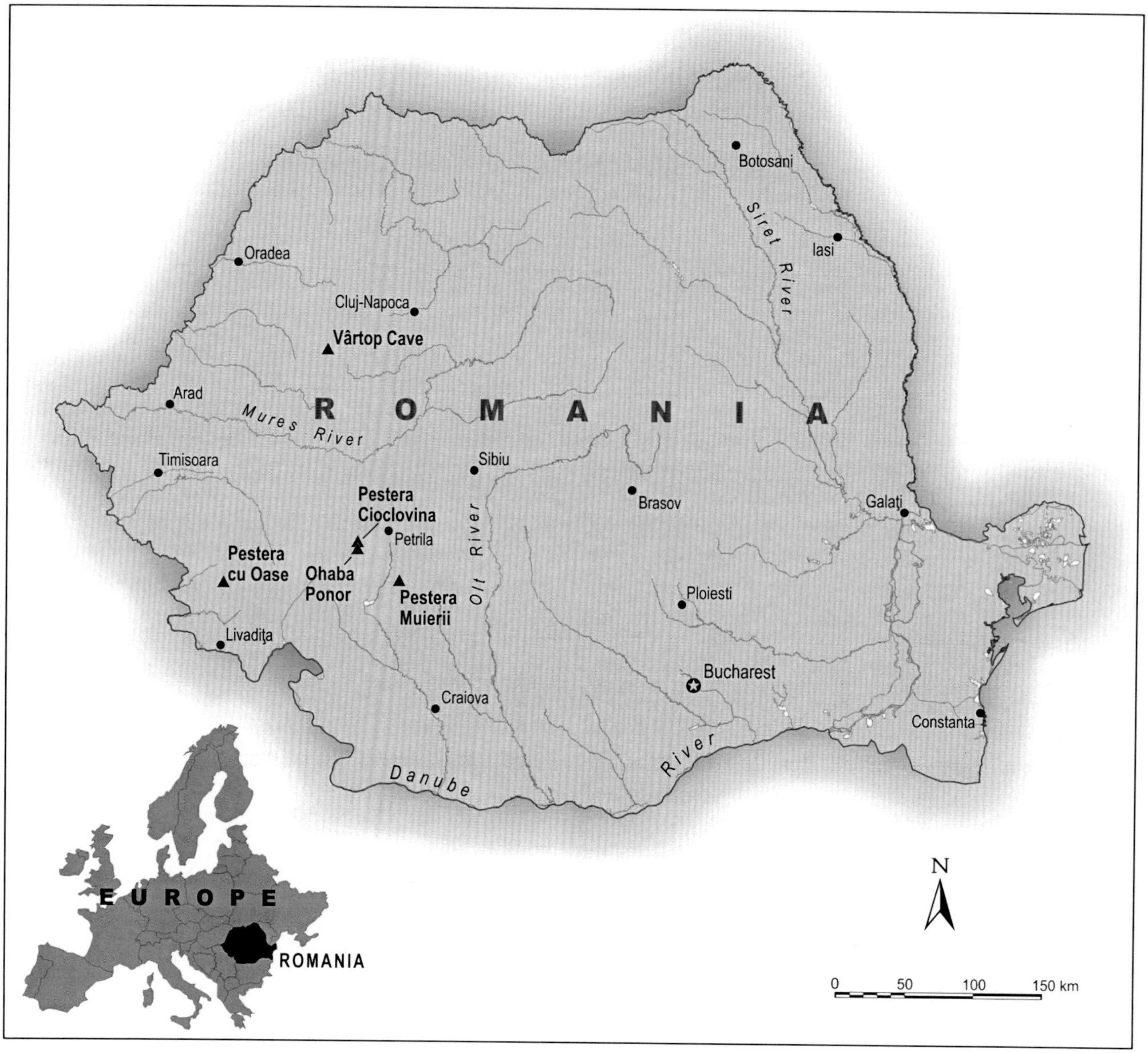

Fig. 4.1 Map of Romania, showing major sites discussed in this chapter

attributed to *H. neanderthalensis* (or *H. primigenius*; Gaál 1928; Necrasov 1971). However, Rainer and Simionescu (1942) caution that most of the artifacts found in the cave belong to the Aurignacian and that the phalanx cannot be confidently assigned to Neanderthals. Recent attempts to locate the specimen(s) were not successful (A. Doboş, personal communication 2014).

Additional Paleolithic human remains include a distal diaphyseal fragment of a right humerus from the loess of Cormani on Dniester in Bessarabia, described by Rainer and Simionescu (1942, p. 490) as relatively gracile and likely late Aurignacian; and a phalanx, found in 1979 in Livadiţa in the Iron Gates Gorge region, purportedly in association with Mousterian artifacts (Terzea 1979; Cârciumaru 1999, p. 70; Pop 2013). No other information exists about these specimens, and their whereabouts are currently unknown.

Human activity during the Middle and Upper Paleolithic in Romania is further documented by a set of three footprints from the Vârtop Cave (Onac et al. 2005) and a large number of footprints from the Ciur-Izbuc Cave (Webb et al. 2014). Vârtop cave was discovered in 1955, and the footprints were found in 1974 in the Room of the Steps, in "a spongy, porous, hardened moonmilk (the cave equivalent of tufa deposited around springs)" (Onac et al. 2005, p. 1151). They appear to have been made by one individual, with the best preserved one measuring 22 cm in length and 10.6 cm in width. These dimensions are proposed to be consistent with a broad Neanderthal foot anatomy

(Onac et al. 2005). Two U–Th dates place the footprints at *ca.* 97–62 ka (Onac et al. 2005). In the Ciur-Izbuc Cave in the Carpathian Mountains, nearly 400 footprints were discovered in 1965. They were originally dated to the end of the Pleistocene and interpreted as having been made by a man, a woman, and a child, based on measurements of 188 of the recorded footprints. In 2012, after the entrance was gated to protect the cave floor from further destruction by tourists, Webb and colleagues re-examined the 51 footprints still in existence (Webb et al. 2014). Based on the range of variation in length (15.7–31.8 cm), the authors established that the footprints were made by at least seven individuals. New AMS ^{14}C dates obtained on bones of *U. spelaeus*, recorded within 2 cm below the surface on which the footprints were made, place the *terminus postquem* at ~36.5 cal kBP. Additional cave bear remains, overlaying the footprints, indicate that the latter had to be made prior to the extinction of the cave bear in the region, estimated to have occurred around ~29.0 cal kBP (Webb et al. 2014). This places the footprints closer to other dated hominin remains from Romania (see below).

All of the well-documented specimens from Romania—from Peştera Cioclovina Uscată, Peştera Muierii, and Peştera cu Oase (Fig. 4.1)—postdate 40 ka and are among the earliest modern humans in Europe. The Cioclovina cave (45°35′ N, 23°07′ E) is part of a large karstic system, known since the 1880s. Rainer and Simionescu (1942) report that it was excavated in 1911 and 1923, yielding Mousterian and Aurignacian artifacts, as well as hearths and charcoal and presumed paleolithic drawings on the cave walls. However, the Cioclovina I calvaria (currently housed at the University of Bucharest, Laboratory of Paleontology) was uncovered much later, along with several artifacts and animal bones, by miners exploiting the cave for phosphates in 1940–1941 (Rainer and Simionescu 1942). The cranium clearly belongs to a modern human (Soficaru et al. 2007; Harvati et al. 2007) and was originally assigned to *Homo sapiens diluvialis* (a common designation for Upper Paleolithic European specimens at the time; Rainer and Simionescu 1942). Although it was considered to date to the early Upper Paleolithic already upon discovery (Rainer and Simionescu 1942; Necrasov 1971; Paunescu 2000), its association with the Aurignacian lithics recovered from the cave could not be demonstrated (Churchill and Smith 2000). The specimen was therefore neglected until Olariu and colleagues reported a direct AMS date of 29,000 ± 700 ^{14}C BP (uncalibrated) on a small portion of the right mastoid, thereby confirming the originally proposed Upper Paleolithic age (Olariu et al. 2002; Olariu et al. 2005). A small piece of the occipital bone, comprising the left posterior border of the foramen magnum, was later dated by Soficaru et al. (2007) with ultrafiltration treatment, providing a similar date of 28,510 ± 170 ^{14}C BP, calibrated at 33,212 ± 693. This age is not only consistent with the later phases of the Aurignacian, but also with the beginning of the Gravettian, in the region.

Peştera cu Muierii (Cave of the Woman) or Muierilor (Cave of the Old Woman), near Baia de Fier (45° 11 N, 23° 46 E), is a karstic cave system in Gorj county, comprising several galleries. The cave was known since the 1870s, and initial archaeological investigations were undertaken in 1929 and again in the early 1950s by Constantin S. Nicolăescu-Plopşor (Alexandrescu et al. 2010). They yielded Middle and Upper Paleolithic, as well as Holocene, archaeological remains in three main galleries (Principală, Secundară, and Musteriană) (Soficaru et al. 2006; Alexandrescu et al. 2010). The Mousterian levels of the Galeria Musteriană—characterized by a Typical Mousterian of non-Levallois debitage (Doboş 2010)—were dated to 42,560 + 1310 – 1120 ^{14}C BP (uncalibrated) through conventional ^{14}C, and to 40,850 ± 450 ^{14}C BP (uncalibrated) through a more recent AMS ultrafiltration ^{14}C date (Soficaru et al. 2006). The Upper Paleolithic lithic material recovered from the entrance to the cave (Galeria Principală,)—including end scrapers, side scrapers, retouched blades, bladelets, burins, raclettes, and one bone projectile—is considered to be consistent with the evolved Aurignacian (Soficaru et al. 2006). Human skeletal remains—a partial cranium, mandible, scapula, and tibia—were discovered in 1952 by Nicolăescu-Plopşor in a surface depression at the back of Galeria Musteriană. The mandible and cranium are thought to represent one individual (Muierii 1): the mandible and the facial skeleton match, and the maxillary and mandibular postcanine dentition are similar in size and in their degree of attrition (Soficaru et al. 2006). Similarly, the modest size of the tibia and the scapula is consistent with the probable female sex of the cranial skeleton, and the bones have been attributed to the same individual. Two additional human skeletal elements of unknown provenance have been reported for this site: a temporal bone (Muierii 2) and a fibular diaphysis (Muierii 3). They were described by Nicolăescu-Plopşor (1968, p. 383) as "*Homo sapiens fossilis* with some archaic traits" (see also Gheorghiu and Haas 1954a, b).

Because of their modern morphology and lack of a firm archaeological context, a Pleistocene age for the Muierii remains could not be ascertained initially. They were therefore largely ignored until direct AMS ^{14}C dating (Olariu et al. 2002) of Muierii 1 produced a date of 30,150 ± 800 ^{14}C BP (uncalibrated) (Olariu et al. 2002, 2005). Soficaru et al. (2006) redated the Muierii 1 cranium and the Muierii 2 temporal bone using AMS ultrafiltration to 29,930 ± 170 ^{14}C BP and 29,110 ± 190 ^{14}C BP (both uncalibrated), respectively. When calibrated, these dates place the Muierii fossils between 34 and 35 ka, thus among the earliest modern human remains known from Europe (Soficaru et al. 2006).

Peştera cu Oase (Cave with Bones) is located in the same general region of the southwestern Carpathian hills

(45°01′ N, 21°50′ E). It was discovered in 2002 by Milota, Bîlgăr, and Sarcina in the course of research on the karstic cave system of Plopa-Ponor (Milota et al. 2013). Mapping and excavation of the fossil-bearing sediments was undertaken in 2003–2004 and 2006 (Trinkaus et al. 2003a, b; Trinkaus 2007; Trinkaus et al. 2013). Skeletal remains of two individuals, comprising a mandible and a partial cranium, were recovered in the surface accumulation of the side gallery (Panta Strămoşilor) among bones of *Ursus spelaeus* and other large mammals (Trinkaus et al. 2003a, b). No lithic artifacts were recovered. Human presence in the Peştera cu Oase, therefore, seems to have been ephemeral and possibly limited to one episode. There are no cut marks or carnivore gnaw marks on either of the specimens, and it is not clear how they were deposited in the cave (Rougier et al. 2007; Trinkaus et al. 2013).

Oase 1 is a large robust adult mandible found on the surface of the Sala Mandibulei in 2002. It is dated to 34,950 + 990 − 890 ^{14}C BP by AMS ^{14}C (calibrated to ~40,400 cal BP) and is therefore one of the oldest early modern human specimens in Europe (Trinkaus et al. 2003a, b; Benazzi et al. 2011). The cranial remains (Oase 2) were found in 2003 among *U. spelaeus* bones during a detailed paleosurface mapping in the Panta Strămoşilor (Trinkaus et al. 2003a, b). The almost complete cranium is that of an adolescent, as evidenced by the unfused spheno-occipital synchondrosis and unerupted M^3s, and therefore did not belong to the same individual as the mandible. Direct AMS ^{14}C dating on the cranium produced a minimum age of 28,890 + ∞ − 170 ^{14}C BP (uncalibrated; Rougier et al. 2007). This individual is therefore considered by the original authors (Rougier et al. 2007, p. 1166) to be either penecontemporaneous with or slightly younger than the Oase 1 specimen.

Given their age and relatively good state of preservation, the Romanian Upper Paleolithic fossils provide a rare glimpse of the earliest modern human populations of Europe. They afford the opportunity to assess hypotheses about the origin of these populations, as well as about possible interbreeding events with Neanderthals. Interbreeding has been postulated from the Neanderthal genomic studies (Green et al. 2010; Prüfer et al. 2014) and is thought to have likely occurred during the initial expansion of modern humans out of Africa in the Near East. Interbreeding events are thought to have been rare, resulting in a minimal contribution of Neanderthals to the recent modern human gene pool (1.5–2.1 %; Prüfer et al. 2014). All three Romanian early Upper Paleolithic samples (Cioclovina, Muierii, and Oase) have been proposed to show a mixture of archaic and modern features and to possibly represent Neanderthal–modern human hybrids on the basis of their anatomy. In the case of Cioclovina, Soficaru et al. (2007) proposed similarities with Neanderthals in the nuchal region, suggesting a hybrid status for this individual. This hypothesis was not supported by a 3D geometric morphometric comparative analysis of the vault shape (Harvati et al. 2007) or a study of the Cioclovina inner ear morphology (Uhl et al. 2016). The Muierii material has been proposed to show several Neanderthal-like features in the face, occipital bone, mandible, and scapula (Soficaru et al. 2006). Finally, the Oase remains have been argued to perhaps reflect admixture with Neanderthals on the basis of the lingual bridging of the mandibular foramen in Oase 1 (Trinkaus et al. 2003a) and several cranial and dental features of Oase 2 (Rougier et al. 2007). A recent genomic analysis of Oase 1 found that this individual indeed possessed a higher proportion of Neanderthal DNA than any recent or early modern human specimen sampled until now, confirming a recent Neanderthal ancestor for this specimen, as recently as 4–6 generations previously (Fu et al. 2015).

In light of these findings, another look at the mandibular sample from the Romanian early Upper Paleolithic is warranted. Although mandibular morphology is generally considered to be greatly influenced by diet and masticatory requirements and to preserve a weaker population history signal than other components of the human skull (see e.g., Smith 2009; von Cramon-Taubadel 2011, 2014), previous work has shown that it does preserve some phylogenetic information (Nicholson and Harvati 2006). Here, we conducted 3D geometric morphometric comparative analyses of the Oase 1 and Muierii 1 mandibles, in order to assess their morphological affinities and explore their potential hybrid status.

The Romanian Upper Paleolithic Mandibles

Oase 1 is a complete adult mandible (Fig. 4.2a). It is very well preserved, lacking only the teeth from I1-P4 on the right and I1-M1 on the left. It is large and robust and was presumed to be male on anatomical grounds, confirmed recently by paleogenetic analysis (Trinkaus et al. 2003a; Fu et al. 2015). It shows a broad ramus and a relatively narrow corpus (both considered plesiomorphic conditions that were lost in Neanderthals). The anterior symphysis presents a bony chin with a prominent tuber symphyseos and minimally developed lateral tubercles—a morphology typical for early modern humans. The absence of a retromolar space, the positioning of the mental foramina under the P_4 alveolus, and the medially placed condyles all align the mandible with modern humans. Trinkaus et al. (2003a) note that a lingual bridging of the mandibular foramen—a trait more commonly present in Neanderthals than in modern humans—is present on the left and absent on the right side of this specimen, suggesting to those authors a Neanderthal contribution to its ancestry. The M_1 and M_2s are worn and therefore of limited

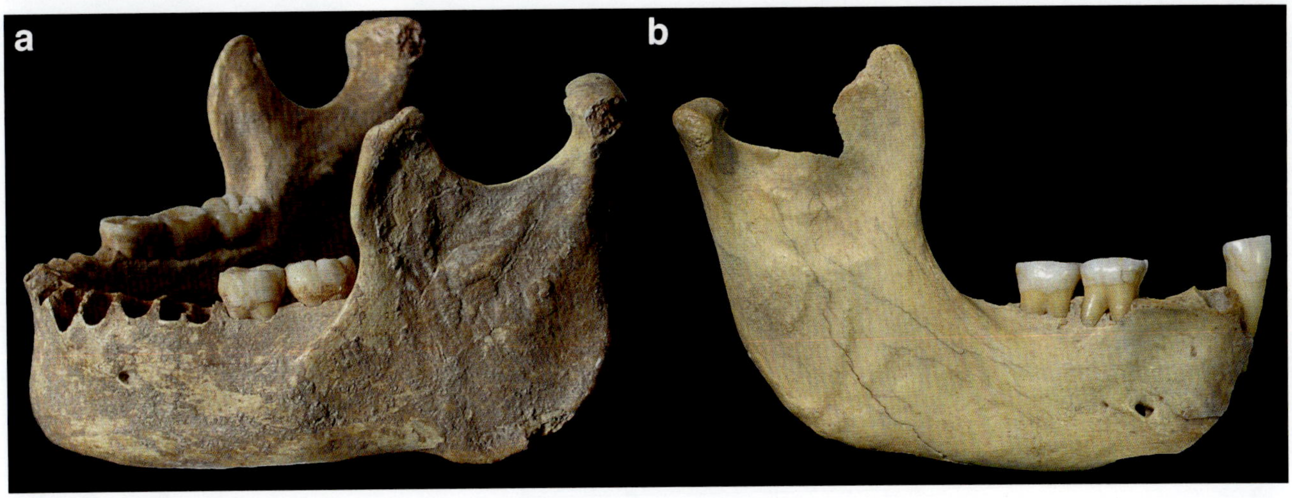

Fig. 4.2 (**a**) Oase 1 mandible, lateral view (photo copyright K. Harvati); (**b**) Muierii 1 mandible, lateral view (Muierii 1 photo kindly provided by and copyright Erik Trinkaus/ISER)

value for morphological analysis. However they exhibit at least five cusps. There are no mid-trigonid crests (a feature that is present in Neanderthals at very high frequencies compared to modern humans; Bailey 2002). The M_3s exhibit small entoconulids and modest anterior foveae, traits more common among early modern humans. Dental metrics are unusual in that the second and third molars fall substantially above mean values reported for both early modern humans (European Upper Paleolithic and the Qafzeh-Skhul sample) and Neanderthals (Trinkaus et al. 2003a).

Muierii 1 is a fragmentary hemimandible (Fig. 4.2b). It preserves a largely complete right ramus and the corpus (with first and second molars) up to the level of the canine. It is lightly built, with a mental foramen situated relatively posteriorly under the P_4. The mandibular notch crest meets the anterior condyle in the lateral half of the middle third of the condyle. The coronoid process is high and the mandibular notch asymmetric, features considered to be Neanderthal traits (Rak 1998; Rak et al. 2002; but see Jabbour et al. 2002; Wolpoff and Frayer 2005). The heavily worn dentition does not preserve diagnostic crown features. It is metrically aligned with more recent Upper Paleolithic humans.

Table 4.1 Recent human (RH) samples included in the comparative analysis

Population	Specimens
Oceania (Australia, New Guinea, and Tasmania)	18
Pacific	18
Southeast Asia (Southeast Asia and China)	14
North Asia (Japan, Korea, Siberia, and Mongolia)	13
East Africa (Masai)	14
South Africa (Khoisan and Bantu)	9
Europe	26
South America	19
Central America (Central America and Mexico)	10
North America Arctic (Alaska, Greenland, and N. Canada)	14
Total	***n*=155**

Additional information in the Appendix

Materials and Methods

Samples: The Oase 1 and Muierii 1 original specimens were digitized by KH at the Institute of Speleology, Cluj, and the Academy of Sciences, Bucharest, respectively. The comparative sample comprised 25 fossil and 155 recent human (RH) mandibles from around the world, digitized by Elisabeth Nicholson Lopez and KH (Tables 4.1 and 4.2, Appendix; see also Nicholson and Harvati 2006). Four Middle Pleistocene European specimens (MP), seven Neanderthal (NEA), twelve Upper Paleolithic/Later Stone Age (UP), and two Late Pleistocene early anatomically modern human (EAM) specimens were included. In cases where we were not able to access the original fossils, high-quality casts were used from the collections of the Division of Anthropology at the American Museum of Natural History, the Department of Anthropology, New York University, and the Department of Human Evolution, Max Plank Institute for Evolutionary Anthropology. Although human mandibular morphology exhibits sexual dimorphism, this was not the focus of the present analysis, and sexes were pooled for both modern and fossil samples. The inclusion of a large modern human comparative sample relative to the much smaller fossil samples in our analyses likely will skew some of the results, for example, the principal components analysis (see below). Nevertheless, we felt that including only the very small fossil modern human sample

Table 4.2 Fossil specimens included in the comparative analysis

Comparative fossil samples	Total: 25
Neanderthals and Early Neanderthals (NEA)	**7**
Amud 1*, Krapina J*, La Ferrassie 1, Shanidar 1*, Tabun C 1*, Zafarraya*, Regourdou	
Middle Pleistocene Hominins (MP)	**4**
Arago 13*, Mauer 1, Montmaurin, Sima de los Huesos 5*	
Early Anatomically Modern Humans (EAM)	**2**
Skhul 5, Qafzeh 9*	
Eurasian Upper Paleolithic (UP)	**12**
Grimaldi-Grotte-Des-Enfants 6*, Isturitz 1950-4-1, Dolni Vestonice 3, 13, 14, 15, 16, Abri Pataud, Ohalo II, Upper Cave 1* and 3*, Oberkassel 2*	

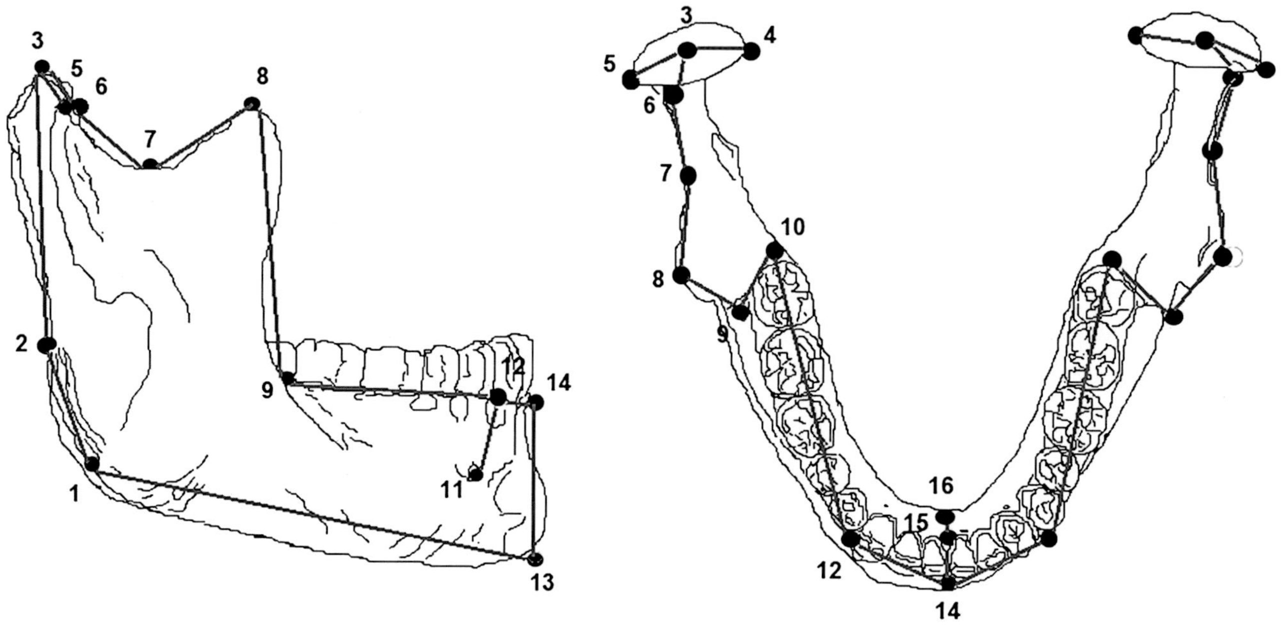

Fig. 4.3 Mandibular landmarks. 1. gonion (*Right* and *Left*), 2. posterior ramus (*Right* and *Left*), 3. condyle tip (*Right* and *Left*), 4. condylion mediale (*Right* and *Left*), 5. condylion laterale (*Right* and *Left*), 6. root of sigmoid process (*Right* and *Left*), 7. sigmoid notch (*Right* and *Left*), 8. coronion (*Right* and *Left*), 9. anterior ramus (*Right* and *Left*), 10. M3 (*Right* and *Left*), 11. mental foramen (*Right* and *Left*), 12. Canine (*Right* and *Left*), 13. gnathion, 14. infradentale, 15. mandibular orale, 16. superior transverse torus (Landmark definitions after Nicholson and Harvati 2006). Figure adapted from Nicholson and Harvati (2006)

would severely limit the range of modern human variation. We therefore decided to include a large and geographically variable modern human sample in order to have an adequate comparative frame for the Romanian specimens.

Data: The data were collected in the form of three-dimensional coordinates of 28 landmarks using a Microscribe 3DX digitizer (for inter- and intraobserver error assessments and landmark definitions see Nicholson and Harvati 2006). Since many of the fossil specimens were incomplete, data reconstruction through reflected relabeling was undertaken (see Nicholson and Harvati 2006). Landmarks were selected not only to represent the overall shape of the mandible but also to quantify as best as possible the described differences among modern human and Neanderthal mandibles (Fig. 4.3; Nicholson and Harvati 2006). Because Muierii 1 is incomplete, the analysis was repeated with only the 12 landmarks preserved by this specimen.

Analysis: Landmark coordinates were superimposed using generalized Procrustes analysis (GPA) in Morphologika (O'Higgins and Jones 2006). GPA superimposes the specimen landmark configurations by translating them to a common origin, scaling them to unit centroid size (the square root of the sum of squared distances of all landmarks to the centroid of the object; the measure of size used here), and rotating them according to a best-fit criterion. This procedure allows for the separate analysis of 'shape' and 'size' (although size-related shape differences may remain; Rohlf 1990; Rohlf and Marcus 1993; Slice 1996; O'Higgins and Jones 1998).

Procrustes methods have been shown to have higher statistical power than alternative geometric morphometric approaches (Rohlf 2000) and have been applied extensively in paleoanthropology (e.g., Harvati 2003; Oettlé et al. 2005; Nicholson and Harvati 2006; Gunz et al. 2009; Bastir et al. 2011; Robinson 2012).

A principal components analysis (hereafter PCA) was conducted on the fitted coordinates to explore the patterns of variation present in the data. An ANOVA was performed on the PCA scores to determine the significance of taxonomic effects on each PC axis. Shape changes along the PC axes were visualized using Morphologika. A discriminant analysis and classification were undertaken. Because the number of variables used for such an analysis should not be greater than the number of specimens in the smallest group, we used only the first four principal components (53.8 % of the total variance for the 28 landmark analysis and 54.7 % for the 12 landmark analysis). Oase 1 and Muierii 1 were treated as unknowns to be classified, as were Skhul 5 and Qafzeh 9. Mean Procrustes distances (hereafter PD) between Oase 1 and Muierii 1, on the one hand, and the comparative samples on the other, as well as interindividual PD between the two Romanian specimens and the fossil sample, were calculated. Statistical analyses were performed in Morphologika, SAS (SAS Institute) and PAST (Hammer et al. 2001), and plots were produced with PAST.

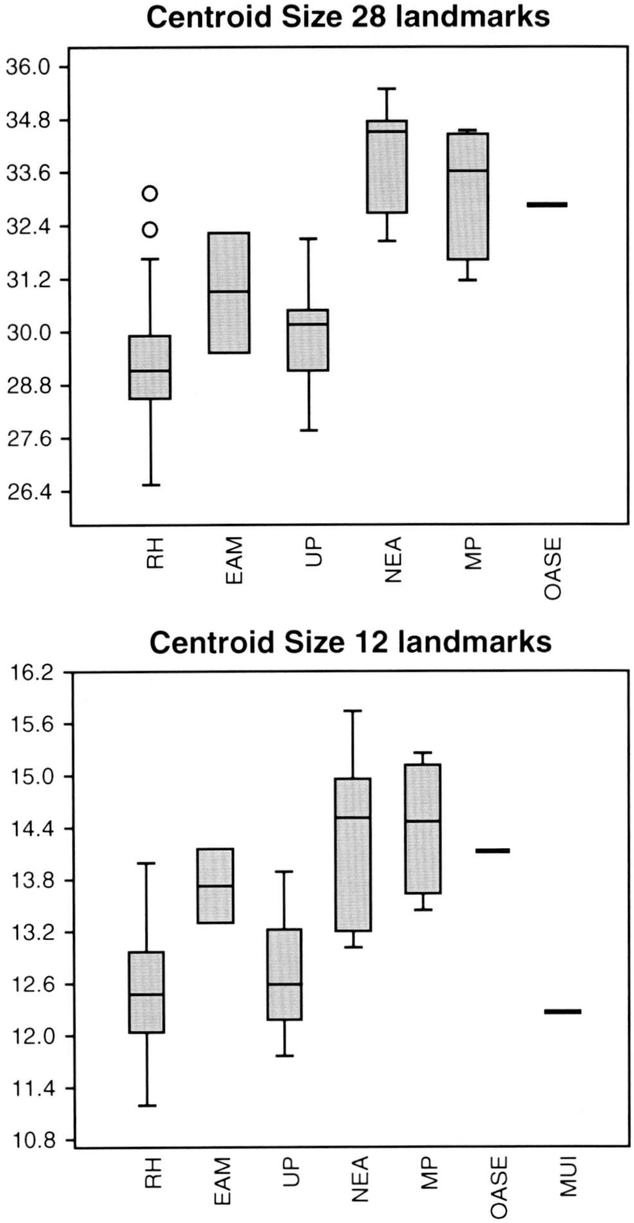

Fig. 4.4 Centroid sizes of the comparative samples used in the geometric morphometric analysis. *Top*: 28 landmark analysis. *Bottom*: 12 landmark analysis

Results

Centroid Size: As previously described, Oase 1 is very large (Fig. 4.4). Its centroid size is slightly above the highest value for the UP and EAM samples. It falls just at the upper end of the recent human variation, and well within the centroid size range of *H. heidelbergensis* and *H. neanderthalensis*. By contrast, Muierii 1 is smaller, and its centroid size falls in the lower part of the UP and recent modern human range of variation, and outside that of the EAM, NEA, or MP samples (Fig. 4.4, bottom).

Twenty-eight Landmark Analysis: In the 28 landmark PCA, PC 1 (23.88 % of the total variance; Fig. 4.5 top) represented intraspecific variation. All taxa overlapped widely along PC1, with short and wide mandibles scoring negatively and long and narrow mandibles scoring positively. Part of the variation along PC1 was driven by Qafzeh 9, which scored very highly on this axis and fell outside the convex hulls of any of the comparative samples. Neanderthals (and Middle Pleistocene *Homo*) were separated from modern humans most clearly along PCs 2 and especially 3 (together accounting for just over 23 % of the total variance; Fig. 4.5, bottom). Both PC 2 and PC 3 were significant for group effects ($p<0.0001$ for both). The recent modern human convex hull is located between the center and the positive side of PC2 and at the positive side of PC3. The UP sample fell nearly completely within the modern human convex hull, but at the center, and thus also close to the NEA and MP. The latter, on the other hand, were separated from the modern human samples, showing more negative PC2 and PC3 scores, and overlapped extensively with each other. Skhul and Qafzeh (EAM) fell within the modern human range, although outside the convex hull of the UP. Oase 1 fell just outside the modern human range and overlapping with the NEA and MP ranges on PC2, but at the border of the UP convex hull along PC2 and 3.

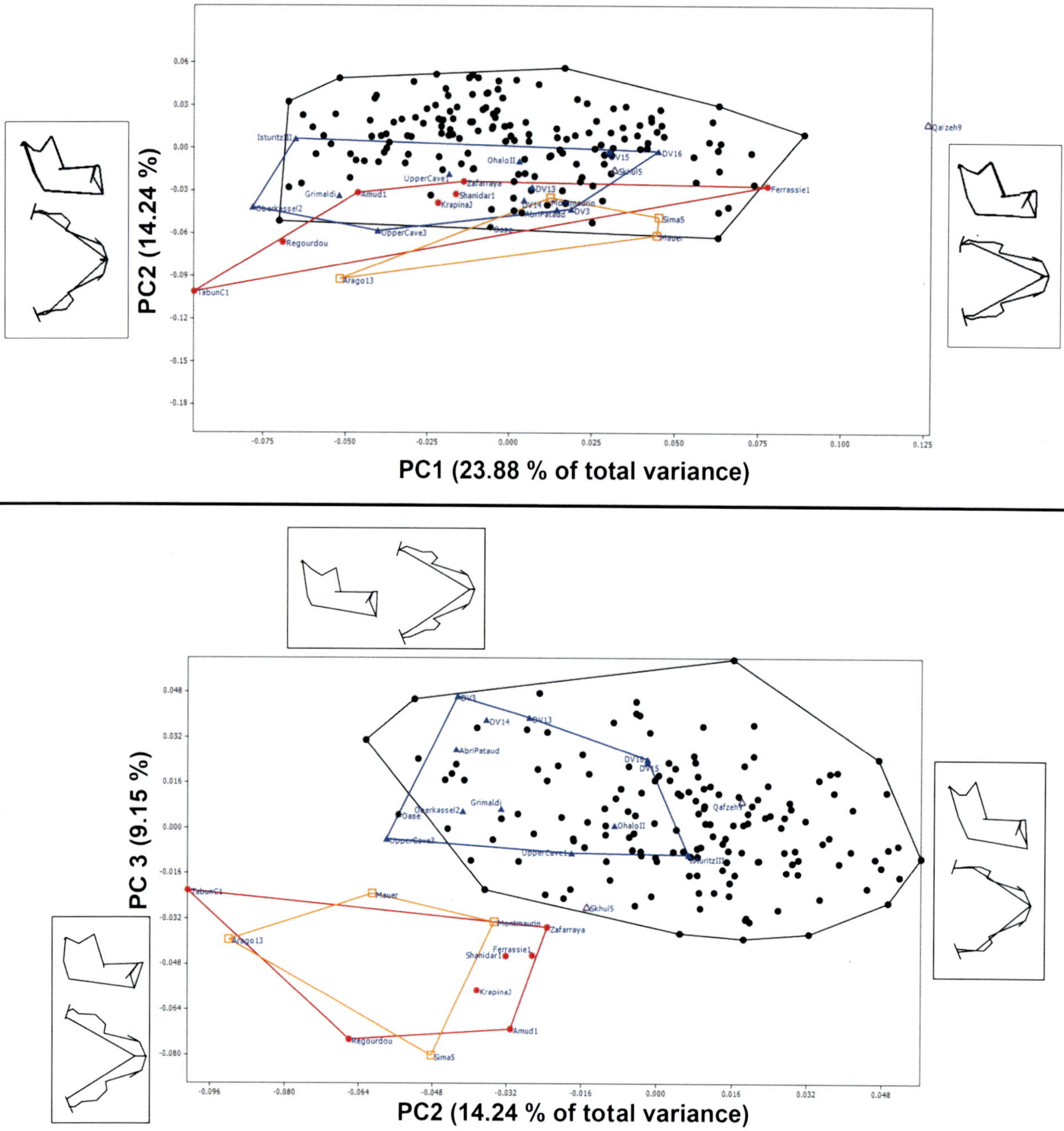

Fig. 4.5 Principal components analysis, 28 landmarks, PCs 1 and 2 (*top*), PCs 2 and 3 (*bottom*). *Red stars*: (NEA), *open yellow squares*: (MP), *black triangles* (EAM), *blue triangles* (UP), *black dots* (modern humans), *black star*: Oase 1. DV stands for Dolni Vestonice in the specimen labels

The discriminant analysis classified Oase 1 as UP with a posterior probability of 0.76. Qafzeh 9 was classified as modern human (probability 0.99), whereas Skhul 5 as MP (0.695). The cross-validation classification classified 81.94 % of modern human mandibles correctly as modern and 18.06 % as UP. 66.67 % of UP were classified correctly, whereas three specimens (25 %) were classified as modern human and one as NEA. None of the NEA or MP specimens were misclassified as modern human or UP. However, they were often misclassified as each other (50 % of MP specimens classified as NEA; 42.86 % of NEA classified as MP). The four closest specimens to Oase 1 in PD, and therefore most similar to it in overall shape, were all UP specimens: Abri Pataud, Upper Cave 101, and Dolni Vestonice 13 and 14 (in that order). Among recent modern humans, Oase 1 was closest to an individual from our Oceania sample (Tasmania). When mean PD values between Oase 1 and all comparative samples were examined, it was by far closest to the UP sample (Table 4.3).

Twelve Landmark Analysis: As in the 28 landmarks analysis, PC1 (20.61 % of total variance, Fig. 4.6, top) reflected

Table 4.3 Mean (top) and interindividual (bottom) Procrustes distances between Oase 1 and Muierii 1 and all comparative samples (top) and individual fossil specimens (bottom)

28 landmarks analysis		12 landmarks analysis			
	Oase 1		Oase 1		Muierii 1
UP	0.0938	UP	0.1252	UP	0.1215
MP	0.1149	EAM	0.1333	EAM	0.1229
RH	0.1152	MP	0.1525	MP	0.1253
NEA	0.1241	RH	0.1613	NEA	0.1272
EAM	0.1353	NEA	0.1667	RH	0.1320
Abri Pataud	0.0702	Abri Pataud	0.0941	Upper Cave 1	0.0922
Upper Cave 1	0.0818	DV15	0.1033	DV16	0.0964
DV13	0.0830	Skhul5	0.1123	Upper Cave 3	0.1049
DV14	0.0837	DV13	0.1157	Zafarraya	0.1054
Ohalo II	0.0875	DV16	0.1164	Ferrassie 1	0.1060
Upper Cave 3	0.0879	Upper Cave 1	0.1211	Montmaurin	0.1078
DV15	0.0908	Oberkassel 2	0.1261	DV15	0.1082
DV3	0.0909	Ohalo II	0.1299	Skhul 5	0.1086
Skhul 5	0.0935	Muierii 1	0.1303	Ohalo II	0.1094
Zafarraya	0.1031	DV3	0.1311	Isturitz III	0.1151
Mauer	0.1067	Upper Cave 3	0.1323	Regourdou	0.1172
DV16	0.1070	DV14	0.1349	Amud 1	0.1193
Montmaurin	0.1077	Isturitz III	0.1388	Krapina J	0.1207
Oberkassel 2	0.1085	Arago 13	0.1388	Oberkassel 2	0.1214
Arago 13	0.1104	Mauer	0.1392	Arago 13	0.1248
Krapina J	0.1124	Montmaurin	0.1516	DV13	0.1298
Isturitz III	0.1128	Qafzeh 9	0.1543	Sima5	0.1301
Amud 1	0.1191	Ferrassie 1	0.1545	Oase 1	0.1303
Grimaldi	0.1212	Tabun C1	0.1584	Abri Pataud	0.1358
Shanidar 1	0.1240	Grimaldi	0.1593	Qafzeh 9	0.1371
Regourdou	0.1271	Regourdou	0.1634	Mauer	0.1386
Ferrassie 1	0.1337	Zafarraya	0.1641	DV14	0.1393
Sima 5	0.1346	Krapina J	0.1676	Grimaldi	0.1464
Tabun C1	0.1493	Amud 1	0.1728	Shanidar 1	0.1492
Qafzeh 9	0.1770	Sima 5	0.1803	DV3	0.1588
		Shanidar 1	0.1858	Tabun C1	0.1725

Values are reported in ascending order

variation within recent modern humans, which overlapped with all other samples on the positive side of this axis. Specimens with negative PC1 scores showed superoinferiorly tall and anteroposteriorly narrow rami, whereas those scoring more positively showed short and wide rami (Fig. 4.6, top). The PCA also separated Neanderthals from modern humans along PCs 2 and 3 (accounting for 14.92 % and 11.27 % of the total variance, respectively; Fig. 4.6 bottom), although their separation was not as clear as in the 28 landmark analysis (with Shanidar 1 and Tabun C1 falling within the modern human convex hull). All three PCs 1–3 were significant for group effects ($p<0.0001$). This greater overlap in the 12 landmarks analysis is likely due to the less complete morphology represented in the dataset. Neanderthals and *H. heidelbergensis* tended to have more negative PC2 scores and more positive PC3 scores than any of the modern human samples, and overlapped widely with each other. These scores reflected a wide ramus, asymmetrical mandibular notch, the presence of a retromolar gap, a posterior placement of the mental foramen relative to the canine, and more lateral position of the condyle relative to the root of the mandibular notch (see also Nicholson and Harvati 2006). The latter trait was reflected by positive PC2 scores and was also shown to some degree by the UP sample, which tended to score positively on PC2. Recent modern humans largely overlapped with the UP sample, although some UP specimens, as well as Skhul 5, showed more positive PC2 scores and fell outside the recent human convex hull. Qafzeh 9 fell well within the range of modern human variation, and outside that of the UP. Oase 1 fell just outside the modern human convex hull, and well within the UP one. Muierii 1, on the other hand, fell outside both recent and UP

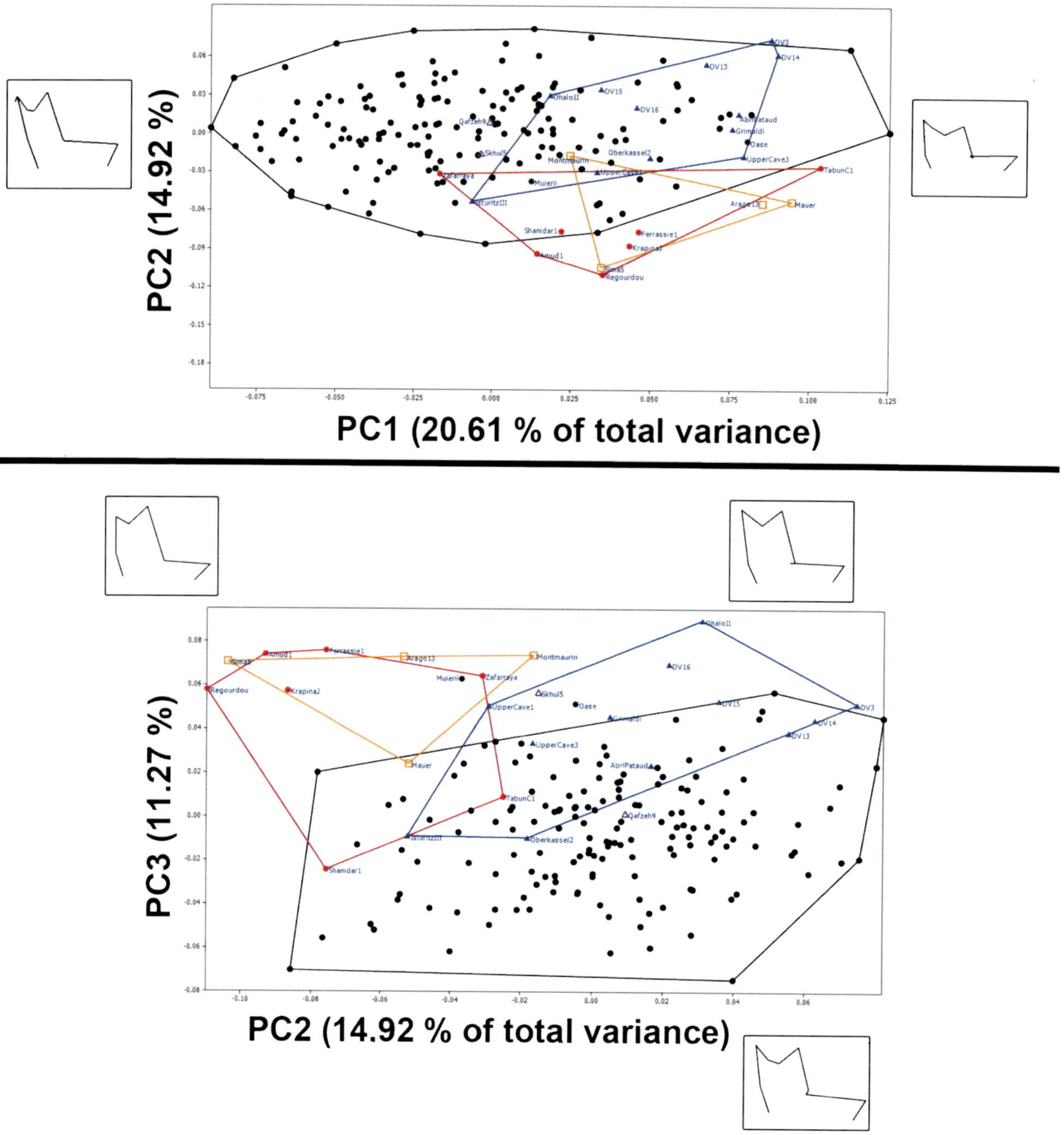

Fig. 4.6 Principal components analysis, 12 landmarks, PCs 1 and 3. Symbols as in Fig. 4.5; *black stars*: Oase 1 and Muierii 1

modern human convex hulls. Its positive PC2 and negative PC3 score placed it within the range of Neanderthals and *H. heidelbergensis*, and closest to the specimens Zafarraya, Montmaurin, and Upper Cave 101.

In the discriminant analysis, Oase 1 was again classified as UP with a posterior probability of 0.98. Muierii 1, on the other hand, was classified as NEA with a probability of 0.53. Qafzeh 9 was classified as recent modern human (0.96) and Skhul 5 as UP (0.86). The cross-validation classification showed more misclassifications among the NEA and MP samples than in the 28 landmarks analysis. 93.55 % of recent modern humans were classified correctly, 9 specimens were classified as UP (5.81 %) and 1 as NEA. Although 66.67 % of UP were classified as UP and 2 specimens (16.67 %) as recent modern humans, 1 was classified as NEA, and 1 as MP. None of the NEA specimens were classified correctly; 5 (71.43 %) were classified as MP, 1 as a recent modern human, and 1 as UP. Only 1 MP was classified correctly, whereas 3

(75 %) were misclassified as NEA. In terms of interindividual PD, Oase 1 was again closest to Abri Pataud, followed by Dolni Vestonice 15, Skhul 5, and a modern human from the Pacific sample (a Melanesian from New Britain). As in the 28 landmark analysis, among the fossil specimens (Table 4.3), Oase 1's nearest neighbors were UP or EAM specimens, and when mean PD values were considered, it was closest to the UP sample. Muierii 1 was closest to a recent modern human (Pacific; again a Melanesian from Nissan, New Britain), followed by Upper Cave 101, an Australian and a South American modern human. Among the fossil specimens, however, and in contrast to Oase 1, Muierii 1's nearest neighbors included UP and EAM, as well as NEA and MP specimens (Table 4.3). In terms of mean PD, it was closest to the UP sample, but this distance was only slightly smaller than its PD to EAM, MP, and NEA (unlike the mean PD distances for Oase 1; Table 4.3).

Discussion and Conclusions

The question of interbreeding between Neanderthals and Upper Paleolithic modern humans, as well as the extent of such admixture and its significance for modern human origins in Europe, has been the subject of long standing debate (e.g., Smith 1992; Bräuer and Broeg 1998; Wolpoff et al. 2001; Stringer 2002; Bräuer et al. 2004; Harvati et al. 2004, 2007; Smith et al. 2005). The study of the European fossil record has been inconclusive in this respect, with different researchers reaching divergent conclusions, sometimes on the same material. Analysis of ancient mtDNA of Neanderthals and early modern human specimens did not support interbreeding (e.g., Krings et al. 1997; Green et al. 2008). However, more recently, the sequencing of the Neanderthal genome indicated a limited contribution of Neanderthals to the gene pool of modern humans (Green et al. 2010; Prüfer et al. 2014). This contribution is estimated to be small, ranging from 1.5 to 2.1 % (Prüfer et al. 2014; although there is still some discussion on whether the observed similarities can be explained at least in part by population substructure of the ancestral modern human population; see Eriksson and Manica 2012; Lowery et al. 2013; Reyes-Centeno et al. 2014, 2015). Surprisingly, Neanderthal alleles are found in all modern people outside Africa, suggesting that the admixture events likely occurred in the Near East before the spread of early modern humans into the rest of Eurasia (Green et al. 2010; Prüfer et al. 2014; Sankararaman et al. 2014; Fu et al. 2014). Additional interbreeding may have occurred in Asia, where admixture levels are observed to be somewhat elevated compared to Europe (Currat and Excoffier 2011; Meyer et al. 2012; Prüfer et al. 2014; see also Vernot et al. 2016). Admixture in South Asia seems to have occurred not only with Neanderthals, but also between Denisovans and the ancestors of present-day Melanesians and native Australians (Rasmussen et al. 2011; Meyer et al. 2012; Vernot et al. 2016). The most recent paleogenomic studies also found that the modern human genome is depleted from Neanderthal-derived alleles in the X chromosome, in genes that are expressed in the testes, and genes affecting specific aspects of the brain (Sankararaman et al. 2014; Vernot and Akey 2014; Vernot et al. 2016), suggesting male infertility for Neanderthal-modern human hybrids and a high level of genetic incompatibility among the two taxa. They have also shed light on the possible time-frame of interbreeding, dated by Sankararaman et al. (2012) to between 37,000 and 86,000 years BP and likely between 47,000 and 65,000 using linkage disequilibrium rates among modern European genomes. This date was further refined to between *ca.* 50 and 60 ka on the basis of an early human specimen from western Siberia, dated to *ca.* 45 ka (Fu et al. 2014). Further admixture events between early modern humans as well as Neanderthals and other Eurasian archaic humans have also been proposed from genomic analyses (Prüfer et al. 2014), suggesting that interbreeding across hominin species, although relatively rare, might have occurred more than once among hominin taxa during the Pleistocene.

Until recently, only two early modern humans had been successively sampled for paleogenetic analysis: Ust'Ishim and Kostenki, dating to *ca.* 45 ka and 36 ka, respectively (Fu et al. 2014; Seguin-Orlando et al. 2014). Both of these possessed similar levels of Neanderthal genetic material as those found among recent humans, although Ust'Ishim also showed longer Neanderthal DNA segments, indicating a Neanderthal ancestor a few thousand years before this individual's lifetime. The recent sequencing of genomic material from Oase 1 showed for the first time an individual with higher Neanderthal genetic contribution than all modern human specimens so far sequenced: 6–9 % of Oase 1's genome was found to derive from Neanderthals, pointing to a recent Neanderthal ancestor (4–6 generations before its lifetime; Fu et al. 2015). At the same time, Oase 1 was found not to have a close genetic affinity with recent Europeans, a result interpreted as indicating that he belonged to an early modern human population which admixed with Neanderthals, but became extinct without contributing genetically to later Europeans (Fu et al. 2015).

In light of these developments, the study of hybridization in the human fossil record has assumed renewed interest. The primary aims of recent research on this subject have been to (a) assess the frequency of admixture across extant primate species and genera, and the relationship of this frequency to evolutionary divergence time and (b) establish criteria for the recognition of potential hybrid individuals or hybrid populations from skeletal morphology. Although information on admixture is not available for all primate taxa, natural hybridization is documented for more than 10 % of recognized primate species (Arnold and Meyer 2006; Zinner et al. 2011). Perhaps unsurprisingly, reproductive isolation among hybridizing mammalian species is correlated with time since divergence, with species that separated less than 2 Ma likely to be interfertile (Holliday 2006; Holliday et al. 2014). Nevertheless, the effects of hybridization on the skeletal phenotype are often difficult to identify. It is sometimes thought that hybrids will exhibit morphology intermediate to that of the two parent populations (e.g., Harvati et al. 2007, 2011). Intermediate phenotypes are indeed documented for pelage coloration and other general morphological features (Bernstein 1966; Froehlich and Supriatna 1996; see also Ackermann 2010; Hamada et al. 2016). Some evidence for intermediate skeletal morphology has been reported for *Papio* and *Macaca* hybrids from hybrid zones (e.g., Froehlich and Supriatna 1996; Frost et al. 2003), as well as intergeneric *Theropithecus–Papio* hybrids (Jolly et al. 1997). However, it is not a necessary consequence of admixture, as hybrids are thought to vary greatly in terms of their phenotype and to also sometimes look identical to one or the other of the parent taxa (Ackermann 2010). This wide range of potential hybrid phenotypes might result in hybrid populations that are more variable relative to their parental groups (Ackermann et al. 2006; Ackermann 2010). Size effects may also be expected, with hybrids being either intermediate in size (Ackermann 2010) or displaying heterosis/dysgenesis (greater or smaller size, respectively, than expected based on the phenotypes of the parental taxa; Cheverud et al. 1993; Schillaci et al. 2005; Ackermann et al. 2006; Ackermann 2010). Among the most tell-tale skeletal markers of hybridization, however, may be various dental and sutural anomalies documented in pedigree *Papio* hybrids by Ackermann et al. (2006; see also Ackermann 2010). These include extremely rare conditions, such as supernumerary molars and canines, as well as extra sutures in the maxilla and parietal bones. Such rare anomalies have been argued to indicate developmental instability, resulting from genetic incompatibility between the parents (Ackermann et al. 2006; Ackermann 2010).

Given varying frequencies of these traits in modern human populations, it is not clear to what degree such criteria are applicable to fossil humans. Nevertheless, on the basis of these criteria, Ackermann (2010) proposed several possible hybrid specimens in the human fossil record, concentrating on the geographical region and temporal framework where hybridization is considered most likely: the Late Pleistocene record of the Near East and Eastern Europe. They include, among Neanderthals: the Krapina sample, where 36 % of the hemimandibles preserving third premolars show rotated premolars (Rougier et al. 2006; Ackermann 2010); and Amud 1, who is thought to show an anomalously small upper M^3 on the right side (Ackermann 2010). Furthermore, Di Vincenzo and colleagues (2012) proposed hybrid status for the Vindija (Croatia) G3 Neanderthals on the basis of intermediate scapular glenoid fossa shape. Among early anatomically modern humans from the Levant: the Skhul sample, where one individual has a rotated upper P^4 (Skhul IV) and another shows pronounced craniofacial asymmetry (Skhul V); and the Qafzeh material, with three individuals showing dental crowding (although the significance of dental crowing in this context is unclear) and one (Qafzeh 11) showing a somewhat rotated lower P_4. Among the Romanian Upper Paleolithic specimens, Ackermann (2010) notes that Oase 2 has been reported to show unusually large upper M^3s that fall outside the range of variation of Upper Paleolithic samples. All molars of this individual are larger than comparative samples used by Rougier et al. (2007) and are also larger than the upper molars of a North African Aterian sample (Bailey, personal communication 2014). As stated above, all three Romanian early Upper Paleolithic samples have been proposed to show hybrid status on the basis of the presence of some presumed Neanderthal-like traits (see Soficaru et al. 2007; also above), although this claim found no support in a 3D comparative analysis of cranial shape of the Cioclovina calvaria (Harvati et al. 2007) and inner ear (Uhl et al. 2016).

In light of the recent genomic results on Oase 1, another look at the morphology of the Romanian Upper Paleolithic mandibles is warranted. In this chapter, we conducted two comparative 3D geometric morphometric analyses: the first on overall mandibular morphology (28 landmarks) including only the more complete Oase 1 specimen; and the second using a reduced dataset (12 landmarks) in order to include both Oase 1 and the less complete Muierii 1 specimen.

In the case of Oase 1, the results of the two analyses were consistent. Oase 1 clustered with Upper Paleolithic modern humans and fell just outside the range of variation of recent modern humans in both analyses. Its position on the PCAs was influenced by its relatively low and broad ramus, tall condyle, and anteroposteriorly long corpus (Figs. 4.5 and 4.6). In both analyses, Oase 1 was classified as UP with high probability. Its nearest neighbors in total shape were UP specimens (although it is interesting to note that its closest recent modern human neighbors in total shape in the two analyses were a specimen from the Oceania (Tasmania) and Pacific (Melanesia) samples; it has been recently found that both groups demonstrate elevated levels of admixture with archaic humans, including Neanderthals and Denisovans; Rasmussen et al. 2011; Meyer et al. 2012; Vernot et al. 2016). Furthermore, in terms of mean Procrustes distances, it was clearly closest to the UP sample in both analyses (Table 4.3).

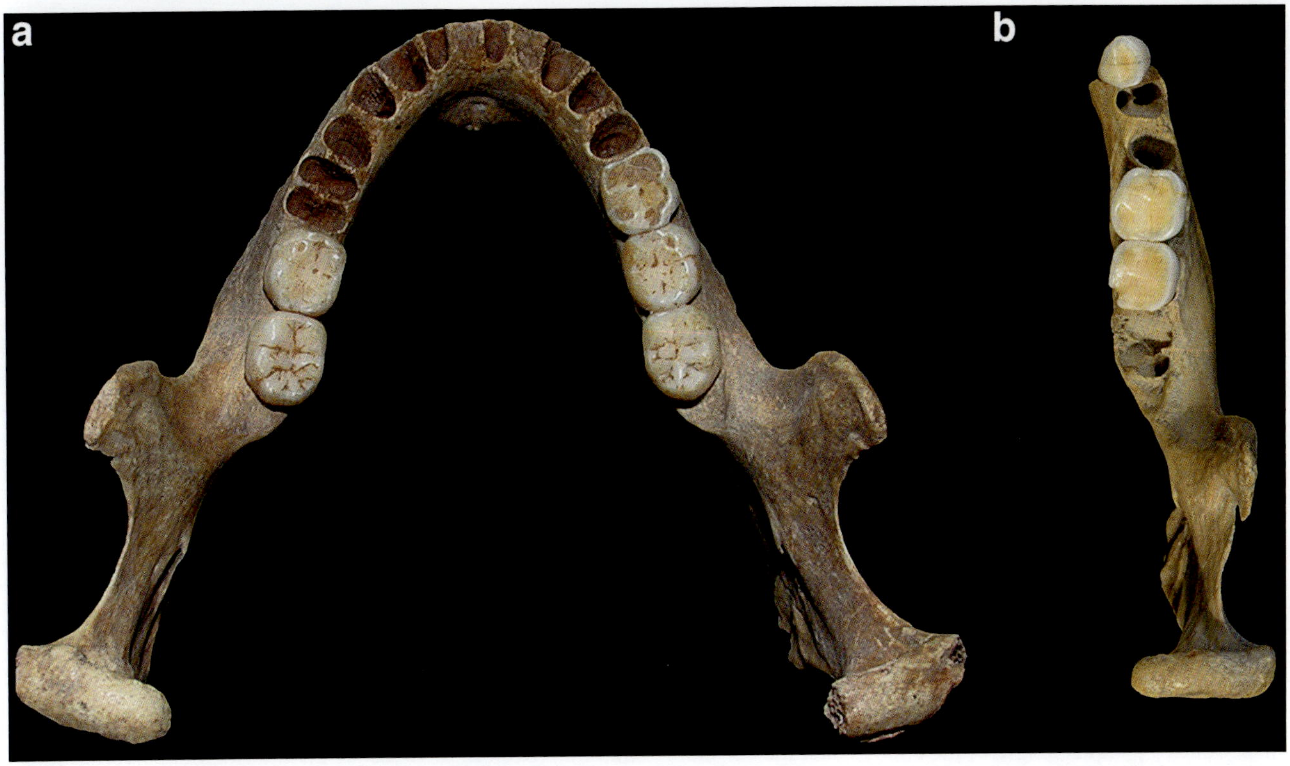

Fig. 4.7 Occlusal view of (**a**) Oase 1 (photo copyright K. Harvati); (**b**) Muierii 1 (Muierii 1 photo kindly provided by and copyright Erik Trinkaus/ISER)

The overall shape of the Oase 1 mandible therefore cannot be characterized as clearly intermediate between Neanderthals and modern humans. Furthermore, this specimen does not show dental anomalies such as supernumerary or rotated teeth (Fig. 4.7). Nevertheless, it was found to be very large in centroid size (Fig. 4.4), falling at the upper end of the recent modern human range of centroid size variation, and slightly above the ranges of the EAM and UP. Exceptionally large size could be interpreted as reflecting admixture (e.g., Ackermann 2010); however, our sample of early modern humans is too small to evaluate whether Oase 1 is exceptionally large relative to the early modern human population. In this case, therefore, the greater proportion of Neanderthal ancestry for the specimen, as revealed by its genetic analysis, is not clearly shown by its overall morphology, and is only suggested by its very large size and by a few nonmetric features (Trinkaus et al. 2003a).

Muierii 1 could only be included in the 12 landmark analysis. Here, the PCA placed it outside the convex hull of either the recent modern human or the UP sample, and within the NEA and MP ranges, close to the area of overlap between NEA, MP, and UP along PC2 and 3 (the axes that separate modern humans from Neanderthals and Middle Pleistocene Europeans). Muierii 1 plotted nearest to Zafarraya and to Upper Cave 101 on the PCA (Fig. 4.6). The discriminant analysis classified it as Neanderthal, albeit with low probability (0.53). It must be noted, however, that although more than 90 % of recent humans were correctly classified in the cross-validation classification, none of the Neanderthals were; instead most were misclassified as MP and two as modern humans (one UP and one recent). In total shape, Muierii 1 was closest to a recent modern human specimen (also from Melanesia) and to the Upper Cave 101 mandible. However, in contrast to Oase 1 in the 12 landmarks analysis, the ten closest specimens to Muierii 1 in the overall sample also include a Neanderthal (Zafarraya). Among the fossil sample (Table 4.3), its closest neighbors include two NEA and 1 MP specimen. Furthermore, the mean Procrustes distances Muierii 1—UP, Muierii 1—EAM, Muierii 1—MP, and Muierii 1—NEA are roughly equivalent in magnitude (Table 4.3). In terms of centroid size, Muierii 1 fell within the lower half of the size ranges of recent and UP modern humans, and below the range of the small EAM sample. Our analysis therefore found an intermediate overall shape for Muierii 1, as reflected by its PCA scores, classification, and by both interindividual and mean Procrustes distances. Although this result might be influenced by the small number of landmarks included in this analysis (which resulted in greater overlap between the samples in the PCA), the results for Oase 1 remain very consistent with those of the complete dataset, suggesting that this influence is limited.

The intermediate position of Muierii 1 in the PCA is driven mainly by its asymmetric mandibular notch and the configuration of its condyle features that have been argued to be Neanderthal derived traits (e.g., Rak 1998; Rak et al. 2002). However, the derived nature of these traits has been questioned. Jabbour et al. (2002) found that although Neanderthals had a significantly higher frequency than modern humans in the least lateral configuration of the crest of the mandibular notch, this feature was not unique to this taxon. These authors also found that the difference in frequencies of this trait between Neanderthals and Middle Pleistocene Europeans was not significant. Wolpoff and Frayer (2005) evaluated the ramal features proposed to be uniquely derived for Neanderthals. They, too, pointed out the large range of variation not only among Neanderthals, but also in Middle Pleistocene hominins and modern humans, in the configuration of the condyle and the crest of the mandibular notch, as well as in other features, such as the depth and position of the deepest point of the mandibular notch and its asymmetric shape (due to unequal height of the condyle and the coronoid process). While our analysis does not evaluate each of these traits separately, our landmark dataset was designed to capture these specific aspects of ramal morphology (see Nicholson and Harvati 2006). The results of our previous work (Nicholson and Harvati 2006), as well as both the analyses presented here, suggest that Neanderthals do differ from modern humans significantly due to the combination of these features. Specifically for the 12 landmarks analysis, these characteristics are among those driving the patterns observed on PCs 2 and 3. These PCs separate Neanderthals from modern humans relatively successfully (especially considering the small number of Neanderthals relative to the large comparative sample of modern humans), although the Neanderthal sample is variable in the expression of these features and some overlap exists. However, PCs 2 and 3 do not separate Neanderthals from earlier Middle Pleistocene specimens, suggesting that at least some of these features may in fact be plesiomorphic in nature. An intermediate position for Muierii 1 could therefore reflect retention of primitive morphology.

Nevertheless, such an intermediate shape could also be interpreted to result from hybridization, as has been claimed previously for this specimen. This claim is consistent with the observation that the occlusal surface of Muierii 1 appears to show a slightly rotated P_4 alveolus (although the distobuccal corner of the P_4 alveolus is partially covered by the outline of the M_1 crown; Fig. 4.7), one of the features listed by Ackermann (2010) as potential indicators of hybrid status. This trait is not uncommon and occurs at variable frequencies among modern human populations (see e.g., Stemm 1971; Gupta et al. 2011), making its significance unclear. Nevertheless, Muierii's intermediate overall shape combined with the rotated premolar is consistent with previous claims of a mixed Neanderthal and modern human like morphology and is suggestive of a possible higher proportion of Neanderthal ancestry for this specimen.

In conclusion, the overall shape of the Oase 1 mandible does not reflect its recent Neanderthal ancestry, as revealed by paleogenomic analysis. The latter is only suggested by its very large size and by nonmetric details of its anatomy. This result not only may be due to the limited phylogenetic signal preserved by mandibular morphology, but also highlights the difficulties of evaluating admixture from skeletal remains, especially in cases where interbreeding occurred several generations previously. On the other hand, the more fragmentary Muierii 1 presents an intermediate overall shape (reflected in the PCA, discriminant analysis, and Procrustes distances), as well as a dental anomaly which is relatively unusual among fossil hominins. These results might be consistent with hybrid status, but, in the case of the intermediate shape, could also reflect primitive retentions. In light of the recent genetic results on Oase 1, however, potential recent Neanderthal ancestry for this individual should be further evaluated by comparative analysis of the cranial remains, as well as by paleogenetic analyses.

Acknowledgments We thank Oana Moldovan and Andre Soficaru for facilitating access to the specimens in Romania. Photographs for the Oase 1 and Muierii 1 mandibles were kindly provided by Prof. E. Trinkaus. The modern human samples, as well as many of the fossil specimens, used in the 3D geometric morphometrics analysis were measured by E. Nicholson-Lopez as part of her undergraduate honors thesis. Tristan Begg helped with the formatting of the chapter. We are immensely grateful to Gisselle Garcia and Eric Delson for their help with the catalogue numbers of the comparative modern human samples. This research was supported by the Max Planck Gesellschaft, the Natural Sciences and Engineering Research Council of Canada (371077-2010), and the European Research Council (ERC STG 283503 'PaGE'). We thank three anonymous reviewers and Eric Delson for helpful comments on earlier versions of the manuscript.

Appendix: Specimens Included in the Comparative Recent Human Samples (Department of Anthropology, American Museum of Natural History, New York; All Specimens are Prefixed AMNH-A)

North American Arctic	Oceania	Central America	East Africa	Europe
Alaska 99/439	Aborigine 99/8153	Huichol 99/2339	Bondei VL/5303	Austria VL/18
Alaska 99.1/440	Aborigine 99/8154	Maya 99/7906	Buera VL/989	Crete VL/4122
Baffin Bay 99/6692	Aborigine 99/8157	Mexico 99/136	Buera VL/990	Czech VL/196
Baffin Bay 99/6693	Aborigine 99/8158	Mexico 99/2237	Maasai VL/4309	Czech VL/1125
Eskimo 99/106	Aborigine 99/8173	Honduras 99/9775	Mgindo Wangindo	Czech VL/1466
Eskimo 99/4103	Aborigine 99/8178	Huichol 99/50	Bantu VL/2797	Czech VL/1467
Eskimo 99/4462	Aborigine 99/8179	Mexico 99/153	Mhehe VL/399	Germany VL/798
Alaska 99.1/220	Australian 99/8217	Mexico 99/2555	Mhuma VL/396	Germany VL/899
Greenland 99/7687	Gundaroo 99/8181	Mexico 99/4028	Mtussi VL/4928b	Germany VL/1435
Greenland 99/7689	Murchison VL/625	Mexico 99/4019	Saramo VL/572	Greece 99.1/42
Alaska 99.1/325	New Guinea 1/1993		Saramo VL/573	Greece 99/9781
Aleut 99.1/696	Tasmania VL/269		Ussandawi VL/1597	Hungary VL/2624
Eskimo 99/6847	Tasmania VL/272		Zanzibar VL/550	Hungary VL/4344
Eskimo 99/8013	Up. Murchison VL/245		Mhuma VL/397	Hungary VL/4348
	Aborigine 99/8155		Schaachi VL/1596	Swiss VL/3274
	Aborigine 99/8175			Bohemia 99.1/835
	Port Darwin VL/1412			Italy VL/1443
	Up. Murchison VL/246			Greece VL/2099
				Czech VL/197
				Sweden VL/2829
				Crete VL/4121
				Germany VL/2346
				Hungary VL/4341
				Hungary VL/4776
				Hungary VL/4783
				Poland VL/629
North Asia	**Pacific**	**South Africa**	**South America**	**South-East Asia**
Chukchi 99/3849	Bismark VL/1416	Bushman VL/8453	Bolivia B/1373	Bangkok VL/596
Japan VL/4671	Bismark VL/1479	Bushman 99/8454	Chile 99.1/758	Bangkok VL/597
Japan VL/4673	Bismark VL/4158	Korana 99/8442	Chile 99/9950	Bangkok VL/2438
Japan VL/4674	Bismark VL/5296	Zulu VL/3461	Churkoni 99/3324	Borneo 'Chinese' VL/1716
Japan VL/4675	Malekula 99/8077	Zulu VL/3573	Churkoni 99/3360	Borneo 'Chinese' VL/1744
Japan VL/4677	Matupi VL/249	Zulu VL/3578	Mundrucan (Brazil) 1/2890	Borneo 'Chinese' VL/1745
Kalmuk 229	Nissan VL/1418	Hottentot VL/77	Peru 99/3677	Borneo 'Chinese' VL/1753
Korea VL/1094	Palau VL/902	Morolong 99/8456	Peru 99/3704	Malaysia VL/1714
Korea VL/1095	Ralum VL/1515	Zulu VL/3462	Peru B/9688	Singapore VL/1295
Korea 99/7748	Ralum VL/1517		Yaghan (Chile) 99.1/762	China 'Malay' VL/1388
Mongolia 99.1/940	Ralum VL/1521		Yaghan 99.1/763	Malay 99/7888
Mongolia 99/8015	Ralum VL/1524		Peru 1/1060	Malay Straits VL/1713
Mongolia 99/8027	Ralum VL/1528		Bolivia 99/385	Bangkok VL/2442
	Ralum VL/1530		Colombia 99/4536	Singapore Malay VL/1385
	Ralum VL/1536		Bolivia B/6551	
	Yap VL/265		Paraguay VL/459	
	Yap VL/906		Peru 1/1059	
	Yap VL/5243		Peru 99/3705	
			Venezuela VL/618	

References

Ackermann, R. R. (2010). Phenotypic traits of primate hybrids: Recognizing admixture in the fossil record. *Evolutionary Anthropology, 19*, 258–270.

Ackermann, R. R., Rogers, J., & Cheverud, J. M. (2006). Identifying the morphological signatures of hybridization in primate and human evolution. *Journal of Human Evolution, 51*, 632–645.

Alexandrescu, E., Olariu, A., Skog, G., Stenström, K., & Hellborg, R. (2010). Os fossiles humains des grottes Muierii et Cioclovina, Roumanie / Human fossil bones from the Muierii Cave and the Cioclovina Cave, Romania. *L'Anthropologie, 114*, 341–353.

Arnold, M. L., & Meyer, A. (2006). Natural hybridization in primates: One evolutionary mechanism. *Zoology, 109*, 261–276.

Bailey, S. E. (2002). A closer look at Neanderthal postcanine dental morphology: The mandibular dentition. *The Anatomical Record, 269*(3), 148–156.

Bastir, M., Rosas, A., Gunz, P., Pena-Melian, A., Manzi, G., Harvati, K., et al. (2011). Evolution of the base of the brain in highly encephalized human species. *Nature Communications, 2*, 588.

Benazzi, S., Douka, K., Fornai, C., Bauer, C. C., Kullmer, O., Svoboda, J., et al. (2011). Early dispersal of modern humans in Europe and implications for Neanderthal behavior. *Nature, 479*, 525–528.

Bernstein, I. S. (1966). Naturally occurring primate hybrids. *Science, 154*, 1559–1560.

Bräuer, G., & Broeg, H. (1998). On the degree of Neandertal-modern continuity in the earliest Upper Paleolithic crania from the Czech Republic: Evidence from non-metrical features. In K. Omoto & P. V. Tobias (Eds.), *Recent advances in human biology. Vol 3: The origins and past of modern humans: Towards reconciliation* (pp. 106–125). Singapore: World Scientific.

Bräuer, G., Collard, M., & Stringer, C. B. (2004). On the reliability of recent tests of the out of Africa hypothesis for modern human origins. *The Anatomical Record, 279A*, 701–707.

Cârciumaru, M. (1999). *Le Paléolithique en Roumanie*. Grenoble: Jérôme Millon.

Cârciumaru, M., Anghelinu, M., & Niţă, L. (2007). The Upper Palaeolithic in the Bistrita Valley (Northeastern Romania). A preliminary review. *Annales d'Université Valahia Târgovişte, section d'Archeologie et d'Histoire, VIII-IX*, 107–124.

Cheverud, J. M., Jacobs, S. C., & Moore, A. J. (1993). Genetic differences among subspecies of the saddle-back tamarin (Saguinus fuscicollis): Evidence from hybrids. *American Journal of Primatology, 31*(1), 23–39.

Churchill, S. E., & Smith, F. H. (2000). Makers of the early Aurignacian in Europe. *Yearbook of Physical Anthropology, 43*, 61–115.

Currat, M., & Excoffier, L. (2011). Strong reproductive isolation between humans and Neanderthals inferred from observed patterns of introgression. *Proceedings of the National Academy of Sciences of the United States of America, 108*(37), 15129–15134.

Di Vincenzo, F., Churchill, S. E., & Manzi, G. (2012). The Vindija Neanderthal scapular glenoid fossa: Comparative shape analysis suggests evo-devo changes among Neanderthals. *Journal of Human Evolution, 62*, 274–285.

Doboş, A. (2010). *Paleoliticul mijlociu de pe teritoriul de la sud şi sud-est de Carpaţi (Oltenia, Muntenia, Dobrogea)*. Ph.D. dissertation, Institutul de Arheologie "Vasile Pârvan", Bucharest.

Doboş, A., & Iovita, R. (2016). The Lower Paleolithic of Romania revisited: New evidence from the site of Dealul Guran. In K. Harvati & M. Roksandic (Eds.), *Paleoanthropology of the Balkans and Anatolia: Human evolution and its context*. Dordrecht: Springer.

Eriksson, A., & Manica, A. (2012). Effect of ancient population structure on the degree of polymorphism shared between modern human populations and ancient hominins. *Proceedings of the National Academy of Sciences of the United States of America, 109*(35), 13956–13960.

Froehlich, J. W., & Supriatna, J. (1996). Secondary intergradation between *Macaca maurus* and *M. tonkeana* in South Sulawesi, and the species status of *M. togeanus*. In J. E. Fa & D. G. Lindburg (Eds.), *Evolution and ecology of macaque societies* (pp. 43–70). Cambridge: Cambridge University Press.

Frost, S. R., Marcus, L. F., Bookstein, F. L., Reddy, D. P., & Delson, E. (2003). Cranial allometry, phylogeography, and systematics of large-bodied papionins (primates: Cercopithecinae) inferred from geometric morphometric analysis of landmark data. *The Anatomical Record, 275*(2), 1048–1072.

Fu, Q., Hajdinjak, M., Moldovan, O. T., Constantin, S., Mallick, S., Skoglund, P., et al. (2015). An early modern human from Romania with a recent Neanderthal ancestor. *Nature, 524*, 216–219.

Fu, Q., Li, H., Moorjani, P., Jay, F., Slepchenko, S. M., Bondarev, A. A., et al. (2014). Genome sequence of a 45,000-year-old modern human from western Siberia. *Nature, 514*, 445–449.

Gaál, I. (1928). Der erste mitteldiluviale Menschenknochen aus Siebernbürgen. *Pub. Muzeuli Jud. Hunedoara, Deva, 3-4*, 61–112.

Gheorghiu, A., & Haas, N. (1954a). Date privind omul primitiv de la Baia de Fier: Consideraţii paleoantropologice. *Sesiunea Secţiunii de Ştiinţe Medicale a Academiei RPR, 41*, 641–663.

Gheorghiu, A., & Haas, N. (1954b). Privind omul primitiv de la Baia de Fier. *Consideraţii paleoantropologice, Sesiunea Secţiunii de Ştiinţe Medicale a Academiei R.P.R., 41*, 641–663.

Green, R. E., Krause, J., Briggs, A. W., Maricic, T., Stenzel, U., Kircher, M., et al. (2010). A draft sequence of the Neandertal genome. *Science, 328*(5979), 710–722.

Green, R. E., Malaspinas, A. S., Krause, J., Briggs, A. W., Johnson, P. L., Uhler, C., et al. (2008). A complete Neandertal mitochondrial genome sequence determined by high-throughput sequencing. *Cell, 134*(3), 416–426.

Gunz, P., Bookstein, F. L., Mitteroecker, P., Stadlmayr, A., Seidler, H., & Weber, G. W. (2009). Early modern human diversity suggests subdivided population structure and a complex out-of-Africa scenario. *Proceedings of the National Academy of Sciences of the United States of America, 106*(15), 6094–6098.

Gupta, S. K., Saxena, P., Jain, S., & Jain, D. (2011). Prevalence and distribution of selected developmental dental anomalies in an Indian population. *Journal of Oral Science, 53*(2), 231–238.

Hamada, Y., San, A. M., & Malaivijitnond, S. (2016). Assessment of the hybridization between Rhesus (*Macaca mulatta*) and Long-Tailed macaques (*M. fascicularis*) based on morphological characters. *American Journal of Physical Anthropology, 159*(2), 189–198. doi:10.1002/ajpa.22862.

Hammer, Ø., Harper, D. A. T., & Ryan, P. D. (2001). PAST: Paleontological statistics software package for education and data analysis. *Palaeontologia Electronica 4*(1), 9. Retrieved from http://palaeo-electronica.org/2001_1/past/issue1_01.htm

Harvati, K. (2003). Quantitative analysis of Neanderthal temporal bone morphology using 3-D geometric morphometrics. *American Journal of Physical Anthropology, 120*, 323–338.

Harvati, K., Frost, S. R., & McNulty, K. P. (2004). Neanderthal taxonomy reconsidered: Implications of 3D primate models of intra- and interspecific differences. *Proceedings of the National Academy of Sciences of the United States of America, 101*, 1147–1152.

Harvati, K., Gunz, P., & Grigorescu, D. (2007). Cioclovina (Romania): Morphological affinities of an early modern European. *Journal of Human Evolution, 53*, 732–746.

Harvati, K., Stringer, C., Grün, R., Aubert, M., Allsworth-Jones, P., & Adebayo Folorunso, C. (2011). The Later Stone Age calvaria from Iwo Eleru, Nigeria: Morphology and chronology. *PLoS One, 6*(9), e24024.

Holliday, T. W. (2006). Neanderthals and modern humans: An example of a mammalian Syngameon? In K. Harvati & T. Harrison (Eds.), *Neanderthals revisited: New approaches and perspectives* (pp. 281–298). Dordrecht: Springer.

Holliday, T. W., Gautney, J. R., & Friedl, L. (2014). Right for the wrong reasons: Reflections on modern human origins in the post-Neanderthal genome era. *Current Anthropology, 55*, 696–724.

Honea, K. (1984). Chronometry of the Romanian Middle and Upper Paleolithic: Implications of current radiocarbon dating results. *Dacia, 28*, 23–39.

Jabbour, R. S., Richards, G. D., & Anderson, J. Y. (2002). Mandibular condyle traits in Neanderthals and other *Homo*: A comparative, correlative, and ontogenetic study. *American Journal of Physical Anthropology, 119*, 144–155.

Jolly, C. J., Woolley-Barker, T., Beyene, S., Disotell, T. R., & Phillips-Conroy, J. E. (1997). Intergeneric hybrid baboons. *International Journal of Primatology, 18*(4), 597–627.

Krings, M., Stone, A., Schmitz, R. W., Krainitzki, H., Stoneking, M., & Pääbo, S. (1997). Neandertal DNA sequences and the origin of modern humans. *Cell, 90*(1), 19–30.

Lowery, R. K., Uribe, G., Jimenez, E. B., Weiss, M. A., Herrera, K. J., Regueiro, M., et al. (2013). Neanderthal and Denisova genetic affinities with contemporary humans: Introgression versus common ancestral polymorphisms. *Gene, 530*(1), 83–94.

Mertens, S. E. (1996). The Middle Paleolithic in Romania. *Current Anthropology, 37*, 515–521.

Meyer, M., Kircher, M., Gansauge, M. T., Li, H., Racimo, F., Mallick, S., et al. (2012). A high-coverage genome sequence from an archaic Denisovan individual. *Science, 338*, 222–226.

Milota, S., Gherase, M., Sarcina, L., Rodrigo, R., Moldovan, O., Trinkaus, E., et al. (2013). Exploration and documentation of the Pestera cu Oase. In E. Trinkaus, S. Constantin, & J. Zilhão (Eds.), *Life and death at the Pestera cu Oase. A setting for modern human emergence in Europe* (pp. 5–15). New York: Oxford University Press.

Mišić, B., Čolić, D., & Dinić, A. (1969). Appendix IIB Ekološko-fitološka istrazivanja. In D. Srejovic (Ed.), *Lepenski Vir* (pp. 207–223). Beograd: Srpska književna zadruga.

Necrasov, O. (1971). Romania. In K. P. Oakley, B. G. Campbell, & T. I. Molleson (Eds.), *Catalogue of fossil hominids part II: Europe* (pp. 283–285). London: British Museum (Natural History).

Necrasov, O., & Cristescu, M. (1965). Données anthropologiques sur les populations de l'âge de la pierre en Roumanie. *Homo, 16*, 129–158.

Nicholson, E., & Harvati, K. (2006). Quantitative analysis of human mandibular shape using 3-D geometric morphometrics. *American Journal of Physical Anthropology, 131*, 368–383.

Nicolăescu-Plopşor, D. (1968). Les hommes fossiles découverts en Roumanie. *VIIème Congrès international des sciences anthropologiques et ethnologiques, III, Moscou, 1964*, 381–386.

O'Higgins, P., & Jones, N. (1998). Facial growth in *Cercocebus torquatus*: An application of three-dimensional geometric morphometric techniques to the study of morphological variation. *Journal of Anatomy, 193*(2), 251–272.

O'Higgins, P., & Jones, N. (2006). *Morphologika 2, v. 2.4*. York: Hull York Medical School.

Oettlé, A. C., Pretorius, E., & Steyn, M. (2005). Geometric morphometric analysis of mandibular ramus flexure. *American Journal of Physical Anthropology, 128*(3), 623–629.

Olariu, A., Hellborg, R., Stenström, K., Skog, G., Faarinen, M., Persson, P., et al. (2002). Dating of some Romanian fossil bones by combined nuclear methods. *Journal of Radioanalytical and Nuclear Chemistry, 253*(2), 307–311.

Olariu, A., Skog, G., Hellborg, R., Stenström, K., Faarinen, M., Persson, P., et al. (2005). Dating of two Paleolithic human fossil bones from Romania by accelerator mass spectrometry. In A. Olariu, K. Stenström, & R. Hellborg (Eds.), *Proceedings of international conference on applications of high precision atomic and nuclear methods, 2–6 September, 2002, Neptun, Romania* (pp. 222–227). Bucuresti: Editura Academiei Romane.

Onac, B. P., Viehmann, I., Lundberg, J., Lauritzen, S. E., Stringer, C., & Popiţă, V. (2005). U–Th ages constraining the Neanderthal footprint at Vârtop Cave, Romania. *Quaternary Science Reviews, 24*(10), 1151–1157.

Paunescu, A. (2000). *Paleoliticul si Mezoliticul din spatiul cuprins intre Carpati si Dunare*. Bucuresti: Asociatia Generala a Inginerilor din Romania (AGIR).

Pop, C. M. (2013). *The Middle Palaeolithic of present day Romania: A critical review*. M.A. Thesis, University of British Columbia.

Prüfer, K., Racimo, F., Patterson, N., Jay, F., Sankararaman, S., Sawyer, S., et al. (2014). The complete genome sequence of a Neanderthal from the Altai Mountains. *Nature, 505*(7481), 43–49.

Rainer, F., & Simionescu, I. (1942). Sur le premier crâne d'homme Paléolithique trouvé en Roumanie. *Analele Academiei Romane, Memoriile Secţiunii Ştiinţifice, Seria III, 17*, 489–503.

Rak, Y. (1998). Does any Mousterian cave present evidence of two hominid species? In T. Akazawa, K. Aoki, & O. Bar-Yosef (Eds.), *Neanderthals and modern humans in Western Asia* (pp. 353–366). New York: Plenum Press.

Rak, Y., Ginzberg, A., & Geffen, E. (2002). Does *Homo neanderthalensis* play a role in modern human ancestry? The mandibular evidence. *American Journal of Physical Anthropology, 119*, 199–204.

Rasmussen, M., Guo, X., Wang, W., Lohmueller, K. E., Rasmussen, S., Albrechtsen, A., et al. (2011). An aboriginal Australian genome reveals separate human dispersals into Asia. *Science, 334*, 94–98.

Reyes-Centeno, H., Ghirotto, S., Détroit, F., Grimaud-Hervé, D., Barbujani, G., & Harvati, K. (2014). Genomic and cranial phenotype data support multiple modern human dispersals from Africa and a southern route into Asia. *Proceedings of the National Academy of Sciences of the United States of America, 111*(20), 7248–7253.

Reyes-Centeno, H., Hubbe, M., Hanihara, T., Stringer, C., & Harvati, K. (2015). Testing modern human out-of-Africa dispersal models and implications for modern human origins. *Journal of Human Evolution, 87*, 95–106.

Robinson, C. (2012). Geometric morphometric analysis of mandibular shape diversity in *Pan*. *Journal of Human Evolution, 63*(1), 191–204.

Rohlf, F. J. (1990). Morphometrics. *Annual Review of Ecology and Systematics, 21*, 299–316.

Rohlf, F. J. (2000). Statistical power comparisons among alternative morphometric methods. *American Journal of Physical Anthropology, 111*(4), 463–478.

Rohlf, F. J., & Marcus, L. F. (1993). A revolution in morphometrics. *Trends in Ecology & Evolution, 8*(4), 129–132.

Rougier, H., Crevecoeur, I., & Wolpoff, M. H. (2006). Lower third premolar rotation in the Krapina sample. *Periodicum Biologorum, 108*, 269–278.

Rougier, H., Milota, S., Rodrigo, R., Gherase, M., Sarcina, L., Moldovan, O., et al. (2007). Peştera cu Oase 2 and the cranial morphology of early modern Europeans. *Proceedings of the National Academy of Sciences of the United States of America, 104*, 1165–1170.

Sankararaman, S., Mallick, S., Dannemann, M., Prüfer, K., Kelso, J., Pääbo, S., et al. (2014). The genomic landscape of Neanderthal ancestry in present-day humans. *Nature, 507*(7492), 354–357.

Sankararaman, S., Patterson, N., Li, H., Pääbo, S., & Reich, D. (2012). The date of interbreeding between Neandertals and modern humans. *PLoS Genetics, 8*(10), e1002947.

Schillaci, M. A., Froehlich, J. W., Supriatna, J., & Jones-Engel, L. (2005). The effects of hybridization on growth allometry and craniofacial form in Sulawesi macaques. *Journal of Human Evolution, 49*(3), 335–369.

Seguin-Orlando, A., Korneliussen, T. S., Sikora, M., Malaspinas, A.-S., Manica, A., Moltke, I., et al. (2014). Genomic structure in Europeans dating back at least 36,200 years. *Science, 346*, 1113–1118.

Slice, D. E. (1996). Three-dimensional generalized resistant fitting and the comparison of least-squares and resistant-fit residuals. In *Advances in morphometrics* (pp. 179–199). New York: Plenum Press.

Smith, F. H. (1992). The role of continuity in modern human origins. In G. Bräuer & F. H. Smith (Eds.), *Continuity or replacement: Neanderthals and their contemporaries controversies in Homo sapiens evolution* (pp. 145–156). Rotterdam: A.A. Balkema.

Smith, H. F. (2009). Which cranial regions reflect molecular distances reliably in humans? Evidence from three-dimensional morphology. *American Journal of Human Biology, 21*, 36–47.

Smith, F. H., Janković, I., & Karavanić, I. (2005). The assimilation model, modern human origins in Europe, and the extinction of Neandertals. *Quaternary International, 137*, 7–19.

Soficaru, A., Doboş, A., & Trinkaus, E. (2006). Early modern humans from the Pestera Muierii, Baia de Fier, Romania. *Proceedings of the National Academy of Sciences of the United States of America, 103*, 17196–17201.

Soficaru, A., Petrea, C., Doboş, A., & Trinkaus, E. (2007). The human cranium from the Peştera Cioclovina Uscată, Romania context, age, taphonomy, morphology, and paleopathology. *Current Anthropology, 48*, 611–619.

Stemm, R. M. (1971). The frequency of malposed unerupted lower premolar teeth. *The Angle Orthodontist, 41*(2), 157–158.

Stringer, C. (2002). Modern human origins: Progress and prospects. *Philosophical Transactions of the Royal Society of London Series B: Biological Sciences, 357*(1420), 563–579.

Terzea, E. (1979). Mamiferele cuaternare din unele peşteri şi adăposturi sub stîncă din zona Porţile de Fier. In T. Orghidan & Ş. Negrea (Eds.), *Speologia* (pp. 104–139). Bucureşti: Editura Academiei Române.

Trinkaus, E. (2007). European early modern humans and the fate of the Neandertals. *Proceedings of the National Academy of Sciences of the United States of America, 104*, 7367–7372.

Trinkaus, E., Constantin, S., & Zilhão, J. (Eds.). (2013). *Life and death at the Peştera cu Oase: A setting for modern human emergence in Europe*. Oxford: Oxford University Press.

Trinkaus, E., Milota, R., Rodrigo, R., Gherase, M., & Moldovan, O. (2003a). Early modern human cranial remains from the Peştera cu Oase, Romania. *Journal of Human Evolution, 45*, 245–253.

Trinkaus, E., Moldovan, O., Milota, S., Bîlgăr, A., Sarcina, L., Athreya, S., et al. (2003b). An early modern human from the Peştera cu Oase, Romania. *Proceedings of the National Academy of Sciences of the United States of America, 100*, 11231–11236.

Uhl, A., Reyes-Centeno, H., Grigorescu, D., Kranioti, E., & Harvati, K. (2016). Inner ear morphology of the Cioclovina early modern European calvaria from Romania. *American Journal of Physical Anthropology, 160*(1), 62–70.

Vernot, B., & Akey, J. M. (2014). Resurrecting surviving Neandertal lineages from modern human genomes. *Science, 343*(6174), 1017–1021.

Vernot, B., Tucci, S., Kelso, J., Schraiber, J. G., Wolf, A. B., Gittelman, R. M., et al. (2016). Excavating Neandertal and Denisovan DNA from the genomes of Melanesian individuals. *Science, 352*(6282), 235–239. doi:10.1126/science.aad9416.

von Cramon-Taubadel, N. (2011). Global human mandibular variation reflects differences in agricultural and hunter-gatherer subsistence strategies. *Proceedings of the National Academy of Sciences of the United States of America, 108*, 19546–19551.

von Cramon-Taubadel, N. (2014). Evolutionary insights into global patterns of human cranial diversity: Population history, climatic and dietary effects. *Journal of Anthropological Sciences, 92*, 43–77.

Webb, D., Robu, M., Moldovan, O., Constantin, S., Tomus, B., & Naeg, I. (2014). Ancient human footprints in Ciur-Izbuc Cave, Romania. *American Journal of Physical Anthropology, 155*, 128–135.

Wolpoff, M. H., & Frayer, D. W. (2005). Unique reamus anatomy for Neandertals? *American Journal of Physical Anthropology, 128*, 245–251.

Wolpoff, M. H., Hawks, J., Frayer, D. W., & Hunley, K. (2001). Modern human ancestry at the peripheries: A test of the replacement theory. *Science, 291*(5502), 293–297.

Zinner, D., Arnold, M. L., & Roos, C. (2011). The strange blood: Natural hybridization in Primates. *Evolutionary Anthropology, 20*, 96–103.

Chapter 5
The Human Fossil Record of Bulgaria and the Formulation of Biogeographic Hypotheses

David S. Strait, Caley M. Orr, Jamie Hodgkins, Nikolai Spassov, Maria Gurova, Christopher Miller, and Tsanko Tzankov

Abstract The location of Bulgaria on the Balkan Peninsula makes it potentially important for evaluating biogeographic hypotheses related to human evolution. The country lies at the crossroads of Europe and Asia Minor and constitutes a key portion of one of the possible dispersal pathways that hominin populations would have employed as they entered and left Europe during the Pleistocene. Unfortunately, the Pleistocene human fossil record of Bulgaria is sparse, and perhaps more importantly, the specific biogeographic hypotheses that human fossil discoveries might address could be more fully articulated. In this chapter, we review the fossil hominins currently known from Bulgaria and discuss the framing of biogeographic hypotheses.

Keywords Balkan • Cave • Climate • Dispersal • Europe • *Homo* • Pleistocene

The Human Fossil Record of Bulgaria

Human fossils recovered from the Pleistocene of Bulgaria are known from only three sites: Bacho Kiro, Temnata Dupka, and Kozarnika. All three are archaeological cave sites in the Balkan Mountains in the north-central and northwest portion of the country (Fig. 5.1). The human fossils derived from these caves are all fragmentary, although the hominins from Bacho Kiro and Kozarnika are potentially important chronologically.

The hominin fossils from Bacho Kiro (Table 5.1; Fig. 5.2) have all been attributed to anatomically modern *Homo sapiens* and most are associated with Aurignacian cultural levels (Glen and Kaczanowski 1982; Ginter and Kozlowski 1982; Mook 1982). The specimens consist of isolated teeth, mandibular fragments, and cranial fragments. The stratigraphic layers from which these specimens derive have been radiocarbon dated to between roughly 29 and 33 ^{14}C k BP (Mook 1982). One specimen (1124), a mandibular fragment with teeth lacking taurodont roots, is notable insofar as it derives from a layer (11/IV) that is associated with the Bachokirian tool industry, considered to be broadly pre-Aurignacian. This layer has been radiocarbon dated to approximately 43 ^{14}C kBP (Mook 1982). In light of recent advances in radiocarbon dating methodology

D.S. Strait (✉)
Department of Anthropology, Washington University, St. Louis, MO, USA
e-mail: dstrait@wustl.edu

C.M. Orr
Department of Cell and Developmental Biology, University of Colorado School of Medicine, Aurora, CO, USA
e-mail: Caley.orr@gmail.com

J. Hodgkins
Department of Anthropology, University of Colorado, Denver, CO 80217-3364, USA
e-mail: Jamie.hodgkins@gmail.com

N. Spassov
National Museum of Natural History, Bulgarian Academy of Sciences, Sofia, Bulgaria
e-mail: nspassov@nmnhs.com

M. Gurova
National Institute of Archaeology and Museum, Bulgarian Academy of Sciences, Sofia, Bulgaria
e-mail: gurovam@yahoo.fr

C. Miller
Institut für Naturwissenschaftliche Archäologie, Eberhard Karls University of Tübingen, Tübingen, Germany
e-mail: Christopher.Miller@uni-tuebingen.de

T. Tzankov
Department of Geography, Ecology and the Preservation of the Environment, South-Western University "Neofit Rilski", Blagoevgrad, Bulgaria

Katerina Harvati and Mirjana Roksandic (eds.), *Paleoanthropology of the Balkans and Anatolia*, Vertebrate Paleobiology and Paleoanthropology, DOI 10.1007/978-94-024-0874-4_5

Fig. 5.1 Satellite image of Bulgaria showing the location of cave sites preserving Paleolithic deposits and human remains

Table 5.1 Hominin fossils from Bacho Kiro Cave

Specimen	Element	Layer	Associated culture/ industry	Age
559	R mand frag., dm2, M1	6a/7	Aurignacian	29,150 kBP[a]
1702	R P4	6a/7	Aurignacian	29,150 kBP[a]
1704	R upper I1	6a/7	Aurignacian	29,150 kBP[a]
2823	R lower I1	7/6b	Aurignacian?	32,700 kBP[a]
3575	R parietal frag.	7	Aurignacian	
2641	R lower I2	7	Aurignacian	
1124	R mand frag., dm1	11/IV	Bachokirian	>43,000 kBP[b]
W-1	Lower di1			

[a]Uncalibrated radiocarbon date on faunal remains from the same stratigraphic layer as the hominin (Mook 1982)
[b]Uncalibrated radiocarbon date on charcoal from the same stratigraphic layer as the hominin (Mook 1982)

(e.g., Conard and Bolus 2003), these dates should be considered approximate at best. Nonetheless, specimen 1124 is arguably among the oldest anatomically modern humans in Europe. Therefore, the redating of level 11/IV with more modern methods should be a research priority. Unfortunately, all of the human remains from Bacho Kiro have been lost.

Three human fossils (Table 5.2; Fig. 5.3) have been recovered from Temnata Dupka (Gambier 1992). These fossils are also fragmentary, consisting of two isolated deciduous incisors and one relatively undiagnostic parietal fragment. They are associated with Gravettian and Epigravettian cultural layers and can reasonably be attributed to anatomically modern *H. sapiens*.

Three hominin fossils have been recovered from Kozarnika (Table 5.3). None have yet been formally described. A phalanx was recovered from level 5b, which has been dated to approximately 26.5 ^{14}C kBP, and a second phalanx was found in level 6/7, which has been dated to approximately 43 ^{14}C kBP (both ages based on uncalibrated radiocarbon dates; Guadelli et al. 2005). The latter is associated with an Early Upper Paleolithic tool industry (Guadelli et al. 2005) that predates the Campanian Ignimbrite eruption (Lowe et al. 2012). If it represents an anatomically modern human, then it is among the earliest representatives of the species in Europe. Guadelli et al. (2005) indicate that a possible hominin deciduous upper central incisor was recovered from level 13 and that this layer has a biochronological date of 1.2–1.4 Ma. Sirakov et al. (2010: table 1) subsequently indicated that the attribution of the specimen to *Homo* was uncertain, and that the age of level 13 was 1.6–1.4 Ma. Thus, there is some uncertainty regarding both the age (see Ivanova 2016; Spassov 2016) and hominin status of the specimen. However, if its hominin status and its proposed biochronological date of >1 Ma are confirmed, then this specimen could be attributable to *H. erectus sensu lato* and is likely one of the very first humans in Europe.

Formulating Biogeographic Hypotheses

One might expect that the discovery of fossil humans and associated archaeological remains in the Balkans might shed light on the patterns by which hominins dispersed into and out of Europe. However, taken in isolation, such discoveries merely document the presence of particular hominin species in a region at particular points in time. These data certainly have biogeographic significance but, for example, it is not obvious that the presence of modern humans at Bacho Kiro prior to 40 ka BP (Mook 1982) is any more significant biogeographically than the presence of modern humans that are as old or older elsewhere in Europe (Benazzi et al. 2011; Higham et al. 2011). In order to maximize the value of paleoanthropological data, it is necessary to formulate explicit biogeographic hypotheses with predictions that are testable (i.e., falsifiable) using the fossil and archaeological record. Here, we outline components of such hypotheses and, as a heuristic exercise, propose a testable hypothesis of hominin biogeography.

It is not especially original to propose hypotheses of hominin biogeography, and many such hypotheses pertaining to the dispersal of hominins into and out of Europe have already been articulated (e.g., Martinón-Torres et al. 2007; Palombo 2010, 2013; van der Made and Mateos 2010; Bermúdez de Castro and Martinón-Torres 2013; Rolland 2013). These overlap in certain respects with our own. However, our purpose is not to propose a fully novel model of hominin dispersals but, rather, to focus attention on how such models might be tested. As a general rule, we argue that it is easier to propose biogeographic hypotheses than it is to test (falsify) them, yet it is only through testing that a hypothesis achieves its full scientific potential.

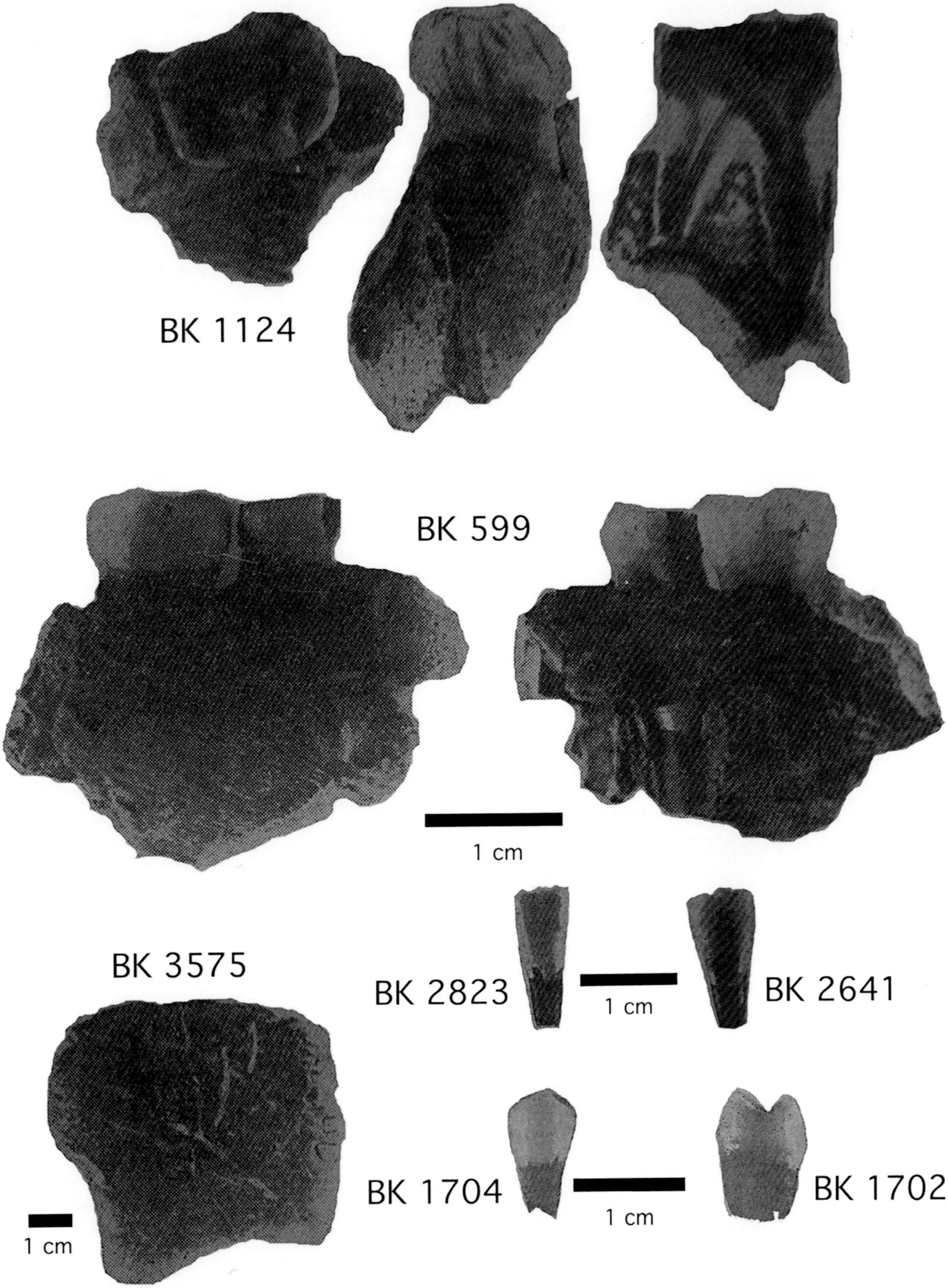

Fig. 5.2 Hominin fossils from Bacho Kiro Cave. Images used with permission from Glen and Kaczanowski (1982)

Table 5.2 Hominin fossils from Temnata Dupka Cave

Specimen	Element	Associated culture/industry
Temnata 1	L lower di2	Epigravettian
Temnata 2	R lower di1	Gravettian
Temnata 3	L parietal frag.	Gravettian

Hypotheses of hominin dispersal should specify *who* (which species) was dispersing, *when* they were dispersing, *the direction* in which they were dispersing, *where* they were dispersing (i.e., the "pathway," loosely defined, along which they traveled), and *why* they were dispersing. When fully formulated, such hypotheses should be accompanied by a discussion of the types of data that could test (falsify) their predictions.

With respect to *who* was dispersing into and out of Europe, the hominin taxa in question plausibly include *Homo erectus sensu lato*, *H. ergaster*, *H. antecessor*, *H. heidelbergensis*, *H. neanderthalensis*, and *H. sapiens*. All of these taxa are within the genus *Homo*, which is unsurprising given that few workers seriously entertain the possibility that australopiths ever left Africa, let alone dispersed into Europe. *Homo habilis* or a similar species is generally not considered a candidate for dispersal into Eurasia, but this view should be tempered by the fact that *Homo floresiensis* exhibits a number of very primitive characteristics that might be broadly construed as being *H. habilis*-like or even australopith-like (e.g., Tocheri et al. 2009; Argue et al. 2009; Brown and Maeda 2009; Jungers et al. 2009), thereby raising the possibility that a relatively primitive *Homo* species might have dispersed out of Africa and been capable of traversing long distances across diverse habitats (e.g., Dennell and Roebroeks 2005).

The questions of *when* and *in which direction* hominins were dispersing is tied to the pattern by which hominin populations were distributed across the Old World at different points in time. Prior to the early Pleistocene, hominins are only known from Africa, so Europe was presumably uninhabited. Thus, given that European hominin fossils date to at least 1.1 Ma (Carbonell et al. 2008; Guadelli et al. 2005; Toro-Moyano et al. 2013; but see Sirakov et al. 2010; Muttoni et al. 2013) and stone artifacts may be even older (Arzarello et al. 2007), it seems likely that hominins dispersed into Europe at some point during the Early Pleistocene. During the Middle Pleistocene, there are closely allied populations of hominins that are found both in and out of Europe. Specifically, archaic humans that may be assigned to *H. heidelbergensis*, *H. rhodesiensis*, and/or archaic *H. sapiens* are found in Africa, Europe, and Asia (e.g., Rightmire 2008; Stringer 2012). If these populations are cospecific or represent sister species, and if they originated in only one region, then a dispersal of these hominins either into or out of Europe is implied, with the direction depending on the region in which these hominins first evolved (e.g., Martinón-Torres et al. 2007). Note, however, that a multiregional origin (broadly speaking) of these hominins would not require any dispersals per se, and that additional dispersals cannot be excluded. In the Late Pleistocene, Neanderthals are known from Europe, Asia, and the Near East (e.g., Schwartz and Tattersall 2002, 2003). Again, if Neanderthals originated in only one region, then another Late Pleistocene dispersal either into or out of Europe is implied. Moreover, anatomically modern humans are known in Africa and the Near East but not in Europe in the early Late Pleistocene. *Homo sapiens* appears in Europe at or around 45 ka BP, based on currently available data (Higham et al. 2011; Benazzi et al. 2011). Thus, it seems probable that modern humans dispersed into Europe in the Late Pleistocene, with the caveat (again) that a multiregional origin of modern humans does not require a dispersal as conventionally defined (i.e., the expansion or relocation of a population); gene flow between adjacent populations is sufficient to explain the pattern.

With respect to *where* hominins may have been entering or leaving Europe, there are relatively few possibilities. The depth of the Gibraltar Strait (−284 m at the Camarinal Sill, its shallowest point; Blanc 2002) strongly suggests that hominins would not have had an opportunity to disperse directly from Northwest Africa to the Iberian Peninsula along a terrestrial pathway during any period in the Pleistocene, even when global sea levels may have been low (e.g., Carbonell et al. 2008). At a minimum, there is not currently any evidence favoring a scenario in which hominins crossed the Strait (Bailey et al. 2008). Indeed, the Strait was evidently a significant (but not insurmountable) barrier to gene flow for humans even after the advent of water transport (Comas et al. 2000). Various island-hopping scenarios could be envisioned that would bring hominins directly from Africa to Europe, but any route involving the crossing of deep water should be viewed as an extraordinary hypothesis requiring extraordinary corroborating evidence. Notably, the Strait of Sicily (between modern Marsala, Sicily and Cape Bon, Tunisia) includes deep troughs that, like the Camarinal Sill, are more than 200 m deep (Maldonado and Stanley 1976). Thus, a dispersal from Africa to the Apennine (Italian) Peninsula seems no more likely than a dispersal across the Strait of Gibraltar. Such dispersals are not impossible, but they seem less likely than terrestrial dispersals. There are three easily identifiable terrestrial (or predominantly terrestrial) possibilities (Fig. 5.4), all of which are abstractions and are merely heuristic ways of thinking about dispersal pathways. There is no reason to expect that hominins followed a "route" as they "marched" into and out of Europe. Nonetheless, the concept of pathways is useful for framing biogeographic hypotheses. The first pathway has been called

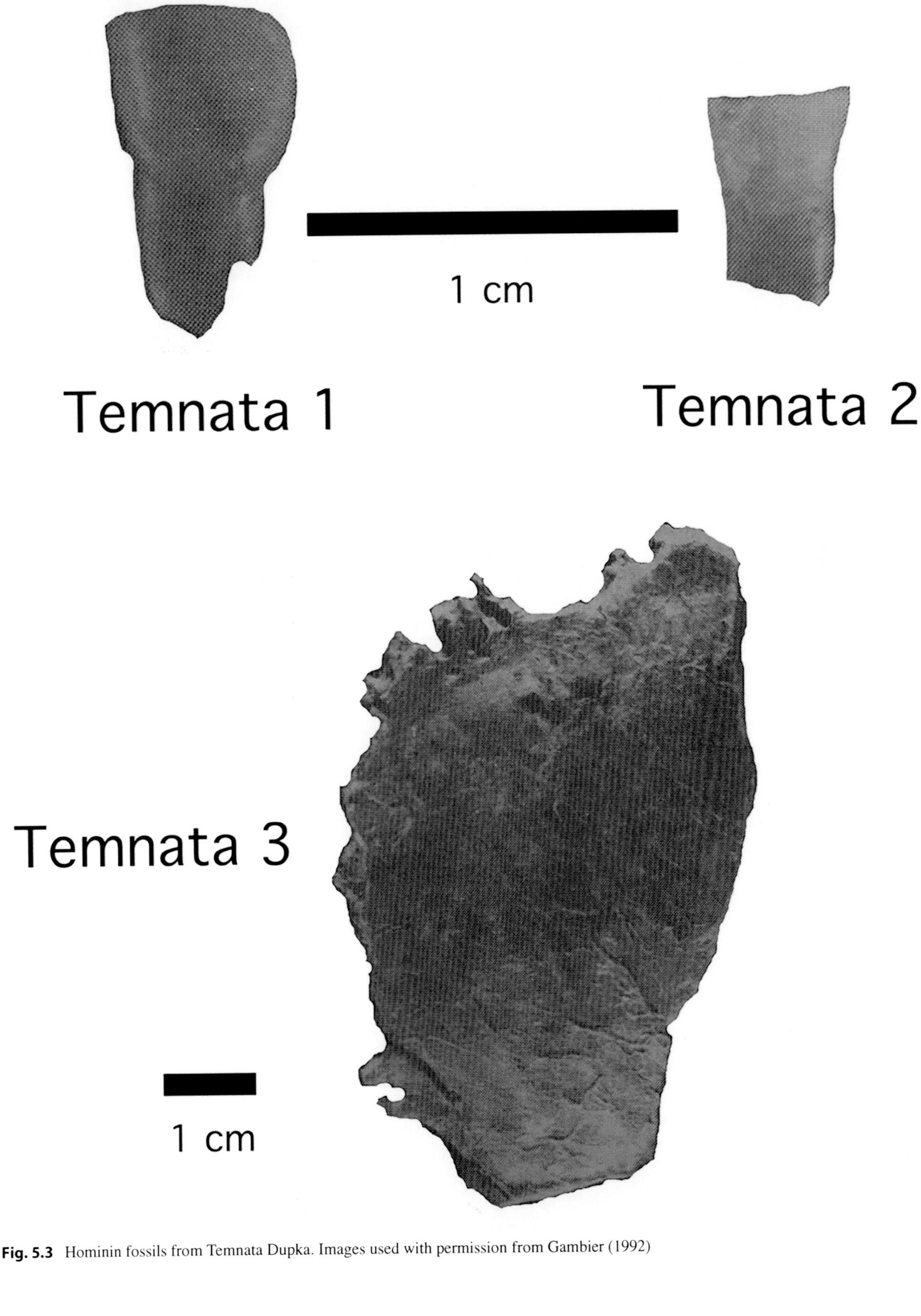

Fig. 5.3 Hominin fossils from Temnata Dupka. Images used with permission from Gambier (1992)

Table 5.3 Hominin fossils from Kozarnika Cave

Element	Level	Age
Phalanx	5b	26,490 kBP[a]
Phalanx	6/7	~43,000 kBP[a]
R upper di1*	13	1.2–1.6 Ma[b]

[a]Uncalibrated radiocarbon date from the same stratigraphic layer as the hominin (Guadelli et al. 2005)
[b]Biostratigraphic date on faunal remains from the layer (Guadelli et al. 2005; Sirakov et al. 2010)
*Hominin status disputed

Peri-Pontic (Spassov 2001, 2016) and involves a dispersal between Europe and central Asia along the northern margins of the Black Sea Basin. We label the second pathway as Trans-Marmaran (also called the Bosphorus pathway [Spassov 2001, 2003, 2016]) insofar as it traverses the landmass surrounding the modern Sea of Marmara between Europe and Asia Minor. Lastly, we identify a Coastal/Trans-Aegean pathway that might have been available had the archipelagoes of the Aegean Sea been joined by land bridges during periods of low sea level in the Pleistocene (e.g., Lykousis 2009). Note that these three pathways are not mutually exclusive; hominins may have used two or more of them simultaneously.

Perhaps the most interesting biogeographic questions concern *why* hominins may have dispersed. These explanations may be classified broadly into several categories. First, one may consider Climate Forcing hypotheses. These explanations suggest that climate change induces change in the distributions of hominin populations. A second category would be Climate Releasing hypotheses in which climate change removes previously existing barriers to dispersal. A third category would be Faunal Wave hypotheses, in which hominins disperse along with other mammalian species, either because entire communities of fauna are dispersing at that time, or because hominins may have a special relationship with one or a few taxa (e.g., a preferred prey species, or a carnivore whose prey could be reliably scavenged). A fourth category would be Culture Releasing hypotheses, in which cultural practices or technology allow hominins to overcome barriers to dispersal that would have been insurmountable in the absence of those behaviors. Our list is not meant to be comprehensive; other categories of hypotheses could certainly be envisioned. Moreover, the categories are not necessarily mutually exclusive.

Testable Biogeographic Hypotheses

Hypotheses are useful only when they can be potentially falsified with data. Here we propose hypotheses about the biogeography of European hominins during the Pleistocene and identify the ways in which they can be falsified or otherwise weakened. We do not claim any special insight into this topic, nor do we feel especially strongly that our hypotheses are correct. Rather, we propose these hypotheses in the belief that, if they are found to be false, then paleoanthropology as a whole will benefit because we will have learned something.

Early Pleistocene Dispersals

We propose that *H. erectus sensu lato* was the first hominin to enter Europe. This species dispersed from the Caucasus along a Peri-Pontic pathway during the Early Pleistocene along with other Asian mammals when steppe-like habitats spread as climates became cooler and more arid (Spassov 2001, 2016). Thus, the proposed dispersal can be categorized as both a Climate Releasing and Faunal Wave hypothesis. The timing of this dispersal must predate the earliest evidence of hominins in Europe. Hominins are dated to 1.1–1.2 Ma (or older; Toro-Moyano et al. 2013; but see Muttoni et al. 2013) in the Iberian Peninsula, and may be as old in the Balkan Peninsula. Trace archaeological evidence from the Apennine Peninsula suggests a possible hominin presence by 1.3–1.7 Ma (Arzarello et al. 2007).

There are a few lines of evidence consistent with this hypothesis. First, it has already been established that Asian steppe mammals disperse into Europe during the Early Pleistocene (e.g., Spassov 2001, 2003, 2016). For example, the Bulgarian paleontological site of Slivnitsa preserves some of the earliest evidence in Europe of Asiatic (*Canis sensu stricto*) and Afro-Asiatic (*Panthera*) carnivorans, the first mass bovid dispersal from Eastern Eurasia, and the first *Ovis* remains in Europe (Spassov 2003). Slivnitsa dates to roughly 2.0 Ma (see Spassov 2016), so these taxa evidently dispersed into Europe during a cooling event at an early stage of the Late Villafranchian. Other southern European Early to late Villafranchian sites (Varshets in Bulgaria; Cernatesti, Tulucesti, and Valea Graunceanului in Romania; and Gerakarou in Greece) also preserve early evidence of faunal dispersals into Europe from Asia (*Nyctereutes*—especially *N.* cf *tingi*) and Afro-Asia (*Mammuthus rumanus*) (Spassov 2003). Subsequent cooling events in the Pleistocene are associated with the arrival in Europe of other eastern fauna such as *Megaceroides* (from Italian sites in the Farnetta unit; Spassov 2003) and *Bison* and *Pontoceros* (from Apollonia in Greece; Koufos et al. 1992; Kostopoulos 1997). Thus, there appears to be a pattern in which arid-adapted Asian and African fauna disperse into southeastern Europe during cooling periods in the Early Pleistocene, and this pattern is consistent with what is known from other paleontological sites throughout Europe (e.g., Maglio and Cooke 1978; van der Made 2001; Koufos and Kostopoulos 2016).

Regarding the taxonomic affinities of the dispersing hominins, we are not yet convinced that the available fossil evi-

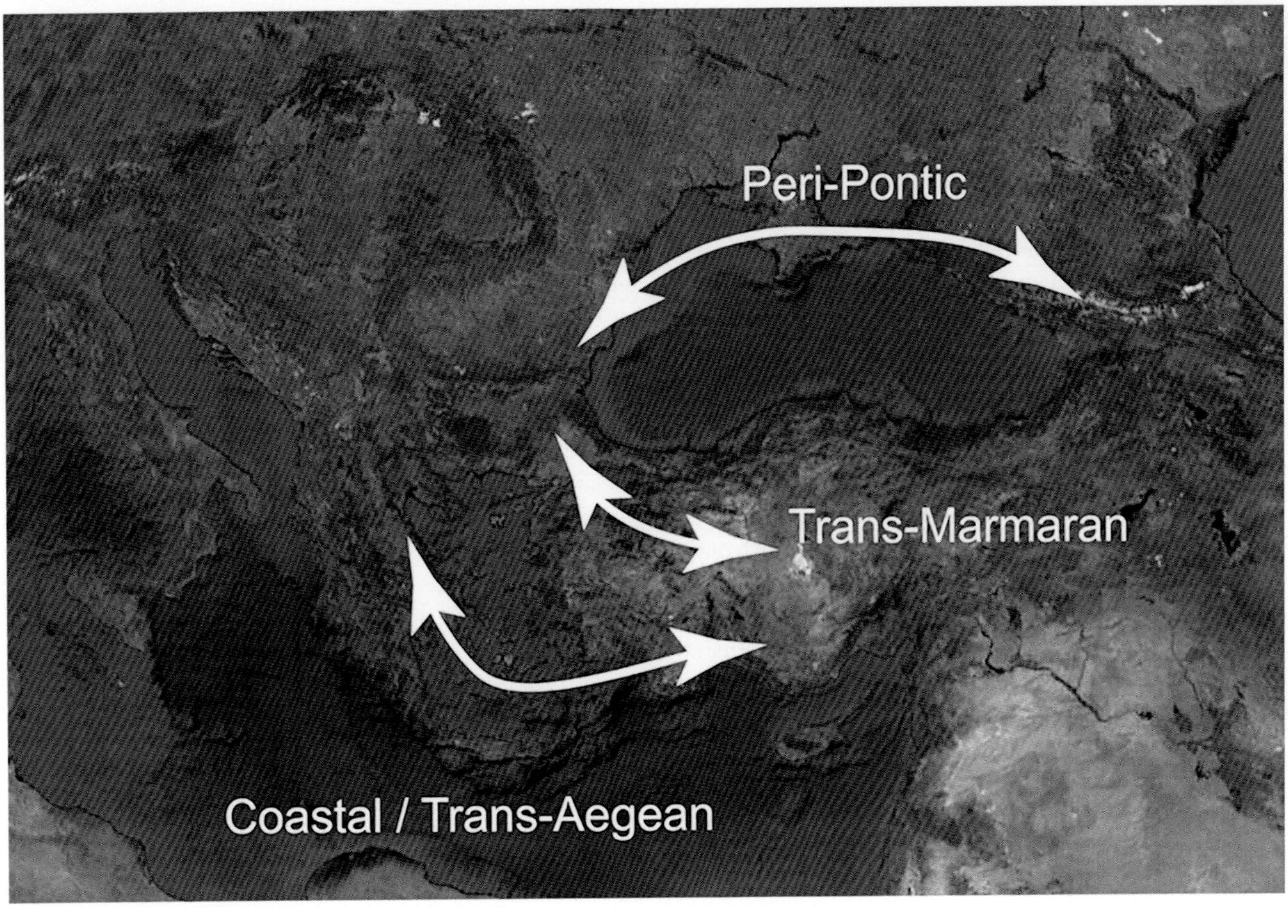

Fig. 5.4 Possible dispersal "pathways" into and out of Europe along which Pleistocene hominins might have traveled. Note that the pathways are merely abstractions that serve heuristically to simplify discussion of what may have been complex and gradual population movements

dence allows us to reject the possibility that the earliest Europeans (including those classified as *H. antecessor*) can be distinguished from *H. erectus*, broadly defined. As to the pathway they may have traversed, evidence from the Crimean site of Sinyaya Balka suggests that the northern Peri-Pontic region was occupied by 1.2 Ma (Shchelinsky et al. 2010).

This hypothesis could be falsified or weakened in a number of ways. First, future fossil hominin discoveries could demonstrate that *H. antecessor* is irrefutably different from *H. erectus sensu lato*, or future discoveries could even demonstrate that the earliest Europeans were *H. habilis*-like. Second, excavations at multiple localities could conceivably demonstrate that the appearance of the first Europeans is consistently associated with warmer rather than cooler periods during the Early Pleistocene. Such evidence would be consistent with the Faunal Wave model of Vrba (1992), known as Habitat Theory, in which low latitude fauna move toward higher latitudes when climates become warmer because the fauna track the changing distributions of vegetational zones. During cool periods, high latitude fauna move toward the equator for the same reason. Finally, the presence of artifacts or hominin fossils anywhere along the Trans-Marmaran or Coastal/Trans-Aegean pathways that are as old or older than Sinyaya Balka would demonstrate those pathways were as likely to have been used as the Peri-Pontic route.

Middle Pleistocene Dispersals

The imperfections of the Middle Pleistocene hominin fossil record make it very difficult to hypothesize about dispersals into and out of Europe during this time period. As a heuristic exercise, we follow Rightmire (2008; see also Stringer 2012; but see Martinón-Torres et al. 2007; Dennel 2009) in hypothesizing that *H. heidelbergensis* originates in Africa at or prior to 600 ka (the date of the Bodo cranium; Clark et al. 1994) and soon afterward disperses into Europe. We cannot presently rule out the possibility that *H. heidelbergensis* evolved *in situ* in Europe prior to 600 ka and then dispersed to Africa and Asia, but we also cannot point to any evidence clearly supporting this possibility. However, the type specimen of *H. heidelbergensis*, the Mauer mandible from Germany, is essentially contemporaneous with Bodo (Wagner et al. 2010) and fossil crania from Yunxian are nearly as old (Chen et al. 1997) and could possibly be attributed to the species. Thus, the location of the first appearance of the species is not obvious. If, in fact, archaic humans like *H. heidelbergensis* dispersed into Europe, then we speculate that such population movements took place during any or all of the periods corresponding to oxygen isotope stages 11, 13, and 15, when climates would have been comparatively warm. Such dispersals might conform to a Climate Releasing model in which African populations disperse into previously inhospitable habitats. If so, hominins may have dispersed along Trans-Marmaran or Trans-Aegean pathways. Subsidence data suggests that up to 50 % of the present Aegean Sea would have been exposed as land even during the relatively warm interglacial during stage 11 when sea levels would have been correspondingly high (Lykousis 2009). One can infer that the same conditions might have applied during older interglacial periods. Once in Europe, *H. heidelbergensis* populations would have gradually evolved *in situ* into Neanderthals (e.g., Hublin 2009), who did not subsequently disperse out of Europe until the Late Pleistocene (see later).

As articulated earlier our hypothesis is highly speculative. It could be rejected by demonstrating clear regional continuity between European *H. heidelbergensis* and European ancestors. If that were the case, then a Middle Pleistocene dispersal out of (rather than into) Europe would be implied, unless one could demonstrate that *H. heidelbergensis* evolved multiregionally across the Old World. It could also be falsified if Neanderthals (or hominins preserving clear Neanderthal apomorphies) are found outside of Europe during the Middle Pleistocene. This would imply the presence of at least one other dispersal either into or out of the continent. It could also be falsified by demonstrating that there is no clear phylogenetic link between Neanderthals and European pre-Neanderthal archaic humans. Mitochondrial DNA from Sima de los Huesos hint at this possibility (insofar those data link a specimen from Sima with Denisovans rather than Neanderthals), although there are many different ways of interpreting the mtDNA evidence (Meyer et al. 2014). Moreover, if the very early dates (Bischoff et al. 2007) obtained for the "pre-Neanderthals" of Sima de los Huesos withstand scrutiny (Endicott et al. 2010; Stringer 2012), then the timing of the dispersals might need to be pushed deeper in time. However, the most recent dates for these fossils point to a younger age (~430 ka; Arsuaga et al. 2014).

Late Pleistocene Dispersals

It is likely that at least two hominin dispersals into or out of Europe took place during the Late Pleistocene. First, it seems likely that Neanderthals dispersed out of Europe and arrived in the Near East by approximately 122 ka (Grün and Stringer 2000) or earlier. This time corresponds roughly to the last interglacial, suggesting that some Neanderthal populations left Europe when it was warm, and may be unrelated to the advance of glaciers (e.g., Bar-Yosef 1992). Thus, the dispersal does not obviously correspond to a Climate Forcing model, although this does not preclude the possibility that glacial maxima later in the Pleistocene may have indeed pushed Neanderthal populations out of Europe at that time. During the last interglacial, a Coastal/Trans-Aegean pathway would likely have been underwater (insofar as oxygen isotope stage 5 is thought to represent a period during which the Black and Mediterranean Seas were connected [Lykousis 2009]) and could not have plausibly led Neanderthals to the Near East, so the Trans-Marmaran pathway seems more likely. The most obvious way in which this dispersal could be challenged would be if Neanderthals or Neanderthal-like hominins were discovered in the Near East prior to the Late Pleistocene. This would raise questions about where Neanderthals first evolved and whether or not they dispersed into or out of Europe. Alternatively, if Near Eastern Neanderthals were found in strata corresponding to the penultimate glacial, then a Climate Forcing model would be an appropriate explanation of why Neanderthals left Europe.

The second Late Pleistocene dispersal corresponds to the movement of anatomically modern *H. sapiens* into Europe at roughly 45 ka (Higham et al. 2011; Benazzi et al. 2011) or older (if the Bohunician lithic industry was produced by modern humans; Richter et al. 2008). This dispersal is remarkable insofar as it represents a northward expansion during a glacial period of a species lacking obvious anatomical adaptations for living in cold environments. The most plausible explanation for such a scenario would be a Culture Releasing model in which technology and culture allowed modern humans to colonize cold habitats that had previously been inhospitable (e.g., Klein 2008). Note that this hypothesis does not necessarily require that the behaviors of non-European modern humans were suddenly transformed so as to allow the dispersal, but could also be compatible with a steady accumulation of behaviors and technology (McBrearty and Brooks 2000) that reached a "critical mass" prior to the dispersal. Interestingly, another northward dispersal of modern humans during the same glacial period may be indicated by the presence of modern humans at approximately 40 ka in Tianyuan Cave near Beijing, China (Shang et al. 2007). If anatomically modern humans entered Europe from the Near East, then it is likely that they traversed a Trans-Marmaran pathway; subsidence data suggest that during the time period in question much of the Coastal / Trans-Aegean pathway would have been underwater, even during glacial maxima when sea levels would have been low (Lykousis 2009).

Conclusion

The human fossil record of Bulgaria is not extensive, but the country's position near pathways into and out of Europe makes it a promising location for paleoanthropological research regarding the biogeography of European hominins. However, this research will be most useful when framed in terms of testable hypotheses. The testing of those hypotheses will ultimately be made possible when researchers from multiple sites and regions in southern Europe, Asia Minor, and the Caucasus cooperate, share data, and adopt a common conceptual framework for addressing research questions. We hope that the volume in which this chapter is found will play a role in stimulating that broad-scale effort. With respect to specific avenues of future research in Bulgaria, priorities include exploration for and excavation of new Paleolithic sites, redating of hominin-bearing stratigraphic layers at existing sites using modern methods, and comparative analysis of preserved putative hominin specimens.

Acknowledgements We thank the symposium organizers for inviting us to participate in the PaGE conference, and the symposium participants for many hours of stimulating conversation. We also thank the government of Bulgaria for permission to excavate at various sites across the country, as well as the anonymous reviewers and Editor.

References

Argue, D., Morwood, M. J., Sutikna, T., Jatmiko, & Saptomo, E. W. (2009). *Homo floresiensis:* A cladistic analysis. *Journal of Human Evolution, 57*, 623–639.

Arsuaga, J. L., Martínez, I., Arnold, L. J., Aranburu, A., Gracia-Téllez, A., Sharp, W. D., et al. (2014). Neanderthal roots: Cranial and chronological evidence from Sima de los Huesos. *Science, 344*, 1358–1363.

Arzarello, M., Marcolini, F., Pavia, G., Pavia, M., Petronio, C., Pretrucci, M., et al. (2007). Evidence of earliest human occurrence in Europe: The site of Pirro Nord (Southern Italy). *Naturwissenschaften, 94*, 107–112.

Bailey, G., Carrion, J. S., Fa, D., Finlayson, C., Finlayson, G., & Rodriguez-Vidal, J. (2008). The coastal shelf of the Mediterranean and beyond: Corridor and refugium for human populations in the Pleistocene. *Quaternary Science Review, 27*, 2095–2099.

Bar-Yosef, O. (1992). The role of Western Asia in modern human origins. *Philosophical Transactions of the Royal Society B Biological Sciences, 337*, 193–200.

Benazzi, S., Douka, K., Fornai, C., Bauer, C. C., Kullmer, O., Svoboda, J., et al. (2011). Early dispersal of modern humans in Europe and implications for Neanderthal behaviour. *Nature, 479*, 525–528.

Bermúdez de Castro, J. M., & Martinón-Torres, M. (2013). A new model for the evolution of the human Pleistocene populations of Europe. *Quaternary International, 295*, 102–112.

Bischoff, J. L., Williams, R. W., & Rosenbauer, J. J. (2007). High-resolution U-series dates from the Sima de los Huesos hominids yields 600 + ∞/−66 kyrs: Implications for the evolution of the early Neanderthal lineage. *Journal Archaeological Science, 34*, 763–770.

Blanc, P.-L. (2002). The opening of the Plio-Quaternary Gibraltar Strait: Assessing the size of a cataclysm. *Geodinamica Acta, 15*, 303–317.

Brown, P., & Maeda, T. (2009). Liang Bua *Homo floresiensis* mandibles and mandibular teeth: A contribution to the comparative morphology of a new hominin species. *Journal of Human Evolution, 57*, 571–596.

Carbonell, E., Bermúdez de Castro, J. M., Parés, J. M., Pérez-González, A., Cuenca-Bescós, G., Ollé, A., et al. (2008). The first hominin of Europe. *Nature, 452*, 465–470.

Chen, T.-M., Yang, Q., Hu, Y.-Q., Bao, W.-B., & Li, T.-Y. (1997). ESR dating of tooth enamel from Yunxian *Homo erectus* site, China. *Quaternary Science Review, 16*, 455–458.

Clark, J. D., de Heinzelin, J., Schick, K. D., Hart, W. K., White, T. D., WoldeGabriel, G., et al. (1994). African *Homo erectus*: Old radiometric ages and young Oldowan assemblages in the Middle Awash valley, Ethiopia. *Science, 264*, 1907–1910.

Comas, D., Calafell, F., Benchemsi, N., Helal, A., Lefrane, G., Stoneking, M., et al. (2000). Alu insertion polymorphisms in NW Africa and the Iberian Peninsula: Evidence for a strong genetic boundary through the Gibraltar Straits. *Human Genetics, 107*, 312–319.

Conard, N., & Bolus, M. (2003). Radiocarbon dating the appearance of modern humans and timing of cultural innovations in Europe: New results and new challenges. *Journal of Human Evolution, 44*, 331–371.

Dennel, R. (2009). *The Paleolithic settlement of Asia*. Cambridge: Cambridge University Press.

Dennell, R., & Roebroeks, W. (2005). An Asian perspective on early human dispersal from Africa. *Nature, 438*, 1099–1104.

Endicott, P., Ho, S., & Stringer, C. (2010). Using genetic evidence to evaluate four palaeoanthropological hypotheses for the timing of Neanderthal and modern human origins. *Journal of Human Evolution, 59*, 87–95.

Gambier, D. (1992). Les vestiges humains des niveaux gravettiens. In J. K. Kozlowski, H. Laville, & B. Ginter (Eds.), *Temnata cave excavations in Karlukovo Karst Area Bulgaria* (Part 1, Vol. 1, pp. 217–222). Krakow: Jagellonian University Press.

Ginter, B., & Kozlowski, J. K. (1982). Excavation and the stratigraphy of the cave. In J. K. Kozlowski (Ed.), *Excavation in the Bacho Kiro Cave (Bulgaria)* (pp. 7–11). Warsaw: Panstowne Wydawnitwo Naukowe.

Glen, E., & Kaczanowski, K. (1982). Human remains. In J. K. Kozlowski (Ed.), *Excavation in the Bacho Kiro Cave (Bulgaria)* (pp. 75–79). Warsaw: Panstowne Wydawnitwo Naukowe.

Grün, R., & Stringer, C. B. (2000). ESR and U-series analyses of dental material from Tabun C1. *Journal of Human Evolution, 39*, 601–612.

Guadelli, J.-L., Sirakov, N., Ivanova, S., Sirakova, S., Anastassova, E., Courtaud, P., et al., (2005). Une séquence du Paléolithique Inférior au Paléolithique Récent dans les Balkans: la grotte Kozarnika à Orechets (nord-ouest de la Bulgarie). In N. Molines, M.-H. Moncel, & J.-L. Monnier (Eds.), *Les Premiers Peuplements en Europe* (British Archaeological Reports, International Series, 164, pp. 87–103).

Higham, T., Compton, T., Stringer, C., Jacobi, R., Shapiro, B., Trinkaus, E., et al. (2011). The earliest evidence for anatomically modern humans in northwestern Europe. *Nature, 479*, 521–524.

Hublin, J. J. (2009). The origin of the Neanderthals. *Proceedings of the National Academy of Sciences of the United States of America, 106*, 16022–16027.

Ivanova, S. (2016). A route through the Balkans and implications for the earliest settlement of Europe. In K. Harvati & M. Roksandic (Eds.), *Paleoanthropology of the Balkans and Anatolia: Human evolution and its context* (pp. 187–211). Dordrecht: Springer.

Jungers, W. L., Harcourt-Smith, W. E. H., Wunderlich, R. E., Tocheri, M. W., Larson, S. G., Sutikna, T., et al. (2009). The foot of *Homo floresiensis*. *Nature, 459*, 81–84.

Klein, R. G. (2008). Out of Africa and the evolution of human behavior. *Evolutionary Anthropology, 17*, 267–281.

Kostopoulos, D. S. (1997). The Plio-Pleistocene artiodactyls (Vertebrata, Mammalia) of Macedonia 1. The fossiliferous site "Apollonia-1", Mygdonia basin of Greece. *Geodiversitas, 19*, 845–875.

Koufos, G. D., & Kostopoulos, D. S. (2016). The Plio-Pleistocene large mammal record of Greece: Implications for early human dispersals into Europe. In K. Harvati & M. Roksandic (Eds.), *Paleoanthropology of the Balkans and Anatolia: Human evolution and its context* (pp. 269–280). Dordrecht: Springer.

Koufos, G. D., Syrides, G. E., Kostopoulos, D. S., & Koliadimou, K. K. (1992). Apollonia, a new vertebrate site in the Pleistocene of the Mygdonia basin (Macedonia, Greece); the first fossil freshwater mollusks in the area. *Comptes Rendus de L'Academie des Sciences, Paris, 315*, 1041–1046.

Lowe, J., Barton, N., Blocklet, S., Ramsey, C. B., Cullen, V. L., Davies, W., et al. (2012). Volcanic ash layers illuminate the resilience of Neanderthals and early modern humans to natural hazards. *Proceedings of the National Academy of Sciences of the United States of America, 109*, 13532–13537.

Lykousis, V. (2009). Sea-level changes and shelf break prograding sequences during the last 400ka in the Aegean margins: Subsidence rates and palaeogeographic implications. *Continental Shelf Research, 29*, 2037–2044.

Maglio, V. J., & Cooke, H. B. S. (Eds.). (1978). *Evolution of African mammals*. Cambridge: Harvard University Press.

Maldonado, A., & Stanley, D. J. (1976). *Late Quaternary sedimentation and stratigraphy in the Strait of Sicily. Smithsonian Contributions to the Earth Sciences, No. 16*. Washington DC: Smithsonian Institution Press.

Martinón-Torres, M., Bermúdez de Castro, J. M., Gómez-Robles, A., Arsuaga, J. L., Carbonell, E., Lordkipanidze, D., et al. (2007). Dental evidence on the hominin dispersals during the Pleistocene. *Proceedings of the National Academy of Sciences of the United States of America, 104*, 13279–13282.

McBrearty, S., & Brooks, A. S. (2000). The revolution that wasn't: A new interpretation of the origin of modern human behavior. *Journal of Human Evolution, 39*, 453–563.

Meyer, M., Fu, Q., Aximu-Petri, A., Glocke, I., Nickel, B., et al. (2014). A mitochondrial genome sequence of a hominin from Sima de los Huesos. *Nature, 505*, 403–406.

Mook, W. G. (1982). Radiocarbon dating. In J. K. Kozlowski (Ed.), *Excavation in the Bacho Kiro Cave (Bulgaria)* (p. 168). Warsaw: Panstowne Wydawnitwo Naukowe.

Muttoni, G., Scardia, G., & Kent, D. V. (2013). A critique of evidence for human occupation of Europe older than the Jaramillo subchron (~1 Ma): Comment on 'The oldest human fossil in Europe from Orce (Spain)' by Toyo-Moyano et al. (2013). *Journal of Human Evolution, 65*, 746–749.

Palombo, M. R. (2010). A scenario of human dispersal in the northwestern Mediterranean throughout the Early to Middle Pleistocene. *Quaternary International, 223–224*, 179–194.

Palombo, M. R. (2013). What about causal mechanisms promoting early hominin dispersal in Eurasia? A research agenda for answering a hotly debated question. *Quaternary International, 295*, 13–27.

Richter, D., Tostevin, G., & Skrdla, P. (2008). Bohunician technology and thermoluminesence dating of the type locality of Brno-Bohunice (Czech Republic). *Journal of Human Evolution, 55*, 871–885.

Rightmire, G. P. (2008). *Homo* in the Middle Pleistocene: Hypodigms, variation and species recognition. *Evolutionary Anthropology, 17*, 8–21.

Rolland, N. (2013). The Early Pleistocene human dispersal in the Circum-Mediterranean Basin and initial peopling of Europe: Single or multiple pathways? *Quaternary International, 316*, 59–72.

Schwartz, J. H., & Tattersall, I. (2002). *The human fossil record Vol. 1: Terminology and craniodental morphology of genus Homo (Europe)*. Hoboken: Wiley-Liss.

Schwartz, J. H., & Tattersall, I. (2003). *The human fossil record Vol. 2: Terminology and craniodental morphology of genus Homo (Africa and Asia)*. Hoboken: Wiley-Liss.

Shang, H., Tong, H., Zhang, S., Chen, F., & Trinkaus, E. (2007). An early modern human from Tianyuan Cave, Zhoukoudian, China. *Proceedings of the National Academy of Sciences of the United States of America, 104*, 6573–6578.

Shchelinsky, V. E., Dodonov, A. E., Baigusheva, V. S., Kulakov, S. A., Simakova, A. N., Tesakov, A. S., et al. (2010). Early Palaeolithic sites on the Taman Peninsula (Southern Azov Sea region, Russia): Bogatyri/Sinyaya Balka and Rodniki. *Quaternary International, 223–224*, 28–35.

Sirakov, N., Guadelli, J.-L., Ivanova, S., Sirakova, S., Boudadi-Maligne, M., Dimitrova, I., et al. (2010). An ancient continuous human presence in the Balkans and the beginnings of human settlement in western Eurasia: A Lower Pleistocene example of the Lower Paleolithic levels in Kozarnika cave (North-western Bulgaria). *Quaternary International, 223–224*, 94–106.

Spassov, N. (2001). The possible time of the first penetrations of *Homo* in Europe in the aspect of the paleoclimate analyses and the mammalian Plio-Pleistocene dispersals from the East. *Annals Institute Museum Archaeology Sofia, 1*, 16–21.

Spassov, N. (2003). The Plio-Pleistocene vertebrate fauna in South-Eastern Europe and the megafaunal migratory waves from the east to Europe. *Revue de Paléobiologie, 22*(1), 197–229.

Spassov, N. (2016). Southeastern Europe as a route for the earliest dispersal of *Homo* towards Europe: Ecological conditions and the timing of the first human occupation of Europe. In K. Harvati & M. Roksandic (Eds.), *Paleoanthropology of the Balkans and Anatolia: Human evolution and its context* (pp. 281–290). Dordrecht: Springer.

Stringer, C. (2012). The status of *Homo heidelbergensis* (Schoetensack 1908). *Evolutionary Anthropology, 21*, 101–107.

Tocheri, M. W., Orr, C. M., Larson, S. G., Sutikna, T., Jatmiko, Saptomo, E. W., et al. (2009). The primitive wrist of *Homo floresiensis* and its implications for hominin evolution. *Science, 317*, 1743–1745.

Toro-Moyano, I., Martínez-Navarro, B., Augustí, J., Souday, C., Bermúdez de Castro, J. M., Matinón-Torres, M., et al. (2013). The oldest human fossil in Europe, from Orce (Spain). *Journal of Human Evolution, 65*, 1–9.

van der Made, J. (2001). Les ongulés d'Atapuerca: stratigraphie et biogéographie. *L'Anthropologie, 105*, 95–113.

van der Made, J., & Mateos, A. (2010). Longstanding biogeographic patterns and the dispersal of early *Homo* out of Africa and into Europe. *Quaternary International, 223–224*, 195–200.

Vrba, E. S. (1992). Mammals as a key to evolutionary history. *Journal of Mammalogy, 73*, 1–28.

Wagner, G. A., Krbetschek, M., Degering, D., Bahain, J.-J., Shao, Q., Falguères, C., et al. (2010). Radiometric dating of the type-site for *Homo heidelbergensis* at Mauer, Germany. *Proceedings of the National Academy of Sciences of the United States of America, 107*, 19726–19730.

Chapter 6
The Human Fossil Record from Turkey

Ahmet İhsan Aytek and Katerina Harvati

Abstract The timing and route of early human dispersals out of the African continent are among the most important issues currently discussed in paleoanthropology. Several questions arise concerning both early and later dispersals: When did migration events happen? From which populations did these dispersing hominins stem? Which routes did they use? One of the likely dispersal corridors passes through Turkey, which is situated between three continents and therefore can be seen as an important bridge between them. Despite its geographic position, paleoanthropological research in Turkey has been limited, and the known fossil human record from this region is small. Although most of the known fossil human remains were found during early investigations, in the last decade new finds have further highlighted the region's potential for paleoanthropological research. This chapter reviews the human fossil record from Turkey, and presents the results of a preliminary geometric morphometric study of the Kocabaş hominin, the oldest and most important fossil human specimen known from the country.

A.İ. Aytek (✉)
Department of Anthropology, Faculty of Arts and Science, Mehmet Akif Ersoy University, Istiklal Campus, Burdur, Turkey

Paleoanthropology, Department of Geosciences, Eberhard Karls Universität Tübingen, Tübingen, Germany

Senckenberg Center for Human Evolution and Paleoenvironment, Eberhard Karls Universität Tübingen, Tübingen, Germany
e-mail: aytek@mehmetakif.edu.tr

K. Harvati
Paleoanthropology, Department of Geosciences, Eberhard Karls Universität Tübingen, Tübingen, Germany

Senckenberg Center for Human Evolution and Paleoenvironment, Eberhard Karls Universität Tübingen, Tübingen, Germany

PhD Program in Anthropology, City University of New York Graduate Center and New York Consortium in Evolutionary Primatology, New York, NY, USA
e-mail: katerina.harvati@ifu.uni-tuebingen.de

Keywords Anatolia • Neanderthal • Upper Paleolithic • Kocabaş • Geometric morphometrics • *Homo erectus*

Introduction

Anatolia lies directly on one of the likely dispersal corridors from Africa and the Near East into Europe. Hominin fossil finds from Turkey can therefore help answer questions about hominin dispersals, both for the dispersal event of early *Homo* species, as well as for the later Out of Africa migration of modern humans (see also Dinçer 2016). Despite its critical geographic position—and similarly to other neighboring countries (see Harvati et al. 2009; Harvati 2016; Roksandic 2016; Strait et al. 2016)—paleoanthropological research in Turkey has been limited, and the known fossil human record from this region is small. Turkey has substantial potential for Paleolithic sites, and the poor fossil record is most likely a consequence of the scarcity of excavations and surveys. According to The Archaeological Settlements of Turkey (www.tayproject.org), there are currently an estimated 448 known Paleolithic sites in the country, and the number is expected to rise as research efforts intensify. However, very few of these sites have been excavated, and, among those excavated, most are poorly documented or damaged.

Beyond the Kocabaş partial cranium (Kappelman et al. 2008), recovered accidentally in a travertine block, Paleolithic hominin remains have been found in just a few excavated localities. These include Karain cave, Merdivenli cave, Beldibi rock shelter, Kanal cave, İncili cave (Big cave), and Üçağızlı cave (Şenyürek 1949; Şenyürek and Bostancı 1956; Bostancı 1963, 1971, 1973; Güleç et al. 2007; see Fig. 6.1). With the exception of Karain and Üçağızlı, publications about these caves are quite old and the information provided is often unclear, especially with regard to their chronology.

Katerina Harvati and Mirjana Roksandic (eds.), *Paleoanthropology of the Balkans and Anatolia*, Vertebrate Paleobiology and Paleoanthropology, DOI 10.1007/978-94-024-0874-4_6

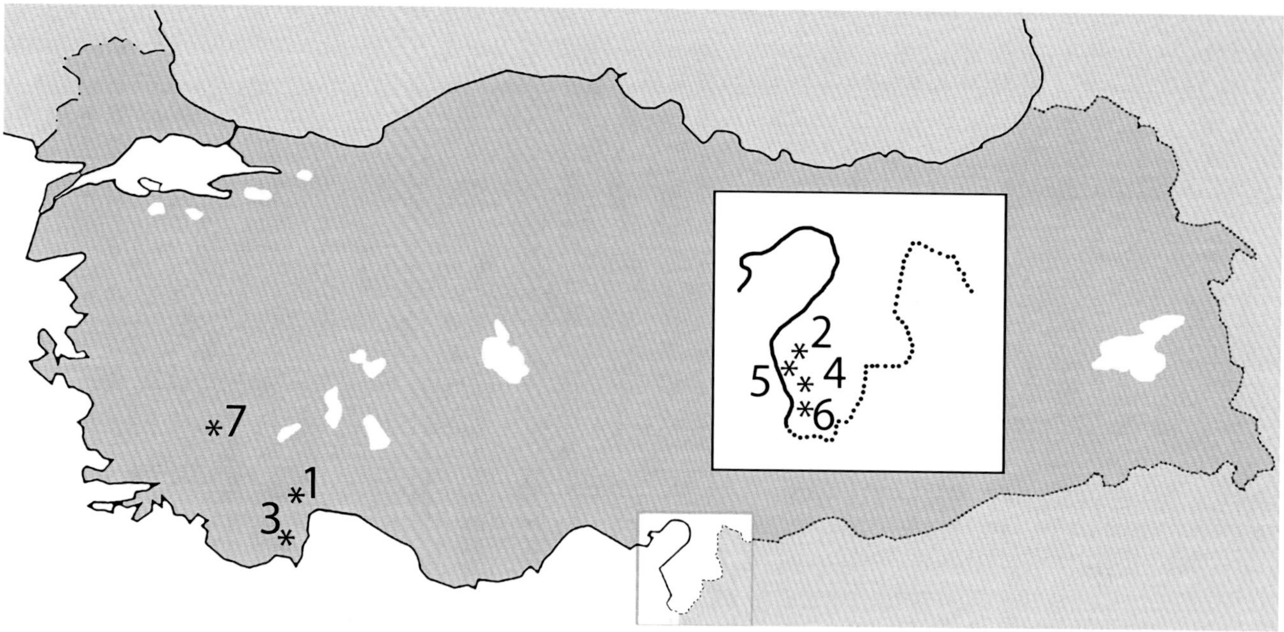

Fig. 6.1 The hominin-bearing Paleolithic sites in Turkey. (1) Karain Cave, (2) Merdivenli Cave, (3) Beldibi Rockshelter, (4) Kanal Cave, (5) İncili Cave, (6) Üçağızlı Cave, (7) Kocabaş

This chapter provides an overview of the known human remains recovered from Paleolithic sites in Turkey, as well as a brief geometric morphometric comparative analysis of the Kocabaş hominin, the earliest human fossil currently known from Anatolia and perhaps the most important human fossil from Turkey (see also Dinçer 2016).

Karain Cave

This cave is located on the southern coast of Turkey (Fig. 6.1), 30 km northwest from the city center of Antalya province (Demirel et al. 2011). It is divided into five chambers, labeled A, B, C, D, and E. Excavations conducted by Prof. Dr. Işın Yalçınkaya are still ongoing in chambers B–E (see also Dinçer 2016). Chamber E contains Lower and Middle Paleolithic layers, yielding abundant lithic assemblages and faunal remains. The extensive use of the cave by humans is thought to span from the Lower Paleolithic to the Roman times. The older levels have produced Acheulean lithic artifacts, including a hand-axe found during the 2007 excavation season, suggested to be at least 400 ka on the basis of its stratigraphic position (Yalçınkaya et al. 2008). Unfortunately, these levels have not yielded human remains yet. Mousterian levels were dated by Thermoluminescence and Electron Spin Resonance to between 160 and 60 kBP (Rink et al. 1994; Otte et al. 1998), suggesting the presence of *Homo neanderthalensis* (Taşkıran 2002), a hominin associated with this industry in Europe. Faunal remains include ibex, fallow deer, roe deer, wild boar, aurochs, equid, hippo, elephant, cave bear, cave hyena, wolf, lynx, wild cat, mustelids, hare, hedgehogs, shrews; as well as birds and freshwater gastropods (Otte et al. 1998). Some of these were intentionally transported to the site by humans; however, others appear to have been accumulated by carnivores that used the cave for shelter or hibernation.

The first hominin remains recovered at Karain were isolated teeth found in 1949 by İ.K. Kökten in chamber D (Şenyürek 1949). They comprise a left upper second deciduous molar (Ldm^2) and a broken root of another tooth. The Ldm^2 was examined by M. S. Şenyürek, who found its morphology and crown dimensions to fall within the Neanderthal range (Şenyürek 1949). The root preserves a small piece of its crown on the upper part. Şenyürek (1949) thought that it likely belonged to a lower incisor, but refrained from further diagnosis. These teeth were proclaimed as the first fossil humans and the first examples of Neanderthals found in Anatolia.

During the 1986 excavation season, three additional human teeth were recovered and were examined by B. Alpagut. In a preliminary report (Yalçınkaya 1988), they were described as a left lower second molar (LM_2), a left upper first molar (LM^1), and a left lower deciduous molar (Ldm_2). Although the LM_2 was described as showing modern-human-like morphological features and measurements (no detail is given), the other teeth were considered to belong to Neanderthals and to show particular similarities with the specimens from Mount-Carmel and Shanidar (Yalçınkaya 1988). Unfortunately, this work was not presented in depth in the article; therefore, the comparative material and methodology used are not clear.

Additional human remains were discovered during the 1996 excavation season. Unlike the previous finds, they included not only teeth, but also mandibular and postcranial remains. The first set of fossils was excavated from the geological layer III.2, archaeological unit F, and dated to *ca.* 200–250 kBP by ESR and TL dating (Otte et al. 1998). It comprises three first phalanges and one third phalanx of the left hand, a radius and a cuboid fragment, a fragment of a fibula (it is not clear if it belongs to a human), and a mandible. Although a detailed analysis of these specimens has not yet been published, some features—including the shape of the mandibular symphysis, the frontal disposition of the incisors, the absence of a bony chin, the posterior position of the mental foramen, and the presence of a retromolar space—all suggest Neanderthal affinities (Otte et al. 1998).

A second set of human remains, including two vertebrae and a fragment of a femoral diaphysis, were excavated in layer III.3 and dated to between *ca.* 250–200 kBP and 350–300 kBP by correlation with oxygen isotope stage OIS 9 (Otte et al. 1998). A possible affinity of these fossils with archaic *Homo sapiens* has been suggested by Otte et al. (1998), principally on the basis of their association with a lithic techno-complex similar to the Acheulo-Yabrudian.

In a recent study, Chevalier et al. (2015) examined the diaphyseal cortical bone thickness of the Karain femur and concluded that it exhibits some features common among Neanderthals, such as a circular outline and a strong posteromedial reinforcement of the cortical thickness on the medial side of the midshaft. Since such posteromedial reinforcements at midshaft are also observed in the Middle Pleistocene specimen from Berg Aukas from Namibia (Grine et al. 1995), Chevalier et al. (2015) point out that larger comparative samples are necessary before reaching a conclusive interpretation of the Karain femur. The human material from Karain cave is currently being analyzed by Işın Yalçınkaya.

A skull found by Kılıç Kökten—currently held in the Anatolian Civilizations Museum—was examined by Güleç (1994). Although the exact provenance of the specimen is not known, it is thought to derive from the Upper Paleolithic or Mesolithic layers of Karain Cave. Unfortunately, the skull is not complete and its state of preservation does not allow for direct dating. The skull consists of a maxilla preserving an incisor, three-fourths of the left orbital region, as well as the frontal and parietal bones. The right side of the skull is covered by hardened sand. The specimen shows some primitive features, including a robust supraorbital torus, some prognathism and great nasal width, as well as a long and low vault. Its measurements (minimum frontal breadth (96?), bistephanion breadth (114?), frontal arch (136?), frontal chord (118?), frontal curve height (28), upper facial height (67), orbital height (33), nasal breadth (16?), nasal height (53), and stephanic index (84,21?)) were found to be consistent with Upper Paleolithic modern humans and Mesolithic people from France, as well as with Dar-es-Soltan and Qafzeh (Güleç 1994).

Merdivenli Cave

This cave is located in the Samandağ region, near the village of Mağracık, in Hatay province (Fig. 6.1). Klaus Hormann, a German geologist, noticed the cave and informed anthropologist Enver Bostancı in 1954 who called it the 'first cave' at the time of discovery, and later renamed it 'Merdivenli cave'. Their first visit to the cave led to the discovery of a flint flake and fossilized animal remains. In 1956, Bostancı conducted test excavations that uncovered a human lower molar and three human bone fragments together with a lithic assemblage and fossil faunal remains (Şenyürek and Bostancı 1958). In the same year, the team proceeded to carry out systematic excavations, recovering cultural remains from the Roman period, but also from the Upper and Middle Paleolithic. The latter include side-scrapers, retouched and unretouched points similar to Levalloiso-Mousterian lithics found in the Near East, and Mousterian artifacts similar to those from Europe (Şenyürek and Bostancı 1956). The fossil fauna includes cave bear, cave lion, and wild boar. During this excavating season, two human upper permanent molars and one lower permanent molar were found in the lower layer of the Middle Paleolithic sediments (Şenyürek and Bostancı 1956). These specimens were considered by the authors to likely represent Neanderthals, but no detailed description and rationale was given. In 1957, a second excavation was carried out but did not yield any further human remains (Şenyürek and Bostancı 1958).

Beldibi Rock Shelter

This rock shelter is located in Beldibi village, 30 km west of Antalya province, and about 25 m above sea level (Fig. 6.1). From the first excavations at Beldibi, conducted in 1959, Bostancı (1963) reported human skull fragments, which were too small to provide meaningful comparative information. More human fossil material—fragments of both right and left femora—were uncovered in test excavations in the following year. These femora were found in the same Upper Paleolithic and Mesolithic layers as the skull fragments. Both were missing proximal and distal ends, limiting the comparative study to examination of the shaft. Bostancı (1963) reported a moderately developed *linea aspera* on both specimens, a condition generally observed among modern humans. The pilaster index (derived from the antero-posterior and transverse diameters of the midshaft) was calculated to be 114.8 for the left and 116.17 for the right femur. Compared to *Homo neanderthalensis*, *Homo erectus*, Skhul III, V and VI, and recent modern humans, the value for the left femur was found to be intermediate between *Homo neanderthalensis* and *Homo sapiens*, while the value for the right femur was found to be similar to present-day modern humans (Bostancı 1963). Paintings and engravings, possibly dating to the Upper Paleolithic, were also found at Beldibi (Bostancı 1959).

Kanal Cave

This cave is located in Çevlik village in Hatay province, 300 m from the coast and also about 600 m from Merdivenli cave (Fig. 6.1). From top to bottom, Kanal cave preserves cultural layers from the Middle Aurignacian, Lower Aurignacian, and Levalloiso-Mousterian (Bostancı 1971). In the course of excavations in 1969, a left upper deciduous canine from a Levalloiso-Mousterian context and a lower molar from the Lower Aurignacian context were found (Bostancı 1971). Crown measurements of the canine, compared to those of Shanidar, Skhul, Pech de L'Azé, and Krapina 3 individuals, were found to be similar to values for Neanderthal juveniles (Bostancı 1971). The second tooth was described as a right M_2. According to Bostancı (1971), the tooth exhibited modern-human morphology, and was assigned to '*Homo sapiens çevlikiyensis*'.

İncili Cave (Big Cave)

Another human fossil-bearing cave is İncili cave (also known as 'Big cave'), located about 300 m south from Kanal cave (Fig. 6.1). According to Bostancı (1973), human remains found during excavation included a mandible, maxilla, femur, tibia, several vertebrae and foot bones, all considered to represent a single male individual of *ca.* 50 years of age. As described by Bostancı (1973), the finds show some primitive characters, such as the absence of bony chin, presence of a simian shelf, large zygomatic process comparable in size to Neanderthals, and larger foot bones than modern humans. Bostancı (1973) suggested that these remains therefore belong to a fossil *Homo* and attributed this specimen to *Homo sapiens çevlikiyensis*, as those from Kanal cave. The lack of stratigraphic context and proper anatomical description makes this attribution problematic.

Üçağızlı Cave

Üçağızlı cave is located in Hatay, the Mediterranean coast of southern Turkey, 6 km from the Syrian border (Fig. 6.1). Excavation was initiated by A. Minzoni-Deroche during the 1980s and directed by E. Güleç from 1997 onwards (Kuhn et al. 2009). The region exhibits many similarities with the Levant with respect to its topography and ecology. Layers preserving Initial Upper Paleolithic and Ahmarian lithic assemblages have been dated through AMS radiocarbon dating on carbonized plant remains and marine shells to between *ca.* 29,000 and 41,000 ^{14}C BP. Later assemblages from the Epipaleolithic period are also preserved (dating to *ca.* 17 ^{14}C 000 BP).

Excavation at Üçağızlı yielded abundant faunal remains, including those of aurochs, red deer, pig, fallow deer, goat, roe deer, carnivores, rodents, fish, tortoises, and shells. In addition to lithics, the material culture remains also comprise bone artifacts and ornaments, mostly produced on shells (Kuhn et al. 2009). Although the cave seems to have been extensively used by humans, as testified by the abundant archaeological remains, human fossils are sparse (Kuhn et al. 2009). Between 1989 and 2012, 14 isolated teeth, a maxillary fragment and a cranial fragment, dated to *ca.* 29.130–41.400 ^{14}C BP, were found in the cave (Güleç et al. 2008). Of these, only ten teeth have been studied: two incisors, two canines, one premolar, and five molars (Güleç et al. 2007). These specimens span from the earliest part of the Initial Upper Paleolithic layers (one specimen) to the end of the Initial Upper Paleolithic layers (three specimens) and to the Ahmarian layers (six specimens). The teeth do not have any pathological conditions and show different wear stages, suggesting that each belongs to a different individual. Güleç et al. (2007) concluded that, while most of them show *Homo sapiens* features, at least one of them possesses some possible Neanderthal traits (Güleç et al. 2007). A detailed description and analysis of these remains is underway by Erksin Güleç.

Kocabaş

The fossil hominin from Kocabaş, also known as the Denizli specimen, was discovered in 2002 in a travertine quarry near the Kocabaş village (S.-W. Turkey, Fig. 6.1; see also Dinçer 2016). Prof. Dr. M. Cihat Alçiçek, a geologist from Pamukkale University, found the fossil during one of his visits to the area in the course of his research on the geological setting of the travertine masses. According to Prof. Alçiçek, it was recovered by workers during the processing of the travertine blocks which were brought from the travertine area to the factory (Dalmersan).

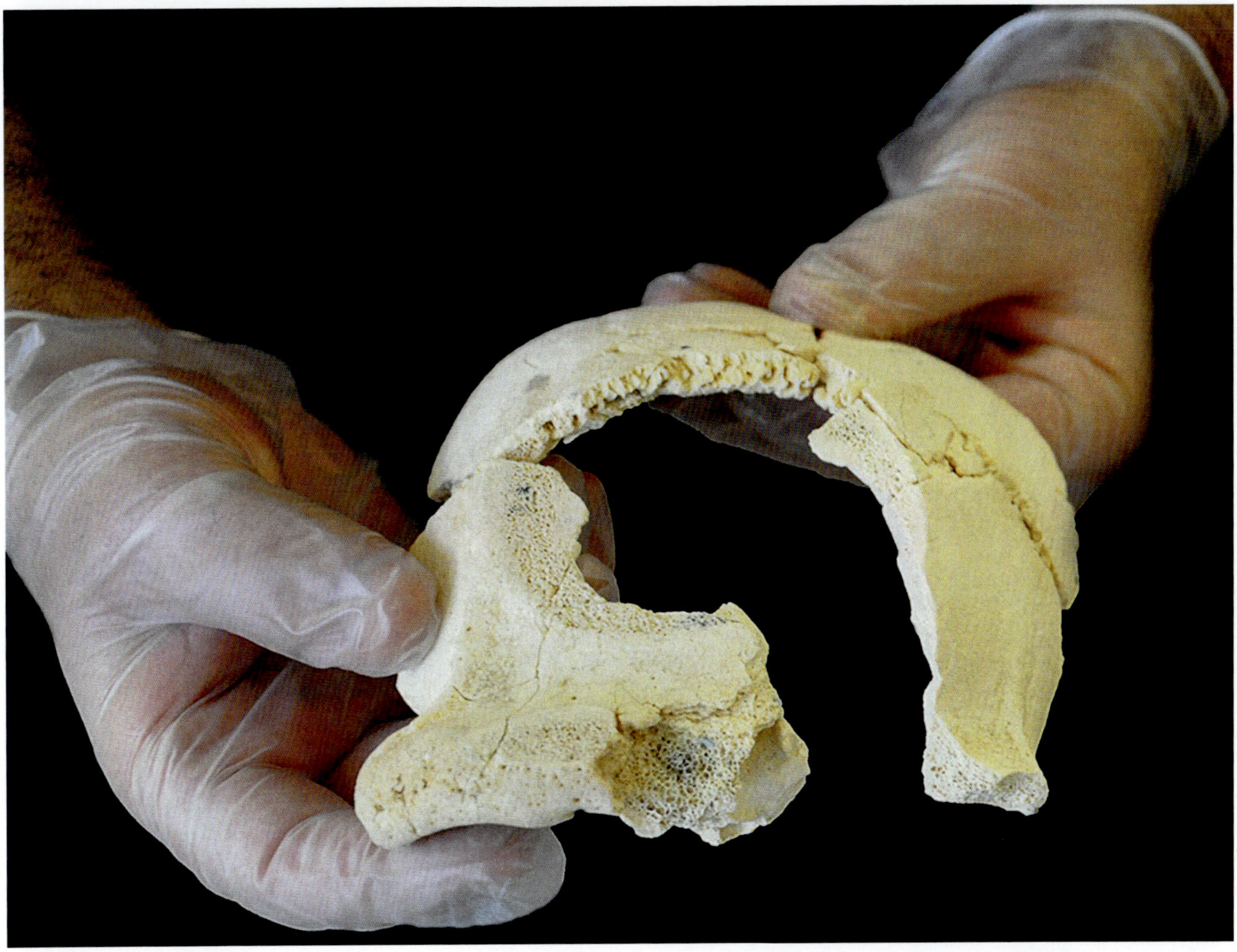

Fig. 6.2 The Kocabaş hominin fossil fragments (Photo: A.İ. Aytek)

The Kocabaş hominin consists of three large bone fragments: a large fragment of the right parietal, a fragment of the right frontal preserving part of the supraorbital torus, and a partial left parietal still articulated with a piece of the left frontal bone (not preserving the supraorbital torus) (Fig. 6.2). Unfortunately, part of the specimen was damaged during the cutting process of the travertine block (Vialet et al. 2012). The endocranial surface of the right frontal bone presents lesion-like traces, interpreted by Kappelman et al. (2008; Fig. 6.3) as a pathological condition. The fossil is now kept in Pamukkale Hierapolis Archaeological Museum in Denizli, Turkey.

The Kocabaş village is situated in south-western Turkey in the Denizli basin, 26 km from the east of Denizli town, in one of the largest river valley systems in Turkey, Büyük Menderes. The Denizli basin is a 70 km long and 50 km wide graben filled with Neogene and Quaternary deposits. The faults, which were generated in the Quaternary, led to the deposition of travertines (Alçiçek et al. 2007). The ancient Denizli travertines are exploited by the marble industry. The Upper Travertine level, from which the fossil likely originates, preserves rich fossiliferous deposits and has yielded various Pleistocene faunal remains, including *Equus*, *Bos*, *Dama*, *Stephanorhinus* sp., cf. *Mammuthus*, *Bison* sp., and *Testudo* sp. (Lebatard et al. 2014; Boulbes et al. 2014).

The exact provenance of the hominin find is not known. Nevertheless, it likely derives from the Upper Travertine level, which was the only one exploited at the time of discovery in 2002 (Vialet et al. 2012; Lebatard et al. 2014). A geological age of 510,000 ± 50,000 to 330,000 ± 30,000 years for the sediments at the specimen's presumed approximate location was originally proposed on the basis of Thermoluminescence dating of the travertines (Kappelman et al. 2008). Because Thermoluminescence has an upper limit of *ca.* 500 ka, a Turkish-French team conducted a paleomagnetic study of a sequence of travertine sediments (Vialet and Alçiçek 2012). These authors proposed an older age of more than 780 kaBP for the fossil-bearing sediments. Further work by Lebatard et al. (2014) combined paleomagnetic measurements with cosmogenic nuclide concentration

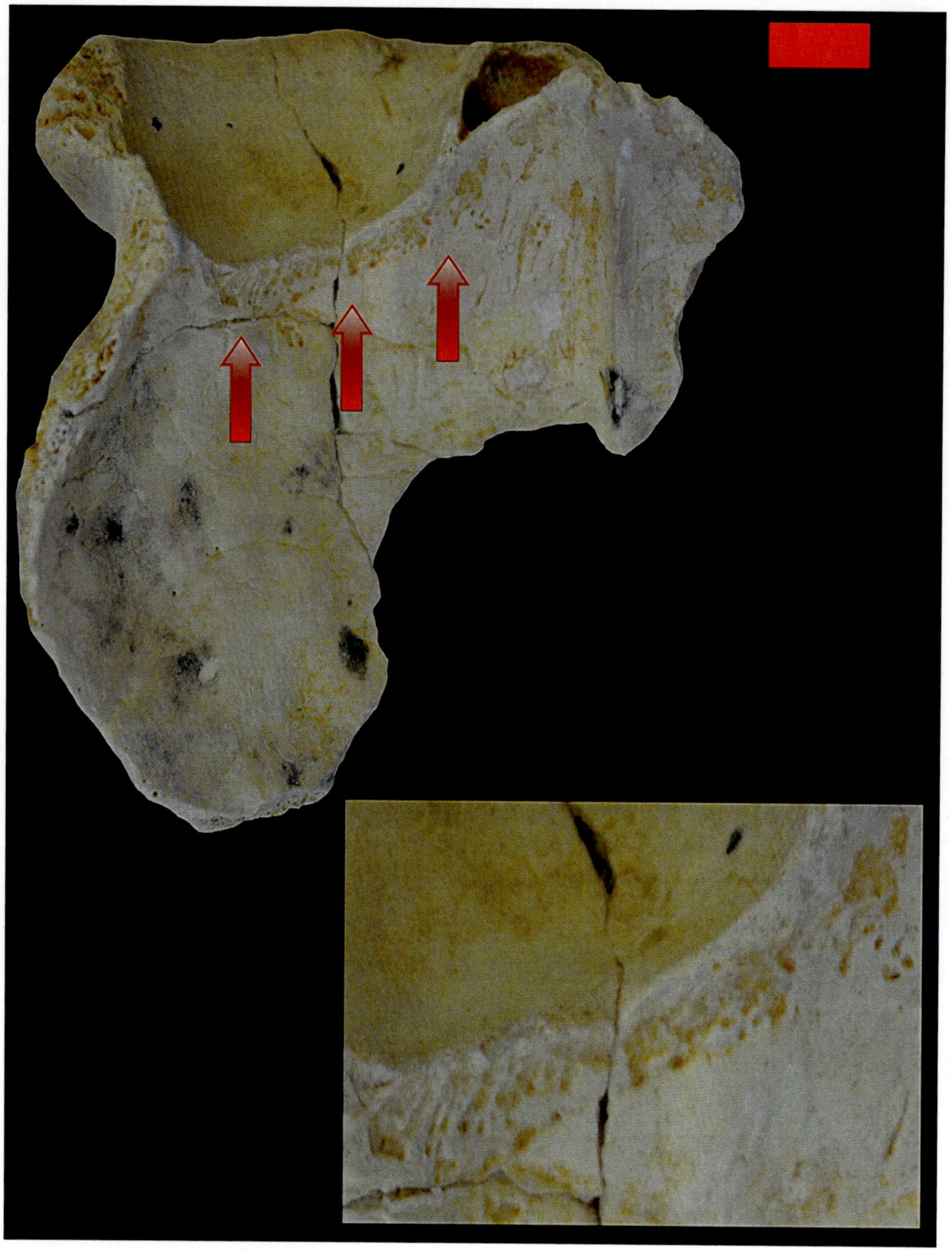

Fig. 6.3 Endocranial aspect of the right frontal bone of the Kocabaş specimen (Scale is 1 cm). *Arrows* point to the lesions on the endocranial surface, shown in magnification in the inset (Photo: A.İ. Aytek)

on travertine sediments in order to refine the chronology of the fossil, and estimated the burial age of ^{26}A/^{10}BE on pebbles from conglomeratic levels which come from the Upper Travertine level. These levels exhibited reverse polarity, suggesting that they were deposited before the Cobb Mountain sub-chron (between 1.22 and *ca.* 1.5 Ma). Lebatard et al. (2014) concluded that the Kocabaş specimen is most likely between 1.1 and 1.3 Ma old. Additional stratigraphic, sedimentological, and paleomagentic study of the region, recently conducted by Khatib et al. (2014), supports an age between 1.2 and 1.6 Ma for the Upper Travertine of Kocabaş (Khatib et al. 2014).

Only a few studies have been published on the morphology and taxonomic position of the Kocabaş specimen itself. A preliminary description and comparative analysis was conducted by Kappelman et al. (2008). These authors provisionally attributed Kocabaş to *Homo erectus sensu lato* on the basis of non-metric features and a few linear measurements. Important traits include the prominent supraorbital torus, which Kappelman et al. (2008) found to resemble Javan and African, rather than Chinese *H. erectus*, and a distinct supraorbital sulcus, commonly considered as a typical *H. erectus* condition (Kappelman et al. 2008). Based on lesions on the endocranial surface of the right frontal bone, Kappelman et al. (2008) proposed that the specimen may represent the most ancient known example of tuberculosis caused by *Leptomeningitis tuberculosa*, or tuberculosis of the meninges. This interpretation, however, has been challenged by other researchers (Roberts et al. 2009).

Kappelman's et al. (2008) conclusions on the taxonomic position of the Kocabaş hominin were recently supported by two other studies. Guipert et al. (2011) reconstructed the specimen on the basis of a CT scan. While the left supraorbital torus was generated by mirroring, Guipert et al. (2011) reconstructed the Kocabaş frontal squama using the frontal bone of Zhoukoudian XI, since they considered this specimen to be most similar to Kocabaş in its proportions and curvatures, as well as in the position of the coronal sutures and temporal lines (Guipert et al. 2011). They then compared their reconstruction with a fossil sample from the middle and late Pleistocene of Asia, Europe, and Africa. Guipert et al. (2011) concluded that Kocabaş is very close to *H. erectus s.l.* in morphology. More recently, Vialet et al. (2012) reconstructed the Kocabaş specimen virtually using the edges of the saw cut as reference planes to bring the bone fragments in connection. The authors then collected a number of linear measurements, which they found to fall within the range of variation of *H. erectus* and to show the strongest resemblance with African and Georgian specimens (Vialet et al. 2012).

In their most recent analysis, Vialet et al. (2014) presented a detailed morphological description, as well as a comparative analysis based on an expanded comparative fossil sample. Additionally, they conducted a geometric morphometric analysis on the basis of 12 3D landmarks and 54 comparative specimens. The linear measurements and indices reported for Kocabaş overlapped with African and Asian *H. erectus* specimens, as well as, in some instances, with Middle Pleistocene individuals such as Kabwe, Petralona, and Bodo. In their geometric morphometric principal components analysis, Kocabaş was found to fall just outside the African *H. erectus* convex hull along PCs 1 and 2 (which comprised specimens quite disparate in time, including KNM ER 3883 and 3733 as well as OH9, Kabwe, and Bodo) and to be relatively removed from the Asian *H. erectus* sample.

The authors concluded, on the basis of their metric, geometric morphometric, and non-metric analysis, that, although Kocabaş shows similarities to Asian *H. erectus*, such as Zhoukoudian, it is in many aspects most similar to ancient African *H. erectus* and *H. ergaster*. Particular similarities were noted to the OH9 and Daka individuals, thought to be contemporaries of the Kocabaş specimen, in the proportions of the frontal bone—although the authors noted that these specimens are quite different in the morphology of the anterior part of the frontal. In light of the recent proposed revision of the Kocabaş geological age to >1 Ma, any morphological affinities with early African *H. erectus* take on particular importance, as they can be interpreted as supporting an early date for this specimen.

Several questions arise regarding the geometric morphometric analysis conducted by Vialet et al. (2014). First, the authors did not include a measure of supraorbital torus thickness in their geometric morphometric shape analysis. Since this aspect of frontal bone morphology is important when considering archaic humans such as *H. erectus*, its lack of representation may have an important influence on their results. For example, the reported PCA does not fully separate modern humans from *H. neanderthalensis*, but also, surprisingly, from *H. erectus* from Java and China. As the above samples show marked differences from one another in their supraorbital morphology, we consider the landmark configuration used by Vialet et al. (2014) to be likely insufficient to fully assess the morphology preserved in Kocabaş. Second, Vialet et al. (2014) were able to include OH9 in their analysis, although this specimen does not preserve bregma. No information is given on how bregma was reconstructed, even though OH9 plays an important role in Vialet's et al. (2014) results and conclusions. Finally, the taxonomic groupings used by Vialet et al. (2014) are unusual and potentially also influence their interpretation of their results and their conclusions: specimens like Bodo and Kabwe, commonly attributed to *H. heidelbergensis s.l.* (see e.g. Rightmire 2009), were assigned by Vialet et al. (2014) to African *H. erectus*, and grouped together with much older specimens, such as KNM-ER 3733 and 3883. Although the authors likely wish to emphasize the geographic origin of these fossils, we consider this unconventional grouping of specimens so far removed in time into one taxonomic unit confusing and potentially misleading.

In order to address these concerns, we conducted a preliminary geometric morphometric comparative analysis of the Kocabaş specimen, using a small comparative modern human and fossil sample. Our goal was to repeat the geometric morphometric analysis of Vialet et al. (2014) using largely overlapping landmark measurements and fossil samples, but incorporating the supraorbital torus morphology and adopting more commonly used taxonomic groupings for the fossil samples, in order to evaluate the degree of similarity between Kocabaş and African *H. erectus*.

In order to digitize landmarks and conduct our analysis, we had to reconstruct the specimen. This was performed virtually, using surface scans of the original individual fragments (made in the Hierapolis Museum, Pamukkale, Denizli, with a portable NextEngine 3D surface scanner (Camera resolution 3 Mp and accuracy 125 μm) by AIA). After three different 3D images of the bones were obtained, they were combined in an anatomical position using the AVIZO software (Avizo 6.3.1.). The parietal bones were merged along the sagittal suture. The frontal fragment preserving the supraorbital torus was merged to the right parietal bone using the temporal line as a guide.

The fragment of the right frontal bone was then mirrored and merged to left side of the specimen, again using the temporal line to align the pieces. The ectocranial splits and fractures were also used to support the accuracy of our reconstruction. Part of the specimen between the frontal and parietal bone is missing. This region was not reconstructed, because the preserved bone was not sufficient for mirroring, and no landmarks were collected in this missing part. Figure 6.4 shows the steps taken to virtually reconstruct the specimen.

Three-dimensional coordinates of 13 landmarks—selected to represent the general shape of the frontal bone and supraorbital torus as best as possible—were digitized on the frontal bone of the virtual reconstruction with AVIZO (Table 6.1). Bregma was not included, in order to maximize the fossil comparative sample and to include, specifically, OH9 in the analysis. Our comparative sample (17 fossil, 20 modern specimens; Table 6.2) was digitized from CT and surface scans obtained from the database of the Tübingen Paleoanthropology section and from the NESPOS online database. All data were collected by AIA. Specimens with missing data were reconstructed through reflected relabeling (Mardia and Bookstein 2000; Harvati 2003) using the software package Morpheus et al. (Slice 1998). A Generalized Procrustes Analysis (GPA) was performed in Morphologika (O'Higgins and Jones 2004)

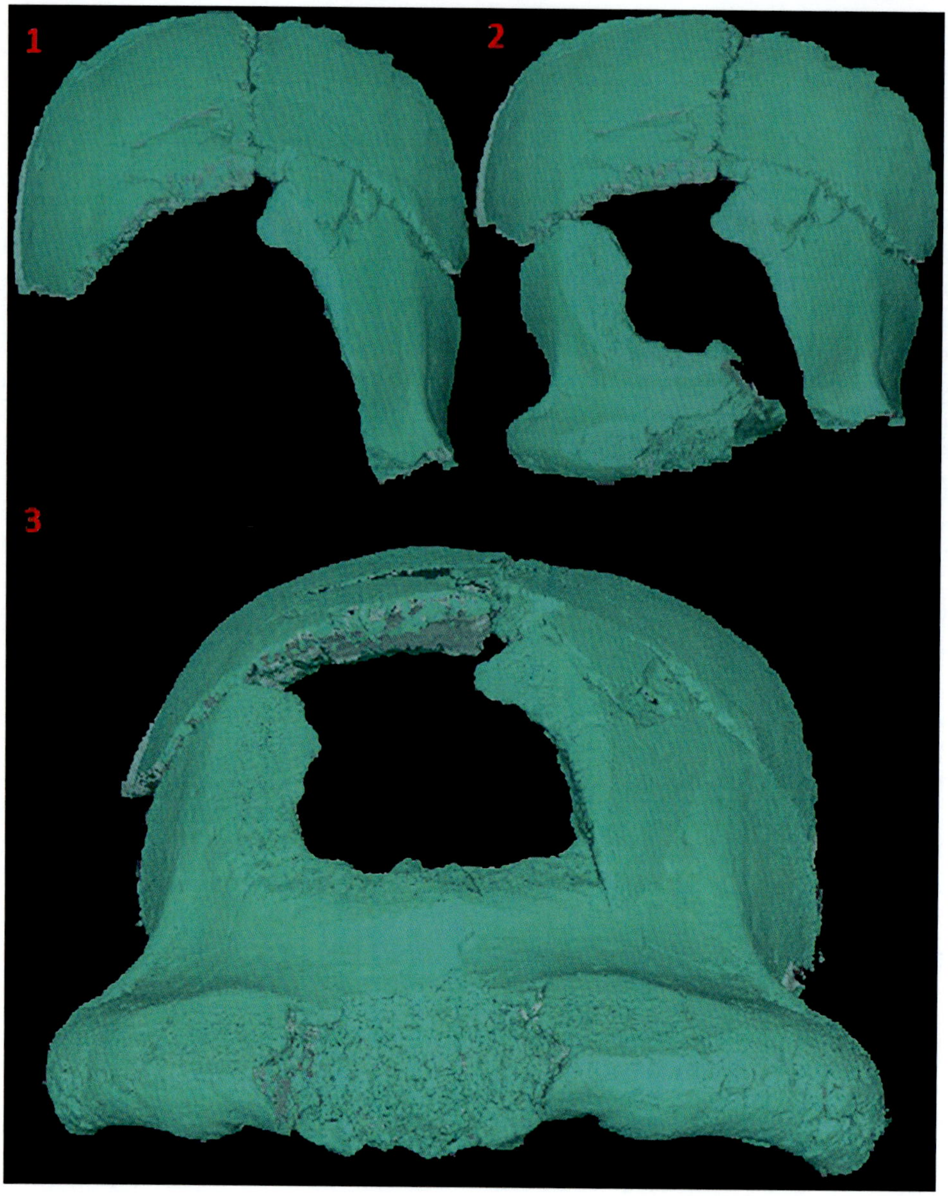

Fig. 6.4 Our virtual reconstruction of the Kocabaş specimen proceeded in the following steps: (1) the parietal bones were merged along the sagittal suture, (2) the frontal fragment preserving the supraorbital torus was merged to the parietal bone using the temporal line as a guide, (3) the fragment of the right frontal bone was then mirrored and merged to the left side of the calvaria, again using the temporal line to align the pieces

Table 6.1 Landmarks used in this study

Landmark No.	Description	Type
1	Supraorbital torus inferior right (The highest point on inferior middle orbit)	3
2	Supraorbital torus superior right (The highest point on superior middle orbit)	3
3	Supraorbital torus inferior left	3
4	Supraorbital torus superior left	3
5	Dacryon right	1
6	Dacryon left	1
7	Deepest point of the postorbital sulcus	2
8	Frontomalare temporale right	1
9	Frontomalare orbitale right	1
10	Frontomolare temporale left	1
11	Frontomolare orbitale left	1
12	Post-orbital constriction right (The deepest point on temporal line)	2
13	Post-orbital constriction left	2

to superimpose the landmark coordinates (Slice 2007). After superimposition, a Principal Components Analysis (PCA) was performed on the fitted coordinates in order to explore the pattern of variation in our samples. Shape changes along the PC axes were visualized using Morphologika. We also examined the inter-individual Procrustes distances (a measure of total shape difference between specimen pairs) between Kocabaş and the other specimens in the sample. All statistical analyses and plots were performed in and produced with Morphologika and PAST.

The first two principal components reflected 64.7 % of the total variance. Modern humans were separated from fossil hominins along PC 1 (52.3 % of the total variance; Fig. 6.5, top panel). Modern humans scored negatively on this axis, while fossil specimens all showed positive PC 1 scores. *H. erectus s.l.* and *H. habilis* showed the most positive PC 1 values, followed by *H. heidelbergensis s.l.* and *H. neanderthalensis* (the latter falling close to 0 on PC 1). All taxa overlapped on PC 2, which reflected within-species variability. *H. neanderthalensis*, however, tended to have more positive PC 2 scores than *H. sapiens*, *H. heidelbergensis*, *H. erectus*, and *H. habilis*. On the other hand, *H. erectus s.l.* showed a very broad range on this axis, with two African specimens falling at each extreme (OH9 showing the most positive and KNM ER 3733 the most negative PC 2 score). Kocabaş fell on the positive side of PC 1, with a negative PC 2 score, plotting with the other fossil specimens and close to the *H. erectus s.l.* convex hull. Although it fell outside the convex hulls of all our samples, it was closest to *H. erectus s.l.* on these two axes. The shape differences along PC 1 and 2 (Fig. 6.5, bottom panel) show that the thickness and shape of the supraorbital torus are very important in driving the patterns seen along both these axes. Indeed PC 1 reflects a supero-inferiorly thicker and more anteriorly projecting supraorbital torus combined with a greater postorbital constriction in specimens with more positive scores, characterizing the divide between modern humans on the negative and fossil humans on the positive side of PC 1. Positive PC 2 scores, on the other hand, indicate a relatively supero-inferiorly thicker supraorbital region which is not projecting laterally, while negative scores suggest a thinner supraorbital torus which is, however, laterally expanded and associated with higher degrees of postorbital constriction. This dichotomy separates Neanderthals and, to a lesser extent, *H. heidelbergensis* from *H. erectus* specimens, although overlap exists. When the inter-individual Procrustes distances, a measure of total shape similarity, were examined, Kocabaş was nearest to Zhoukoudian III, Ceprano, and Arago (in that order; Table 6.3). It was farthest from OH9.

Our results broadly agree with those previously reported. Kocabaş grouped nearest to the *H. erectus s.l.* sample in the PCA and showed the smallest Procrustes distance, and therefore the greatest overall shape similarity, to a *H. erectus* specimen (Zhoukoudian III). However, it also showed relatively small Procrustes distances to *H. heidelbergensis* specimens (Ceprano, Arago) suggesting affinities also with this taxon. Our findings differ from those of Vialet et al. (2014) in some important ways. Although those authors confirmed the previously proposed affinity of Kocabaş with *H. erectus s.l.*, as do we, they also found significant similarities between Kocabaş and African *H. erectus* specimens. Vialet et al. (2014) do not report Procrustes distances or a classification analysis. However, on the basis of their PCA plot (PC 1-2), as well as of their linear measurements and anatomical observations, they suggest that Kocabaş is in many respects most similar to African *H. erectus*, and particularly to specimens like OH9 and Daka. Although our fossil comparative sample was relatively limited, we did not observe any particular similarity with African rather than Asian *H. erectus* specimens. In fact the three specimens with overall smaller Procrustes distances to Kocabaş (and therefore the most similar in overall shape) in our analysis were all Eurasian and considered significantly younger in geological age than the >1 ma recently proposed for Kocabaş. Although Daka was not included in our analysis, OH9 showed the largest Procrustes distance to Kocabaş than any other specimen in our comparative sample.

The differences in the results of our study compared to the findings of Vialet et al. (2014) may stem from several factors. First, two different reconstructions were produced for these two studies, and therefore slight differences and inconsistencies in how the specimen was put together in each of them might have produced differences in the results. Secondly, a larger fossil comparative sample was used by Vialet et al. (2014), which comprised an expanded *H. erectus s.l.* sample compared to our own. By comparison, our study's fossil sample was rather limited, and consisted overwhelmingly of casts rather than original specimens. Nevertheless, we feel that an

Table 6.2 Comparative samples used in this study

Specimen	Taxon	Origin	Missing landmarks	Scan
Kocabaş	?	Turkey	None	Original
Sangiran 17	*Homo erectus s.l.*	Asia	None	Cast
Zhoukoudian XII	*Homo erectus s.l.*	Asia	6, 8, 9	Cast
Zhoukoudian III (Locus E)	*Homo erectus s.l.*	Asia	6	Cast
ER 3883	*Homo erectus s.l.*	Africa	None	Cast
ER 3733	*Homo erectus s.l.*	Africa	None	Cast
OH9	*Homo erectus s.l.*	Africa	8, 9	Cast
ER 1813	*Homo habilis*	Africa	None	Cast
Petralona	*H. heidelbergensis s.l.*	Europe	None	Original
Ceprano	*H. heidelbergensis s.l.*	Europe	None	Cast
Arago	*H. heidelbergensis s.l.*	Europe	None	Cast
Dali	*H. heidelbergensis s.l.*	Asia	None	Cast
Bodo	*H. heidelbergensis s.l.*	Africa	None	Original
Kabwe	*H. heidelbergensis s.l.*	Africa	None	Cast
Feldhofer	*H,. neanderthalensis*	Europe	None	Cast
Guattari	*H. neanderthalensis*	Europe	None	Cast
Gibraltar	*H. neanderthalensis*	Europe	None	Cast
Krapina 3	*H. neanderthalensis*	Europe	13	Original
African 1	*H. sapiens* (VA-014)		None	Original
African 2	*H. sapiens* (VA-019)		None	Original
African 3	*H. sapiens* (VA-023)		None	Original
African 4	*H. sapiens* (VA-024)		None	Original
African 5	*H. sapiens* (VA-025)		None	Original
Asian 1	*H. sapiens* (VA-026)		None	Original
Asian 2	*H. sapiens* (VA-027)		None	Original
Asian 3	*H. sapiens* (LIA-835)		None	Original
Australian 1	*H. sapiens* (VA-013)		None	Original
Australian 2	*H. sapiens* (VA-016)		None	Original
Australian 3	*H. sapiens* (VA-017)		None	Original
Australian 4	*H. sapiens* (VA-020)		None	Original
European 1	*H. sapiens* (VA-003)		None	Original
European 2	*H. sapiens* (VA-004)		None	Original
European 3	*H. sapiens* (VA-005)		None	Original
European 4	*H. sapiens* (VA-006)		None	Original
European 5	*H. sapiens* (VA-008)		None	Original
European 6	*H. sapiens* (VA-009)		None	Original
European 7	*H. sapiens* (VA-010)		None	Original
European 8	*H. sapiens* (VA-011)		None	Original

VA Virtual Anthropology; University of Vienna, Department of Anthropology
LIA Leipzig Institute of Anatomy

important factor driving the differences in our results is the different landmark dataset employed by us vs. Vialet et al. (2014). We made an effort to quantify the shape and thickness of the supraorbital torus with our landmarks, even though this region is only partially preserved and has been affected by the postmortem damage. As is seen by our PCA results, this morphology is important in driving the patterns seen along the first two PC axes (Fig. 6.5, bottom panel). On the other hand, we excluded bregma in order to expand our fossil sample and to be able to include OH9 in particular, which does not preserve this region. Vialet et al. (2014) left the supraorbital torus unrepresented in their shape analysis, but included bregma, and were therefore able to assess the height of the frontal bone in their analysis. They were also able to include OH9, although no information is provided about how bregma was reconstructed in that specimen. Both these aspects of frontal bone morphology (shape and thickness of supraorbital torus, frontal bone height) are important when considering archaic humans such as *H. erectus*; their absence likely has a strong influence on results and could influence the observed

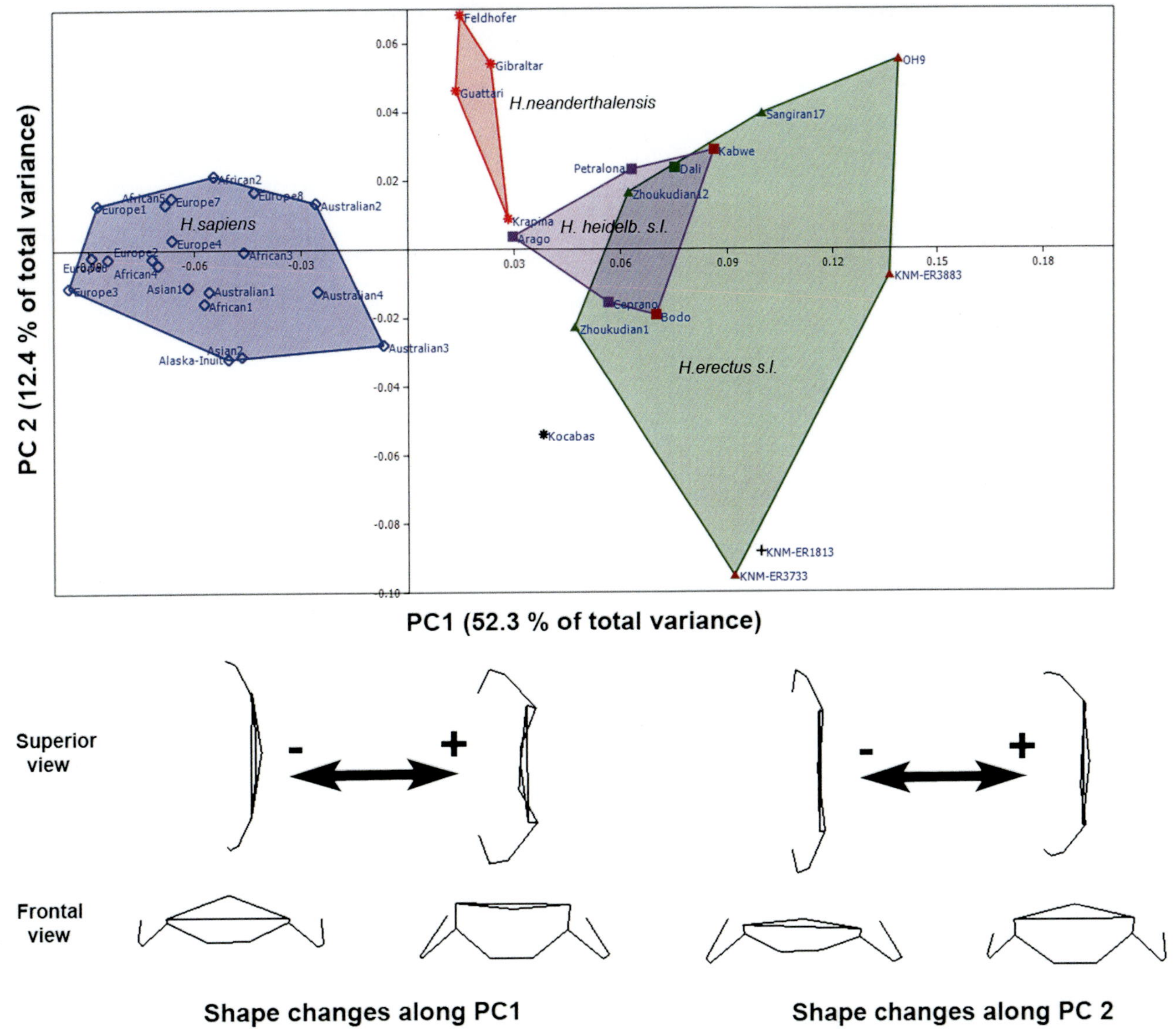

Fig. 6.5 *Top panel*: Principal Components Analysis. PC1 plotted against PC2. *Blue diamonds*: *H. sapiens*; *Triangles*: *H. erectus s.l.* (*Brown*: African, *Green*: Asian); *Squares*: *H. heidelbergensis s.l.* (*Brown*: African, *Lavender*: European, *Green*: Asian); *Red stars*: *H. neanderthalensis*; *Cross*: *H. habilis*; *Black star*: Kocabaş. *Bottom panel*: Shape changes for PC1 and PC2, superior and anterior views

overlap among modern human and fossil samples in the PCA reported by Vialet et al. (2014). We therefore consider the landmark configuration used by Vialet et al. (2014) to be insufficient to fully assess the affinities of Kocabaş.

Conclusions

In summary, the known fossil human record from Turkey is sparse, but still substantial when considering the lack of Paleolithic and paleoanthropological research in this country. Both Upper, as well as Middle and Lower Paleolithic human remains are likely represented. Nevertheless, this record often suffers from insufficient description and documentation, as well as a dearth of information regarding its geological and chronological context. Perhaps the best documented specimen is the Kocabaş partial cranium, the oldest and most important known hominin from Turkey. Our preliminary analysis of this specimen supports previous attributions to *H. erectus s.l.*, but does not confirm similarity with African as opposed to Asian *H. erectus*. Kocabaş was instead found to be most similar to Eurasian specimens, including Zhoukoudian III, as well as Ceprano and Arago, suggesting some affinity with younger Middle Pleistocene specimens commonly assigned to *H. heidelbergensis*. This result may be driven by our limited fossil samples, but may also in part reflect the inclusion of taxonomically important features (i.e. the morphology of the supraorbital torus) and exclusion of others (frontal bone height) in our analysis.

Table 6.3 Inter-individual Procrustes distances for Kocabaş, reported in ascending order

	Kocabaş
Zhoukoudian III	0.0949
Ceprano	0.0963
Arago	0.1013
Krapina 3	0.1179
Bodo	0.1241
Petralona	0.1316
Zhoukudian XII	0.1363
KNM-ER3733	0.1389
Europe8	0.1414
Australian3	0.1423
Europe7	0.1447
African1	0.1460
Dali	0.1465
Australian4	0.1471
Asian2	0.1488
Alaska-Inuit	0.1490
Europe2	0.1491
Guattari	0.1504
African3	0.1514
Gibraltar	0.1537
Australian2	0.1580
Feldhofer	0.1594
African4	0.1594
Asian1	0.1595
African5	0.1595
Kabwe	0.1613
KNM-ER1813	0.1624
Europe6	0.1641
Europe4	0.1650
Europe3	0.1677
Australian1	0.1692
Sangiran17	0.1736
Europe5	0.1752
African2	0.1762
Europe1	0.1812
KNM-ER3883	0.1815
OH9	0.1879

Given these limitations, the results reported here should be considered preliminary, and further work is necessary to support or reject them, especially with regard to Kocabaş' possible relationship with *H. heidelbergensis*, a finding that is especially relevant given the specimen's uncertain provenance and recently revised chronology. Future research should aim to represent as much of the preserved morphology of this fossil as possible, including the parietals, possibly using semilandmarks; to assess the effects of size on the expression of the relevant morphology; and to incorporate a larger fossil sample, in order to further clarify the taxonomic status and phylogenetic relationships of this important hominin.

Acknowledgments We thank The Ministry of Culture and Tourism of Turkey and the Pamukkale Hierapolis Archaeological Museum for allowing access to the Kocabaş specimen for the purposes of this study. We are grateful to Prof. Mehmet Cihat Alçiçek for his comments on the geology of the area and sharing with us information about the fossil and the site. The scanning equipment used in our study of the Kocabaş specimen was borrowed from the Tübingen Paleoanthropology Imaging Laboratory (Senckenberg Center for Human Evolution and Paleoecology, Eberhard Karls University of Tübingen). We also thank Amelie Vialet for helpful comments on a previous draft of this manuscript, as well as Mirjana Roksandic and two anonymous reviewers whose comments and suggestions greatly improved this paper. This work was supported by the ERC Starting Grant project 'Paleoanthropology at the Gates of Europe' (PaGE) Nr. 283503.

References

Alçiçek, H., Varol, B., & Özkul, M. (2007). Sedimentary facies, depositional environments and palaeogeographic evolution of the Neogene Denizli Basin, SW Anatolia, Turkey. *Sedimentary Geology, 202*, 596–637.

Bostancı, E. (1959). Research on the Mediterranean coast of Anatolia: A new Paleolithic site at Beldibi near Antalya, Anatolia. *Anadolu (Anatolia), 4*, 129–178.

Bostancı, E. Y. (1963). Human fossil remains in Beldibi and Belbaşı rock shelters on the Mediterranean coast of Anatolia. *Antropoloji, 1*, 17–36.

Bostancı, E. Y. (1971). Kanal Mağarası"nda Levalloise-Mousterian Seviyede Keşfedilen Bir Üst Süt Canine ile Alt Aurignacien Seviyede Bulunan Bir Mandibulae Molar Hakkında İnceleme. *Antropoloji, 5*, 45–82.

Bostancı, E. Y. (1973). Homo sapiens çevlikiensis in the Canal and Big caves of Çevlik near Samandağ of the province of Antakya on the Mediterranean coast of Anatolia. *Antropoloji, 6*, 29–56.

Boulbes, N., Mayda, S., Titov, V. V., & Alçiçek, M. C. (2014). Les grands mammifères du Villafranchien supérieur des travertins du Bassin de Denizli (Sud-Ouest Anatolie, Turquie). *L'Anthropologie, 118*(1), 44–73.

Chevalier, T., Özçelik, K., De Lumley, M.-A., Kosem, B., De Lumley, H., Yalçınkaya, I., et al. (2015). The endostructural pattern of a Middle Pleistocene human femoral diaphysis from the Karain E site (Southern Anatolia, Turkey). *American Journal of Physical Anthropology, 157*, 648–658.

Demirel, A., Andrews, P., Yalçınkaya, I., & Ersoy, A. (2011). The taphonomy and palaeoenvironmental implications of the small mammals from Karain Cave, Turkey. *Journal of Archaeological Science, 38*, 3048–3059.

Dinçer, B. (2016). The Lower Paleolithic in Turkey: Anatolia and hominin dispersals out of Africa. In K. Harvati & M. Roksandic (Eds.), *Paleoanthropology of the Balkans and Anatolia: Human evolution and its context*. Dordrecht: Springer.

Grine, F. E., Jungers, W. L., Tobias, P. V., & Pearson, O. M. (1995). Fossil *Homo* femur from Berg Aukas, northern Namibia. *American Journal of Physical Anthropology, 97*, 151–185.

Guipert, G., Vialet, A., & Alçiçek, M. C. (2011). The *Homo erectus* from Kocabaş in Turkey and the first settlements in Eurasia. *American Journal of Physical Anthropology, 144*(S52), 151.

Güleç, E. (1994). Akdeniz Bölgesi'nde Bulunan Üç Fosil Kafatası. *Belleten, 223*, 531–552.

Güleç, E., Kuhn, S. L., Stiner, M. C., Özer, İ., Sağır, M., Erkman, A. C., et al. (2008). *2006-2007 Yılı Üçağızlı Mağarası Kazısı. 30.* Kazı Sonuçları Toplantısı, Vol. 4, pp. 233–242.

Güleç, E., Özer, İ., Sağır, M., & Kuhn, S. (2007). Early Upper Paleolithic human dental remains from Üçağızlı Cave (Hatay, Turkey). *American Journal Physical Anthropology, 132*(S44), 122.

Harvati, K. (2003). The Neanderthal taxonomic position: Models of intra- and inter-specific craniofacial variation. *Journal of Human Evolution, 44*, 107–132.

Harvati, K. (2016). Paleoanthropology in Greece: Recent findings and interpretations. In K. Harvati & M. Roksandic (Eds.), *Paleoanthropology of the Balkans and Anatolia: Human evolution and its context*. Dordrecht: Springer.

Harvati, K., Panagopoulou, E., & Runnels, C. (2009). The paleoanthropology of Greece. *Evolutionary Anthropology, 18*, 131–143.

Kappelman, J., Alçiçek, M. C., Kazancı, N., Schultz, M., Özkul, M., & Şen, Ş. (2008). Brief communication: First Homo erectus from Turkey and implications for migrations into temperate Eurasia. *American Journal of Physical Anthropology, 135*(1), 110–116.

Khatib, S., Rochette, P., Alçiçek, M. C., Lebatard, A. E., Demory, F., & Saos, T. (2014). Études stratigraphique, sédimentologique et paléomagnétique des travertins de Kocabaş, Bassin de Denizli, Anatolie, Turquie, contenant des restes fossiles quaternaires. *L'Anthropologie, 118*(1), 16–33.

Kuhn, S. L., Stiner, M. C., Güleç, E., Özer, İ., Yılmaz, H., Baykara, İ., et al. (2009). The early Upper Paleolithic occupations at Üçağızlı Cave (Hatay, Turkey). *Journal of Human Evolution, 56*, 87–113.

Lebatard, A. E., Alçiçek, M. C., Rochette, P., Khatib, S., Vialet, A., Boulbes, N., et al. (2014). Dating the *Homo erectus* bearing travertine from Kocabaş, (Denizli, Turkey) at at least 1.1 Ma. *Earth and Planetary Science Letters, 390*, 8–18.

Mardia, K. V., & Bookstein, F. L. (2000). Statistical assessment of bilateral symmetry of shapes. *Biometrika, 87*, 285–300.

O'Higgins, P., & Jones, N. (2004). *Tools for statistical shape analysis*. Heslington: Functional Morphology and Evolution Research Group, Hull York Medical School.

Otte, M., Yalçınkaya, I., Kozlowski, J., Bar-Yosef, O., Bayon, I. L., & Taşkıran, H. (1998). Long-term technical evolution and human remains in the Anatolian Paleolithic. *Journal of Human Evolution, 34*, 413–431.

Rightmire, G. P. (2009). Middle and later Pleistocene hominins in Africa and Southwest Asia. *Proceedings of the National Academy of Sciences, USA, 106*(38), 16046–16050.

Rink, W. J., Schwarz, H. P., Grün, R., Yalçınkaya, I., Taşkıran, H., Otte, M., et al. (1994). E.S.R. dating of the last interglacial Mousterian at Karain Cave, Southern Turkey. *Journal of Archaeological Science, 21*, 839–849.

Roberts, C. A., Pfister, L., & Mays, S. (2009). Letter to the editor: Was tuberculosis present in *Homo erectus* in Turkey? *American Journal of Physical Anthropology, 139*, 442–444.

Roksandic, M. (2016). The role of the Central Balkans in the peopling of Europe: The paleoanthropological evidence. In K. Harvati & M. Roksandic (Eds.), *Paleoanthropology of the Balkans and Anatolia: Human evolution and its context*. Dordrecht: Springer.

Şenyürek, M. (1949). Türk Tarih Kurumu Adına Yapılan Karain Kazısında Bulunan İki Fosil Dişe Dair Kısa Ön Rapor. *Belleten, 52*, 833–837.

Şenyürek, M., & Bostancı, E. (1956). The excavation of a cave near the village Mağracık in the Vilayet of the Hatay. *Anatolia, 1*, 81–83.

Şenyürek, M., & Bostancı, E. (1958). Hatay Vilayetinde Prehistorya Araştırmaları. *Belleten, 86*, 147–156.

Slice, D. E. (1998). *Morpheus et al.: Software for morphometric research*. Revision 01-30-98. Department of Ecology and Evolution, State University of New York, Stony Brook, New York.

Slice, D. E. (2007). Geometric morphometrics. *Annual Review of Anthropolgy, 36*, 261–281.

Strait, D. S., Orr, C. M., Hodgkins, J., Spassov, N., Gurova, M., Miller, C., et al. (2016). The human fossil record of Bulgaria, and the formulation of biogeographic hypotheses. In K. Harvati & M. Roksandic (Eds.), *Paleoanthropology of the Balkans and Anatolia: Human evolution and its context*. Dordrecht: Springer.

Taşkıran, H. (2002). *Karain Mağarası. Arkeo Atlas* (Vol. 1, p. 50). İstanbul: DBR Yayıncılık.

Vialet, A., & Alçiçek, C. (2012). *A Lower Pleistocene human fossil from Kocabaş (Denizli, Turkey) pushing Homo erectus far to the west of Asia*. European Society for the Study of Human Evolution, 2nd Annual Meeting, Vol. 34, p. 187.

Vialet, A., Guipert, G., Alçiçek, M. C., & de Lumley, M. A. (2014). La calotte crânienne *d'Homo erectus* archaïque de Kocabaş, Bassin de Denizli, Anatolie, Turquie (The archaic *Homo erectus* skullcap from Kocabaş, Denizli Basin, Anatolia, Turkey). *L'Anthropologie, 118*, 74–107.

Vialet, A., Guipert, G., & Alçiçek, M. C. (2012). *Homo erectus* found still further West: Reconstruction of the Kocabaş cranium (Denizli, Turkey). *Comptes Rendus Palevol, 11*, 89–95.

Yalçınkaya, I. (1988). *1986 Yılı Karain Kazıları. 9.* Kazı Sonuçları Toplantısı, pp. 15–37.

Yalçınkaya, I., Taşkıran, H., Kartal, M., Özçelik, K., Kösem, M. B., & Kartal, G. (2008). *2007 Yılı Karain Mağarası Kazıları. 30.* Kazı Sonuçları Toplantısı, Vol. 2, pp. 285–300.

Part II
The Archaeological Record

Chapter 7
The Middle and Upper Paleolithic on the Western Coast of the Mani Peninsula (Southern Greece)

Andreas Darlas and Eleni Psathi

Abstract Along the western coast of the Mani peninsula (S. Greece) numerous caves with Upper Pleistocene and Early Holocene deposits, preserving cultural remains from the Middle Paleolithic to the end of the Neolithic, form an important group of archaeological sites located in a restricted geographic area. Excavations have been carried out in seven of these caves. The excavation of Kalamakia yielded data about the Middle Paleolithic, while the other six caves have yielded remains of all Upper Paleolithic phases. Of particular interest is the discovery of transitional Middle-Upper Paleolithic layers in Kolominitsa cave. Although preliminary, this evidence demonstrates the importance of systematic research on a regional scale through the comparative study of neighboring and contemporaneous sites. Finally, these sites enable us to date the arrival of anatomically modern humans in this area and to study subsequent ecological and cultural changes.

Keywords Upper Pleistocene • Peloponnese • Cave sites • Environment • Lithic industry • Human diet • Middle-Upper Paleolithic transition

Introduction

The Paleolithic of Greece is still poorly known, since the relevant research remains in the margin of the overall archaeological research in the country, which focuses almost exclusively on the study of later periods. Although Paleolithic projects have multiplied during the last decades, most of them are field surveys, usually yielding finds without stratigraphic context, which are therefore not informative enough (if not questionable). Excavations are comparatively rare and reliable data remain sparse. To date, less than a dozen excavations have been carried out. Petralona cave was, until recently, the only excavated Middle Pleistocene site (Poulianos 1971; Darlas 2014; but see Panagopoulou et al. 2015; Galanidou et al. 2016). The remaining excavated sites date to the Upper Pleistocene (Middle and Upper Paleolithic). Earlier excavations, carried out between 1940s and 1970s, include those of Seidi (Schmidt 1965), Asprochaliko and Kastritsa (Higgs and Vita-Finzi 1966; Bailey et al. 1983), Franchthi (Perlès 1987), and Kefalari (Reisch 1980); while those of Klithi (Bailey 1997), Boila (Kotjabopoulou et al. 1997), Theopetra (Kyparissi-Apostolika 2000), Klisoura (Koumouzelis et al. 2001; Kaczanowska et al. 2010), and Maara (Trantalidou and Darlas 1992) are more recent.

In this context, the presence of numerous caves with Pleistocene deposits containing Paleolithic remains in the Mani peninsula becomes very important for Paleolithic research in Greece. Apart from Lakonis (Panagopoulou et al. 2002–2004), located on the northeastern end of the peninsula, all other caves mentioned here are situated along the western coast. A few contain deposits with Middle Paleolithic remains, while most of them preserve deposits containing Upper Paleolithic remains, providing a sequence of Middle Paleolithic and all phases of the Upper Paleolithic in a restricted area. Because of its rich record and limited geographic area, the Mani peninsula is particularly suitable for systematic research and represents one of the richest areas of the Greek Paleolithic record.

The Mani Peninsula

Mani is the middle of the three peninsulas formed in the Southern Peloponnese. It constitutes the extension of the Taygetos mountain range, which begins at the center of

A. Darlas (✉)
Ephoreia of Paleoanthropology and Speleology, Ardittou 34b, Athens 11636, Greece
e-mail: adarlas@culture.gr

E. Psathi
Ephoreia of Antiquities of Arcadia, Kalavryton 211, Tripolis 22100, Greece
e-mail: epsathi@culture.gr

Katerina Harvati and Mirjana Roksandic (eds.), *Paleoanthropology of the Balkans and Anatolia*, Vertebrate Paleobiology and Paleoanthropology, DOI 10.1007/978-94-024-0874-4_7

Peloponnese and ends at cape Tenaron, and is mostly formed of Upper Cretaceous–Lower Eocene metamorphic limestones (Thiebault 1982). Throughout the entire western coast and especially on the vertical cliffs overhanging the sea—several tens of meters high—there are numerous caves belonging to an extended karstic system (Bassiakos 1993). The size of these caves varies: the majority are small, some tens of meters deep, opened at the current sea level or some meters above it; others are very large with impressive stalagmitic formations (e.g. Diros caves, Aghios Dimitrios cave). Most have been used by humans from the Paleolithic until the recent past. In many cases, the Pleistocene fillings have been eroded; however, nearly all of them still preserve Paleolithic evidence (see also Tourloukis et al. 2014). Today, several caves preserve Paleolithic deposits suitable for excavation.

Geomorphological Evolution and Human Use of Caves During the Upper Pleistocene

The western side of the peninsula presents a stepped morphology, due to the successive horizontal surfaces and the overhanging vertical cliffs. At the current sea level, and slightly above, a Tyrrhenian terrace forms a horizontal zone several kilometers long and 10–100 m wide. Two marine deposits (marine crusting and beach rock) are locally deposited on its surface, attributed to the MIS 5e and 5c transgressions (based on similar formations in Crete; Keraudren et al. 2000). This zone is dominated by a vertical cliff face, several tens of meters high, onto which most of the caves open. Almost all of these caves, especially those situated 0–20 m above sea level (asl), have been eroded by the marine transgression of MIS 5e and 5c and have completely lost their sediment. As a consequence, although the caves could have been inhabited in the Middle Pleistocene, or even earlier, their current infill dates to the Upper Pleistocene and contains archaeological remains of the Middle and Upper Paleolithic as well as of subsequent periods.

The Tyrrhenian terrace (as well as other lower plateaus, currently submerged by the sea) must have been completely free at the beginning of the last glacial, since screes had not yet accumulated. The caves overlooking these plateaus were the most favorable for habitation at that time, with the lower ones inhabited first, since access to the higher caves would have been difficult or even impossible.

Screes were gradually accumulated on the bottom of the cliffs blocking the lower caves first and pushing the human occupation to the higher ones, with the sloping surface of screes facilitating the access. Therefore, the lower caves usually contain Middle Paleolithic remains, while those opened at a higher level contain mostly Upper Paleolithic and younger remains (Darlas 2012). It is worth mentioning that the scree "sealed" the majority of the caves, protecting their filling from erosion.

A similar process took place at the lower level, on the bottom of the cliff in front of the Tyrrhenian terrace. The marine regression in the beginning of MIS 5b and mainly MIS 4 revealed numerous caves, which were consequently inhabited by both Neanderthals of the Middle Paleolithic and *Homo sapiens* of the Upper Paleolithic.

The Holocene marine transgression dramatically limited the living space of the peninsula inhabitants. Most of the previously inhabited caves became submerged and only a small proportion was still available for habitation. On the other hand, after erosion of the scree, the lower caves lost their "protective wall" and erosion of their deposits began. As a result, many caves have lost their entire filling (or a part of it), and are currently empty.

The Archaeological Research

Previous paleoanthropological research at Apidima caves from 1980 to 1984 (Fig. 7.1) recovered two crania attributed to Middle Pleistocene hominins, and a headless burial of a female individual of a possible Upper Paleolithic age (Coutselinis et al. 1991; Pitsios 2000; Harvati et al. 2011).

Kalamakia cave (Fig. 7.1) was excavated from 1993 to 2006 and yielded Middle Paleolithic remains (De Lumley and Darlas 1994; Darlas and De Lumley 1999, 2004). A broader project on the caves of the Mani peninsula was carried out from 1999 to 2005 in the course of which 103 caves were explored. Most are small cavities with archeological remains, and more than 50 of them contain Pleistocene deposits. Small test pits opened in six caves—Kolominitsa, Kastanis, Skini 4, Skini 3, Tripsana, and Melitzia (Fig. 7.1)—yielded Upper Paleolithic material.

The above research demonstrated the great density of Paleolithic caves in Mani and yielded valuable, if preliminary, information about the Upper Paleolithic (Darlas and Psathi 2008). Given the great number of caves with preserved remains of this period, successive restricted excavations were no longer sufficient and the research was organized around systematic exploration of three caves—Melitzia, Kolominitsa, and Skini 2; Fig. 7.1—which contained deposits from all phases of the Upper Paleolithic. The main excavation was undertaken at Melitzia cave, with research in the other two caves complementing the findings. At Kolominitsa, the concentration is on Early Upper Paleolithic phases, which may not be represented at Melitzia cave. At Skini, 2 a series of samples will be collected for paleoenvironmental analysis.

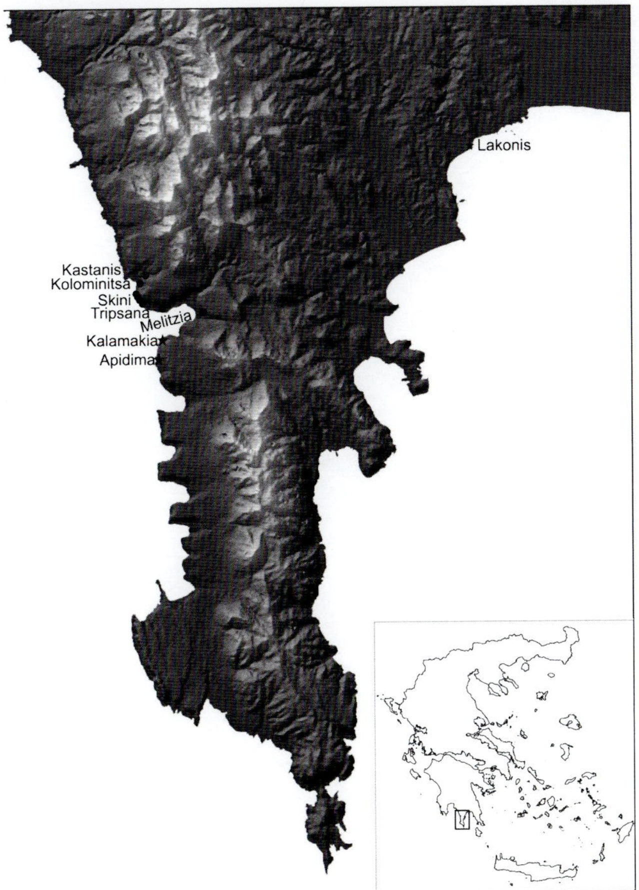

Fig. 7.1 Geographic position of excavated Paleolithic sites in the Mani peninsula (*stars* represent Middle Paleolithic sites; *circles* Upper Paleolithic sites)

Fig. 7.2 View of the entrance of the Kalamakia cave, showing the excavation trenches

The systematic excavation of Melitzia started in 2009; the initial test pit in Kolominitsa was reopened in 2011 in order to examine the deposits in greater depth; while the exploration of Skini 2 has not yet been carried out. Complementing the research in western Mani, the excavation of Lakonis cave at the NE border of the peninsula (Panagopoulou et al. 2002–2004; Harvati et al. 2003), as well as recent work on the northern end of the western coast (Tourloukis et al. 2014), have yielded Middle Paleolithic material.

The Middle Paleolithic

Kalamakia Cave

Kalamakia cave (Fig. 7.1) is located at the entrance of Itylo bay, approximately 2.5 km North-West of Areopolis (36° 40′ 33.68″ N, 22° 21′ 51.45″E). It opens 10 m inland from the current sea shore, at 2.5 m asl, directly on the Tyrrhenian terrace. The cave is 20 m deep, with a 7 m wide and 8 m high entrance. Throughout their entire height, the walls are perforated by *Lithophaga* sp., indicating that during the Pleistocene the cave was submerged for a long period of time. The Pleistocene filling of the cave is more than 7 m thick (Figs. 7.2 and 7.3). At the bottom, two marine deposits (Units 0 [marine crusting] and II [beach rock]) are attributed to the marine transgressions of MIS 5e and 5c respectively. Above these two layers, there are 7 m of accumulated continental deposits, more than 4 m of which are rich in Middle Paleolithic remains (Units III and IV), while the uppermost 2.5 m are practically culturally sterile (Unit VI; De Lumley and Darlas 1994; Darlas and De Lumley 2004).

A thorough horizontal excavation was conducted at the site. The sediments of Unit IV were excavated throughout their vertical expanse in an area of 4–10 m^2. In Unit III, due to the extremely hard, lithified sediments, only the top layers in an area of 8 m^2 were excavated (Darlas and De Lumley 2004). The attribution of the beach rock (Unit II) to the MIS 5c transgression seems to be confirmed by the U/Th dating of a marine shell (Institut de Paléontologie Humaine in Paris, IPH Kal 9304: 109,000 + 14,000/−13,000 kBP); De Lumley

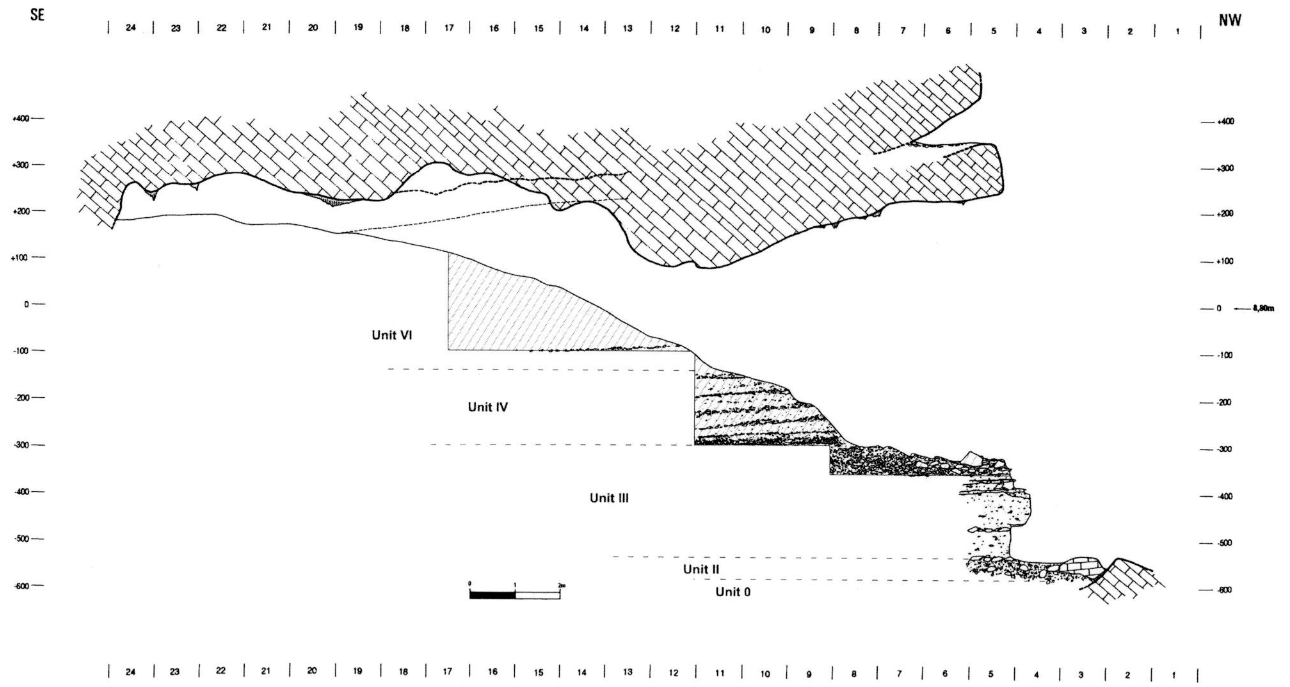

Fig. 7.3 Synthetic schematic representation of the Kalamakia cave stratigraphy. From bottom to top: Unit 0: marine crusting; Unit II: beach rock; Unit III: lithified angular gravel in a reddish sandy clay matrix; Unit IV: loose angular gravel in a reddish sandy clay matrix; Unit VI: layered clayey silts. Adapted from Darlas and De Lumley 2004

and Darlas 1994. The top of Unit IV has been dated to >39,000 years BP (^{14}C AMS dating on charcoal at Gif-sur-Yvette in France, GifA 94592). Therefore, the archaeological deposits of the cave are considered to date between 100,000 and >39,000 years BP. Additionally, one coprolite from the culturally sterile Unit VI has been dated by ^{14}C AMS to 22,410 ± 120 (27,770–26,330 cal BP; Beta-245334).

Environmental data: The large mammal fauna of Kalamakia comprises 17 taxa (Table 7.1). Fallow deer dominates the assemblage, followed by ibex and wild boar (Table 7.2). A few remains belong to elephants and rhinoceros. Carnivores are also present, albeit in low numbers, throughout the stratigraphic sequence, with the red fox as the most common species. The site has also yielded the remains of 60 small vertebrate taxa, including abundant land tortoise remains collected mostly from the lower half of Unit IV. These belong principally to *Testudo marginata* and, to a lesser degree, to *Testudo hermanni*.

On the basis of both pollen and faunal data, during at least the first half of the last glacial, the climate in Kalamakia was mild. This was due both to its geographic position at the southern edge of the Peloponnese, as well as to its coastal location. The surrounding landscape was covered with maquis vegetation and some Mediterranean presteppic forest taxa, such as *Quercus ilex-coccifera*, *Artemisia*, and *Ephedra*. The Mediterranean taxa *Olea* and *Phillyrea* are also relatively well represented (Lebreton et al. 2008). This combination of biotopes would have been able to support substantial faunal populations with various ecological restrictions. The study of micro-vertebrates, especially rodents, indicates a generally open landscape surrounding the cave with dry and relatively warm climatic conditions (Roger and Darlas 2008).

Human Remains: The excavation has yielded 14 isolated human remains, mostly teeth, attributed to Neandertals (Harvati et al. 2013; Harvati 2016).

Human versus *carnivore use of the cave*: Humans seem to have occupied the cave periodically. Zooarchaeological data (processed for publication by E.P.) suggest that humans were responsible for the formation and modification of the larger part of the mammal and tortoise assemblages. The taphonomic analysis—with emphasis on body part representation, fragmentation, and cortex alteration of the bone material—indicates systematic and complete processing of medium-sized ungulates at the site (especially fallow deer and ibex), but also of tortoises. However, carnivores scavenged and inflicted damage both on mammal and tortoise remains in most archaeological layers. Furthermore, carnivores contributed to the formation of some short-term layers in Units IV and VI. Most of the observed carnivore marks are consistent with canid activity (i.e. fox, wolf).

Table 7.1 Faunal list from Kalamakia

Carnivora	***Rodentia***	***Reptilia***	***Aves***
Ursus arctos	*Sciurus vulgaris*	*Testudo marginata*	*Puffinus puffinus*
Panthera pardus	*Myoxus glis*	*Testudo hermani*	*Accipiter nisus*
Lynx lynx	*Apodemus* sp.	*Scincidae indet.*	*Falco* cf. *vespertinus*
Felis silvestris	*Apodemus mystacinus*	cf. *Tarentola* sp.	*Alectoris graeca*
Canis lupus	*Cricetulus migratorius*	*Lacertidae indet.*	*Coturnix coturnix*
Vulpes vulpes	*Microtus arvalis*	*Lacerta* sp.	*Eudromias morinellus*
Martes sp.	*Microtus guentheri*	*Anguis fragilis*	*Chlidonias* sp.
Mustela sp.	*Microtus thomasi*	*Pseudopus* cf. *apodus*	*Columba livia/oenas*
	Chionomys nivalis	*Pseudopus* sp.	*Otus scops*
Proboscidea		*Eryx jaculus*	*Athene noctua*
Palaeoloxodon antiquus	***Insectivora***	*Hierophis gemonensis*	*Strix aluco*
	Erinaceus sp.	cf. *Dolichophis caspius*	*Apus apus*
Perissodactyla	*Talpa* sp.	*Malpolon monspessulanus*	*Apus* cf. *pallidus*
Stephanorhinus sp.	*Crocidura suaveolens*	*Malpolon* sp.	*Hirundo rustica*
		Coronella sp.	*Certhia* sp.
Artiodactyla	***Chiroptera***	*Coronella* cf. *austriaca*	*Turdus* cf. *philomenos*
Sus scrofa	*Myotis* sp.	*Elaphe quatuorlineata*	*Corvus corone*
Cervus elaphus	*Myotis blythii*	*Zamenis longissima*	cf. *Corvus monedula*
Dama dama	*Rhinolophus hipposideros*	*Zamenis* cf. *situla*	cf. *Pica pica*
Capreolus capreolus		cf. *Telescopus* sp.	*Pyrrhocorax pyrrhocorax*
Bos primigenius	***Amphibia***	*Natrix natrix*	*Emberiza citrinella*
Capra ibex	*Bufo bufo*	*Vipera* sp.	
	Rana sp.		
Lagomorpha			
Lepus europaeus			

Human occupation of the cave: *The excavation brought to light significant information concerning the occupation of the site and its spatial organization* (*Darlas and De Lumley* 2004):

(a) Living floors: 17 consecutive living floors have been revealed within Unit IV. The composition and density of the remains in these floors demonstrate that the cave served mostly as a short-term site and sometimes as a longer-term camp.
(b) Hearths: Several hearths have been uncovered. Three different types can be distinguished: (1) simple accumulations of ashes on the ground, (2) accumulations of ashes with stones, and (3) a basin-like hearth bounded by a circle of stones.
(c) Stone structures: Two dallages and a circle of stones have been uncovered.

The evidence from mammal taxa association, as well as from the demographic and taphonomic analysis of the faunal remains from the lower half of Unit IV, indicates prolonged or intense human occupation of the cave in this period. On the other hand, the upper half of this unit clearly shows an alternation of human and carnivore occupations, with carnivores mostly scavenging the animal bone remains accumulated by humans.

Lithic Assemblages: The lithics constitute Mousterian assemblages marked by an elevated frequency of the Levallois method (Table 7.3A). Levallois flakes represent 14.7 % of the total flake component, reaching 19.7 % among the flint flakes and 24.1 % among those made on andesite (Darlas and De Lumley 2004). The main raw materials are flint (obtained at a distance of 15 km) quartz and quartzite (found at a distance of 10 km) and andesite, which comes from a distance of 30 km. The features of lithic industry vary from one living floor to another, thus displaying a good example of variability. The observed differences mainly concern the choice of raw materials and the tool-kit and to a lesser degree, the technological features.

However, it must be emphasized that the main technological, as well as typological, characteristics remain unchanged throughout the entire stratigraphic sequence. These are: very small dimensions of artifacts; very small number of cores, which are extremely wasted (small-sized); total absence of cortical flakes; small number of "debitage by-products"; abundant microflakes ("retouch by-products"); and a high percentage of retouched tools. These features are obviously due to the distant origin and the scarcity of raw materials, which arrived at the cave in an advanced stage of processing, or in the form of already finished tools.

Table 7.2 Frequency of large mammal and tortoise remains in Kalamakia

Taxa/NISP[a]	Units III + IV	% NISP
Ursus arctos	17	0.42
Panthera pardus	13	0.32
Lynx lynx	7	0.17
Felis silvestris	29	0.71
Canis lupus	6	0.15
Vulpes vulpes	85	2.09
Martes sp.	17	0.42
Mustela sp.	1	0.02
Paleoloxodon antiquus	40	0.98
Stephanorhinus sp.	8	0.20
Sus scrofa	234	5.76
Bos primigenius	36	0.89
Capra ibex	568	13.98
Cervus elaphus	124	3.05
Dama dama	2671	65.76
Capreolus capreolus	64	1.58
Lepus europaeus	142	3.50
Mammal NISP	**4062**	**100.00**
Testudo (*marginata* + *hermanni*)	11,140	
Total NISP	**15,202**	
Unidentified *Cervidae*	115	
Unidentified *Artiodactyla*	2517	
Unidentified *Carnivora*	61	
Unidentified mammal bone fragments[b]	6934	
Total NS[c]	**24,829**	

[a]Number of identified specimens
[b]Unidentified fragments longer than 2 cm
[c]Number of specimens

Table 7.3 General composition (A) and tool types (B) of lithic assemblages from Kalamakia

A		
	Nb	%
Flakes	687	73.9
Blades	18	1.9
Cores	18	1.9
Debris	81	8.8
Pseudolevallois points	3	0.3
Levallois	122	13.2
Total	**929**	**100.0**
Debitage < 15 mm	9052	

B		
	Nb	%
Lateral scraper	85	46.8
Transversal scrapers	12	6.6
Double scrapers	24	13.2
Dejete scrapers	7	3.8
Convergent scrapers	3	1.7
Mousterian points	8	4.4
Limace	1	0.5
Notches	14	7.8
Bec	2	1.1
Denticulates	7	3.8
Endscraper	6	3.3
Boreer	7	3.8
Spines	4	2.2
Burin	1	0.5
Bifacial piece	1	0.5
Total	**182**	**100.0**

Levallois products are mostly flakes that display centripetal and unipolar (rarely bipolar) negatives. Levallois blades are very rare, while points are absent. A high percentage (22 %) of debitage products are retouched tools (Table 7.3B). The most common tools are scrapers (77 %), well-shaped and of small dimensions. Converging tools, especially points, are very well shaped. Points are usually very finely shaped (Fig. 7.4).

It is also worth mentioning that the uppermost living floor contains marine shells of the species *Callista chione*, which were retouched into tools in the same way as the lithic artifacts. This is a good example of human adaptation to the environment and exploitation of available natural resources.

The Upper Paleolithic

Kolominitsa Cave

Kolominitsa is located at about 1 km North of Itylo bay (36° 42′ 16.00″N, 22° 20′ 54.96″E; Fig. 7.1). It opens 100 m from the current sea shore and on the top of a talus, at 22 m asl (Fig. 7.5). It is about 40 m deep, 10 m wide, and 12 m high. Kolominitsa is a large cave that served as a major occupation site during both the Paleolithic and subsequent periods, as testified by its very thick stratigraphic sequence with dense cultural remains. The Pleistocene layers date to the Middle and Upper Paleolithic. In most parts of the cave, the uppermost layers have been eroded; they are preserved only at the back of the cave where they are approximately 3 m thick (Figs. 7.5 and 7.6) and contain archaeological remains. Two dates are available from these layers (Table 7.4). The Holocene occupations followed the erosion of the deposits.

A small test pit (1.30 × 1.30 m), opened at the entrance of the cave, was initially excavated to the depth of 95 cm and, in 2011, extended to the depth of 1.70 m. So far, 19 spits have been excavated (Fig. 7.7). From these layers four dates are available (Table 7.5). The two upper spits (1 and 2) yielded a lithic assemblage with backed bladelets (Gravettian). Spits 3–8 yielded an Aurignacian lithic assemblage (Table 7.6; Fig. 7.8). However, the assemblages are very small, and further analysis is not possible. The dating of a sample from the sixth spit (33,870 ± 550 ^{14}C BP; see Table 7.5) points to an early Aurignacian phase.

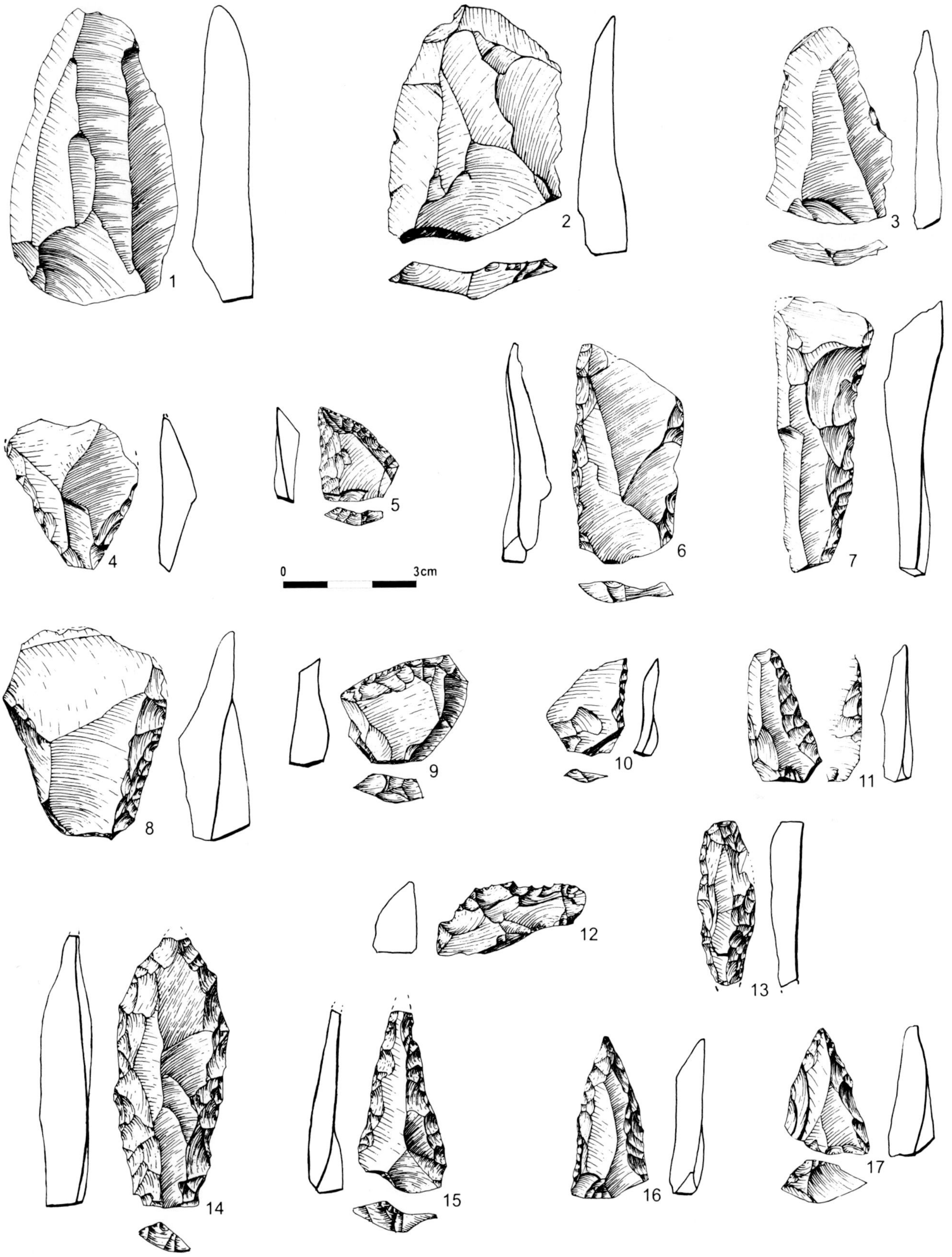

Fig. 7.4 Middle Paleolithic artifacts from Kalamakia. 1, 2, 3: Levallois flakes; 4, 6, 7, 8, 10, 11: scrapers; 9, 12: transversal scrapers; 13: *limace*; 5, 14, 15, 16, 17: Mousterian points. Adapted from Darlas and De Lumley 2004

Spits 9 and 10 were very poor in finds. It is worth mentioning that they have not yielded typical Upper Paleolithic lithic tool types but only scrapers. However, these are not diagnostic of the Middle Paleolithic either. In general, the whole cultural material raises the question of the existence of a transitional Middle-Upper Paleolithic phase (Darlas and Psathi 2008). This question was the main reason for the recent resumption of the excavations. The sediments of spits 11–13 (30 cm thick) have been extremely lithified and contain abundant large stones. However, artifacts are rare and not diagnostic. Nevertheless, the dating of a burnt bone from spit 11 (see Table 7.5), places this layer chronologically in the broader transitional Middle-Upper Paleolithic period.

Further down, spits 14–19 (60 cm thick) contained loose sandy-clayed sediments with very dense archaeological remains, especially lithic and bone material (Tables 7.6 and 7.7). The lithic assemblage displays an unquestionable mixture of Middle and Upper Paleolithic elements, primarily Levallois products and convergent scrapers, but also bladelets extracted from "cores of volumetric reduction" (Figs. 7.9 and 7.10), as well as typical Aurignacian carinated end scrapers (Fig. 7.10: 14, 15). Although a thorough stratigraphic-sedimentological analysis is not yet available, stratigraphic perturbation is not macroscopically visible.

The dating of two charcoal samples from these spits produced ages which are not far from the transitional period (despite the questions arising from the inversion of the ages; Table 7.5). At the same time, faunal data from the same layers are reminiscent of the Middle Paleolithic pattern known from Kalamakia, with fallow deer dominating over red deer and the sudden appearance of land tortoises (Table 7.7). In sum, according to the cultural and faunal material as well as the radiocarbon dates, the 90 cm thick spits 9–19 of Kolominitsa likely correspond to the Middle-Upper Paleolithic transition

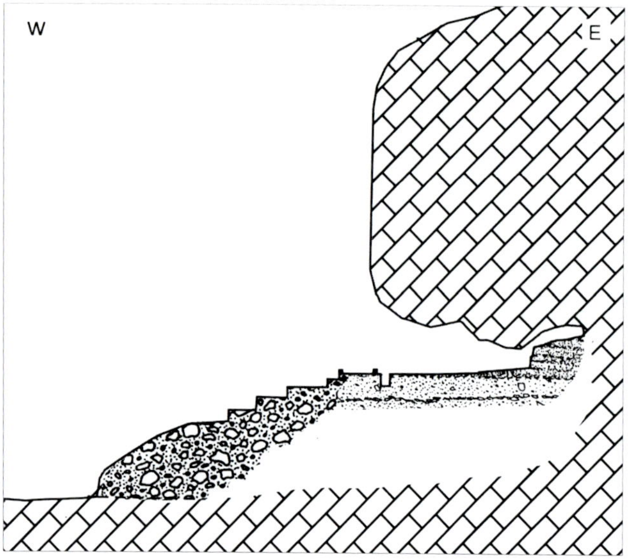

Fig. 7.5 Profile of Kolominitsa and schematic representation of its deposits. In the back of the cave all layers are preserved up to the original top of the filling. The upper layers have been eroded in the rest of the cave. In front of the cave, a thick scree accumulation forms a sloping surface leading from the Tyrrhenian terrace to the entrance of the cave

Fig. 7.6 General view of the interior of Kolominitsa. The paved floor has been constructed on the surface created by the erosion of the upper layers of the deposits. The latter can be seen in the background, where they are preserved up to their original height

Table 7.4 Radiocarbon dates from Kolominitsa (upper layers, preserved at the back of the cave)

Depth	Laboratory code	Material	Method	Conventional age	cal years BP
140	Beta-237175	Charcoal	AMS	19,560 ± 120	23,840–22,690
210	Beta-237176	Charcoal	AMS	21,940 ± 140	26,850–25,930

(*sensu lato*). Pure Middle Paleolithic layers have not yet been reached.

Large mammal remains: Abundant faunal material is attributed mainly to *Dama dama* and *Capra* sp., and, to a lesser degree, to the following taxa: *Cervus elaphus*, *Capreolus capreolus*, *Sus scrofa*, *Bos primigenius*, and *Lepus europaeus*. Present but very rare are *Ursus arctos* and *Canis lupus*. Land tortoises, especially *Testudo marginata*, become very common toward the deeper layers (Table 7.7).

Other Finds: Noteworthy is the presence of pieces of ferrous mineral (hematite) in all layers, from spit 16 to the top.

Fig. 7.7 East profile of the excavation trench at Kolominitsa. From bottom to top: loose reddish sandy clay with large stones (MPL/UPL); strongly lithified reddish sandy clay with stones (IUP?); loose reddish sandy clay with rare small stones (UPL)

Melitzia Cave

Melitzia is located on the eastern coast of Itylo bay (36° 41′ 31.95″N, 22° 23′ 35.66″E; Fig. 7.11). It opens at 350 m inland from the current sea-shore at 64 m asl. It is a 20 m wide, 20 m deep, and 4 m high spacious cavity.

After an initial test pit, a more extended excavation over 8 m^2 and reaching 1.5 m in depth has been carried out since 2009. Since the material of the systematic excavation has not yet been studied, the following presentation is based on the results of the initial test excavation. The upper 70 cm yielded reworked sediments containing remains of both prehistoric and historic times (i.e. pottery). The underlying layers date to the Pleistocene and contain Upper Paleolithic material, though very eroded and disturbed (Fig. 7.12). The intense disturbance of the sediments, in addition to the very wet plastic clay (mud), did not allow a clear identification of different layers during the excavation, or the separation of the material uncovered from these layers. Only the lowest excavated layers, around 1.5 m of depth, appear completely or nearly undisturbed.

The excavated layers date to the Upper Paleolithic and, more specifically, between *ca.* 24,000 and 11,000 cal BP (Table 7.8). However, a hiatus appears between 21,000 and 13,500 cal BP. This hiatus is probably due to the erosion of the lower ("Gravettian") layers, the truncation of which can be seen on the profile of the trench (Fig. 7.12).

The dense archaeological material, together with the strong presence of burnt remains, testifies to the intense occupation of the cave. Both large mammals and small vertebrates were uncovered. Red deer heavily dominates the faunal assemblage, followed by wild goat and wild boar (Table 7.9). Large bovids, probably aurochs, are rare and so is the red fox. Among the small-sized species, hares and birds are very abundant. Marine shells and land snails are abundant, while land tortoises are sporadic. The dominance of red deer remains suggests a specific use of the site, linked to the hunting of this animal. Entire carcasses, belonging mostly to adult animals, were brought to the cave.

The lithic assemblage is characterized by the strong presence of projectiles: backed bladelets and points (Table 7.10; see below: Fig. 7.13a). Noteworthy is the complete absence of geometric microliths and microburins. The

Table 7.5 Radiocarbon dates from Kolominitsa (test pit)

Spits	Laboratory code	Material	Method	Conventional age	cal years BP
6	Beta-193416	Charcoal	AMS	33,870 ± 550	40,390–37,180
11	Beta-307820	Burnt bone	AMS	34,320 ± 250	40,040–38,730
16	Beta-333515	Charcoal	AMS	37,840 ± 300	42,800–42,020
18	Beta-333516	Charcoal	AMS	34,150 ± 280	39,650–38,610

Table 7.6 General composition and tool types of lithic assemblages from Kolominitsa

	Gravettian (spits 1 and 2)		Aurignacian (spits 3–8)		IUP? (spits 9 and 10)		MPL-UPL (spits 11–19)	
	Nb	%	Nb	%	Nb	%	Nb	%
Debitage <20 mm	2	2.0	382	39.3	97	48.5	1329	70.5
Debitage <20 mm	69	69.0	460	47.4	79	39.5	383	20.3
Levallois flakes	0		0		0		19	1.0
Pebbles	0		1	0.1	2	1.0	0	
Bladelet cores	0		5	0.5	0		0	
Prismatic cores	1	1.0	7	0.7	1	0.5	2	0.1
Divers cores	3	3.0	12	1.2	5	2.5	15	0.8
Crested blades	1	1.0	0		0		1	0.1
Blades	0		1	0.1	0		2	0.1
Bladelets	14	14.0	51	5.2	4	2.0	21	1.1
Burin spall	1	1.0	1	0.1	1	0.5	6	0.3
Retouched tools	8	8.0	52	5.4	11	5.5	107	5.7
Total	**99**	**100.0**	**972**	**100.0**	**200**	**100.0**	**1885**	**100.0**
Tools								
End scrapers	1	12.5	10	19.2	1	9.1	3	2.8
Carinated end scrapers	1	12.5	2	3.8	0		2	1.9
Retouched blades	1	12.5	1	1.9	0		1	0.9
Backed bladelets	1	12.5	1	1.9	0		0	
Truncated blades	0		0		0		1	0.9
Splintered pieces	3	37.5	12	23.2	1	9.1	9	8.4
Burins	0		7	13.5	2	18.2	6	5.6
Scrapers	1	12.5	2	3.8	2	18.2	48	44.9
Mousterian points	0		0		0		4	3.7
Misc. (fragments incl.)	0		17	32.7	5	45.4	33	30.9
Total	**8**	**100.0**	**52**	**100.0**	**11**	**100.0**	**107**	**100.0**

excavation also yielded bone artifacts (awls, needles, points, and rounded blades), a few ornaments (perforated canines and *Dentalium* beads), as well as pieces of ferrous mineral (hematite) found in all layers.

Kastanis Cave

The erosion in Kastanis, a small shallow cave that opens at about 300 m North of Kolominitsa cave (36° 42′ 35.16″N, 22° 21′ 03.16″E; Fig. 7.1), revealed Pleistocene deposits with archaeological material. A very small test pit (30 × 40 cm and only 20 cm deep) yielded charcoal, animal bones, and stone tools. A charcoal sample from the bottom of the pit was dated to 12,390 ± 70 ^{14}C BP (14,910–14,070 cal BP; Beta-237174). The lithic assemblage is marked by the presence of thin backed bladelets and is attributed to the Epigravettian (Table 7.11). The small sample of faunal remains contains *Canis* sp., *Vulpes vulpes*, *Cervus elaphus*, *Capra* sp., and *Lepus europaeus*. Noteworthy is the abundance of hares and birds (Darlas and Psathi 2008).

Skini 4 Cave

Four shallow caves open at cape Skini, on the mouth of Itylo bay, about 500 m South of Kolominitsa (36° 41′ 54.50″N, 22° 20′ 57.01″E; Figs. 7.1 and 7.14). Skini 1 has been eroded by the sea, while Skini 2, 3, and 4 preserve nearly their entire filling. Skini 4, the northernmost of these caves, opens at a distance of 60 m from the current sea-line at 27 m asl. It is 4 m deep, 6 m wide, and 2.5 m high. A small trench (1.5 × 1.5 m) excavated to the depth of 85 cm (Fig. 7.15) produced the following results.

Successive ash layers and very dense archeological material imply an intensive use of this small cave. The dating (^{14}C AMS) of a charcoal from the bottom of the trench gave an age of 26,240 ± 200 ^{14}C BP (31,210–30,540 cal BP; Beta-193419). The lithic industry appears homogenous and

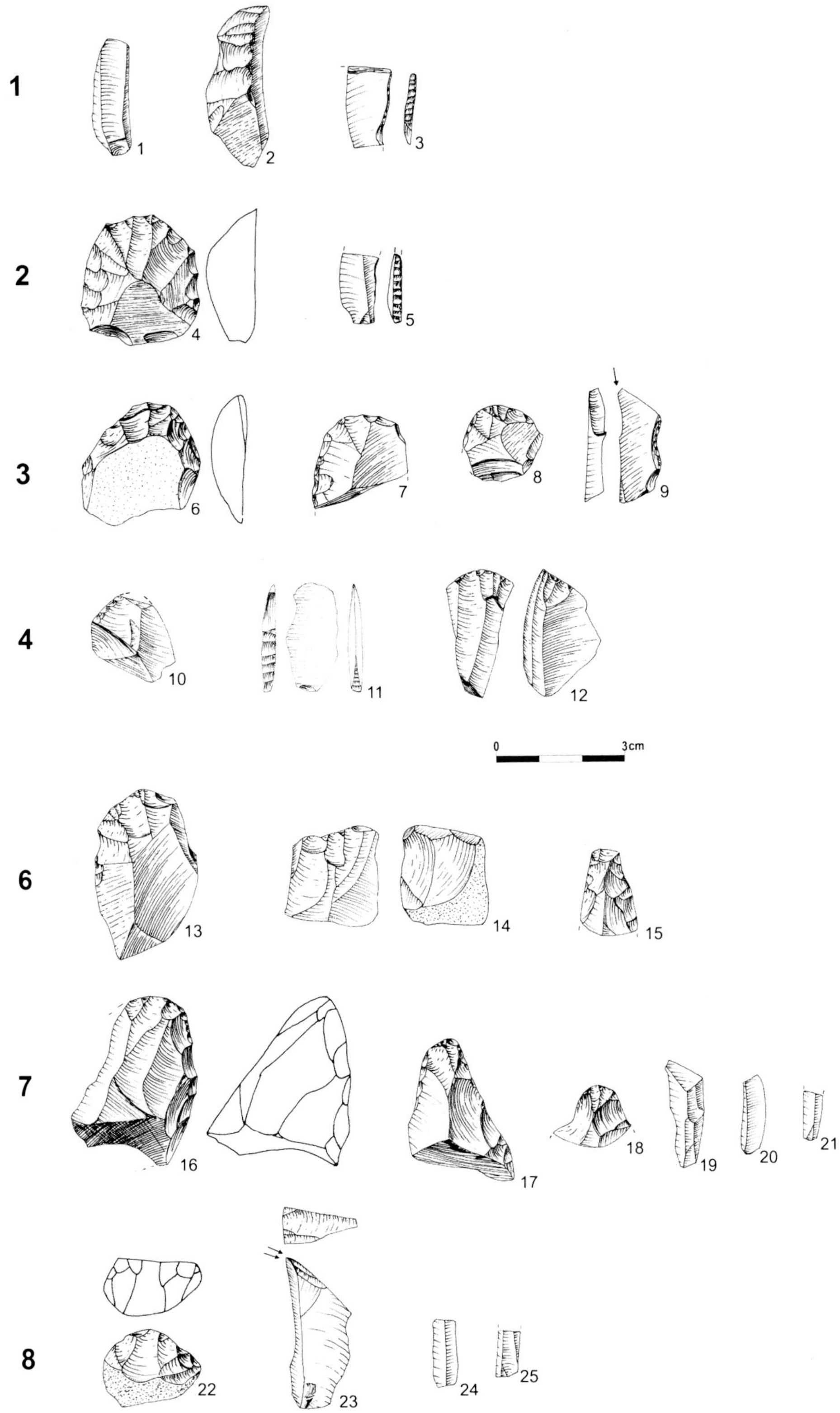

Fig. 7.8 Upper Paleolithic artifacts from Kolominitsa (spits 1–8). 1, 19, 20, 21, 24, 25: bladelets; 2: crested blade; 3, 5, 11: backed bladelets; 4, 6, 7, 8, 10, 12, 13, 15, 16, 17, 18, 22: various end scrapers; 9, 23: burins; 14: core. Adapted from Darlas and De Lumley 2004

Table 7.7 Frequency of large mammal and tortoise remains in Kolominitsa

Taxa/NISP[a]	Spits 1 and 2	Spits 3–8	Spits 9 and 10	Spits 11–19	Total	% NISP
Ursus cf. *arctos*		1			1	0.19
Canis lupus			1		1	0.19
Sus scrofa		1	1		2	0.37
Bos primigenius			3	2	5	0.94
Capra sp.	3	35	13	77	128	23.97
Cervus elaphus	5	18	1	11	35	6.55
Dama dama	12	74	24	140	250	46.82
Capreolus capreolus	1	14	3	10	28	5.24
Lepus europaeus	1	2		4	7	1.31
Testudo (*marginata* and sp.)		2	1	74	77	14.42
Total NISP	**22**	**147**	**47**	**318**	**534**	100.00
Unidentified *Carnivora*	2	2	1		5	
Unidentified *Cervidae*	16	58	8	45	127	
Unidentified *Artiodactyla*	23	116	31	135	305	
Unidentified[b]	236	997	123	2661	4017	
Total NS[c]	**299**	**1320**	**210**	**3159**	**4988**	

[a]Number of identified specimens
[b]Unidentified fragments longer than 2 cm
[c]Number of specimens

unchanged from the bottom to the top of the stratigraphic sequence. It is assigned to an assemblage with backed bladelets (Gravettian). Noteworthy is the presence of shouldered points (see below: Fig. 7.13b). Nearly all steps of the lithic reduction are present in the assemblage: crested blades, other technical pieces, flakes, numerous cores, blades, and bladelets (Table 7.12).

Large mammal bone remains belong to the following taxa: *Vulpes vulpes*, *Felis silvestris*, *Sus scrofa*, *Cervus elaphus*, *Dama dama*, *Bos* sp., *Capra* sp., and *Lepus europaeus* (Table 7.13). Red deer dominates the faunal assemblage, while carnivores are very rare. Several fragments of antler tips have been recorded, including one which had been transformed into a point (see below: Fig. 7.16: 1). All layers yielded pieces of ferrous mineral (hematite).

Skini 3 Cave

Skini 3 opens about 20 m to the South of the Skini 4 cave at the same altitude (36° 41′ 53.42″N, 22° 20′ 57.82″E; Fig. 7.14). A test pit brought to light a double Late Neolithic burial at just 15 cm below the current ground level. The Paleolithic layers immediately underlying the burial were extremely poor in remains, indicating a sporadic use of this cave in comparison with the neighboring Skini 4. Few lithic artifacts were collected (only 32 are longer than 20 mm), including backed bladelets. The dating of a charcoal gave an age of 25,560 ± 190 ^{14}CBP, (30,890–29,700 cal BP; Beta-193418), indicating that the cave was occupied at approximately the same age as Skini 4.

Tripsana Cave

On the north coast of Itylo bay (36° 41′ 36.80″N, 22° 21′ 57.91″E; Figs. 7.1, 7.11, and 7.17) at the Tripsana location, a rescue excavation has been carried out in a small cave. The natural substratum was reached in both opened trenches: in the interior, Trench A reached 1.60 m depth (Fig. 7.18), while near the mouth of the cave Trench B reached 1.28 m in depth. A charcoal sample from the 14th spit (depth 1.40 m) of trench A gave a date of 28,060 ± 250 ^{14}C BP (33,025–31,550 cal BP; Beta-237180). Archaeological remains testify to the ephemeral use of the site during the Gravettian. Both excavation trenches yielded few stone artifacts attributed to the Gravettian (Table 7.14; Fig. 7.13C). In addition to lithic artifacts, five fragments of bone tools have also been collected (four points and one retouched splinter; Fig. 7.16), along with hematite pieces. The identified remains of large mammals belong to the following taxa: *Canis lupus*, *Vulpes vulpes*, *Felis silvestris*, *Martes* sp., cf. *Mustela*, *Sus scrofa*, *Cervus elaphus*, *Dama dama*, *Bos* sp., *Capra* sp., and *Lepus europaeus*. *Cervus* and *Capra* dominate the faunal assemblage (Fig. 7.15); fallow deer remain quite common, while numerous hare remains have been recorded in the top layers.

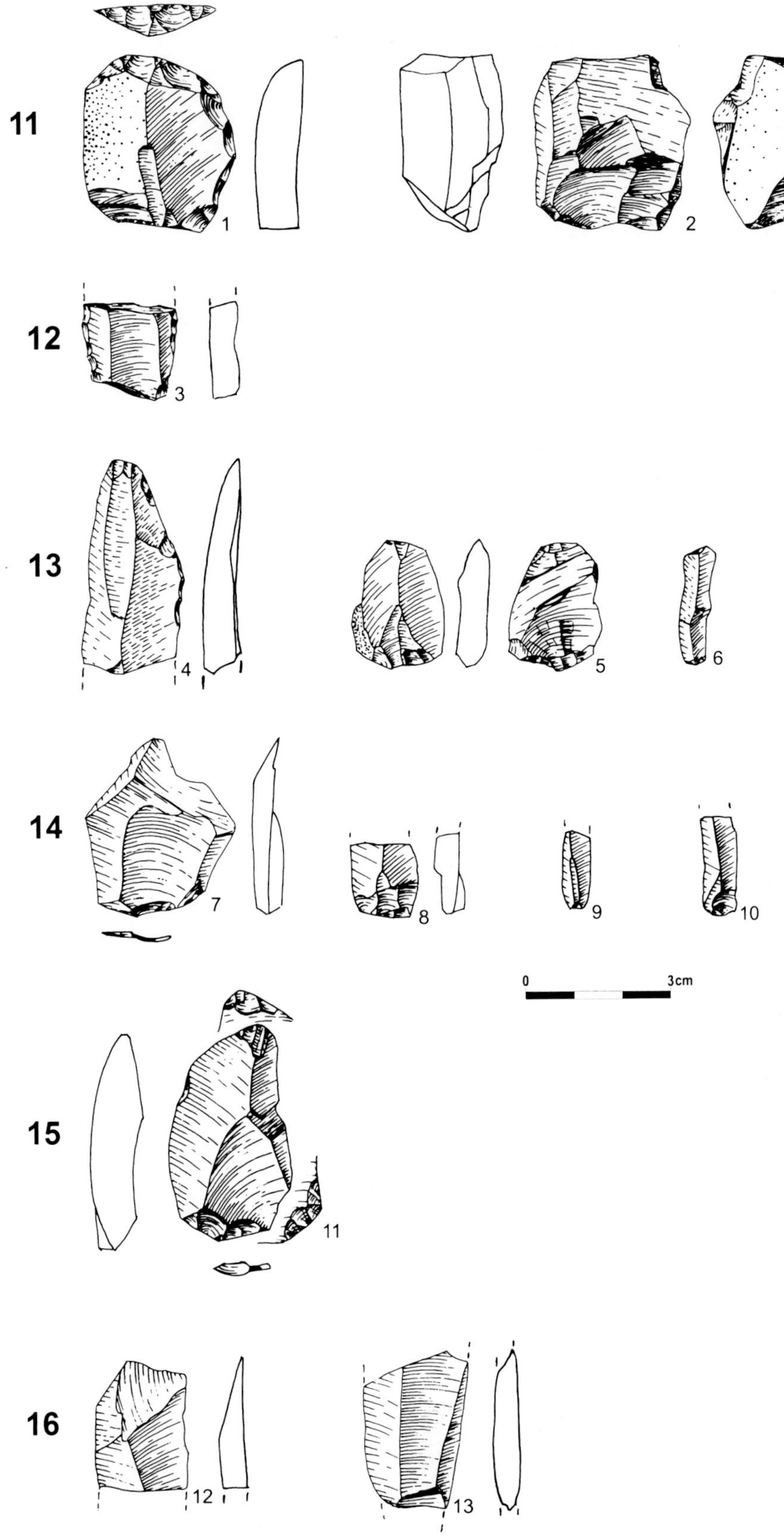

Fig. 7.9 Kolominitsa. Lithic artifacts from the Middle-Upper Paleolithic transition phase (spits 11–16). 1: end scraper; 2: unipolar core; 3: double scraper; 4: end scraper on blade; 5, 11: splintered pieces; 6, 8, 9, 10: bladelets; 7, 12, 13: Levallois flakes

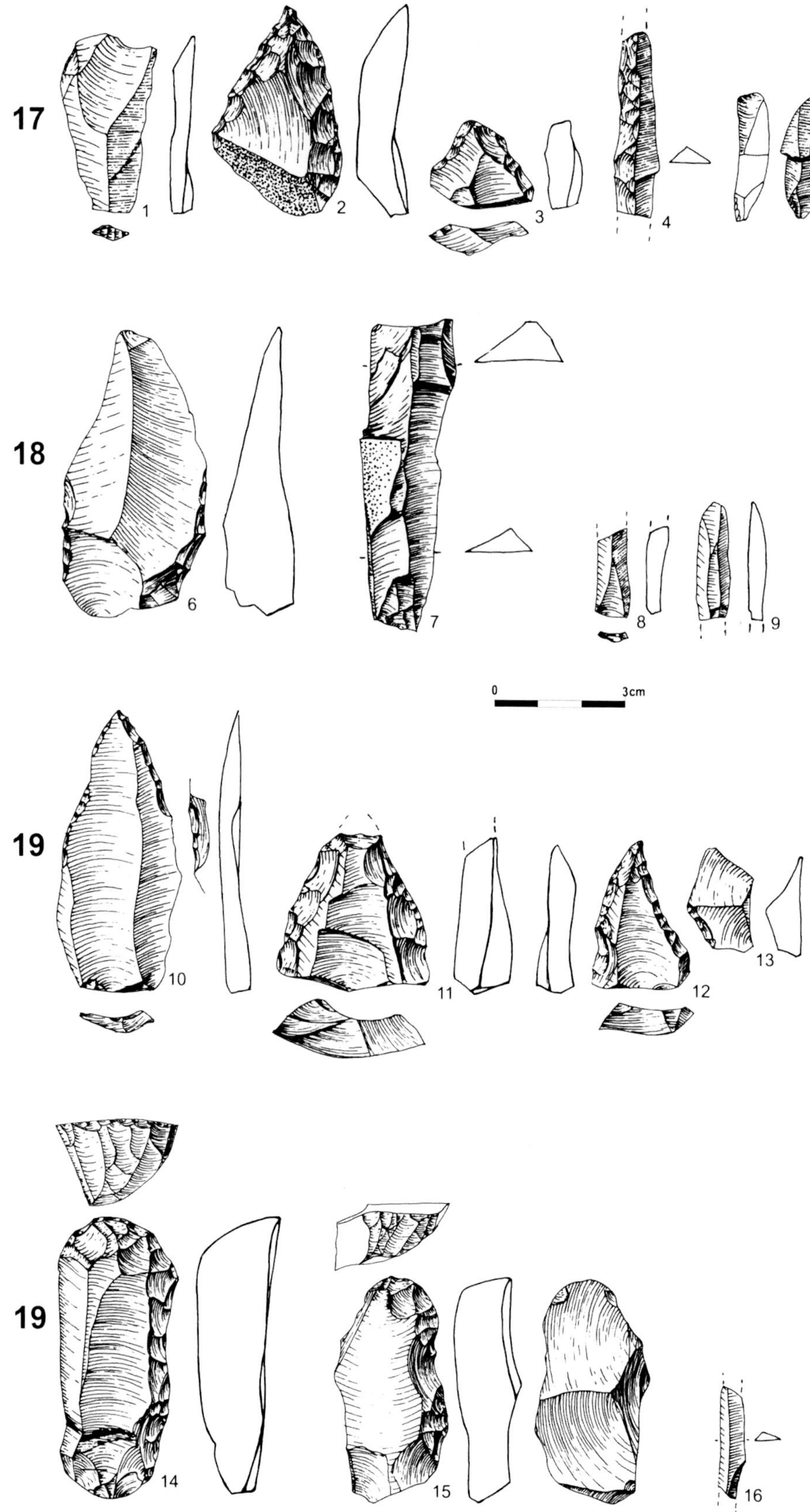

Fig. 7.10 Kolominitsa. Lithic artifacts from the Middle-Upper Paleolithic transition phase (spits 17–19). 1: Levallois flake; 2, 11, 12: Mousterian points; 3: convergent scraper; 4: crested blade; 5: burin; 6 double scraper; 7; blade; 8, 9, 16: bladelets; 10: elongated Mousterian point on a Levallois blade; 13: pseudo-Levallois point; 14: carinated end scraper on retouched blade; 15: carinated end scraper on retouched blade (blank: pseudo-levallois point)

Fig. 7.11 Location of Melitzia on the eastern coast of Itylon bay (view from SE). Tripsana is also indicated at the northern coast as well as Cape Skini at the northern end of the bay

Data Synthesis on the Upper Paleolithic of the Mani Peninsula

Test excavations in the six caves described above brought to light significant cultural remains dating from the beginning to the end of the Upper Paleolithic. The Middle-Upper Paleolithic transition and the Aurignacian are represented only in Kolominitsa. Gravettian has been attested in Kolominitsa, Skini 4, Skini 3, and Tripsana caves, as well as in the "lower layers" of Melitzia, while the phase corresponding to the Epigravettian has been identified in Melitzia ("middle layers") and Kastanis caves.

Environmental Data and Human Diet

In the absence of completed laboratory analyses, the environmental evidence from the Upper Paleolithic of the Mani peninsula is poor compared to that of the Middle Paleolithic. So far, only the preliminary study of the large mammal fauna provides some evidence of the change toward drier and colder climatic conditions that led to the restriction of the Mediterranean forest. The latter is mainly attested through the progressive replacement of the fallow deer, the typical eastern Mediterranean cervid, by the red deer, better adapted to a sparser vegetation cover.

Rather than reflecting significant climate change, extinctions and/or oscillations in the frequency of several species (e.g. cervids versus *Capra*, increasing rarity of land tortoises), as well as new faunal associations (e.g. *Lepus* and avian species) present during the first half of the Upper Pleistocene (Kalamakia cave), and up to the second half of the Upper Pleistocene (the above-mentioned six Upper Paleolithic cave sites), might reflect changes in subsistence strategies adopted by humans in each of these Upper Paleolithic sites. This was possibly combined with a rise in human population size. The Upper Paleolithic large mammal fauna contains significantly fewer taxa. Several carnivore species, as well as the very large elephants and rhinos disappear, while the frequency of smaller species, such as hares and birds, sharply rises. Land tortoises,

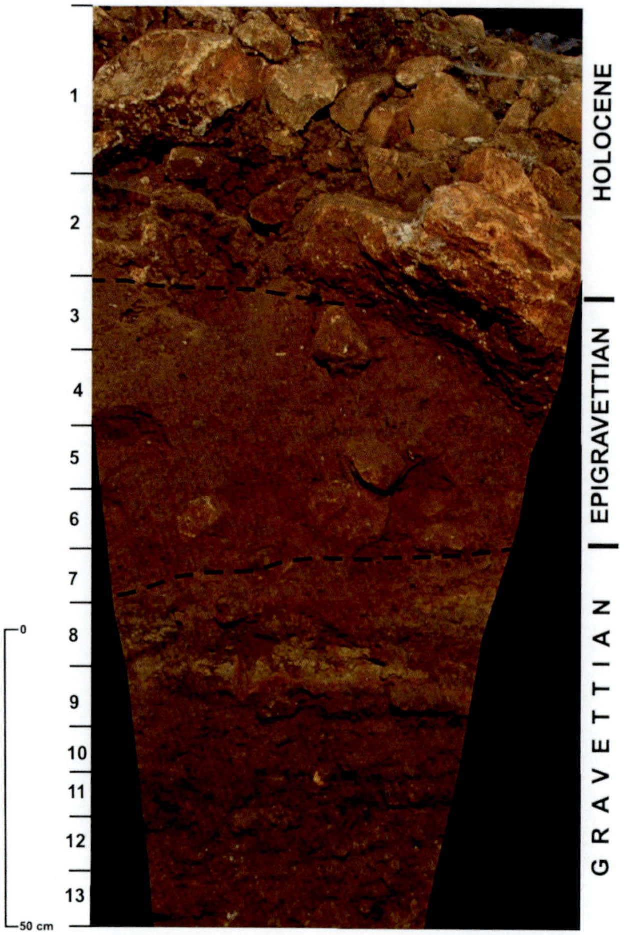

Fig. 7.12 West profile of the excavation trench at Melitzia. From bottom to top: successive layers of ashes, burnt remains and clay, truncated at the top ("Gravettian"); very plastic red clay ("Epigravettian"); disturbed layer of stones in plastic red clay matrix ("Holocene")

very abundant at least in the lower half of the stratigraphic sequence at Kalamakia, become rapidly very rare, while the consumption of edible sea shells and land snails—which are not well represented in Kalamakia and Kolominitsa cave (at least in the lower layers of the latter)—intensifies toward the end of the Paleolithic, at least in Melitzia cave.

Lithics

Despite the small scale of the excavations and low number of recovered artifacts, it was possible to define the main characteristics of the lithic assemblages.

The *Middle-Upper Paleolithic transition* appears in the record of Kolominitsa cave. The lithic assemblage of spits 19–14 is marked by a mixture of Middle and Upper Paleolithic elements (Figs. 7.9 and 7.10). Middle Paleolithic markers are Levallois products (quite rare) as well as the products of discoid debitage and, finally, some characteristic tool-types, such as Mousterian points and convergent scrapers. Among Levallois products, most characteristic are the very elongated flakes or blades with completely parallel ridges (products of the unipolar recurrent method). On the other hand, the presence of several blades and bladelets, extracted without any doubt from "cores of volumetric reduction", is considered as evidence for the Upper Paleolithic. These laminar pieces are not retouched. Their high frequency implies that the presence of these artifacts cannot be random or coincidental. Moreover, the carinated end scrapers are the most characteristic Upper Paleolithic tools (Fig. 7.10: 14, 5). The Kolominitsa spits 14–19 probably correspond to layer VI at Klissoura Cave 1 in Argolid,

Table 7.8 Radiocarbon dates from Melitzia

Depth	Laboratory code	Material	Method	Conventional age	cal years BP
87	Beta-269603	Charcoal	AMS	11,670 ± 60	13,680–13,380
97	Beta-269604	Charcoal	AMS	9330 ± 60	10,700–10,390
91	Beta-269605	Charcoal	AMS	9550 ± 60	11,150–10,680
98	Beta-286709	Charcoal	AMS	19,300 ± 80	23,270–22,600
108	Beta-286710	Charcoal	AMS	11,270 ± 50	13,240–13,090
116	Beta-307818	Charcoal	AMS	19,460 ± 80	23,450–23,230
122	Beta-307819	Charcoal	AMS	18,870 ± 80	22,530–22,330
144	Beta-333517	Charcoal	AMS	17,970 ± 80	21,530–21,340
133	Beta-359676	Charcoal	AMS	19,120 ± 80	22,990–22,480
144	Beta-359677	Charcoal	AMS	20,460 ± 90	24,560–24,310
134	Beta-359678	Charcoal	AMS	20,160 ± 90	24,320–23,870

Table 7.9 Frequency of large mammal remains in Melitzia

Taxa/NISP[a]	Spits 1 and 2	Spits 3–7	Spits 8–13	Total	% NISP
Canis lupus	0	2	1	3	1.97
Vulpes vulpes	0	3	3	6	3.95
Mustelidae	0	0	1	1	0.66
Sus scrofa	5	2	1	8	5.26
Cervus elaphus	11	63	21	95	62.50
Capra sp.	1	21	3	25	16.45
Lepus europaeus	1	11	2	14	9.21
Total NISP	**18**	**102**	**32**	**152**	**100.00**
Unidentified *Artiodactyla*	34	151	101	286	
Unidentified[b]	39	350	139	528	
Total NS[c]	**91**	**603**	**272**	**966**	

[a]Number of identified specimens
[b]Unidentified fragments longer than 2 cm
[c]Number of specimens

Table 7.10 General composition and tool types of lithic assemblages from Melitzia

	Disturbed (spits 1 and 2)		Epigravettian (spits 3–7)		Gravettian (spits 8–13)	
	Nb	%	Nb	%	Nb	%
Debitage <20 mm			47	18.8	5	11.6
Debitage >20 mm	5	62.5	124	49.6	25	58.2
Pebbles	1	12.5	0		0	
Prismatic cores	0		0		1	2.3
Divers cores	0		10	4.0	7	16.3
Crested blades	0		1	0.4	0	
Blades	0		6	2.4	0	
Bladelets	0		17	6.8	0	
Burin spalls	0		4	1.6	0	
Retouched tools	2	25.0	41	16.4	5	11.6
Total	**8**	**100.0**	**250**	**100.0**	**43**	**100.0**
Tools						
End scrapers	0		0		1	20.0
Retouched blades	0		1	2.4	1	20.0
Backed bladelets	0		27	65.8	1	20.0
Burins	0		1	2.4	0	
Splintered pieces	0		4	9.9	1	20.0
Misc. (fragments incl.)	2	100.0	8	19.5	1	20.0
Total	**2**	**100.0**	**41**	**100.0**	**5**	**100.0**

which presents a mixture of Middle and Upper Paleolithic elements (detailed analysis on the lithic material of this layer is not yet available; Kaczanowska et al. 2010). At Kolominitsa, the overlying layers (spits 11–13) did not yield any diagnostic finds. More Uluzzian specifically, they do not contain any new types of tools, characteristic of the Initial Upper Paleolithic, as for example, the curved backed points of "uluzzian" type found in the layer V of Klissoura (Koumouzelis et al. 2001; Kaczanowska et al. 2010). It seems highly probable that the corresponding layers of Kolominitsa have been eroded, at least in the area of the test pit. We hope that they have been preserved in another area of the cave.

Both the Kolominitsa stratigraphic sequence and the radiocarbon dates obtained from that site present great analogies with those of Klissoura Cave 1 (Kuhn et al. 2010). On the other hand, there are no analogies with Lakonis, which does not seem to contain any layer of the Middle-Upper Paleolithic transition. The excavators of Lakonis have argued for the presence of this phase (Panagopoulou et al. 2002–2004; Elefanti et al. 2008). Nonetheless, the lithic assemblages coming from the deposits claimed as dating from the "Initial Upper Paleolithic", do not show any characteristic attributes of the Upper Paleolithic artifacts (neither in technological nor in typological terms), and therefore could not justify the above arguments (see Kozlowski and Otte 2009). In any case, with caution due to the limited test-character of the excavation, it could be suggested that the mixture of Middle and Upper Paleolithic elements in spits 19–14 of Kolominitsa could indicate the coexistence and cultural interaction between Neanderthals and anatomically modern humans.

From the sites discussed here, the *Aurignacian* is represented only in Kolominista (spits 3–8). However, the finds, especially the lithic ones, are very rare and do not allow any detailed description. We note only the relatively high frequency of carinated end scrapers and splintered pieces (Table 7.6; Fig. 7.8).

Contrary to the very sparse Aurignacian evidence, the presence of industries with backed bladelets (*Gravettian*) is very strong and found in most caves discussed in this chapter. While blades generally represent only a small part of the debitage products, these assemblages are marked by the high frequency of backed bladelets and points, particularly the most characteristic shouldered points. The Gravettian layers from these sites are dated between 28,260 ± 250 ^{14}C BP (33,025–31,550 cal BP) and 19,580 ± 120 BP (23,835–22,692 cal BP).

The lithic industry from the middle layers of Melitzia and the small sample from Kastanis can be attributed to the *Epigravettian*. Their age ranges between 12,390 ± 70 ^{14}C BP (14,910–14,070 cal BP) and 9350 ± 60 ^{14}C BP (10,320–10,310 cal BP). In Melitzia, we note the abundance of extremely small backed bladelets and points (Table 7.10). Splintered pieces are present in all Epigravettian layers, while geometric microliths and microburins are completely absent.

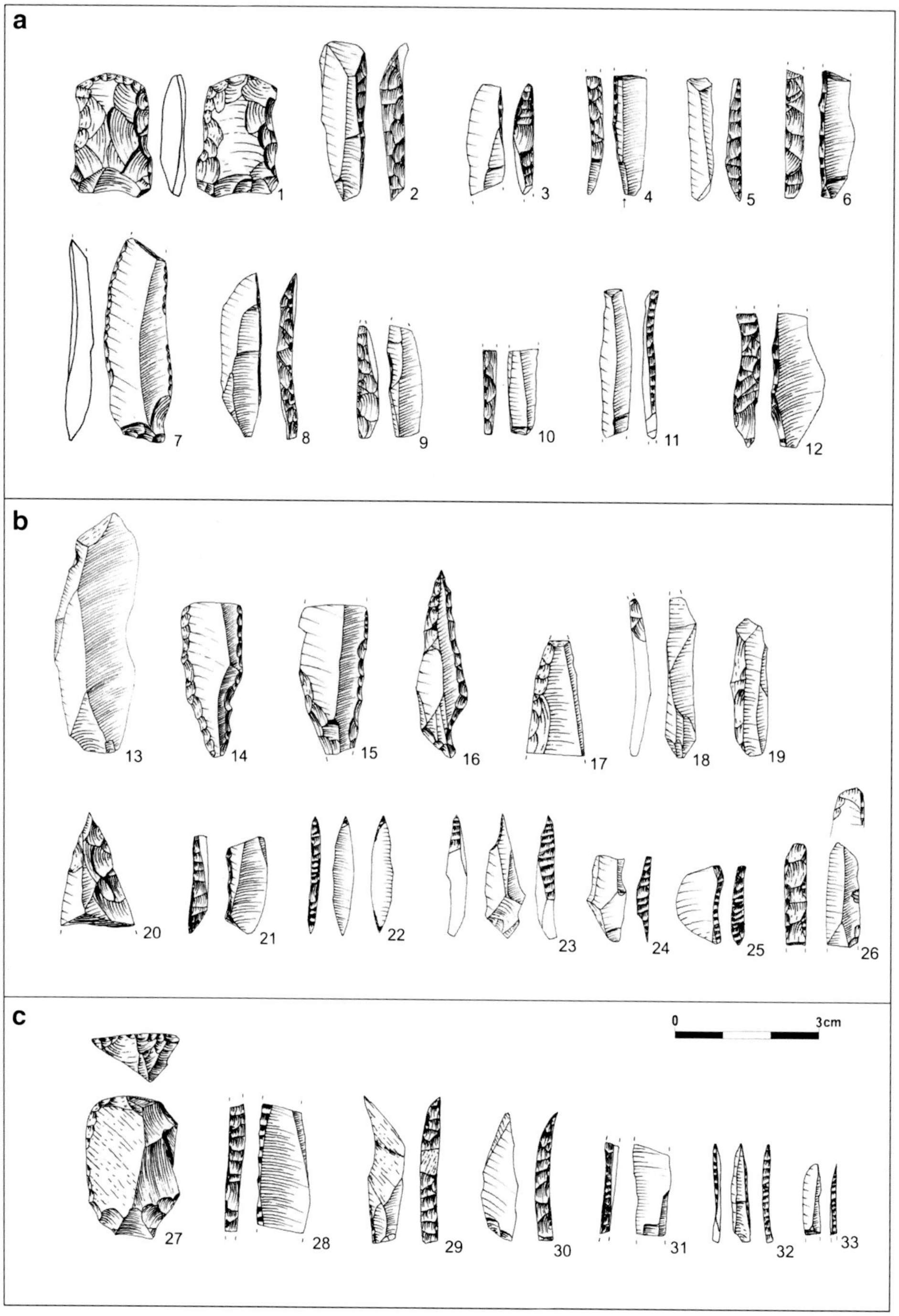

Fig. 7.13 Upper Paleolithic artifacts from Melitzia (**a**) Skini 4 (**b**) and Tripsana (**c**). 1: bifacial tool; 2, 3, 4, 5, 6, 8, 9, 10, 11, 12, 21, 24, 25, 26, 28, 29, 30, 31, 33: backed bladelets; 7, 13: blades; 14, 15: shouldered pieces; 16: shouldered point; 17, 18, 20, 22, 30, 32: various points; 19: crested blade; 23: borer. Adapted from Darlas and De Lumley (2004)

Table 7.11 General composition and tool types of the lithic assemblage from Kastanis

	Nb	%
Debitage <20 mm	63	53.0
Debitage >20 mm	37	31.1
Prismatic cores	1	0.8
Crested blades	1	0.8
Bladelets	6	5.1
Burin spalls	1	0.8
Retouched tools	10	8.4
Total	**119**	**100.0**
Tools		
End scrapers	1	10.0
Backed bladelets	5	50.0
Splintered pieces	1	10.0
Misc. (fragments incl.)	3	30.0
Total	**10**	**100.0**

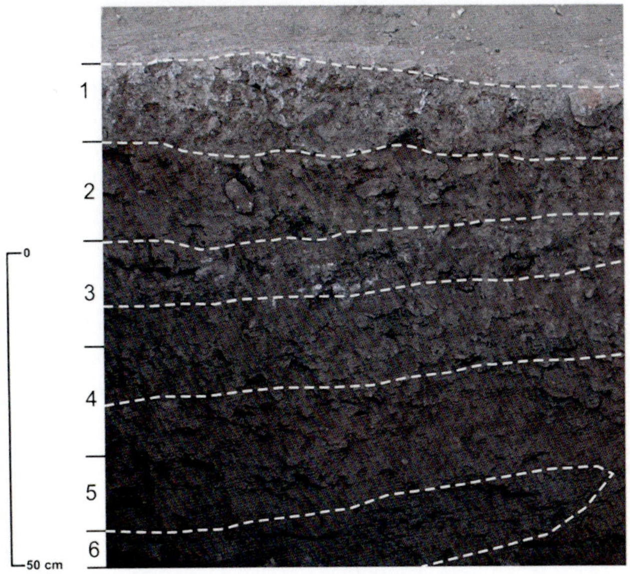

Fig. 7.15 East profile of the excavation trench at Skini 4. Loose sandy clay sediments with successive ash layers

Fig. 7.14 General view of Cape Skini; Skini 1, 2, 3, and 4 can be seen from right to left (S–N)

The main raw materials used for the lithic tools are flints of various colors, quartz, quartzite, and andesite. The latter, known also as "stone of Krokees", is not a local rock, and must have been transported from a relatively great distance (30 km). After the Aurignacian, the red jasper—a rock of very good quality and distant origin—appears and becomes very common.

Table 7.12 General composition and tool types of the lithic assemblage from Skini 4

	Nb	%
Debitage <20 mm	244	24.4
Debitage >20 mm	508	50.9
Prismatic cores	23	2.3
Divers cores	21	2.1
Pebble	1	0.1
Crested blades	5	0.5
Blades	20	2.0
Bladelets	49	4.9
Burin spall	1	0.1
Retouched tools	127	12.7
Total	**999**	**100.0**
Tools		
End scraper	13	10.2
Retouched blades	18	14.2
Backed bladelets	19	15.0
Shouldered point	1	0.8
Splintered pieces	10	7.8
Burin	2	1.6
Scrapers	1	0.8
Misc. (fragments incl.)	63	49.6
Total	**127**	**100.0**

Bone Tools

In addition to lithic artifacts, bone tools appear from the beginning of the Upper Paleolithic, although not in great quantities. These are fragmented bone and awls, rounded blades and, finally, needles, the latter discovered in Melitzia cave (Fig. 7.16).

Pieces of Ferrous Minerals

A special mention should be made of the presence of pieces of ferrous minerals, especially hematite. This material, although naturally present in the broader area, is completely absent from Kalamakia and other Middle Paleolithic sites, as well

Table 7.13 Frequency of large mammal remains in Skini 4

Taxa/NISP[a]	Total	% NISP
Vulpes vulpes	7	5.34
Felis silvestris	1	0.76
Sus scrofa	2	1.53
Cervus elaphus	46	35.11
Dama dama	10	7.63
Cervus/*Dama*	42	32.06
Capra sp.	13	9.92
Lepus europaeus	10	7.63
Total NISP	**131**	**100.00**
Unidentified *Artiodactyla*	101	
Unidentified[b]	585	
Total NS[c]	**817**	

[a]Number of identified specimens
[b]Unidentified fragments longer than 2 cm
[c]Number of specimens

Fig. 7.16 Fragments of bone points and tools from Skini 4 (1), Tripsana (2, 3, 4), Kolominitsa (5, 6) and Melitzia (7)

Fig. 7.17 The entrance of Tripsana, at the northern coast of Itylon bay, opens in the middle of the vertical cliff, at the top of the scree talus

Fig. 7.18 Stratigraphy of Tripsana deposits. From bottom to top: light brown sandy clay; dark brown sandy clay with large blocks of stones; two successive thin layers of humus

Table 7.14 General composition and tool types of the lithic assemblage from Tripsana

	Nb	%
Debitage <20 mm	44	28.9
Debitage >20 mm	81	53.4
Prismatic cores	2	1.3
Divers cores	5	3.3
Blades	2	1.3
Bladelets	7	4.6
Burin spalls	1	0.6
Retouched tools	10	6.6
Total	**152**	**100.0**
Tools		
End scrapers	2	10.0
Retouched blades	1	10.0
Backed bladelets	3	30.0
Splintered pieces	2	20.0
Scrapers	1	10.0
Misc. (fragments incl.)	1	10.0
Total	**10**	**100.0**

Table 7.15 Frequency of large mammal remains in Tripsana

Taxa/NISP[a]	Total	% NISP
Canis lupus	1	0.21
Vulpes vulpes	13	2.75
Felis silvestris	8	1.69
Martes sp.	2	0.42
Mustela sp.	1	0.21
Sus scrofa	10	2.12
Cervus elaphus	122	25.85
Dama dama	26	5.51
Cervus/Dama	56	11.86
Capra sp.	56	11.86
Lepus europaeus	177	37.50
Total NISP	**472**	**100.00**
Unidentified Carnivora	20	
Unidentified Artiodactyla	395	
Unidentified[b]	841	
Total NS[c]	**1728**	

[a]Number of identified specimens
[b]Unidentified fragments longer than 2 cm
[c]Number of specimens

as from the current lowest layers of Kolominitsa cave (spits 17–19). In contrast, it is consistently present in the Upper Paleolithic layers of the caves discussed here. It seems that this material might have been used by *Homo sapiens* as a colorant, while it was ignored by the Neanderthals. This material can be considered as local and it is found in the form of irregular "iron pebbles" of various sizes (often 2–10 cm long). The finds yielded by excavation are mostly intact, but they can also be slightly processed, in the form of "pebble tools" or "cortical flakes". Their surface is usually well preserved; sometimes it is corroded and occurs as "rusty". Traces of use could not be established macroscopically on any of the pieces. Their future analysis will undoubtedly yield more information.

Conclusion

The above brief synthesis demonstrates the wealth of information that can be obtained by our research project. The dense cluster of the western Mani cave sites, located in a very restricted geographic zone and sharing very similar formation and occupation histories, makes it possible for us to carry out a detailed regional study. With data spanning the whole Upper Pleistocene, and a very good resolution, the regional approach allows detailed comparisons among contemporary sites of the same type. Finally, such regional studies possibly constitute the best way to study, describe and define the Greek Paleolithic, which is still poorly known in comparison to the state of research in most European and Mediterranean areas.

Acknowledgements The investigations in the Mani peninsula have been funded by the Hellenic Ministry of Culture (Ephoreia of Paleoanthopology-Speleology), the J. F. Costopoulos Foundation, the Institute for Aegean Prehistory, the Psychas Foundation, and the Municipality of Eastern Mani. All the above institutions are gratefully acknowledged. We also thank the three anonymous reviewers and the Editors for their very useful comments on the manuscript.

References

Bailey, G. N. (Ed.). (1997). *Klithi: Palaeolithic settlement and quaternary landscapes in Northwest Greece*. Cambridge: McDonald Institute for Archaeological Research.

Bailey, G. N., Carter, P. L., Gamble, C. S., & Higgs, H. P. (1983). Asprochaliko and Kastritsa: Further investigations of Palaeolithic settlement and economy in Epirus (North-West Greece). *Proceedings of the Prehistoric Society, 49*, 15–42.

Bassiakos, E. I. (1993). Chronologissi apolithomaton ke spilaeothematon me ti methodo tou syntonismou tis ilektronikis stroformis kai meleti morfologias ypogeiou karst kai ton schetikon radiometrikon ke geologikon synthikon se spilaeoperivallonta tis periohis Dirou Manis (Dating of Fossils from Caves and Speleothems: Evidence from Electron Spin Resonance (ESR) Technique, the Study of Underground Karst Morphology and the Relevant Radiometric and Geological Conditions in Speleoenvironments of Dyros, Mani). Ph.D. Dissertation, University of Athens.

Coutselinis, A., Dritsas, C., & Pitsios, T. K. (1991). Expertise médico-légale du crâne pléistocène LAO1/S2 (Apidima II), Apidima, Laconie, Grèce. *L'Anthropologie, 95*(2-3), 401–408.

Darlas, A. (2012). Gheomorfologhiki exelixi ke katikissi ton spilaeon tis dytikis aktis tis Manis kata to anotero Plistokeno ke to Olokeno (Geomorphological evolution and use of the caves of the Western coast of Mani during the Upper Pleistocene and the Holocene). In N. Zacharias, et al. (Eds.), *Proceedings of 5th Symposium of the Hellenic Society of Archaeometry* (pp. 237–253). Athens: Papazissis.

Darlas, A. (2014). Anazitontas tin taftotita kai ti spoudeotita tou spilaiou Petralonon gia tin elliniki kai evropaiki proistoria (in search of the identity of Petralona cave and its importance for the Greek and European Prehistory). In E. Stefani, et al. (Eds.), *1912–2012. A century of research in prehistoric Macedonia*, International Conference Proceedings, Archaeological Museum of Thessaloniki 22–24 November 2012 (pp. 195–204). Thessaloniki: Ziti.

Darlas, A., & De Lumley, H. (1999). Palaeolithic research in Kalamakia Cave, Areopolis, Peloponnese. In G. N. Bailey et al. (Eds.), *The Palaeolithic archaeology of Greece and adjacent areas* (pp. 293–302). Athens: British School at Athens Studies 3.

Darlas, A., & De Lumley, H. (2004). La grotte de Kalamakia (Aréopolis, Grèce). Sa contribution à la connaissance du Paléolithique moyen de Grèce. In *XIVème Congrès UISPP. 5ème Section: Le Paléolithique moyen*, BAR International Series 1239 (pp. 225–233). Oxford: Archaeopress.

Darlas, A., & Psathi, E. (2008). Le Paléolithique supérieur dans la péninsule du Mani (Péloponnèse, Grèce). In A. Darlas & D. Mihailovic (Eds.), *The Palaeolithic of the Balkans*, BAR International Series 1819 (pp. 51–59). Oxford: Archaeopress.

De Lumley, H., & Darlas, A. (1994). Grotte de Kalamakia (Aréopolis, Péloponnèse). *Bulletin de Correspondance Hellenique, 118*(2), 535–559.

Elefanti, P., Panagopoulou, E., & Karkanas, P. (2008). The transition from the Middle to the Upper Paleolithic in the Southern Balkans: The evidence from the Lakonis I Cave, Greece. *Eurasian Prehistory, 5*(2), 85–95.

Galanidou, N., Athanassas, C., Cole, J., Iliopoulos, G., Katerinopoulos, A., Magganas, A., et al. (2016). The Acheulean site at Rodafnidia, Lisvori on Lesbos, Greece: Part I. In K. Harvati & M. Roksandic (Eds.), *Paleoanthropology in the Balkans and Anatolia: Human evolution and its context*. Dordrecht: Springer.

Harvati, K. (2016). Paleoanthropology in Greece: Recent findings and interpretations. In K. Harvati & M. Roksandic (Eds.), *Paleoanthropology in the Balkans and Anatolia: Human evolution and its context*. Dordrecht: Springer.

Harvati, K., Darlas, A., Bailey, S. E., Rein, T. R., El Zaatari, S., Fiorenza, L., et al. (2013). New Neanderthal remains from Mani peninsula, Southern Greece: The Kalamakia Middle Palaeolithic cave site. *Journal of Human Evolution, 64*, 486–499.

Harvati, K., Panagopoulou, E., & Karkanas, P. (2003). First Neanderthal remains from Greece: The evidence from Lakonis. *Journal of Human Evolution, 45*, 465–473.

Harvati, K., Stringer, C., & Karkanas, P. (2011). Multivariate analysis and classification of the Apidima 2 cranium from Mani, Southern Greece. *Journal of Human Evolution, 60*, 246–250.

Higgs, E. S., & Vita-Finzi, C. (1966). The climate, environment and industries of stone age Greece: Part II. *Proceedings of the Prehistoric Society, 32*, 1–29.

Kaczanowska, M., Kozlowski, J. K., & Sobczyk, K. (2010). Upper Palaeolithic human occupations and material culture at Klissoura Cave 1. *Eurasian Prehistory, 7*(2), 133–285.

Keraudren, B., Dallongeville, R., Bernier, P., Caron, V., & Renault-Miskovsky, J. (2000). Le Pléistocène supérieur marin (Tyrrénien) en Crète nord-occidentale (Grèce). *Géomorphologie: Relief, Processus, Environnement, 3*, 177–190.

Kotjabopoulou, E., Panagopoulou, E., & Adam, E. (1997). The Boila Rockshelter: A preliminary report. In G. N. Bailey (Ed.), *Klithi: Palaeolithic settlement and quaternary landscapes in Northwest Greece* (pp. 427–437). Cambridge: McDonald Institute for Archaeological Research.

Koumouzelis, M., Ginter, B., Kozlowski, J. K., Pawlikowski, M., Bar-Yosef, O., Albert, R. M., et al. (2001). The Early Upper Palaeolithic in Greece: The excavations in Klissoura Cave. *Journal of Archaeological Science, 28*, 515–539.

Kozlowski, J. K., & Otte, M. (2009). La fin du Paléolithique Moyen et le début du Paléolithique Supérieur dans l'est des Balkans. In I. Gatsov & J. L. Guadelli (Eds.), *Saxa loquuntur* (Honour of the 65th Anniversary of Nikolai Sirakov, pp. 183–196). Sofia: Avalon.

Kuhn, S. L., Pigati, J., Karkanas, P., Koumouzelis, M., Kozlowski, J. K., Ntinou, M., et al. (2010). Radiocarbon dating results for the Early Upper Paleolithic of Klissoura Cave 1. *Eurasian Prehistory, 7*(2), 37–46.

Kyparissi-Apostolika, N. (Ed.). (2000). *Theopetra Cave. Twelve years of excavation and research 1987–1998*. Athens: Institute for Aegean Prehistory.

Lebreton, V., Psathi, E., & Darlas, A. (2008). Environnement végétal des néandertaliens de la grotte de Kalamakia (Aréopolis, Grèce). In A. Darlas & D. Mihailovic (Eds.), *The Palaeolithic of the Balkans* (BAR International Series 1819, pp. 61–68). Oxford: Archaeopress.

Panagopoulou, E., Karkanas, P., Tsartsidou, G., Kotjabopoulou, E., Harvati, K., & Ntinou, M. (2002–2004). Late Pleistocene archaeological and fossil human evidence from Lakonis Cave, Southern Greece. *Journal of Field Archaeology, 29*, 323–349.

Panagopoulou, E., Tourloukis, V., Thompson, N., Athanassiou, A., Tsartsidou, G., Konidaris, G. E., et al. (2015). Marathousa 1—A new Middle Pleistocene archaeological site from Greece. *Antiquity Project Gallery, 89*(343).

Perlès, C. (1987). *Les industries lithiques taillées de Franchthi (Argolide, Grèce): Tome I, Présentation générale et industries paléolithiques*. Bloominghton: Indiana University Press.

Pitsios, T. K. (2000). *O Tenarios anthropos (The Tainarian man)*. Athens: Ellinika Grammata.

Poulianos, A. (1971). Petralona. A Middle Pleistocene Cave in Greece. *Archaeology, 24*, 6–11.

Reisch, L. (1980). Pleistozän und Urgeschichte des Peloponnes (unpublished). Habilitationschrift, Friedriech-Alexander Universität Erlangen, Nürberg.

Roger, T., & Darlas, A. (2008). Microvertébrés, paléo environnement et paléoclimat de la grotte de Kalamakia (Péloponnèse, Grèce). In A. Darlas & D. Mihailovic (Eds.), *The Palaeolithic of the Balkans* (BAR International Series 1819, pp. 77–84). Oxford: Archaeopress.

Schmidt, E. (1965). Die Seïdi-Höhle eine jungpaläolitische Station in Griechenland. *Proceedings of IV Colloque International de Spéléologie, 1er en Grèce, Athens 1963*, pp. 163–174.

Thiebault, F. (1982). *Évolution géodynamique des Hellénides externes en Péloponnèse méridional (Grèce)*. Publications de la Société Géologique du Nord 6.

Tourloukis, V., Thompson, N., Garefalakis, C., Karkanas, P., Panagopoulou, E., Harvati, K. (2014). *New Middle Paleolithic sites from the Mani peninsula, Southern Greece*. Abstracts book, XVII World UISPP Congress (pp. 491–492). Burgos: Union International de Science Préhistoriques et Protohistoriques.

Trantalidou, K., & Darlas, A. (1992). Erevnes sta spilea tou nomou Dramas (researches in the caves of Prefecture of Drama) 1992. *To archeologhiko ergo sti Makedonia ke Thraki, 6*, 587–603.

Chapter 8
The Acheulian Site at Rodafnidia, Lisvori, on Lesbos, Greece: 2010–2012

Nena Galanidou, Constantin Athanassas , James Cole, Giorgos Iliopoulos, Athanasios Katerinopoulos, Andreas Magganas, and John McNabb

Abstract Rodafnidia is an Acheulian site on Lesbos Island, in the north-east Aegean Sea. This chapter presents the model that guided Paleolithic investigations on the island, the history of research, and the results of the 2012 expedition of systematic work in the field, which consisted of surface survey and excavation. The typology and technology of lithic artifacts from the surface and the uppermost Unit 1, as well as the first cluster of luminescence dates, firmly place the early component of the site in the Middle Pleistocene. The Acheulian industry derives from fluvio-lacustrine deposits at a locale with abundant fresh-water and lithic resources. Situated in the north-east Mediterranean Basin, an area where research on early hominin prehistory is intensifying, Rodafnidia holds the potential to contribute to Eurasian Lower Paleolithic archaeology and fill the gap in our understanding of early hominin presence and activity where Asia meets Europe.

Keywords Lower Paleolithic • Large cutting tools • Middle Pleistocene • West Asia • pIRIR dating

Introduction

Rodafnidia, at Lisvori on Lesbos Island in the north-east Aegean Sea (Fig. 8.1), is a new open-air site with a distinctive Lower Paleolithic component. It lies at 26°11′54.58″ E and 39°6′15.42″ N, in a volcanic setting, near the local thermal springs, and 2 km from the south-west shore of the Kalloni Gulf (Fig. 8.2). The site has produced compelling evidence for the presence of groups who used Acheulian tools on Lesbos (Galanidou 2013; Galanidou et al. 2013). By virtue of its content and position at the junction between west Anatolia, the Aegean Archipelago and the Balkan Peninsula, Rodafnidia links the early archaeology of south-east Europe with that of west Asia. In this chapter, we report the key geographic features of Lesbos that guided research on early hominin archaeology of the island, the history of site discovery, the background work, the objectives of our project, and the results of the 2012 campaign, including the first cluster of pIRIR dating results obtained for the excavated sediments.

N. Galanidou (✉)
Department of History and Archaeology, University of Crete, 74100 Rethymno, Greece
e-mail: galanidou@uoc.gr

C. Athanassas
Centre de Recherche and d'Enseignement de Géosciences de l'Environnement (C.E.R.E.G.E.), Europôle Méditerranéen de l'Arbois, Avenue Louis PHILIBERT BP 80-13545, Aix-en-Provence cedex 04, France
e-mail: athanassas@cerege.fr

J. Cole
School of Environment and Technology, University of Brighton, Brighton BN2 4GJ, UK
e-mail: J.N.Cole@brighton.ac.uk

G. Iliopoulos
Department of Geology, University of Patras, 26504 Rio Patras, Greece
e-mail: iliopoulosg@upatras.gr

A. Katerinopoulos • A. Magganas
Faculty of Geology and GeoEnvironment, University of Athens, 15784 Athens, Greece
e-mail: akaterin@geol.uoa.gr; amagganas@geol.uoa.gr

J. McNabb
Department of Archaeology, University of Southampton, Southampton SO17 1BF, UK
e-mail: J.McNabb@soton.ac.uk

Katerina Harvati and Mirjana Roksandic (eds.), *Paleoanthropology of the Balkans and Anatolia*, Vertebrate Paleobiology and Paleoanthropology, DOI 10.1007/978-94-024-0874-4_8

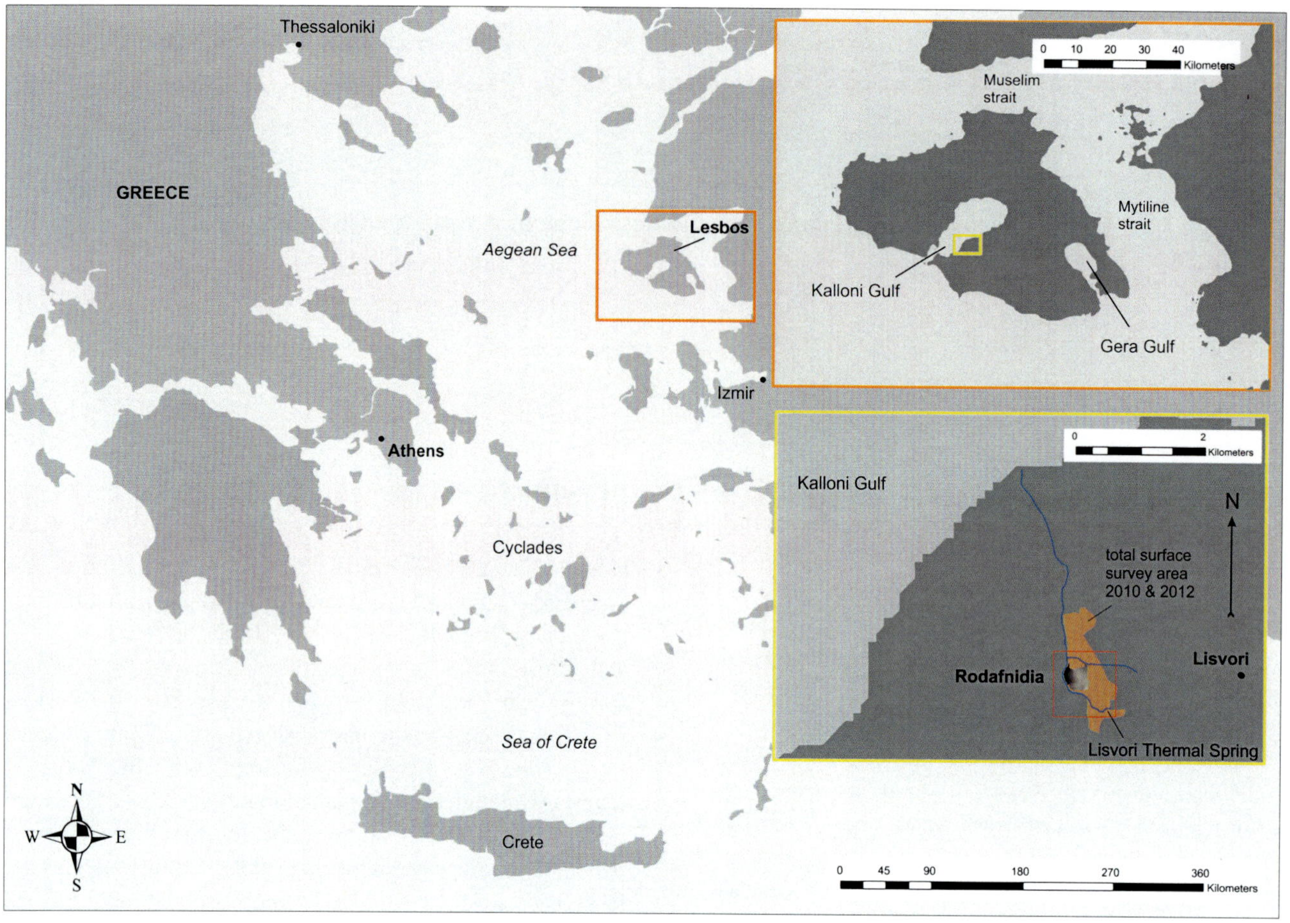

Fig. 8.1 Location map for the island of Lesbos (*left*), the Kalloni Gulf (*upper right*) and the Rodafnidia archaeological site (*bottom right*)

The Key Geographic Features of Lesbos Guiding the Project

Lesbos is the third largest of the Greek islands, measuring some 1600 km^2. Its topography and landscape have been significantly affected by volcanism, sedimentation, tectonism, eustasy, and isostasy. Around Lisvori, the island's Cenozoic volcanic and sedimentary history is mainly manifested by volcanic rocks, volcaniclastic deposits including large ignimbrite bodies and tuffs, siliceous and marly limestones, and geothermal springs (Hecht 1974; Pe-Piper 1978; Pe-Piper and Piper 1993, 2002; Lamera 2004; Kouli and Seymour 2006; Lambrakis and Stamatis 2008; Thomaidou 2009). Within the same area, various hard, mostly siliceous rocks of volcanic and/or diagenetic origin, are commonly outcropped as layers, nodules, or fracture fillings, which can be easily used as raw materials for knapping. These rocks are also found as clastic constituents of the Quaternary strata of the area.

Quaternary deposits are not widespread on Lesbos. However, they are fairly abundant in the south and east part of the island, consisting mainly of clastic fluvial and alluvial deposits (Soulakellis et al. 2006). Lesbos is separated from the Asian coast by two sea straits. The north strait, Lamna, is a faulted trough more than 150m-deep, lying along a major splay of the south branch of the North Anatolian Fault. The east strait, Mytilene, is mostly shallow, less than 50 m deep, with a flat, smooth seafloor (Fig. 8.1). Thus, a glacial sea-level drop of only 50 m would have been enough to expose the latter, connect the island with the Anatolian mainland, and allow the migration of hominins and terrestrial animals. Such a Pleistocene movement can be attested to by the presence of several fossiliferous sites at Vatera that have yielded a rich Early Pleistocene paleontological record with over 15 mammal taxa, including the giant macaque *Paradolichopithecus arvernensis*. This evidence represents a fauna that can be characterised as continental (De Vos et al. 2002; Lyras and van der Geer 2007),

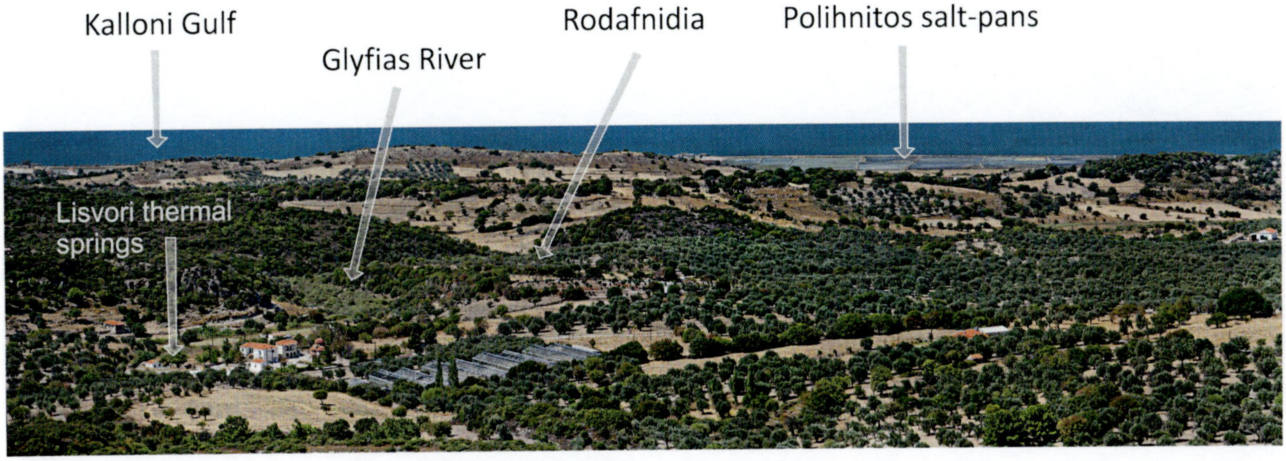

Fig. 8.2 Panoramic view of the Kalloni Gulf area where Rodafnidia is situated, looking north

reflecting the long-term, close association of Lesbos with Asia (Fig. 8.1).

Glacial periods with accompanying low sea levels were the times for terrestrial animals to disperse onto what are today the islands of the east Aegean Sea. These faunal dispersals likely also encompassed hominin population movements, and Rodafnidia at Lisvori offers archaeological evidence to support this hypothesis, adding a human component to the rich paleontology of Lesbos. Interglacial periods with accompanying high sea levels were key periods that cut Lesbos off from the Asian mainland, producing the insular picture that one sees today (Sakellariou and Galanidou 2016). Such events of land fragmentation occurred several times during the Pleistocene, isolating animal and hominin populations from the large expanses of Anatolia and limiting them to the islands mentioned earlier.

A further important feature of Lesbos is the presence of two shallow and enclosed gulfs, the Kalloni Gulf and the Gera Gulf (Fig. 8.1). Both embayments are connected to the open sea through shallow straits. During Pleistocene glacial periods, both gulfs would have lain well above sea level. However, it is not certain that they were dry. Our null hypothesis, which is still only supported by a small body of marine geological data, is that, during past low sea-level periods, these gulfs may have been shallow, initially semi-salted but eventually fresh-water lakes. If this were the case, then, in addition to abundant lithic raw materials of chert composition, hominins on Lesbos would have had a variety of survival possibilities associated with fresh-water resources. Envisioning the Kalloni Basin as a large, resource-rich, Pleistocene lake suggests that it might have been a point of attraction and persistent occupation for hominins in west Anatolia and the larger Aegean landmass during glacial periods (Lykousis 2009; Sakellariou and Galanidou 2016).

The Site, Its Discovery, and Objectives of Research

Rodafnidia is situated on a spur of a low hill, bordered to the north by a small stream and to the west by the Glyfias stream (Figs. 8.1 and 8.2). The Glyfias receives its brackish water from the local geothermal spring that lies less than 400 m to the south-east of the hill. It joins the little stream at the north-west of the hill to debouch into the Kalloni Gulf east of the Polichnitos salt pans. The south and west sides of the hill, being made up of ignimbrites, are rather steep and rocky, forming a small gorge, whereas the north side presents a smooth relief with a gentle slope, covered now with olive groves. The toponym 'Rodafnidia' refers to the oleanders, which once used to grow in the area where a large olive grove, segmented into numerous properties, stretches today (Fig. 8.3). The hill is divided into a south and a north part by a narrow farm track; its west end slopes down smoothly and meets the Glyfias. Some 100 m north of this point, on the lowermost terrace, a nineteenth-century watermill represents the only standing historical monument on the hill, apart from the stone installation with a now-dried-up fresh-water spring. The task of recording the watermill's plan brought to the area two medical doctors with an interest in the cultural heritage of Lesbos. They identified an extensive scatter of knapped stone artifacts, the greater portion of which had Levallois, proto-Levallois, and Acheulian affinities, and belonged to the Middle Paleolithic and the end of the Lower Paleolithic (Harisis et al. 2000).

Against the sparse background of early hominin sites within mainland Greece, the Aegean Archipelago and west Turkey (Jöris 2014; Otte et al. 1999; Galanidou 2004; Harvati et al. 2009), this earlier brief report, coupled with an evaluation of the island's key geographic features, led to an initial visit to the site by NG in 2009. A subsequent surface survey in 2010 with

Fig. 8.3 View of Rodafnidia, looking east. In the foreground, the site prior to excavations. In the background, the village of Lisvori

a small team[1] established the boundaries, character, and affinities of the lithic scatter (Fig. 8.1).

From the results of the initial investigation it was determined that the distribution of knapped stone artifacts was extensive, ceramic finds were almost completely absent, and a component of the lithic assemblage was Lower Paleolithic in character, including several Large Cutting Tools (LCTs) (as described by Kleindienst 1962; McNabb et al. 2004). Although the highest concentration of lithic finds was indeed near the watermill, the Lower Paleolithic component was not located in its immediate vicinity. A good number of the surface finds belonged to later Prehistory, namely the Late Neolithic and the Bronze Age.

Having ascertained that Rodafnidia had significant potential for systematically exploring Middle Pleistocene hominin presence at the junction between Anatolia and the Aegean Archipelago, the University of Crete obtained a permit in 2011 to undertake a 5-year (2012–2016) program of on-site and off-site research in order to:

a) conduct archaeological excavation, surface survey, and geophysical survey;
b) establish a chronological framework for the archaeological record based on relative and absolute dating; and
c) evaluate and correlate existing and new regional paleontological, paleoclimatic, geomorphological, and oceanographic evidence.

Put together, this work sheds new light on the history of hominin movements and dispersals between Africa and Eurasia, and on the early occupation of Europe, covering the current lacuna of early sites in south-east Europe and the west Anatolian coast (Dennell et al. 2011; Jöris 2014). Through an extensive program of reconstructing the site catchment and landscape evolution of the Aegean Archipelago, it further explores the attractions that the Kalloni basin, Lesbos, and the north-east Aegean basin offered to early humans during the Middle Pleistocene (Sakellariou and Galanidou 2015).

Investigative Methodology

The investigation strategy of the first season in the field conducted in August and September 2012, comprised archaeological surface, sub-surface, geological, and paleogeographic

[1]The members of the 2010 campaign team were Christina Papoulia, Elli Karkazi, Aggeliki Garidi, and Mihalis Spyridakis.

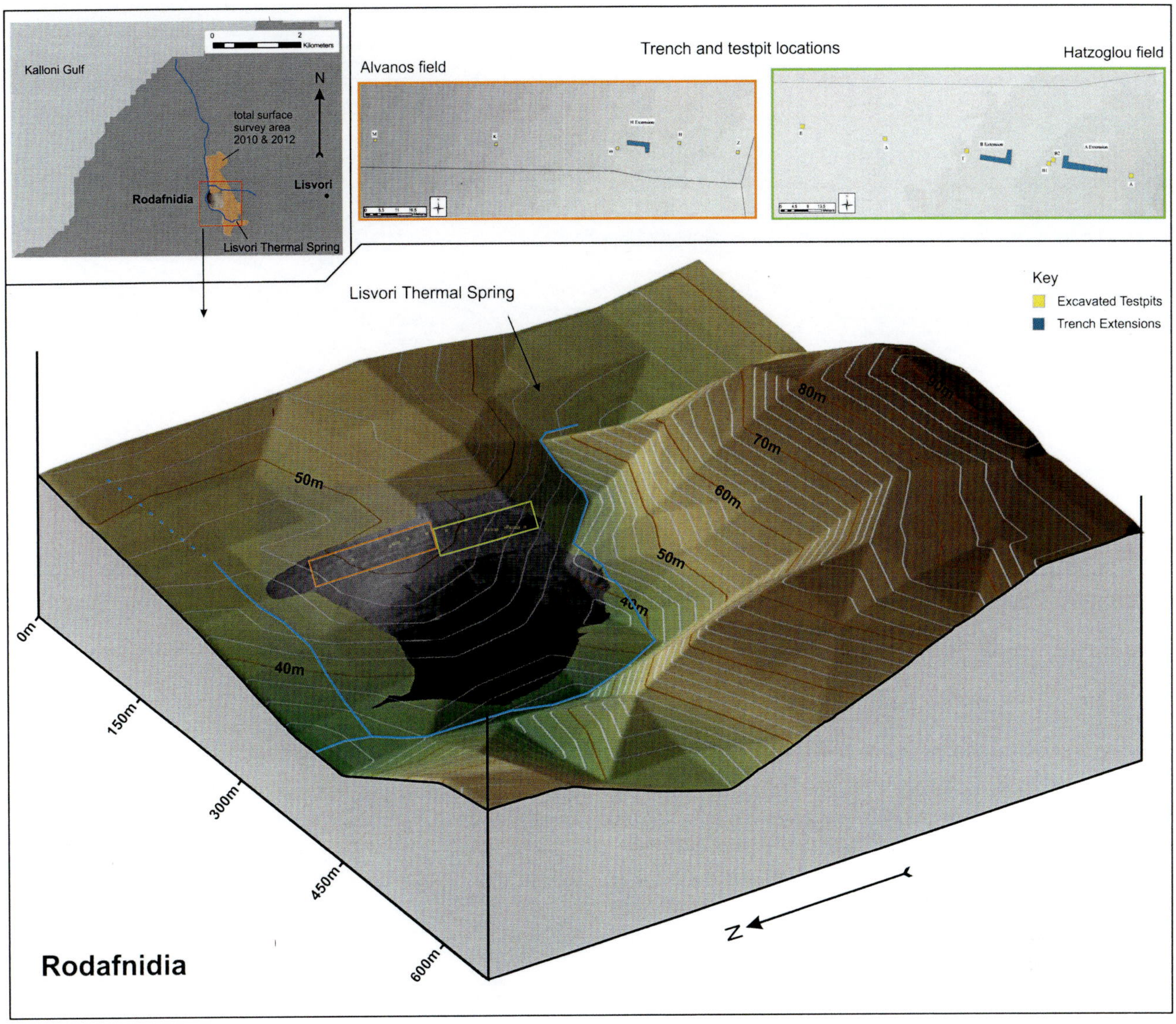

Fig. 8.4 DEM showing the location of Rodafnidia, the 2010 and 2012 survey areas marked in gray, and the two successive properties where excavations were conducted during the 2012 expedition *(bottom)*. Plans of the trench and test pit locations in the two successive properties, namely Alvanos and Hatzoglou (*upper middle and right*)

work.[2] A detailed topographic GPS survey was conducted over the extent of the hill, with test pit, trench, and find locations also being recorded (Fig. 8.4).

The excavation was guided by the initial 2010 surface survey work that had identified areas containing concentrations of finds from different periods. Given this assessment of spatial variation present on the site, the 2012 field season focussed on areas associated with LCTs in order to gain insight into the stratigraphy, and to understand the geological background to the Paleolithic remains. A major question, therefore, was the origin of the surface scatters.

Intrusive investigation took place in two adjoining plots on a north–south axis across the top of the Rodafnidia knoll (Fig. 8.4), which form a continuous strip of land, a transect across the top of the spur. The plots are named after their owners, Hatzoglou to the south of the dirt track and Alvanos to the north. Their surface survey during 2010 yielded numerous LCTs. The location of these two plots was crucial in that they allowed a geological assessment of the knoll at its widest point. Along this transect we laid out a series of fourteen

[2]The members of the 2012 campaign scientific team were as follows: James Cole, Giorgos Iliopoulos, Athanasios Katerinopoulos, Geoff King (Institut de Physique du Globe, Paris), Andreas Magganas, John McNabb, Ageliki Theodoropoulou (Institut de Paléontologie Humaine, Paris), Chronis Tzedakis (University College London), Katerina Vasileiadou; graduate students were as follows: Elli Karkazi, Thanos Rousis, Lena Kouklamani, Eleni Zervaki, Stefanos Fotinis (Univ. of Crete); and the undergraduate students were as follows: Ageliki Garidi, Elina Latsou, Eirini Saloustrou, Vaso Kourkouli (Univ. of Crete), Jeanine Curvers (Katholik Univ. of Leuven), and Roy Waterston (Univ. of York).

Fig. 8.5 Rodafnidia: view of the Hatzoglou property, test pits BI and B2 in the foreground and Trench A Extension, looking south-west

1×1 m test pits at regular 20 m intervals and gave them the Greek alphabet letters A to Ξ; of these we opened ten. Based on the results of the smaller test pit sampling strategy, we also opened three longer L-shaped trenches: Trench A Extension (11×1–3×1 m), Trench B Extension (7.5×1–3×1 m), and Trench H Extension (7×1–3×1 m) (Fig. 8.4). These were dug in order to expose large geological sections at key locations (Fig. 8.5), to allow sedimentological and geological sampling and to refine the preliminary geological interpretation of the site established via the test pits.

In September 2012, before leaving the site the majority of trenches and test pits were backfilled so that the plots could be returned to their owners as they were prior to excavations. Test pits Γ, Θ and M, the most significant in terms of stratigraphy, to which we wanted to have immediate access in the future for the purpose of stratigraphic referencing, paleomagnetic sampling and dating, were backfilled using geotextile and polystyrene blocks, topped by cobbles and loose earth from the excavation debris (Fig. 8.6).

We conducted additional surface surveys in both the excavated plots and in the surrounding areas (Fig. 8.7). Artifacts lying on the topsoil (the top of Unit 0, see below) were collected and their positions plotted using an RTK GPS (Fig. 8.8).

Stratigraphy and Date

The sedimentary series were exposed to a maximum depth of approximately 2.5–2.7 m. These stratigraphic sequences from different trenches excavated in the north and south

Fig. 8.6 Picture showing test pit backfilling in the Alvanos plot. In the foreground, test pit M backfilling using geo-textile and polystyrene blocks, topped by cobbles and loose earth from the excavation debris (September 2012)

Fig. 8.7 Surface surveying at the Alvanos plot (August 2012)

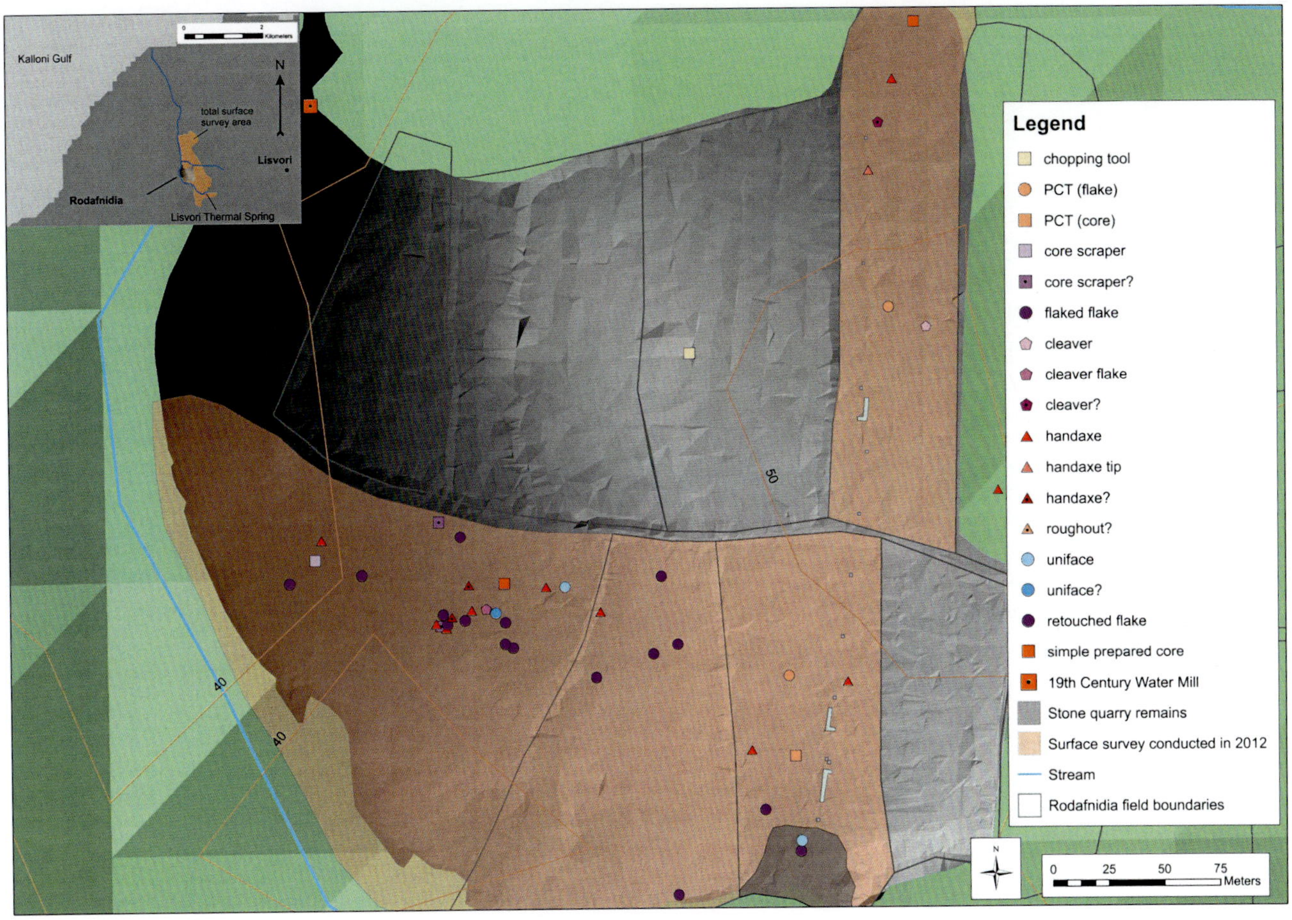

Fig. 8.8 Map showing the Rodafnidia knoll; in beige the 2012 surface survey area and location of surface finds collected

parts of the site have been correlated and are described as follows. Four sedimentary units (Units 0–3) were identified above both the weathered bedrock and the unaffected bedrock proper (Fig. 8.9).

- Unit 0 is the topsoil. It consists of brown silts with scattered rounded pebbles of relatively small sizes, and its upper part is loose due to farming activity.
- Unit 1 is a matrix-supported conglomerate, with red/brown silt as the matrix. In some places the matrix contains more sand, as well as rounded to sub-rounded pebbles and cobbles of various sizes. The larger cobbles have a diameter of about 20 cm.
- Unit 2 is a red-brown mud with calcitic nodules and numerous mud-cracks filled with calcium carbonate, particularly to the north of the site where the unit gets thicker.
- Unit 3, of which at least the top 20–30 cm were exposed, is a matrix-supported conglomerate with red silt as the matrix, accompanied by pebble-sized clasts.

Units 0 and 1 contained archaeological finds, whilst Unit 2 was barren. Unit 3 yielded no artifacts in 2012. The recovery of any archaeological finds in it will have to await fuller and deeper excavation. The lithology suggests a relatively small alluvial plain, which represents the depositional environment for the artifacts. Two types of deposit can be distinguished: floodplain and fluvial. The floodplain sediments, Unit 2, are red to red-brown muds with mud-cracks filled with carbonates (cracks were formed when the muds were exposed and dried, and during soil formation they were filled with carbonates). The fluvial deposits are the conglomerate accumulations of Unit 1 that characterise a fluvial network, be it river or stream, that shifted its course over time, eroding and forming new river beds that cut through the floodplain sediments deposited during older flood events.

This picture can be made out in most trenches. The exception is the northernmost excavated test pit M, the one closest to the present-day Kalloni Gulf shore. Here, green clay below Unit 0 indicates the presence of a still fresh-water deposit: a pond, a marsh, or a small lake (possibly an oxbow). The green color of the clay is a result of its reducing conditions. The presence of such still fresh-water bodies is a common feature found alongside fluvial systems developing across alluvial plains (Marriott 2006).

Sediment samples were collected for luminescence and TCN dating (conducted by Constantin Athanassas), for micro-

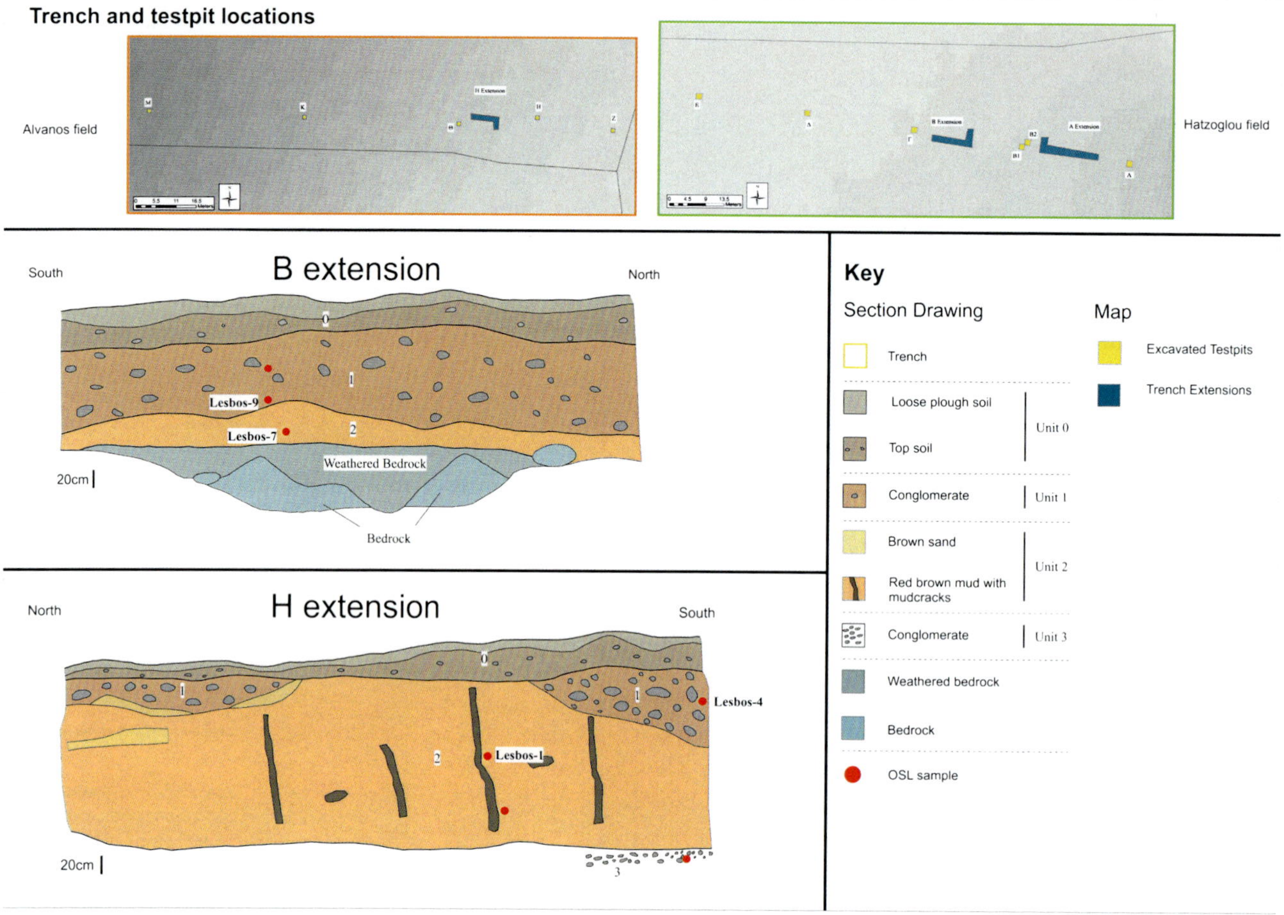

Fig. 8.9 Showing the stratigraphy of the Rodafnidia site exemplified through Trench B Extension (*middle*) and Trench H Extension (*bottom*). The origin of luminescence dating samples is denoted by red dots

fossil preparation (conducted by Katerina Vasileiadou), and from test pit M for palynological preparation (conducted by Chronis Tzedakis). The sediments from M were highly oxidised and did not preserve any fossil content apart from a few algae. More promising is the study of micro-fossil remains. Amongst the finds from the first sample (H extension), the presence of charophyte gyrogonite was noted; this stonewort calcareous spore, if contemporary with the sampled sediments, indicates a freshwater depositional environment.

For the luminescence dating method, the samples were collected from depths well below the ground surface and from soil profiles exposed in the trenches (Fig. 8.9). Cohesive layers were sampled in the daytime by inserting aluminium tubes into the sections, while loose sediments were sampled at night under dimmed-red portable lighting, by delving into the soil profile with a spade and sealing the extricated sediment into light-tight wrapping. Four of the samples were submitted to the Luminescence Dating suite of the Laboratory of Archaeometry at N.C.S.R. 'Demokritos', Athens.

The speculated Mid-Pleistocene age of the artifacts necessitated the employment of extended-range luminescence dating methods instead of conventional optically stimulated luminescence from quartz (OSL), as the latter was expected to be saturated on the time scales considered here. Examination of the material revealed a predominance of tectosilicate mineralogy, abundantly supplied by the extensive volcanic contexts of the wider area.

Quartz from volcanic environments has been proven unsuitable with respect to its luminescence properties (e.g. Bonde et al. 2001). For that reason thermally transferred OSL from quartz (e.g. Wang et al. 2007) was avoided; the abundance of feldspars instead dictated an infrared stimulated luminescence (IRSL) approach. For preliminary chronological evidence, fast track runs of the elevated-temperature IRSL protocol of Thiel et al. (2011) were carried out on the feldspars. This method, established as 'post infrared infrared stimulated luminescence' (pIRIR) dating, allows estimation of the paleodose by measuring the IRSL at 290°C. Furthermore, it has the purported advantage that it circumvents underestimations potentially induced by loss of signal from feldspars in ambient conditions, a phenomenon known as 'anomalous fading' (Wintle 1973; Spooner 1994).

Table 8.1 pIRIR dates obtained for the excavated sediments at Rodafnifia Unit 1 and Unit 2

Sample code	Trench	Unit	Depth below surface (m)	Age	MIS stage
Lesbos-4	H Extension	1	0.8–0.9	164 ± 33 ka	6
Lesbos-9	B Extension	1	1.2–1.4	258 ± 48 ka	8
Lesbos-1	H Extension	2	1.3	272 ± 25 ka (475 ± 48 ka)	9 (13)
Lesbos-7	B Extension	2	1.6–1.7	476 ± 62 ka	13

In the absence of *in situ* γ-dose rate measurements, dosimetry was limited to radio-elemental analyses by inductively coupled plasma mass spectrometry (ICP-MS).

Even though $pIRIR_{290}$ signal response was found not to be close to saturation, two of the delivered ages (Lesbos-4: 164 ± 33 kBP and Lesbos-9: 258 ± 48 kBP) appeared broadly spread (Table 8.1). Additionally, the age distribution of Lesbos-1 brought forth two clusters: one centred at 272 ± 25 kBP and a second at 475 ± 48 kBP. The latter shows, at least, consistency with the age of the sample Lesbos-7 (476 ± 62 kBP) (Table 8.1).

The bimodality seen in the distribution of these preliminary results raises the obvious question as to the origin of this behavior. It remains uncertain whether it is caused by environmental, anthropological, microdosimetric, or laboratory measurement conditions. In situations where adequate signal resetting can be evidenced by the environmental conditions, a series of post-depositional processes may be responsible for altering the paleodose of some grains, leading to skewed and multimodal age dispersal (Lomax et al. 2007). Employment of age modelling is therefore necessary here to explore different approaches to establishing the luminescence age.

The possibility that the observed scatter is a laboratory artifact due to the acceleration of measurement procedure cannot be ruled out in this case. Current ages were calculated using default settings proposed by Thiel et al. (2011), but it is recommended that the paleodose be measured over a range of temperatures to establish optimum measurement conditions of IRSL, and the maximum reproducibility of the paleodose. Additionally, the ages should be further tested for the presence of any anomalous fading in the $pIRIR_{290}$ signal.

Despite the spread and the methodological challenges, all $pIRIR_{290}$ results suggest a Middle Pleistocene age for Rodafnidia. The delivered ages for the samples from Unit 2 (Table 8.1) indicate that this unit might have been deposited during MIS stage 13 and thus during an interglacial period (interstadial). Conversely, the delivered ages for the samples from Unit 1, Lesbos-4 and Lesbos-9 (Table 8.1), suggest that Unit 1 in Trench H Extension was possibly deposited during MIS 6 (164 ± 33 kBP) and Unit 1 in Trench B Extension during MIS 8 (258 ± 48 kBP). Hence, despite the age difference, the sediments in both cases appear to have been deposited during glacial periods (stadials).

It is important to note that these observations are in agreement with the lithological character of the sampled units (Fig. 8.9). The fine grained Unit 2 must have been deposited during an interglacial period when sea level was significantly higher and the climate was wet enough with increased precipitation. On the other hand, the coarse grained deposits of Unit 1, as well as of Unit 3, which is located below Unit 2 in Trench H Extension, represent sediments deposited during glacial periods when sea level was lower and the climate drier. It should be noted that the deposits of Unit 1 appear mostly as lenses undercutting the sediments of Unit 2, for reasons explained later. We assume that during the stadial MIS 13 Rodafnidia was located closer to the sea shore (the paleo-Kalloni Gulf), thus representing a floodplain environment where marshes and temporary ponds would develop, allowing the presence of fresh water dwellers (charophytes, gastropods). Conversely, during glacial periods Rodafnidia became an elevated inland area where erosional processes would dominate. Hence fluvial systems would first develop, eroding the substrate, which in this case would be the sediments of Unit 2; these fluvial channels would be subsequently filled by fluvial coarse-grained deposits of Unit 1. Unit 1 might also represent coarse-grained fluvial deposits that were deposited in different fluvial networks formed during two different glacial periods, MIS 6 and 8, respectively. These are the find-bearing sediments. The artifacts must have accumulated originally in older sediments (units), possibly older than MIS 13, that were eroded upstream and were carried downstream through the fluvial channels to where they were finally deposited. Similarly, the coarse grained Unit 3 represents fluvial deposits formed during a glacial period before MIS 13.

In summary, based upon the surface survey collections and excavations, a working hypothesis has emerged: that the present surface material may have originated from the subsurface geological features. The presence of a buried channel, or possibly network of channels, across the knoll is suggested by the nature of the geological deposits with a spatial variation in the stratigraphy towards the Kalloni Gulf shore.

The Lithic Finds

The vast majority of Rodafnidia lithic artifacts recovered in 2012 were produced on chert of a wide range of colors. The most common hues are light brown and beige, while dark red-brown is occasionally present. Rarely, black, white, or translucent samples occur. The majority of these cherts are fossiliferous; macro and microfossils are included. Many of them present wood tissue and could be characterised as fossilised remains of plants. Others present faunal (mainly gastropod) macrofossils; these are endocasts of the original gastropod shells and provide determinations only to the genus level. In order to use fossils for accurate biostratigraphic datings, determinations to the species level are needed. In the future, thin sections need to be produced in order to identify and determine possible microfossils that will provide us with relative dating of the age of the rocks. This in turn would guide research into lithic raw material provenance.

Petrological analyses on siliceous raw materials recovered from Rodafnidia and the wider Lisvori—Polichnitos area suggest that cherts may have been formed either through chemical precipitation of SiO_2 from silica-rich fluids within a hydrothermal, possibly geyser-type environment, connected to the volcanism of the past; or by thermally induced diagenesis of the lacustrine siliceous limestones of Pliocene date and the siliceous marly limestones of the fresh-water swamp that occur close to Rodafnidia. Within both environments, biogenic (opal-A) and/or non-biogenic (opal-A′) silica was transformed to chert with microcrystalline quartz and chalcedony, through an intermediate stage of opal-CT (Stamatakis and Magganas 1988). This change is mostly due to the existing high heat flow in the area, while compaction and probably alkaline pore waters played a subordinate role (Kelepertsis 1993).

The presence of handaxes and cleavers indicates a Lower Paleolithic component to the Rodafnidia assemblage. Initial observations on the lithic finds collected during the 2010 surface survey suggested a broad similarity between artifacts from Rodafnidia and those from Kaletepe Deresi 3 in Cappadocia, central Anatolia (Slimak et al. 2008), Gesher Benot Ya'aqov in north Israel (Goren-Inbar and Saragusti 1996) and even from certain African assemblages, for example at Olduvai Gorge and elsewhere (Leakey and Roe 1994; Sharon 2007). In light of this, a variation of the methodology applied at the South African Acheulian sites of the Cave of Hearths and Canteen Koppie (McNabb et al. 2004; McNabb and Sinclair 2009; McNabb and Beaumont 2012) was used to conduct a preliminary study of the artifacts. The major technogroups identified in Rodafnidia lithics are (Table 8.2) as follows: (1) Large Cutting Tools, (2) Prepared Core Technology (PCT), (3) Non-PCT Flake Cores, (4) Flakes and Detached Pieces, and (5) Retouched Flakes.

Table 8.2 Main artifact types found at Rodafnidia in the 2010 and the 2012 campaigns

Artifact type	Artifact provenance			
	2012 excavation	2010 surface survey	2012 surface survey	Total
Blade	2	–	–	**2**
Core	74	20	111	**205**
Core (tool)	–	1	2	**3**
Discoidal core	–	–	1	**1**
Flake	240	30	164	**434**
Flaked flake	–	1	11	**12**
LCT	5	5	20	**30**
PCT (core)	2	–	2	**4**
PCT (flake)	–	–	2	**2**
Retouched Flake	–	1	7	**8**
Scraper	–	–	2	**2**
Simple prepared core	–	–	2	**2**
Total	**323**	**58**	**324**	**705**

Table 8.3 Counts of LCT types recovered from Rodafnidia in 2010 and 2012

LCT type	Artifact provenance			
	2012 excavation	2010 surface survey	2012 surface survey	Total
Cleaver	–	–	1	**1**
Cleaver flake	1	–	–	**1**
Cleaver?	–	1	1	**2**
Handaxe	2	3	12	**17**
Handaxe tip	–	–	1	**1**
Handaxe?	–	–	2	**2**
Rough-out	–	1	–	**1**
Trihedral	1	–	–	**1**
Uniface	1	–	2	**3**
Uniface?	–	–	1	**1**
Total	**5**	**5**	**20**	**30**

The Large Cutting Tools Technogroup

The 2012 database (for the preliminary investigations and the first systematic survey and excavation season) records a total of 30 Large Cutting Tools (LCT) (Table 8.3).

Handaxes are tools with a converging tip that have been wholly or partially made by bifacial thinning and shaping. They lack the flat/guillotine-shaped cutting edge (cleaver bit), which is characteristic of cleavers. The database records 16 whole handaxes, two broken tip fragments, and two further examples whose identification is less certain (a selection is shown in Fig. 8.10).

Unifaces represent handaxes with converging tips where all, or virtually all, of the thinning and shaping is confined to one face of the tool. There are three unifaces and a possible fourth in the database (Table 8.3). Trihedrals are tools that have a triangular shape in cross-section. This is either a result of the original cobble/nodule form, or of flaking on an unusually thick nodule/cobble. In some instances trihedrals may result from flaking on an unusually thick natural or struck flake. One clear example of a trihedral was found at Rodafnidia (Table 8.3; Fig. 8.11c). Rough-outs are large flat artifacts with a small number of flake removals on each face. They tend to be flat in cross-section. It is often difficult to distinguish these from ordinary cores, which happen to be

Fig. 8.10 (**a–e**) Five examples of the handaxe component for the Rodafnidia LCTs

made on a flat nodule. Their status as an unfinished LCT is therefore a subjective call. Two examples were identified at Rodafnidia (Table 8.3; Fig. 8.11d).

Cleavers are defined on a number of criteria (Mourre 2003). (1) The presence of a clear flat cleaver edge or bit. (2) They are made on flake blanks. (3) The flake blanks show evidence of preparation of the core prior to the detachment of the flake. This evidence takes the form of large primary flake scars whose point of origin, where discernable, originates well beyond the current margins of the cleaver. (4) Adjacent to the cleaver bit, on the dorsal face, is a large flat flake scar, which represents one of the original blank scars just noted. With regard to criteria two to four, these cleavers conform to Sharon's Large Flake Acheulian (2007). (5) Where it is possible to observe, the flake blanks are side-struck and the lateral margins of the cleaver (i.e. the proximal and distal of the original flake-blank) have been removed. Criteria two to five represent a common pattern in the African Acheulian and one of us (JM) has seen numerous identical examples from the Middle Pleistocene of South Africa. So much so, that the cleavers at Rodafnidia could be said to conform precisely to the 'Acheulian package' noted elsewhere (Sharon 2008, 2009; McNabb and Sinclair 2009). The flaking away of the proximal area, accompanied by some thinning and shaping of the former distal end of the blank, to form the cleaver sides, is particularly diagnostic. The definition of cleavers adopted here is, primarily, a technological one. The database at the end of the 2012 season[3] recorded one certain (Fig. 8.12a) and one possible example of these cleavers (Fig. 8.12c), though the latter may be a broken handaxe. A number of large flakes from Rodafnidia would be suitable for LCT blanks. One in particular is highly suggestive of a flake blank from the prepared surface of a core or boulder (Fig. 8.12b). It was recovered from the excavated Unit 1 of Trench B Extension.

Cleavers represent one of the most interesting aspects of the Rodafnidia Acheulian lithic assemblage. In that a number

[3] As the paper goes to print in 2016, four additional seasons in the field have brought to light a larger sample of LCTs and cleavers.

Fig. 8.11 Four examples of Lower Paleolithic artifacts from Rodafnidia. (**a**) Scraper; (**b**) Single platform core or massive scraper; (**c**) Trihedral; (**d**) Rough-out

Fig. 8.12 Three examples of the cleaver component for the Rodafnidia LCTs. (**a**) Cleaver on a side struck flake; (**b**) Cleaver flake; (**c**) Cleaver broken at tip or handaxe broken medially

of cleavers are made on pre-prepared flake blanks, they are closer in concept to the technologically defined Gesher Benot Ya'aqov (GBY) cleavers (Goren-Inbar and Saragusti 1996), than to other non-technologically defined assemblages elsewhere in the Near East. Cleavers may be defined on morphometric (Leakey and Roe 1994), or on typological grounds (Kleindienst 1962; Wymer 1961). Even a brief perusal of some of the Acheulian literature from Near Eastern sites suggests, from the illustrations, that many of the LCTs described as cleavers are in fact ovate handaxes with transverse tranchet blows resulting in square-ended handaxes (the tips are plainly convergent and are made by bifacial thinning and shaping prior to the tranchet blow). We would restrict the definition of cleavers to the purely technological definition set forth here, and so see all convergent tips with square ends, whether made by thinning and shaping detachments, or by tranchet finish, as handaxes (narrow square-ended). Irrespective of this, the technologically defined cleavers from Rodafnidia present an interesting problem. The large flake-blank cleaver, or Acheulian package as described here (Sharon 2007; McNabb 2009), is normally associated with intractable lithologies such as andesite in South Africa or the volcanics of East Africa and GBY. On more tractable rock types, particularly those that knap like siliceous rocks, the preforming of flake blanks is usually unnecessary. A future research question for our project will be to examine why Acheulian knappers at Rodafnidia resorted to this package, when the chert, which is a common lithology at the site, knaps so well and did not require it?

While there is a strong African flavor to the Acheulian assemblage at Rodafnidia originating from Units O and 1, it is important to remember that any Acheulian settlement of Lesbos will have originated from mainland Anatolia. Kaletepe Deresi 3 is the only excavated Acheulian site in Turkey (Slimak et al. 2008; Dinçer 2016). Located on a bank of a seasonal drainage in the Göllüdağ region of central Anatolia (well known for its obsidian sources), it has revealed assemblages manufactured on obsidian and andesite with a strong affinity to the Large Flake Acheulian described by Sharon (2007). Since Rodafnidia also falls within this group of assemblages, a key future research goal will be to compare these two assemblages.

The Prepared Core Technology Technogroup

From Unit 1 we also recovered artifacts, smaller than LCT flake-blanks, which demonstrate clear preparation of a surface prior to flaking. The Prepared Core Technology (PCT) technogroup can encompass Levallois, as well as other forms of PCT such as Victoria West (McNabb and Beaumont 2011, 2012), Kombewa and simple prepared cores (McNabb and Sinclair 2009), as well as the Tabelbala Tachengit technique noted at Kaletepe Deresi 3 (Slimak et al. 2008; Dinçer 2016). What unites them as a group is that one surface on the core or flake will be considered more important than the other, and from this preferential surface flakes or a single flake will be removed. This is irrespective of whether preparation or flaking of that surface has been conducted or not. Thus, the common thread here is that all these artifacts are conceived of as possessing a hierarchical relationship between their upper and lower halves.

The two forms of classic Levallois present at Rodafnidia are as follows:

- Radial/centripetal. In practice the cores need not always be circular in their plan form. One example of a radial core in a worn state was found on the surface during field walking (Fig. 8.13a).
- Convergent/point. Two examples of Levallois convergent cores were found (Table 8.3), both worn, the latter a surface find possibly made on a flake. A single example of an atypical Levallois convergent point was recovered from one of the test pits, and a second atypical point was found on the surface during field walking (Fig. 8.13b, c).

Additionally, there exists the artifact category known as 'Simple Prepared Cores' that represent a form of 'Stripped Down Levallois' (White and Ashton 2003). They conform to a number of the rules for Levallois as identified by Boëda (Boëda 1995). However, a carefully prepared surface and careful maintenance of lateral and distal convexities is not practised (White and Ashton 2003). Two such simple prepared cores were discovered during surface survey collection (Table 8.2).

For many archaeologists the presence of PCT, and Levallois in particular, signals the Middle Paleolithic (Clark 1994, 1999). However, the temporal boundary between these periods based on tool typology is becoming blurred (McBrearty 2001, 2003; Shea 2006; Beaumont 2011). There are a number of sites from Africa where Acheulian artifacts are clearly contemporary with Levallois and other forms of PCT, e.g. the Kapthurin Formation, Kenya (Tryon et al. 2005), and Canteen Koppie, South Africa (McNabb and Beaumont 2012). This has been noted elsewhere. Currently, the presence of Levallois may signal an Acheulian with PCT or the presence of a Middle Paleolithic assemblage. A major question for future research will be determining the relationship between the LCTs and the PCTs.

The Non-PCT Flake Cores Technogroup

A number of flake core morphologies persist throughout the African Early Stone Age (Leakey 1971; Kuman 2007) and

Fig. 8.13 Examples of PCT artifacts from Rodafnidia. (**a**) Radial Levallois core; (**b**) Convergent Levallois point (atypical); (**c**) Convergent Levallois point

can be found in other parts of the Old World where similar ranges of raw materials occur. These forms are choppers/chopping tools, discoids/discoidal cores, single platform cores (including the typological category of core scraper), and polyhedrons and irregular polyhedrons (McNabb and Sinclair 2009; McNabb and Beaumont 2011). How culturally diagnostic, or indicative of a specific period (say Lower Paleolithic but not Middle Paleolithic) they are, is open to debate. Raw material considerations may weigh heavily in the choice of knapping procedures. Furthermore, enigmatic types such as spheroids/sub-spheroids are also reported from Early Stone Age sites (Kleindienst 1962; Leakey 1971, 1979).

A single example of a chopping tool (Table 8.2) is present in the Rodafnidia database, made of a coarse-grained lava and found on the surface during field walking. A number of other cores resemble chopping tools, but their morphology is not sufficiently diagnostic for a confident interpretation. There were two typological discoids (Table 8.2), both made by alternate flaking. There is one example of a single platform core, which would class typologically as a core scraper (Table 8.2; Fig. 8.11b); as well as a spheroid made on lava (Table 8.2).

The Flakes and Detached Pieces Technogroup

A large number of flakes were recovered during field walking and during the excavation of the test pits and the extension trenches. The majority are un-diagnostic waste flakes. Some possess dihedral butts, but this need not be an indicator of the Middle Paleolithic alone. Some of the larger ones may have been used as cores.

The Retouched Flake Technogroup

The Retouched and Modified Flake technogroup is divided into two broad sub-groups. The first are the flaked flakes. These are a common Lower Paleolithic tool type in which a flake of any size is flaked again by one or more removals. They are not cores, since the intent appears to be the modification of the edge (Ashton et al. 1991). Our database records 12 examples (Table 8.2), all having the razor-sharp edges produced by this technique. Technologically they are similar to those found at other Lower Paleolithic/ESA sites, with removals being single or multiple, direct or inverse, proximal, distal or from the laterals. The Retouched Flake group encompasses scrapers, made on flakes (Fig. 8.11a) as well as on unworked pieces, three denticulates, an un-diagnostic retouched point, a possible wedge, and a potential awl.

Summary of Lithic Artifact Analysis

The artifacts recovered from controlled excavations (Unit 1) and from systematic field walking around the excavated fields (Unit 0) clearly demonstrate the presence of the Acheulian at Rodafnidia. The Trench B Extension confirmed that diagnostic Lower Paleolithic artifacts were present within the channel fills of Unit 1. It is possible that Middle Paleolithic groups associated with Levallois technology were also present there. Given the very small numbers of diagnostic artifacts, the presence of the latter remains to be established by further research.

Discussion

Lower Paleolithic sites in the north-east Mediterranean Basin are sparse and discontinuous due to archaeological research traditions and priorities in the countries involved (but see Tourloukis 2016 for a geoarchaeological perspective). They are found in a variety of settings, in caves or in the open air, and associated with good quality lithic raw materials and larger or smaller bodies of fresh water. Together they produce a fragmentary picture of a number of hominin dispersal episodes at different times of the Early and the Middle Pleistocene. In Turkey and in Greece, bifacial technology is better known from material recovered on the surface rather than through excavation (Galanidou et al. 2016, appendix 1; Dinçer 2016). Working with material deriving from open contexts presents us with a number of problems in discussing the character and presence of Acheulian groups.

Recently, Kuhn (2010a) set out the research questions and current status of Lower Paleolithic research in Anatolia, and in this volume Dinçer offers an updated comprehensive account of the evidence available. The issues here concern the quality and the affordances of the record, which stem from recovery conditions and procedures, rather than the absolute numbers of the sites and the finds reported in the large expanses of Anatolia. Out of a total of 170 Lower Paleolithic sites documented in the Archaeological Settlements of Turkey Project database (www.tayproject.org), only a handful can be included in detailed paleoanthropological discussion. In central Anatolia, the two major reference sites are situated at altitudes higher than 1000 m (Dinçer 2016). Kaletepe Deresi 3 near Cappadocia is the only excavated site with what is described as a 'geologically *in situ* Acheulian component' and a Middle Paleolithic component overlying it (Slimak et al. 2008; Dinçer 2016). Its Acheulian component provides the nearest published comparanda to the Rodafnidia Acheulian assemblage. Further west, Dursunlu, located in a lignite quarry, has yielded less than 30 quartz artifacts, mostly flakes and flake tools, and associated faunal remains. These, coupled with paleomagnetic dating, document a hominin presence here in the Early Pleistocene, probably sometime around or post 1 Ma (Güleç et al. 2009).

In Aegean Turkey, the Lower Paleolithic inventory includes part of a *Homo erectus* skull found embedded in a travertine block in a quarry near Kocabaş in the province of Denizli, recently dated to around 1.1 Ma (Kappelman et al. 2008; Lebatard et al. 2014; Aytek and Harvati 2016), and the odd site containing bifaces, big flakes, or chopping tools (Dinçer 2016). The picture of Lower Paleolithic in Turkey is completed with two reference cave sites in the far south and the far north of the country. The earliest component of the 11-m-long and impressive Karain Cave sequence (Otte et al. 1998, 1999) on the Mediterranean coast south of the Taurus Mountain consists of Clactonian flakes, denticulates, and three bifaces manufactured on a variety of radiolarite, flint, and calcareous stones. In European Turkey, Yarimburgaz Cave contains Middle Pleistocene deposits and a lithic assemblage with Middle Paleolithic affinities (Arsebük 1993; Arsebük and Özbaşaran 1999; Kuhn 2003, 2010b). Yarimburgaz Cave may be combined with open-air sites of low chronological resolution that contain choppers, chopping tools, and other lithics (Dinçer 2016) to make up the patchy and enigmatic record of Turkish Thrace.

In the southern part of the Balkan peninsula, the Lower Paleolithic inventory numbers less than a handful of sites or find spots, presenting fewer than a dozen Large Cutting Tools (Harvati et al. 2009; Galanidou 2004, 2014a, b; Panagopoulou et al. 2015). Three are key sites. First is Petralona cave in Macedonia, yielding a precious *Homo heidelbergensis* cranium (Henning et al. 1982; Grün 1996; Harvati 2009) and early artifacts, though neither can be directly associated with the other (Darlas 2014).

The second key site is Kokkinopilos, an ancient wetland site in the karstic landscape of Epirus with eroding *terra rossa* deposits out of which three impressive flint LCTs originate: an elongated 'Micoquian handaxe' (Runnels and Van Andel 1993a) and two more bifaces (Tourloukis 2009, 2016). Of these latter pair, one may be considered to have Acheulian affinities, but the other, originating from a stratified context, has affinities with the Keilmesser group, and so perhaps may be part of the Kokkinopilos Middle Paleolithic component (Galanidou et al. 2016). At Kokkinopilos an overlap, in chronostratigraphic terms, between the Acheulian and an early Mousterian may be envisaged. This is concordant with Tourloukis and Karkanas' (2012) description of the site as 'a low energy depositional environment of a shallow lake formed in a tectonic basin, at times drying out' (2012: 4). Serious attempts to come to grips with its stratigraphy and dating have taken place, especially in the undisturbed localities. Luminescence dating of the biface-bearing sediments suggests minimum ages between 207 and 220 kBP (ibid.). These match the Runnels and van Andel (1993a, 198) date calculated from the rate of sedimentation, namely 250 kBP, proposed for the context of the first handaxe. The two lines of evidence combined give us confidence that hominin presence at Kokkinopilos began during the late Middle Pleistocene. The question that remains open is who these hominins were (Galanidou 2016).

The third Lower Paleolithic site is Marathousa 1 in the Megalopolis basin, an area long-known for its rich paleontological yield. Archaeological excavation on this Middle Pleistocene site has produced the remains of *Elephas (Palaeoloxodon) antiquus* and lithics (mainly flakes and various kinds of fragments though no bifacially worked specimens) in a fine-grained geological matrix (Panagopoulou et al. 2015). Ongoing research is expected to clarify the depositional history of these remains.

Beyond these three instances, if one adds in Lenormant's (1867) nineteenth-century reference to a biface claimed to originate from the Megalopolis basin in the Peloponnese, the handaxe discovered by Eric Higgs (1964) in west Macedonia during his first expedition to Greece in the 1960s, as well as two more sites in the Peneios River, Thessaly (Runnels and van Andel (1993b) and Nea Artaki, Euboea (Sarantea 1986) whose assignment to the Lower Paleolithic sites is possible yet lacks secure chronostratigraphic confirmation, one has enumerated the whole of the scanty Lower Paleolithic record of mainland Greece.

Claims for Lower Paleolithic finds from a handful of island sites that include Milos, Aegean Sea (Chelidonio 2001), Kefallonia, Ionian Sea (Foss 2002), Loutro and Plakias in south Crete, Libyan Sea (Mortensen 2008; Strasser et al. 2010), and Gavdos, Libyan Sea (Kopaka and Matzanas 2009, 2011), are based on artifact typology, as most finds derive from low-resolution surface collections. At Plakias they are also based on dating the associated geological contexts (Strasser et al. 2011); unequivocal evidence, however, is absent, since association of the claimed early finds with the better dated geological contexts has not been adequately shown (Galanidou 2014a, b).

Rodafnidia is unique for the richness of an Acheulian lithic assemblage, which has to date no counterpart within the Lower Paleolithic of the Aegean Turkey or Greece (Galanidou 2013; Galanidou et al. 2013). The site therefore stands out as an exciting target for enriching the Lower Paleolithic record of the north-east Mediterranean and for obtaining dates for Acheulian activity. The importance of this site and thence of our project lies in (i) *the time* and duration of the hominin and human presence, with pIRSL results suggesting that the upper part of the excavated sequence dates to the Middle Pleistocene. (ii) *The size* of the entire site—the Acheulian site may be extensive: explaining the archaeological assemblage distribution forms a central focus for future research. And (iii) *the geography* of the site which has a critical element to it: both on a local scale, in a fluvio-lacustrine environment of the Kalloni basin right by geothermal springs, and on a regional scale, with its central geographical position on the border of two continents and the heart of Eurasia. The proximity of Lesbos to Anatolia makes Rodafnidia a key site in the attempt to comprehend both hominin migration into Europe (Bar-Yosef and Belfer-Cohen 2001; Moncel 2010), as well as Acheulian occupation northwards of the Jordanian Rift Valley (Dennell et al. 2011; Goren-Inbar et al. 2000; Otte et al. 1999; Lordkipanidze et al. 2000). Further systematic exploration of the site will furnish research into human origins with archaeological data to address the role of these two key Eurasian regions, either as areas of occupation and stasis, or as mere passageways in hominin dispersals during the Middle Pleistocene.

Acknowledgments The first year of systematic research at Rodafnidia, Lisvori was supported by the University of Crete, the Secretariat General for the Aegean and Island Policy and the authority of the North East Aegean Region. We are grateful to the Municipality of Lesbos for its support and lab space granted. Doukas and Toula Alvanou and Theodoros Hatzoglou granted us permission to excavate their properties in 2012, and for this they deserve our special thanks. We thank the Editors and anonymous reviewers for the helpful comments and suggestions.

References

Arsebük, G. (1993). Yarımburgaz, a Lower Paleolithic cave site near Istanbul. In M. Frangipane, H. Hauptman, M. Liverani, P. Matthiac, & M. Mellink (Eds.), *Between the rivers and over the mountains, archaeologica Anatolica et Mesopotamica* (pp. 23–36). Roma: Gruppo Editoriale Internazionale.

Arsebük, G., & Özbaşaran, M. (1999). Pleistocene archaeology at the cave of Yarımburgaz in Eastern Thrace/Turkey: Preliminary results. In G. N. Bailey, E. Adam, E. Panagopoulou, C. Perlès, & K. Zachos (Eds.), *The Palaeolithic archaeology of Greece and adjacent areas* (pp. 59–72). Nottingham: The British School at Athens.

Ashton, N., Dean, P., & McNabb, J. (1991). Flaked flakes: What, Where, When and Why? *Lithics, 12*, 1–11.

Aytek, I. A., & Harvati, K. (2016). The human fossil record from Turkey. In K. Harvati, & M. Roksandic (Eds.), *Paleoanthropology in the Balkans and Anatolia: Human evolution and its context*. Dordrecht: Springer.

Bar-Yosef, O., & Belfer-Cohen, A. (2001). From Africa to Eurasia—Early dispersals. *Quaternary International, 75*, 19–28.

Beaumont, P. (2011). The edge: More on fire making by about 1.7 million years ago at Wonderwerk Cave in South Africa. *Current Anthropology, 52*, 585–594.

Boëda, E. (1995). Levallois: A volumetric construction, methods, a technique. In H. L. Dibble & O. Bar-Yosef (Eds.), *The definition and interpretation of Levallois technology* (pp. 41–68). Madison, Wisconsin: Prehistory Press.

Bonde, A., Murray, A. S., & Friedrich, W. L. (2001). Santorini: Luminescence dating of a volcanic province using quartz. *Quaternary Science Reviews, 20*, 789–793.

Chelidonio, G. (2001). Manufatti litici su ciottolo da Milos (isole cicladi): Nota preliminare. *Pegaso: Rivista annuale di cultura mediterranea, 1*, 117–144.

Clark, J. D. (1994). The Acheulean industrial complex in Africa and elsewhere. In R. S. Corruccini & R. L. Ciochon (Eds.), *Integrative paths to the past. Paleoanthropological advances in honour of F. Clark Howell* (pp. 451–469). New Jersey: Prentice Hall.

Clark, J. D. (1999). Cultural continuity and change in hominid behavior in Africa during the Middle to Upper Pleistocene transition. In H. Ullrich (Ed.), *Hominid evolution. Lifestyles and survival strategies* (pp. 277–292). Archaea: Gelsenkirchen.

Darlas, A. (2014). *Researching the identity and importance of the Petralona cave in Greek and European prehistory*. In A century of Research in prehistoric Macedonia 1912–2012, Proceedings of International Conference Archaeological Museum of Thessaloniki November 2012 (pp. 195–204) (Original in Greek).

De Vos, J., Van der Made, J., Athanassiou, A., Lyras, G., Sondaar, P., & Dermitzakis, M. D. (2002). Preliminary note on the Late Pliocene fauna from Vatera (Lesvos, Greece). *Annales Géologiques des Pays Hélléniques, 1e Série, 39*(A), 37–70.

Dennell, R. W., Marinón-Torres, M., & Bermúdez de Castro, J. M. (2011). Hominin variability, climatic instability and population demography in Middle Pleistocene Europe. *Quaternary Science Reviews, 30*, 1511–1524.

Dinçer, B. (2016). The Lower Paleolithic in Turkey: Anatolia and hominin dispersals. In K. Harvati, & M. Roksandic (Eds.), *Paleoanthropology of the Balkans and Anatollia: Human evolution and its context* (pp. 213–228). Dordrecht: Springer.

Foss, P. (2002). Kephallenia: Archaeology and history: The ancient Greek cities. In K. Randsbourg (Ed.), *Acta Archaelogica* (Vol. 73). Copenhagen: Blackwell Munksgaard.

Galanidou, N. (2004). Human presence in the Balkans during the Palaeolithic. In H. J. Griffiths, J. Reed, & B. Kristufek (Eds.), *Balkan biodiversity: Pattern and process in the European hotspot* (pp. 157–175). Dordrecht: Kluwer Academic Publishers.

Galanidou, N. (2013). Looking for the earliest occupants of the Aegean—Palaeolithic excavations at Rodafinidia, Lisvori, Lesvos. In M. Alvanou (Ed.), *Island identities* (pp. 15–17). Mytilene: Secretariat General for the Aegean and Island Policy.

Galanidou, N. (2014a). Advances in the Palaeolithic and Mesolithic archaeology of Greece for the new millenium. *Pharos, 20*(1), 1–40.

Galanidou, N. (2014b). Archaic hominins on Crete: Fact or fiction. *Journal of Mediterranean Archaeology, 27*(2), 260–267.

Galanidou, N. (2016). Before ancient Nericos, What? Palaeolithic finds from Karyotes Lefkas. In C. Papadatou-Yannopoulou (Ed.), *Nericos, Lefkas, Castro. The longest-living capital of Lefkas*. Proceedings of a Conference held at Lefkas in August 2010. Lefkas: Cultural Centre of the Lefkas Municipality 83–111.

Galanidou, N., Cole, J., Iliopoulos, G., & McNabb, J. (2013). East meets West: The Middle Pleistocene site of Rodafnidia on Lesvos, Greece. *Antiquity Project Gallery, 87*(336). http://antiquity.ac.uk/projgall/galanidou336/.

Galanidou, N, Papoulia, C., & Ligovanlis, S. (2016). The Middle Palaeolithic bifacial tools from Megalo Karvounari. In B. Forsén, N. Galanidou, & E. Tikkala (Eds.), *The Thesprotia Expedition vol. III*. Helsinki: Papers and Monographs of the Finnish Institute at Athens.

Goren-Inbar, N., Feibel, C. S., Versoub, K. L., Melamed, Y., Kislev, M. E., Tchernov, E., & Saragusti, I. (2000). Pleistocene milestones on the Out-of-Africa corridor at Gesher Ya'aqov, Israel. *Science, 289*(5481), 944–947.

Goren-Inbar, N., & Saragusti, I. (1996). An Acheulian biface assemblage from the site of Gesher Benot Ya'aqov, Israel: Indications of African affinities. *Journal of Field Archaeology, 23*, 15–30.

Grün, R. (1996). A re-analysis of electron spin resonance dating results associated with the Petralona hominid. *Journal of Human Evolution, 30*, 227–241.

Güleç, E., White, T., Kuhn, S., Özer, I., Sağır, M., Yılmaz, H., & Howell, F. C. (2009). The Lower Pleistocene lithic assemblage from Dursunlu (Konya), central Anatolia, Turkey. *Antiquity, 83*(319), 11–22.

Harisis, H. B., Durand, P., Axiotis, M., & Harisis, T. B. (2000). Traits of Palaeolithic settlement on Lesbos. *Archaeology and Arts, 76*, 83–87. Original in Greek.

Harvati, K. (2009). Petralona: Link between Africa and Europe? In L. A. Schepartz, S. C. Fox, & C. Bourbou (Eds.), *New directions in the skeletal biology of Greece* (Hesperia Supplement, Vol. 43, pp. 31–47). Princeton: American School of Classical Studies at Athens.

Harvati, K., Panagopoulou, E., & Runnels, C. (2009). The paleoanthropology of Greece. *Evolutionary Anthropology, 18*(4), 131–143.

Hecht, J. (1974). *Geological map of Greece, 1/50Π000, 'Polichnitos' sheet*. Athens: Institute of Geology and Mineral Exploration. Original in Greek.

Henning, G. J., Herr, W., Weber, E., & Xirotiris, N. I. (1982). Petralona cave dating controversy. *Nature, 299*, 281–282.

Higgs, E. S. (1964). A hand axe from Greece. *Antiquity, 38*, 54–55.

Jöris, O. (2014). Early Palaeolithic Europe. In C. Renfrew & P. Bahn (Eds.), *The Cambridge world prehistory* (Vol. III, pp. 1703–1746). Cambridge: Cambridge University Press.

Kappelman, J., Alçiçek, M. C., Kazanci, N., Schultz, M., Özkul, M., & Sen, S. (2008). First *Homo erectus* from Turkey and implications for migrations into temperate Eurasia. *American Journal of Physical Anthropology, 135*(1), 110–116.

Kelepertsis, A. (1993). The mineralogy—Geochemistry of pliocene lacustrine sediments of the polichnitos area, Lesvos Island, Greece and the formation of silica polymorphs. *Mineral Wealth, 83*, 25–34.

Kleindienst, M. R. (1962). Components of the East African Acheulean assemblage: An analytic approach. In G. Mortelmans (Ed.), *Actes du IVème Congrès Panafricain de Préhistoire et de l'Étude du Quaternaire* (Vol. 40, pp. 81–105). Tervuren: Musée royal de l'Afrique centrale.

Kopaka, K., & Matzanas, C. (2009). Palaeolithic industries from the island of Gavdos, near neighbour to Crete in Greece. *Antiquity Project Gallery, 83*(321). http://www.antiquity.ac.uk/antiquitynew/projgall/kopaka321/.

Kopaka, K., & Matzanas, C. (2011). Early marine travels in the Aegean and Crete. Reflections departing from the knapped-stone industries from Gavdos island. In M. Andreadaki-Vlazaki (ed.) Proceedings of the 10th International Cretological Conference, Chania, 1-8 October 2006, A1, 43-82. (in Greek). Chania, Chrysostomos Scholar Society.

Kouli, M., & Seymour, S. K. (2006). Contribution of remote sensing techniques to the identification and characterization of Miocene calderas, Lesvos Island, Aegean Sea, Hellas. *Geomorphology, 77*(1–2), 1–16.

Kuhn, S. L. (2003). Flexibility and variation in the Lower Paleolithic: A view from Yarımburgaz Cave. In M. Özbaşaran, O. Tanındı, & A. Boratav (Eds.), *Archaeological essays in honour of Homo amatus: Güven Arsebük* (pp. 149–157). İstanbul: Ege Yayınları.

Kuhn, S. L. (2010a). Was Anatolia a bridge or barrier to early hominin dispersals? *Quaternary International, 223–224*, 434–435.

Kuhn, S. L. (2010b). The Yarımburgaz lithic assemblage. In F. C. Howell, G. Arsebük, S. L. Kuhn, M. Özbaşaran, & M. C. Stiner (Eds.), *Culture and biology at the crossroads: The Middle Pleistocene record of Yarımburgaz Cave (Thrace, Turkey)* (pp. 93–129). İstanbul: Ege Yayınları.

Kuman, K. (2007). The earlier Stone Age in southern Africa: Site context and the influence of cave studies. In T. Pickering, K. Schick, & N. Toth (Eds.), *Breathing life into fossils. Taphonomic studies in honour of C.K. (Bob) Brain* (pp. 181–198). Bloomington, Indiana: Stone Age Institute Press.

Lambrakis, N. J., & Stamatis, G. N. (2008). Contribution to the study of thermal waters in Greece: Chemical patterns and origin of thermal water in the thermal springs of Lesvos. *Hydrological Processes, 22*, 171–180.

Lamera, S. (2004). *The Polichnitos ignimbrite of Lesvos island*. Ph.D. Dissertation, University of Patras. (Original in Greek).

Leakey, M. D. (1971). *Olduvai Gorge: Excavations in Beds I and II 1960–1963*. Cambridge: Cambridge University Press.

Leakey, M. D. (1979). *Olduvai Gorge. My search for early man*. Glasgow: Book Club Associates for William Collins Sons and Co.

Leakey, M. D., & Roe, D. A. (1994). *Olduvai Gorge Volume V; Excavations in Beds II, IV and the Msek Beds 1968–1971*. Cambridge: Cambridge University Press.

Lebatard, A. E., Alçiçek, M. C., Rochette, P., Khatib, S., Vialet, A., Boulbes, N., et al. (2014). Dating the *Homo erectus* bearing travertine from Kocabaş, (Denizli, Turkey) at least 1.1 Ma. *Earth and Planetary Science Letters, 390*, 8–18.

Lenormant, F. (1867). L' âge de la pierre en Grèce. *Revue Archéologique, I*, 16–19.

Lomax, J., Hilgers, A., Twidale, C. R., Bourne, J. A., & Radtke, U. (2007). Treatment of broad palaeodose distributions in OSL dating of dune sands from the western Murray Basin, South Australia. *Quaternary Geochronology, 2*, 51–56.

Lordkipanidze, D., Bar-Yosef, O., & Otte, M. (2000). *Early humans at the gates of Europe*. Proceedings of the First International Symposium, Dmanisi, Tbilisi (Georgia) September 1998. Liège: Etudes et Recherches Archéologiques de l'Université de Liège (E.R.A.U.L.).

Lykousis, V. (2009). Sea-level changes and shelf break prograding sequences during the last 400 ka in the Aegean margins: Subsidence rates and palaeogeographic implications. *Continental Shelf Research, 29*(16), 2037–2044.

Lyras, G. A., & van der Geer, A. A. E. (2007). The Late Pliocene vertebrate fauna of Vatera (Lesvos Island, Greece). *Cranium, 24*(2), 11–24.

Marriott, S. B. (2006). Floodplain. In A. S. Goodie (Ed.), *Encyclopedia of geomorphology* (Vol. 1, pp. 381–384). New York: Routledge.

McBrearty, S. (2001). The Middle Pleistocene of East Africa. In L. S. Barham & K. Robson-Brown (Eds.), *Human roots. Africa and Asia in the Middle Pleistocene* (pp. 81–98). Bristol: Western Academic and Specialist Press.

McBrearty, S. (2003). Patterns of technological change at the origin of *Homo sapiens. Before Farming, 3*(9), 1–5.

McNabb, J. (2009). The ESA stone tool assemblage from the Cave of Hearths, Beds 1–3. In J. McNabb & A. Sinclair (Eds.), *The cave of hearths: Makapan middle pleistocene research project: Field research by Anthony Sinclair and Patrick Quinney* (University of Southampton Series in Archaeology, Vol. 1, pp. 75–94). Oxford: Archaeopress.

McNabb, J., & Beaumont, P. (2011). *A report on the archaeological assemblages from excavations by Peter Beaumont at Canteen Koppie, Northern Cape, South Africa* (University of Southampton Series in Archaeology, Vol. 4). Oxford: Archaeopress.

McNabb, J., & Beaumont, P. (2012). Excavations in the Acheulean levels at the earlier Stone Age site of Canteen Koppie, Northern Province, South Africa. *Proceedings of the Prehistoric Society, 78*, 51–71.

McNabb, J., Binyon, F., & Hazelwood, L. (2004). The large cutting tools from the South African Acheulean and the questions of social traditions. *Current Anthropology, 45*(5), 653–677.

McNabb, J., & Sinclair, A. (2009). *The cave of hearths: Makapan middle pleistocene research project: Field research by Anthony Sinclair and Patrick Quinney, 1996–2001* (University of Southampton Series in Archaeology, Vol. 1). Oxford: Archaeopress.

Moncel, M.-H. (2010). Oldest human expansions in Eurasia: Favouring and limiting factors. *Quaternary International, 223–224*, 1–9.

Mortensen, P. (2008). Lower to Middle Palaeolithic artifacts from Loutró on the south coast of Crete. *Antiquity Project Gallery, 82*(317). http://www.antiquity.ac.uk/projgall/mortensen/.

Mourre, V. (2003). *Implications culturelles de la technologie des hachereaux*. Ph.D. Dissertation, Université de Paris X – Nanterre. (Original in French).

Otte, M., Yalçınkaya, I., Bar-Yosef, O., Kozlowski, J. K., Léotard, J.-M., Taşkıran, H., Noiret, P., & Kartal, M. (1999). The Anatolian Palaeolithic: Data and reflections. In G. N. Bailey, E. Adam, E. Panagopoulou, C. Perlès, & K. Zachos (Eds.), *The Palaeolithic archaeology of Greece and adjacent areas* (pp. 73–85). Nottingham: The British School at Athens.

Otte, M., Yalçınkaya, I., Kozlowski, J., Bar-Yosef, O., Bayón, I. L., & Taşkıran, H. (1998). Long-term technical evolution and human remains in the Anatolian Palaeolithic. *Journal of Human Evolution, 34*, 413–431.

Panagopoulou, E., Tourloukis, V., Thompson, N., Athanassiou, A., Tsartsidou, G., Konidaris, G. E., et al. (2015). Marathousa 1: A new Middle Pleistocene archaeological site from Greece. *Antiquity Project Gallery, 89*, 343.

Pe-Piper G. (1978). *The Cenozoic volcanic rocks of Lesbos*. Ph.D. Dissertation, University of Patras. (Original in Greek).

Pe-Piper, G., & Piper, D. J. W. (1993). Revised stratigraphy of the Miocene volcanic rocks of Lesbos, Greece. *Neues Jahrbuch für Geologie und Paläontologie Monatshefte, 2*, 97–110.

Pe-Piper, G., & Piper, D. J. W. (2002). The igneous rocks of Greece. *Beiträge zur Regionalen Geologie der Erde, 30*, 1–573.

Runnels, C., & Van Andel, T. H. (1993a). A Handaxe from Kokkinopilos, Epirus, and its implications for the Paleolithic of Greece. *Journal of Field Archaeology, 20*, 191–203.

Runnels, C., & Van Andel, T. H. (1993b). The lower and middle paleolithic of thessaly (Greece). *Journal of Field Archaeology, 20*, 299–318.

Sakellariou, D., & Galanidou, N. (2016). Pleistocene submerged landscapes and Palaeolithic archaeology in the tectonically active Aegean region. In J. Harff, G. N. Bailey, & F. Lüth (Eds.), *Geology and archaeology: Submerged landscapes of the continental shelf* (Special Publications, Vol. 411). London: Geological Society 145–178. doi:10.1144/SP411.9.

Sarantea, E. (1986). *Prehistoric finds from Nea Artaki Euboea*. Athens: Arsenides Editions. Original in Greek.

Sharon, G. (2007). *Acheulian large flake industries: Technology, chronology and significance* (British Archaeological Reports International Series, Vol. 1701). Oxford: Archaeopress.

Sharon, G. (2008). The impact of raw material on Acheulian large flake production. *Journal of Archaeological Science, 35*, 1329–1344.

Sharon, G. (2009). Acheulian giant core technology: A worldwide perspective. *Current Anthropology, 50*, 335–367.

Shea, J. J. (2006). The origins of lithic projectile point technology: Evidence from Africa, the Levant, and Europe. *Journal of Archaeological Science, 33*, 823–846.

Slimak, L., Kuhn, S. L., Roche, H., Mouralis, D., Buitenhuis, H., Balkan-Atli, N., et al. (2008). Kaletepe Deresi 3 (Turkey): Archaeological evidence for early human settlement in Central Anatolia. *Journal of Human Evolution, 54*(1), 99–111.

Soulakellis, N., Novak, I., Zouros, N., Lowman, P., & Yates, J. (2006). Fusing Landsat 5/TM imagery and shaded relief maps in tectonic and geomorphic mapping: Lesvos Island Greece. *Photogrammetric Engineering and Remote Sensing, 72*(6), 693–700.

Spooner, N. A. (1994). The anomalous fading of infrared-stimulated luminescence from feldspars. *Radiation Measurements, 23*, 625–632.

Stamatakis, M., & Magganas, A. (1988). Thermally induced silica transformation of Pliocene diatomaceous layers from Aegina Island, Greece. In J. R. Hein & J. Obradovic (Eds.), *Siliceous deposits of the Tethys and Pacific regions* (pp. 141–150). New York: Springer Verlag.

Strasser, T. F., Panagopoulou, E., Runnels, C., Murray, P., Thompson, N., Karkanas, P., McCoy, F., & Wegmann, K. (2010). Stone age seafaring in the Mediterranean: Evidence from the Plakias region for Lower Palaeolithic and Mesolithic habitation of Crete. *Hesperia, 79*(2), 145–190.

Strasser, T. F., Runnels, C., Wegmann, K., Panagopoulou, E., McCoy, F., Digregorio, C., Karkanas, P., & Thompson, N. (2011). Dating Palaeolithic sites in southwestern Crete, Greece. *Journal of Quaternary Science, 26*(5), 553–560.

Thiel, C., Buylaert, J.-P., Murray, A., Terhorst, B., Hofer, I., Tsukamoto, S., & Frechen, M. (2011). Luminescence dating of the Stratzing loess profile (Austria). Testing the potential of an elevated temperature post-IR IRSL protocol. *Quaternary International, 234*, 23–31.

Thomaidou, E.L. (2009). *Geological structure of Lesbos Island*. Ph.D. Dissertation, Aristotelian University of Thesaloniki. (Original in Greek).

Tourloukis, V. (2009). New bifaces from the Palaeolithic site of Kokkinopilos, Greece and their stratigraphic significance. *Antiquity Project Gallery, 83*(320). http://antiquity.ac.uk/projgall/tourloukis/.

Tourloukis, V. (2016). On the spatio-temporal distribution of Mediterranean Lower Paleolithic sites: A geoarchaeological perspective. In K. Harvati, & M. Roksandic (Eds.), *Paleoanthropology of the Balkans and Anatolia: Human evolution and its context* (pp. 303–323). Dordrecht: Springer.

Tourloukis, V., & Karkanas, P. (2012). The Middle Pleistocene archaeological record of Greece and the role of the Aegean in hominin dispersals: New data and interpretations. *Quaternary Science Reviews, 43*, 1–15.

Tryon, C. A., McBrearty, S., & Texier, P.-J. (2005). Levallois lithic technology from the Kapthurin Formation, Kenya: Acheulian origin and Middle Stone Age diversity. *African Archaeological Review, 22*(4), 199–229.

Wang, X. L., Wintle, A. G., & Lu, Y. C. (2007). Testing a single-aliquot protocol for recuperated OSL dating. *Radiation Measurements, 42*, 380–391.

White, M. J., & Ashton, N. (2003). Lower palaeolithic core technology and the origins of the Levallois method in North-Western Europe. *Current Anthropology, 44*, 598–609.

Wintle, A. G. (1973). Anomalous fading of thermoluminescence in mineral samples. *Nature, 245*, 143–144.

Wymer, J. (1961). The Lower Palaeolithic succession in the Thames Valley and the date of the ancient channel between Caversham and Henley, Oxon. *Proceedings of the Prehistoric Society, 27*, 1–27.

Chapter 9
Technological Changes and Population Movements in the Late Lower and Early Middle Paleolithic of the Central Balkans

Dušan Mihailović and Katarina Bogićević

Abstract Recent archaeological investigations have enabled preliminary insight into the Lower to Middle Paleolithic transition in the Central Balkans. Industries containing tools made from pebbles and flakes, within which Levallois artifacts were present to a lesser (Kosovska Kosa) or greater (Samaila) extent, have been encountered at the sites in the Zapadna Morava valley. The Charentian, likely dating to the Middle Pleistocene (possibly MIS 7) on the basis of microfaunal remains, has been reported in Velika and Mala Balanica in Sićevo. With regard to later (MIS 5–4) industries, assemblages of Typical Mousterian (Crvena Stijena, Hadži Prodanova cave), Charentian (Pešturina) and assemblages where Taubachian–Charentian component, Charentian elements, and backed bifaces are combined (Petrovaradin fortress) are encountered in the Central Balkans. After examining all available data, we propose the hypothesis that in addition to climatic, ecological, and behavioral factors, demographic factors also probably had considerable impact on the variability of lithic assemblages. Migrations and cultural transmission could have resulted in the appearance of Near Eastern elements in the Central Balkans as well as Balkan elements in the Near East. The homogeneity and/or variability of industries could be considerably influenced by the degree of isolation of human groups living in this region.

Keywords Charentian • Mousterian • Balanica • Pešturina • Lithic assemblages • Small mammals

D. Mihailović (✉)
Department of Archaeology, Faculty of Philosophy, University of Belgrade, Čika Ljubina 18-20, 11000 Belgrade, Serbia
e-mail: dmihailo@f.bg.ac.rs

K. Bogićević
Department of Paleontology, Faculty of Mining and Geology, University of Belgrade, Kamenička 6, 11000 Belgrade, Serbia
e-mail: k.bogicevic@rgf.bg.ac.rs

Introduction

Technological and behavioral changes during the transition from the Lower to the Middle Paleolithic have been well documented in Western Europe as well as in the Near East. However, they remain very poorly understood in Southeastern Europe. This is partly due to the fact that this period has been very poorly investigated in the region. Recently, however, the situation has changed, as multiple areas in the Balkans have been subjected to detailed surveying, and several relevant multilayered cave sites have now been systematically excavated (e.g., Kozarnika, Balanica, and Crvena Stijena). New preliminary dates, as well as paleoecological and geoarchaeological information have been obtained, and the results of earlier investigations have been reassessed (Morley 2007; Sirakov et al. 2010; Tourloukis 2010; Mihailović et al. 2011; Iovita et al. 2012). In this chapter, we will briefly present the chronology and artifact assemblages for several key new sites in the Central Balkans (Fig. 9.1), and then we will discuss the contribution of these new data to our understanding of cultural developments and population movements in the Balkan Paleolithic.

The Sites

Kremenac

Until a few years ago, the only site in Serbia yielding material with Lower Paleolithic affinities was the open-air site of Kremenac near Niš, Southern Serbia. Kremenac is situated on the northern fringes of the Niš basin, on an old lake terrace of Miocene age, at a location where secondary deposits of opal and chert were recovered. During many decades of surface collection at the site, archaeologists have recorded hundreds of artifacts and pseudo-artifacts—primarily resulting from the mechanical disintegration of chert nodules on the ground

Katerina Harvati and Mirjana Roksandic (eds.), *Paleoanthropology of the Balkans and Anatolia*, Vertebrate Paleobiology and Paleoanthropology, DOI 10.1007/978-94-024-0874-4_9

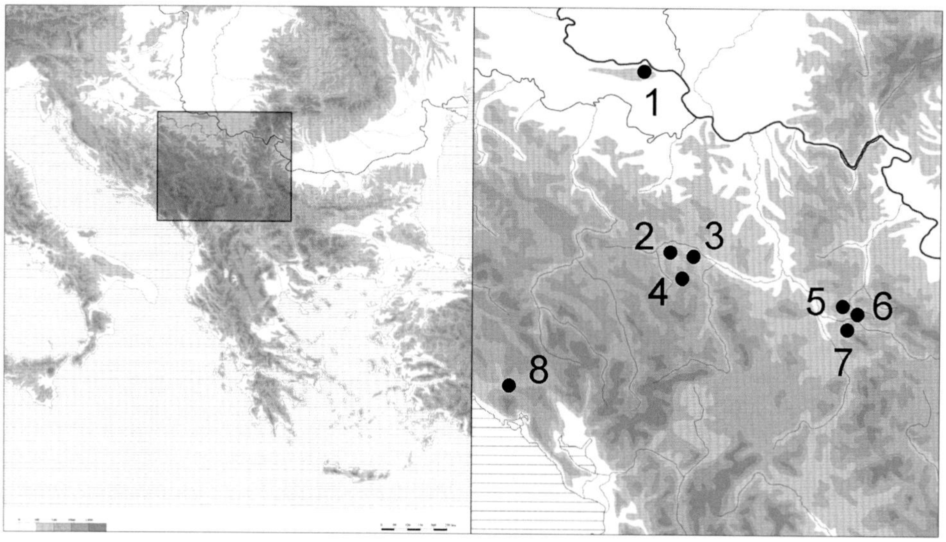

Fig. 9.1 Paleolithic sites in the Central Balkans, mentioned in the text: *1*—Petrovaradin fortress; *2*—Kosovska Kosa; *3*—Samaila; *4*—Hadži Prodanova Cave; *5*—Kremenac, *6*—Mala and Velika Balanica; *7*—Pešturina; *8*—Crvena Stijena

surface. By the mid 1990s, when Kaluđerović conducted test excavations at Kremenac, there were indications that there may be Lower Paleolithic artifacts present in this material (Kaluđerović 1996). Surveying continued in 2005, when it was confirmed that there were abundant tools manufactured on asymmetrical flakes and partially flaked pebbles including minimally modified pseudo-bifaces (Mihailović 2008a, 2009a). However, a precise determination of this industry was not possible since the artifacts were not recovered from an undisturbed stratigraphic context that could be securely dated by absolute dating techniques (Mihailović 2008a). Independent of that project, J. Šaric from the Institute of Archaeology in Belgrade visited the site on two occasions (2008 and 2009) and gathered 27 artifacts from the surface. As a result of this, he subsequently published an article where he concluded that finds from Kremenac could be related to the earliest phase of settlement in the Balkans (Šarić 2011).

Kosovska Kosa and Samaila

A much clearer situation is evident in the Čačak-Kraljevo basin, in the western part of Central Serbia. During a survey between 2010 and 2012, we recorded over 30 findspots, as well as a number of sites yielding between 200 and 300 artifacts from both the Lower and Middle Paleolithic (e.g., Samaila, Viljuša, Ježevica, Kosovska Kosa), on the highest river terrace of the Zapadna Morava (Mihailović et al. 2014). These artifacts were not recovered from a secure stratigraphic context, but rather collected on the ground surface. However, the technological and typological homogeneity of the assemblages, as well as the fact that no traces of natural transportation (e.g., within a colluvial or fluvial environment) were observed on the surfaces of the artifacts, suggest spatial, chronological, and cultural/technological integrity of the assemblages.

Finds at the site of Kosovska Kosa near Čačak are concentrated on both sides of a paleochannel that bisects one of the northern foothills of Mount Jelica. An assemblage of 162 artifacts was gathered on the right-bank of the channel in 2012. These artifacts are made on chert and quartz pebbles, which were most likely gathered from secondary deposits in the vicinity of the settlement. The assemblage comprises 13.6 % cores, choppers and chopping tools, 46.3 % unretouched artifacts, 10.5 % large chunks, 19.1 % tools, and 7.4 % chips and small fragments (Figs. 9.2 and 9.3). Among the cores, apart from choppers and chopping tools, polyhedral cores and one preferential core formed on a pebble were also identified. Flakes are asymmetrical, often with cortex present and exhibiting obliquely oriented platforms that are plain and rarely faceted. In addition, a single Levallois flake was also found. Denticulates, sidescrapers, endscrapers, and retouched flakes prevail among the tools, while other types are represented by only one or two specimens. Sidescrapers on thick flakes and on pebble fragments are characteristic of the assemblage. One Quinson point was made on a thick flake, triangular in section, and exhibits Quina-like coarse retouch. Denticulated endscrapers and flakes thinned at the proximal end by inverse retouch were also recovered.

A similar industry has been identified at nearby Samaila, near Kraljevo (Fig. 9.4), but the frequency of choppers is

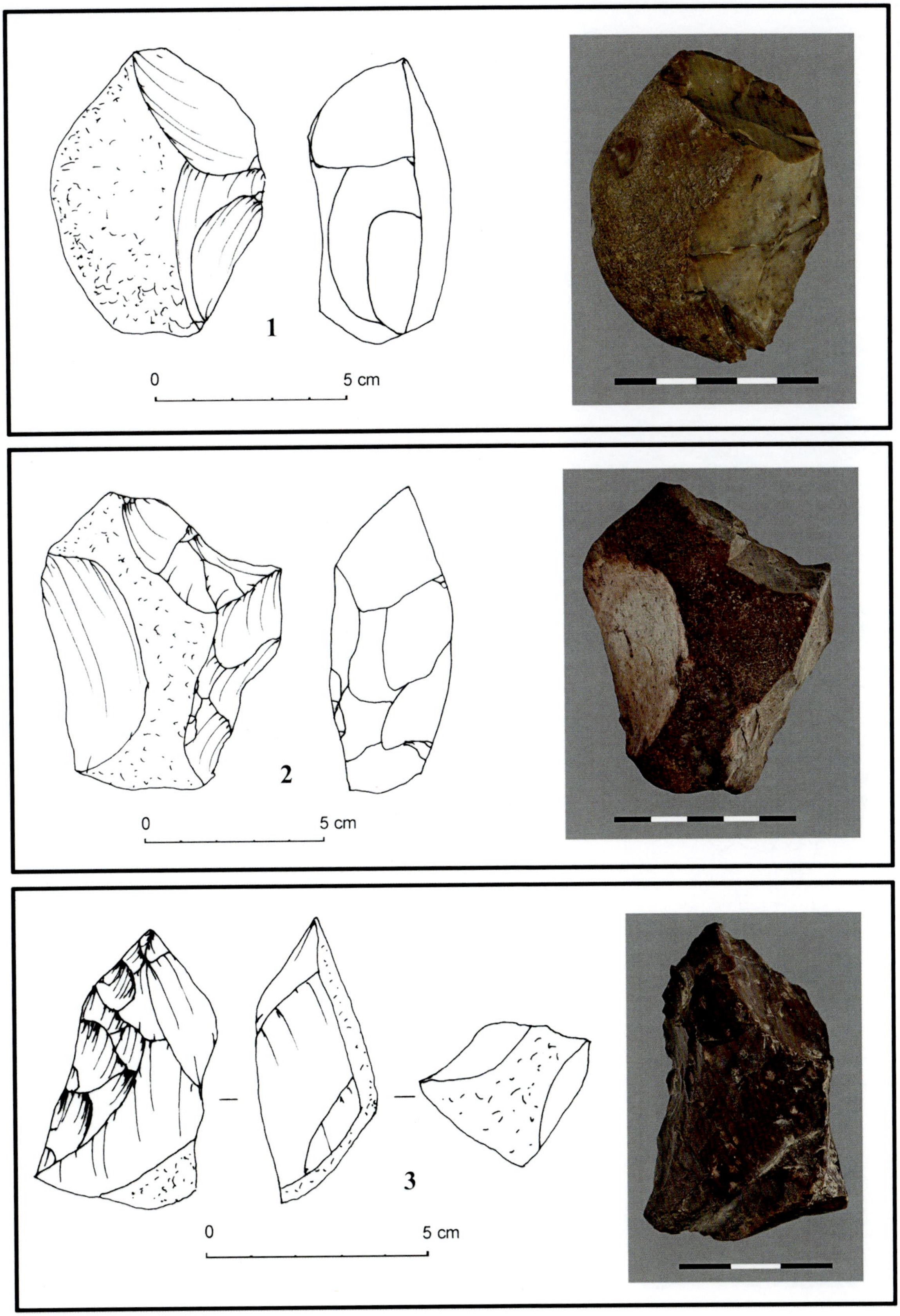

Fig. 9.2 Kosovska Kosa: side-choppers (*1*, *2*) and Quinson point (*3*)

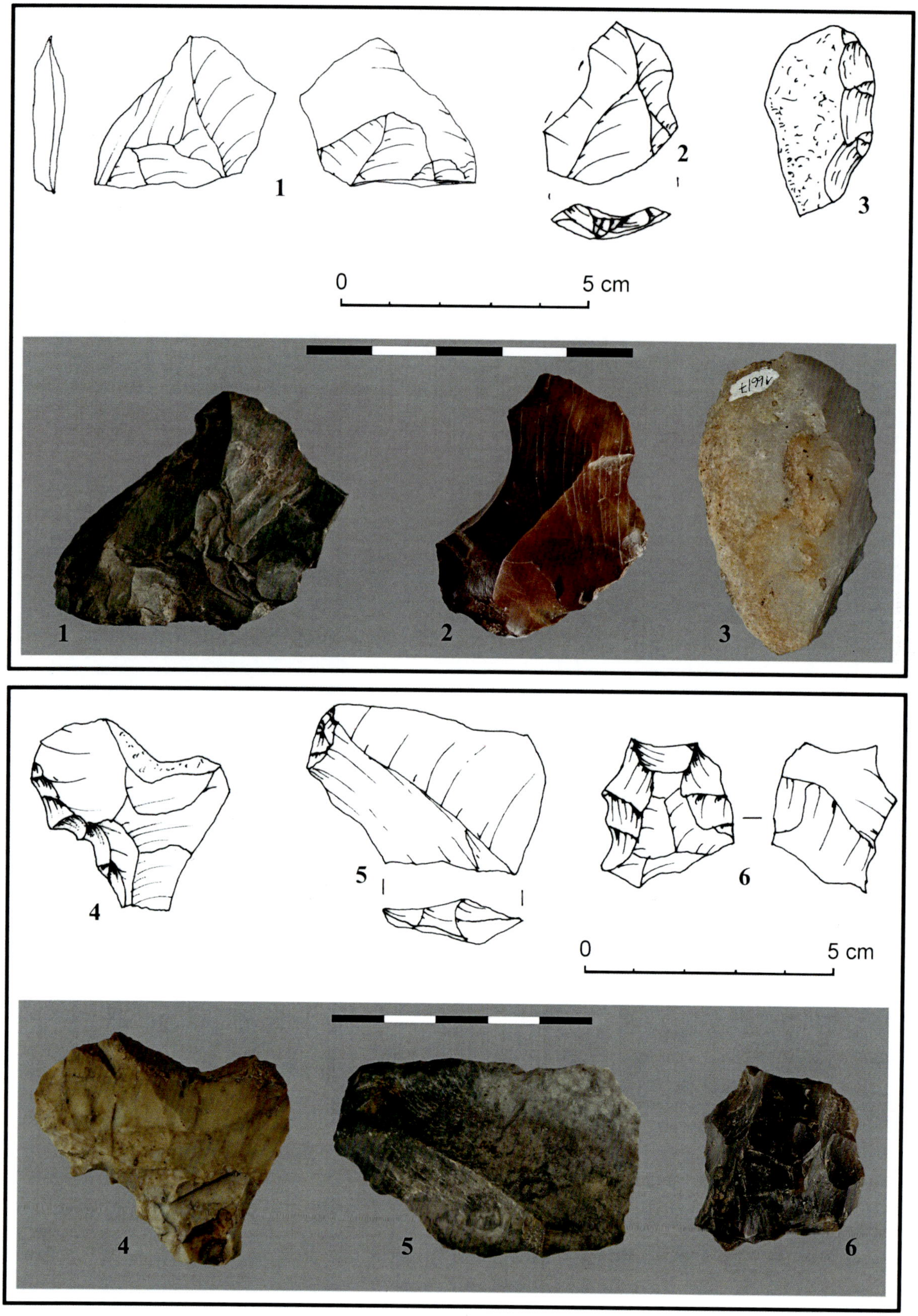

Fig. 9.3 Stone artifacts from Kosovska Kosa: ventrally thinned flake (*1*), Levallois flake (*2*), denticulate tools (*4*, *6*), endscraper (*5*)

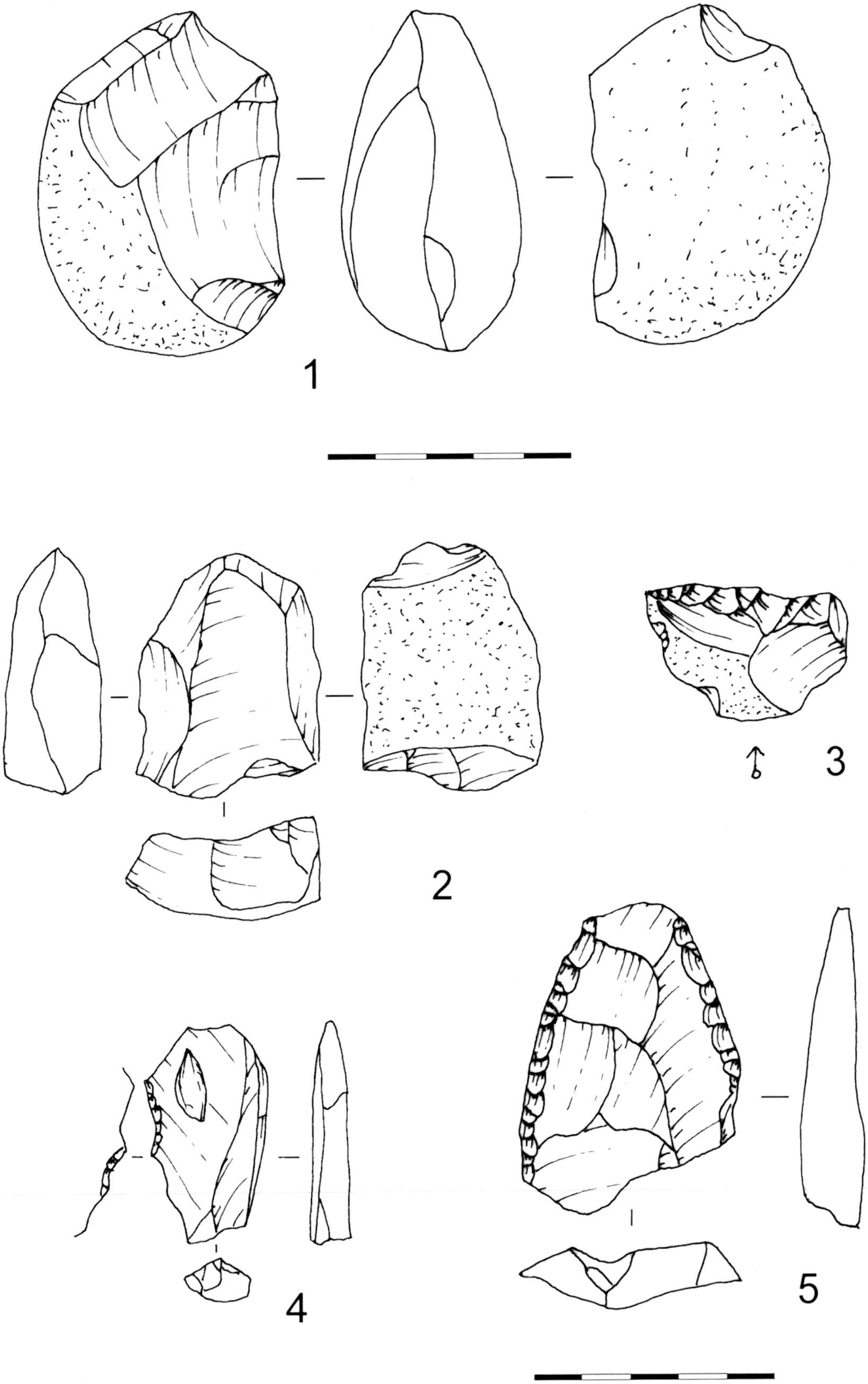

Fig. 9.4 Stone artifacts from Samaila: side-chopper (*1*), preferential core (*2*), transverse scraper (*3*), naturally backed knife (*4*), double scraper on Levallois flake (*5*)

considerably lower and the frequency of Levallois artifacts greater (Mihailović and Bogosavljević-Petrović 2010). The preferential "proto-Levallois" cores of variable sizes appear among the cores, and cores of Kombewa type were also documented. The assemblage contains a number of sidescrapers, including macrolithic sidescrapers, lateral sidescrapers, and sidescrapers thinned on the ventral side, as well as one transversal sidescraper. Endscrapers on asymmetrical flakes, naturally backed knives and one combined tool (endscraper-perforator) made on a thick flake were also recorded.

Velika and Mala Balanica

The Balanica cave complex, which comprises the Velika and Mala Balanica caves, is situated 10 km east of Niš at the exit of the Sićevo gorge. A hominin mandible attributed to *Homo erectus s.l.* (Roksandic et al. 2011; Roksandic 2016), recently dated to a period prior to 397–525 ka BP (Rink et al. 2013), was discovered in the lowest excavated layer (3b) at Mala Balanica in 2006. In addition to the mandible, a Charentian-type industry was confirmed from layer 2a–2c in Mala Balanica and layers 3a–3b in Velika Balanica (Fig. 9.5), while Typical Mousterian was recorded in the upper layers of Velika Balanica (2a–2c) (Mihailović 2008b, 2009a). Microfaunal remains suggest a Middle Pleistocene date for the Charentian layers. Forest species predominate (*Apodemus sylvaticus/flavicollis, Apodemus mystacinus, Muscardinus* sp., *Dryomys nitedula, Rhinolophus ferrumequinum*), but we have recorded meadow dwellers (*Sorex minutes*), species with wider distribution (*Microtus subterraneus* and *Arvicola* sp.), as well as steppe species (*Ochotona pusilla, Allocricetus bursae* and *Lagurus* sp.). The warm-loving *Apodemus sylvaticus/flavicollis* and *mystacinus* are consistent with interglacial environments. This is consistent with the geoarchaeological record, which also indicates climatic amelioration during the deposition of the Charentian layers at Mala Balanica.

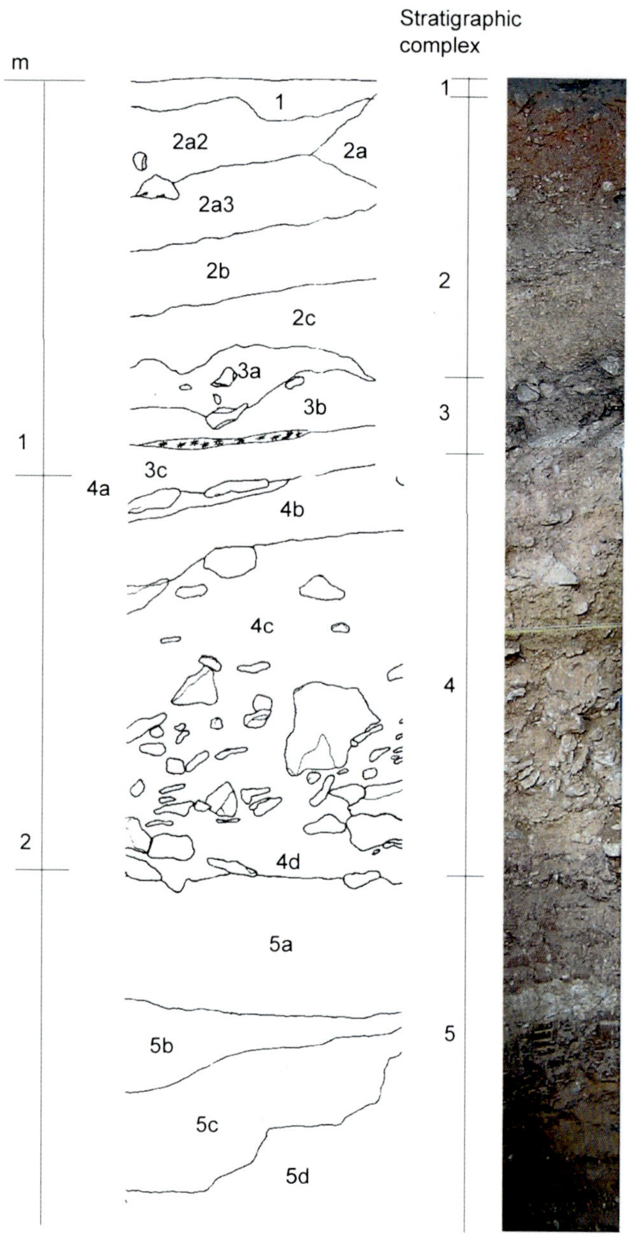

Fig. 9.5 Photograph and stratigraphic section of the north profile of square M22

The lithic assemblage recovered from layers 3a–3c at Velika Balanica and layers 2a–2c in Mala Balanica is characterized by a combination of predominantly expedient tools made on quartz pebbles collected from the river bank, and a smaller proportion of tools made on high-quality raw material. Most of the artifacts were produced by centripetal and "cortical backed" methods (Bourguignon 1997; Hiscock et al. 2009), aimed at the production of flakes with a thick platform or lateral back opposite the working edge. Scrapers predominate in both caves, comprising 62.1 and 33.8 % of the tool assemblages in Mala Balanica and in Velika, respectively. This is followed by denticulate tools, representing 24.1 % in Mala and 25 % in Velika Balanica, with other types of tools less well represented. Levallois artifacts, blades, and artifacts with faceted platform are completely absent. Intensively used Quina-type scrapers were documented in both caves, and one limace and a Mousterian point were recorded in Velika Balanica (Fig. 9.6). Quina elements are somewhat less well represented in the Balanica industry than in a classical Quina Mousterian assemblage (Bordes 1953). In terms of both typological and technological aspects, the Balanica industry exhibits a strong Charentian character. The assemblage from the upper layers in Velika Balanica (2a–2c) could be ascribed to Typical Mousterian, with Levallois artifacts present and lateral sidescrapers of diverse types prevailing among the tools.

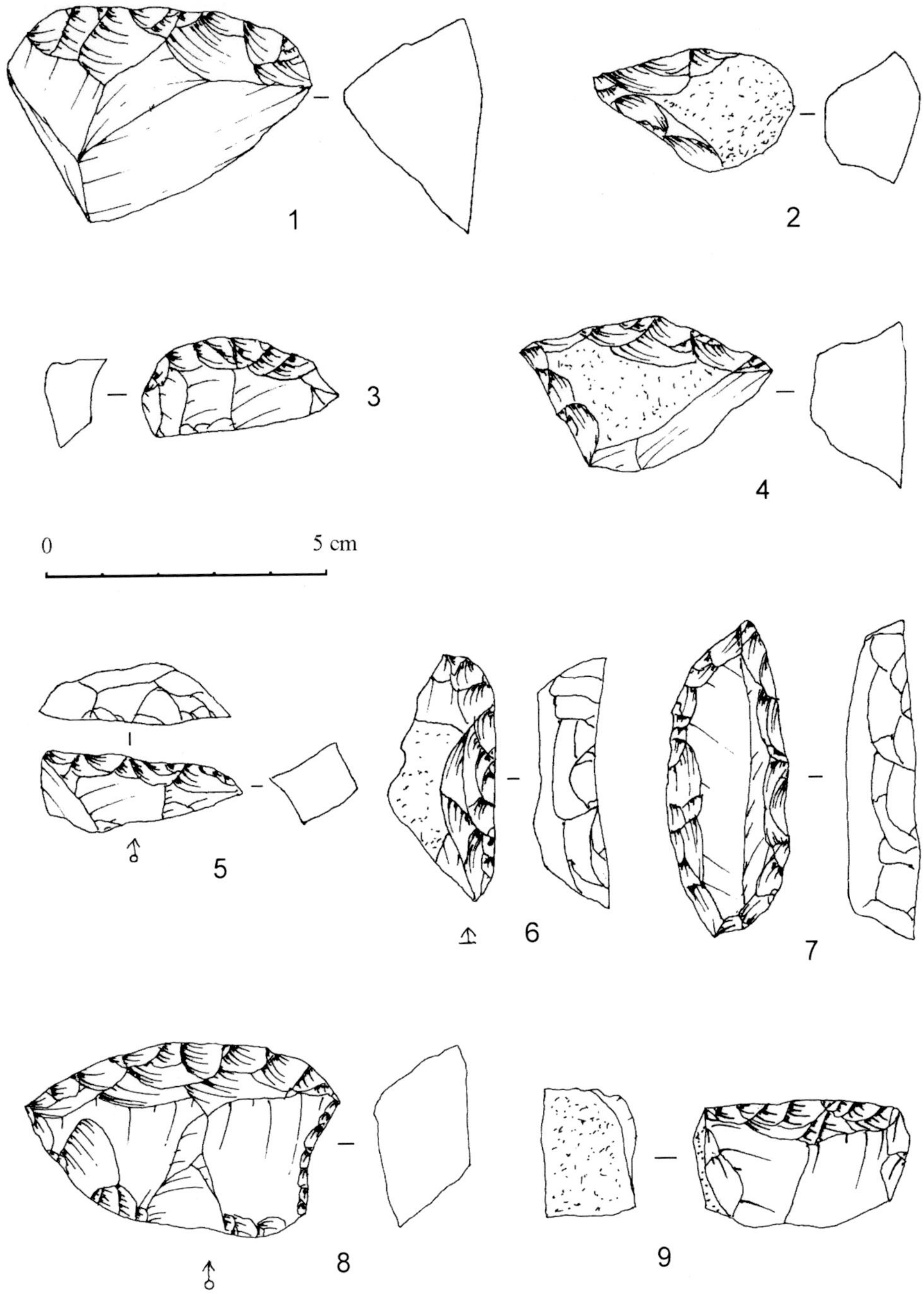

Fig. 9.6 Scrapers (*1–5*) and limace (*6*) from Velika Balanica and scrapers from Mala Balanica (*8* and *9*)

Petrovaradin Fortress

Many thousands of artifacts were found within an area of 92 square meters during rescue archaeological excavations at Petrovaradin Fortress, near Novi Sad, Northern Serbia (Mihailović 2009b). The finds were encountered within loessic sediments (layers 2a, 2b), deposited directly above the overlying bedrock. On the basis of preliminary examination of the site stratigraphy, the lower layer was related to the phase L1L2 (after the Chinese loess stratigraphic system—Kukla 1987), i.e., to MIS 4 (Marković et al. 2004). The upper layer was interpreted as a paleosoil within the initial phase of formation (Marković et al. 2004). Subsequent investigations, however, and the preliminary results of aminoacid dating of the lower layer (T. Gaudenyi, pers. comm.) suggest the possibility that layer 2b could correspond to

MIS 6 instead. Artifacts were predominantly made on white chert from nearby deposits and on quartz pebbles from the banks of the Danube. Diverse flaking techniques were employed, including a simplified Levallois technique of preferential type, the "salami slice" (or "cobble wedge") technique and the Kombewa method. The average length of recorded artifacts is relatively small, ranging from 24 to 25.5 mm. Tool structures are characterized by a predominance of sidescrapers in various forms. Most numerous are straight and arched lateral sidescrapers, but transversal specimens and massive bifacially flaked backed sidescrapers have also been recovered (Fig. 9.7). Denticulated and notched tools, as well as non-standardized tools of Upper Paleolithic type (endscrapers and burins), mostly made of quartz, are also very frequent. Mousterian points have not been found and only a few tools were made from Levallois flakes. On the whole, both Charentian and Levallois components were combined in the industry from Petrovaradin fortress, while bifacial backed sidescrapers of somewhat larger size (10–15 cm) superficially resemble specimens of Micoquian type (Bosinski 1967).

Hadži Prodanova Cave

Investigations in Hadži Prodanova cave, Southwest Serbia, were conducted in 2003 and 2004 (Mihailović and Mihailović 2006). At least three Middle Paleolithic layers were confirmed in the cave, with the two basal layers (5b–5c) probably dating to MIS 5 on the basis of stratigraphic position and microfauna. A relatively mild climate at the time of deposition of these layers is indicated by frequent occurrences of the rodent species *Microtus subterraneus* (25–57 %). This conclusion is corroborated by the lack of remains of cricetidae, genus *Lagurus*, and other typical steppe forms that appear during the later Upper Pleistocene in the Balkans (Dimitrijević 1997; Kowalski 2001). Quartz artifacts and tools made of flint (Levallois blades, endscrapers, sidescrapers, retouched flakes) that were probably introduced from outside the catchment area were recovered from layers 5b and 5c. Abundant animal remains were recorded in the Middle Paleolithic layers. Some of these animals inhabited the cave (cave bear, wolf), while others were part of the prey fauna, dominated by ibex remains (Milošević 2010).

Pešturina

The Pešturina cave has been investigated since 2005 and is situated in the immediate vicinity of Balanica, on the east fringes of the Niš basin (Mihailović and Milošević 2012). Two Middle Paleolithic layers (3 and 4) were recorded in the cave and small quantities of artifacts are present in these layers associated with numerous Pleistocene faunal remains. Following a program of ^{14}C and ESR dating, it has been established that layer 3 formed 38–45 ka BP, corresponding to MIS 3 (Alex and Boaretto 2014; Blackwell et al. 2014), while most of the dates for layer 4, the average age of which is 95 ka BP, suggest an MIS 5c origin for this layer (Blackwell et al. 2014).

The assemblage from layer 4 has Charentian characteristics from both a technological and typological point of view. The assemblage is dominated by centripetal cores on pebble truncations, as well as by broad flakes with laterally oriented cortex exhibiting damage on the opposite edge, suggesting their use as naturally backed knives. A large quantity of sidescrapers and denticulated tools were also found. Transverse sidescrapers prevail in relation to lateral specimens. Sidescrapers with Quina and demi-Quina retouch are also present. Levallois artifacts were also recorded in the assemblage, including one core and three blades flaked by the recurrent technique.

Lithic Industries

Before considering the technological grouping of assemblages from these sites some remarks should be made that are essential to understanding of the Paleolithic of the Balkans. First, it should be noted that the precise assignment of assemblages to specific industries is not currently possible because most of these assemblages have not yet been analyzed in detail and, as a result, there are no quantitative data. Second, the terminology used in the study of the Lower and Middle Paleolithic in Western Europe is not quite adequate to describe technological developments occurring in Southeast Europe and Southwest Asia, not just because of differences in the incidence of distinct classes and types of tools (Shea 2013), but also because of their spatial and temporal discontinuity in the west of the continent. This has resulted in terminological confusion, which is manifested in the relativization of western European terms ("Typical Mousterian," Charentoid, etc.), as well as in the use of distinct terms for defining regional phenomena. It is not possible to synchronize these terms before gaining insights into the mechanisms driving the variability seen in industries of the Lower and Middle Paleolithic of this region.

Core-Flake-Tool Industries

The earliest industries with tools made on pebbles and flakes in Central and Southeast Europe are dated to the Lower Pleistocene (Kozarnika), while sites in Hungary (Vértesszőlős),

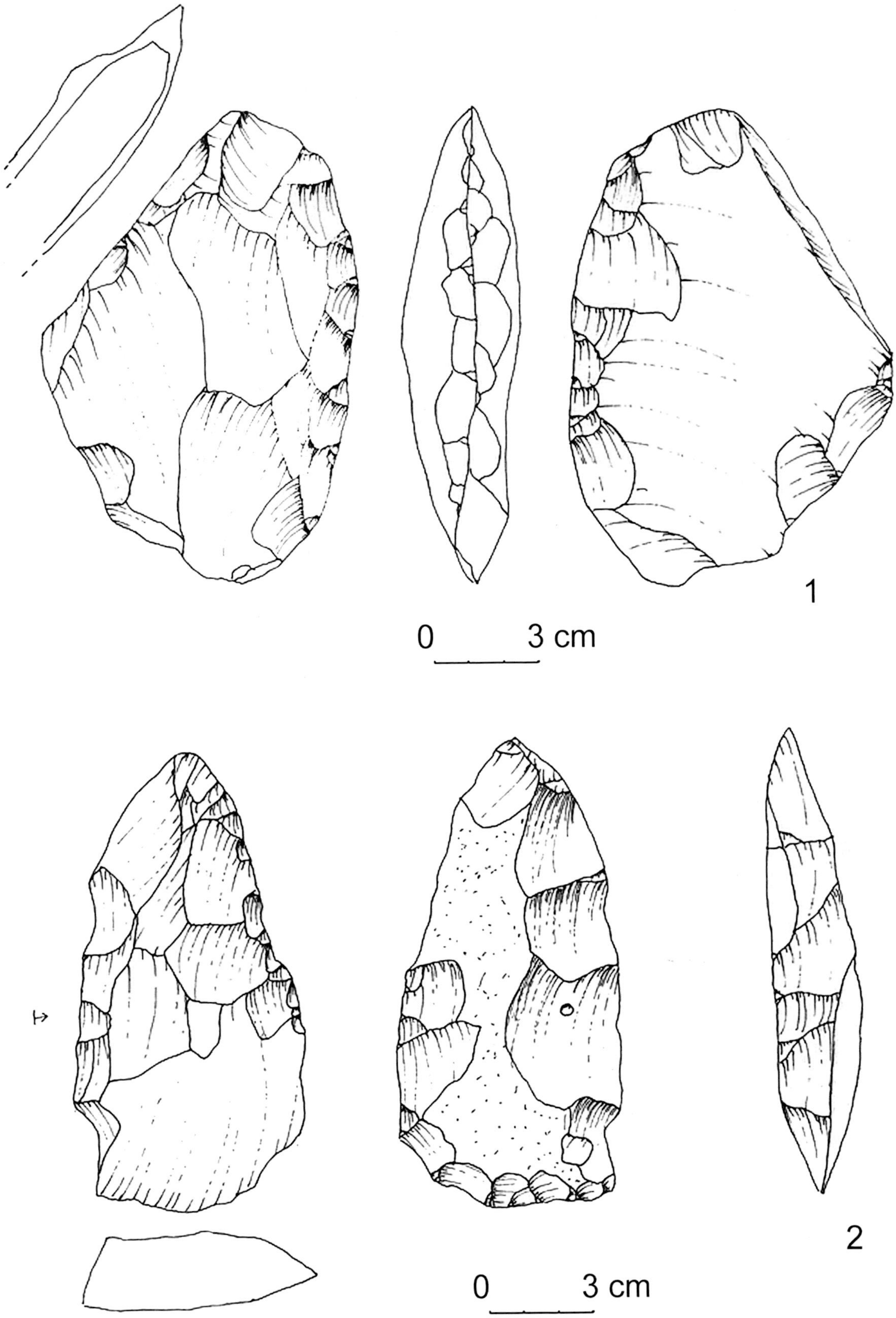

Fig. 9.7 Bifacial backed scrapers from Petrovaradin fortress

Romania (Dealui Guran), Greece (Rodia), and Western Turkey (Yarimburgaz, Karain) date from the Middle Pleistocene (Kretzoi and Dobosi 1990; Otte et al. 1998; Clark Howell et al. 2010; Sirakov et al. 2010; Tourloukis 2010; Iovita et al. 2012). Within this entire area, also including Eastern Europe, there is first the appearance of the Clactonian technique of flake production (e.g., Kozarnika, Karain, Treugol'naya) towards the end of the Middle Pleistocene (Otte et al. 1998; Doronichev 2008; Sirakov et al. 2010; Doronichev and Golovanova 2010), and later including branching or "ramifying" pattern of core knapping (Goren-Inbar and Belfer-Cohen 2002; Kuhn and Stiner 2010). Certainly at this point in time, the earliest occurrence of the Levallois technique in the Balkans is recorded in layers 10a–10b in Kozarnika (Guadelli et al, 2005) and the lowest layers at Crvena Stijena (Kozłowski 2002). Acheulian finds have not been confirmed with certainty in any locality and only isolated bifaces have been encountered at sites in Croatia and Greece (Tourloukis 2010; Tourloukis et al. 2015). However, recently a somewhat larger concentration of bifaces has been confirmed at the island of Lesvos in the Aegean (Galanidou et al. 2013, 2016).

How does the material from the Central Balkans fit into this picture? The finds from Kremenac include asymmetrical endscrapers and denticulated tools on massive flakes, as well as preferential cores that could not have been produced naturally (Mihailović 2008a). Therefore, it could be assumed that the Lower Paleolithic really can be recognized at this site. Nevertheless, pseudo-artifacts resulting from mechanical rock disintegration do appear on the surface together with genuine artifacts. "Choppers," pseudo-bifaces and flakes only exhibit a minimal amount of diagnostic scars, making the recognition of technological patterns particularly difficult. Bearing in mind the above, and the biased selection strategy of the collection approach along with the lack of contextual data, Šaric's claims for parallels with the Acheulean (Šarić 2011) cannot be accepted without reservation.

In contrast to Kremenac, the technological patterns at Kosovska Kosa are clearly recognizable and the context of the finds is much less ambiguous. Settlement at the site is confirmed by the presence of artifacts grouped on the channel bank made from raw material gathered from local sources, and by the fact that there are no pseudo-tools recorded at the site. Among the choppers, massive examples were encountered that were either flaked on the side or on the terminal end and were certainly used as cores. Biconvex specimens with a cutting edge in the plane of the intersection could have been used as tools, while smaller pieces retouched along the edge could be classified as sidescrapers. The entire suite of technological and typological indicators described above suggest that Kosovska Kosa dates from the Lower Paleolithic, rather than representing the functional Levallois-Mousterian facies confirmed at other sites in the valley. If this is so—and only new investigations will confirm this—it might be proposed that the assemblage has similarities with the Tayacoid and Charentoid industries of Middle Pleistocene age recorded across certain areas of Europe (Doronichev 2008; Doronichev and Golovanova 2010; Moncel 2011).

Levallois-Mousterian of Samaila Type

The assemblages from Samaila and other sites in the Zapadna Morava valley are very distinctive because Lower Paleolithic elements in those assemblages are more frequent than at many other Middle Paleolithic open-air sites in Southeast Europe, Northern Bosnia, or in the Pineios valley for example (Baumler 1987; Tourloukis 2010). Despite the somewhat dubious context and the lack of quantitative data, it seems that the appearance of elements from both periods is probably not accidental as it is possible to follow the technological progression from the flaking of choppers, through "proto-Levallois" and Levallois cores on pebbles and massive flakes, to Kombewa cores on massive flakes (Mihailović and Bogosavljević-Petrović 2010). Finds from Samaila and other sites in the Zapadna Morava valley are unlikely to be older than MIS 7 or MIS 6 when the Levallois technique appears in the Balkans (e.g., Kozarnika, Crvena Stijena), but also not younger than MIS 5, the period from which similar industries from Asprochaliko, Zobište, and the lower layers at Crvena Stijena have been recorded (Basler 1975; Bailey et al. 1992; Baumler 1987; Sirakov et al. 2010). Evidence of a dense population in the valley speaks in favor of an interglacial date for the assemblage (MIS 7 or 5) considering that other evidence suggests that the Balkans were intensively populated during that period. For example, most of the layers from the middle part of the stratigraphic sequence at Crvena Stijena date to this time (Basler 1975; Morley 2007).

Charentian at Balanica

Even though we still do not have at our disposal a complete quantitative dataset, we could say that the industry in the lower layers at Velika and Mala Balanica also have a well-defined Charentian character in terms of technological and typological characteristics. This is an exciting discovery, as the Charentian has hitherto been recognized only at sites to the north of the Sava and the Danube (Gábori 1976). It was also generally considered that the industry does not predate the last interglacial and Early glacial, i.e., the age which was obtained for Krapina and Charentian sites in the Carpathian basin (Mihailović 2008b, 2009a). Microfaunal data, however, show that the Charentian layers in both caves are probably older, dating from the Middle Pleistocene (MIS 7?). If this observation proves to be true, it would mean that Balanica is one of the earliest Charentian sites in Europe. It is apparent, however, that Quina retouch is also present on tools from earlier periods, including Arago and La Micoque (Turq 1989; Bourguignon 1997; Geneste et al. 1997), but it is also known that Charentian facies in Western and Central Europe appear only from MIS 6 or MIS 5e (Simek and Smith 1997; Moncel 2011). On the other hand, the earliest manifestations in the East date from a much earlier time. The beginning of the Yabrudian in the Levant dates from 400 to 350 ka BP (Rink et al. 2004), the Acheulo-Yabrudian from 350 to 200 ka BP (Mercier et al. 1995), and Kozłowski (2002) has proposed a date of 300–330 ka BP for the Proto-Charentian of Karain in Anatolia. Charentian elements (scrapers and rare limaces) are noted in the Caucasus as early as MIS 10 (Treugol'naya), and they are present in late Acheulian (Kudarian) industries associated with MIS 9–7 (Doronichev 2008; Doronichev and Golovanova 2010).

Regardless of the issues in dating some of these sites, it seems that Charentian industries in the East predate those in the West and that there is a temporal trend in the spread of Charentian technologies from Southwest Asia to Southeast Europe. As we have seen, the similarities between the Yabrudian and the Charentian of Western Europe exist not only on the typological (Bordes 1953) but also the technological level (Bourguignon 1997). It is becoming clear that the "eastern Charentian" represents a unique entity, an integral part of the Acheulo-Yabrudian and Kudarian traditions (Doronichev 2008), with a strong Acheulian component. For this reason, the Charentian of Karain, where bifacial techniques are only very rarely present, was most commonly associated with Acheulo-Yabrudian sequence rather than the Charentian of Southeast Europe (Gábori 1976). The evidence from Balanica shows that there is a third "Balkan-Anatolian" group where Acheulian elements are absent and local Lower Paleolithic and Mousterian elements (especially in Balanica) dominate.

Charentian and Mousterian in the Early Glacial

Only two facies, the Charentian and Typical (i.e., Levallois) Mousterian, can be distinguished in the early Middle Paleolithic of the Central Balkans, and it is only in that period that they are clearly distinguished. The Charentoid character of the later early glacial industries (Kozłowski 1992) is best shown by the finds from Pešturina Cave and Petrovaradin fortress. Charentian elements are well represented in the material from Pešturina but the recovered assemblage is at present rather small, so it is possible that this situation will change when more artifacts are gathered. However, substantial amounts of material have been recorded at Petrovaradin fortress, so it is quite clear that elements usually related to diverse cultural traditions are present at this site. The Taubachian–Charentian component is apparent in the microlithic character of the industry, use of quartz, high incidence of denticulated and notched tools, and high Charentian index. The Levallois technique is confirmed by cores and flakes produced by the preferential and recurrent method, while backed bifaces reveal a Central European affinity. Nevertheless, it would not be justified to insist on the attribution of the industry from Petrovaradin fortress to any distinct technological facies at this time. A Taubachian component is present only at a general technological level, while Quina retouch is not recorded on any tools. Levallois artifacts are scarce, while bifacial scrapers resemble Micoquian specimens in terms of their concept rather than specific flaking technique. Instead of attempting to precisely define a "cultural" affiliation of the industry, it seems more important to emphasize that the eclectic character of the industry from Petrovaradin is probably indicative of a high level of social and cultural connection between Middle Paleolithic communities inhabiting the Carpathian basin.

When considering the distribution of different techniques, the practical purpose of specific tool types should also be taken into account. For example, the connection between the Charentian technique and the technique of flaking backed bifaces is suggested among others by the fact that in both instances of sidescraper manufacture blanks of the "déjeté" type were used exhibiting thick and broad platforms representing distinct "talon-dos" (Turq 1989), opposite the transversal working edge. In contrast to this, in the Levallois (Typical Mousterian) a different technique of hafting, based on reducing the ventral side of the tool, was practiced. This technique is well documented in MIS 6/5 and in the somewhat later Mousterian of Karain and the Mousterian in the Zagros region (Otte et al. 1998). The expansion of a "Crvena Stijena-type Mousterian" out to the East is—according to some—one of the primary sources of evidence for an "eastward expansion" of the Neanderthals (Kozłowski 1992, 2002; Hublin 2002).

Concluding Remarks: Demographic Factors of Technological Variability

Data from the Central and Eastern Balkans, though meager at present, indicate that industries with tools made on pebbles and flakes appear in the earlier phase of the Lower Paleolithic, followed first by Clactonian industries with or without (proto) Levallois elements, and later by the Charentian. Therefore, although perhaps premature, the question arises: to what degree have the similarities and differences in technological development in this part of Europe been influenced by common regional traditions and similar climatic and ecological conditions, rather than by other factors? It has recently been established that demographic factors, including migrations and extinction rates within local communities, could have had a significant impact on the variability of industries (Premo and Kuhn 2010), and such a scenario may account for the patterning of sites in the Balkans.

The two-way movement and mixing of populations, including cultural transmission, between Southwest Asia and Southeast Europe could probably be explained by the geographic position of the Balkans and the fact that it is open to the south. In other words, the Balkan peninsula was probably situated on the northwestern fringes of a "central area of dispersals" (Dennell et al. 2010, 2011), whose border shifted to the rhythm of climatic oscillations. The Balkans may have played a significant role as a transitional zone between a permanently settled area in the southeast and areas in the northwest that were not permanently (or densely) populated during glacial periods. In any case, the southern and western areas of the Balkans could have served as a refugium for human groups retreating south during glacial periods, as it could be assumed to have been intensively populated in the interglacials. Gradually, a "pumping in" of populations and cultural packages from the East into Europe (and vice versa) and their intermixing in the Balkan region might well have occurred.

It is neither necessary nor helpful to explain the demographic reasons for technological changes and regional differentiation during the Lower and Middle Paleolithic using a conventional model of migration and cultural transmission. A similar conclusion could be drawn if we take into account the possibility that demographic stress also had a significant impact. According to the most recent studies, it is assumed that elements of continuity and homogeneity of phenomena in material culture are considerably more prominent in communities that inhabited geographically isolated or recolonized areas (Premo and Kuhn 2010). In this context, the fact that elements of continuity and homogeneity present in Middle Paleolithic assemblages of the Balkans are much more prominent at Crvena Stijena (in the mountainous hinterland of the narrow coastal region) than at Petrovaradin fortress and other sites in densely populated areas is quite understandable.

Acknowledgments This work was supported by grants from the Ministry of Education and Science of the Republic of Serbia, projects no. 177023 and 176015. We thank the Editors and anonymous reviewers for their comments.

References

Alex, B., & Boaretto, E. (2014). Radiocarbon chronology of Pešturina Cave. In D. Mihailović (Ed.), *Paleolithic and Mesolithic research in the Central Balkans* (pp. 39–47). Belgrade: Serbian Archaeological Society.

Bailey, G., Papakonstantinou, V., & Sturdy, D. (1992). Asprochaliko and Kokkinopilos: TL dating and reinterpretation of Middle Paleolithic sites in Epirus, north-west Greece. *Cambridge Archaeological Journal, 2*, 136–144.

Basler, Đ. (1975). Stariji litički periodi u Crvenoj stijeni. In Đ. Basler (Ed.), *Crvena stijena—zbornik radova* (pp. 11–120). Nikšić: Zajednica kulturnih ustanova.

Baumler, M.F. (1987). Core reduction sequences, an analysis of blank production in the Middle Paleolithic of Northern Bosnia (Yugoslavia). Ph.D. Dissertation, The University of Arizona. Tucson.

Blackwell, B., Chu, S., Chaity, I., Huang, Y. E. W., Mihailović, D., Roksandic, M., et al. (2014). ESR dating ungulate tooth enamel from the Mousterian Layers at Pešturina, Serbia. In D. Mihailović (Ed.), *Paleolithic and Mesolithic research in the Central Balkans* (pp. 21–38). Belgrade: Serbian Archaeological Society.

Bordes, F. (1953). Essai de classification des industries "moustériennes". *Bulletin de la Société Préhistorique Française, 7–8*, 457–466.

Bosinski, G. (1967). *Die Mittelpaläolithischen Funde im Westlichen Mitteleuropa*. Köln: Böhlau Verlag.

Bourguignon, L. (1997). Le Moustérien de type Quina: Nouvelle définition d'une entité technique. Ph.D. dissertation, Université de Paris X, Paris.

Howell, F. C., Arsebük, G., Kuhn, S. L., Özbaşaran, M., & Stiner, M. C. (Eds.). (2010). *Culture and biology at a crossroads: The Middle Pleistocene record of Yarımburgaz Cave (Thrace, Turkey)*. Istanbul: Zero Books/Ege Publications.

Dennell, R. W., Martinon-Torres, M., & Bermudez de Castro, J. M. (2010). Out of Asia: The initial colonisation of Europe in the Early and Middle Pleistocene. *Quaternary International, 223–224*, 439.

Dennell, R. W., Martinón-Torres, M., & Bermúdez de Castro, J. M. (2011). Hominin variability, climatic instability and population demography in Middle Pleistocene Europe. *Quaternary Science Reviews, 30*, 1511–1524.

Dimitrijević, V. (1997). Upper Pleistocene mammals from cave deposits of Serbia. *Geološki anali Balkanskog poluostrva, 61*(2), 179–370.

Doronichev, V. (2008). The Lower Paleolithic in Eastern Europe and the Caucasus: A reappraisal of the data and new approaches. *PaleoAnthropology, 2008*, 107–157.

Doronichev, V., & Golovanova, L. (2010). Beyond the Acheulean: A view on the Lower Paleolithic occupation of Western Eurasia. *Quaternary International, 223–224*, 327–344.

Gábori, M. (1976). *Les civilisations du Paléolithique moyen entre les Alpes et l'Oural*. Budapest: Akadémiai Kiadó.

Galanidou, N., Cole, J., Iliopoulos, G., & McNabb, J. (2013). East meets West: The Middle Pleistocene site of Rodafnidia on Lesvos, Greece. *Antiquity, 87*, 336.

Galanidou, N., Athanassas, C., Cole, J., Iliopoulos, G., Katerinopoulos, A., Magganas, A., McNabb, J. (2016). The Acheulean site at Rodafnidia, Lisvori on Lesbos, Greece: Part I. In K. Harvati, & M. Roksandic (Eds.), Paleoanthropology of the Balkans and Anatolia: Human evolution and its context. Dordrecht: Springer.

Geneste, J.-M., Jaubert, J., Lenoir, M., Meignen, L., & Turq, A. (1997). Approche technologique des Moustériens charentiens du sud-ouest de la France et du Languedoc oriental. *Paléo, 9*, 101–142.

Goren-Inbar, N., & Belfer-Cohen, A. (2002). The technological abilities of the Levantine Mousterians: Cultural and mental capacities. In T. Akazawa, K. Aoki, & O. Bar-Yosef (Eds.), *Neanderthals and modern humans in West* (pp. 205–221). New York: Kluwer Academic Publishers.

Guadelli, J.-L., Sirakov, N., Ivanova, S., Sirakova, S., Anastassova, E., Courtaud, P., et al. (2005). Une séquence du Paléolithique inférieur au Paléolithique récent dans les Balkans: La grotte Kozarnika à Orechets (Nord-Ouest de la Bulgarie). In N. Molines, M.-H. Moncel, & J.-L. Monnier (Eds.), *Données récentes sur les modalités de peuplement et sur le cadre chronostratigraphique des industries du Paléolithique inférieur et moyen en Europe* (BAR International Series, Vol. 1364, pp. 87–104). Oxford: Archaeopress.

Hiscock, P., Turq, A., Faivre, J.-P., & Bourguignon, L. (2009). Quina procurement and tools production. In B. Adams & B. S. Blades (Eds.), *Lithic materials and Paleolithic societies* (pp. 232–246). Singapore: Wiley-Blackwell.

Hublin, J.-J. (2002). Climatic changes, paleogeography, and the evolution of the Neandertals. In T. Akazawa, K. Aoki, & O. Bar-Yosef (Eds.), *Neanderthals and modern humans in Western Asia* (pp. 295–310). New York: Kluwer Academic Publishers.

Iovita, R., Fitzsimmons, K. E., Dobos, A., Hambach, U., Hilgers, A., & Zander, A. (2012). Dealul Guran: Evidence for Lower Paleolithic (MIS 11) occupation of the Lower Danube loess steppe. *Antiquity, 86*, 973–989.

Kaluđerović, Z. (1996). Kremenac kod Niša—Paleolitsko nalazište. *Starinar, XLVII*, 289–290.

Kowalski, K. (2001). Pleistocene rodents of Europe. *Folia Quaternaria, 72*, 1–389.

Kozłowski, J. K. (1992). The Balkans in the Middle and Upper Paleolithic: The gate to Europe or a cul-de-sac? *Proceedings of the Prehistoric Society, 58*, 1–20.

Kozłowski, J. K. (2002). The Middle and the Early Upper Paleolithic around the Black Sea. In T. Akazawa, K. Aoki, & O. Bar-Yosef (Eds.), *Neanderthals and modern humans in Western Asia* (pp. 461–482). New York: Kluwer Academic Publishers.

Kretzoi, M., & Dobosi, V. T. (1990). *Vértesszőlős: Site, man and culture*. Budapest: Akadémiai Kiadó.

Kuhn, S. L., & Stiner, M. C. (2010). Culture and biology in transition: Summary of findings and conclusion. In F. C. Howell, G. Arsebük, S. L. Kuhn, M. Özbaşaran, & M. C. Stiner (Eds.), *Culture and biology at a crossroads: the Middle Pleistocene record of Yarımburgaz Cave (Thrace, Turkey)* (pp. 301–319). Istanbul: Zero Books/Ege Publications.

Kukla, G. J. (1987). Loess stratigraphy in Central China. *Quaternary Science Reviews, 6*, 191–219.

Mercier, N., Valladas, H., Valladas, G., Reyss, J. L., Jelinek, A., Meignen, L., & Joron, J. L. (1995). TL dates of burnt flints from Jelinek's excavations at Tabun and their implications. *Journal of Archaeological Science, 22*, 495–509.

Marković, S. B., Mihailović, D., Oches, E. A., Jovanović, M., & Gaudenyi, T. (2004). The Last Glacial climate, environment and the evidence of Paleolithic occupation in Vojvodina province, Serbia: An overview. *Antaeus, 27*, 147–152.

Mihailović, D. (2008a). Istraživanja paleolita na području Niške kotline između 2004. i 2006. godine. *Zbornik radova Odbora za kras i speleologiju, IX*, 105–118.

Mihailović, D. (2008b). New data about the Middle Paleolithic of Serbia. In A. Darlas & D. Mihailović (Eds.), *The Paleolithic of the Balkans* (BAR International Series, Vol. 1819, pp. 93–100). Oxford: Archaeopress.

Mihailović, D. (2009a). Pećinski kompleks Balanica i paleolit Niške kotline u regionalnom kontekstu. *Arhaika, 2*, 3–26.

Mihailović, D. (2009b). Middle Paleolithic settlement at Petrovaradin Fortress. Petrovaradin edition, Vol. II, Novi Sad: The City Museum of Novi Sad.

Mihailović, D., & Bogosavljević-Petrović, V. (2010). Samaila—Vlaška glava, paleolitsko nalazište na otvorenom prostoru. *Naša prošlost, 10*, 21–43.

Mihailović, D., & Mihailović, B. (2006). Paleolitsko nalazište Hadži Prodanova pećina kod Ivanjice. *Arheološki pregled Srpskog arheološkog društva, 1*(2003), 13–16.

Mihailović, D., Mihailović, B., Milošević, S., & Lopičić, M. (2011). Tehnološka varijabilnost u ranom srednjem paleolitu centralnog Balkana. *Glasnik Srpskog arheološkog društva, 25*, 7–23.

Mihailović, D., & Milošević, S. (2012). Istraživanja paleolitskog nalazišta Pešturina kod Niša. *Glasnik Srpskog arheološkog društva, 28*, 87–106.

Mihailović, D., Milošević, S., & Radović, P. (2014). New data about the Lower and Middle Paleolithic in the Western Morava valley. In D. Mihailović (Ed.), *Paleolithic and Mesolithic research in the Central Balkans* (pp. 57–67). Belgrade: Serbian Archaeological Society.

Milošević, S. (2010). Reading of human on-site activities on the basis of faunal taphonomy at the Paleolithic site of Hadži Prodanova cave (Serbia). Bone Commons, 1390. Retrieved January 20, 2013 from http://alexandriaarchive.org/bonecommons/items/show/1390.

Moncel, M.-H. (2011). Technological behavior and mobility of human groups deduced from lithic assemblages in the late middle and early late Pleistocene of the middle Rhône Valley (France). In N. Conard & J. Richter (Eds.), *Neanderthal lifeways, subsistence and technology* (pp. 261–287). Dodrecht: Springer.

Morley, M. (2007). Mediterranean Quaternary rockshelter sediment records: A multi-proxy approach to environmental reconstruction. Ph.D. Dissertation, The University of Manchester.

Otte, M., Yalcinkaya, I., Kozłowski, J. K., Bar-Yosef, O., Lopez Bayon, I., & Taskiran, H. (1998). Long-term technical evolution and human remains in the Anatolian Paleolithic. *Journal of Human Evolution, 34*, 413–431.

Premo, L. S., & Kuhn, S. L. (2010). Modeling effects of local extinctions on culture change and diversity in the Paleolithic. *PLoS One, 5*(12), e15582. doi:10.1371/journal.pone.0015582.

Rink, W. J., Schwarcz, H. P., Ronen, A., & Tsatskin, A. (2004). Confirming a near 400 ka age for the Yabrudian industry of Tabun Cave, Israel. *Journal of Archaeological Science, 31*, 15–20.

Rink, W. J., Mercier, N., Mihailović, D., Morley, M. W., Thompson, J. W., & Roksandic, M. (2013). New radiometric ages For the BH-1 hominin from Balanica (Serbia): Implications for understanding the role of the Balkans in Middle Pleistocene human evolution. *PLoS One, 8*(2), e54608. doi:10.1371/journal.pone.0054608.

Roksandic, M., Mihailović, D., Mercier, N., Dimitrijević, V., Morley, M. W., Rakočević, Z., Mihailović, B., Guibert, P., & Babb, J. (2011). A human mandible (BH-1) from the Pleistocene deposits of the Mala Balanica cave (Sićevo Gorge, Niš, Serbia). *Journal of Human Evolution, 61*, 186–196.

Roksandic, M. (2016). The role of the Central Balkans in the peopling of Europe: Paleoanthropological evidence. In K. Harvati, & M. Roksandic (Eds.), Paleoanthropology of the Balkans and Anatolia: Human evolution and its context (pp. 15–33). Dordrecht: Springer.

Shea, J. (2013). *Stone tools in the Paleolithic and Neolithic near East*. Cambridge: Cambridge University Press.

Simek, J. F., & Smith, F. H. (1997). Chronological changes in stone tool Assemblages from Krapina (Croatia). *Journal of Human Evolution, 32*, 561–575.

Sirakov, N., Guadelli, J. L., Ivanova, S., Sirakova, S., Boudadi-Maligne, M., Dimitrova, I., et al. (2010). An ancient continuous human presence in the Balkans and the beginnings of human settlement in western Eurasia: A Lower Pleistocene example of the Lower Paleolithic levels in Kozarnika Cave (North-western Bulgaria). *Quaternary International, 223–224*, 94–106.

Šarić, J. (2011). Lower Paleolithic site Kremenac near the village Rujnik (Serbia). *Starinar, LXI*, 7–31.

Tourloukis, V. (2010). *The Early and Middle Pleistocene archaeological record of Greece—Current status and future prospects*. Leiden: Leiden University Press.

Tourloukis, V., Karkanas, P., & Wallinga, J. (2015). Revisiting Kokkinopilos: Middle Pleistocene radiometric dates for stratified archaeological material in Greece. *Journal of Archaeological Science, 57*, 355–369.

Turq, A. (1989). Approche technologique et économique du faciès Moustérien de type Quina: étude préliminaire. *Bulletin de la Société préhistorique française, 86*(8), 244–256.

Chapter 10
Recent Research on the Croatian Middle/Upper Paleolithic Interface in the Context of Central and Southeast Europe

Ivor Karavanić, Rajna Šošić-Klindžić, James C.M. Ahern, Natalija Čondić, Ivor Janković, Krunoslav Zubčić, and Fred H. Smith

Abstract This chapter presents some new data on, and interpretations of the Croatian Middle and Early Upper Paleolithic. Alternative interpretations of the Middle/Upper Paleolithic interface in Vindija cave (situated in the Zagorje region of northwestern Croatia) are reported, together with preliminary results of research on the early Upper Paleolithic site of Bukovac pećina (situated in the region of Gorski kotar), and the late Dalmatian Middle Paleolithic sites of Mujina pećina, Velika pećina in Kličevica and Kaštel Štafilić—Resnik. The archaeological assemblage (Mousterian industry) and the results of chronometric dating make the sequences of these Dalmatian sites contemporary with late Neandertals and with the earliest known anatomically modern human groups in Europe. This recent research greatly contributes to our understanding of the distribution of Neandertals and the complexity of the Middle/Upper Paleolithic interface.

Keywords Mousterian • Aurignacian • Neandertals • Early modern humans

Introduction

Paleoanthropological, archeological, and genetic evidence from the Croatian Middle Paleolithic sites has played an important role in scientific debates about later human evolution, Neandertal adaptation, and the origins of anatomically modern humans. Despite the importance and relative abundance of the Croatian Paleolithic record, several gaps still remain. This chapter presents alternative interpretations of the Middle/Upper Paleolithic interface in Vindija, as well as preliminary results of the research conducted at Bukovac pećina in the Gorski kotar region and from three Dalmatian Middle Paleolithic sites. These sites are important for the reconstruction and comparison of behavioral processes between Central and Southeast (SE) Europe during the late Middle Paleolithic and/or early Upper Paleolithic.

The Paleolithic sites of Croatia are generally situated in two main geographic regions: continental (Hrvatsko zagorje,

I. Karavanić (✉)
Department of Archaeology, Faculty of Humanities and Social Sciences, University of Zagreb, Ivana Lučića 3, HR-10000 Zagreb, Croatia

Department of Anthropology, University of Wyoming, 1000 E. University Ave, Laramie, WY, 82071 USA
e-mail: ikaravan@ffzg.hr

R. Šošić-Klindžić
Department of Archaeology, Faculty of Humanities and Social Sciences, University of Zagreb, Ivana Lučića 3, HR-10000 Zagreb, Croatia

J.C.M. Ahern
Department of Anthropology, University of Wyoming, Laramie, WY 82071, USA
e-mail: rsosic@ffzg.hr; JAhern@uwyo.edu

N. Čondić
Archaeological Museum Zadar, Trg opatice Čike 1, HR-23000 Zadar, Croatia
e-mail: ncondic@amzd.hr

I. Janković
Institute for Anthropological Research, Gajeva 32 HR-10000 Zagreb, Croatia

Department of Anthropology, University of Wyoming, Laramie, WY 82071, USA
e-mail: ivor@inantro.hr

K. Zubčić
Croatian Conservation Institute, Mesnička 49, HR-10000 Zagreb, Croatia
e-mail: kruno.zubcic@gmail.com

F.H. Smith
Department of Sociology and Anthropology, Illinois State University, Bloomington-Normal, IL 61790, USA
e-mail: fsmith@ilstu.edu

Katerina Harvati and Mirjana Roksandic (eds.), *Paleoanthropology of the Balkans and Anatolia*, Vertebrate Paleobiology and Paleoanthropology, DOI 10.1007/978-94-024-0874-4_10

Gorski kotar, Lika) and Adriatic (Istria, Kvarner, Dalmatia). The most famous sites are Krapina and Vindija, located in the continental region of the Hrvatsko zagorje (northwestern Croatia), which differ geographically and ecologically from the Mediterranean sites found farther south on the Adriatic coast and its hinterland (Fig. 10.1). Human fossil remains and Paleolithic industries from these two sites have been analyzed and described in many publications (see Smith 1976; Simek and Smith 1997; Wolpoff 1999; Cartmill and Smith 2009 and references therein; also Janković et al. 2016). The Vindija cave, in particular, has yielded both Middle and Upper Paleolithic stratigraphic units that have had an important role in the debate surrounding the European Middle/Upper Paleolithic transition.

In addition to these famous sites, a few other localities are known from continental Croatia. The site of Velika pećina, also in the Zagorje region, was initially best known for a human frontal bone thought to be associated with the early Upper Paleolithic at the site (see Smith 1984), but later shown to be intrusive into the Upper Paleolithic strata (Smith et al. 1999). However, the site has yielded a small series of artifacts, including bone points, that are clearly derived from the early Upper Paleolithic (Malez and Vogel 1970; Karavanić and Smith 1998). About 100 years ago a single bone point was found at Bukovac pećina situated in the continental region of Gorski kotar, located between the Hrvatsko zagorje and the Adriatic (Malez 1979; Fig. 10.1). Based solely on that bone point, this site was designated an early Upper Paleolithic locality. However, the lack of corroborating finds makes this attribution questionable. Recent excavations carried out in 2010–2012 in this cave aimed to determine the layer from which the bone point originated and to obtain samples from that level for dating (Janković et al. 2011b, 2016).

In contrast to Hrvatsko zagorje, the cultural and paleoecological situation in Dalmatia (southern Croatia) is not as extensively known. Until recently, Paleolithic research in this region was rare. Archaeological material was mainly collected from the surface of open-air sites and determination was based solely on typology (Batović 1965, 1973, 1988; Vujević 2007). Many pseudoartifacts, pseudotools, and naturally fragmented pieces were found together with artifacts and tools, sometimes in mixed cultural contexts. The only site in Dalmatia with a clear and homogenous Mousterian stratigraphic sequence that was excavated systematically (1995–2003) is Mujina pećina near the city of Kaštela. Radiocarbon AMS and ESR dates obtained from Mujina pećina are the first chronometric dates for the Mousterian industry in Dalmatia (Rink et al. 2002). A test excavation of another Dalmatian site, Velika pećina in Kličevica near Benkovac, was conducted in 2006 (Karavanić et al. 2007; Karavanić 2008). More extensive excavation was carried out in 2012 and 2013, establishing a short stratigraphic sequence, with several layers yielding numerous Mousterian finds. Furthermore, small-scale excavation at the underwater open-air Mousterian site of Kaštel Štafilić—Resnik, using a grid, was conducted in 2008 and continued through 2010–2013, when only surface finds were collected over a larger area (Karavanić et al. 2009).

The Istrian peninsula is home to several Paleolithic sites, but these have yielded mostly later Upper Paleolithic occurrences. Exceptions are lower layers (H—E) from the site of Šandalja II which have produced Aurignacian artifacts (see Malez 1979; Karavanić 2009). Except for Šandalja II, only one other possible Aurignacian (Ivšišće) and two Mousterian sites (Romualdova pećina and Campanož) are known from the Istrian region of Croatia (Komšo et al. 2007; Komšo 2012).

Sites

Vindija

Vindija cave is a Middle and Upper Paleolithic site (with Holocene archaeological deposits as well), in which Neandertal skeletal remains were found (Malez 1975; Malez et al. 1980; Wolpoff et al. 1981; Janković et al. 2016). The site is situated 2 km west of the village of Donja Voća, and 20 km west of Varaždin. Its entrance lies in a narrow gorge 275 m above sea level. The cave is more than 50 m deep, up to 28 m wide and more than 10 m high at some places (Fig. 10.2). Vuković (1950), who first visited the site in 1928, excavated the cave for more than 30 years, with some interruptions. Malez started systematic excavations at Vindija in 1974, and fieldwork continued every season until 1986. Most of the lithic and faunal material, as well as all of the fossil human remains known from the site, were recovered during this latter period (Ahern et al 2004; Janković et al. 2006, 2011a). The stratigraphic profile, which is about 9 m high, comprises some 20 strata that, according to Malez and Rukavina (1979), covered the period from the onset of the Riss glaciation (oxygen isotope stage 6 or earlier) through the Holocene. The G complex, comprising five stratigraphic levels numbered G_1 (top) through G_5, produced all of the Neandertal skeletal remains from the site (although one or two fragmentary pieces may derive from earlier levels, cf., Ahern et al. 2004). Level G_3 contained approximately 100 fragmentary Neandertal skeletal remains associated with a late Mousterian industry. These remains were directly dated to >42 kBP (uncalibrated) by radiocarbon AMS (Krings et al. 2000) and 4 years later to *ca.* 38 kBP (uncalibrated) by the same method (Serre et al. 2004). An additional AMS radiocarbon date on a Neandertal bone from unit G (level unknown) has yielded results of about 44 kBP (uncalibrated; Green et al. 2010; for other dates see Wild et al. 2001; Ahern et al. 2004: Table 1).

Fig. 10.1 Map with most important Croatian sites mentioned in the text. Downloaded from GinkgoMaps-project, http://www.ginkgomaps.com licensed under CC-BY-3.0, modified by: M. Vuković

A series of human skeletal remains derive from level G_1 and diagnostic morphology from these specimens identifies them as Neandertals (Smith and Ahern 1994; Smith et al. 1999; Ahern et al. 2004). Several different radiocarbon dates on bone samples from this level have been obtained (see Ahern et al. 2004: Table 1). The most important are direct AMS dates from Neandertal skeletal remains, specifically the Vi 207 mandible and 208 parietal. These bones were first dated to 28 and 29 ^{14}C kBP, respectively (Smith et al. 1999). More recently, however, the same samples were redated, using ultrafiltration pretreatment, to 32.4±0.8 ^{14}C kBP, 31.4±0.2 ^{14}C kBP, and 32.4±1.8 ^{14}C kBP, respectively (Higham et al. 2006). Since these dates are uncalibrated, the calibrated age would be older.

Neandertal skeletal remains from level G_3 show distinct changes in facial morphology compared to earlier Neandertals;

Fig. 10.2 View from inside Vindija cave. Photo I. Karavanić

these differences characterize the entire G_3 Vindija sample, not just selected specimens (see Smith 1984; Wolpoff 1999; Ahern et al. 2004; Cartmill and Smith 2009; Janković et al. 2016). The small sample of Neandertals from level G_1 shows the same basic morphological characteristics as those from comparable elements in the G_3 sample (Wolpoff et al. 1981; Smith and Ahern 1994; Smith et al. 1999). In these features, the Vindija G_3 and G_1 specimens are intermediate between the geologically earlier Krapina (and most other) Neandertals and early modern Europeans, although still closer overall to the former group (Smith 1994; Karavanić and Smith 1998; Cartmill and Smith 2009).

The Vindija faunal remains were studied on several occasions (Miracle 1991; Brajković 2005; Brajković and Miracle 2008; Karavanić and Patou-Mathis 2009). Results from both faunal and stable isotope analysis show that the Vindija Neandertals were top-level carnivores, obtaining almost all of their dietary protein from animal sources (Richards et al. 2000; Karavanić and Patou-Mathis 2009). In this respect, the Vindija people are similar to Neandertals from other parts of Europe (e.g., Bocherens and Drucker 2006; Bocherens 2011).

The Vindija stratigraphy contains levels with both Middle (Mousterian) and Upper Paleolithic industries. In the lower Mousterian levels, tools were produced on local raw materials (Kurtanjek and Marci 1990; Blaser et al. 2002) using the Levallois method (Montet-White 1996). In contrast, the Levallois method was not used in level G_3, but local raw materials (chert, quartz, tuff, etc.) continued to be used. Seventeen percent of lithic items from G_3 were transformed into tools. This Late Mousterian industry is dominated by sidescrapers, notched pieces, and denticulates, but also contains some Upper Paleolithic types (e.g., endscrapers, see Karavanić and Smith 1998). Some endscrapers might have come from the Upper Paleolithic levels as a result of sediment mixing. However, in addition to flake technology, level G_3 includes evidence of bifacial technology and blade technology (Karavanić and Smith 1998). New analyses (Karavanić and Patou-Mathis 2009) show that some "retouchers" from the G layer (Karavanić and Šokec 2003;

Ahern et al. 2004) are in fact pseudoartifacts. Markings on a cave bear baculum (Karavanić and Smith 1998) could also be the result of natural processes and not of human activity (Karavanić and Patou-Mathis 2009).

As in level G_3, a combination of Middle and Upper Paleolithic typological characteristics is also present in the stone tool assemblage from level G_1, where various lithics, bone points, and Neandertal fossils were found. Some of the lithic items from this level, previously identified as tools, probably represent pseudotools (see Zilhão and d'Errico 1999; Janković et al. 2006: 4; Zilhão 2009). While relatively meager, the lithic industry of this level suggests continuation of the Mousterian technological and typological tradition (with the absence of the Levallois method). In contrast, the bone tools from the same level are typical of the Upper Paleolithic, and therefore, this industry was attributed to the Aurignacian (Karavanić 1995). This unusual association of Neandertal skeletal remains and Upper Paleolithic bone points in level G_1 has been explained either as a result of mixing of different strata (Kozlowski 1996; Zilhão and D'Errico 1999; Bruner 2009; Zilhão 2009), or as a true cultural assemblage (Montet-White 1996; Karavanić 1995, 2000b, 2007; Karavanić and Smith 1998, 2000; Janković et al. 2006; Karavanić and Patou-Mathis 2009).

A number of interpretations have been given for the G_1 lithic industry (e.g., Karavanić 1995, 2000b; Kozlowski 1996; Montet-White 1996; Karavanić and Smith 1998; Miracle 1998; Zilhão 2009). Kozlowski (1996) sees it as Mousterian; Svoboda (2001, 2006) has suggested affinities to the Szeletian; while Montet-White (1996) used the term Olschewian (see also Karavanić and Smith 1998). Karavanić (2000b; 2007) used the Olschewian to designate a possible regionally specific "transitional" industry. Recently, Zilhão (2009) also claimed that this industry is Szeletian. More generally, Straus (1999), Montet-White (1996), Karavanić and Smith (1998; 2011), Karavanić and Patou-Mathis (2009), Ahern et al. (2004), and Janković et al. (2006, 2011a, b) see the unusual G_1 associations in the context of a more complex pattern that characterizes the Middle/Upper Paleolithic transition in this region, as some of the so-called transitional industries that combine Mousterian and certain Upper Paleolithic technological and typological aspects are found at many localities, especially in Central Europe (for a more detailed insight and references see Janković and colleagues 2006, 2011a; Karavanić and Smith 2011).

Bukovac Pećina

Bukovac pećina is located in Croatia's Gorski kotar region, southeast of the town of Lokve on the northwestern slopes of Sleme Hill (Malez 1979). It is situated in a mountain region in the border zone between the Mediterranean and continental zones of Croatia, closer to the Adriatic than to the Hrvatsko zagorje sites (Fig. 10.1). The cave was first test excavated by Kormos (1912) and Szilágy in 1911 (Malez 1979). A trench excavated in the front of the cave yielded no significant discoveries, but a test pit deeper inside the cave resulted in the recovery of faunal remains and a bone point. The point was assigned to different cultures (Malez 1979), but today the overriding view is that it belongs to the Aurignacian or Olschewian (Malez 1979; Montet-White 1996; Horusitzky 2004). The base of the point is missing, but based on the sudden thinning of the widest part it can be argued that it was a so-called Mladeč point. No additional artifacts were recovered during the 1970s excavations by Malez (1979). Therefore, the assignment of the industry to the Upper Paleolithic on the basis of this single point might be questionable. One of the major aims of the excavations under the direction of I. Janković from 2010 to 2014 (Janković et al. 2011b), was to determine the layer from which this find originated, based on the stratigraphy provided by Kormos (1912), and to obtain material for dating (Fig. 10.3). Unpublished radiocarbon dates confirm the Aurignacian timeframe. In addition, a second artifact (a stone core) was found in a trench in front of the cave in 2013.

Velika Pećina in Kličevica

Velika pećina is located in the canyon surrounding the Kličevica creek near the town of Benkovac in Dalmatia, Southern Croatia. Savić (1984) collected several lithics from the cave and its surroundings. Malez visited the site, collected several artifacts and conducted a small-scale excavation in the cave (Savić, personal communication). Božićević (1987) published the layout of the cave and a longitudinal cross-section. Karavanić and Čondić (2006), visited the site with a small team in 2003 and collected several artifacts from the surface of the cave floor. A test excavation was conducted in 2006 (Fig. 10.4). In a small trench (1 × 2 m initially, somewhat expanded during the excavation in order to reach the cave wall) several Mousterian levels were established. A total of 105 finds were found *in situ*, among which stone artifacts dominate, while animal bones and teeth are less abundant. Additionally, a number of items were found in the sieve. Animal bones from level D were dated by radiocarbon AMS to *ca.* 39 ^{14}C kBP (Karavanić et al. 2007). Recently, an animal bone from level D was cut in two pieces and sent for AMS radiocarbon dating (Karavanić et al. 2014). Half of the bone was prepared for the AMS in the standard way, while the other half was prepared by ultrafiltration. The first sample, prepared in standard way was dated to *ca.* 35 ^{14}C kBP. The other, prepared by ultrafiltration, was dated to *ca.* 32 ^{14}C kBP (Karavanić

Fig. 10.3 Excavation at Bukovac pećina. Photo I. Karavanić

Fig. 10.4 Excavation at Velika pećina in Kličevica. Photo I. Karavanić

et al. 2014). Comparing with an earlier date from Velika pećina and the dates from Mujina pećina (Rink et al. 2002), these new dates are too recent for late Mousterian in Dalmatia (for further discussion of these dates see Karavanić et al. 2014). The tools (Fig. 10.5) are small (similar to the so-called Micromousterian), and made on local chert. Based on typology (most tools are scrapers, some of which are transversal), the artifacts represent the Late Mousterian (or Balkan Charentian according to the terminology of Kozlowski).

Excavation squares from the earlier excavations were expanded, and two additional squares were opened in 2012. In one of them the basal rock was soon unearthed, while in the other a layer yielding Mousterian artifacts and animal bones was found after a layer of mixed sediment was removed. Additional squares were opened in 2013 and further excavation of the site is planned.

Mujina Pećina

Mujina pećina is situated in the hills north of Trogir and west of Split (Fig. 10.1). The cave is about 10 m deep and 8 m wide, located at about 280 m above sea level. Finds were initially collected in 1977 from the surface inside and outside the cave (Malez 1979), and the first test excavation took place in 1978 (Petrić 1979). In 1995, a joint project of the Department of Archaeology at the University of Zagreb and the Museum of the Town of Kaštela launched systematic excavations. The last year of excavation was 2003. Following standard archaeological methodology for Paleolithic cave sites, all artifacts and ecofacts with dimensions of 2 cm or more in size were recorded in three dimensions on site plans, and all sediments were sieved (Fig. 10.6). The northern stratigraphic profile inside the cave is only about 1.5 m deep and comprises poorly sorted Quaternary sediments composed of large fragments of

Fig. 10.5 Scrapers from Velika pećina in Kličevica. Photo I. Karavanić

Fig. 10.6 Excavation at Mujina pećina. Photo S. Burić

carbonate rock, gravel and sand grains, rarely silt, and some clay (for further discussion of the stratigraphy of the site see Karavanić and Bilich-Kamenjarin 1997; Rink et al. 2002). The interface between Level E2 and E1 was dated by AMS to 45 ^{14}C kBP, while the AMS age of overlying levels, calculated as the mean of 5 dates from these levels, is about 39 ^{14}C kBP (for discussion on these and ESR dates see Rink et al. 2002).

Two localized areas of burning, probably representing open, unconstructed and unpaved Mousterian hearths, were found in the occupation level D2. Anthracotomical analysis shows that *Juniperus sp.* was used for fuel at both hearths (Culiberg, personal communication; Karavanić et al. 2008b).

All lithic finds are attributable to the Mousterian industry (Karavanić et al. 2008a, b). No human skeletal remains were recovered. However, given the nature of the lithic assemblage and the radiometric dates, it is assumed that Neandertals were responsible for the evidence of human occupation at the site. Presence of Levallois debitage was detected in levels D1 and D2. In levels B and C, tools make up 1/3 of the lithic assemblage. Of these, flakes are the dominant technological product. The most frequent tool types in these levels are denticulates and notched pieces. Tools are generally small in size (around 3 cm in length) and strongly resemble the so-called Micromousterian. Of the total lithic material from levels D1 and D2 only about 1/5 are definite tools. The most frequent tool types are simply retouched flakes, made on local chert pebbles and nodules, which are often small. It seems more likely that the use of small pebbles available near the cave, as well as the low flaking quality of larger pieces of some local cherts (rather than the intentional selection of small pebbles for production of small tools) dictated the small tool size in the Mousterian of Dalmatia (Karavanić et al. 2008a, b).

Faunal remains from Mujina pećina also show differences in dominance of animal species between these two stratigraphic complexes, especially in their frequency. The relative frequency of chamois/ibex, equids, and large-sized carnivores increases dramatically from the lower levels D1/D2 to levels B and C, while the relative frequency of hare and red deer decreases significantly (Miracle 2005). Red deer and hare are often regarded as indicative of temperate conditions, and their decrease could be interpreted as evidence of a shift towards cooler and drier climates in levels B and C. However, we believe sedimentary analyses to be a more reliable indicator of local climate. While levels D1 and D2 contain cryoclastic stone debris indicative of cold climate with some or no gravel and/or fine sediment, levels B and C contain brown sandy sediment with stone debris indicative of a relatively warm climate (Karavanić and Bilich-Kamenjarin 1997; Rink et al. 2002). Data from the fossil plant remains agree with the climatic conditions ascertained on the basis of sediment data (Karavanić et al. 2008b). The discordance between sediments and faunal assemblages therefore most likely reflects prey selection by humans and/or other bone collectors (Miracle 2005).

The frequency of large carnivores, especially bears, that used the cave for hibernation and as a nursery, suggests their more regular occupation of Mujina pećina in levels B and C relative to levels D1 and D2 (Miracle 2005). During the accumulation of levels B and C people visited the site, but their cultural remains are less numerous than in some earlier levels (E1, E2 and E3). Impact scars and cut marks are present in all analyzed levels (B+C and D1+D2) but are found only on faunal long-bone shafts, suggesting first defleshing and then cracking long bones for marrow extraction by humans. Alongside the evidence of carcass processing, the dominance of prime-age adults among red deer, chamois/ibex, and large bovid assemblages suggests hunting activities by the Mujina pećina inhabitants (Miracle 2005). There is a difference between levels B and C, and D1 and D2 reflected in animal activities in the cave. In levels D1 and D2 carnivores were scavenging human food refuse, while in the levels B and C bear activity is noticeable. The assemblage with evidence of human processing does not indicate targeting particular prey to the exclusion of other species or specialized procurement (Miracle 2005; Karavanić at al. 2008b).

The northern niche, which provided a good shelter from bad weather conditions, was the most intensively used area of the cave during the formation of stratigraphic units B, D1, D2, and E2 (Nizek and Karavanić 2012). On the other hand, most of the material from the level E1 was concentrated along the southern edge of the excavation area, while another extensively used area for levels E2 and E3 was the entrance to the cave. The oldest levels (E3, E2, E1) at Mujina pećina are richest in anthropogenic finds, indicating much more intensive human activity than in younger levels. The richest levels may suggest long-term occupation (Karavanić 2000a), but may also result from the repeated use of the site for brief occupations (see Conard 1996). The lower density of finds in the upper levels (B, D1 and D2) suggests that the site was used as an occasional hunting camp during the formation of these levels (Nizek and Karavanić 2012).

There is strong evidence that people used Mujina pećina during the autumn throughout the sequence (at least in the analyzed layers), as well as for spring visits in level B (Miracle 2005; Karavanić et al. 2008b). There is no evidence of human activities in Mujina during the summer and winter, while bears were active at the cave during the winter in level B. These observations bring up the question of where the Mujina pećina people lived during the summer and the winter. One distinct possibility is that they were closer to the coast during the winter to take advantage of seasonally migrating game and relatively warmer and more sheltered locations. If so, such locations are most likely under sea level at the present time, or were damaged and washed away by subsequent changes in sea level.

Kaštel Štafilić: Resnik

The site of Resnik is a well-known locality from the Hellenistic and late Roman periods, and finds have been collected both on land and under water (Brusić 1990, 2004). Neolithic finds also have been collected from an underwater site, but at a different location from the Hellenistic and late Roman finds (Brusić 2004). Of particular importance is the discovery of an underwater site that yielded Paleolithic artifacts. The site is located at a depth of about 4 m, and the discovery was reported by I. Svilan (Karavanić et al. 2009).

Small-scale excavation at the site of Kaštel Štafilić using a grid was conducted in 2008 (Karavanić et al. 2009) and continued in 2010–2015, when only surface finds were collected (Fig. 10.7). The methodology used is described in detail elsewhere (Karavanić 2015). The locality itself represents an open air site dating to the time when the sea level was much lower than today. Although the finds are somewhat disturbed (due to the action of waves and other factors) it seems that their accumulation is not a result of displacement from another locality as was reported earlier (Karavanić et al. 2009).

Among the tools, several pseudotools and numerous naturally broken pieces of chert were found. The excavations ascertained the presence of the centripetal method and confirmed that the artifacts (side scrapers are most abundant) belong to the Mousterian industry. The finds are not numerous enough to allow a more detailed determination of the type of Mousterian, and the question whether the site is contemporaneous to, or older than the occupation at nearby Mujina pećina remains open. There is a possibility that the same group of hunters used both sites during different seasons.

This site is important for several reasons. It adds to the overall picture of the area that was once land and connects it to other sites. It also allows for the development of a methodology for underwater excavation of Paleolithic sites, which is one of the important directions Paleolithic archaeology will take in the near future. Additionally, it opens up a whole set of questions related to the processes of formation of underwater sites.

Interpretative Summary

Late Middle and early Upper Paleolithic sites in Croatia are found in two geographical regions: continental and Adriatic. This enables us to study the adaptation of late Neandertals and early modern humans in two different paleoenvironmental settings. The most important site for the study of the Middle/Upper Paleolithic interface in northwestern Croatia is the Vindija cave, as it contains fossil remains of late Neandertals associated with artifacts. Lately, it has been claimed (Zilhão 2009) that the most recently published date of 32.4 ^{14}C kBP (Higham et al. 2006) for the Vindija G_1 layer Neandertals is likely a minimum date, and a recent study by Higham and colleagues (2014) implies the same. Zilhão (2009) further claimed that the actual age of these remains must be older in order for Vindija to support the assimilation model of modern human origins (see Smith et al. 1989, 2005; Cartmill and Smith 2009). Zilhão (2011) holds that all Neandertals and the Mousterian predate all early modern humans and the Upper Paleolithic. Thus for him, Vindija must predate any occurrence of modern humans or the Upper Paleolithic to constitute evidence of a Neandertal contribution to early modern populations.

The assimilation model posits that archaic Eurasians, including Neandertals, made small, but not insignificant, contributions to early modern human populations as the latter spread throughout Eurasia (Smith et al. 1989, 2005; Ahern et al. 2013). Interbreeding between early modern humans and Neandertals, as well as other archaic humans, has been suggested for some time based on morphological studies (see reviews in Wolpoff 1999; Smith 1994; Cartmill and Smith 2009). More recently, genetic studies have also supported interbreeding (Green et al. 2010; Sankararaman et al. 2012, 2014; Prüfer et al. 2014), although they have also shown that the Neandertal (and other archaic human) contributions to the modern human gene pool were uniformly small. Initially, Green and colleagues (2010) estimated that interbreeding between Neandertals and early moderns must have occurred before Asian and European modern populations diverged from one another, at *ca.* 100 ka. More recently, however, Sankararaman et al. (2012) found that the last genetic exchange occurred most likely between 37 and 86 ka. This range overlaps with the dates for Vindija G_3. Zilhão's (2009) assertion that Vindija has to date earlier that the first modern humans in Central Europe for the assimilation model to apply, is simply not the case. We believe that even if the Vindija G_1 dates were slightly older when calibrated, they still overlap with early modern dates such as those from the Grotta del Cavallo (Benazzi et al. 2011) and Oase (Trinkaus et al. 2003). Thus, as we explain in detail elsewhere (Karavanić and Smith 2011; Janković et al. 2011a, 2016; Ahern et al. 2013), we interpret the Vindija morphology, not as Zilhão does, but rather as an indication of modern human gene flow into a late Neandertal population. In the context of that interpretation, the younger age of Vindija makes perfect sense (contra Zilhão). It is important to reemphasize that the assertion that Vindija reflects modern human gene flow into late Neandertals is a morphological argument, not demonstrated by the current genetic evidence. Still, it seems highly unlikely that gene flow occurred in only one direction, particularly given the 6–9 % contribution of Neandertals to early modern Central Europeans (Fu et al. 2015).

Much of the debate concerning the possibility of Neandertal–early modern human interaction at Vindija is

Fig. 10.7 Collecting material from the surface of underwater site Kaštel Štafilić. Photo K. Zubčić, Croatian Conservation Institute

based on the archaeological industry from level G_1. In this layer, various lithics, bone points and Neandertal skeletal remains were found, and a mixture of Middle and Upper Paleolithic typological characteristics is present in the stone tool assemblage. It is likely that some of the lithics (e.g., Vi 1061, Vi 3383) are pseudotools, as argued recently by Zilhão (2009). The presence of pseudotools and the results of refitting (Bruner 2009; Zilhão 2009) confirms that there was some mixing of different layers, and that the presence of certain Upper Paleolithic lithic tool types made on high quality silex from G_1 and G_3 layers might be explained as a result of this mixing (Karavanić and Smith 2011). Different authors have long recognized that both bioturbation and cryoturbation occurred at Vindija and likely resulted in mixing of elements from different layers in some parts of the cave (Malez and Rukavina 1975; Smith 1984; Kozlowski 1996; d'Errico et al. 1998; Karavanić and Smith 1998). However, these phenomena are not seen uniformly throughout the site, and the area where many of the relevant finds were found does not show evidence of disturbance (Karavanić and Smith 1998, 2011). Furthermore, the change in the raw material seen from early Middle Paleolithic levels to late Upper Paleolithic levels (increase in chert and decrease in quartz; see Blaser et al. 2002; Ahern et al. 2004: Table 9) is more easily explained as a reflection of behavioral change.

In light of the documented disturbance of layers, the Olschewian hypothesis regarding the transitional industry of the G_1 layer (Karavanić 2000b, 2007) is not likely. While Pacher (2010) correctly pointed out the lack of attributable elements required to define Olschewian as an Initial Upper Paleolithic industry, her suggestion that fossil human remains from Vindija level G_1 are not Neandertals has no foundation. Even though the human remains are very fragmented, as she properly noted, their anatomical features clearly indicate attribution to Neandertals with some modern human characteristics (see Smith and Ahern 1994; Karavanić and Smith 1998; Wolpoff 1999; Cartmill and Smith 2009). The attribution of the G_1 industry to the Szeletian was first proposed by Malez (1979) more than 30 years ago, although it is unclear whether he was referring to the G_1 unit specifically, or to some other G unit layer. Likewise, Svoboda (2001, 2006) noted some similarities between the G_1 layer of Vindija and the Szeletian, and, recently, Vindija G_1 was attributed to the Szeletian by Zilhão (2009). However, the evidence for the presence of the Szeletian industry in G_1 is based solely on one tool, a nicely shaped bifacial point. There is no evidence of *in situ* production of this tool, and it was made on nonlocal raw material (red radiolarite) that was imported from Hungary (Montet-White 1996; Karavanić and Smith 1998; Biró and Markó 2007).

Therefore, the best determination for the G_1 lithic industry is Mousterian (Karavanić and Smith 2011, see also Kozlowski 1996), while the Szeletian bifacial stone point should be seen as an import, the result of the contact among various Neandertal groups (if the Szeletian was produced by Neandertals) or a contact between Neandertals and early modern humans (if the Szeletian was produced by early modern humans) between northwestern Croatia and Hungary. Although most Szeletian assemblages and sites from Hungary are older than Vindija G_1 (Adams 2009), a contemporary late phase of the Szeletian is known in western Slovakia (Kaminská et al. 2011). Even though we cannot completely rule out the possibility of disturbed contexts, we argue that the Upper Paleolithic elements in the same level, especially the bone points and possibly some lithic types, may well be a result of contact (exchange or acculturation) between Neandertals and anatomically modern groups (Karavanić and Smith 2011).

Although direct dating of the bone points from Vindija and Velika pećina (both in the Hrvatsko zagorje, NW Croatia) failed (Smith et al. 1999), an age of 34 ^{14}C kBP was determined for the "i" layer of that site (Malez and Vogel 1970). Thus, the same age can be assumed for the bone points (most likely with split bases) from the same layer of the same site (Malez and Vogel 1970). A bone point (most likely with a split base) from Divje babe I (Slovenia) comes from a layer that has been dated to about 35 kBP (Nelson 1997). This point was directly dated to *ca.* 30 ^{14}C kBP (Moreau et al. 2015) while points from Potočka zijalka (Slovenia) are dated to between 35 and 29 kBP (Hofreiter and Pacher 2004; Moreau et al. 2015). The oldest bone projectile points from Hungary are dated to 37–38 kBP (Davies and Hedges 2008–2009). All of these dates are uncalibrated. Although we do not have direct dates on the points, a date from a comparable archaeological layer suggests that the bone points from Velika pećina (Hrvatsko zagorje) are older than, or contemporaneous with, the Vindija Neandertals. If we adhere to the generally accepted view that such points are associated with modern humans, this also raises the question of possible interactions between these groups.

Upper Paleolithic bone points have also been found at other Croatian sites, such as the presumed Aurignacian specimen from Bukovac cave discussed earlier. From the eastern Adriatic, only a single bone point has been found, and it comes from the layer H at the site of Šandalja II in Istria. It is relatively small compared to the points from Central Europe and has a split-base and rounded cross section. It is similar to points from the Franco-Cantabrian Magdalenian (Straus, personal communication); and based on the recent date for the layer F at Šandalja II, it should be older than 32 kBP (Richards et al. 2015), if it did not originally come from one of the Epigravettian layers. The Dalmatian area has several known Mousterian sites: open air sites with surface finds at the area of Ravni Kotari, north of Zadar; the open air site Veli Rat at the island of Dugi; the Giljanovići open air site north of Kaštela; Velika pećina in

Kličevica near Benkovac and Mujina pećina near Kaštela cave sites; and the Kaštel Štafilić—Resnik underwater site. However, to date these have not been fully investigated. Only the site of Mujina pećina has been systematically excavated. Systematic research at Kaštel Štafilić is in progress and systematic excavations of Velika pećina in Kličevica started in 2013. To date no bone points have been recovered from these localities.

It is clear that mixture of artifacts from different levels occurred at Vindija, and this fact alone casts a cloud of suspicion on the nature of the level G_1 cultural assemblage. However, given the fact that these potential examples of interaction are rare and often ephemeral, it seems wise not to entirely dismiss the Vindija evidence. For example, another site offered as evidence for Neandertal acculturation, the Gotte du Renne at Arcy-sur-Cure, has been argued, most recently by Higham and colleagues (2010) and Bar-Yosef and Bordes (2010), to show effects of disturbance resulting in mixing material from different levels. The evidence from Grotte du Renne, in the form of the Initial Upper Paleolithic Châtelperronian assemblage, extends over several archaeological levels, making extensive mixing seem unlikely (see Hublin et al. 2012). However, Vindija level G_1 is a very different story. It is a relatively thin level that is not found in all parts of the cave, and there is a reason to suspect considerable erosion of deposits from caves in Central Europe during this time span (Malez and Rukavina 1979). Thus, the nature of G_1 as an archaeological level, plus the obvious presence of cave bear in the cave, makes it difficult to conclusively demonstrate that mixing of layers did not occur.

Still there is some evidence against the argument that mixing explains all the interesting associations in Vindija level G_1. First of all, there is no evidence that the Neandertal skeletal material in G_1 originates from another level. The fragmentary cranial material from the younger F complex is basically modern (Smith et al. 1985) and the G_1 remains are clearly Neandertal, as discussed previously. Moreover, the direct AMS dates on two of the G_1 Neandertals are significantly younger than the dates obtained from the Vindija G_3 Neandertals. Additionally, the Vi 207 mandible, as well as other specimens such as the Vi 307 zygomatic and Vi 308 supraorbital torus fragment, have the distinctive red clayey/loam sediment of level G_1 embedded in crevices and spaces in the bones, and lack the distinctly different sediments of stratigraphically adjacent layers. The Vi 3437 split-based bone point also had the same distinctive red sediment and was found directly next to Vi 207 (Radovčić, personal communication). Furthermore, the same distinctive red sediment infiltrated the Vi 3439 massive based (Mladeč) point. Of course these factors do not prove that the bone point could not be in level G_1 as the result of mixture of the layers, but it makes it less likely. It should also be noted that the F complex does not have other examples of split-based bone points. Thus, there is not an assemblage of such points from which one ended up artificially mixed into G_1.

Bruner's (2009) study of refitting shows a relatively high percentage of refit among pieces from different stratigraphic levels. She points out that refitted pieces come from levels presumably separated by another level, which would suggest particularly poor stratigraphic control. However, in many cases, levels of the G complex are not continuous in the cave, so that refitted pieces from say G_1 and G_3 may actually reflect mixing between contiguous levels. Thus, the extent of the problem is likely not as great as she suggests.

In discussing the Châtelperronian, Klein (2009) indicates that more than one or two sites with possible evidence of Neandertal–early modern human interaction are needed to rule out coincidence of other factors. Zilhão (2011) is skeptical of claims of such interaction for another reason; as explained previously, he believes all Initial Upper Paleolithic (like the Châtelperronian) is earlier than the appearance of modern humans and their cultural manifestations in Europe. Because of the problems with Vindija, we know that it will never convince skeptics, regardless of the basis for their skepticism. But we believe that there is a strong case to be made that enough evidence exists to suggest the real possibility of a culturally based interaction, to go along with the indications of biological interaction, at Vindija. However, the Vindija case also demonstrates the difficulties inherent in separating a culturally mixed circumstance from one of natural mixture and thus serves as a reminder how carefully these ephemeral manifestations must be excavated in the future.

Compared to Vindija, the Adriatic region offers little to aid in understanding the Middle-Upper Paleolithic transition, but it does offer some important insights. There is evidence that people used Mujina pećina during the autumn and spring while there is no evidence for hominin activity at this site during the summer and winter (Miracle 2005) which raises a question where these hominins lived during these periods (Karavanić et al. 2008b). They might have moved closer to the coast and one of the locations on their trail might be the Kaštel Štafilić underwater site, while other locations are most likely also below see level at present time, or destroyed by subsequent changes in sea level. Although no diagnostic fossil hominin remains have been found at Dalmatian Middle Paleolithic sites, the archaeological assemblage (Mousterian industry) and the results of chronometric dating indicate that their sequences are contemporary with the late Neandertals and earliest known anatomically modern human groups in Europe.

Sites dated to the early Upper Paleolithic are rare in this area, as well as in the whole eastern Adriatic (Karavanić 2009; Mihailović 2009), and there is a chronological gap between the late Middle and early Upper Paleolithic (see Karavanić 2009; Papagianni 2009; Papagianni and Morse 2013). Further, no industry from a single site of the eastern

Adriatic region shows a progressive or transitional nature, and there is no evidence of an *in situ* transition at any site in this region. Possible reasons for this situation are as follows: insufficient level of research, flooding or abrasion as a result of rising sea levels, and/or low population density in the eastern Adriatic during the Middle/Upper Paleolithic transition and early Upper Paleolithic (Karavanić 2009). It is also possible that Neandertal populations had disappeared from this region before the arrival of the first anatomically modern humans (see Papagianni 2009; Papagianni and Morse 2013), or Neandertals where late inhabitants of several niches in eastern Adriatic (Šošić Klindžić et al. 2014) which were avoided by early modern humans.

Although it is not clear why no site in the eastern Adriatic region thus far documents the Middle/Upper Paleolithic transition, and why early Upper Paleolithic sites are very rare, new research on Dalmatian Mousterian sites enables a better comparison with other Adriatic and continental sites. Furthermore, this new research makes a contribution towards our understanding of the distribution of Mousterian people, the complexity of the processes that underlie the interactions between Middle and Upper Paleolithic populations in the late Pleistocene of Central and SE Europe, and the reconstruction of the mobility patterns of Paleolithic populations. Therefore, it is of crucial importance to continue research that will include mapping and test excavations of both cave and open-air sites, as well as underwater research at the Kaštel Štafilić site and underwater survey for Paleolithic sites.

Acknowledgments We are grateful to the organizers of the Symposium (Human Evolution in the Southern Balkans), particulary to our colleague Katerina Harvati for inviting us to contribute to both the conference and to the edited volume. The results presented in this chapter in part stem from the projects of the Ministry of Science, Education and Sports (130-0000000-0871), Ministry of Culture of Republic of Croatia (Kaštel Štafilić and Velika pećina in Kličevica) and Croatian Science Foundation. Field work was also supported by the University of Wyoming, Illinois State University, Faculty of Humanities and Social Sciences—University of Zagreb and Zadar County. We are also indebted to our colleagues Katerina Harvati, Mirjana Roksandic and three anonymous reviewers for their suggestions and comments, making this a better chapter.

References

Adams, B. (2009). The Bükk Mountain Szeletian: Old and new views on "transitional" material from the eponymous site of the Szeletian. In M. Camps & P. Chauhan (Eds.), *Sourcebook of Paleolithic transitions: Methods, theories and interpretations* (pp. 427–440). New York: Springer.

Ahern, J., Janković, I., Voisin, J., & Smith, F. H. (2013). Modern human origins in Central Europe. In F. H. Smith & J. C. M. Ahern (Eds.), *The origins of modern humans: Biology reconsidered* (pp. 151–221). Hoboken: Wiley-Blackwell.

Ahern, J. C. M., Karavanić, I., Paunović, M., Janković, I., & Smith, F. H. (2004). New discoveries and interpretations of hominid fossils and artifacts from Vindija cave, Croatia. *Journal of Human Evolution, 46*, 25–65.

Bar-Yosef, O., & Bordes, J.-G. (2010). Who were the makers of the Châtelperronian culture? *Journal of Human Evolution, 59*, 586–593.

Batović, Š. (1965). Prvi paleolitski nalazi u srednjoj Dalmaciji. *Diadora, 3*, 205–209.

Batović, Š. (1973). Prapovijesni ostaci na zadarskom otočju. *Diadora, 6*, 5–139.

Batović, Š. (1988). Paleolitički i mezolitički ostaci s Dugog otoka. *Poročilo o raziskovanju paleolita, neolita in eneolita v Sloveniji, 16*, 7–54.

Benazzi, S., Douka, K., Fornai, C., Bauer, C., Kullmer, O., Svoboda, J., et al. (2011). The early dispersal of modern humans in Europe and implications for Neanderthal behaviour. *Nature, 479*, 525–528.

Biró, K., & Markó, A. (2007). Stones "hit the road". In T. Kovács (Ed.), *Highways and byways in space and time* (pp. 14–19). Budapest: Hungarian National Museum.

Blaser, F., Kurtanjek, D., & Paunović, M. (2002). L'industrie du site néandertalien de la grotte de Vindija (Croatie): Une révision des matières premières lithiques. *L'Anthropologie, 106*, 387–398.

Bocherens, H. (2011). Diet and ecology of Neanderthals: Implications from C and N isotopes. Implications from bone and tooth biochemistry. In N. Conard & J. Richter (Eds.), *Neanderthal lifeways, subsistence and technology* (pp. 73–85). Dordrecht: Springer.

Bocherens, H., & Drucker, D. (2006). Dietary competition between Neanderthals and modern humans: Insights from stable isotopes. In N. Conard (Ed.), *When Neanderthals and modern humans met* (pp. 129–143). Tübingen: Kerns Verlag.

Božićević, S. (1987). Speleološke pojave benkovačkog kraja i njihovo značenje. In J. Medini, J. Petricioli, & M. Pekić (Eds.), *Benkovački kraj kroz vjekove, zbornik 1* (pp. 29–35). Benkovac: Narodni list.

Brajković, D. (2005). *Korelacija tafodema skupine Ungulata iz gornjopleistocenskih špilja Vindija, Velika pećina i Veternica u sjeverozapadnoj Hrvatskoj*. Unpublished Ph.D. Dissertation. University of Zagreb.

Brajković, D., & Miracle, P. T. (2008). Middle Palaeolithic and early Upper Palaeolithic subsistence practices at Vindija cave, Croatia. In A. Darlas & D. Mihailović (Eds.), *The Palaeolithic of the Balkans* (pp. 107–116). Oxford: Archaeopress.

Bruner, K. M. (2009). *Testing stratigraphic integrity of Upper and Middle Paleolithic deposits in Vindija cave (Croatia): A chipped stone refitting analysis*. MA Thesis, University of Kansas, Lawrence.

Brusić, Z. (1990). Resnik kod Kaštel Novog: Helenističko pristanište. *Arheološki Pregled, 29*(1988), 117–119.

Brusić, Z. (2004). *Resnik—Hydroarchaeological research*. Kaštela: Muzej grada Kaštela.

Cartmill, M., & Smith, F. (2009). *The human lineage*. Wiley-Blackwell: Hoboken.

Conard, N. J. (1996). *Middle Paleolithic settlement in Rhineland. In N. J. Conard (Ed.) Middle Paleolithic and Middle Stone Age settlement system*. U.I.S.P.P.—XIII Congrés (pp. 255–268). Forlì: Abaco.

d'Errico, F., Zilhão, J., Julien, M., Baffier, D., & Pelegrin, J. (1998). Neanderthal acculturation in western Europe? A critical review of the evidence and its interpretation. *Current Anthropology, 39*(Suppl.), S1–S44.

Davies, W., & Hedges, R. (2008–2009). Dating a type site: Fitting Szeleta cave into its regionalchronometric context. *Praehistoria, 9–10*, 35–45.

Fu, Q., Hajdinjak, M., Moldovan, O. T., Constantin, S., Mallick, S., Skoglund, P., et al. (2015). An early modern human from Romania with a recent Neanderthal ancestor. *Nature, 524*, 216–219.

Green, R., Krause, J., Briggs, A., Maricic, T., Stenzel, U., Kircher, M., et al. (2010). A draft sequence of the Neandertal genome. *Science, 328*, 710–725.

Higham, T., Douka, K., Wood, R., Bronk Ramsey, C., Brock, F., Basell, L., et al. (2014). The timing and spatiotemporal patterning of Neanderthal disappearance. *Nature, 512*, 306–309.

Higham, T., Jacobi, R., Julien, M., David, F., Basell, L., Wood, R., et al. (2010). Chronology of the Grotte du Renne (France) and implications for the context of ornaments and human remains within the Châtelperronian. *Proceedings of the National Academy of Sciences of the United States of America, 107*, 20234–20239.

Higham, T., Ramsey, C. B., Karavanić, I., Smith, F. H., & Trinkaus, E. (2006). Revised direct radiocarbon dating of the Vindija G1 Upper Paleolithic Neandertals. *Proceedings of the National Academy of Sciences of the United States of America, 103*, 553–557.

Hofreiter, M., & Pacher, M. (2004). Using ancient DNA to elucidate raw material origin of bone points from Potočka zijalka (Slovenia): Preliminary results. In M. Pacher, V. Pohar, & G. Rabeder (Eds.), *Potočka zijalka: Palaeolontological and archaeological results of the campaigns 1997–2000* (pp. 201–210). Vienna: Österreichischen Akademie der Wissenschaften.

Horusitzky, F. Z. (2004). Les artefacts en os et bois de cerf à Bukovac, Lokve (Croatie): Une seconde flûte possible? Relations entre les chasseurs de Lokve et les montagnards d'Olcheva au début du Paléolithique supérieur. *Arheološki vestnik, 55*, 9–37.

Hublin, J.-J., Talamo, S., Julien, M., David, F., Connet, N., Bodu, N., et al. (2012). Radiocarbon dates from the Grotte du Renne and Saint-Césaire support a Neandertal origin for the Châtelperronian. *Proceedings of the National Academy of Sciences of the United States of America, 109*, 18743–18748.

Janković, I., Ahern, J. C. M., Karavanić, I., & Smith, F. H. (2016). The importance of Croatian Pleistocene hominin finds in the study of human evolution. In K. Harvati, & M. Roksandic (Eds.), *Paleoanthropology of the Balkans and Anatolia: Human evolution and its context*. Dordrecht: Springer.

Janković, I., Karavanić, I., Ahern, J. C. M., Brajković, D., Mauch Lenardić, J., & Smith, F. H. (2006). Vindija cave and the modern human peopling of Europe. *Collegium Antropologicum, 30*, 457–466.

Janković, I., Karavanić, I., Ahern, J. C. M., Brajković, D., Mauch Lenardić, J., & Smith, F. H. (2011a). Archaeological, paleontological and genomic perspectives on late European Neandertals at Vindija cave, Croatia. In S. Condemni & G.-C. Weniger (Eds.), *Continuity and discontinuity in the peopling of Europe: One hundred fifty years of Neanderthal study* (pp. 299–313). Dordrecht: Springer.

Janković, I., Karavanić, I., Ahern, J. C. M., & Mihelić, S. (2011b). Pećina Bukovac. *Hrvatski arheološki godišnjak, 7*(2010), 477–479.

Kaminská, Ľ., Kozlowski, J. K., & Škrdla, P. (2011). New approach to the Szeletian—Chronology and cultural variability. *Eurasian Prehistory, 8*(1–2), 29–49.

Karavanić, I. (1995). Upper Paleolithic occupation levels and late-occurring Neandertal at Vindija cave (Croatia) in the context of Central Europe and the Balkans. *Journal of Anthropological Research, 51*, 9–35.

Karavanić, I. (2000a). Research on the Middle Palaeolithic in Dalmatia, Croatia. *Antiquity, 74*, 777–778.

Karavanić, I. (2000b). Olschewian and appearance of bone technology in Croatia and Slovenia. In J. Orschiedt & G.-C. Weniger (Eds.), *Neanderthals and modern humans—Discussing transition* (Wissenschaftliche Schriften des Neanderthal Museums, Bd. 2, pp. 159–168). Mettmann: Neanderthal Museum.

Karavanić, I. (2007). The Middle/Upper Paleolithic interface in Croatia. In J. Riel-Salvatore & G. A. Clark (Eds.), *New approaches to the study of Early Upper Paleolithic "Transitional" industries in Western Eurasia: Transitions great and small* (BAR International Series, Vol. 1620, pp. 75–89). Oxford: Archaeopress.

Karavanić, I. (2008). Le Paléolithique moyen en Dalmatie (Croatie): Résultats des nouvelles recherches. *Études balkaniques, 15*, 19–32.

Karavanić, I. (2009). Adriatic coast of Croatia and its hinterland from 50,000 to 25,000 BP. In M. Camps & C. Szmidt (Eds.), *The Mediterranean from 50,000 to 25,000 BP: Turning points and new directions* (pp. 163–178). Oxford: Oxbow Books.

Karavanić, I. (2015). Research on underwater Mousterian: The site of Resnik—Kaštel Štafilić, Dalmatia, Croatia. In S. Sázelova, M. Novák, & A. Mizerová (Eds.), *Forgoten times and spaces: New perspectives in Paleoanthropological, Paleontological and Archaeological studies* (pp. 73–79). Brno: Institute of Archaeology of the CAS, Brno & Masaryk University.

Karavanić, I., & Bilich-Kamenjarin, I. (1997). Musterijensko nalazište Mujina pećina kod Trogira: Rezultati trogodišnjih iskopavanja. *Opuscula Archaeologica, 21*, 195–204.

Karavanić, I., & Čondić, N. (2006). Probno sondiranje Velike pećine u Kličevici kod Benkovca. *Obavijesti Hrvatskog arheološkog društva, 38*(2), 45–50.

Karavanić, I., Čondić, N., & Vukosavljević, N. (2007). Velika pećina u Kličevici. *Hrvatski arheološki godišnjak, 3*, 345–347.

Karavanić, I., Golubić, V., Kurtanjek, D., Šošić, R., & Zupanič, J. (2008a). Litička analiza materijala iz Mujine pećine. *Vjesnik za arheologiju i povijest dalmatinsku, 101*, 29–58.

Karavanić, I., Miracle, P. T., Culiberg, M., Kurtanjek, D., Zupanič, J., Golubić, V., et al. (2008b). The Middle Paleolithic from Mujina Pećina, Dalmatia, Croatia. *Journal of Field Archaeology, 33*, 259–277.

Karavanić, I., & Patou-Mathis, M. (2009). Middle/Upper Paleolithic interface in Vindija cave (Croatia): New results and interpretations. In M. Camps & P. Chauhan (Eds.), *Sourcebook of Paleolithic transitions: Methods, theories and interpretations* (pp. 397–405). New York: Springer.

Karavanić, I., & Smith, F. H. (1998). The Middle/Upper Paleolithic interface and the relationship of Neanderthals and early modern humans in the Hrvatsko Zagorje, Croatia. *Journal of Human Evolution, 34*, 223–248.

Karavanić, I., & Smith, F. H. (2000). More on the Neanderthal problem: The Vindija case. *Current Anthropology, 41*, 838–840.

Karavanić, I., & Smith, F. H. (2011). Middle/Upper Paleolithic interface at Vindija cave (Croatia) in the context of central Europe and the Adriatic. In A. P. Derevianko & M. V. Shunkov (Eds.), *Characteristic features of the Middle to Upper Paleolithic transition in Eurasia* (pp. 124–134). Novosibirsk: Institute of Archaeology and Ethnography SB RAS.

Karavanić, I., & Šokec, T. (2003). The Middle Paleolithic percussion or pressure flaking tools? The comparison of experimental and archaeological material from Croatia. *Prilozi Instituta za arheologiju u Zagrebu, 20*, 5–14.

Karavanić, I., Vukosavljević, N., Šošić Klindžić, R., Ahern, J. C. M., Čondić, N., Becker, R., et al. (2014). Projekt *Kasni musterijen na istočnom Jadranu—temelj za razumijevanje identiteta kasnih neandertalaca i njihovog nestanka*: Sažetak 1. godine istraživanja/*The Late Mousterian in eastern Adriatic—towards understanding of late Neanderthals' identity and their demise Project:* A summary of 1st year of research. *Prilozi Instituta za arheologiju u Zagrebu, 31*, 139–157.

Karavanić, I., Zupčić, K., Pešić, M., Parica, M., & Šošić Klindžić, R. (2009). Kaštel Štafilić—Podvodno paleolitičko nalazište. *Hrvatski arheološki godišnjak, 5*(2008), 549–551.

Klein, R. (2009). *The human career* (3rd ed.). Chicago: University of Chicago Press.

Komšo, D. (2012). Srednji paleolitik u Istri: Romualdova pećina i Campanož/Middle Palaeolithic in Istria: Romualdova pećina and Campano. In *Put neandertalca/The Neandertal Trail* (pp. 192–205). Zagreb: Arheološki muzej.

Komšo, D., Balbo, A. I., & Miracle, P. T. (2007). Čepićko polje. *Hrvatski arheološki godišnjak, 3*(2006), 225–228.

Kormos, T. (1912). Die ersten Spuren des Urmenschen in Karst-Gebirge. *Foldtani Kozlony, 52*, 97–104.

Kozlowski, J. K. (1996). Cultural context of the last Neanderthals and early modern humans in central-eastern Europe. In O. Bar-Yosef, L. L. Cavalli-Sforza, R. J. March, & M. Piperno (Eds.), Proceedings of the XIII UISPP Congress. *The Lower and the Middle Palaeolithic*, Vol. 5 (pp. 205–218). Forlì: Abaco.

Krings, M., Capelli, C., Tschentscher, F., Geisert, H., Meyer, S., Von Haeseler, A., et al. (2000). A view of Neandertal genetic diversity. *Nature Genetics, 26*, 144–146.

Kurtanjek, D., & Marci, V. (1990). Petrografska istraživanja paleolitskih artefakata spilje Vindije. *Rad Jugoslavenske akademije znanosti i umjetnosti, 449*, 227–238.

Malez, M. (1975). Die Höhle Vindija—eine neue Fundstelle fossiler Hominiden in Kroatien. *Bulletin scientifique Conseil des Academies des Sciences et des Arts de la RSF de Yougoslavie, 20*(5–6), 139–141.

Malez, M. (1979). Nalazišta paleolitskog i mezolitskog doba u Hrvatskoj. In A. Benac (Ed.), *Praistorija jugoslavenskih zemalja* (Vol. I, pp. 227–276). Sarajevo: Svjetlost.

Malez, M., & Rukavina, D. (1975). Krioturbacijske pojave u gornjopleistocenskim naslagama pećine Vindije kod Donje Voće u sjeverozapadnoj Hrvatskoj. *Rad Jugoslavenske akademije znanosti i umjetnosti, 371*(17), 245–265.

Malez, M., & Rukavina, D. (1979). Položaj naslaga spilje Vindije u sustavu članjenja kvartara šireg područja Alpa. *Rad Jugoslavenske akademije znanosti i umjetnosti, 383*(18), 187–218.

Malez, M., Smith, F. H., Radovčić, J., & Rukavina, D. (1980). Upper Pleistocene hominids from Vindija, Croatia, Yugoslavia. *Current Anthropology, 21*, 365–367.

Malez, M., & Vogel, J. C. (1970). Die Ergebnisse der Radiokarbonanalisen der quartären Schichten der Velika Pećina in Nordwest-Kroatien. *Bulletin scientifique Conseil des Academies des Sciences et des Arts de la RSF de Yougoslavie, 15*(11–12), 390–391.

Mihailović, D. (2009). *Upper Paleolithic and Mesolithic Chipped Stone Industries from Crvena Stijena*. Prehistoric settlements in caves and rock-shelters of Serbia and Montenegro. Fascicule II. Belgrade: University of Belgrade.

Miracle, P. T. (1991). Carnivore dens or carnivore hunts? A review of Upper Pleistocene mammalian assemblages in Croatia and Slovenia. *Rad Hrvatske akademije znanosti i umjetnosti, 458*(25), 193–219.

Miracle, P. T. (1998). The spread of modernity in Europe. In K. Omoto & P. V. Tobias (Eds.), *The origins and past of modern humans—Towards reconciliation* (pp. 171–187). Singapore: World Scientific.

Miracle, P. T. (2005). Late Mousterian subsistence and cave-use in Dalmatia: The zooarchaeology of Mujina Pećina, Croatia. *International Journal of Osteoarchaeology, 15*, 84–105.

Montet-White, A. (1996). *Le Paléolithique en ancienne Yougoslavie*. Grenoble: Jérôme Millon.

Moreau, L., Odar, B., Higham, T., Horvat, A., Pirkmajer, D., & Turk, P. (2015). Reassessing Aurignacian of Slovenia: Techno-economic behaviour and direct dating of osseous projectile points. *Journal of Human Evolution, 78*, 158–180.

Nelson, D. E. (1997). Radiocarbon dating of bone and charcoal from Divje babe I cave. In I. Turk (Ed.), *Mousterian "Bone Flute" and other finds from Divje babe I cave site in Slovenia* (pp. 51–64). Ljubljana: Slovenian Academy of Sciences and Arts.

Nizek, R., & Karavanić, I. (2012). Prostorna analiza nalaza musterijenskih razina D2, E1, E2, i E3 Mujine pećine/The spatial analysis of finds from Mousterian levels D2, E1, E2, and E3 of Mujina pećina. *Prilozi Instituta za arheologiju u Zagrebu, 29*, 25–56.

Pacher, M. (2010). Raw material analysis of Upper Paleolithic bone points and the invention of the Olschewian. In C. Neugebauer-Maresch & L. R. Owen (Eds.), *New aspects of the central and eastern European Upper Palaeolithic—Methods, chronology, technology and subsistence* (pp. 319–325). Vienna: Verlag der Österreichischen Akademie der Wissenschaften.

Papagianni, D. (2009). Mediterranean southeast Europe in the Late Middle and Early Upper Palaeolithic: Modern human route to Europe or Neanderthal refugium? In M. Camps & C. Szmidt (Eds.), *The Mediterranean from 50,000 to 25,000 BP: Turning points and new directions* (pp. 115–136). Oxford: Oxbow Books.

Papagianni, D., & Morse, M. A. (2013). *The Neanderthals rediscovered: How modern science is rewriting their story*. New York: Thames & Hudson.

Petrić, N. (1979). Mujina pećina, Trogir—Paleolitičko nalazište. *Arheološki pregled, 20*(1978), 9.

Prüfer, K., Racimo, F., Patterson, N., Jay, F., Sankararaman, S., Sawyer, S., et al. (2014). The complete genome sequence of a Neanderthal from the Altai Mountains. *Nature, 505*, 43–49.

Richards, M. P., Karavanić, I., Pettitt, P., & Miracle, P. (2015). Isotope and faunal evidence for high levels of freshwater fish consumption by Late Glacial humans at the Late Upper Palaeolithic site of Šandalja II, Istria, Croatia. *Journal of Archaeological Science, 61*, 204–212.

Richards, M. P., Pettitt, P. B., Trinkaus, E., Smith, F. H., Paunović, M., & Karavanić, I. (2000). Neanderthal diet at Vindija and Neanderthal predation: The evidence from stable isotopes. *Proceedings of the National Academy of Sciences of the United States of America, 97*, 7663–7666.

Rink, W. J., Karavanić, I., Pettitt, P. B., van der Plicht, J., Smith, F. H., & Bartoll, J. (2002). ESR and AMS based ^{14}C dating of Mousterian levels at Mujina Pećina, Dalmatia, Croatia. *Journal of Archaeological Science, 29*, 943–952.

Sankararaman, S., Mallick, S., Dannemann, M., Prüfer, K., Kelso, J., Pääbo, S., et al. (2014). The genomic landscape of Neanderthal ancestry in present-day humans. *Nature, 507*, 354–357.

Sankararaman, S., Patterson, N., Li, H., Pääbo, S., & Reich, D. (2012). The date of interbreeding between Neandertals and modern humans. *PLoS Genetics, 8*(10), e1002947. doi:10.1371/journal.pgen.1002947.

Savić, M. (1984). *Perspektive muzejske službe u Benkovcu*. Unpublished report, Benkovac.

Serre, D., Langaney, A., Chech, M., Teschler-Nicola, M., Paunović, M., Hofreiter, M., et al. (2004). No evidence of Neandertal mtDNA contribution to early modern humans. *PLoS Biology, 2*, 313–317.

Simek, J. F., & Smith, F. H. (1997). Chronological changes in stone tool assemblages from Krapina (Croatia). *Journal of Human Evolution, 32*, 561–575.

Smith, F. H. (1976). *The Neandertal remains from Krapina: A descriptive and comparative study*. University of Tennessee Department of Anthropology Reports of Investigation 15. Knoxville: University of Tennessee.

Smith, F. H. (1984). Fossil hominids from the Upper Pleistocene of central Europe and the origin of modern Europeans. In F. H. Smith & F. Spencer (Eds.), *The origin of modern humans: A world survey of the fossil evidence* (pp. 137–209). New York: Alan R. Liss.

Smith, F. H. (1994). Samples, species and speculations in the study of modern human origins. In M. Nitecki & D. Nitecki (Eds.), *Origins of anatomically modern humans* (pp. 227–249). New York: Plenum.

Smith, F. H., & Ahern, J. C. (1994). Brief communication: Additional cranial remains from Vindija cave, Croatia. *American Journal of Physical Anthropology, 93*, 275–285.

Smith, F. H., Boyd, C. D., & Malez, M. (1985). Additional Upper Pleistocene human remains from Vindija cave, Croatia, Yugoslavia. *American Journal of Physical Anthropology, 68*, 375–388.

Smith, F. H., Falsetti, A. B., & Donnelly, S. M. (1989). Modern human origins. *Yearbook of Physical Anthropology, 32*, 35–68.

Smith, F. H., Janković, I., & Karavanić, I. (2005). The assimilation model, modern human origins in Europe, and the extinction of Neandertals. *Quaternary International, 137*, 7–19.

Smith, F. H., Trinkaus, E., Pettitt, P. B., Karavanić, I., & Paunović, M. (1999). Direct radiocarbon dates for Vindija G1 and Velika Pećina Late Pleistocene hominid remains. *Proceedings of the National Academy of Sciences of the United States of America, 96*, 12281–12286.

Šošić Klindžić, R., Karavanić, I., & Vukosavljević, N. (2014). *Neandertal niches on the eastern edge of the great Adriatic plain*. Abstracts. 20th Annual Meeting of the European Association of Archaeologists (p. 334). Istanbul: EAA.

Straus, L. G. (1999). The Neanderthal problem continued: Reply. *Current Anthropology, 40*, 352–355.

Svoboda, J. (2001). La question szélétienne. In D. Cliquet (Ed.), *Les industries a outils bifaciaux du paléolithique moyen d'Europe occidentale* (pp. 221–330). Liège: ERAUL 98.

Svoboda, J. (2006). The Danubian gate to Europe: Patterns of chronology, settlement archaeology, and demography of late Neandertals and early modern humans on the Middle Danube. In N. Conard (Ed.), *When Neanderthals and modern humans met* (pp. 233–267). Tübingen: Kerns Verlag.

Trinkaus, E., Moldovan, O., Milota, S., Bîlgăr, A., Sarcina, L., Athreya, S., et al. (2003). An early modern human from the Peştera cu Oase, Romania. *Proceedings of the National Academy of Sciences of the United States of America, 100*, 11231–11236.

Vujević, D. (2007). *Srednji paleolitik na području južno od Ražanca*. Unpublished MA Thesis, University of Zadar, Zadar.

Vuković, S. (1950). Paleolitska kamena industrija spilje Vindije. *Historijski zbornik, 1–4*, 241–256.

Wild, E. M., Paunović, M., Rabeder, G., Steffan, I., & Steier, P. (2001). Age determination of fossil bones from the Vindija Neanderthal site in Croatia. *Radiocarbon, 43*, 1021–1028.

Wolpoff, M. H. (1999). *Paleoanthropology* (2nd ed.). New York: McGraw-Hill.

Wolpoff, M. H., Smith, F. H., Malez, M., Radovčić, J., & Rukavina, D. (1981). Upper Pleistocene human remains from Vindija cave, Croatia, Yugoslavia. *American Journal of Physical Anthropology, 54*, 499–545.

Zilhão, J. (2009). Szeletian, not Aurignacian: A review of the chronology and cultural associations of the Vindija G1 Neandertals. In M. Camps & P. Chauhan (Eds.), *Sourcebook of Paleolithic transitions* (pp. 407–426). New York: Springer.

Zilhão, J. (2011). Aliens from outer time? Why the "human revolution" is wrong and where do we go from here? In S. Condemi & G.-C. Weniger (Eds.), *Continuity and discontinuity in the peopling of Europe. One hundred fifty years of Neanderthal study* (pp. 331–366). Dordrecht: Springer.

Zilhão, J., & D'Errico, F. (1999). The Neanderthal problem continued: Reply. *Current Anthropology, 40*, 355–364.

Chapter 11
The Lower Paleolithic of Romania Revisited: New Evidence from the Site of Dealul Guran

Adrian Doboş and Radu Iovita

Abstract Southeastern Europe represents a key area in investigating hominin dispersals during the Pleistocene. However, the understanding of these phenomena is hampered by the scarcity of data, especially for the Lower and Middle Pleistocene. The discoveries from Romania assigned to these periods (either credited as *in situ* or from disturbed contexts) are rather doubtful. After reviewing the state of the art, our paper presents the site of Dealul Guran, discovered in 2010 during a systematic survey carried out in the province of Dobrogea, southeastern Romania. The site is a collapsed rockshelter located on a limestone hill, very rich in flint nodules. Three archaeological layers were identified, and absolute ages indicate that the two oldest archaeological units correspond to an OIS 11 occupation of the site. The assemblages consist mostly of cortical flakes and there are many tested blocks from these units, likely reflecting flint quarrying activities.

Keywords Core-and-flake industries • OIS 11 • Osteodontokeratic • Pseudo-tools • Taphonomy • *Trés Ancien Paléolithique*

A. Doboş (✉)
Department of Paleolithic Archaeology, Institute of Archaeology "Vasile Parvan", 11 Henri Coanda Street, Sector 1, Bucharest 010667, Romania
e-mail: addobos@yahoo.co.uk

R. Iovita
MONREPOS Archaeological Research Centre and Museum for Human Behavioural Evolution, Römisch-Germanisches Zentralmuseum, Leibniz-Forschungsinstitut für Archäologie, Schloss Monrepos, 56567 Neuwied, Germany
e-mail: iovita@rgzm.de

Introduction

Southeastern Europe represents a key area in investigating Pleistocene hominin dispersals. A good understanding of these phenomena, however, is hampered by the scarcity of data, especially for the Lower and Middle Pleistocene. Scenarios for the hominin colonization of Eurasia can be grouped into two categories based on where the explanatory emphasis lies. Theories of the first group are centered on factors extrinsic to hominin evolution, such as the expansion of grassland habitats into northern latitudes and the migration of mammalian guilds (Turner 1992; Martínez-Navarro 2010; Van der Made and Mateos 2010; see also Spassov 2016). Others favor causes intrinsic to the development of hominin groups, ranging from technological innovations to changes in brain and body size and increased life expectancy (Carbonell et al. 1999; Tappen 2009; Bar-Yosef and Belmaker 2011). Whichever factor, or combination of factors, might have prevailed, it is certain that the geographic location of the Balkans place this region on most colonization itineraries (e.g., Korisettar and Petraglia 1998; Moncel 2010). During the harsh conditions of glacial periods, the southern parts of Eurasia, such as the Balkans, Iberia, and the Italian peninsula, were probably used as refugia by hominin populations (Dennell et al. 2011; Macdonald et al. 2011; Harvati 2016). At the end of the Middle Pleistocene Climate Transition (roughly between 920 and 640 ka), the duration of glacial cycles shifted from *ca.* 41 to 100 ka, thus allowing for longer human occupations at higher latitudes (Mudelsee and Schulz 1997; Schmeider et al. 2000). Furthermore, sedimentary records in southeastern Europe were never interrupted by ice sheet advances, a situation which makes the Balkans a good candidate area for adding viable landmarks on the migration routes from Africa and/or Asia, as well as for shedding light on small-scale population dynamics throughout the Pleistocene.

Our paper, presenting the data from Romania, is divided into three main parts. In the first part, we present a critical

Katerina Harvati and Mirjana Roksandic (eds.), *Paleoanthropology of the Balkans and Anatolia*, Vertebrate Paleobiology and Paleoanthropology, DOI 10.1007/978-94-024-0874-4_11

review of the Lower Paleolithic research in Romania, focusing on twentieth century records. In the second, we present the newly discovered site of Dealul Guran, discovered during a systematic field survey in the province of Dobrogea, southeastern Romania (Lower Danube Survey for Paleolithic Sites—LoDanS http://lodans.wordpress.com) (Iovita et al. 2014). Three Paleolithic layers were identified, and luminescence dating indicates an OIS 11 occupation for the oldest layers, which represents the first securely dated, stratified Lower Paleolithic site in Romania. Finally, in the third part we discuss the position of Dealul Guran within the context of the Lower Paleolithic of Eastern Europe.

Romania: Landmarks in the Research on the Lower Paleolithic

The research on the Lower Paleolithic in Romania is over a century old and can be divided into three stages on both chronological and methodological grounds (Doboş 2008). The pioneering phase, which evolved between the two world wars, was mainly connected with the activity of archaeologists specializing in various historic periods, geologists, and paleontologists who had a variable interest in the study of the Paleolithic. A prominent figure of Paleolithic research in Transylvania was Márton Roska, who discovered numerous Paleolithic open air and cave sites, some of which (Iosăşel, Căpuşu Mic, Zimbru, etc.) were assigned to the Lower Paleolithic, mostly according to tool types regarded as typical for the period. Among the stone tools found by Roska, the bifacially worked ones were regarded as potentially Chellean, Acheulean, and Micoquian (Roska 1928, 1931a, b, 1933). Similarly, Lower Paleolithic industries were reported from sites on the right bank of Dniester River, on today's territory of the Republic of Moldova (Ambrojevici 1926). These discoveries generated a vivid debate among the scholars of the time regarding the role of human agency in producing these lithic remains. In many ways anticipating later twentieth-century trends in Paleolithic research elsewhere, the authenticity of some of the alleged stone tools was primarily challenged on taphonomic grounds: some were regarded as products of natural agents, such as the influence of frost or water transportation (Nicolăescu-Plopşor 1929, 1930, 1931). A further critique focused on the lack of accurate stratigraphic information, which could lead to the incorrect cultural attribution of the artifacts (Moroşan 1931).

M. Moga was the first to publish a regional review of the Lower Paleolithic. In his critical analysis of the archaeological record from Transylvania, he showed that no reliable data could be found for proving the existence of Lower Paleolithic sites in Romania (Moga 1938). After World War II, research on the Paleolithic changed as a result of political developments. The new communist authorities provided extensive funding for archaeological research in order to support the nationalist discourse, and the Paleolithic was no exception. One of the prominent figures of the period was C.S. Nicolăescu-Plopşor, whose extensive activity covered all of Romania's territory and all Paleolithic periods (Doboş 2005). Some of the discoveries made in this period were interpreted as being of Lower Paleolithic age. Most of these were found in river terrace gravels and, given their general aspect resembling pebble tools, they were assigned to the Acheulean and Abbevillian. Rolled flakes were generally assigned to the Clactonian (Nicolăescu-Plopşor 1956, 1957; Nicolăescu-Plopşor and Moroşan 1959; Dicu 1972, 1973, 1979; Nania 1972; Păunescu 1980).

The perspective on the Lower Paleolithic took a new turn after the of Plio-Pleistocene[1] paleontological sites in the Olteţ River valley, where bone fragments interpreted as tools were found among mammal remains. Given their presumed Villafranchian age, the discovery of Osteodontokeratic industries was advocated in scientific journals (Nicolăescu-Plopşor and Nicolăescu-Plopşor 1963; Nicolăescu-Plopşor 1964b), as well as in popular literature (Nicolăescu-Plopşor 1965, 1970; Roşu 1987).

Finally, the third research stage was characterized by efforts to create a geochronological framework for the Lower and Middle Pleistocene in Romania, which was meant to accurately systematize the archaeological discoveries on the one hand, and, on the other hand, to correlate them to the European sequences. The prominent figures of this period were the paleontologists C. Rădulescu and P. Samson, who contributed important data to the understanding of Quaternary environments in Romania (Păunescu et al. 1982; Rădulescu et al. 1998). From an archaeological point of view, the work of Al. Păunescu was of the greatest importance for this period. Păunescu studied and published all the available lithic collections, providing the first modern synthesis of the Paleolithic record of Romania (Păunescu 1999a, b, 2000, 2001).

The Lower Paleolithic Record of Romania

In this section, we will present a synthesis of the archaeological record assigned to the Lower Paleolithic. During the century-old research on the topic, the accumulated evidence can be grouped into two main categories: the *in situ* discoveries and discoveries from disturbed contexts. Issues regarding the definitions of Lower Paleolithic divisions in Romanian archaeology, extensively discussed by one of us (Doboş 2008) will only be briefly mentioned.

[1]The Pliocene/Pleistocene boundary was at 1.8 ma when these discoveries were made.

In Situ Discoveries

This category includes the sites credited as yielding *in situ* discoveries in the Romanian literature. Wherever appropriate, we mention the issues that challenge their allegedly undisturbed character. The oldest discoveries are situated in and near the Olteţ River Valley (Fig. 11.1), namely in the vicinity of Tetoiu village, Vâlcea county (the name of the village until 1968 was Bugiuleşti, thus some of the sites are also known by this name). During the Plio-Pleistocene, the area included the shoreline of the Getic Lake (Samson and Rădulescu 1973). A total of eight paleontological and/or archaeological find spots were reported (Păunescu 2000), of which the most important three are presented below.

Tetoiu—Pietrişu Vijoieşti. This paleontological site was investigated through a 126 m^2 and 7.27 m deep excavation in 1960–1961. The fossil concentration was identified over an area of *ca.* 50 m^2 in a sandy layer, at *ca.* 6 m depth. The following faunal taxa were identified: *Archidiskodon meridionalis, Nyctereutes megamastoides, Lynx issiodorensis, Eucladoceros* sp., *Pliotragus ardeus, Stephanorhinus etruscus, Plessipus athanasiui, Beremendia* cf. *fissidens, Trogontherium dacicum*, and *Vulpes alopecoides*; the faunal association was dated to the Upper Pliocene—Tiglian (Rădulescu et al. 1998; Rădulescu et al. 2003; Fig. 11.2). Many of the fossil bones were found in anatomical position, a situation which suggested, according to C.S. Nicolăescu-Plopşor, that the mammals were "sinking" in the muddy lakeshore, thus becoming an easy target for predators (Nicolăescu-Plopşor 1964a). Recent excavations were carried out in 2004 by a team from the Institute of Speleology "Emile Racovitza" in Bucharest and Baylor University, Waco, Texas. No fossil fauna was recovered, suggesting a possible exhaustion of the fossiliferous deposit (Petculescu pers. comm.)

Tetoiu—Dealul Mijlociu. In 1960, a chopping tool and a protobiface (Fig. 11.3: 1, 2) were recovered from a slope of

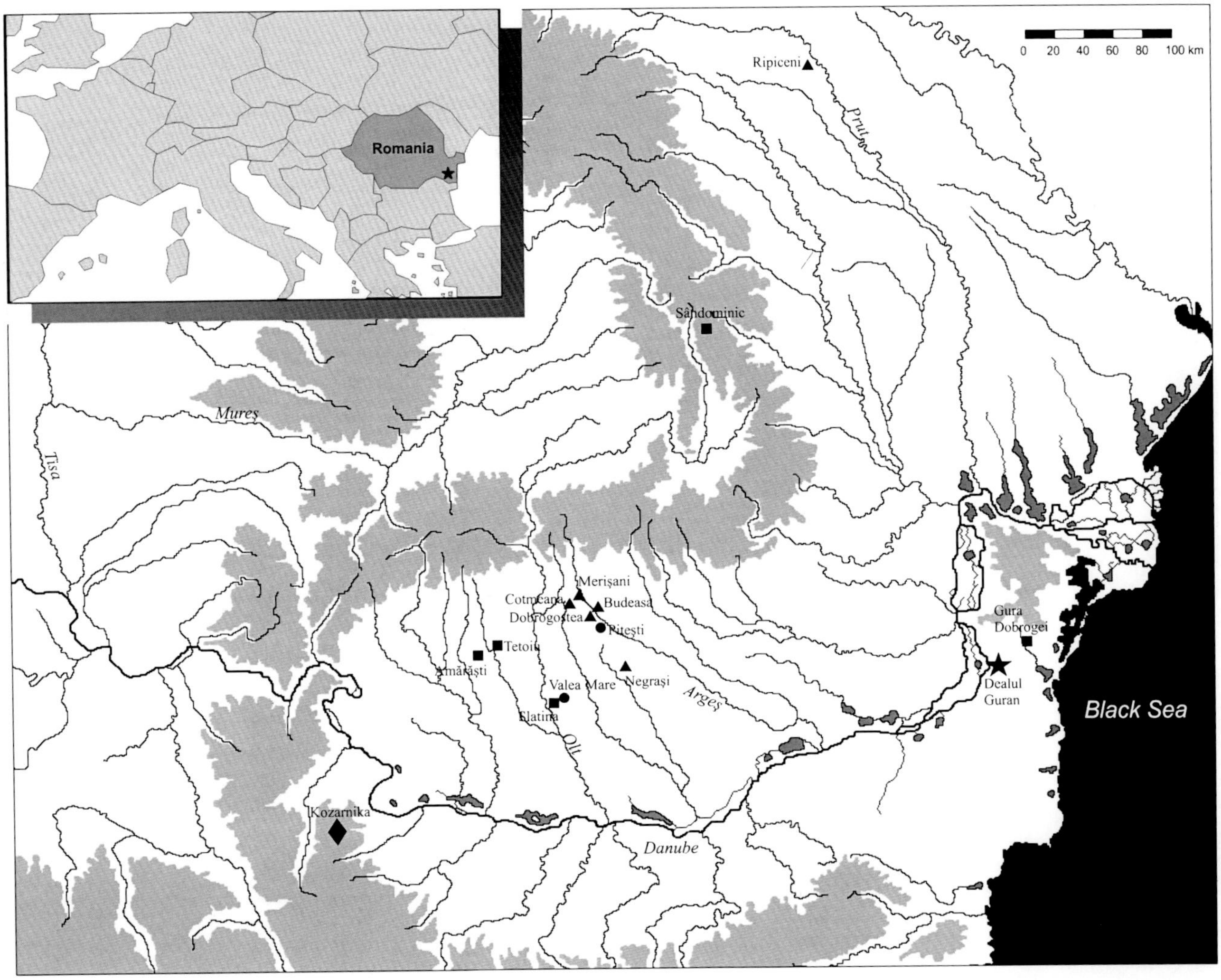

Fig. 11.1 Map of the most important sites mentioned in the text. Star = Dealul Guran; Squares = *in situ* discoveries; Triangles = disturbed context locales with 40–100 pieces; Circles = disturbed context locales with more than 100 pieces

MA	*OIS*	Stage	Guiding taxon	Romania	France, Italy, Greece	Germany, Hungary	Georgia, Russia
0.125	*5e*	Saalian					
0.38			*S gregalis martelensis*	**Sândominic - 2(LP)**	La Fage	Steinheim	
			Arvicola terrestris *Pliomys relictus*	**Sândominic - 1(LP)**		Castellum	Khazarian
						Uppony	Singlian
0.4	*11*	Holsteinian					
							Okaian
		Elsterian		**Slatina (terrace) (LP)**			
						Vértesszölös(LP)	
0.5	*14*		*Arvicola cantianus*	**Gura Dobrogei (2)(LP)**			
				Gura Dobrogei (1)	L'Escale	Mosbach - 2 Tarko	
0.86	*21*	"Cromerian"		Casian - 1	**Isernia (LP)**		Tiraspolian
1.12	*32*	Bavelian			**Vallonet (LP)**		
1.32	*41*	Menapian					
		Waalian					Tamanian
		Eburonian	*Canis etruscus*	Fântâna Alortiţei	Olviola		Odessan
1.76	*60*						**Dmanisi (LP)**
		Tiglian	*Paradolicopithecus* *Mitilanotherium* *Manis*	Valea lui Grăunceanu	Tiglian 5C	Villany - 3	
				Dealul Mijlociu (LP)			
			Trogontherium dacicum	Pietrişu-Vijoieşti			
					Chilhac (LP) Volakas St. Vallier		Khaprovian

Fig. 11.2 Correlation of the fossil sites of Romania assigned to the Lower Paleolithic (LP), together with their principal biochronological equivalents (redrawn from Rădulescu et al. 1998; the Oxygen Isotopic Stages and absolute ages taken from (Gibbard and Cohen 2008)

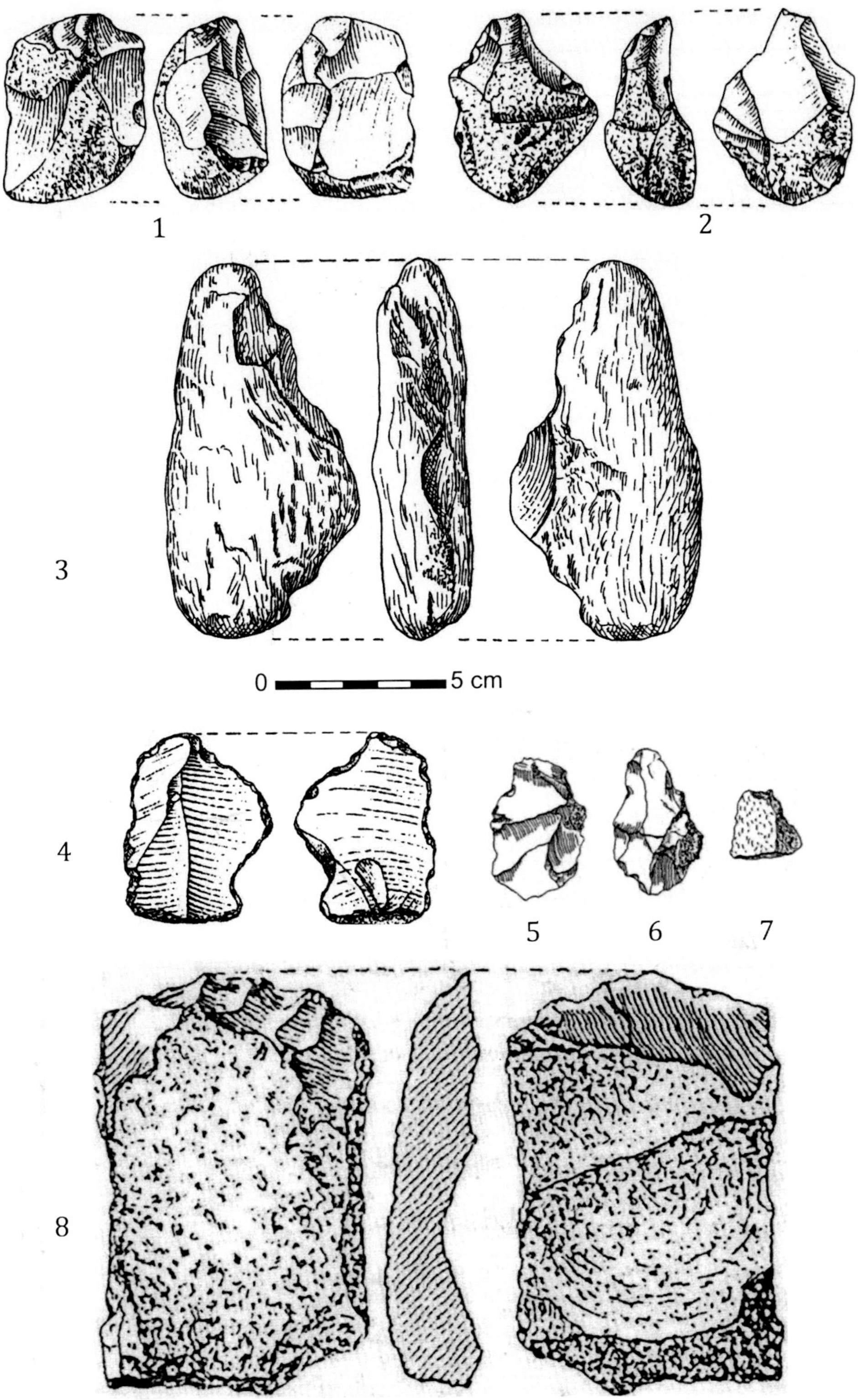

Fig. 11.3 Tetoiu—Dealul Mijlociu: *1*—Protobiface; *2*—Chopping tool (Păunescu 2000); Tetoiu—Valea lui Grăunceanu: *3*—Chopping tool (Păunescu 2000); Slatina—Terrace: *4*—Levallois retouched flake (Păunescu 2000); Gura Dobrogei—Peştera Liliecilor: *5*—Flake with retouched edge; *6*—Sidescraper; *7*—Flake; *8*—Chopping tool (Rădulescu et al. 1998)

the Dealul Mijlociu (Mijlociu Hill), in a layer consisting of gravel and sand mixture. No faunal remains were found. Although the sediment was not in primary context, it was argued that, given the absence of rolling traces on the edges of the tools, they originated from a nearby locality. The age of the deposit was estimated at *ca.* 1.7 Ma (Rădulescu and Samson 1991; Păunescu 2000; Spassov 2016).

Tetoiu—Valea lui Grăunceanu. The site was excavated in the 1960s over a surface of 200 m^2. The faunal fossil remains were concentrated in an area of *ca.* 90 m^2 and at a depth between 4.77 and 5.6 m, in sediments composed of clay mixed with sand, overlying a sterile layer of lacustrine deposits. The paleontological record included the following taxa: *Paradolichopithecus arvernensis geticus, Archidiskodon meridionalis, Equus stenonis, Gazellospira troticornis, Pliotragus ardeus, Macedontherium martini, Dicerorhinus* sp., *Cervus philisi, Croizetoceros ramosus, Castor plicidens, Trogontherium cuvieri, Nyctereutes megamastoides, Ursus etruscus, Crocuta perrieri, Homotherium crenatidens, Megantereon megantereon, Felis issiodorensis, Felis toscana*, and *Meles* sp. The faunal assemblage was placed in the Upper Villafranchian (Rădulescu et al. 1998), with an estimated age of *ca.* 1.8 Ma (Upper Tiglian). Among the *ca.* 20,000 bones present in this layer, a number of fragments were interpreted as bone tools, and, given the age of the site, the existence of Osteodontokeratic industries was postulated. The anthropic origin of these industries was supposed to be corroborated by the presence of three rocks found in the same deposits, which were interpreted as manuports originating from over 40 km away (Nicolăescu-Plopşor and Nicolăescu-Plopşor 1963, 1965; Nicolăescu-Plopşor 1964a, b). A chopping tool was likewise reported from the layer superposing the fossil concentration (Păunescu 2000), but the piece looks rather like a natural accident (Fig. 11.3: 1, 2).

The three Tetoiu sites raise serious doubts on their status as Lower Paleolithic sites. As elsewhere in the world (Singer 1956; Wolberg 1970; Brain 1981), the alleged Osteodontokeratic industries most likely reflect taphonomic processes rather than intentional modification (see a more detailed presentation in Doboş 2005, 2008). The relative chronological position of the Tetoiu sites is unconvincing, since a site with two choppers and no fauna—Dealul Mijlociu—is interpreted as older than a site with presumed Osteodontokeratic industries—Valea lui Grăunceanu. The choppers from Dealul Mijlociu, although presented as *in situ*, actually originate from a different spot and hence should be counted with the disturbed context discoveries (Spassov 2016).

Gura Dobrogei (Constanţa County). Gura Dobrogei is a cave site, also known as Peştera Liliecilor (Bats' cave). In 1971, the excavation in a sector of the cave called "Secondary Gallery" yielded artifacts interpreted to be of Lower Paleolithic age. The important chronologic landmark is a silt deposit where the following rodent taxa were found: *Allactaga orghidani, Apodemus sylvaticus, Cricetulus gr. migratorius, Mesocricetus newtoni, Cricetus cricetus praeglacialis, Ellobius calabaei, Spermophilus gr. nogaici, Clethrionomys glareolus, Lagurus transiens dacicus, Eolagurus gromovi vistornensis, Arvicola cantianus, Microtus guentheri, Microtus arvalis, Pitymis arvaloides, Stenocranius gregalis*, and *Ochotona pussila*. Based on these microfaunal remains, the layer was estimated to have a late Cromerian age (OIS 13). A chopping tool and a retouched flake (Fig. 11.3: 5, 8) were found in a loess layer overlying the silt deposit; from another loess layer, below the silt, a side-scraper and a quartzite flake were reported (Fig. 11.3: 6, 7) (Păunescu et al. 1982; Rădulescu et al. 1998; Păunescu 1999a). Given the scarcity of the lithic inventory, and the absence of published plans and profiles from the excavation, caution is recommended when counting Gura Dobrogei among the multi-layered Lower Paleolithic sites.

Slatina-terrace (Olt County). In the vicinity of Slatina, five sites of archaeological and/or paleontological interest were reported. Slatina-terrace is an important landmark in the geochronological framework of Romania. The stratigraphic sequence of the river Olt terrace is *ca.* 45 m high and shows the existence of several layers with fossil fauna. Of great significance is layer 37, which yielded an interesting faunal association, including *Trogontherium dacicum, Mimomys* sp., *Unio aspcheronicus, Unio bozdagiensis, Anodonta* sp., *Euphrata* sp., *Corbicula* sp., and *Viviparus lineatus*. The layer was dated through paleomagnetism at *ca.* 1.8–1.6 Ma (Andreescu et al. 1981). The archaeology of the site, on the other hand, is represented by a single retouched Levallois flake (Fig. 11.3: 4) found in the terrace gravels, that was potentially assigned to the Lower Paleolithic (Rădulescu et al. 1998; Cârciumaru 1999; Păunescu 2000), although, in the rest of Europe, the appearance of Levallois technology is considered one of the characteristics of the Middle Paleolithic. Most importantly, however, since no association could be made between the flake and any of the paleontological layers, there is no basis on which to consider Slatina-terrace a Lower Paleolithic site.

Amărăşti (Dolj County). The site was discovered when a water dam was built near the river. A small excavation (of unknown size) led to the discovery—in a layer of clay at a depth between 2.7 and 4.05 m—of parts of an *Elephas trogontherii* skeleton and a few lithics: two interpreted as knapped manuports, two cortical flakes, three denticulates, and a *tranche de citron* flake (Fig. 11.4: 1–4). The interpretation of the finds as remnants of a hunting party (Cârciumaru 1999; Păunescu 2000) can be challenged for several reasons: most of the lithics are so rudimentary that they can also be natural; no faunal analysis was made to find any cutmarks on the bones; and the very few finds ($n = 8$) were scattered over a depth of *ca.* 1.3 m. The overall situation raises questions

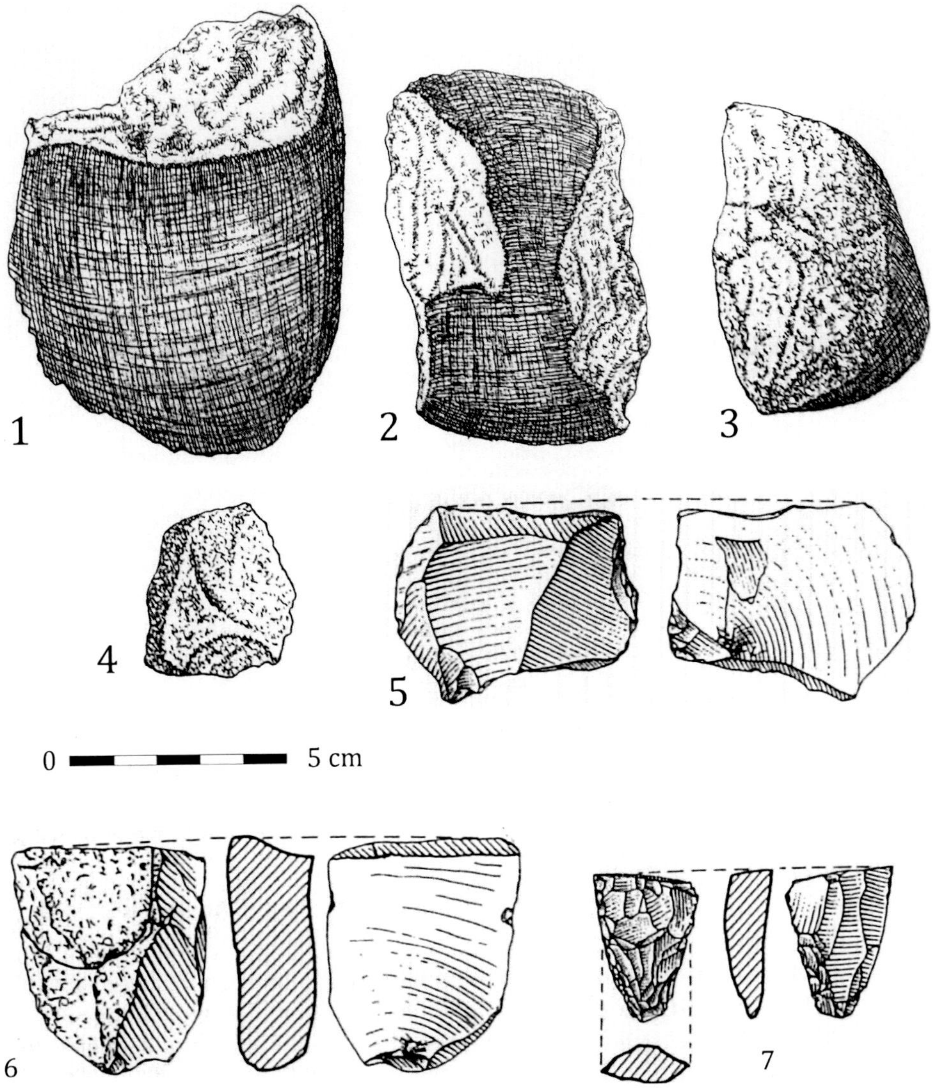

Fig. 11.4 Amărăşti—Baraj: *1, 2*—Pebbles with knapping negatives; *3*—Flake; *4*—*Tranche de citron* flakes (Păunescu 2000); Sândominic—Travertine quarry: *5*—Scraper on Levallois flake; *6*—Flake fragment; *7*—Biface fragment (Păunescu 2000)

about the *in situ* character of the site, much less warranting its interpretation as a Lower Paleolithic hunting episode.

Sândominic (Harghita County). This site is located in a travertine quarry. Lower Paleolithic finds were reported from a rock fissure where sediment accumulated. According to the authors of the excavation (Rădulescu et al. 1998), two sedimentary deposits were identified. The lower one, a *terra rosa*, was dated to the late Holsteinian (OIS 11) based on the presence of *Arvicola terrestris* and *Pliomys relictus*. Three quartzite lithics (two flakes and a piece of shatter) and a sandstone biface fragment (Fig. 11.4: 7) were found. The upper layer was dated to the early Saalian (OIS 10/8) based on the discovery of *Stenocranius gregalis martelensis*. The lithic inventory (Fig. 11.4: 5, 6) is composed of a sidescraper on a Levallois blank and a flake fragment (Cârciumaru 1999; Păunescu 2000). Similar to the situation from the site of Gura Dobrogei, the archaeological evidence is fairly poor for a multi-layer stratified Lower Paleolithic site.

The Disturbed Contexts

Most of the lithics assigned to the Lower Paleolithic come from disturbed contexts, especially from river terrace gravels. According to their typological features, the pebble tools were generally assigned to *Trés Ancien Paléolithique* (TAP) industries. Bifacially worked tools were integrated into post-TAP Lower Paleolithic industries, for which no specific term was used, although sometimes the Acheulean was tacitly implied. Finally, flakes were mainly classified as Premousterian.

A total of 65 locations related to the Lower Paleolithic are known from the Romanian literature. They were associated

Fig. 11.5 Olt River valley: *1–4* “Choppers” (*1*—Valea Muierii; *2, 4*—Unknown provenience; *3*—Valea Dârjovului)

either with a single Lower Paleolithic division, or with several, depending on the typology of the recovered material. The number of discovered pieces ranges from one to over one hundred. Most of the locations are in Walachia and Oltenia (53), followed by five in Moldavia, four in Transylvania and three in Dobrogea (Păunescu 1999a, b, 2000, 2001). Only those locations where more than 40 lithics were found are included on the map in Fig. 11.1. The total number of lithic pieces reported is around 1100.

The understanding of the Lower Paleolithic record based on these surface collections is hampered by several issues. First, as with the stratified finds, the anthropic origin of some of the lithics is doubtful. As mentioned above, they were mainly collected from river gravels, which makes natural transport by water action a likely agent in the creation of these lithics. During the transportation process, cobbles may break in countless ways, resulting in some which may resemble real archaeological choppers and chopping tools (Fig. 11.5). In many of the published accounts, the drawings show clearer knapping features than are visible on the originals (Fig. 11.6). Second, even the true artifacts from the disturbed contexts are problematic. According to the *fossiles directeurs* principle, they were considered cultural markers for the Lower Paleolithic. However, while some forms (e.g., choppers, chopping tools, bifaces) are more common in the Lower Paleolithic, they are not completely absent from the later periods, so again, caution is recommended.

The data presented above show that the presence of the Lower Paleolithic in Romania based on the twentieth century finds is uncertain. The discoveries credited as *in situ* were fairly few and poorly documented, whereas the discoveries from disturbed contexts were doubtful because of their disputable artifactual character and/or their chronological position. Moreover, some of the lithics, including those credited as *in situ*, have been lost, and any reassessment would have to rely on the drawings which may not accurately reflect the attributes of the lithic tools.

The Site of Dealul Guran

The LoDanS Project

The Lower Danube Survey for Paleolithic Sites (LoDanS) project was aimed at discovering new sites in the Romanian province of Dobrogea, between the Danube and the Black Sea. Dobrogea occupies a central geographic point of potential importance for several dispersal routes in and out of Europe, for both fauna and humans. On the one hand, most of the

Fig. 11.6 Slatina—Valea Muierii: *1*—Chopper; Valea Mare: *2*—Chopping tool; Brebeni: *3*—Chopping tool (Drawings after Păunescu 2000)

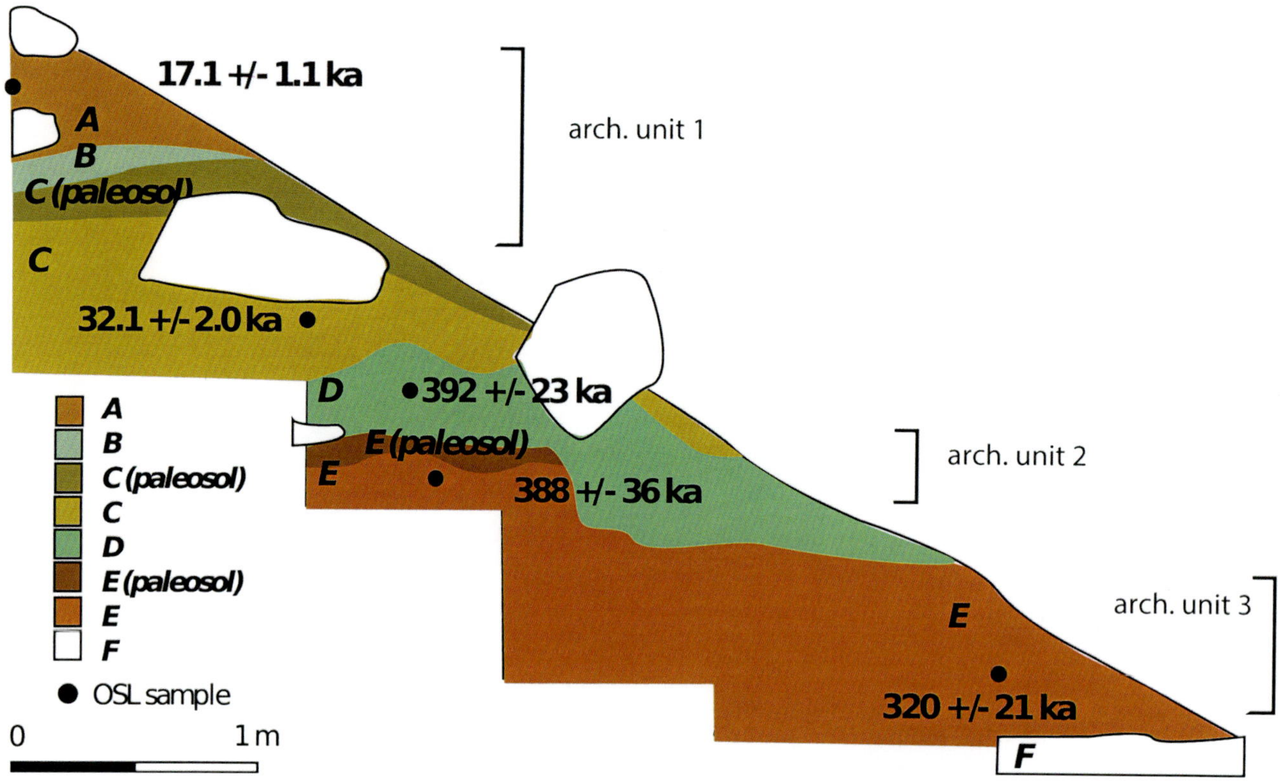

Fig. 11.7 Dealul Guran: Schematic drawing of the stratigraphic sequence, after Iovita et al. 2012, Fig. 3B (original drawing by K. Fitzsimmons, changed to reflect the 2012 excavation season)

Danube tributaries here flow south to north, acting as a conduit for migrations from the Balkans along the Danube toward Central Europe or northeastwards to the Pontic steppes. Here again, Dobrogea is directly connected to the great Eurasian loess steppe (Haase et al. 2007). Faunal guild similarities between this region and the Crimea have previously been noted (Petculescu and Ştiucă 2008). This connection has not been investigated very much, despite the fact that loess archives provide an important record for the paleoenvironmental changes within the last 1 Myr period (Fitzsimmons et al. 2012). The field surveys carried out in 2010 and 2012 (Iovita et al. 2014) in caves and rockshelters, as well as in the open air, identified 61 prospective sites, 32 of which yielded surface lithic material; of those, eight locations were chosen for test excavations, with the most promising one so far being Dealul Guran (Iovita et al. 2012).

Stratigraphy and Age

The site is a collapsed rockshelter facing north-east; it lies close to the top of the Guran Hill, *ca.* 30 m above the Peştera, a tributary of the river Danube. The geology of the valley can be best observed on the slope opposite the Dealul Guran, known as Dealul Peşterica (Small Cave Hill). There, contemporary quarrying activities have revealed a Cretaceous-Tertiary sequence of alternating limestone and sandstone layers, some of them containing flint nodules, which made the area attractive for Paleolithic communities searching for raw material (Iovita et al. 2012). The rockshelter from Dealul Guran was formed by differential erosion of glauconitic clays and limestone/sandstone units, with the fossiliferous limestone acting as the overhanging roof and the kaolinitic sandstone as the rockshelter floor, on which the glauconitic clays were redeposited during the Quaternary occupation of the site. An aeolian loess component is present only in microscopic quantities, a situation explained by the site's position in the landscape, in the lee of dominant aeolian transport direction (Buggle et al. 2008; see also Iovita et al. 2012).

The archaeological units follow relatively closely the geological layers (Fig. 11.7), with the Upper Paleolithic unit 1 separated from the Lower Paleolithic units 2 and 3 by a series of large boulders (probably from a collapsed rockshelter roof), which sealed these older units. Both units 2 and 3 lie in geological layer E, with the overlying D being essentially sterile. Due to the nature of the excavation trench, unit 2 was sampled over a smaller area, and consequently, the lithic assemblage from this unit is also less well represented.

Three luminescence ages were obtained by K. Fitzsimmons (Luminescence Laboratory, Max Planck Institute for Evolutionary Anthropology, Leipzig) using post-IR-IRSL$_{225}$ measurements on the polymineral fine-grained fraction (4–11 μm) (for further details see Iovita et al. 2012, including online supplement). All three determinations suggest an age corresponding to OIS 11 (420–360 ka), which is corroborated by the slumping of rockshelter sediments and clay weathering, both of which require relatively humid conditions for deposition, in contrast to the OIS 10 glacial which was relatively arid in this region (Fitzsimmons et al. 2012).

Lithic Assemblage

Only lithics larger than 2.5 cm were individually piece-provenienced with the total station and analyzed in the lab. The lithic assemblage from the Lower Paleolithic archaeological units (2 and 3) contains a total of 512 artifacts >2.5 cm (of which 77 are in unit 2), and 1164 lithics <2.5 cm (216 for unit 2), which were recovered after wet sieving. The difference between Unit 2 and Unit 3 regarding the proportions of small and large lithics is not significant ($\chi^2 = 3.05, p = 0.81$). In terms of site preservation, the absence of size sorting among the artifacts shows that no major post-depositional processes have affected the site (Bertran et al. 2012). This assumption is corroborated by the very high percentage of lithics (89 %) with no edge damage.

Only five formal tools could be identified, two denticulates, two scrapers (pictured, Fig. 11.8: 7), and a notch. The assemblage composition can be described as follows: 300 complete flakes, 5 complete tools, 73 cores and 10 core fragments, as well as 118 broken flakes (see Fig. 11.9). Seven cores have a single platform and an adjacent flaking surface; three multiple-platform cores were also found (Fig. 11.8: 5), as well as two Kombewa cores and an additional core-on-flake. The production of flakes on flake-cores is also demonstrated by five Kombewa flakes and one Kombewa core. The low blank to core ratio (4), as well as the cortex ratio (after Dibble et al. 2005; Douglass et al. 2008) CR = 1.07, suggest that slightly more cortex is present than expected under a neutral reduction model. This indicates that the main activity at the site was cobble-testing, with some of the non-cortical products being taken out of the site.

Discussion and Conclusions

The evidence for the Lower Paleolithic in the Balkans, although fragmentary, has been growing in the last few decades, leading to a more accurate and realistic understanding of the period. On the one hand, the current evidence has been critically reassessed (e.g., Kuhn 2002; Doboş 2008; Doronichev 2008, 2015; Sirakov et al. 2010; Tourloukis 2010; Ling 2012; Dinçer 2016; Harvati 2016), and on the other, newly discovered sites (Iovita et al. 2012; Panagopoulou et al. 2015; Darlas and Psathi 2016; Galanidou et al. 2016) and human fossils (Roksandic et al. 2011; Rink et al. 2013) have brought valuable additional data.

When reviewing the alleged Lower Pleistocene lithic collections coming from Romania, two major issues occur: some of them are natural, and the others are not valid chrono-cultural markers. Thus, speaking of sites of such an age based on the available evidence is a dangerous venture. On the other hand, in the 1.5–1.2 Ma time interval, there is a scatter of several Mode 1 sites with secure context in the southern parts of Europe: Sima del Elefante (Rosas et al. 2006), Barranco de León and Fuente Nueva (Oms et al. 2000; Toro-Moyano et al. 2013), Pirro Nord (Arzarello et al. 2007), and Kozarnika (Sirakov et al. 2010). Although an earliest colonization of even the southern fringes of the European continent that pre-dates the Jaramillo subchron (~1 Ma) has been recently challenged (Muttoni et al. 2013), the evidence at *ca.* 1 Ma is currently accepted by most. As such, given its geographic position, we estimate that the future discovery of sites significantly older than Dealul Guran in Romania is quite likely. The discoveries of Sândominic, Gura Dobrogei and Amărăşti, which were previously assigned a Middle Pleistocene age based on the faunal remains and were presented as *in situ*, however, do not stand up to a critical evaluation, because of the doubtful anthropogenic character of the small number of lithics and the poor documentation of their provenience.

Dealul Guran is the first securely dated Lower Paleolithic site in Romania, and at the same time the oldest site currently known in the country. While there is no reason to assume that the antiquity of hominin settlement in Romania could not date earlier than the Middle Pleistocene, there are several factors to keep in mind. First, the progressive drying-up of the Pannonian and Getic Lakes and the eventual creation of the modern Danube took up most of the early Pleistocene (Olteanu and Jipa 2006). The open steppic landscape which characterizes the region today, and which we believe would have been conducive to larger-scale land-use patterns, and hence, migration, dates only to the beginning of the deposition of loess, which is currently dated to 700 ka to 1 Ma. Therefore, we speculate that it is easier to imagine hominins settling in the region after 1 Ma, although this hypothesis must be tested through systematic surveys. Such surveys could target either known Pleistocene fossiliferous sites, as was previously done, or use the stratigraphy of loess/paleosol sequences to search for *in situ* artifacts, which can be dated and subsequently excavated. In Dobrogea, the recent construction of the A2 highway and other infrastructure projects

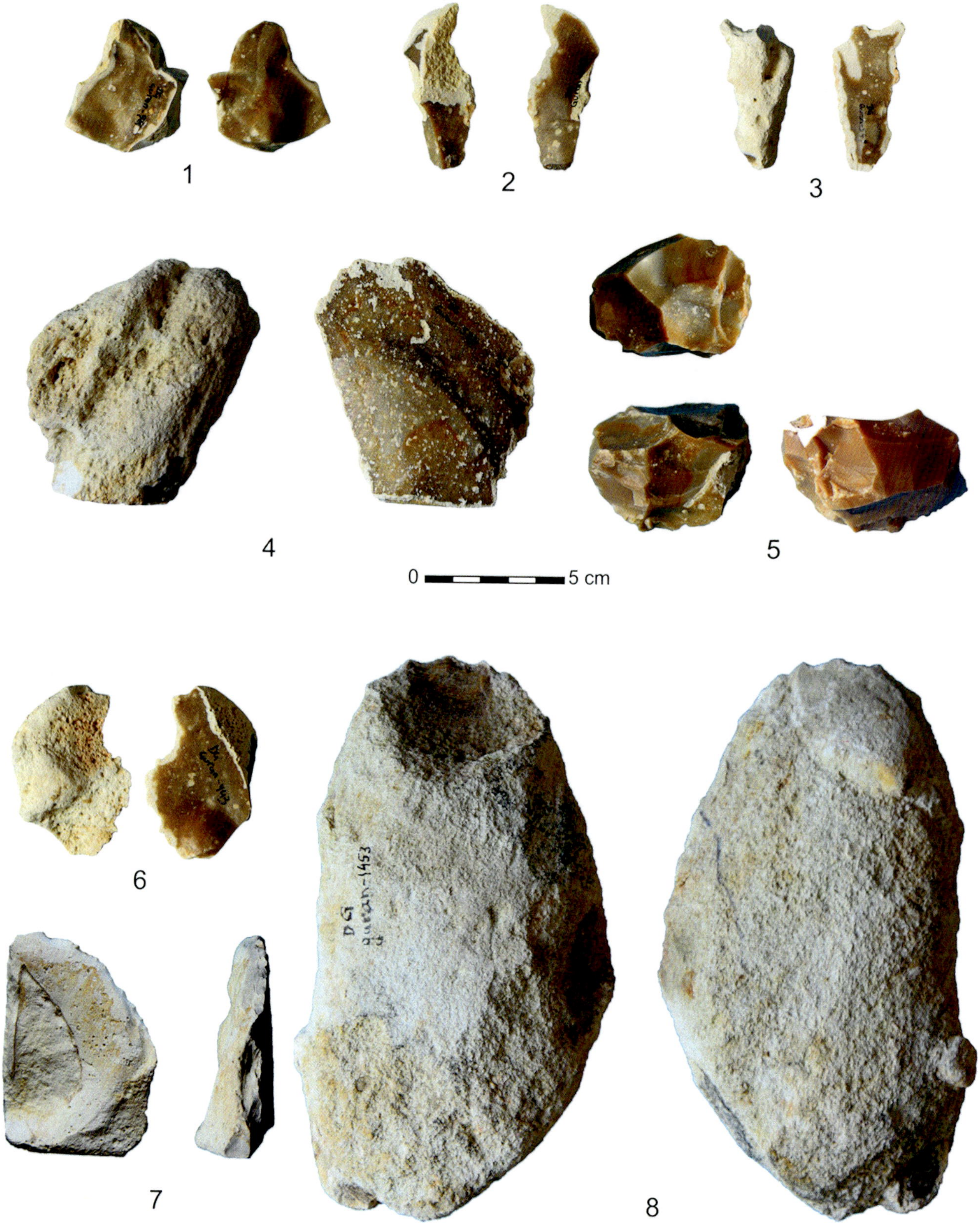

Fig. 11.8 Dealul Guran: *1*, *2*, *3*, *4*, *6*—Complete flakes; *5*—Core; *7*—Sidescraper; *8*—Tested block

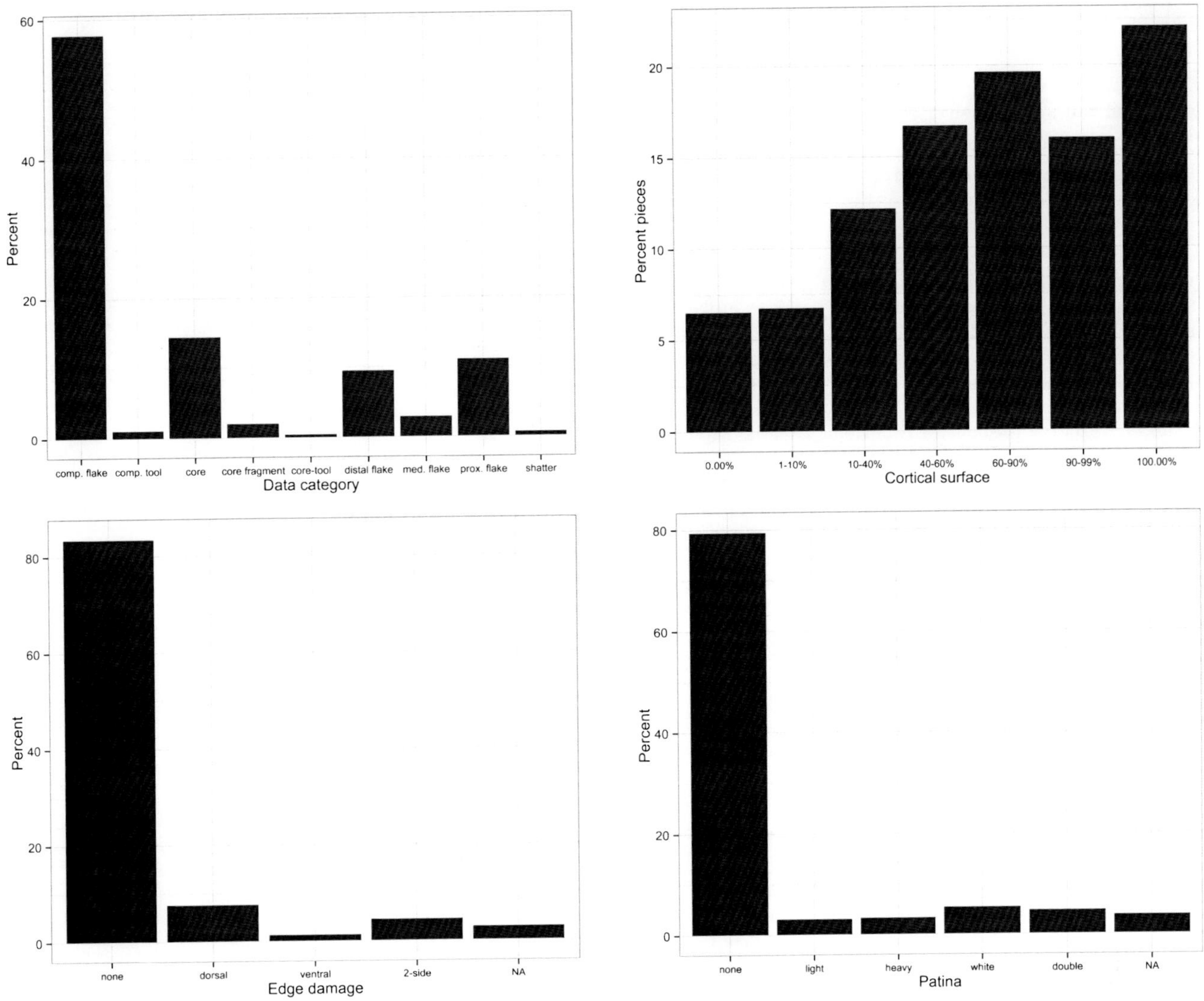

Fig. 11.9 Dealul Guran: Summary of the lithic assemblage

has opened a large number of visible sections, some of which we have studied and sampled for dating, so that, in the near future, a better database of the Paleolithic settlement of the region can be built (Iovita et al. 2014).

As to the significance of the site for reconstructing hominin dispersals at the Gates of Europe, the stone industry at Dealul Guran does not offer many clues that would invite speculation about cultural links with sites of a similar age in the region. Rather, it brings a new perspective in the understanding of the Middle Pleistocene economic behavior, i.e., raw material exploitation. In the OIS 13-OIS 11 time bracket, numerous lithic industries from Central and Eastern Europe are dominated by small-sized artifacts, as is the case for Vértesszölös, OIS 13 (Dobosi 1998, 2003), Bilzingsleben, OIS 11 (Haidle and Pawlik 2010), Schöningen, OIS 11 (Thieme 2003), and Treugol'naya, OIS 11 (Doronichev 2011). This common feature was sometimes explained through the predilection toward knapping small-sized tools, driven by the so-called "small tool tradition" (Burdukiewicz and Ronen 2003), although an alternative explanation relates the size of artifacts to the size of raw material.

In the case of Dealul Guran, the character of the site was clearly influenced by raw material abundance: the numerous cortical flakes, tested blocks and almost complete lack of retouched tools indicate a quarrying site. This situation adds a new pattern to the Lower Paleolithic industries of European sites. The particularities of Dealul Guran indicate that it was just a small part of a wider landscape-exploitation system employed by the hominins that settled here during OIS 11, and that other sites, perhaps displaying other functional characteristics, are needed before an accurate reconstruction of Middle Pleistocene behavior at the western margin of the loess steppe can be attempted.

In conclusion, the data presented in this paper suggest that Dealul Guran is the only reliable Lower Paleolithic site in Romania. The *in situ* discoveries are poorly documented, and the over 60 find spots with lithics in river gravels are hardly reliable indicators for the presence of Lower Paleolithic sites. On the other hand, the discovery of the site of Dealul Guran through a systematic field survey shows that such research projects are appropriate undertakings for answering some of the numerous questions about the earliest hominin occupation of Europe.

Acknowledgements The LoDanS project was funded by Max Planck Institute for Evolutionary Anthropology, Leipzig and was carried out in concordance with Romanian law and within the terms of the permits nos. 1/2011 and 107/2012 granted by the Ministry of Culture of Romania. The Romanian research collective was headed by Dr. Roxana Dobrescu (2010) and Dr. Vlad V. Zirra (2011, 2012). Tiberiu Tecar, from National Museum of Transylvanian History, Cluj provided prints. Professor A. Barnea (Institute of Archaeology "Vasile Parvan") and Mariana Petrut (History Museum, Adamclisi) provided logistical assistance during the excavation season. We wish to thank the mayor of Peştera, Valentin Vrabie, the vice-mayor, Dorin Grosu as well as the people of the villages of Peştera and Adamclisi, who welcomed and helped us through the duration of the field seasons. Thanks to G. Heinz (Fachhochschule Mainz/RGZM) for assistance with the total station, as well as N. Schlösser and K. Weber (University of Mainz), along with several teams of dedicated student volunteers, who made the project possible. We thank the Editors and anonymous reviewers for their help in improving this chapter.

References

Ambrojevici, C. (1926). Urmele omului diluvial în Basarabia. *Buletinul Muzeului Național de Istorie Naturală, 1*, 67–76.

Andreescu, I., Rădulescu, C., Samson, P., Tschepalzyga, A., & Troubikhin, I. (1981). Chronologie (Mollusques, Mammiferes, Paleomagnetism) des formations Plio-Pléistocenes de la zone de Slatina (Bassin Dacique). *Travaux de l'Institut de Spéleologie "Emile Racovitza", 20*, 127–137.

Arzarello, M., Marcolini, F., Pavia, G., Pavia, M., Petronio, C., Petrucci, M., et al. (2007). Evidence of earliest human occurrence in Europe: The site of Pirro Nord (Southern Italy). *Naturwissenschaften, 94*, 107–112.

Bar-Yosef, O., & Belmaker, M. (2011). Early and Middle Pleistocene faunal and hominins dispersals through Southwestern Asia. *Quaternary Science Reviews, 30*, 1318–1337.

Bertran, P., Lenoble, A., Todisco, D., Desrosiers, P., & Sørensen, M. (2012). Particle size distribution of lithic assemblages and taphonomy of Palaeolithic sites. *Journal of Archaeological Science, 39*, 3148–3166.

Brain, C. K. (1981). *The hunters or the hunted? An introduction to African cave taphonomy*. Chicago: University of Chicago Press.

Buggle, B., Hambach, U., Glaser, B., Marcović, S., Glaser, I., & Zöller, L. (2008). Long-term Paleoclimate records in SE-Europe—The loess paleosol sequences Batajnica/Stari Slankamen (Serbia) and Mircea Voda (Romania). Hauptversammlung der DEUQUA, Wien, 31 August–6 September 2008, *62*, 15–19.

Burdukiewicz, J., & Ronen, A. (Eds.). (2003). *Lower Palaeolithic small tools in Europe and the Levant*. Oxford: British Archaeological Reports International Series 1115.

Carbonell, E., Mosquera, M., Rodriguez, X. P., Sala, R., & van der Made, J. (1999). Out of Africa: the dispersal of the earliest technical systems reconsidered. *Journal of Anthropological Archaeology, 18*, 119–136.

Cârciumaru, M. (1999). *Le Paléolithique en Roumanie*. Grenoble: Jérome Millon.

Darlas, A., & Psathi, E. (2016). The Middle and Upper Paleolithic on the Western coast of the Mani Peninsula (Southern Greece). In K. Harvati, & M. Roksandic (Eds.), *Paleoanthropology of the Balkans and Anatolia: Human evolution and its context*. Dordrecht: Springer.

Dennell, R., Martinon-Torres, M., & Bermudez de Castro, J. M. (2011). Hominin variability, climatic instability and population demography in Middle Pleistocene Europe. *Quaternary Science Reviews, 30*, 1511–1524.

Dibble, H., Schurmans, U., Iovita, R., & McLaughlin, M. (2005). The measurement and interpretation of cortex in lithic assemblages. *American Antiquity, 70*(3), 545–560.

Dicu, P. (1972). Descoperiri atribuite paleoliticului inferior pe cursul mijlociu al Argeşului (jud. Argeş). *Studii şi Cercetări de Istorie Veche, 23*(2), 223–234.

Dicu, P. (1973). Noi descoperiri de unelte atribuite paleoliticului inferior pe cursul mediu al Argeşului. *Studii şi Cercetări de Istorie Veche, 24*(4), 653–638.

Dicu, P. (1979). Paleoliticul inferior în Piemontul Cotmeana şi depresiunea Sibiu. *Studii şi Cercetări de Istorie Veche, 30*(4), 575–589.

Dinçer, B. (2016). The Lower Paleolithic in Turkey: Is Anatolia on the route for early hominin dispersals out of Africa? In K. Harvati, & M. Roksandic (Eds.), *Paleoanthropology of the Balkans and Anatolia: Human evolution and its context*. Dordrecht: Springer.

Doboş, A. (2005). C. S. Nicolăescu-Plopşor şi arheologia paleoliticului. *Studii de Preistorie, 2*, 233–247.

Doboş, A. (2008). The Lower Paleolithic of Romania: A critical review. *PaleoAnthropology*, 218–233.

Dobosi, V. (1998). Le site paléolithique inferieur de Vértesszölös, Hongrie. *L'Anthropologie, 92*(4), 1041–1050.

Dobosi, V. (2003). Changing environment—Unchanged culture at Vértesszölös, Hungary. In J. Burdukiewicz & A. Ronen (Eds.), *Lower Palaeolithic small tools in Europe and the Levant* (pp. 101–111). Oxford: British Archaeological Reports International Series 1115.

Doronichev, V. (2008). The Lower Paleolithic in eastern Europe and the Caucasus: A reappraisal of the data and new approaches. *PaleoAnthropology*, 107–157.

Doronichev, V. (2011). Le Paléolithique ancien de l'Europe orientale et du Caucase. *L'Anthropologie, 115*, 197–246.

Doronichev, V. (2015). The Pre-Mousterian industrial complex in Europe between 400 and 300 ka: Interpreting its origin and spatiotemporal variability. *Quaternary International*. doi:10.1016/j.quaint.2015.05.063.

Douglass, M., Holdaway, S., Fanning, P., & Shiner, J. (2008). An assessment and archaeological application of cortex measurement in lithic assemblages. *American Antiquity, 73*(3), 513–526.

Fitzsimmons, K., Marković, S., & Hambach, U. (2012). Pleistocene environmental dynamics recorded in the loess of the middle and lower Danube basin. *Quaternary Science Reviews, 41*, 104–118.

Galanidou, N., Athanassas, C., Cole, J., Iliopoulos, G., Katerinopoulos, A., Magganas, A., & McNabb, J. (2016). The Acheulean site at Rodafnidia, Lisvori on Lesbos, Greece: Part I. In K. Harvati & M. Roksandic (Eds.), *Paleoanthropology of the Balkans and Anatolia: Human evolution and its context*. Dordrecht: Springer

Gibbard, P., & Cohen, K. (2008). Global chronostratigraphical correlation table for the last 2.7 million years. *Episodes, 31*, 243–247.

Haase, D., Fink, J., Haase, G., Ruske, R., Pécsi, M., Richter, H., et al. (2007). Loess in Europe—Its spatial distribution based on a European Loess Map, scale 1:2,500,000. *Quaternary Science Reviews, 26*, 1301–1312.

Haidle, M., & Pawlik, A. (2010). The earliest settlement of Germany: Is there anything out there? *Quaternary International, 223–224*, 143–153.

Harvati, K. (2016). Paleoanthropology in Greece: Recent findings and interpretations. In K. Harvati, & M. Roksandic (Eds.), *Paleoanthropology of the Balkans and Anatolia: Human evolution and its context*. Dordrecht: Springer.

Iovita, R., Doboş, A., Fitzsimmons, K., Probst, M., Hambach, U., Robu, M., et al. (2014). Geoarchaeological prospection in the loess steppe: Preliminary results from the Lower Danube Survey for Paleolithic Sites. *Quaternary International, 351*, 98–114. doi:10.1016/j.quaint.2013.05.08.

Iovita, R., Fitzsimmons, K., Doboş, A., Hambach, U., Hilgers, A., & Zander, A. (2012). Dealul Guran: Evidence for Lower Palaeolithic (MIS 11) occupation of the Lower Danube loess steppe. *Antiquity, 86*, 973–989.

Korisettar, R., & Petraglia, M. D. (1998). The archaeology of the Lower Paleolithic: Background and overview. In M. D. Petraglia & R. Korisettar (Eds.), *Early human behavior in global context. The rise and diversity of the Lower Paleolithic record* (pp. 1–22). London: Routledge.

Kuhn, S. (2002). Paleolithic archeology in Turkey. *Evolutionary Anthropology, 11*, 198–210.

Ling, V. (2012). *The Lower Palaeolithic colonisation of Europe. Antiquity, magnitude, permanency and cognition*. Oxford: British Archaeological Reports International Series 2316/Archaeopress.

Macdonald, K., Martinon-Torres, M., Dennell, R., & Bermudez de Castro, J. M. (2011). Discontinuity in the record for hominin occupation in south-western Europe: Implications for occupation of the middle latitudes of Europe. *Quaternary International, 271*, 84–97.

Martínez-Navarro, B. (2010). Early Pleistocene faunas of Eurasia and hominin dispersals. In J. Fleagle, J. Shea, F. Grine, A. Baden, & R. Leakey (Eds.), *Out of Africa I: The first hominin colonization of Eurasia* (pp. 207–224). Dordrecht: Springer.

Moga, M. (1938). Paleoliticul inferior în Transilvania. *Anuarul Comisiunii Monumentelor Istorice, Secţiunea pentru Transilvani, 4*, 3–21.

Moncel, M.-H. (2010). Oldest human expansions in Eurasia: Favouring and limiting factors. *Quaternary International, 223–224*, 1–9.

Moroşan, N. N. (1931). Existe-t-il du micoquien en Bessarabie et quelle serait sa place dans la chronologie du Pléistocene? *Bulletin de la Societe Prehistorique Francaise, 28*(4), 223–233.

Mudelsee, M., & Schulz, M. (1997). The Mid-Pleistocene climate transition: Onset of 100 ka cycle lags ice volume build-up by 280 ka. *Earth and Planetary Science Letters, 151*, 117–123.

Muttoni, G., Scardia, G., & Kent, D. (2013). A critique of evidence for human occupation of Europe older than the Jaramillo subchron (~1 Ma): Comment on 'The oldest human fossil in Europe from Orce (Spain)' by Toro-Moyano et al. (2013). *Journal of Human Evolution, 65*, 746–749.

Nania, I. (1972). Unelte ale paleoliticului inferior descoperite pe văile Dâmbovnicului şi *Mozacului. Studii şi Cercetări de Istorie Veche şi Arheologie, 23*(2), 235–244.

Nicolăescu-Plopşor, C. S. (1929). Cultura şeleană în Romania? *Arhivele Olteniei, 8*(45–46), 469–473.

Nicolăescu-Plopşor, C. S. (1930). Iarăsi cultura şeleană în România. *Arhivele Olteniei, 9*(49–50), 211–213.

Nicolăescu-Plopşor, C. S. (1931). Asupra culturii acheuleene şi micocquiene în România. *Arhivele Olteniei, 10*(53), 46–52.

Nicolăescu-Plopşor, C. S. (1956). Rezultatele principale ale cercetărilor paleolitice în ultimii 4 ani în RPR. *Studii şi Cercetări de Istorie Veche, 7*(1–2), 7–39.

Nicolăescu-Plopşor, C. S. (1957). Cercetări asupra paleoliticului timpuriu. *Materiale si Cercetari Arheologice, 3*, 281–290.

Nicolăescu-Plopşor, C. S. (1964a). Date noi cu privire la cunoaşterea începuturilor şi sfârşitului paleoliticului României. *Studii şi Cercetări de Istorie Veche, 15*(3), 307–320.

Nicolăescu-Plopşor, C. S. (1964b). Nouvelles donées sur la possibilité de l'existence de protohominiens dans le villafranchien de Roumanie. *Dacia N.S., 8*, 47–52.

Nicolăescu-Plopşor, C. S. (1965). *Oamenii din vârsta veche a pietrei*. Bucureşti: Editura Ştiinḑifică.

Nicolăescu-Plopşor, D. (1970). Vârsta omului. *Magazin Istoric, 4*(36), 80–84.

Nicolăescu-Plopşor, C. S., & Moroşan, I. N. (1959). Sur le commencement du paléolithique en Roumanie. *Dacia N.S., 3*, 9–33.

Nicolăescu-Plopşor, C. S., & Nicolăescu-Plopşor, D. (1963). The possible existence of protohominids in Romania's villafranchean. *Dacia N.S., 7*, 8–25.

Nicolăescu-Plopşor, C. S., & Nicolăescu-Plopşor, D. (1965). *Cu privire la începuturile istoriei.României* (Omagiu lui P. Constantinescu-Iaşi, pp. 31–37). Bucureşti: Editura Academiei Republicii Populare Române.

Olteanu, R., & Jipa, D. (2006). Dacian basin environmental evolution during Upper Neogene within the Paratethys domain. *Geo-Eco-Marina, 12*, 91–105.

Oms, O., Parés, J., Martínez-Navarro, B., Agustí, J., Toro, I., Martínez-Fernández, G., et al. (2000). Early human occupation of Western Europe: Paleomagnetic dates for two Paleolithic sites in Spain. *Proceeding of the National Academy of Sciences United States of America, 97*, 10666–10670.

Panagopoulou, E., Tourloukis, V., Thompson, N., Athanassiou, A., Tsartsidou, G., Konidaris, G. D., Karkanas, P., & Harvati, K. (2015). Marathousa 1—A new Middle Pleistocene archaeological site from Greece. *Antiquity Project Gallery, 89*(343).

Păunescu, A. (1980). Evoluţia istorică pe teritoriul României din paleolitic până la începutul neoliticului. *Studii şi Cercetări de Istorie Veche şi Arheologie, 31*(4), 519–545.

Păunescu, A. (1999a). *Paleoliticul şi mezoliticul de pe teritoriul Dobrogei*. Bucureşti: Satyia Sai.

Păunescu, A. (1999b). *Paleoliticul şi mezoliticul de pe teritoriul Moldovei cuprins între Siret şi Prut*. Bucureşti: Satya Sai.

Păunescu, A. (2000). *Paleoliticul şi mezoliticul de pe teritoriul cuprins între Carpaţi şi Dunăre*. Bucureşti: AGIR.

Păunescu, A. (2001). *Paleoliticul şi mezoliticul din spaţiul transilvan*. Bucureşti: AGIR.

Păunescu, A., Rădulescu, C., & Samson, P. (1982). Decouvertes du Paleolithique Inferieur en Roumanie. *Travaux de l'Institut de Spéleologie "Emile Racovitza", 22*, 53–62.

Petculescu, A., & Ştiucă, E. (2008). Peculiarity of the mammal associations from the Upper Pleistocene (Dobrogea, Romania). *Quaternary International, 179*(1), 79–82.

Rădulescu, C., & Samson, P. (1991). Traces d'activite humaine a la limite Pliocène/Pléistocène dans le Bassin Dacique (Roumanie). In E. Bonifay & B. Vandermeersch (Eds.), *Les premiers Européens* (pp. 203–207). Paris: Comité des Travaux Historiques et Scientifiques.

Rădulescu, C., Samson, P., Petculescu, A., & Ştiucă, E. (2003). Pliocene Large Mammals of Romania. *Coloquios de Paleontología, 53*(1), 549–558.

Rădulescu, C., Samson, P., & Ştiucă, E. (1998). Cadre biostratigraphique du paléolithique inferieur en Roumanie. *Quaternaire, 9*(4), 283–290.

Rink, W., Mercier, N., Mihailović, D., Morley, M., Thompson, J., & Roksandic, M. (2013). New radiometric ages for the BH-1 hominin from Balanica (Serbia): Implications for understanding the role of the Balkans in Middle Pleistocene human evolution. *PLoS One, 8*(2), e54608. doi:10.1371/journal.pone.0054608.

Roksandic, M., Mihailović, D., Mercier, N., Dimitrijević, V., Morley, M.W., Rakočević, Z., et al. (2011). A human mandible (BH-1) from the Pleistocene deposits of Mala Balanica cave (Sićevo Gorge, Nis, Serbia). Journal of Human Evolution, 61 (2), 186–196.

Rosas, A., Huguet, R., Pérez-González, A., Carbonell, E., Bermúdez de Castro, J. M., Vallverdú, J., et al. (2006). The "Sima del Elefante" cave site at Atapuerca (Spain). *Estudios Geológicos, 62*(1), 327–348.

Roska, M. (1928). Le paléolithique inferieur de Zimbru (Zimbro, judeţul Arad). *Buletinul Societăţii de Ştiinţe din Cluj, 4*(2), 35–37.

Roska, M. (1931a). Notă preliminară asupra cercetărilor paleolitice făcute în Ardeal în cursul anului 1928. *Anuarul Institutului Geologic al României, 14*, 79–97.

Roska, M. (1931b). Paleoliticul ardealului. *Anuarul Institutului Geologic al României, 14*, 99–126.

Roska, M. (1933). Recherches paleolithiques en Transylvanie en 1927. *Dacia, 3–4*, 8–23.

Roşu, L. (1987). *Treptele antropogenezei*. Bucureşti: Editura Ştiinţifică şi Enciclopedică.

Samson, P., & Rădulescu, C. (1973). Les faunes de Mamiféres et la limite Pliocéne-Pleistocéne en Roumanie. *Travaux de l'Institut de Spéleologie "Emile Racovitza", 12*, 191–228.

Schmeider, F., von Dobeneck, T., & Bleil, U. (2000). The Mid-Pleistocene climate transition as documented in the deep South Atlantic Ocean: Initiation, interim state and terminal event. *Earth and Planetary Science Letters, 179*, 539–549.

Singer, R. (1956). The "Bone Tools" from hopefield. *American Anthropologist, 58*(6), 1127–1134.

Sirakov, N., Guadelli, J.-L., Ivanova, S., Sirakova, S., Boudadi-Maligne, M., Dimitrova, I., et al. (2010). An ancient continuous human presence in the Balkans and the beginnings of human settlement in western Eurasia: A Lower Pleistocene example of the Lower Palaeolithic levels in Kozarnika cave (North-western Bulgaria). *Quaternary International, 223–224*, 94–106.

Spassov, N. (2016). Southeastern Europe as a route for the earliest dispersal of *Homo* towards Europe: Ecological conditions and the timing of the first human occupation of Europe. In K. Harvati, & M. Roksandic (Eds.), *Paleoanthropology of the Balkans and Anatolia: Human evolution and its context*. Dordrecht: Springer.

Tappen, M. (2009). The wisdom of the aged and out of Africa I. In J. Shea & D. Lieberman (Eds.), *Transitions in prehistory: Essays in honor of Ofer Bar-Yosef* (pp. 33–53). Oakville: Oxbow Books.

Thieme, H. (2003). Lower Paleolithic sites at Schöningen, Lower Saxony, Germany. In J. Burdukiewicz & A. Ronen (Eds.), *Lower Palaeolithic small tools in Europe and the Levant* (pp. 9–27). Oxford: British Archaeological Reports International Series 1115.

Toro-Moyano, I., Bienvenido Martínez-Navarro, B., Agustí, J., Souday, C., Bermúdez de Castro, J. M., Martinón-Torres, M., et al. (2013). The oldest human fossil in Europe dated to ca. 1.4 Ma at Orce (Spain). *Journal of Human Evolution, 65*, 1–9.

Tourloukis, E. (2010). *The Early and Middle Pleistocene archaeological record of Greece*. Leiden: Leiden University Press.

Turner, A. (1992). Large carnivores and earliest European hominids: Changing determinants of resource availability during the Lower and Middle Pleistocene. *Journal of Human Evolution, 22*, 109–126.

Van der Made, J., & Mateos, A. (2010). Longstanding biogeographic patterns and the dispersal of early Homo out of Africa and into Europe. *Quaternary International, 223–224*, 195–200.

Wolberg, D. (1970). The hypothesized osteodontokeratic culture of the Australopithecinae: A look at the evidence and the opinions. *Current Anthropology, 11*(1), 23–37.

Chapter 12
A Route Through the Balkans and Implications for the Earliest Settlement of Europe

Stefanka Ivanova

Abstract Traces of settlement possibly dating to the Lower Paleolithic have recently been discovered in Bulgaria, including Kozarnika cave and surface sites from the Rhodope Mountains. Chopping tools, cores, flakes, and other stone tools are present in some flint assemblages. In rare cases, bifacial forms have been found. Based on their biostratigraphic position, the assemblages from Kozarnika are estimated to date from the period ranging between 1.6 Ma and 400 ka (Sirakov et al. 2010). This suggested dating is discussed here. The age of the surface sites in the Rhodope Mountains is estimated on the basis of the typo-technological characteristics and the stratigraphic location of the artifacts. The surface sites in the Western Rhodopes may date from the Middle Pleistocene, while the surface sites in the Eastern Rhodopes might be even older. The results of the research on Lower Paleolithic sites in Bulgaria are discussed in the framework of hypothesized repeated waves of dispersal towards Europe.

Keywords Lower Paleolithic • Bifaces • Chopping tools • Dispersal

Introduction

The earliest human dispersal to Eurasia is among the most actively debated topics in archaeology and paleoanthropology today, and numerous recent studies have examined the timing and routes of migration of this event. Experts have discussed several possible routes: across the Strait of Gibraltar, across the Strait of the Bosporus, along the northern shores of the Black Sea, across the Sicilian Channel, and across the Bab el Mandeb Straits (see Spassov 2016; Strait et al. 2016). Faunal, geomorphological, and climatic evidence has been brought forth to support each of these possible routes. However, sufficient archaeological evidence to track the paths of movement of early hominins from Africa to Europe is still lacking, and each new discovery complicates the chronological framework of the different stages of the colonization of Eurasia. Several proposed migration corridors pass through the Balkan Peninsula in southeastern Europe (Fig. 12.1:1). However, since research on the Lower Paleolithic of this region started relatively late, the number of such early sites is limited and there is still a lack of sufficient evidence to conduct comparative analyses. The objective of this chapter is to provide an overview of the relatively little-known Lower Paleolithic sites in the territory of Bulgaria (southeastern part of the Balkan Peninsula).

Evidence for human presence in Bulgaria during the Lower Paleolithic comes from Kozarnika Cave, located at the foothills of the Balkan Mountains (Northwestern Bulgaria); from surface sites in the Western Rhodopes (Kremenete, Shiroka Polyana); and from surface sites in the Eastern Rhodopes—Benkovski (Ivanova 2006, 2009; Sirakov et al. 2010; see also Spassov 2016) (Fig. 12.1:2, 3).

The Early Paleolithic at Kozarnika

Kozarnika is located in northwestern Bulgaria, in the northern part of the western Pre-Balkans, close to the Danube Valley. The cave is situated at 480 m above sea level and 85 m above the valley bottom of the river Skomlia. Its Pleistocene sequence is characterized by rich lithic assemblages and bone remains (Sirakov et al. 2010). The entrance of Kozarnika faces south. It measures 8 m wide at the base and has a height of 3.5 m (Fig. 12.2:1, 2). The length of the cave is 210 m. Trenches I, II, and III are located near the entrance, while

S. Ivanova (✉)
National Institute of Archaeology and Museum, Bulgarian Academy of Sciences, Saborna Str., Sofia 1000, Bulgaria
e-mail: ivanova.stefanka@gmail.com

Katerina Harvati and Mirjana Roksandic (eds.), *Paleoanthropology of the Balkans and Anatolia*, Vertebrate Paleobiology and Paleoanthropology, DOI 10.1007/978-94-024-0874-4_12

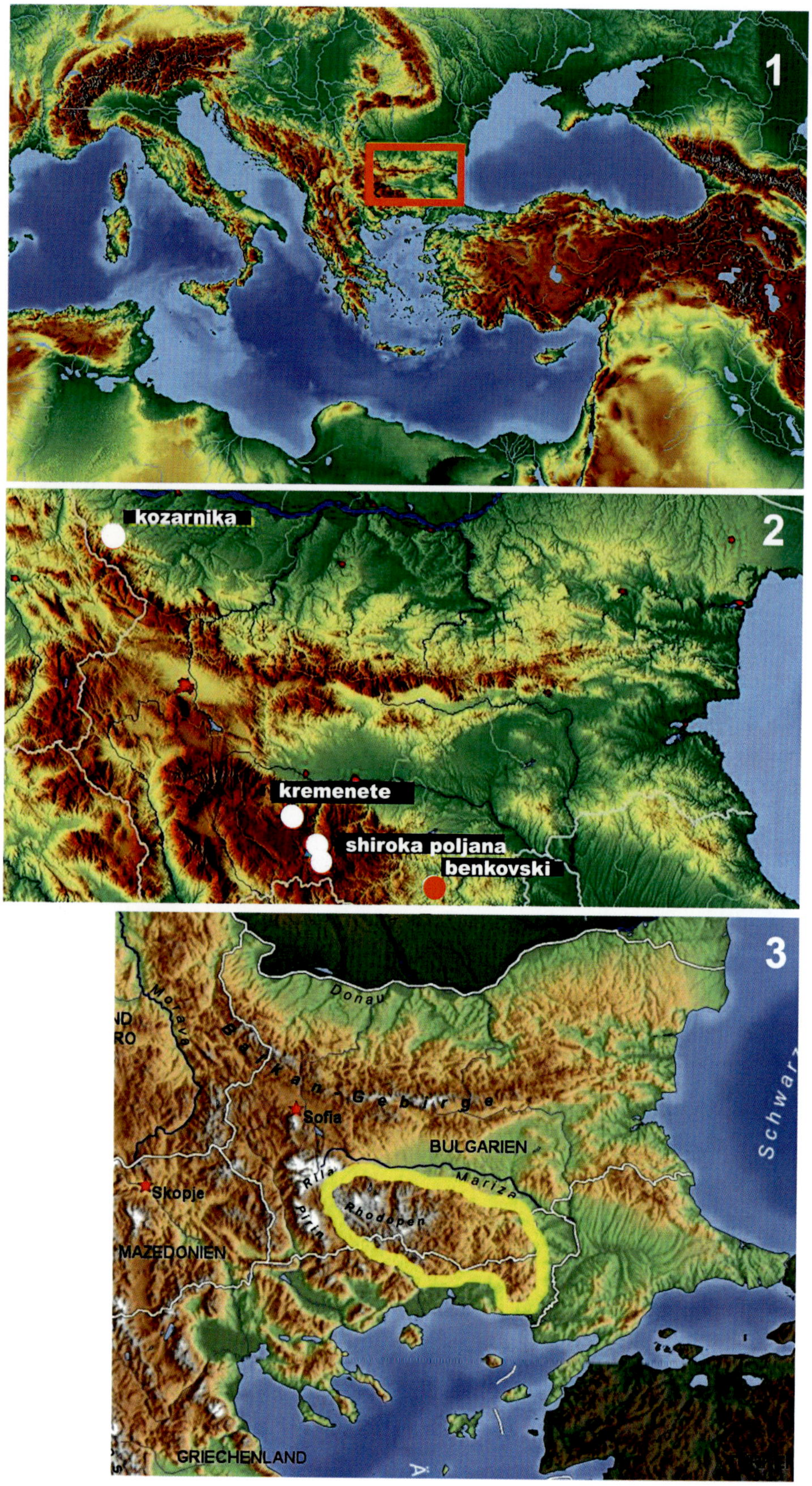

Fig. 12.1 Bulgaria at the crossroads of three continents. (*1*) The location of Bulgaria; (*2*) Early Paleolithic sites from Bulgaria; (*3*) The location of the Rhodope Mountains

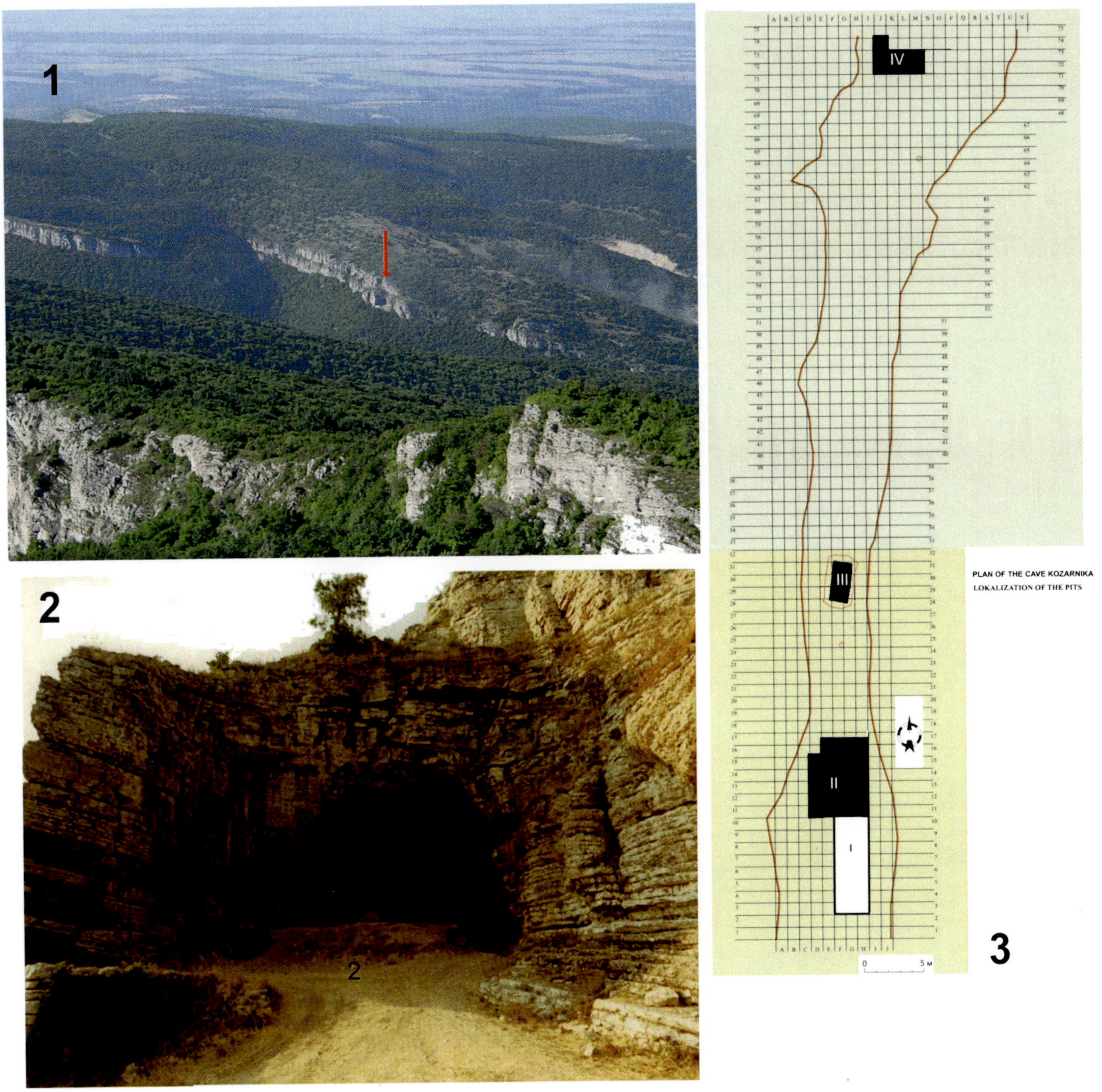

Fig. 12.2 Kozarnika cave. (*1*) A view to the cave; (*2*) A view to the entrance; (*3*) Location of the trenches in Kozarnika cave

trench IV is further inside the cave, *ca.* 72 m from the entrance. The surface and the upper part of the Kozarnika sediments have been destroyed by modern activities in the cave. However, the sediments that we studied (at around 10 m depth in trench II) preserved traces of long-term human activity in the cave dating to the Lower, Middle, and Upper Paleolithic, as well as the Holocene. The Kozarnika Pleistocene sequence is characterized by rich lithic assemblages and bone remains. The Upper Paleolithic period is represented by the Epigravettian and Gravettian. The Middle Paleolithic assemblages are part of the Levallois-Mousterian with leaf points. The Lower Paleolithic assemblages from the trenches II, III, and IV in Kozarnika (Fig. 12.2:3) were studied by the author during the period 1996–2011, in the framework of the Bulgarian-French research project "Les plus anciennes Manifestations de la présence Humaine en Bulgarie du nord."

Stratigraphic Notes

The sediments in Kozarnika have been separated into three groups based on their location in trench II (Sirakov et al. 2010). The first group includes the sterile layers 16–14. The second group comprises layers 13–11, which have yielded Early Paleolithic assemblages (in trenches II and III; Fig. 12.3:1, 3) and k, l, and m in trench IV (Fig. 12.3:2). A more detailed description of the lithostratigraphic units in this group is discussed in Sirakov et al. (2010). The third sediment group includes layers 10–3, which have yielded Middle and Upper Paleolithic assemblages.

The Early Paleolithic layers have been dated on the basis of their faunal composition. Based on the distinctive biostratigraphic zones in the cave a date between 1.6 Ma–0.5 ka BP has been proposed (Sirakov et al. 2010:12). The different layers are dated as follows: 13–11c: 1.4–0.9 Ma BP; 11b: 800–600 ka BP; 11a: 600–400 ka BP (Guadelli and Guadelli 2004). However, this chronological assessment is subject to discussion (see later). Within each layer several facies are distinguished. Because of the small differences in the characteristics of the lithic assemblages of the various facies, and due to the difficulties in distinguishing among facies in certain situations, here I shall describe and compare combined assemblages from several facies within a given layer.

Raw Materials

The raw material used in the Kozarnika Early Paleolithic assemblages is local. Flint nodules are oval shaped and usually small (most frequently *ca.* 3 cm, although larger nodules between 8 and 12 cm are also encountered). They are distributed in a limestone deposit and have been broken into fragments with natural break surfaces due to rockfall from the cave ceiling. The quality of the raw material used is relatively poor. Pieces have been split along planes parallel to the surface. A surface perpendicular to the axis of the concretion was also utilized. Concretions were frequently split into segments. In layers 13 and 12, we observe preparations of some of the nodules for testing or knapping. There appears to be a specific, rather rare, technique for breaking them. The nodules were struck on a hard surface, while being rotated around their axes. Little fragments were chipped off, and a strip (a "band") on the surface of the nodules was formed. Then, after the nodule was prepared in this way, it was broken by a blow on the surface of the band. Strikes were delivered almost at the tangent of the circumference of the surface. Thus, a large point was formed in the central part of the intersecting plane of the concretion (Ivanova 2003:12).

The flint artifacts are relatively small (most are 3–4 cm long), due to size constraints and particularities of the local raw materials. Some larger artifacts were made from larger, better quality pieces of raw material, as well as, alternatively, from quartzite pebbles.

Lower Paleolithic Assemblages of Trenches II and III

Characteristics of the lithic assemblages in Layer 13: This lithostratigraphic unit includes artifacts from the earliest human occupation that has been discovered so far in this part of the cave. Its lithic assemblages are divided into "13 upper" and "13 lower."

Lithic assemblage of Layer 13 lower: Artifacts were recovered in sediment that is preserved on a large limestone block (its surface is about 1.5 m^2). The assemblage includes 28 artifacts (2 pieces and nodules, 7 cores, 17 tools, and 2 flakes). The cores vary in size from 2 to 4 cm, as well as in exploitation techniques (some are spheroids). One can distinguish a group of tools in Layer 13 lower, which includes artifacts with similar techno-typological and metric characteristics. These tools are flat, prismatic, and elongated. Their shapes are predetermined by the nature of the raw material used. The ratio of length and width is close to 2:1 and lengths are in the range of 6–7 cm. The largest artifact of the group is a blade-like specimen with an asymmetric, elongated proximal part forming a solid, exposed tip, with abrupt negatives across the distal part (Fig. 12.4:1). On the edges of some of the artifacts there is partial, irregular retouch, but this is not always possible to distinguish from pseudo-retouch. One example is a scraper showing coarse, abrupt retouch of the edges (Fig. 12.4:2). The ventral and the dorsal surfaces of the artifacts can be either natural or intentional. One specimen shows flat, partial surface negatives (Fig. 12.4:3). There is also evidence of frost activity on the surface. The other artifacts have separate negatives on the sides or on the edges and proximal parts, and can be defined as notched tools. Most of the artifacts in the oldest assemblage from Kozarnika are made of raw materials rarely used in the upper layers. The sources of this type of raw material are found about 5 km away from the cave.

Lithic assemblage of Layer 13 upper: The artifacts were found on a surface of 4 m^2. The assemblage includes the following groups: 3 specimens belonging to the group of preliminary knapping, 16 cores, 45 tools, and 76 flakes (140 pieces in total). Flakes are the largest component of this assemblage, which is dominated by specimens with multidirectional negatives on the dorsal surface and a size of 3–4 cm.

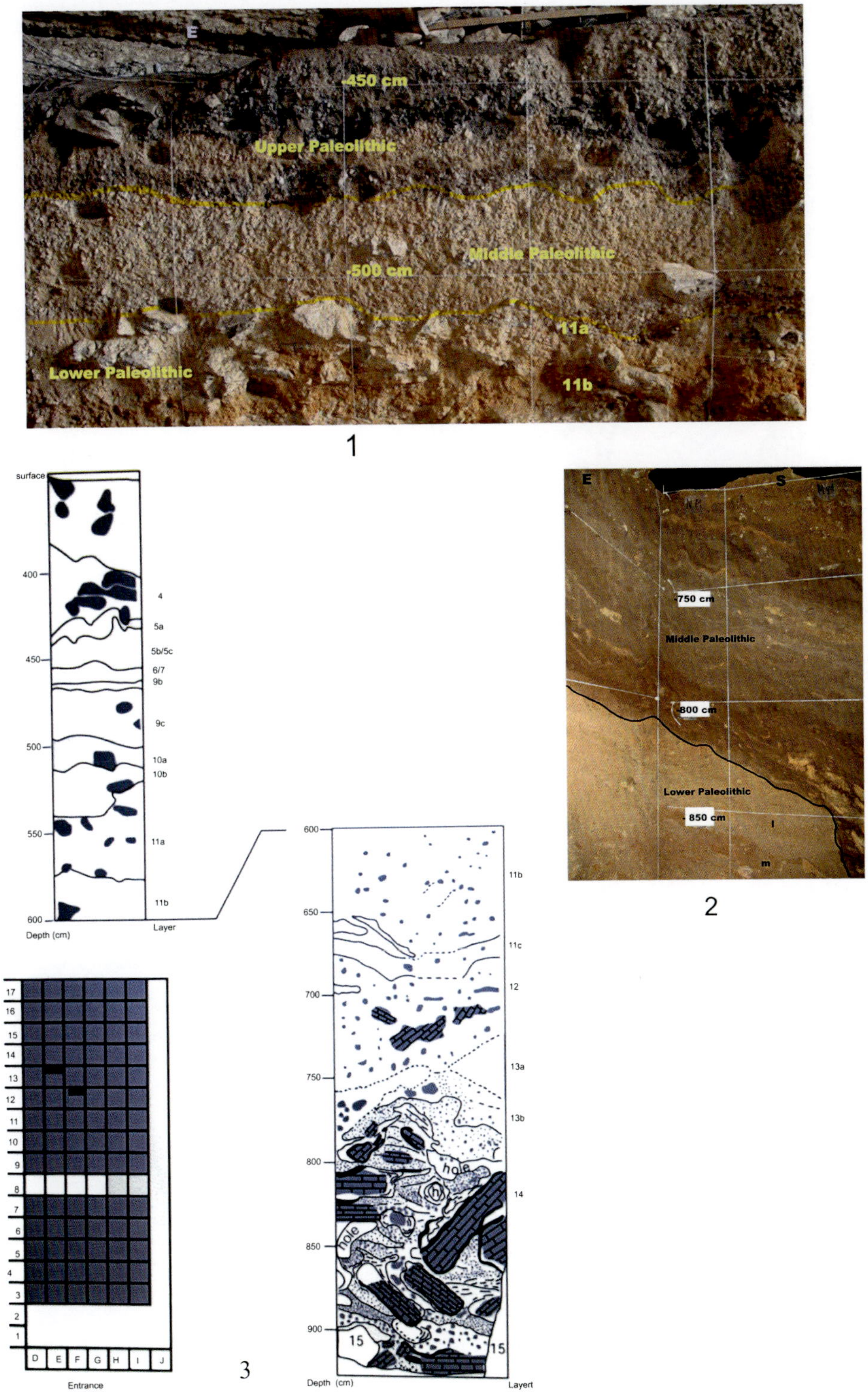

Fig. 12.3 Kozarnika cave. (*1*) Trench II, upper part of the East profile; (*2*) Trench IV, NE profile and position of layers l and m; (*3*) Kozarnika cave, position of the stratigraphic columns in trench II. Courtesy of V. Popov (Popov 2009)

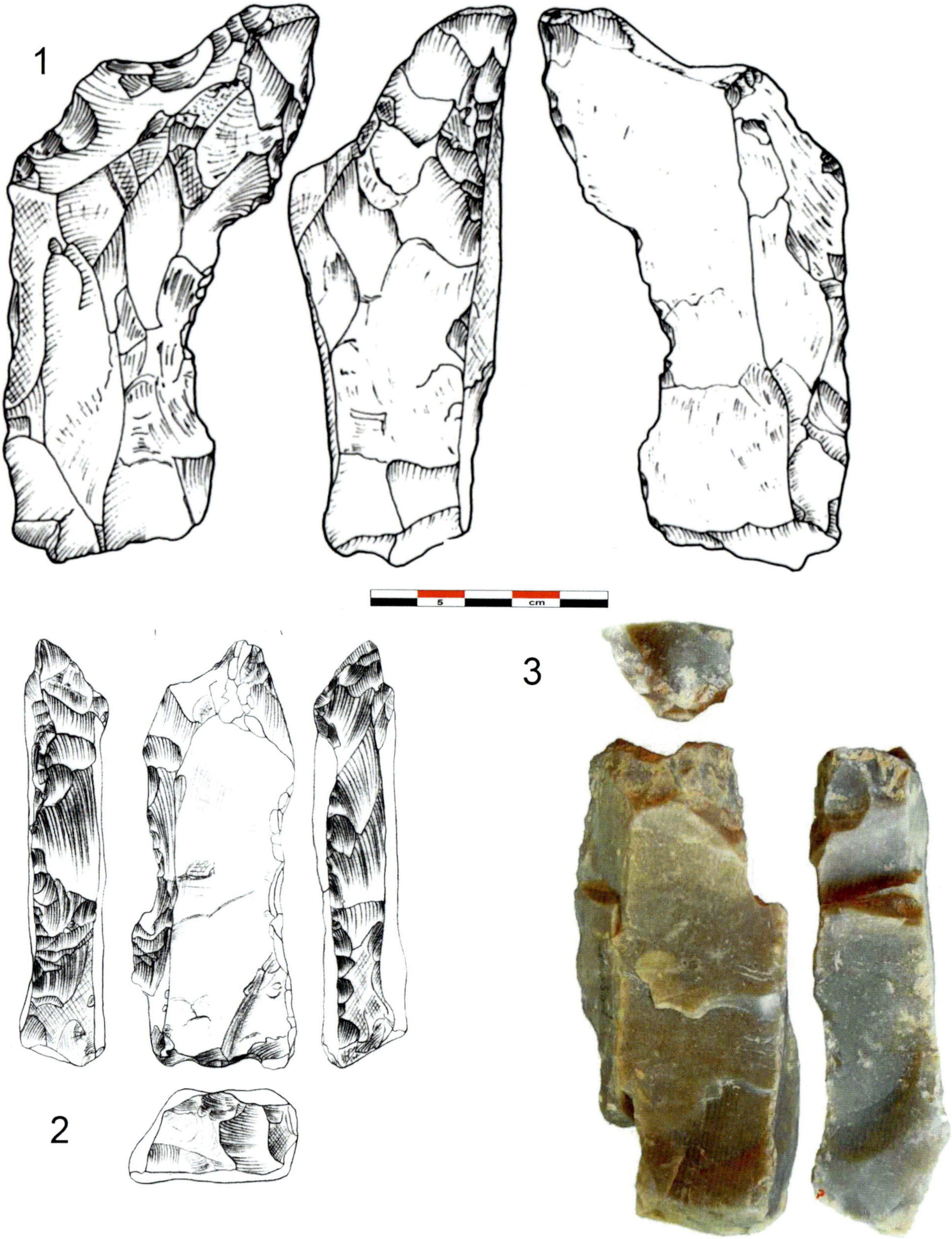

Fig. 12.4 Kozarnika cave, artifacts from Layer 13 lower. (*1*) Asymmetric pointed artifacts; (*2*) Scraper; (*3*) Artifact with coarse negatives

In order to break one of the nodules, the technique of first chipping the surface to form "a band," around the circumference of the nodules was used. Then the prepared nodules were broken with blows. Most cores vary in size from 2 to 4 cm. They were produced through a knapping technique which exploited bipolar or multidirectional cores by removing surfaces at a 90° angle. Removals are broad and short (Fig. 12.5:4, 8). Larger specimens with broad and short removal surfaces were exploited by rotation in one direction. Small flakes were separated from small cores and micro cores. There are several cores on quartzite, which were exploited by direct knapping without preparation.

Retouched flakes are the most abundant tool group. Retouch is partial, abrupt, and semiabrupt. The edges are often uneven (Fig. 12.5:2, 7, 10). There is a preference for retouch on smaller flakes, with almost half of the retouched flakes being small (1.5–2.0 cm). Borers, scrapers, notched tools, and core tools are also present. Borers and scrapers are found in almost equal numbers. The dimensions of the borers vary, and some are small fragments (about 2 cm). The scrapers also vary in size, with some showing cortex on the dorsal surface (Fig. 12.5:1). Retouch is abrupt and semiabrupt, and often stepped. Tools are formed on flakes with large distal parts, flat butts and wide flaking angles, and their dimensions appear almost standardized (3–4 cm). The retouch is abrupt and semiabrupt. The core tools are larger. It is often difficult to determine whether the separation of flakes was planned as exploitation, or whether it was aimed to give some form to the artifacts.

Lithic assemblage from Layer 12: This assemblage includes more than 400 artifacts. The group of preliminary preparation (concretions, nodules, pieces with traces of attempts at breaking, cortex removal, and flake removals) comprises 28 % of the total. Artifacts of relatively small size (3–5 cm) predominate, and debitage forms the largest group of this assemblage (52 % of the total). Among the latter, there are many flint pieces with separate intentional negatives on some of the surfaces, but the typical intentional flakes are few. The size of the flakes varies (1.0–3.5 cm), but most measure between 1.5 and 2.0 cm. Short artifacts with flat platforms, convex bulbs, and wide platform angles dominate. Short fragments also dominate the group of artifacts of larger size (3–5 cm). The group of flakes with sizes 6–7 cm is small but noticeable. Some large artifacts are Pontinian flakes. Among cores, specimens between 2.5 and 3.5 cm in size dominate. Cores are mainly single platform, or exhibit attempts to change the orientation. On some cores, the change of the direction of exploitation was applied only on one side, but there are also specimens on which it was applied on several sides (Fig. 12.6:2, 4, 5). Double-platform cores are also well represented. They were used to produce small flakes of around 1.5 cm in size. Flakes of up to 3 cm were obtained from the exploitation of multidirectional cores. Several larger quartzite pebbles have been exploited without preliminary preparation (Fig. 12.7:2). A knapping technique, consisting of striking off flakes from opposite directions and evidenced by several polyhedral cores, was used during advanced exploitation. Knapping on the shorter axis of the core was preferred for all the cores. In the assemblage of Layer 12, for the first time among cores, one can find specimens with traces of preparation reminiscent of Levalloisian and discoid cores. These elements can be also observed in the debitage.

The percentage of the tools in the assemblage is 13.3 %. Among tools, the most abundant group is that of retouched flakes (Fig. 12.6:3, 6). These vary in size and in the types of retouch, which include abrupt, as well as flat, surface retouch. The most common other tools are denticulates and notches. Scrapers are also abundant, often atypical, and of different sizes. Retouch is usually abrupt or semiabrupt, and the retouched edge is often wavy or concave. Several large specimens were formed from massive quartzite pebbles or fragments (Fig. 12.7:1). The core tools are very large and their surfaces are covered with single flat negatives. Partial retouch on their edges forms large, projected tips (Fig. 12.6:7). Some of the core tools are solid scrapers.

A unique finding originates from Layer 12. This is a fragment of bone preserving what appear to be intentional incisions, grouped in 4 series of 4 incisions (Fig. 12.7:3; Guadelli and Guadelli 2004; Guadelli et al. 2005). This specimen could represent an early example of the human ability for abstract thinking. In the same layer, evidence for precision work on animal skin has also been recovered in the form of a first phalanx of a marmot with a cut mark, most likely the result of skinning (Sirakov et al. 2010:12). Such an operation on such a small bone (length of only 2 cm) requires precise work with a tool of very small dimensions. This could explain the numerous little flakes in the assemblage (1 cm and smaller). Working with flakes of such small size furthermore requires the use of some sort of a handle. This finding increases our knowledge about the manual abilities of these early inhabitants of Kozarnika and their efficient utilization of natural resources.

Lithic assemblages from Layer 11: For the purpose of this chapter, assemblages from the various facies of Layer 11 (11a, 11b, 11c) will be examined together. This layer contains the youngest Lower Paleolithic assemblages in Kozarnika. The total number of artifacts here exceeds 1000. Flakes account for the greatest portion of this assemblage (between 40 and 50 %), followed by pebbles, nodules with traces of testing and removal attempts, and separate negatives of detached flakes (15–20 %). The percentage of tools is also high (20 %). Among the flakes, the number of specimens with elongated shapes with flat platforms is higher than that observed in other layers. However, short, broad flakes predominate and small flakes (<1.5 cm) are most common. More than 70 % have convex bulbs and vis-

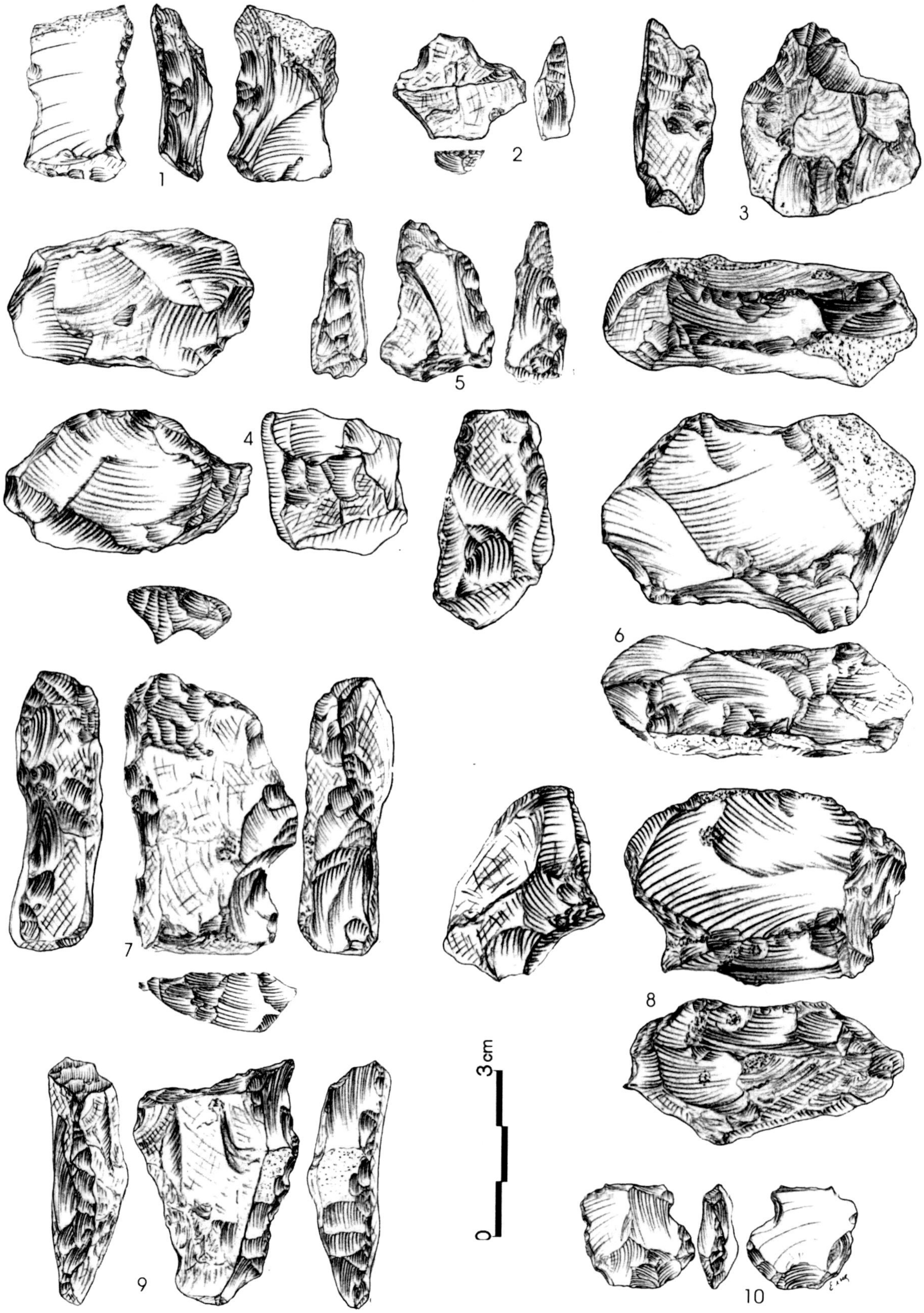

Fig. 12.5 Kozarnika cave, artifacts from layer 13 upper. (*1*) Scraper; (*2*, *7*, *10*) Retouched flakes; (*3*, *5*) Borers; (*4*, *8*) Cores; (*6*) Core tool. Borer scraper

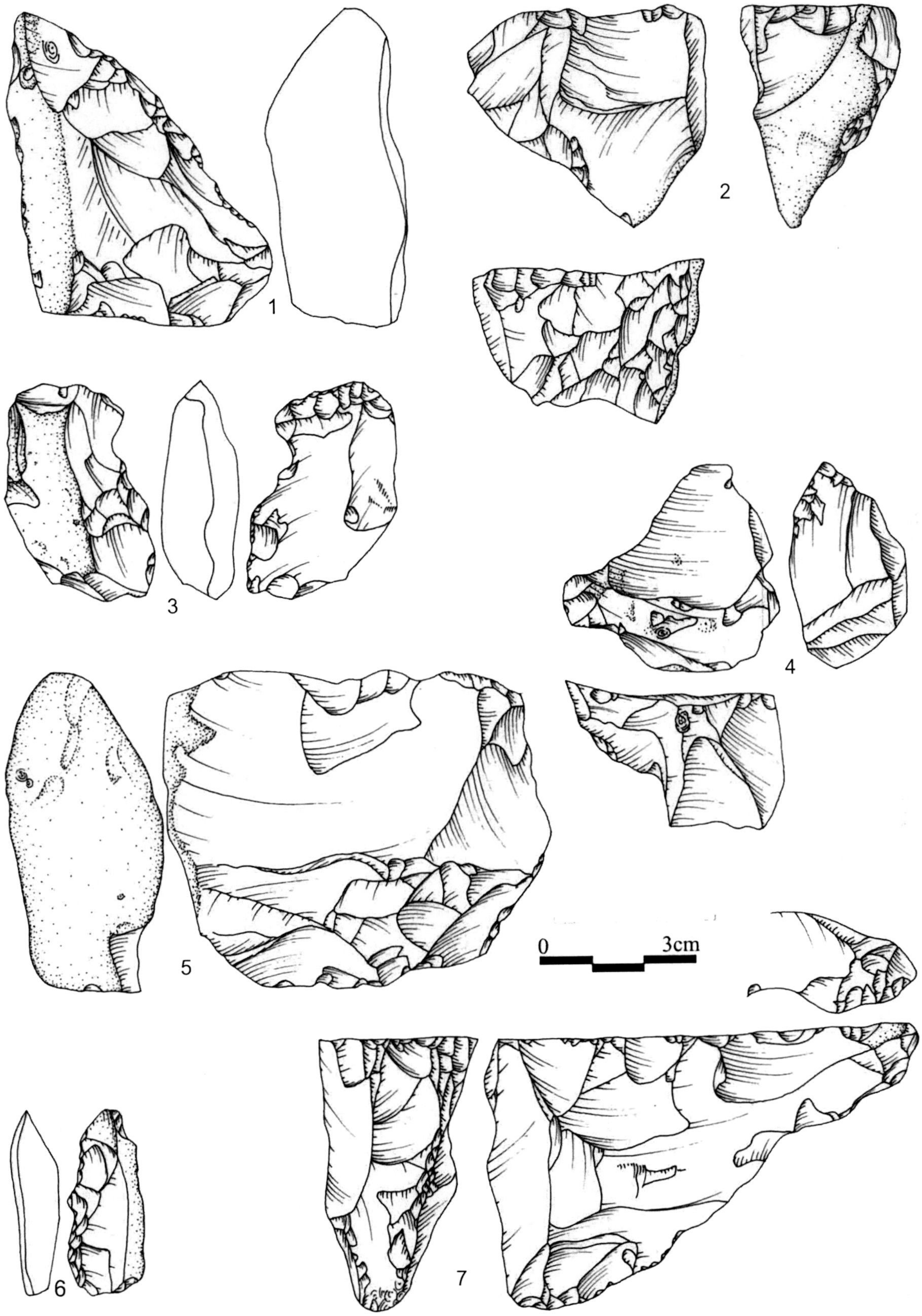

Fig. 12.6 Kozarnika cave, artifacts from layer 12. (*1*) Flake; Cores (*2*, *4*, *5*); Retouch flakes (*3*, *6*); Core tool (*7*)

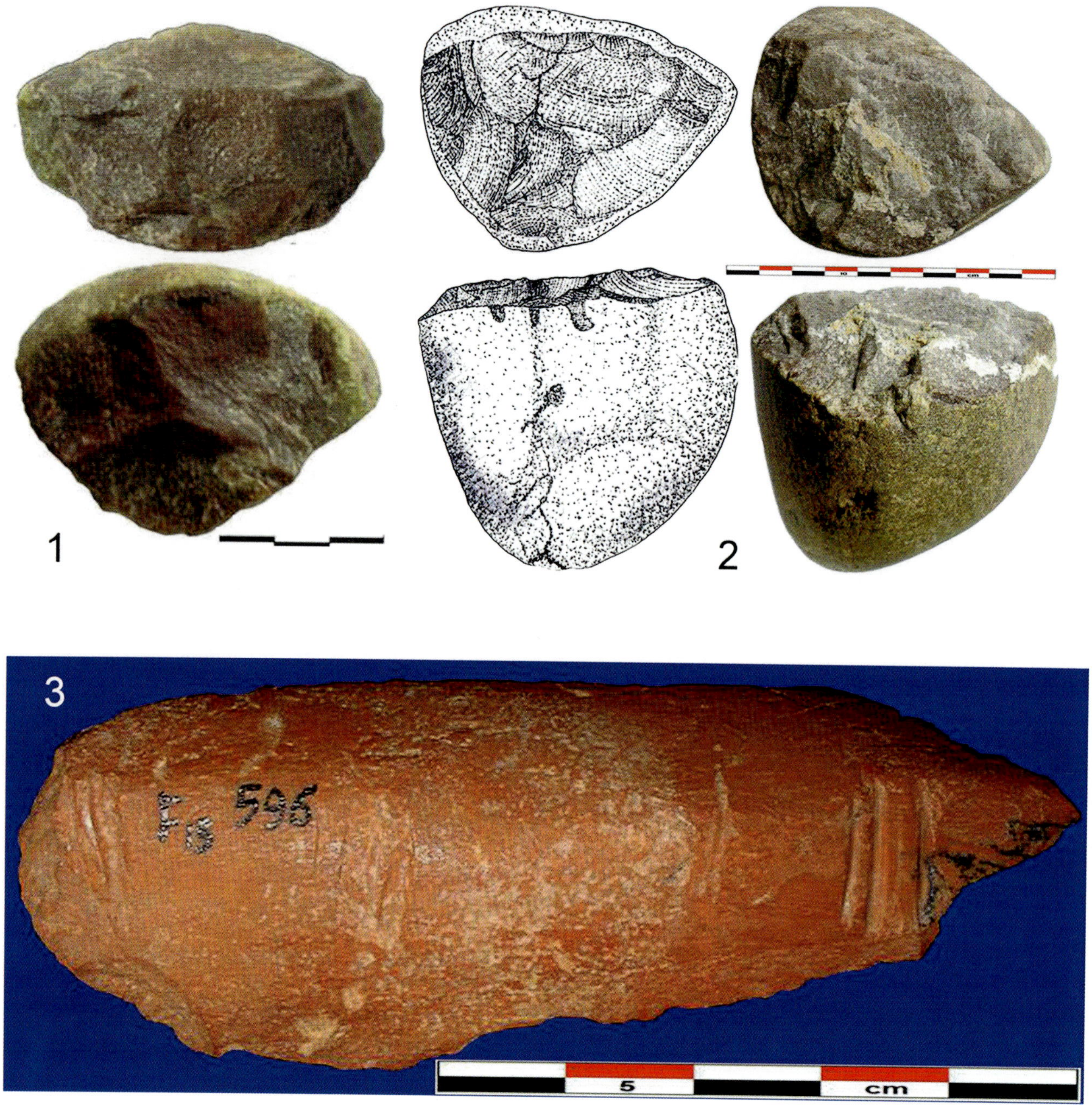

Fig. 12.7 Kozarnika cave, artifact from layer 12. (*1*) Quartzite pebble tool; (*2*) Quartzite pebble core; (*3*) Fragment of bone with intentional cutting

ible impact points. Flakes with wide platform angles predominate. Most flakes have multidirectional negatives on a dorsal surface. Among cores, single platform ones are the most frequent; cores with multidirectional exploitation and flat or oval shape, as well as cores with unprepared platforms, are also present (Fig. 12.8:1). Some artifacts were exploited as atypical bifacial discoid cores. Some cores are polyhedral.

The most numerous tool categories are retouched flakes and notched flakes. Some of the flakes are large (5–6 cm) with coarse retouch, while flakes with irregular sizes and partial retouch were also found. Among the scrapers, elongated specimens predominate. The retouch is semiabrupt or abrupt, mostly coarse and, in some cases, stepped. Usually it is not continuous. Retouched edges are often wavy, convex, or concave. Fragments of scrapers from knapping accidents are

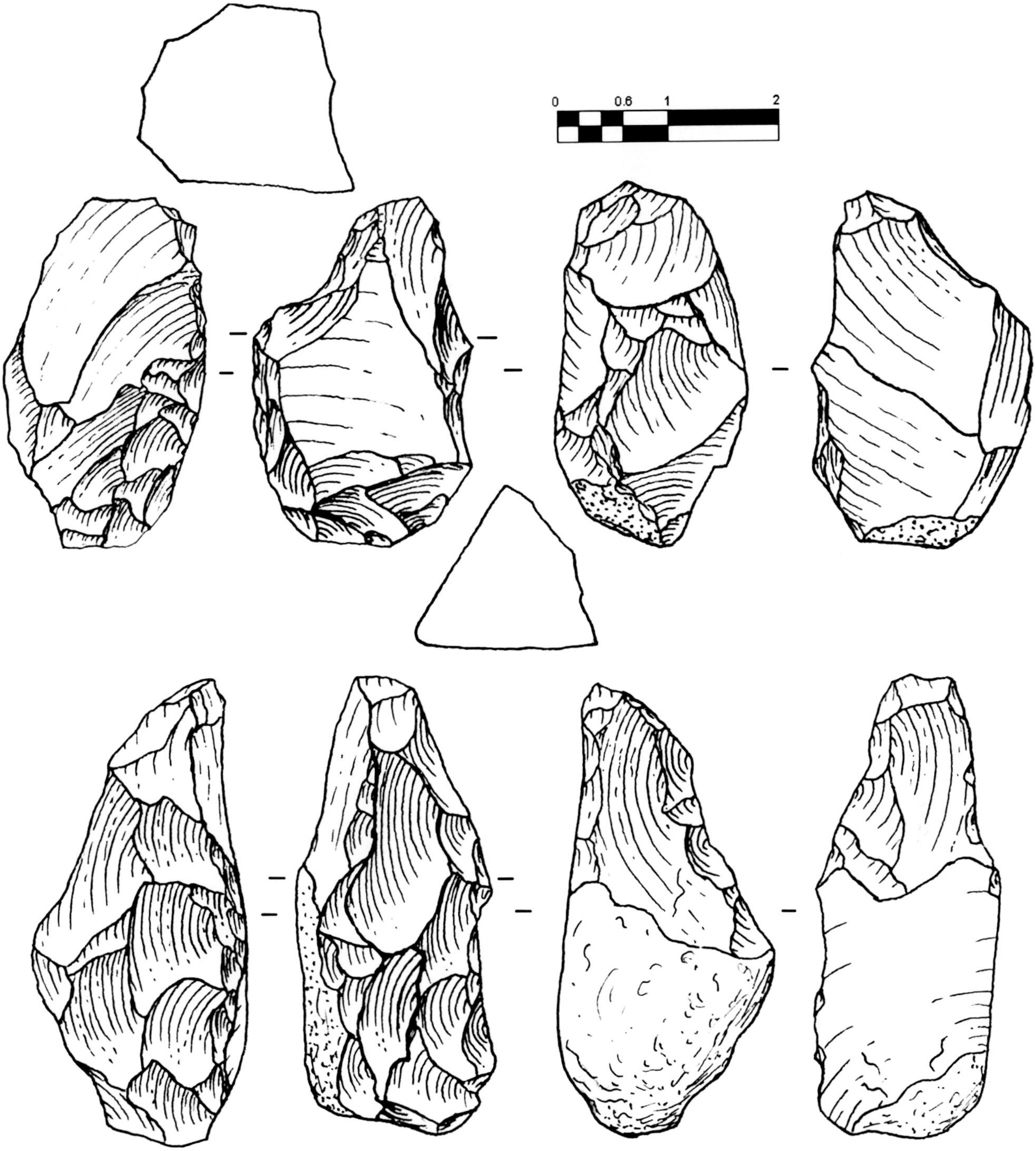

Fig. 12.8 Kozarnika cave, artifacts from layer 11. (*1*) Core; (*2*) Pic

the majority. Borers are small in size. Among the core-like tools the majority have well-formed solid tips. There are also core-like scrapers, notched core tools, and a pic (Fig. 12.8:2).

Bifacial tools and artifacts with bifacial retouch are a diagnostic group of high interpretational value. Although rare (14 specimens), they demonstrate the existence of bifaces in the Early Paleolithic of Kozarnika. These artifacts are small, oval, or slightly elongated. Some of them are in the initial stages of shaping. Formally, they could be referred to as "atypical." The bifaces are formed with coarse, flat, surface retouch, sometimes with preserved fragments of cortex over parts of their surfaces (Fig. 12.9:1–4). The bifaces of Kozarnika differ significantly from the single findings of bifaces in the Rhodope Mountains.

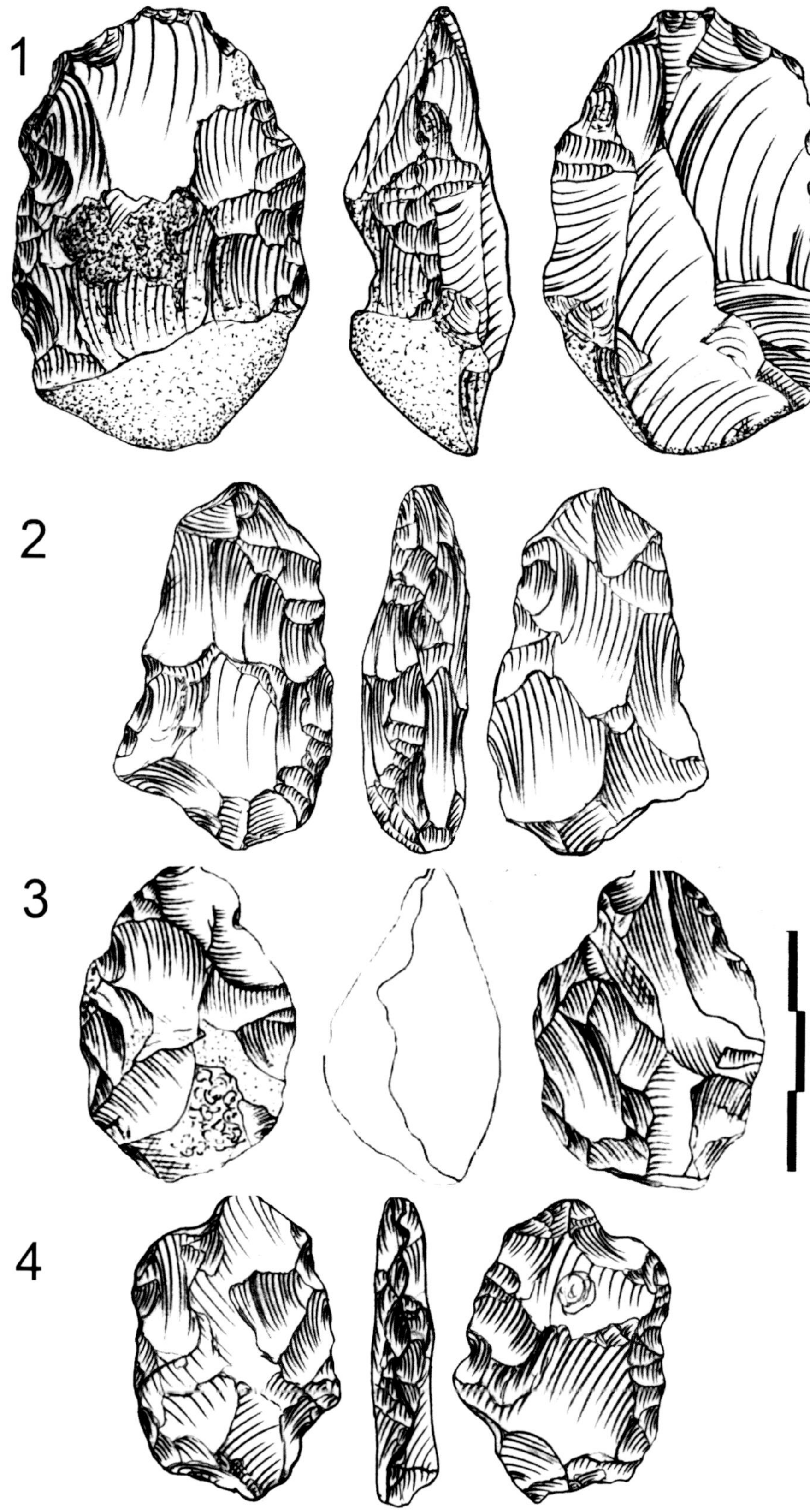

Fig. 12.9 Kozarnika cave, artifacts from layer 11. (*1–4*) Bifaces

The typical core in Layer 11c is a single platform core with a wide short removal surface with a triangular shape. Characteristic of the assemblage of 11c is the larger size of the artifacts, the deep and coarse negatives, and the typological variety of the scrapers.

In the assemblage of Layer 11b, we can find some flakes produced with the "Citrus," or Pontinian, technique, which consists of the splitting of smaller pebbles into thick halves and using these "citrus slices" as blanks (Debenath and Dibble 1994:33). An increase in the number of borers is also observed. The distal parts are formed with coarse, single negatives, fine smooth retouch, or steep alternative retouch. Both large specimens, as well as micro borers, have been recovered. Four end scrapers, made on flakes with abruptly retouched edges, were also found. Among the debitage, there are a large number of specimens with elongated, blade-like shapes. Some of the retouched flakes are micro-flakes.

Lower Paleolithic Artifacts of Trench IV

Trench IV is located *ca.* 72 m from the entrance of the Kozarnika cave (Fig. 12.2:3). Despite this relatively large distance, the location is convenient for human habitation. Due to the height of the cave's entrance, sunlight reaches this location even nowadays when the sun is low, despite the thick sediments between the entrance and trench IV. In this trench three lithostratigraphic units include artifacts with Lower Paleolithic characteristics: layers k, l, and m (Fig. 12.3:2). The technological characteristics of the assemblage of layer k are similar to the assemblages of Layer 11 in trench II (see above). The difference is in the increased number of larger (*ca.* 6 cm) quartzite flakes and quartzite pebbles with traces of exploitation without preparation, as well as in the smaller number of artifacts resulting from knapping accidents.

Lithic assemblage from layers l and m: In the eastern part of trench IV, a concentration (9 pieces) of large quartzite artifacts was discovered in a series of very fine, dark and light gray, powdery sediments (layers I, m; Fig. 12.3:2). These sediments differ from the lithostratigraphic units of the trenches near the entrance. Only a small surface area has been excavated so far, and no in-depth research has been carried out yet. The artifacts are made on relatively large quartzite pebbles (>12 cm). No such pebbles have been found in the sediments researched so far and no traces of water flow that could have moved artifacts of that size are evident. It is therefore hypothesized that they might have been brought into the cave by people. A large (14 × 11.5 × 9.0 cm) chopper, made on a quartzite pebble, has no analog in the assemblages discussed so far (Fig. 12.10:1). The remaining quartzite artifacts are from pebbles of approximately uniform sizes (11.0–11.2 cm) × (7.0–9.0 cm) (Fig. 12.10:2, 3). Their characteristics suggest that these artifacts are older than the ones from Layer 13 "lower." Excavation of a larger surface area and in greater depth would help provide more evidence for the early occupation of Kozarnika.

Summary

In summary, Kozarnika contains clear evidence of occupation during the Lower Paleolithic. The lithostratigraphic units 11 (a, b, c) contain assemblages with "atypical" bifaces and bifacial forms. They are small in size and, although rare, their presence is constant. The bifaces from Kozarnika differ entirely from the known biface assemblages of the Middle East and Europe, as well as from the bifaces found at Lower Paleolithic surface sites in the Rhodope Mountains. The group of core tools is dominated by artifacts with projected tips. The technique for forming the tips of the artifacts in assemblages 12 and 13 upper is similar to the burin blow technique from later periods.

However, it must be emphasized that there are questions about these findings that are still the subject of discussion. Most notably, the opinion of researchers regarding the dating of the Kozarnika Lower Paleolithic layers is divided. Some authors (e.g., Fernandez and Cregut-Bonnoure 2007) have correlated the faunal assemblage from the lower levels of Kozarnika to MNQ18. However, others have pointed out that "…the obvious uncertainty of some of the determinations makes this association relatively unreliable in terms of biochronology" (Kahlke et al. 2011:11). Several publications (Guadelli and Guadelli 2004; Guadelli et al. 2005; Sirakov et al. 2010) have correlated the earlier layers 12–13 to the biostratigraphic zone B2-2 (1.6 Ma–0.9 ka) on the basis of the large mammalian remains from these layers. However, N. Spassov is of the opinion that "Although Guadelli et al. (2005) place the large mammal assemblage of the Kozarnika lower level (B2-2) at an approximate age of 1.4 Ma, it is more likely that the assemblage is closer to the beginning of the Epivillafranchian (N.S. and D.K., pers. comment)," (Kahlke et al. 2011:11; see also Spassov 2016). Based on the analysis of small mammals, V. Popov also suggested a later chronology and concluded that "the geographical location of Kozarnika in SE Europe probably accounts for the occurrence of species characteristic for the Villanyian and Early Biharian in Central and Western Europe in assemblages dominated by Late Biharian elements. Their presence here is of a relic character which should be taken into

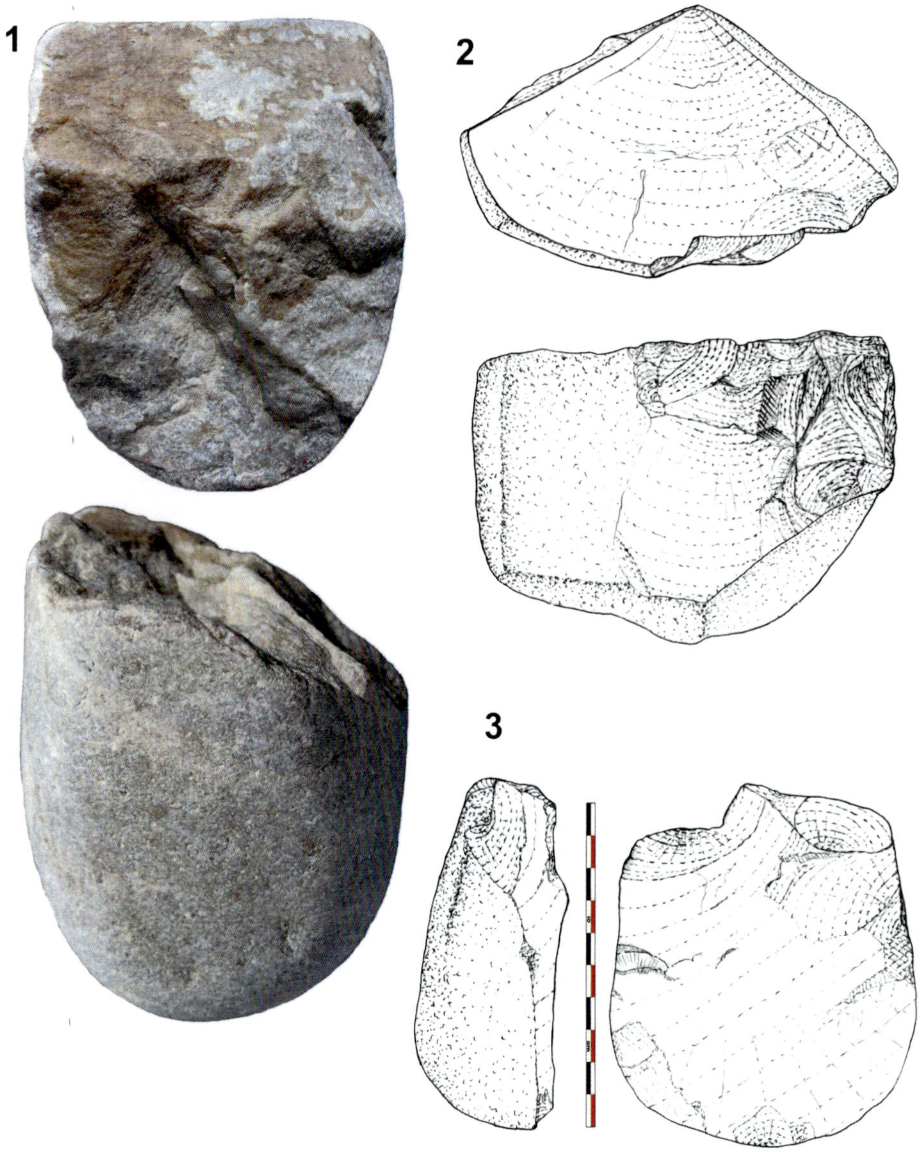

Fig. 12.10 Kozarnika cave, artifacts from layers l and m. (*1*) Chopper; (*2, 3*) Quartzite pebble tools

account in biostratigraphic correlations" (Popov and Marinska 2007; Popov 2009).

The analysis of the archaeological assemblages of the different layers reveals a very slow technological evolution and long-term use of almost identical knapping methods in the Lower Paleolithic sequence. There is also no significant difference in the typological characteristics of the assemblages. Elements of Levallois and discoid concepts can be observed already in the technological characteristics of Layer 12 and continue in the assemblages of Layer 11. The difference is in the presence of bifaces in the assemblages of Layer 11. Such a minor change in the methods of knapping and the typology of the artifacts is difficult to reconcile with the extremely long period of time (900–400 ka) that these Kozarnika layers are proposed to represent. I therefore share the opinion that the earliest assemblages in Kozarnika that have been researched so far are no older than 1 Ma. Nevertheless, the completely different character of the lithic assemblage from Layer 13 "lower" in trench II and the artifacts of layers l and m in trench IV must also be taken into consideration. The stratigraphic position of the artifacts in trench IV and their archaic typological characteristics might indicate an earlier settlement.

The Lower Paleolithic Sites of the Rhodope Mountains

Characteristics of the Rhodope Mountain Region

Early Paleolithic surface sites were also discovered in Southern Bulgaria in the Rhodope Mountains. There is no evidence of glacial valleys or mountain glaciation in the Rhodopes. Throughout the entire Rhodopes massif there are about ten peaks higher than 2000 m and no peaks higher than 2200 m asl (Mihnevski and Genkova 1989). The Western Rhodopes have an average altitude of 1098 m. No Quaternary terraces were formed due to the extensive uplift (almost 500 m) of these mountains in the late Pliocene (Yaranov 1939). This region is extremely rich in diverse and easily accessible flint. The presence of opal and chalcedony tabular pieces (some very large, up to 80 × 60 cm) on the surface in the area of Shiroka Polyana and Kremenete is derived from very early magmatic activity (Bozhkov et al. 1978). The Eastern Rhodopes comprise a smaller part of the Rhodope Mountains. This region is characterized by low mountainous and hilly relief with an average altitude of only 320 m asl. Unlike the deep canyons typical of the Western Rhodopes, the Eastern Rhodopes show broad valleys with gentle slopes. Volcanic rocks are typical for this region.

Early Paleolithic Collection from Western Rhodopes: The surface Sites of Shiroka Polyana

The area studied is located at 1500 m above sea level on a large flat hilltop that is dated to the Early Miocene denudation levels. Here, an area of 4 km^2 is covered by opal-chalcedony raw material and artifacts. The area is smoothly sided, cut by widely meandering brooks, which form shallow micro-valleys (30–80 cm deep). These valleys are the main source of the artifacts. The artifacts were collected from the bare surface and from these micro-valleys (Fig. 12.11).

The separate concentrations of artifacts are considered to represent distinct phases of settlement characterized by different technological and typological features, and originating from different time periods. The majority of the collected artifacts are attributed to Middle Paleolithic on the basis of their typological characteristics. However, some of the artifacts could be attributed to an earlier phase of the Paleolithic. These have an archaic appearance with amorphous specimens predominating. None of the specimens included in this group has parallels in the Middle Paleolithic assemblages of this part of the Balkan Peninsula (Ivanova 1979). This Lower Paleolithic collection includes several bifacial tools, large cores with bifacial exploitation and solid discoid core. Traces of cortex or natural surfaces are preserved on the bifacial forms. A biface on a large flat piece (12.0 × 6.5 × 4.7 cm; Fig. 12.12:1) shows retouch only on its proximal part. Bifacial retouch was also used on thin, flat artifacts and on a bifacial scraper made from a large tabular piece (Fig. 12.11:4). The assemblage also includes a chopping tool, made on a flat, naturally short piece and exhibiting semiabrupt surface retouch forming a slightly convex edge at the widest part of the specimen (Fig. 12.12:2).

The collection from Shiroka Polyana is characterized by the large size of the artifacts and by bifacial retouch on some tools. The cores have been exploited in a multidirectional manner, through direct knapping without preparation, or with only limited preparation of the striking platform, sometimes bifacially. There are no cores with only one striking platform, whether without preparation or with limited preparation. Despite the small number of artifacts (13) the collection can be tentatively assigned to the Acheulean.

Early Paleolithic Collection from the Western Rhodopes: The Surface Site of Kremenete

Kremenete is located about 10 km north of Shiroka Polyana and is situated on a broad flat ridge at *ca.* 1600 m above sea level. The artifacts come from the sediments (underlying the humus layer) consisting of materials weathered from the basalt rock. The raw material consists of large blocks of flint, which are dispersed on the surface. More than 1200 artifacts have been collected, and most can be assigned to the Middle Paleolithic. However, a group of 11 chopping tools with Lower Paleolithic techno-typological characteristics differs from the general characteristics of the collection. They are variable in size (9–15 cm) and are characterized by flat-oval shapes, bifacial retouch of one edge and natural surfaces, or blunted, with single steep retouch scars on the opposite edge. Natural or cortical surfaces are visible on the sides of the artifacts. In some cases, the surface retouch forms flat dorsal and ventral sides when the raw material is not naturally flat (Fig. 12.13:1–4).

In summary, the surface collections from the Western Rhodopes include bifaces, choppers, and tools with bifacial retouch, demonstrating that the Pleistocene inhabitants of the Rhodopes used the bifacial technique. The lack of stratigraphic data hampers the chronological assessment

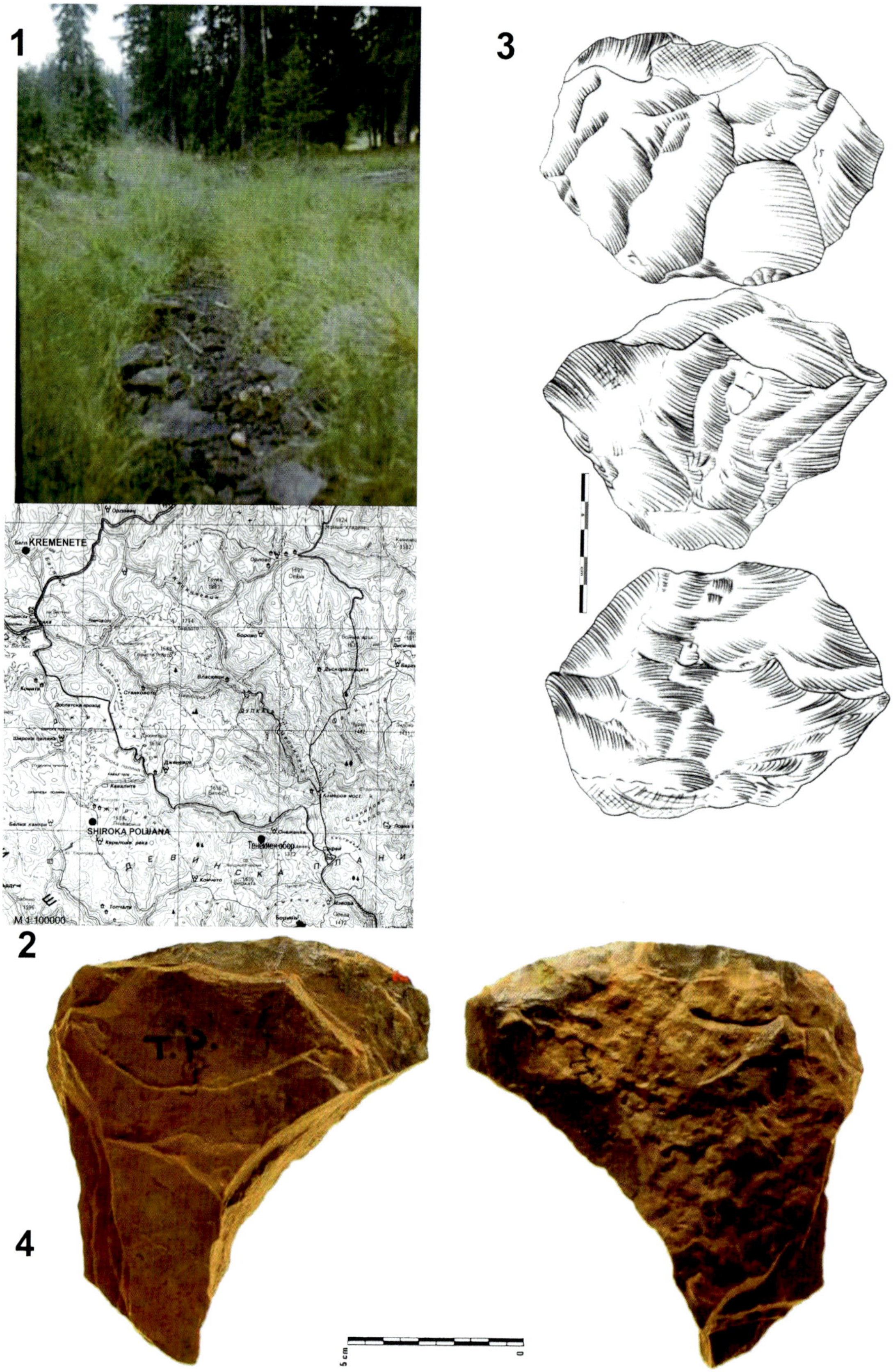

Fig. 12.11 Surface site Shiroka Polyana (West Rhodope Mountains). (*1*) View; (*2*) The location of Lower Paleolithic sites; (*3*) Solid discoid core; (*4*) Bifacial scraper

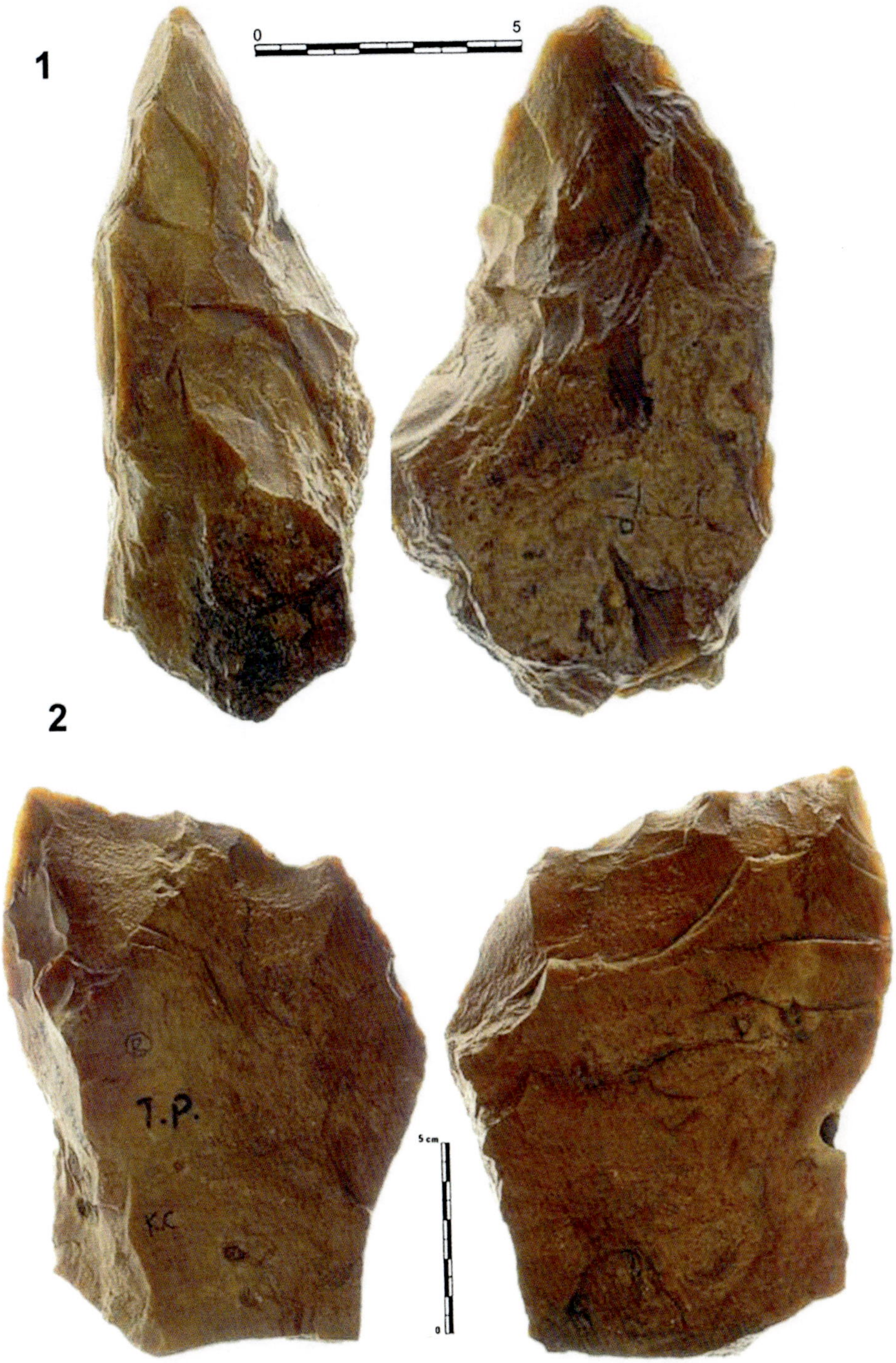

Fig. 12.12 Surface site Shiroka Polyana (West Rhodope Mountains). (*1*) Biface; (*2*) Chopping tool

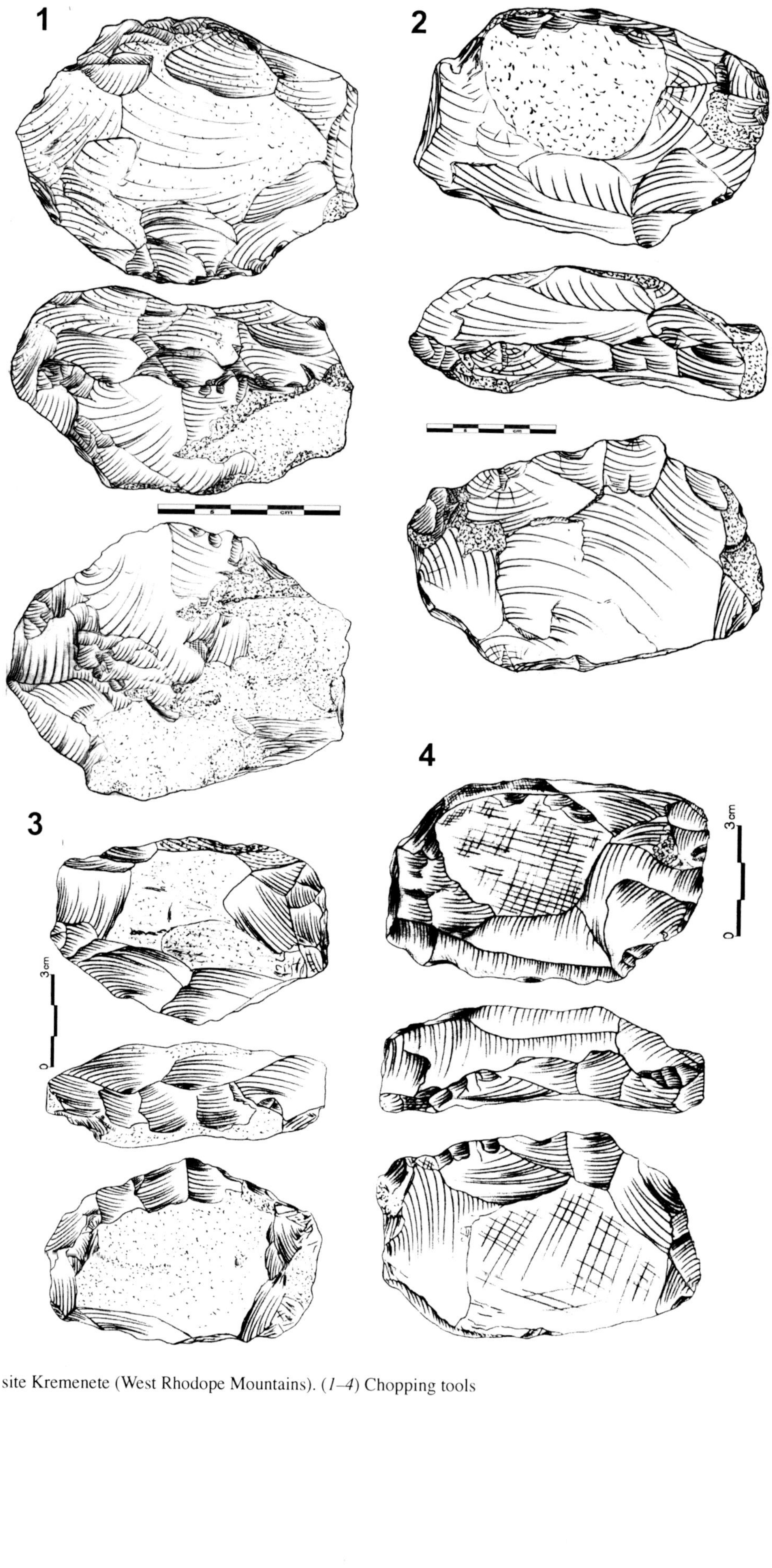

Fig. 12.13 Surface site Kremenete (West Rhodope Mountains). (*1–4*) Chopping tools

of these finds. However, their typological characteristics suggest a Middle Pleistocene age, the period when most of the biface cultures in Europe emerged. The bifacial forms from the Western Rhodopes have no similarities in their technological and typological characteristics to those of Kozarnika. They are also probably younger than the Kozarnika bifaces. However, there is a close similarity between the tools from Kremenete and those found in the Caucasus in Tsona cave (Georgia). The Acheulean layers in Tsona cave are dated based on biostratigraphic and geochronological data to the Middle Pleistocene faunal complex (Tiraspol), which has been dated by ESR to 583,000 ± 112,000 (Lybin and Belyaeva 2006).

Early Paleolithic Collections from the Eastern Rhodopes: Surface Site of Benkovski

Benkovski is located east of the town Zlatograd, near the border between Bulgaria and Greece. The region is low and hilly, with large areas of bare volcanic rocks. The primary location of the flint artifacts was on a small plateau. Although the Rhodopes massif was uplifted during the Early Pleistocene, uplift was much more limited in the Eastern than in the Western Rhodopes. Gradually, new river valleys were formed and incised their way through the plateau. Many of the artifacts were found in such "young" valleys (Fig. 12.14:1). They were transported from the plateau through the process of erosion.

The raw materials are silicified volcanic rocks of very good quality, which are widespread throughout the area in the form of large or very large (more than 90 cm) pieces. The artifacts were moved a short distance during landslides, sliding together with sediments from the edge of the plateau into the newly formed valleys. However, there is no trace of water transport movement over a long period of time. The artifact edges have traces of pseudo-retouch and their surfaces are covered with patina. Artifacts were found in separate concentrations located at relatively short distances from each other. They appear not to have been moved significantly in the valley. It is possible that their arrangement corresponds to their original position on the plateau. These concentrations suggest that the area was regularly visited during different periods of the Early and Middle Pleistocene. Unlike the situation in the Western Rhodopes, no Middle Paleolithic artifacts were recovered.

More than 60 artifact concentrations were discovered along about 1 km in the valley of Marasi Dere. Several artifacts were found *in situ* on the plateau near the edge of the valley. Several hundred artifacts were collected, the most numerous of which are the pieces with evidence of testing. Several very large (42 × 18 × 10 cm), elongated artifacts are unique, with very large tips and with a base part formed by coarse surface retouch (Fig. 12.14:2). A large number of pieces of differing sizes only show negatives of one or few removals. Cores vary, both in size and in the methods of knapping. Among the most common core types are single-platform cores from pebbles. There are cores with change of orientation, without repetitive schemes, and bifacial core tools (Fig. 12.15), as well as a very large spheroid core (23 × 22 × 16 cm—Fig. 12.16). In some cases the artifacts cannot be defined precisely. Some specimens can be defined as core tools or as unfinished bifaces.

Several scrapers and a borer (Fig. 12.17:1, 2) are also present. Removal flakes are extremely diverse. They are elongated and large (13–17 cm), and some are plate shaped. This elongation is due to the intrinsic qualities of the raw material. Some of the flakes show irregular partial retouch and several specimens exhibit flat abrupt retouch on the ventral surface. Flat or natural platforms dominate. Platform angles are wide. Along with the very large sized flakes, numerous flakes of smaller sizes have also been found. The very small sized flakes are most likely the product of abrupt retouch.

Of particular interest are two tools: A short and wide biface (18 × 12.5 × 8 cm) with large surface negatives. Several negatives form the distal part of this specimen, and finer negatives form its edges (Fig. 12.16:1). A chopping tool made on a jasper pebble is also noteworthy. The distal part is a natural part of the pebble, but the proximal part is formed with bifacial retouch, and long negatives form a convex sharp edge (Fig. 12.17:1).

Discussion and Conclusions

Possible Routes for the Earliest Colonization of Europe

One of the most interesting questions in Paleolithic archaeology today relates to the timing and routes of early hominin dispersal into the European continent (e.g., Villa 1994; Dennell and Roebroeks 1996; Bar-Yosef 1998; Turner 1999; Bar-Yosef and Belfer-Cohen 2000; Dennell 2000; Roebroeks 2001; Petraglia 2003; Derricourt 2005; Ronen 2006; de Lumley et al. 2009; Abbate and Sagri 2012; Arzarello and Peretto 2010). Recent investigations of the Early Paleolithic in the Balkans provide grounds to consider this region as a possible dispersal corridor. Evidence for mammalian dispersals from the East provides strong

1

2

Fig. 12.14 Surface site Benkovski (East Rhodope Mountains). (*1*) View to the valley; (*2*) Solid tool with elongated tip

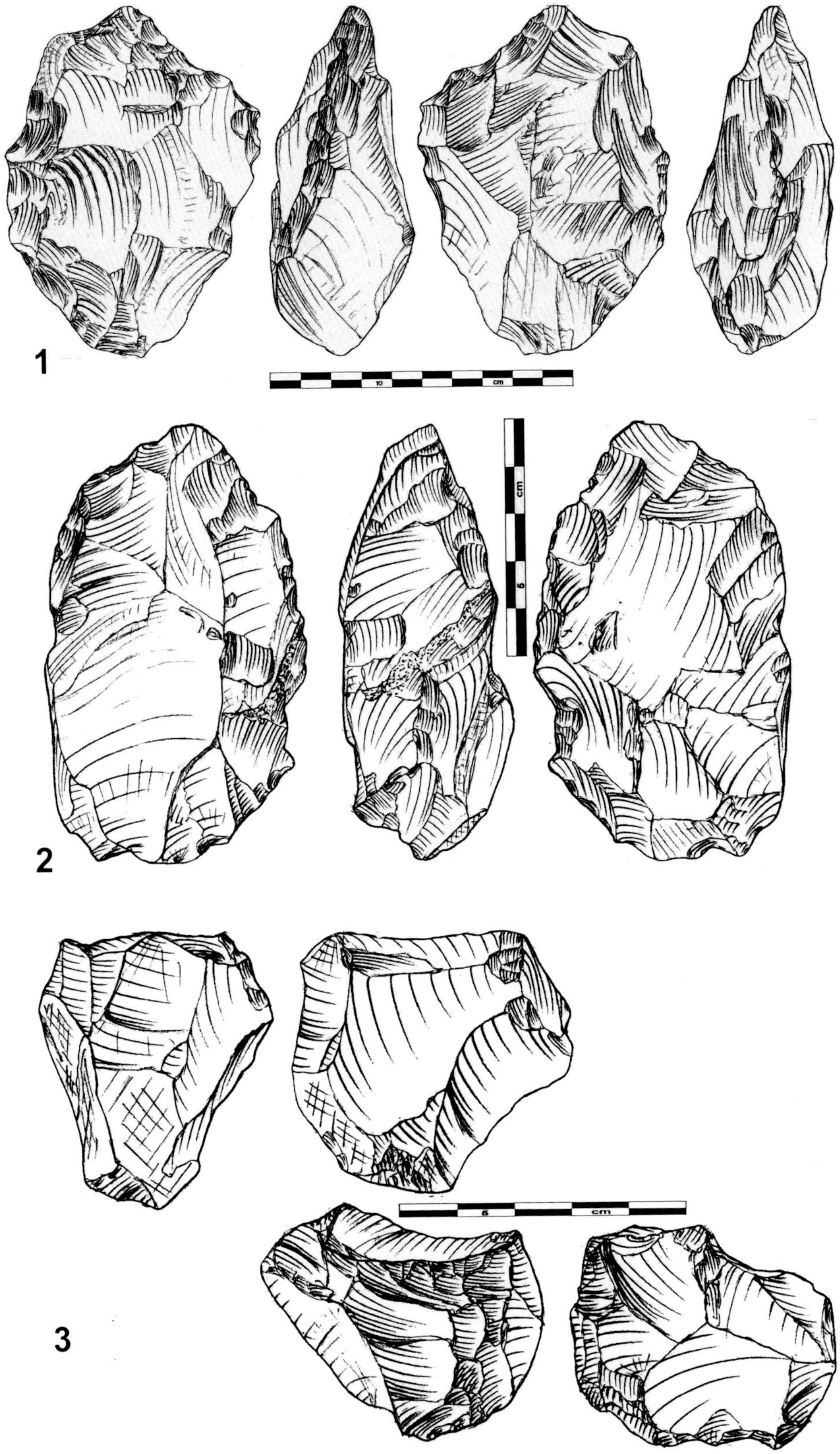

Fig. 12.15 Surface site Benkovski (East Rhodope Mountains). (*1*, *2*) Bifacial artifacts; (*3*) Core

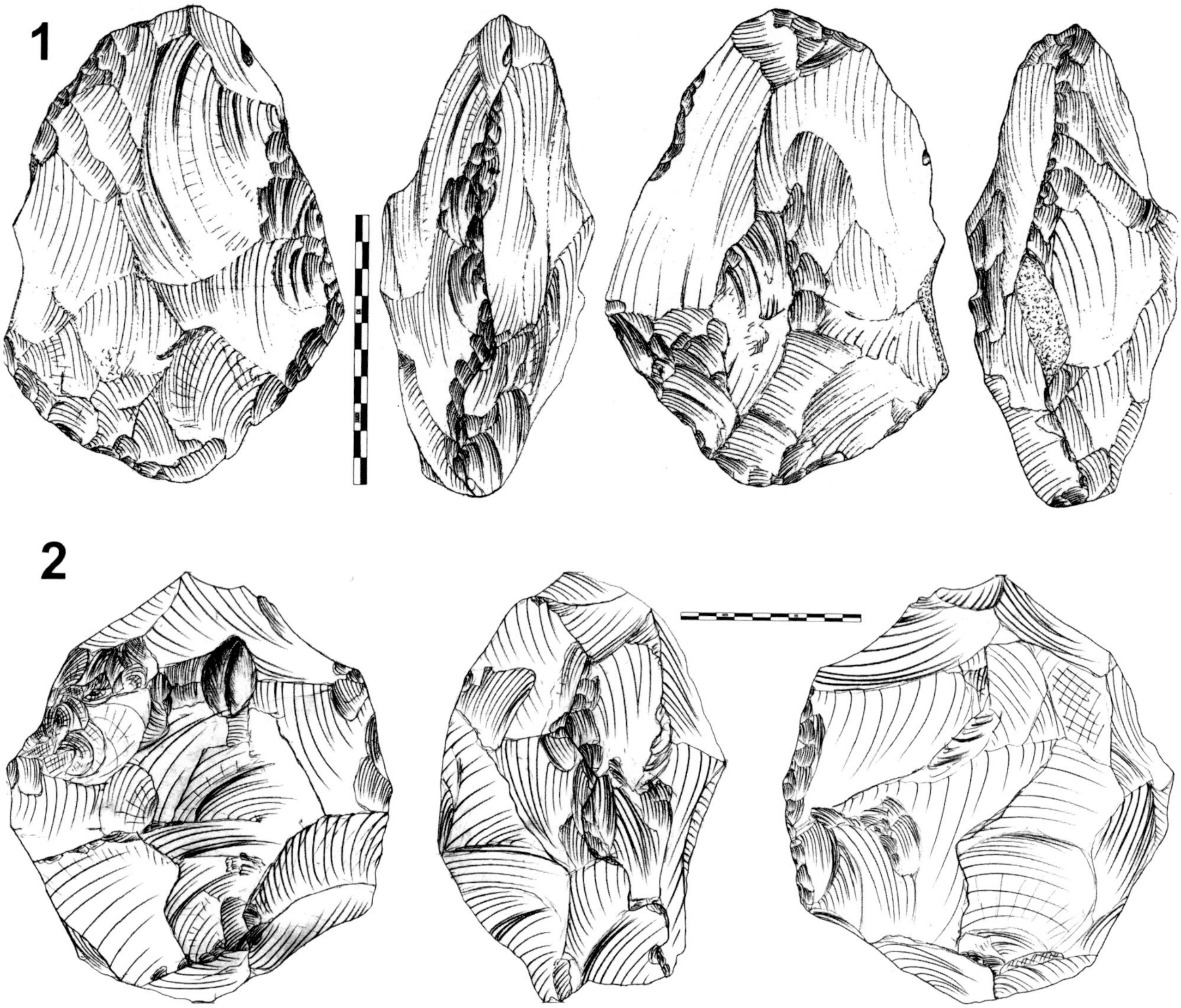

Fig. 12.16 Surface site Benkovski (East Rhodope Mountains). (*1*) Biface; (*2*) Spheroid core

arguments for the chronology of the earliest possible human migration to Europe (see Koufos and Kostopoulos 2016; Spassov 2016). A temporary closure of the Bosporus in the Early Pleistocene suggests that the earliest possible migration dates could be between 2 and 1.9 Ma BP. The earliest migration of some mammals from Asia to Europe (*Canis, Panthera*) is thought to have occurred at this time. The mass migration of bovids took place during the Villafranchian. The first human groups likely entered Europe following the path of migrating herds (Spassov 2001; Spassov 2016), possibly as early as 1.8–1.0 Ma BP. The earliest Paleolithic finds, currently known from Europe, date to this broad time period. Some of the most discussed routes of migration to the European continent (across the Bosporus and along the northern coast of the Black Sea) cross the eastern part of the Balkan Peninsula (through the territory of modern-day Bulgaria). The Lower Paleolithic sites from Bulgaria, therefore, can play a crucial role in understanding the proposed routes and timing of human dispersals into Europe. Further field research, as well as more in depth comparative analysis of the known material, will be necessary to shed light on these issues.

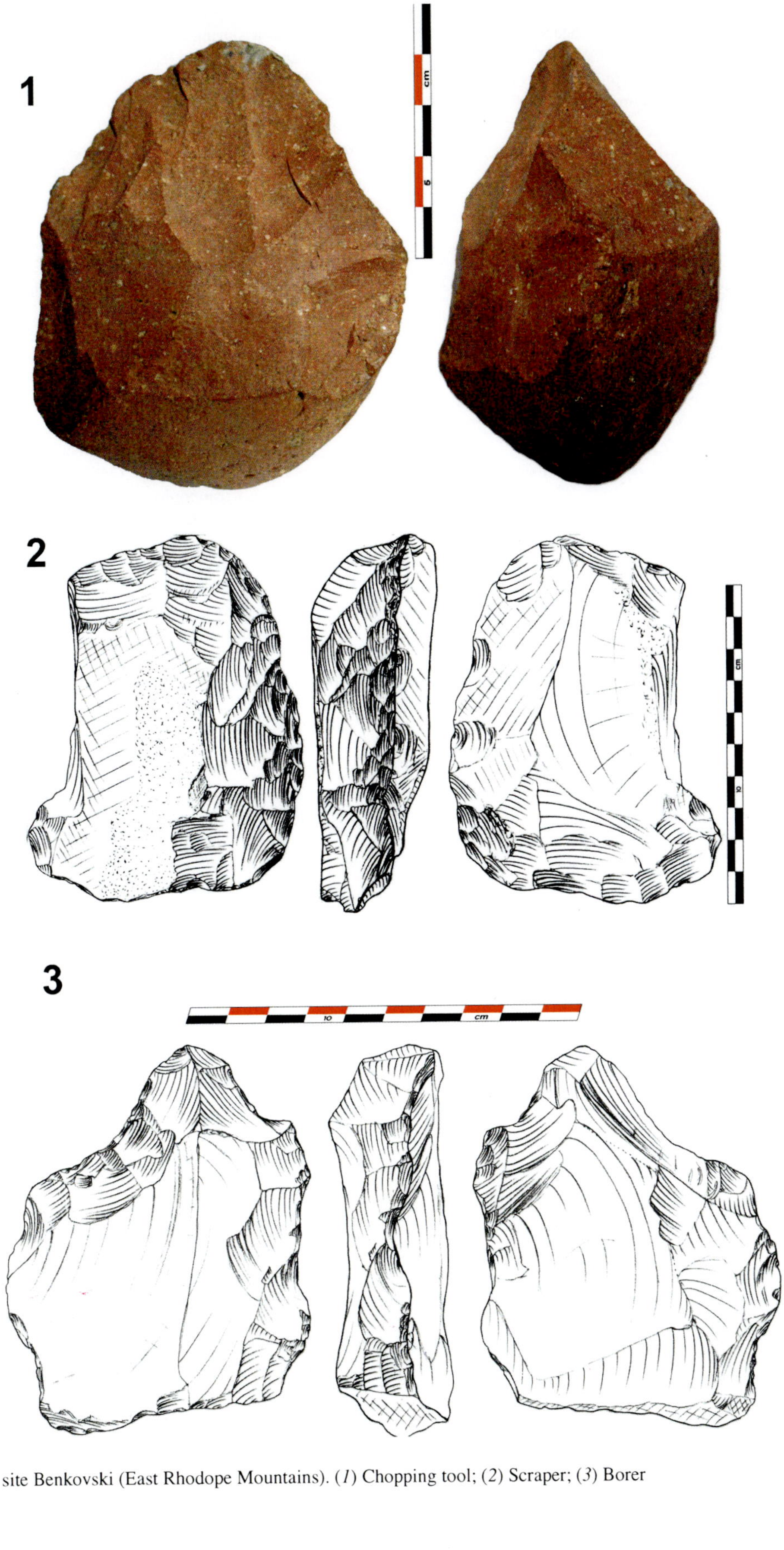

Fig. 12.17 Surface site Benkovski (East Rhodope Mountains). (*1*) Chopping tool; (*2*) Scraper; (*3*) Borer

Acknowledgments This study was made possible with financial supported by the Advisory Committee of the Archaeological Researches abroad (MAEE, France) - DGRCST, by the CNRS (Centre National de la Recherche Scientifique), by the Bulgarian Academy of Sciences. The comments and suggestions of two anonymous reviewers and K. Harvati greatly improved this chapter.

References

Abbate, E., & Sagri, M. (2012). Early to Middle Pleistocene *Homo* dispersal from Africa to Eurasia: Geological, climatic and environmental constraints. *Quaternary International, 267*, 3–19.

Arzarello, M., & Peretto, C. (2010). Out of Africa: The first evidence of Italian peninsula occupation. *Quaternary International, 223–224*, 65–70.

Bar-Yosef, O. (1998). Early colonizations and cultural continuities in the Lower Palaeolithic of western Asia. Early human behaviour in global context. In M. Petraglia & R. Korisettar (Eds.), *The rise and diversity of the Lower Palaeolithic record. One world archaeology* (Vol. 28, pp. 221–279).

Bar-Yosef, O., & Belfer-Cohen, A. (2000). Early human dispersal: The unexplored constraint of African diseases. *ERAUL, 92*, 79–86.

Bozhkov, I., Plotnikova, I., & Raynova, M. (1978). On the development of garlovinnite facies of rhyolite on Bratsigovo-Dospatska depression. *Bulgarian Geological Society, 39*(3), 344–348 (in Bulgarian).

de Lumley, H., Barsky, D., & Cauche, D. (2009). Les premières étapes de la colonization de l'Europe et l'arrivée de l'Homme sur les rives de la Méditerranée. *L'Anthropologie, 113*, 1–46.

Debenath, A., & Dibble, H. L. (1994). *Handbook of Paleolithic typology: Lower and Middle Paleolithic of Europe*. Philadelphia: University of Pennsylvania Museum of Archaeology and Anthropology.

Dennell, R. (2000). Dispersal and colonization, long and short chronologies: How continuous is the Early Pleistocene record for hominids outside East Africa. *Journal of Human Evolution, 45*, 421–440.

Dennell, R., & Roebroeks, W. (1996). The earliest colonization of Europe: The short chronology revisited. *Antiquity, 70*, 535–542.

Derricourt, R. (2005). Getting "Out of Africa": Sea crossings, land crossing and culture in the hominin migrations. *Journal of World Prehistory, 19*, 119–132.

Fernandez, P., & Cregut-Bonnoure, E. (2007). Les Caprinae (Rupicaprini, Ovibovini, Ovini et Caprini) de la séquence Pleistocene de Kozarnika (Bulgarie du Nord): Morphometrie, biochronologie et implications phylogeniques. *Revue de Paléobiologie, 26*, 425–503.

Guadelli, A., & Guadelli, J.-L. (2004). *Une expression "symbolique" sur os dans le Paléolithique inférieur de la grotte Kozarnika, Bulgarie du Nord-Ouest. Colloque international "La spiritualite."* U.I.S.P.P, VIIIeme Commission: Paléolithique supérieur Liège, 10-11-12 Decembre 2003, vol. 106 (pp. 87–95). *ERAUL*.

Guadelli, J.-L., Sirakov, N., Ivanova, S., Sirakova, S., Anastassova, E., Courtaud, P., et al. (2005). Une séquence du Paléolithique inférieur au Paléolithique récent dans les Balkans: La grotte Kozarnika à Oreshets (Nord-Ouest de la Bulgarie). In N. Molines, M.-H. Moncel, & J.-L. Monnier (Eds.), *Les premiers peuplements en Europe. Actes du colloque international "Données récentes sur les modalités de peuplement et sur le cadre chronostratigraphique, géologique et paléogéographique des industries du Paléolithique inférieur et moyen en Europe"* (BAR International Series, Vol. 1364, pp. 87–103). Oxford: Archaeopress.

Ivanova, S. (1979). Cultural differentiation in the Middle Paleolithic on the Balkan Penninsula. In J. K. Kozlowski (Ed.), *Middle and Early Upper Palaeolithic in Balkans* (Prace Archeologiczne, Vol. 28, pp. 13–33). Krakow: Zeszyty Naukowe Uniwersytetu Jagiellonskiego.

Ivanova, S. (2003). Raw materials exploitation strategy on the territory of Bulgaria during early Palaeolithic period. In T. Tsonev, & E. Montagnari Kokelj (Eds.), *The humanized mineral world: Towards social and symbolic evaluation of prehistoric technologies in South Eastern Europe*. Proceedings of the ESF workshop, September 2003, Vol. 103 (pp. 9–16). *ERAUL*.

Ivanova, S. (2006). Early and Middle Paleolithic flint assemblages from the region of the Shiroka Poljana, Western Rhodopes. *Excavations and surveys, 36*, 7–137. Sofia ISSN 0205–0722 (in Bulgarian).

Ivanova, S. (2009). *Early Paleolithic assemblages from Kozarnika cave*. Saxa Loquuntur Collection in honor of the 65th anniversary of Nikolay Sirakov (pp. 29–57). Sofia: Avalon (in Bulgarian).

Kahlke, R.-D., García, N., Kostopoulos, D., Lacombat, F., Lister, A., Mazza, P., et al. (2011). Western Palaearctic palaeoenvironmental conditions during the early and early Middle Pleistocene inferred from large mammal communities, and implications for hominin dispersal in Europe. In J. Carrión, J. Rose, & C. Stringer (Eds.), *Ecological scenarios for human evolution during the early and early Middle Pleistocene in the western Palaearctic* (Quaternary science reviews, Vol. 30, pp. 1368–1395).

Koufos, G. D., & Kostopoulos, D. S. (2016). The Plio-Pleistocene large mammal record of Greece: Implications for early human dispersals into Europe. In K. Harvati, & M. Roksandic (Eds.), *Paleoanthropology of the Balkans and Anatolia: Human evolution and its context*. Dordrecht: Springer.

Lyubin, V.P. & Belyaeva, E.V. (2006). Cleavers and handaxes with transverse cutting edge in the Acheulian of Caucasus. In N. Goren-Inbar and S. Gonen (Eds), Axe age: Acheulian tool-making from quarry to discard. London: Equinox pp.347–364.

Mihnevski, N., & Genkova, A. (1989). Snow avalanches in the Rhodopes and conditions for their formation. *Geographica Rhodopica, 1*, 129–132.

Petraglia, M. D. (2003). The Lower Paleolithic of the Arabian peninsula: Occupations, adaptations and dispersal. *Journal of World Prehistory, 17*, 141–179.

Popov, V. (2009). The small mammals from sectors 7, 8, and 12 of Kozarnika cave: Results from the excavations during the field season of 2008 and some comments of the biostratigraphy of the deposits. Projet de recherche conjoint: "Les plus anciennes manifestations de la présence humaine dans les Balkans" Mission Préhistorique Française en Bulgarie du Nord (MAE) (pp. 211–217).

Popov, V., & Marinska, M. (2007). An almost one million year long (Early to Late Pleistocene) small mammal succession from the archaeological layers of Kozarnika Cave in Northern Bulgaria. *Courier Forschungsinstitut Senckenberg, 259*, 79–92.

Roebroeks, W. (2001). Hominid behaviour and the earliest occupation of Europe: An exploration. *Journal of Human Evolution, 41*, 437–461.

Ronen, A. (2006). The oldest groups in the Levant. *Comptes Rendus Palevol, 5*, 343–351.

Sirakov, N., Guadelli, J.-L., Ivanova, S., Sirakova, S., Boudadi-Maligne, M., Dimitrova, I., et al. (2010). An ancient continuous human presence in the Balkans and the beginnings of human settlement in western Eurasia: A Lower Pleistocene example of the Lower Palaeolithic levels in Kozarnika cave (North-western Bulgaria). *Quaternary International, 223 224*, 94 106.

Spassov, N. (2001). The possible time of the first penetrations of *Homo* in Europe in the aspect of the paleoclimate analyses and the mammalian Plio-Pleistocene dispersals from the East. *Annual of the Institute and Museum of Archaeology, Sofia, 1*, 16–21.

Spassov, N. (2016). Southeastern Europe as a route for the earliest dispersal of *Homo* towards Europe: Ecological conditions and the timing of the first human occupation of Europe. In K. Harvati, &

M. Roksandic (Eds.), *Paleoanthropology of the Balkans and Anatolia: Human evolution and its context*. Dordrecht: Springer.

Strait, D. S., Orr, C. M., Hodgkins, J., Spassov, N., Gurova, M., Miller, C., & Tzankov, T. (2016). The human fossil record of Bulgaria, and the formulation of biogeographic hypotheses. In K. Harvati, & M. Roksandic (Eds.), *Paleoanthropology of the Balkans and Anatolia: Human Evolution and its context*. Dordrecht: Springer.

Turner, A. (1999). Assessing the earliest human settlement of Eurasia: Late Pliocene dispersions from Africa. *Antiquity, 73*, 563–570.

Villa, P. (1994). Lower and Middle Pleistocene archaeology. In S. J. de Laet (Ed.), *History of humanity* (Prehistory and the beginnings of civilization, Vol. 1, pp. 44–61). London: Routledge.

Yaranov, D. (1939). Contributions to the morphology of the Western Rhodopes. *Proceedings of the Bulgarian Geographical Society, 7*, 197–240 (in Bulgarian).

Chapter 13
The Lower Paleolithic in Turkey: Anatolia and Hominin Dispersals Out of Africa

Berkay Dinçer

Abstract Modern-day Turkey covers a vast area and includes many different ecological regions. Based on its geographic position, the Asian portion of Turkey, Anatolia, is accepted as a major route of early hominin dispersals. While it represents a reasonably direct route, Anatolia should not be conceptualized as a convenient land bridge for hominins originating in Africa, one that could be traversed without anatomical and/or technological adaptations. The available data show, at minimum, the presence of hominins in Anatolia at various times in the Pleistocene and the presence of various lithic traditions in the region. These are not sufficient for clarifying Anatolia's role as a passage in the earliest occupations of Eastern Europe, and theories suggesting Anatolia as a major hominin dispersal route must remain preliminary.

Keywords Lithic technology • Migratory routes • Oldowan • Acheulian • Bifaces

Introduction

Modern Turkey covers an area of 783,562 km^2 at the junction between Asia and Europe. The Asian portion of Turkey is Anatolia (Asia Minor) and the European portion is Eastern Thrace. The country can be divided into seven distinct geographical regions: Southeastern, Eastern, and Central Anatolia; the Mediterranean, the Black Sea, Marmara, and Aegean regions. Within these seven regions, there are 21 sub-regions separated by different climatological characteristics and topographic features, making it difficult to conceptualize Turkey as a single homogeneous geographical entity. Over 55 % of Turkey's surface is higher than 1000 m above sea level (asl). This mountainous landscape creates many closed basins with hundreds of separated ecological micro-regions, (like archipelagos) providing a high degree of biodiversity and endemism (Avcı 2005).

In many publications where human dispersals are depicted through maps bearing arrows with hypothetical human migratory routes, there is often one arrow route through Anatolia and Thrace toward Europe. Due to their location at the junction between the European and Asian continents, Anatolia and Thrace are regarded as the most likely, if not the only, routes by which African hominins could have reached Europe (e.g. Bar-Yosef and Belfer-Cohen 2001). For many authors, modern Turkey is a simple east–west oriented land bridge for early hominins. Widespread acceptance of this simplified model is probably a result of a Eurocentric point of view in which "destination," rather than the "source" represents the starting point in the search for the origins of the first Europeans and their migratory routes. While the route through Anatolia is actually the shortest land route from Africa to Europe, it is important to remember that the distance between the southeastern and northwestern ends of the Anatolian peninsula is more than 1000 km as the crow flies, and that it presents the migrants with an equally imposing range of ecological challenges. Regarding it simply as a trajectory does not recognize the cultural (and perhaps anatomical) adaptations that needed to take place in order to successfully inhabit the area.

As dispersals out of Africa most likely occurred several times, Anatolia's contribution to human evolution can only be fully appreciated by relying on larger and better contextualized physical evidence of early occupation of the region, with an understanding that Europe was not the sole "destination" of migrating human groups. Since Anatolia would have been the first place where hominins originating in Africa would have faced cold and snowy winters, this land could have been a barrier for various dispersal events. While the presence of a few sites could be understood as evidence of

B. Dinçer (✉)
Prehistory Department, Faculty of Letters, Istanbul University, Istanbul, Turkey
e-mail: berkaydincer@gmail.com

Katerina Harvati and Mirjana Roksandic (eds.), *Paleoanthropology of the Balkans and Anatolia*, Vertebrate Paleobiology and Paleoanthropology, DOI 10.1007/978-94-024-0874-4_13

successful occupation episodes, it could equally well represent unsuccessful attempts at colonizing new habitats.

However, the greatest limitation to our understanding of the role Anatolia played in hominin dispersals is the scarcity of primary evidence in the form of fossil hominins and archaeological sites. The low level of interest in Paleolithic archaeology among archaeologists in Turkey might be the result of the history of archaeology in the country, where large mounds of later prehistory and prosperous classical period sites drew more attention than Paleolithic sites (Arsebük 1998a), leaving the country with very few Paleolithic archaeologists (but see Tourloukis 2016 for a geoarchaeological perspective on a similar situation in Greece). With the currently available data, it is not possible to draw any firm conclusions about the role Anatolia might have played in early hominin dispersals. The number of systematically excavated Paleolithic sites is very small and those that are identified are geographically dispersed over large areas. The distribution of excavated sites does not even permit the construction of a reasonable regional stratigraphic sequence (Fig. 13.1).

According to The Archaeological Settlements of Turkey Project database (www.tayproject.org), the number of Paleolithic sites in Turkey increased from 210 to 459 over the last 15 years. The total number of Lower Paleolithic sites in this database has increased from 86 to 170 (Harmankaya and Tanındı 1997). Most of the recently discovered sites ($n=75$) are located in Southeastern Turkey, where large-scale, systematic surveys have been conducted in areas slated for inundation after dam constructions. This database does not contain some of the most recent discoveries, and the number of recorded sites is growing.

The known Paleolithic sites are not distributed evenly across Turkey due to unequal coverage of the country by systematic surveys (Harmankaya 1997). Results of many earlier studies were not published in detail: in many cases only the name of the sites and the presence of Paleolithic artifacts are recorded. Detailed descriptions or standard drawings are often missing, and very summary typological assessments by the authors are the only available evidence. These are often imprecise and use terms interchangeably. For example, the use of terms "handaxe" and "biface" interchangeably in this text is the result of the practice in the available literature. While many bifaces and handaxes were reported, it is not clear if all handaxes were bifacially produced. Since many of those artifacts are lost or not available for further research, the biface/handaxe terminology was kept here as used by the cited authors. These terms possibly indicate the same types of artifacts, but it is not possible to confirm this assumption. Furthermore, the exact locations of many sites are missing and most of the statements cannot be verified (Arsebük 1998a; Kuhn 2002).

Even with such poor data, it is still interesting to explore both the (1) distribution information, in order to understand the choice of environment of early inhabitants; and (2) the technology, in order to build a relative chronology of the Lower Paleolithic in Anatolia. In this chapter, I will summarize data from available literature, present a general picture

Fig. 13.1 Map showing sites and provinces (italic) mentioned in the text. Shaded areas indicate regions with concentrated Paleolithic research. *1*—Şehremuz, *2*—Dülük, *3*—Helale, *4*—AktaŞ, EŞkini Sefine and Madler, *5*—KD3 and Göllüdağ, *6*—Dursunlu, *7*—Karain, *8*—Şenköy, *9*—Gez Alanı, *10*—KocabaŞ, *11*—Belentepe, *12*—MenekŞe Kayalar, *13*—Kuzfındık, *14*—Bozyer, *15*—Yarımburgaz, Göksu and Eskice Sırtı, *16*—Yatak, Balıtepe, KuŞtepe and AkçeŞme

of the Lower Paleolithic in Turkey, and evaluate the validity of possible migratory route hypotheses for Anatolia toward Europe taking into account lithic tool technologies.

Southeastern Anatolia

Southeastern Anatolia could be considered as the northern part of Mesopotamia. It is the most intensively surveyed region in Turkey with regards to the Paleolithic, due mainly to the salvage projects that preceded the construction of dams on the Euphrates and Tigris rivers. Atatürk, Karakaya, Kargamış, and Ilısu are the largest artificial lakes in the region and research has been concentrated in and around their reservoir areas (Taşkıran 2008). Many of the Paleolithic sites recorded in the survey of the region are currently inundated and not available for further research.

High-quality flint sources are abundant in Southeastern Anatolia, and Lower Paleolithic sites in the region cover large areas and bear many artifacts due to the easy availability of raw materials (Dinçer 2010a). Information on the distribution of flint sources and their properties, however, has not been reported in detail.

The only excavated Lower Paleolithic site in Southeastern Anatolia is Şehremuz (Albrecht and Müller-Beck 1988). This is an artifact-bearing gravel locality attributed to the Late Acheulian. No radiometric dates are available from the site. Şehremuz has yielded 236 definite artifacts, 197 of which are handaxes or handaxe-like tools. Morphological analyses supported by microscopic use-wear analyses showed that the cutting edges of these bifacial handaxes are generally on their tips. End-retouched long flakes, side scrapers, and semi-triangular and triangular Levallois flakes have also been recovered from the site (Albrecht and Müller-Beck 1988).

Many open-air sites were discovered on the terraces of the Euphrates river. In neighboring Syria, sites were geologically correlated with the Euphrates terraces (Copeland 2004; Sanlaville 2004). On the Turkish side, this kind of research has been conducted only in the area around Gaziantep (SE Anatolia). Geological and archaeological data are comparable with those reported for Syria. The oldest stone tools (only three flakes) were found in the QfIV (*Quaternaire fluviatile*) formation (Minzoni-Déroche 1987). While the QfIII formation in Turkey has not yielded many stone tools, in Syria, especially in Balikh Valley near the border with Turkey, stone tools found within this formation have been attributed to the Middle Acheulian and dated to roughly 700 ka (Copeland 2004). At Nizip (Gaziantep), 177 stone tools have been found in this formation (Minzoni-Déroche 1987), including five picks, seven bifaces, seven cores and 158 flakes and blades. QfII, the richest artifact-bearing formation, probably dates to approximately 600 ka (Sanlaville 2004), with assemblages attributed to the Upper Acheulian. A total of 739 stone tools have been found in this formation, including seven picks, 158 bifaces, 48 cores and 526 flakes and blades. All handaxes are made on flint nodules. It is also important to note the presence of Levallois cores in the QfII (Minzoni-Déroche 1987; Minzoni-Déroche and Sanlaville 1988). Since those finds were recovered during survey, the relationship between Levallois products and handaxes is not clear.

In 1998, prior to the construction of the Kargamış and Ilısu dams, large areas of Southeastern Anatolia were systematically surveyed. More than 70 Lower Paleolithic sites were discovered in the Euphrates and Tigris valleys (the Tigris Valley has fewer Paleolithic sites than the Euphrates Valley; Algaze and Rosenberg 1991; Taşkıran 2008) and elaborately published by Taşkıran and Kartal (1999, 2001, 2004). The majority of the artifacts collected and reported are bifaces. Some of them are partial bifaces on flakes. There are some cleavers and picks, while choppers and chopping tools are very rare. Levallois products are often associated with Lower Paleolithic assemblages in the region (Taşkıran and Kartal 1999, 2001, 2004; Taşkıran 2002a, b, 2008), lending support to their purported association at surface sites from the QfII formation in the Nizip area.

Another systematic survey in Southeastern Anatolia took place in the Sakçagözü area between 1995 and 1996. During this survey by A. Garrard et al. (2004) eight Lower Paleolithic sites were discovered, two of which were cave sites. Bifaces, small and made on flint nodules, dominate these assemblages. Symmetry and standardization are at a very low level and the bases of the handaxes are generally cortical. Raw material sources are located at higher altitudes than the sites. There is also a remarkable decrease in the number of Paleolithic sites at the distance of 4–5 km from raw material sources (Garrard et al. 2004).

Many of the studies of Paleolithic sites conducted in Southeastern Anatolia since the end of nineteenth century were unsystematic and results were published only partially. One example is that of the surface collections from Dülük. This is a very large high-quality flint source covering an area of more than one square kilometer. After its discovery in 1938, many archaeologists have visited Dülük and collected artifacts (Atasayan 1939; Çambel 1947; Kökten 1947a, 1952; Bostancı 1962). However, only typical artifacts such as bifaces, Levallois cores, and retouched flakes were collected and, as a result, most of these collections do not reflect the actual artifact assemblages. One of the most distinctive aspects of the Dülük assemblage, beside the presence of bifaces, Levallois cores and retouched flake tools, is the presence of blade-like long flakes (Fig. 13.2; Dinçer 2010a). A very similar assemblage has been recently found at Helale near Mardin (exhibition of Mardin Museum).

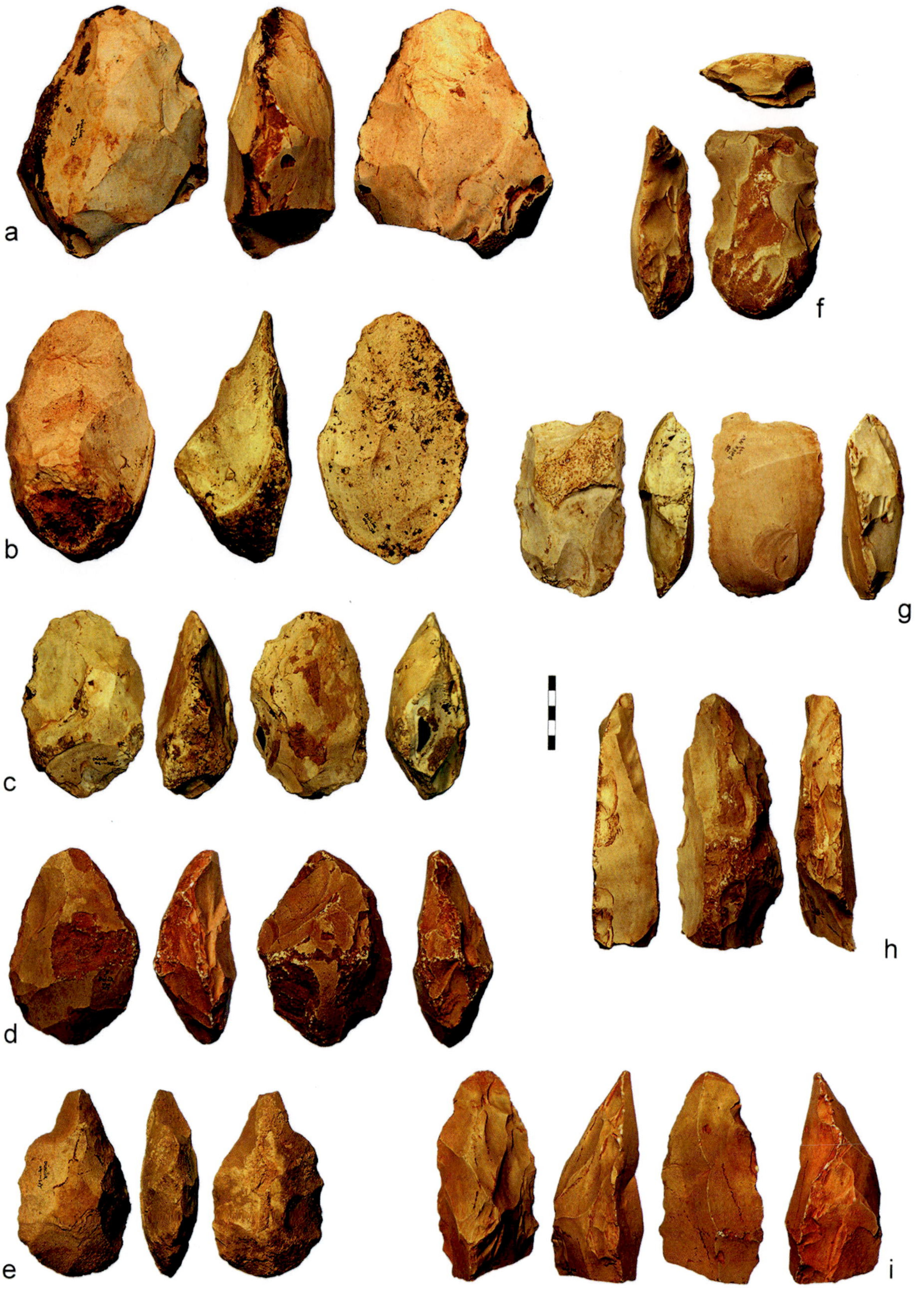

Fig. 13.2 Bifaces (**a–e**), cleavers (**f**, **g**), a retouched long flake (**h**), and a trihedral pick (**i**) from Dülük

In summary, bifaces (especially handaxes) are characteristic of the Lower Paleolithic of Southeastern Anatolia. Apart from Şehremuz, all sites in this region are known solely from surveys and surface collection (Yalçınkaya 1986). For most of these sites, geomorphological settings are not very well known. Many of the sites are surface scatters associated with raw material sources. The presence of rich raw material sources made possible the production of large bifaces and Levallois production accompanies bifacial artifacts at many of these sites. This could be evidence of either the use of the Levallois method in the Lower Paleolithic and/or the use of the same raw material sources in later time periods. Since there is no stratigraphical evidence and all Levallois and bifacial tools were recovered from surface sites, further elaboration is not possible. The scarcity of core tool assemblages without bifaces (i.e. chopper/chopping tool assemblages) in Southeastern Anatolia is noteworthy. Furthermore, given that this region is closest to Africa and directly lies on the route to Dmanisi, the scarcity of Lower Paleolithic core tool/flake assemblages in the region requires further examination.

As mentioned above, Paleolithic sites in the Euphrates Valley are more numerous than those in Tigris Valley. It is not clear whether this represents a clue to population densities in these respective regions, or if it reflects geomorphological differences between the two river courses (see Tourloukis 2016 for the discussion of geomorphological influences on site distribution patterns).

Eastern Anatolia

Eastern Anatolia is a mountainous highland landscape with very cold winters. Paleolithic research in this region is mainly concentrated in two areas: the Keban dam reservoir area and the Kars area. Important obsidian sources are known around the Lake Van (Marro and Özfırat 2004; Baykara et al. 2016) and Erzurum-Kars areas (Koyabashi and Sagona 2008).

A recent survey near the Lake Van in 2014 revealed many Lower and Middle Paleolithic sites related to the Erciş and Gürgürbaba obsidian sources. The Lower Paleolithic sites bear many bifacial and unifacial handaxes made on large flakes. The preliminary results of this survey are published by Baykara et al. (2016).

The most remarkable finds of the Keban dam lake area are the pebble tools found at the sites of Aktaş, Eşkini Sefine, and Madler (Kökten 1974). These are accepted as typologically the oldest artifacts from this area. The assemblages from these sites are characterized by the presence of bifacial artifacts made of flint (Kökten 1974), while tools are made of basalt in the Kars area (Yalçınkaya 1981, 1985).

Eastern Anatolia is probably the most promising area for the understanding of the earliest hominin dispersals in Turkey. The earliest known hominin occupation out of Africa for the moment is documented at Dmanisi (Gabunia et al. 2001; Mgeladze et al. 2011), which is very close to the border between Turkey and Georgia. If one accepts that the Dmanisi hominins originated in Africa, Eastern Anatolia should play a key role in documenting the evidence of the earliest hominin dispersals in Turkey. However, there have been no stratified and dated sites found in the region that are as old as Dmanisi. As this is the largest geographical sub-region in Turkey and a *terra incognita* for the Paleolithic, current knowledge does not reflect its actual potential.

Central Anatolia

Central Anatolia is a high plateau with many closed basins. Research in the region is concentrated in two areas: in the vicinity of Ankara and in Cappadocia. Research on the Paleolithic around Ankara was mainly conducted in the first half of the twentieth century and was not systematic. In contrast with the number of reported Lower Paleolithic sites ($n=11$), the number of reported artifacts is limited ($n=62$); furthermore, some of these artifacts have been lost. The Lower Paleolithic of this area is characterized by bifacial artifacts made on flint (Kökten 1947b; Yalçınkaya 1981; Kartal 2005). However, it is probable that this is in part due to selective collection practices by early twentieth century researchers.

Göllüdağ is a large obsidian source in Cappadocia. Research at Göllüdağ began in 1995 with the purpose of investigating the Neolithic obsidian trade. In 2000, a Paleolithic site, Kaletepe Deresi 3 (KD3), was found on the bank of a seasonal drainage close to the Kaletepe Neolithic obsidian workshop excavation (Slimak and Dinçer 2007). KD3 was excavated between 2000 and 2008. In two trenches (*aval* and *amont*) each covering an area of nearly 25 m^2, 19 Lower and Middle Paleolithic levels were uncovered over a depth of nearly 8 m. A total of 6354 artifacts were collected, 977 of these coming from the Middle Paleolithic levels (I-II). Faunal remains from KD3 are restricted to Middle Paleolithic levels due to the acidic soil; only two equid mandibles and a few isolated equid teeth were found in the entire sequence (Slimak and Dinçer 2007; Slimak et al. 2007; Kuhn et al. 2009).

Kaletepe Deresi 3 is the only excavated Acheulian site in Turkey, which is geologically *in situ* (Slimak 2004; Slimak et al. 2008). Basal levels of KD3 lie on top of a 1.1 ± 0.02 Ma rhyolithic bedrock dated by K/Ar. Higher in the sequence, at the top of level II, there is a tephra layer dated to 160 ka (Mouralis 2003), providing a chronological frame. Recent tephrachronological analysis suggested that the age of the basal levels might be Middle and possibly Lower Pleistocene (Tryon et al. 2009). Still, the ages of KD3 assemblages are not certain (Kuhn 2010a).

Below level IV, the stratigraphic sequences of the *aval* and the *amont* trenches are different from each other. Acheulian levels (IV–XII) in the *aval* trench were deposited in a *cuvette*-like natural basin. Gravels of natural rhyolite and obsidian in this basin suggest deposition by water or colluvial action. Outside the basin, the artifact density is very low. Raw materials at these levels indicate differentiation in raw material use. Obsidian is mainly used for bifaces and fragments related to bifacial tool making, whereas coarse grained rhyolitic or andesitic stones are mainly used for polyhedral pieces, flakes, choppers, and chopping tools. Cleavers are made of both obsidian and andesite and some of them are of the "Tabelbala Tachengit" type from North Africa (Slimak et al. 2008). Assemblages from these levels are variants of the "large flake Acheulian" tradition. The presence of large obsidian Levallois flakes and cores in Lower Paleolithic levels is also important to note (Slimak et al. 2005, 2008; Kuhn et al. 2009).

The Acheulian levels of the *amont* trench (IV, Vam,[1] VIam, and VI'am) are characterized by higher frequencies of the use of rhyolite/andesite. With the exception of the VI'am level, where the use of rhyolite/andesite accounts for more than 40 % of the artifacts, in all other levels rhyolite/andesite comprise 90 % or more of the assemblages. In the VI'am level, frequencies of obsidian and basalt are almost equal. Since VI'am is the oldest level and obsidian is a commonly used raw material, it is difficult to explain the low frequency of obsidian use in later levels. The explanation could be cultural rather than geological, since the obsidian was abundant. Frequencies of choppers and chopping tools made of rocks other than obsidian are higher than 25 % of tool types at all levels. It is important to note that polyhedral pieces are numerous (Slimak et al. 2007). For the moment it is not clear if the Acheulian levels of the *aval* and *amont* trenches are contemporary.

Systematic survey of the Göllüdağ region began in 2007. In an investigation area of slightly over 100 km^2, more than 200 Lower and Middle Paleolithic sites were recorded (Balkan-Atlı et al. 2011). This high density is the result of both the concentration of hominin occupation in the Paleolithic due to rich raw material sources and intensive survey strategies. However, there are only two *in situ* stratified Paleolithic sites and both date to the Middle Paleolithic. The Lower Paleolithic sites found during survey were limited to dispersed surface finds and concentrations in lag deposits (Kuhn et al. 2015).

The Lower Paleolithic of Göllüdağ is characterized by bifacial handaxes, with large flakes frequently used as biface blanks. A few cleavers have also been found. Unlike the Acheulian levels of KD3, basalt handaxes comprise nearly one-third of the bifaces from Göllüdağ (Fig. 13.3). Choppers/chopping tools and polyhedral pieces were not recovered in high numbers (Balkan-Atlı et al. 2008, 2009, 2010, 2011), while they were numerous at KD3. This might be the result of the collection strategy applied during the survey, since "typical" tools such as bifaces and Levallois cores are easy to recognize, while core tools and polyhedral pieces are difficult to recognize in the field, especially when they are made of materials other than obsidian.

Another Lower Paleolithic site in Central Anatolia, Dursunlu, is located in the western part of the region, nearly 250 km to the west of KD3. Stone tools and faunal remains were found 10–12 m below the modern surface in a lignite quarry. Because of the mining activities, only blocks of soils from tip-heaps could be excavated. Dursunlu has been dated to 0.78–0.99 Ma by paleomagnetism and to 0.85–0.90 Ma by microfaunal evidence (Güleç et al. 2009). Paleomagnetic data suggest that the Dursunlu assemblage could postdate the Jaramillo event and predate the Brunhes/Matuyama reversal. Faunal remains are taxonomically very diverse and include large mammals, birds, amphibians, and tortoises. Avifauna indicates a lacustrine or marshy environment, and terrestrial bird species also indicate steppe conditions around wetland. Human modifications on faunal remains are very restricted: there is only one bone with probable cutmarks (Güleç et al. 1999, 2009).

The lithic assemblage of Dursunlu is very small, only 135 pieces, 95 % of which are made of quartz. Quartz flaking properties sometimes make it difficult to identify intentional modification by hominins; however, the kind of quartz used for artifacts is not naturally present in the sediments at Dursunlu. Güleç et al. (2009) developed a system for providing "certainty scores" for artifacts and found 26 artifacts with low, 81 with moderate, and 28 with high certainty score. The assemblage consists mainly of flakes and flake fragments. Flake and flake tool platforms preserved on 49 artifacts are mainly plain ($n = 27$). The presence of crushed platforms ($n = 13$) indicates the use of bipolar (hammer on anvil) percussion. Only 12 pieces have secondary modifications: three modified flakes, three polyhedral cores, two notches, one chopper, one polyhedron, one *pièce esquille*, and one tested chunk. There is no evidence of biface or Levallois production at Dursunlu (Güleç et al. 2009).

Central Anatolia presented a serious adaptive challenge for early hominins. The two excavated sites are located at altitudes higher than 1000 m asl, while the surveyed sites at Göllüdağ are located at altitudes between 1300 and 2000 m asl. Due to its continental climate, winters in central Anatolia are harsh. Faunal remains at Dursunlu found in

[1]Since the stratigraphic sequence differs in the aval (av) and amont (am) tranches, the av and am designations are used to indicate that those levels are in one or the other trench. The apostrophes in level names indicate levels in the same geological context but with different finds.

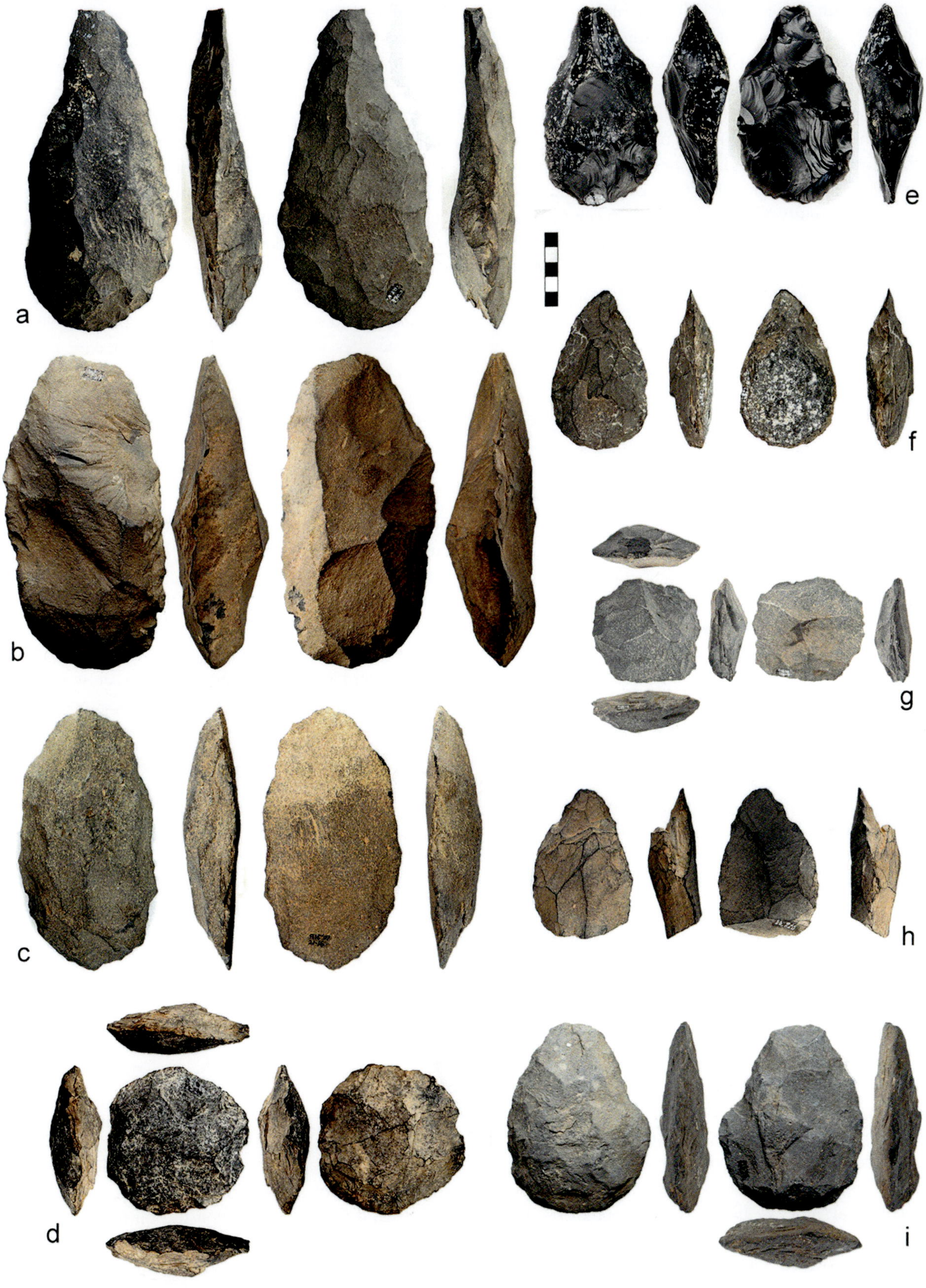

Fig. 13.3 Bifaces (**a**, **e**, **f**, **h**, **i**), retouched large flakes (**b**, **c**), and bifacial disks (**d**, **g**) from Göllüdağ survey

association with stone tools, help describe environmental conditions for the first documented hominin occupation of Turkey in the Lower Pleistocene (Güleç et al. 2009), but the small sample of artifacts does not permit a full account of the tool-making technologies. However, it is not clear whether these earliest hominin occupations of Central Anatolia took place under glacial or interglacial conditions, and it is currently not possible to draw conclusions about the level of continuity of human presence in central Anatolia in the Lower Pleistocene.

Kaletepe Deresi 3 shows a different picture than Dursunlu. With a deep stratigraphy including both Lower and Middle Paleolithic levels and spanning a time interval of at least a half million years, KD3 is clearly a key site for the early Paleolithic in central Turkey, and reflects human occupations of Central Anatolia at various times during the Lower and Middle Paleolithic. The main problems concerning KD3 are the uncertain dating and lack of faunal remains.

Mediterranean Region

Located south of the Taurus Mountain range along the Mediterranean Sea, this part of present-day Turkey has a relatively mild climate. Due to karstic bedrock geology, caves are abundant in the region. Research is concentrated in the eastern and western parts of the region, which correspond to Hatay and Antalya provinces respectively. The African Rift Valley system extends into Hatay via the Levantine Corridor.

Karain cave is the most extensively researched Paleolithic site in Turkey (see also Aytek and Harvati 2016). The first excavations were conducted between 1946 and 1973; and have been ongoing continuously since 1985. The deposits at Karain—divided into five major geological units—are 11 m deep and cover all periods of the Paleolithic, representing the key chronological sequence for Turkey (Yalçınkaya 1992). The Lower Paleolithic was ascertained in the geological unit V, with the upper layers of the unit dated to 370–400 ka (Otte et al. 1998a). The material consists of thick Clactonian flakes with pronounced bulbs of percussion, and with notches and denticulates as main tool types. Cores are mainly polyhedral (Otte et al. 1998a, b, 1999). Common raw materials are radiolarite, flint, and calcareous stones. In earlier excavations two bifaces were reported (Taşkıran 1998) but only one biface made of calcareous stone has been found in later excavations. This biface was found below the *tayacian* layers in the geological unit V and thus likely predates 400 ka (Yalçınkaya et al. 2007, 2008, 2009a, b).

In Hatay province, the Lower Paleolithic is known only from surveys. The flint sources in the area are of almost the same quality and richness as those in Southeastern Anatolia. Since there has been no systematic research, only typical tools have been collected in surveys. The most commonly reported artifacts are bifaces (Şenyürek and Bostancı 1958; Şenyürek 1960, 1961), with one cleaver and one chopping tool (Özçelik 2003). The most important Lower Paleolithic site in Hatay is Şenköy (Bostancı 1971), a large flint source with associated artifacts. Bifaces—most of which are small and crude—are well-represented in surface assemblages. A few chopping tools and preferential and laminar Levallois cores were also reported (Özçelik 2003). Most of the flakes show the Clactonian kind of a developed bulb of percussion and open platform angles. It is important to note the presence of long flakes. The Şenköy assemblage most likely reflects a workshop associated with a flint source (author's study).

Three sites with bifaces, Clactonian flakes, scrapers on blades, and Levallois cores and flakes were identified in the province of Kahramanmaraş (Erek 2011).

The assemblages from the eastern part of the Mediterranean region have striking resemblances with those from Southeastern Anatolia. The abundance of similar high-quality flint raw material in the eastern Mediterranean and Southeastern Anatolia regions might have played an important role in the resemblances of these industries. The most important site in the western part of the Mediterranean region, on the other hand, is Karain cave, whose Lower Paleolithic assemblage is characterized by a *tayacian*-like flake industry, contrasting with the Acheulian assemblages of the eastern Mediterranean and Southeastern Anatolia.

Black Sea Region

Prehistoric research in the Black Sea region is generally very limited, due in part to the high density of vegetation. Only sporadic discoveries of single bifaces have been reported so far: one biface made of andesite was found near Samsun (Kökten 1951); other bifaces were found in the Ordu (Kökten 1963) and Kastamonu provinces (Bostancı 1952). Only one assemblage—consisting of basalt handaxes, retouched flake tools, points and one obsidian blade—was recovered at Gez Alanı, near Bayburt (Gündüzalp 1986).

Aegean Region

Characterized by parallel mountain ranges running in an east–west direction, this region lies on the eastern coast of the Aegean Sea. Its inland areas—usually seen as an obligatory route from Central Anatolia toward the west—are characterized by high altitude and a relatively continental climate. Before the 1990s, evidence for Lower Paleolithic occupation here was limited to three bifaces: two from Izmir (Kansu

1963; Yalçınkaya 1981) and one from Eskişehir (Chaput 1941; Tomsky 1982). In the last two decades, however, the number of Paleolithic sites from this region has increased, with systematic research in areas like southern Bursa, Kuzfındık,[2] Kureyşler, and Aizanoi (Dinçer and Türkcan 2011; Dinçer 2014a, b, 2015; Dinçer et al. 2014).

The only *Homo erectus* fossil from Turkey comes from the Kocabaş village in the province of Denizli in the Aegean region (Aytek and Harvati 2016). It was recovered from a travertine block quarried for construction material. The block had been cut, reducing the Kocabaş fossil fragment to a frontal bone with pronounced browridges. The first age proposed for this specimen was 1.11±0.11 Ma, based on the ESR dating of the travertine (Engin et al. 1999); later revised to 510–490 ka based on thermoluminescence dating (Kappelman et al. 2008). However, a recent multidisciplinary study confirmed the age of at least 1.1 Ma (Lebatard et al. 2014). Kappelman et al. (2008) also claimed that the Kocabaş fossil exhibited a pathological condition consistent with tuberculosis, purported to have resulted from a weakened immune system related to vitamin D deficiency; a condition observed in dark-skinned people living in northern latitudes (Kappelman et al. 2008). This interpretation has been debated (e.g. Roberts et al. 2009) and, more recently, rejected (Lebatard et al. 2014). No stone tools were found in association with the fossil.

One single stone tool claimed to be as old as the Kocabaş fossil has been reported recently (Maddy et al. 2015). This artifact was found in one of the terraces of Gediz River and dated to 1.24 and 1.17 Ma. It is a quartz hard hammer flake with one flake removal on its dorsal face (Maddy et al. 2015). Since it is a single, isolated flake, it is not possible to fully understand the technological properties of the earliest occupation of the region.

Belen Tepe is an artifact-bearing primary flint source situated south of the Uludağ volcano in Bursa. The site is located on a hilltop, at the elevation of 1050 m asl. The surface assemblage is mainly composed of cortical flakes, which numerous Clactonian-like flakes; two bifaces, one of which was left unfinished, while the other was broken; one large chopping tool made of flint; and two preferential Levallois cores, one of which is made on the ventral face of a Clactonian flake. One bifacial disk bears signs of heavy battering, pointing to its use as an anvil or *percuteur dormant* (Fig. 13.4; Dinçer 2010b, 2014a).

Surface sites around the Kütahya province have yielded many bifaces and Clactonian flakes (Efe 1990). A single handaxe made of flint and shaped by soft hammer percussions (Taşkıran and Taşkıran 2011), found near Afyon at the site of Menekşe Kayalar, resembles Upper Acheulian specimens. One isolated chopping tool made of quartzite was reported at the Kuzfındık valley near Eskişehir (Dinçer and Türkcan 2011). A probable Lower Paleolithic site with core tools and Levallois cores has been reported at Bozyer in the province of Manisa, but details have not yet been published (Roosevelt and Luke 2010).

Sites in the Aegean region are few and dispersed over a large distance. At present, they do not provide a clear picture of the Lower Paleolithic in the region. Most of the surveys were conducted by archaeologists interested in later periods, causing irregular reporting that might have dramatically affected assemblage compositions. Poor stone tool raw material resources might also account for some characteristics of the assemblages, with bifacial tools found only when high-quality raw material was present, as in the case of Belen Tepe and Menekşe Kayalar (Dinçer 2014b; Taşkıran and Taşkıran 2011).

Marmara Region

The Marmara region lies at the junction between Asian Anatolia and the European Balkans, and accordingly, this area has always been considered as the passage between the two regions (Özdoğan 1998). Research in the Marmara region is mainly limited to the vicinity of Bosporus.

Located west of Bosporus, Yarımburgaz cave is the only excavated Paleolithic site in the region. It is a two-chamber (upper and lower) cave. Between 1988 and 1990, over 130 m^2 were excavated in the lower chamber, in nine trenches (Kuhn et al. 1996). The Yarımburgaz assemblage is attributed to the Lower Paleolithic. Electron spin resonance dating of 17 *Ursus deningeri* teeth from upper layers suggests a date of 270–390±40–60 kBP for the recent uptake model, and between 200–220±20–30 kBP for the linear uptake model (Arsebük 1998b; Arsebük and Özbaşaran 1999). According to ESR results, the most recent Paleolithic layers were deposited between 226±24 and 211±22 kBP, during the MIS7d (Blackwell et al. 2010). Analysis of the small mammal fauna places the Paleolithic occupation of the cave during a cold phase in the middle of the Middle Pleistocene (Santel and Koenigswald 1998; Koenigswald et al. 2010).

Different species of *Ursus* constitute 93 % of the identified large mammals from the site. Most specimens are attributed to the Middle Pleistocene cave bear, *Ursus deningeri*. Non-ursid carnivores constitute 3 % and ungulates 4 % of the faunal assemblage (Stiner et al. 1996, 1998; Kuhn et al. 1998). Human modifications on animal bones are restricted to less than 1 % of ungulate bones, while gnawing damage is found on 23 % of the ungulates, 10 % of the ursids, and 18 % of the non-ursid carnivores (Arsebük 1998b; Kuhn et al. 1998).

[2] Southern Bursa and western Eskişehir are located in Marmara and Central Anatolia regions respectively. Lower Paleolithic finds from these areas are discussed here due to their close proximity to the eastern Aegean region.

Fig. 13.4 Bifaces (**a**, **b**), Levallois core on Clacton flake (**c**), bifacial disk (**d**), and chopping tool (**e**) from Belen Tepe

The Paleolithic deposits in Yarımburgaz cave are almost 1.2 m thick. Sedimentary infilling of the cave occurred in three cycles separated by two hiatuses (Arsebük 1998b). Nonetheless, the lithic industry of Yarımburgaz is homogeneous throughout these different cycles (Arsebük 1993). Flint dominates the lithic industry represented by 1675 artifacts collected from controlled excavations. Quartz and quartzite are two of the other most frequently used raw materials. A high frequency of retouched flake tools ($n=538$) dominated by denticulates and sidescrapers is the most notable characteristic of the lithic industry, with abundant flakes—broken or whole—($n=366$), notches and *bec/perçoirs*. Levallois production is absent in the Yarımburgaz assemblage and there is no evidence of bifaces/handaxes or their by-products (Kuhn et al. 1996; Arsebük 1998b).

There are clear differences in the use of different kinds of raw material. More than 70 % of the retouched tools and flakes are made of flint, while more than 75 % of the core tools are made of quartzite, which, in turn, comprises only 14 % of the assemblage. The number of core tools ($n=64$) is strikingly low (Kuhn et al. 1996; Kuhn 2010b). Due to the high frequency of retouched flake tools, the Yarımburgaz assemblage could easily be attributed to the Middle rather than the Lower Paleolithic (Kuhn 2003).

Many Lower and Middle Paleolithic sites have been found in surveys of the Istanbul area. Only a few bifaces were found on the east side of the Bosporus, including one "typical Late Acheulian" biface from the site of Göksu, as well as many choppers and chopping tools (Jelinek 1980). On the western side of the Bosporus, choppers and chopping tools characterize open-air sites such as Eskice Sırtı (Runnels and Özdoğan 2001). This is also the case further west in Turkish Thrace, where choppers and chopping tools comprise nearly half of the assemblages at the open-air sites of Yatak, Akçeşme, Kuştepe, and Balıtepe (Fig. 13.5; Dinçer and Slimak 2007). In contrast to the Yarımburgaz assemblage, open-air sites in Turkish Thrace do not bear many flakes and retouched flake tools. Only two very atypical bifaces have been found in Turkish Thrace and neither can be accepted as a real handaxe. The presence of anvils bearing pronounced signs of heavy battering is important to note (Dinçer 2001; Dinçer and Slimak 2007). Lower Paleolithic sites in this area are concentrated on old terraces and closely associated with quartz and quartzite gravel deposits.

Lower Paleolithic sites in the Marmara region are commonly accepted as evidence of the earliest human dispersals since they lie on the most direct route from Africa toward Europe. Abundant choppers and chopping tools, namely the "Mode 1" industries, seem to support these ideas; however, the dates of these open-air sites remain uncertain.

The stratigraphic sequence of the Paleolithic in the Marmara region is not clear. The available data from Yarımburgaz, which has only one cultural component, are insufficient for understanding the Paleolithic in a regional perspective. The lithic industry shows no change over time and the occupation of the cave reflects only a relatively short period of time in the Middle Pleistocene. With a predominance of retouched flake tools and low frequency of chopper/chopping tools, the Yarımburgaz lithic assemblage could be classified as Middle Paleolithic. There is no clear correlation amongst the assemblages of the Lower Paleolithic open-air sites in Marmara region and Yarımburgaz. The state of research (mostly limited to surveys) is far from clarifying the actual occupation of the Marmara region in the Lower Paleolithic.

Conclusions: A Synthesis of the Lower Paleolithic in Turkey

Most of the Lower Paleolithic localities, including the ones that were excavated (Şehremuz, KD3, and Dursunlu), are open-air sites. With the exception of Dursunlu, no faunal remains have been found in association with stone tools at any of them. Only two Lower Paleolithic cave sites have been systematically excavated: Karain and Yarımburgaz. These caves are situated very far from each other: one located at the southern, and the other at the northern end of the country.

At the current state of research, it appears that one of the characteristics of Lower Paleolithic sites in Turkey could be their close association with primary sources of raw material: flint in Southeastern Anatolia and obsidian in Central Anatolia. Raw material could also help explain the relative abundance and distribution of bifacial tools, since the production of bifaces requires large pieces of high-quality homogenous raw material such as those found in Southeast Anatolia. On the other hand, in areas such as Thrace, where primary sources of good quality raw material are not available, sites are usually concentrated on old river terraces that bear quartz and quartzite pebbles. Thus, technical adaptations to the available raw materials may explain some of the particular features of local Lower Paleolithic industries.

Runnels (2003) proposed a boundary in the distribution of Lower Paleolithic assemblages in Turkey (Runnels 2003), with bifaces abundant in the eastern part of the country, and less frequent and more "atypical" in the western parts. Since the oldest Paleolithic assemblages are thought to approach 1 Ma in age, it is important to question how long this boundary might have persisted. Did it last the whole duration of the Lower Paleolithic or did it move over time? An even more important question is why did such a boundary exist? As

Fig. 13.5 Choppers from AkçeŞme (**f**), Balıtepe (**a**, **e**), KuŞtepe (**b**), and Yatak (**c**, **d**, **g**)

mentioned above, bifaces have a close relationship with raw material sources. There are very limited high-quality raw material sources at most of western and northwestern Turkey. As evidenced by Menekşe Kayalar (Taşkıran and Taşkıran 2011), Belen Tepe (Dinçer 2010b, 2014a, b), and the Kütahya region (Efe 1990), when raw materials were available, there seems to be no cultural boundary in the distribution of bifaces. Bifaces have recently been found even further west, on the island of Lesvos (Galanidou et al. 2016). Thus, this hypothetical boundary seems to be related to raw material sources rather than cultural boundaries.

Lower Paleolithic occupations (documented so far in Turkey) occur at a period that significantly postdates Dmanisi. Excluding Dursunlu, all of the Lower Paleolithic sites could be dated to the Middle Pleistocene. The scarcity of sites dated to the Lower Pleistocene could reflect a low density of occupation in the early Lower Paleolithic, low level of research, or geological factors that cause reduced preservation and accessibility of sites dated to the Lower Pleistocene (see Tourloukis 2016).

The presence of only three sites (e.g. Dursunlu, Kocabaş, and probably Kaletepe Deresi 3) dated to nearly 1 Ma does not provide evidence for a stable hominin occupation. It is important to recognize that there may have been many unsuccessful hominin attempts to occupy the higher-elevation regions under harsh glacial conditions. Turkey might have been the first place encountered by dispersing African hominins in which they could not survive, given their existing anatomical and cultural adaptations during these first attempts at occupation. Sites with technologies that seem to be earlier than the Acheulian are rare, which could be evidence of unsuccessful attempts at occupations, and as far as many different stone tool-making traditions are evidenced, it is not possible to conclude that the earliest occupation of Anatolia at Dursunlu led to a continuous occupation. Successful colonization of the inland parts of Anatolia might have required several unsuccessful attempts and probably a very long time for early hominins to adapt to the climate and to the environmental conditions of Anatolia.

References

Albrecht, G., & Müller-Beck, H. (1988). The Palaeolithic of Şehremuz near Samsat on the Euphrates River. Summary of the excavation findings and a morphology of the handaxes. *Paléorient, 14*(2), 76–86.

Algaze, G., & Rosenberg, M. (1991). The Tigris-Euphrates archaeological reconnaissance project, 1989. *Araştırma Sonuçları, XIII*, 137–161.

Arsebük, G. (1993). Yarımburgaz, a Lower Paleolithic cave site near Istanbul. In M. Frangipane, H. Hauptman, M. Liverani, P. Matthiac, & M. Mellink (Eds.), *Between the rivers and over the mountains, archaeologica Anatolica et Mesopotamica* (pp. 23–36). Roma: Gruppo Editoriale Internazionale.

Arsebük, G. (1998a). A review of the current status of Pleistocene archaeology in Turkey. In G. Arsebük, M. Mellink, & W. Schirmer (Eds.), *Light on top of the black hill* (pp. 71–76). İstanbul: Ege Yayınları.

Arsebük, G. (1998b). Yarımburgaz Mağarası, Pleistosen arkeolojisi ile ilgili son çalışmalara 1997 gözüyle özet bir bakış. *Turkish Academy of Sciences Journal of Archaeology (TÜBA-Ar), 1*, 9–25.

Arsebük, G., & Özbaşaran, M. (1999). Pleistocene archaeology at the cave of Yarımburgaz in Eastern Thrace/Turkey: Preliminary results. In G. N. Bailey, E. Adam, E. Panagopoulou, C. Perlès, & K. Zachos (Eds.), *The Palaeolithic archaeology of Greece and adjacent areas* (pp. 59–72). Nottingham: The British School at Athens.

Atasayan, M. (1939). 1938 yılında Gaziantep köylerinden Dülük ve civarında bulunan çakmaktaşı aletler üzerine bir not. *Türk Arkeoloji Mecmuası, 19–22*, 314–317.

Avcı, M. (2005). Çeşitlilik ve Endemizm Açısından Türkiye'nin Bitki Örtüsü. *İstanbul Üniversitesi Coğrafya Dergisi, 13*, 27–55.

Aytek, A.I., & Harvati, K. (2016). The human fossil record from Turkey. In K. Harvati, & M. Roksandic (Eds.), *Paleoanthropology of the Balkans and Anatolia: Human evolution and its context evolution and its context* (pp. 79–91). Dordrecht: Springer.

Balkan-Atlı, N., Kuhn, S., Astruc, L., Çakan, G., Dinçer, B., & Kayacan, N. (2008). Göllüdağ 2007 survey. *Anatolia Antiqua, XVI*, 293–312.

Balkan-Atlı, N., Kuhn, S., Astruc, L., Kayacan, N., Dinçer, B., Balcı, S., Erturaç, K., & Grenet, M. (2011). Göllü Dağ survey 2010. *Anatolia Antiqua, XIX*, 259–278.

Balkan-Atlı, N., Kuhn, S., Astruc, L., Kayacan, N., Dinçer, B., & Çakan, G. (2009). Göllü Dağ 2008 survey. *Anatolia Antiqua, XVII*, 301–315.

Balkan-Atlı, N., Kuhn, S., Astruc, L., Kayacan, N., Dinçer, B., & Çakan, G. (2010). Göllüdağ survey 2009. *Anatolia Antiqua, XVIII*, 191–202.

Bar-Yosef, O., & Belfer-Cohen, A. (2001). From Africa to Eurasia—Early dispersals. *Quaternary International, 75*, 19–28.

Baykara, İ., Dinçer, B., Şahin, S., Koç, E., Silibolatlaz-Baykara, D., Özer, İ., Sağır, M. (2016). 2014 yılı Van ili Neojen ve Pleistosen Dönemleri Yüzey Araştırması. *Araştırma Sonuçları Toplantısı* 33/2: 539–552.

Blackwell, B. A. B., Schwarcz, H. P., Farrand, W. R., Lundburg, J., Skinner, A. R., Divjak, M. N., & Blickstein, J. I. B. (2010). Electron spin resonance (ESR) and ^{230}TH/^{234}U dating in the lower chamber at Yarımburgaz, Turkey. In F. C. Howell, G. Arsebük, S. L. Kuhn, M. Özbaşaran, & M. C. Stiner (Eds.), *Culture and biology at the crossroads: The Middle Pleistocene record of Yarımburgaz cave (Thrace, Turkey)* (pp. 51–72). İstanbul: Ege Yayınları.

Bostancı, E. Y. (1952). Gök Irmak vadisinde prehistorya araştırmaları. *Dil Tarih ve Coğrafya Fakültesi Dergisi, X*(2), 137–142.

Bostancı, E. Y. (1962). Researches in south-east Anatolia, the Chellean and Acheulean industry of Dülük and Kartal. *Anadolu/Anatolia, 6*, 111–162.

Bostancı, E. Y. (1971). Antakya; Şenköy'de yeni Orta ve Üst Pleistosen insanına ait kültürler üzerinde bir araştırma. *Antropoloji, 5*, 83–99.

Çambel, H. (1947). Achäologischer Bericht über Anatolien. *Orientalia, 16*, 263–270.

Chaput, E. (1941). *Phrygie: Exploration archéologique, Tome 1: Géologie et géographie*. Paris: de Boccard.

Copeland, L. (2004). The Paleolithic of the Euphrates Valley in Syria. In O. Aurenche, M. Le Meère, & P. Sanlaville (Eds.), *From the river to the sea* (pp. 19–114). Oxford: BAR International Series.

Dinçer, B. (2001). Trakya'da yeni bulunan Paleolitik Çağ buluntu yeri: Yatak. *Bilim ve Ütopya, 81*, 72–77.

Dinçer, B. (2010a). Dülük (Gaziantep) Paleolitik Çağ buluntuları. MA Thesis, Uludağ University.

Dinçer, B. (2010a). Bursa ve çevresi yüzey araştırmaları 2008–2009 tarihöncesi buluntuları. *Arkeoloji ve Sanat, 134*, 1–16.

Dinçer, B. (2014a). The Paleolithic of Kocasu Basin (NW Anatolia). In D. B. Erciyas & E. Sökmen (Eds.), *Regional studies in archaeology symposium proceedings* (pp. 23–50). İstanbul: Ege Yayınları.

Dinçer, B. (2014b). Dağlık Bursa Paleolitiği. In M. Şahin (Ed.), *Bursa ve İlçeleri Arkeolojik Kültür Envanteri-1: Olympos Araştırmaları* (pp. 159–177). Bursa: Bursa Kültür AŞ.

Dinçer, B. (2015). Kureyşler Baraj Gölü Havzasında 2014 Yılı Paleolitik Araştırmaları. In M. Türktüzün, S. Ünan (Eds.), *Kureyşler Barajı Kurtarma Kazıları-* 2014 (191–210). Ankara: Bilgin Kültür Sanat Yayıncılık.

Dinçer, B., & Slimak, L. (2007). The Paleolithic of Turkish Thrace: Synthesis and recent results. *Turkish Academy of Sciences Journal of Archaeology (TÜBA-Ar), 10*, 49–61.

Dinçer, B., & Türkcan, A. U. (2011). Frigya'da ilk insanın izleri: Kuzfındık vadisi Paleolitik dönem bulguları (Eskişehir). *Arkeoloji ve Sanat, 137*, 45–52.

Dinçer, B., Türkcan, A. U., & Erikan, F. (2014). Aizanoi 2012 Yılı Paleolitik Buluntuları. In E. Özer (Ed.), *Aizanoi—1* (pp. 2–8). Ankara: Bilgin Kültür Yayınları.

Efe, T. (1990). 1988 yılında Kütahya, Bilecik, Eskişehir illerinde yapılan yüzey araştırmaları. *Araştırma Sonuçları Toplantısı, VII*, 405–424.

Engin, B., Güven, O., & Köksal, F. (1999). Electron spin resonance age determination of a travertine sample from the southwestern part of Turkey. *Applied Radiation and Isotopes, 51*(6), 689–699.

Erek, M. C. (2011). Kahramanmaraş ili prehistoryası. In H. Taşkıran, M. Kartal, K. Özçelik, M. B. Kösem, & G. Kartal (Eds.), *Işın Yalçınkaya'ya armağan* (pp. 113–126). Ankara: Bilgin Kültür Yayınları.

Gabunia, L., Antón, S. C., Lordkipanidze, D., Vekua, A., Justus, A., & Swisher, C. C., III. (2001). Dmanisi and dispersal. *Evolutionary Anthropology, 10*, 158–170.

Galanidou, N., Athanassas, C., Cole, J., Iliopoulos, G., Katerinopoulos, A., Magganas, A., & McNabb, J. (2016). The Acheulean site at Rodafnidia, Lisvori on Lesbos, Greece: Part I. In K. Harvati, & M. Roksandic (Eds.), *Paleoanthropology of the Balkans and Anatolia: Human evolution and its context* (pp. 119–138). Dordrecht: Springer.

Garrard, A., Conolly, J., McNabb, J., & Moloney, N. (2004). Palaeolithic-Neolithic survey in the Sakçagözü region of the north Levantine rift valley. In O. Aurenche, M. Le Meère, & P. Sanlaville (Eds.), *From the river to the sea* (pp. 145–164). Oxford: BAR International Series.

Güleç, E., Howell, F. C., & White, T. (1999). Dursunlu—A new Lower Pleistocene artifact-bearing locality in southern Anatolia. In H. Ullrich (Ed.), *Hominid evolution: Lifestyles and survival strategies* (pp. 349–364). Gelsenkirchen: Archaea.

Güleç, E., White, T., Kuhn, S., Özer, İ., Sağır, M., Yılmaz, H., & Howell, F. C. (2009). The Lower Pleistocene lithic assemblage from Dursunlu (Konya), central Anatolia, Turkey. *Antiquity, 83*, 11–22.

Gündüzalp, N. (1986). İçdoğu Karadeniz bölgesinden prehistorik buluntular. *Türk Tarih Kongresi, IX*(1), 49–54.

Harmankaya, S. (1997). Türkiye Paleolitik araştırmaları üzerine bir değerlendirme. In S. Harmankaya & O. Tanındı (Eds.), *Türkiye Arkeolojik Yerleşmeleri-I, Paleolitik-Epipaleolitik* (pp. 7–19). İstanbul: Ege Yayınları.

Harmankaya, S., & Tanındı, O. (1997). *Türkiye Arkeolojik Yerleşmeleri-I, Paleolitik-Epipaleolitik*. İstanbul: Ege Yayınları.

Jelinek, A. J. (1980). Collection of Paleolithic materials from valleys on the east side of the Bosporus. In H. Çambel & R. J. Braidwood (Eds.), *Prehistoric research in Southeastern Anatolia* (pp. 309–327). İstanbul: Edebiyat Fakültesi Yayınları.

Kansu, Ş. A. (1963). Note préliminaire sur le Paléolithique de la région egéenne. *Belleten, XXVII*(107), 485–490.

Kappelman, J., Alçiçek, M. C., Kazancı, N., Schultz, M., Özkul, M., & Şen, Ş. (2008). First *Homo erectus* from Turkey and implications for migrations into temperate Eurasia. *American Journal of Physical Anthropology, 135*, 110–116.

Kartal, M. (2005). Yontmataş buluntu toplulukları ışığında Ankara: Neyi biliyoruz? Neyi bilmiyoruz? Yeni değerlendirmeler ve sonuçlar. *Anadolu/Anatolia, 28*, 49–72.

Koenigswald, W., Lindenau, C., & Santel, W. T. (2010). Ecological significance of the small mammal fauna from Yarımburgaz cave (Turkish Thrace). In F. C. Howell, G. Arsebük, S. L. Kuhn, M. Özbaşaran, & M. C. Stiner (Eds.), *Culture and biology at the crossroads: The Middle Pleistocene record of Yarımburgaz cave (Thrace, Turkey)* (pp. 73–92). İstanbul: Ege Yayınları.

Kökten, İ. K. (1947a). Bazı prehistorik istasyonlar hakkında yeni gözlemler. *Dil Tarih ve Coğrafya Fakültesi Dergisi, V*(1–5), 223–236.

Kökten, İ. K. (1947b). 1945 yılında Türk Tarih Kurumu adına yapılan tarihöncesi araştırmaları. *Belleten, XI*(43), 431–472.

Kökten, İ. K. (1951). Kuzeybatı Anadolu'nun tarih öncesi hakkında yeni gözlemler. *Dil Tarih ve Coğrafya Fakültesi Dergisi, IX*(3), 201–2014.

Kökten, İ. K. (1952). Anadolu'da prehistorik yerleşme yerlerinin dağılışı üzerine bir araştırma. *Dil Tarih ve Coğrafya Fakültesi Dergisi, X*(3–4), 167–207.

Kökten, İ. K. (1963). Anadolu Ünye'de Eskitaş devrine (Paleolitik) ait yeni buluntular. *Dil Tarih ve Coğrafya Fakültesi Dergisi, XX*(3–4), 277–278.

Kökten, İ. K. (1974). *Prehistoric research in the Keban dam lake area; 1971* (Keban Projesi 1971 Çalışmaları, pp. 7–11). Ankara: ODTÜ Keban Projesi Yayınları.

Koyabashi, K., & Sagona, A. (2008). A survey of obsidian sources in the provinces of Erzurum, Erzincan, Rize and Bitlis, 2006. *Araştırma Sonuçları Toplantısı, XXV*(2), 185–196.

Kuhn, S. L. (2002). Paleolithic archaeology in Turkey. *Evolutionary Anthropology, 11*, 198–210.

Kuhn, S. L. (2003). Flexibility and variation in the Lower Paleolithic: A view from Yarımburgaz cave. In M. Özbaşaran, O. Tanındı, & A. Boratav (Eds.), *Archaeological essays in honour of Homo amatus: Güven Arsebük* (pp. 149–157). İstanbul: Ege Yayınları.

Kuhn, S. L. (2010a). Was Anatolia a bridge or barrier to early hominin dispersals? *Quaternary International, 223–224*, 434–435.

Kuhn, S. L. (2010b). The Yarımburgaz lithic assemblage. In F. C. Howell, G. Arsebük, S. L. Kuhn, M. Özbaşaran, & M. C. Stiner (Eds.), *Culture and biology at the crossroads: The Middle Pleistocene record of Yarımburgaz cave (Thrace, Turkey)* (pp. 93–129). İstanbul: Ege Yayınları.

Kuhn, S. L., Arsebük, G., & Howell, F. C. (1996). The Middle Pleistocene lithic assemblage from Yarımburgaz cave, Turkey. *Paléorient, 22*(1), 31–49.

Kuhn, S., Balkan-Atlı, N., & Dinçer, B. (2009). 2008 excavations at Kaletepe Deresi 3. *Anatolia Antiqua, XVII*, 291–299.

Kuhn, S. L., Dinçer, B., Balkan-Atlı, N., & Erturaç, K. (2015). Paleolithic occupations of the Göllü Dağ, Central Anatolia, Turkey. *Journal of Field Archaeology, 40*(5), 581–602.

Kuhn, S. L., Stiner, M. C., & Howell, F. C. (1998). Middle Pleistocene hominids and bears at Yarımburgaz cave (Thrace, Turkey). In M. Otte (Ed.), *Préhistoire d'Anatolie, genèse de deux mondes* (pp. 579–598). Liège: Eraul.

Lebatard, A.-E., Alçiçek, M. Ç., Rochette, P., Khatib, S., Vialet, A., Boulbes, N., Bourlés, D. L., Demory, F., Guipert, G., Mayda, S., Titov, V. V., Vidal, L., & de Lumley, H. (2014). Dating the *Homo erectus* bearing travertine from Kocabas (Denizli, Turkey) at least 1.1 Ma. *Earth and Planetary Science Letters, 390*, 8–18.

Maddy, D., Schreve, D., Demir, T., Veldkamp, A., Wijbrans, J. R., van Gorp, W., van Hinsbergen, D. J. J., Dekkers, M. J., Scaife, R., Schoorl, J. M., Stemerdink, C., & van der Schriek, T. (2015). The earliest securely-dated hominin artefact in Anatolia? *Quaternary Science Reviews, 109*, 68–75.

Marro, C., & Özfırat, A. (2004). Pre-classical survey in Eastern Turkey. Second preliminary report: The Erciş region. *Anatolia Antiqua, XII*, 227–265.

Mgeladze, A., Lordkipanidze, D., Moncel, M.-H., Despriee, J., Chagelishvili, R., Nioradze, M., & Nioradze, G. (2011). Hominin occupations at the Dmanisi site, Georgia, Southern Caucasus: Raw materials and technical behaviours of Europe's first hominins. *Journal of Human Evolution, 60*, 571–596.

Minzoni-Déroche, A. (Ed.). (1987). *Le Paléolithique du bassin du Nizip (Gaziantep)*. Istanbul: Institut français d'études anatoliennes.

Minzoni-Déroche, A., & Sanlaville, P. (1988). Le Paléolithique inférieur de la région de Gaziantep. *Paléorient, 14*(2), 87–98.

Mouralis, D. (2003). Les complexes volcaniques quaternaires sur les paysages de Cappadoce (Göllüdağ et Acıgöl-Turquie): Évolutions morphodinamiques et implications environnementales. Ph.D. Dissertation, Université Paris.

Otte, M., Yalçınkaya, I., Bar-Yosef, O., Kozlowski, J. K., Léotard, J.-M., Taşkıran, H., Noiret, P., & Kartal, M. (1999). The Anatolian Palaeolithic: Data and reflections. In G. N. Bailey, E. Adam, E. Panagopoulou, C. Perlès, & K. Zachos (Eds.), *The Palaeolithic archaeology of Greece and adjacent areas* (pp. 73–85). Nottingham: The British School at Athens.

Otte, M., Yalçınkaya, I., Kozlowski, J., Bar-Yosef, O., Bayón, I. L., & Taşkıran, H. (1998a). Long-term technical evolution and human remains in the Anatolian Palaeolithic. *Journal of Human Evolution, 34*, 413–431.

Otte, M., Yalçınkaya, I., Kozlowski, J., Bar-Yosef, O., Bayón, I. L., & Taşkıran, H. (1998b). *Évolution technique á long terme dans le Paléolithique anatolien* (Préhistoire d'Anatolie, Genèse de deux mondes, pp. 463–477). Liège: ERAUL.

Özçelik, K. (2003). 1998 Yılı Hatay yüzey araştırması buluntularının tekno-tipolojik değerlendirilmesi. *Anadolu/Anatolia, 24*, 79–92.

Özdoğan, M. (1998). Anatolia from the last glacial maximum to the Holocene climatic optimum: Cultural formations and the impact of the environmental setting. *Paléorient, 23*(2), 25–38.

Roberts, C. A., Pfister, L.-A., & Mays, S. (2009). Letter to the editor: Was tuberculosis present in *Homo erectus* in Turkey? *American Journal of Physical Anthropology, 139*, 442–444.

Roosevelt, C. H., & Luke, C. (2010). Central Lydia archaeological survey: 2008 results. *Araştırma Sonuçları Toplantısı, XXVII*(2), 1–24.

Runnels, C. (2003). The Lower Paleolithic of Greece and NW Turkey. In M. Özbaşaran, O. Tanındı, & A. Boratav (Eds.), *Archaeological essays in honour of Homo amatus: Güven Arsebük* (pp. 195–202). İstanbul: Ege Yayınları.

Runnels, C., & Özdoğan, M. (2001). The Palaeolithic of the Bosphorus region, NW Turkey. *Journal of Field Archaeology, 28*(1–2), 69–92.

Sanlaville, P. (2004). Les terrasses Pléistocènes de la vallée de l'Ephrate en Syrie et dans l'extrême sud de la Turquie. In O. Aurenche, M. Le Meère, & P. Sanlaville (Eds.), *From the river to the sea* (pp. 115–133). Oxford: BAR International Series.

Santel, W. T., & Koenigswald, W. (1998). Preliminary report on the Middle Pleistocene small mammal fauna from Yarımburgaz cave in Turkish Thrace. *Eiszeitalter und Gegenwart, 48*, 162–169.

Şenyürek, M. (1960). Test excavations made in a cave in the vicinity of Samandağ in 1958, preliminary report. *Anadolu/Anatolia, 3*, 64–70.

Şenyürek, M. (1961). Altındere'nin üst Acheuléen endüstrisi. *Belleten, XXV*(98), 149–185.

Şenyürek, M., & Bostancı, E. (1958). The Palaeolithic cultures of the Hatay province. *Belleten, XXII*(86), 171–209.

Slimak, L. (2004). Implantations humaines et exploitation des obsidiennes en Anatolie centrale durant le Pléistocène. *Paléorient, 30*(2), 7–20.

Slimak, L., Balkan-Atlı, N., Binder, B., & Dinçer, B. (2005). Installations paléolithiques en Cappadoce. État des connaissances de cinq années de recherche sur les premiers peuplements humains en Anatolie centrale. *Anatolia Antiqua, XIII*, 287–294.

Slimak, L., & Dinçer, B. (2007). Kaletepe Deresi 3, Orta Anadolu'da tabakalanma veren bir ilk Paleolitik Çağ yerleşmesi. *Turkish Academy of Sciences Journal of Archaeology (TÜBA-Ar), 10*, 33–47.

Slimak, L., Kuhn, S., Balkan-Atlı, N., Binder, B., Grenet, M., & Dinçer, B. (2007). Kaletepe Deresi 3: de l'Acheulléen au Moustérien en Anatolie centrale. *Anatolia Antiqua, XV*, 257–273.

Slimak, L., Kuhn, S. L., Roche, H., Mouralis, D., Buitenhuis, H., Balkan-Atlı, N., Binder, D., Kuzucuoğlu, C., & Guillou, H. (2008). Kaletepe Deresi 3 (Turkey): Archaeological evidence for early human settlement in Central Anatolia. *Journal of Human Evolution, 54*, 99–111.

Stiner, M., Achyuthan, H., Arsebük, G., Howell, F. C., Josephson, S. C., Juell, K. E., Pigati, J., & Quade, J. (1998). Reconstructing cave bear paleoecology from skeletons: A cross-disciplinary study of Middle Pleistocene bears from Yarımburgaz cave, Turkey. *Paleobiology, 24*(1), 74–98.

Stiner, M. C., Arsebük, G., & Howell, F. C. (1996). Cave bears and Paleolithic artifacts in Yarımburgaz cave, Turkey: Dissecting a palimpsest. *Geoarchaeology: An International Journal, 11*(4), 279–327.

Taşkıran, H. (1998). *The distribution of bifaces in Anatolia* (Préhistoire d'Anatolie, Genèse de deux mondes, pp. 569–577). Liège: ERAUL.

Taşkıran, H. (2002a). Karkamış baraj gölü alanında yapılan Paleolitik Çağ yüzey araştırması üzerine genel bir değerlendirme. *İdol, 13*, 8–10.

Taşkıran, H. (2002b). The Palaeolithic survey in the Carchemish dam reservoir region: 2000 season. In N. Tuna & J. Velibeyoğlu (Eds.), *Ilısu ve Karkamış Baraj Gölleri Altında Kalacak Arkeolojik Kültür Varlıklarını Kurtarma Projesi 2000 Yılı Çalışmaları* (pp. 413–429). Ankara: ODTÜ TAÇDAM Yayınları.

Taşkıran, H. (2008). Réflexions sur l'Acheuléen d'Anatolie. *L'Anthropologie, 112*, 140–158.

Taşkıran, H., & Kartal, M. (1999). Palaeolithic survey in the Carchemish dam reservoir region: Preliminary observations. In N. Tuna & J. Öztürk (Eds.), *Ilısu ve Karkamış Baraj Gölleri Altında Kalacak Akeolojik Kültür Varlıklarını Kurtarma Projesi 1998 Yılı Çalışmaları* (pp. 57–63). Ankara: ODTÜ-TAÇDAM Yayınları.

Taşkıran, H., & Kartal, M. (2001). Palaeolithic survey in the Carchemish dam reservoir region: 1999 Season. In N. Tuna, J. Öztürk, & J. Velibeyoğlu (Eds.), *Ilısu ve Karkamış Baraj Gölleri Altında Kalacak Arkeolojik Kültür Varlıklarını Kurtarma Projesi 1999 Yılı Çalışmaları* (pp. 515–528). Ankara: ODTÜ TAÇDAM Yayınları.

Taşkıran, H., & Kartal, M. (2004). 2001 Palaeolithic survey in the Ilısu dam region. In N. Tuna, J. Greenhalgh, & J. Velibeyoğlu (Eds.), *Ilısu ve Karkamış Baraj Gölleri Altında Kalacak Arkeolojik Kültür Varlıklarını Kurtarma Projesi 2001 Yılı Çalışmaları* (pp. 725–737). Ankara: ODTÜ TAÇDAM Yayınları.

Taşkıran, H., & Taşkıran, Z. F. (2011). İki yüzeyli aletlerin Anadolu'daki dağılımında yeni bir nokta: Afyonkarahisar. In H. Taşkıran, M. Kartal, K. Özçelik, M. B. Kösem, & G. Kartal (Eds.), *Işın Yalçınkaya'ya Armağan* (pp. 235–244). Ankara: Bilgin Kültür Yayınları.

Tomsky, J. (1982). *Das Altpalaeolithikum in vorderen Orient*. Wiesbaden: Dr. Ludwig Reichert.

Tourloukis, V. (2016). On the spatio-temporal distribution of Mediterranean Lower Paleolithic sites: A geoarchaeological perspective. In K. Harvati, & M. Roksandic (Eds.), *Paleoanthropology of the Balkans and Anatolia: Human evolution and its context* (pp. 303–323). Dordrecht: Springer.

Tryon, C. A., Amelia, M., Logan, V., Mouralis, D., Kuhn, S., Slimak, L., & Balkan-Atlı, N. (2009). Building a tephrostratigraphic framework for the Paleolithic of Central Anatolia, Turkey. *Journal of Archaeological Science, 36*, 637–652.

Yalçınkaya, I. (1981). Le Paléolithique inférieur de Turquie. In J. Cauvin & P. Sanlaville (Eds.), *Préhistorie du Levant* (pp. 207–218). Paris: Editions du CNRS.

Yalçınkaya, I. (1985). Araştırmaların ışığında Anadolu Alt Paleolitiği ve sorunlarına genel bir bakış. *Antropoloji, 12*, 395–435.

Yalçınkaya, I. (1986). Güneydoğu Anadolu'da yapılan Paleolitik Çağ araştırmaları üzerinde genel bir değerlendirme. *Türk Tarih Kongresi, X*(1), 35–42.

Yalçınkaya, I. (1992). 1990 yılı Karain Kazıları. *Kazı Sonuçları Toplantısı, XIII*(1), 33–54.

Yalçınkaya, I., Özçelik, K., Kartal, M., & Taşkıran, H. (2009a). Diffusion des cultures à bifaces en Turquie. *Anadolu/Anatolia, 35*, 1–38.

Yalçınkaya, I., Taşkıran, H., Kartal, M., Özçelik, K., Kösem, M. B., & Kartal, G. (2007). 2005 yılı Karain Mağarası kazıları. *Kazı Sonuçları Toplantısı, XXVIII*(1), 539–558.

Yalçınkaya, I., Taşkıran, H., Kartal, M., Özçelik, K., Kösem, M. B., & Kartal, G. (2008). 2006 Yılı Karain Mağarası kazıları. *Kazı Sonuçları Toplantısı, XXIX*(1), 467–482.

Yalçınkaya, I., Taşkıran, H., Kartal, M., Özçelik, K., Kösem, M. B., & Kartal, G. (2009b). 2007 yılı Karain Mağarası kazıları. *Kazı Sonuçları Toplantısı, XXX*(2), 285–300.

Chapter 14
Technological Variability of the Middle-to-Upper Paleolithic Transition: Examples from the Balkans and Neighbouring Regions

Valéry Sitlivy

Abstract This chapter attempts to track the technological variability, innovations, and changes in lithic technologies that occurred in several regions of the Balkans and its neighbouring areas during the late Middle and Early Upper Paleolithic (EUP). A diversity of knapping methods and techniques has been recorded by analysing debitage, reduction sequences, and refitting data of selected key assemblages. The Middle Paleolithic (MP) technological structure appears to include a series of independent methods of blank production: Levallois, non-Levallois, and the volumetric laminar production. Similarly, the EUP technological structure also follows several models: (a) exclusive/dominant Upper Paleolithic (UP) laminar strategies; (b) a combination of UP strategies with different MP technologies; and (c) a fusion in one reduction sequence of volumetric blade and "flat" Levallois point concepts. Originally identified by refitting, the latter innovative model is central to understanding the marked technological transition and the emergence of the UP technological package.

Keywords Mousterian • Early Upper Paleolithic • Lithic Technology • Reduction sequences • Refits

Introduction

The origin of anatomically modern humans in Europe and their relationship with Neandertals is one of the key topics in Paleolithic archaeology. Archaeologists have long focused on changes in the material record across the Middle/Upper Paleolithic boundary, and the end of the Middle Paleolithic (MP), the transition, and the onset of the Upper Paleolithic (UP) have been extensively debated in Western Europe and the Near East. With few exceptions, the rich archaeological record from other parts of the Old World has rarely been incorporated into an overall picture. Independent of geographic area, both continuity and discontinuity in technological development during the MP to UP transition have been described in the literature stemming from substantial research conducted in recent decades within well-controlled contexts throughout Eurasia, especially concerning the period just prior to the onset of the UP. Recently, published data suggest that the combination of patterned behaviors is even more complex than commonly depicted (e.g. Brantingham et al. 2004; Derevianko 2005; Camps and Szmidt 2009), showing that a universal model of the "Big Transition" does not always fit regional patterns, which do not follow a unique scenario, but rather support the complexity of changes that—depending on local conditions—led to the establishment of the UP following several different pathways. A number of hypotheses have been introduced concerning the origins of anatomically modern humans, modern behavior, and the expansion of populations and/or technological innovations. In some instances, it is suggested that laminar and Levallois strategies influenced or developed directly into Initial Upper Paleolithic (IUP) technologies, documenting a local transformation/transition. In others, cultural diffusion from adjacent or distant regions is purported. Central to understanding this process are Levallois-based industries, very often equipped with innovative volumetric blade production technologies both in support of local continuity (e.g. Marks and Volkman 1983; Meignen and Bar-Yosef 2002; Derevianko 2010), discontinuity (e.g. Chabai 2004), or diffusion of ideas or populations (e.g. Škrdla 1996; Svoboda 2007; Bar-Yosef 2006).

This chapter focuses on technological variability of lithics in order to track the changes, innovations, and trends that occurred during the transition from the Middle to Upper Paleolithic in several key regions of the Balkans and its neighbouring regions of Central and Southeast Europe. The analysis is restricted to core reduction and blank production of Late Middle Paleolithic ("non-Micoquian"), Initial ("transitional"),

V. Sitlivy (✉)
Rue de Berghes 6/41, 4020 Liège, Belgium
e-mail: vsitlivy@voo.be

Katerina Harvati and Mirjana Roksandic (eds.), *Paleoanthropology of the Balkans and Anatolia*,
Vertebrate Paleobiology and Paleoanthropology, DOI 10.1007/978-94-024-0874-4_14

and Early Upper Paleolithic industries from secure archaeological contexts with long cultural sequences. In order to properly reconstruct reduction sequences, special attention is given to refitting, a fundamental process that provides undisputable information on technology. While proposed technological variability is restricted to qualitative analysis, extensive, already published quantitative data stands behind notions as "exclusive", "dominant", "prevailed", "rare", or "absent". Finally, this proposed synthesis is based on lithic assemblages excavated and studied by the author in Eastern, Central, and Southeast Europe (Ukraine, Poland, Romania, and Greece) as is indicated by references and Tables 14.1, 14.2, and 14.3. Data from Czech, Bulgarian, Moldavian, Russian, and Greek sites (excluding Klissoura Cave 1) are taken from publications.

Transition or Mechanical Mixing?

The progress of several new field projects in Southeast, Central, and Eastern Europe, innovative technological studies, refitting and attribute analyses, as well as recent programmes of radiometric dating have allowed for a more precise evaluation of the variability encountered during this period in different geographic areas (e.g. Svoboda and Bar-Yosef 2003; Tostevin and Škrdla 2006; Chabai et al. 2004, 2006; Sitlivy et al. 2008a, b, 2009a, b). The transitional period in these regions shows significant variability and the co-occurrence, overlapping, and persistence of different "cultural" entities (Fig. 14.1), knapping methods, and debitage/detachment techniques (e.g. Chabai et al. 2004; Kozłowski 2000; Sitlivy and Zięba 2006). Recent re-examination of seemingly "certain" classical industries have resulted in their dramatic reconsideration (Rigaud and Lucas 2006; Tsanova 2006; Teyssandier 2003), supporting the need for more detailed studies on both old and new lithic assemblages. Additionally, the uncritical use of rich and sometimes "attractive" lithic assemblages lacking a proper taphonomical understanding and secure stratigraphic context obstructs or complicates the interpretation of change.

While the possibility of finding different *fossiles directeurs* in the same occupation layer due to natural (e.g. frost) or anthropic (e.g. trampling) post-depositional processes is frequently cited (e.g. Hoffecker 2009), they are not always taken into consideration during assemblage analysis and are often ignored in final interpretations. Problematic stratigraphic integrity and doubtful radiocarbon sampling are examples that feed scepticism regarding the supposed Late Mousterian in Southern Carpathian caves (Anghelinu et al. 2012), while "transitional" industries deserve similar criticism, as they typically display stratigraphic mixing (e.g. Mitoc-Valea Izvorului—Tuffreau et al. 2009; Anghelinu et al. 2012 for the Romanian record or Theopetra, layer II4 in Greece—Valladas et al. 2007). The frequently cited candidate for the Bohunician/Emiran—layer III-d at Kulichivka, Western Ukraine, with a single tentative date of *ca.* 31 ^{14}C kBP (Savich 1975) raises doubts on the site's homogeneity (i.e. a possible mixing with the uppermost UP). Recent studies, based on the undisturbed part of Kulichivka, indicate Moustero-Levallois features (for details, see Meignen et al. 2004; Sitlivy and Zięba 2006) contradicting previous claims (Stepanchuk and Cohen 2000–2001). In comparison with the Levantine site Boker Tachtit (reference assemblage for this peculiar IUP innovative technology), MP features in level 1 at Kulichivka are significant: more flat cores, short blanks, and Mousterian retouched tools (Meignen et al. 2004), while, on the other hand, the industry of the lowermost layer was previously described as having the "UP bidirectional (pointed) Levallois blade producing technique" (Demidenko and Usik 1993; for details, see Usik et al. 2006b). In sum, the divergence in interpretations is related to the "clarity" of the samples studied and to the degree of contamination from the overlaying UP industry. Mechanical mixing of both (not necessarily late) Middle and (not necessarily early) Upper Paleolithic artifacts, producing various "transitional industries" was and still remains a real danger for the correct assessment of the Middle-to-Upper Paleolithic transition.

It has recently become evident that MP technological structure may include independent non-Levallois laminar technologies, which were used in parallel with other "flat-core" strategies, including Levallois. This is not the case for the Emiran or Bohunician, where UP blade and Levallois methods were mixed in one long reduction sequence (Marks and Volkman 1983; Škrdla 1996), clearly showing the technological transformation (transition). Confirmed through refitting analysis, this unique innovative technological trend became a good marker illustrating a shift to the UP model of raw material exploitation and persistence of "archaic" knapping habits (Sitlivy et al. 2014c). Unfortunately, numerous other candidates for the "new behavior" which lack refitting data, support an alternative interpretation within the context of the "archaic behavior", that is, either a non-fusion of UP laminar and Levallois approaches, which maintained their independent technological status, and/or a mechanical mixture of artifacts representing different technologies and archaeological periods (e.g. Mitoc-Valea Izvorului, Kulichivka, etc., see below). These alternatives emphasise yet again the importance of analysing suitable lithic assemblages for an accurate reconstruction of reduction sequences. The problem may be resolved by conjoins of various debitage products together in a single block or separately into several refits, including MP (e.g. Levallois, discoidal cores, and corresponding blanks with fine butt faceting and developed bulbs) and UP (e.g. prismatic cores and corresponding

Table 14.1 Central Europe. Late middle and early upper Paleolithic blank production

Site	Raw materials	Blade methods						Other methods	Tool-kit	Cultural and chronological attribution
		Direct	Prepared	Working surface(s), localization	Direction of debitage	Platform(s)	Technique			
Piekary IIa, layer 7c	Medium-sized nodules of good quality local flint	Rare: by cortical and *débordant* blades	Two-sloped central crests, secondary, lateral neo-crests, bifacial shaping	Narrow with extension to sides, partially turned	Bidirectional and unidirectional	Facetted and plain, tablets, some abrasion	Hard stone percussion	Lineal Levallois for flakes and discoidal	Sidescrapers, notches, retouched blades and flakes, truncated facetted pieces, backed blades, burin	Blade Levallois-Mousterian, OIS 3, 61–48 kBP
Piekary IIa, layer 7b	*Idem*	By cortical and *débordant* blades	Lateral crests	Partially turned (from narrow to wide side)?	Unidirectional	Plain and facetted, abrasion	Hard stone percussion	Levallois (recurrent centripetal for flakes and lineal convergent for points), discoidal, Kombewa	*Idem* + denticulates, backed knifes, retouched flake and blade	Blade Levallois-Mousterian, 39 kBP
Piekary IIa, layer 7a	*Idem*	By cortical and *débordant* blades, bidirectional debitage, changes in orientation	Lateral and central crests, bifacial pre-form, neo-crests	Partially turned (from narrow to wide side; from wide to narrow surfaces)	Bidirectional prevailed	Plain, several scars, tablets, abrasion	Soft and hard hammer percussion	Recurrent centripetal Levallois (decreasing of Levallois debitage), discoidal	Sidescrapers > notches, retouched flakes, splintered pieces, *raclette*, rare tools on blades	Blade Levallois-Mousterian, 46–33 kBP
Piekary IIa, layer 6	Medium-sized nodules of good quality local flint		1. Lateral and central crests (prevailing of one-sloped)	1. Turned (from the narrow opposing flaking surfaces to the large opposing sides) and partially turned (from the narrow flaking surface to the large sides)	1. Bidirectional	Plain, several scars, tablets, abrasion	Soft and hard hammer percussion		Retouched blades and flakes prevail over burins, endscrapers, rare sidescrapers	EUP, 32–26 kBP
			2. Several crest positions during reduction, neo-crests	2. Partially turned (from wide to narrow side; from narrow to wide side)	2.Unidirectional					
			3. By cortical and *débordant* blades	3. Turned	3. Unidirectional					

(continued)

Table 14.1 (continued)

Site	Raw materials	Blade methods						Other methods	Tool-kit	Cultural and chronological attribution
		Direct	Prepared	Working surface(s), localization	Direction of debitage	Platform(s)	Technique			
Sitlivy et al. (2008b) and Valladas et al. (2008)										
Księcia Józefa, layer III	Large and medium-sized nodules of good and mediocre quality local flint	1. On nodules by means of naturally backed and plunging blades, distal trimming		1. Partially turned (from narrow to wide side and inversely)	1. Unidirectional	1. Plain, rejuvenation by several scars	Hard stone percussion	Polyhedral (flakes/blades), discoidal, Kombewa, unidirectional (flakes), Levallois for points	Sidescrapers, notched, denticulates, naturally backed knives, retouched flakes and blades, *raclettes*, rare endscrapers	Mousterian, ~44 kBP
		2. On nodule by means of naturally backed blades		2. Partially turned (from wide to narrow side and inversely) + convergent for point at the end of reduction	2. Bidirectional with isolated surfaces	2. Dihedral				
		3. On flake		3. Partially turned (from narrow to large dorsal surface and back)	3. Unidirectional	3. Plain (use of flake butt for core reduction)				
		4. On nodule with central two-sloped crest		4. Partially turned (from narrow to wide surface) + elongated point and bidirectional debitage	4.Bidirectional	4. Facetted; *chapeau de gendarme* platform for point removal				

Księcia Józefa, layer II	*Idem*		1. On nodules, frontal crests (one-sloped, partial) on narrow part, lateral and postero-lateral crests	1. Partially turned: narrow via wide side (s), wide with extension to narrow part (s)	1. Bidirectional	1. Prepared by several scars, facetted, partial tablets, abrasion	1. Hard stone percussion	Bladelet (core on flake/burin-like)	Non-specific isolated types: sidescraper, retouched flake and blade, *raclette*, notch, splintered piece and borer	EUP, ~40 kBP
			2. On flakes, lateral crest, neo-crest	2. Partially turned (from narrow to wide dorsal flake face)	2. Bidirectional	2. *Idem*	2. Hard and soft stone percussion			
			3. On nodules, frontal crests (complete or partial two-sloped), postero-lateral	3. Partially turned (narrow with slight extension to sides; from one ridge to another and back)	3. Unidirectional	3. Plain, abrasion, partial tablet	3. Hard stone percussion			
Sitlivy et al. (2004, 2009a, b, 2014c)										
Korolevo II, layer II	Mostly local large andesite blocks	Rare (mostly on pebbles)	Bifacial shaping (narrow wedge-shaped bifacial pre-core with central crest), one- and two-sloped crests, neo-crests	Narrow part was reduced to flat cores with wide surface, narrow wedge-like surface, partially turned, rare turned	Bidirectional, unidirectional (mostly exhausted)	Plain and facetted before blade removal, abrasion	Mostly hard stone percussion		UP (endscrapers, retouched blades) prevail over MP (scrapers etc.) + bifacial leaf-points	EUP, >38.5 kBP

(continued)

Table 14.1 (continued)

Site	Raw materials	Blade methods						Other methods	Tool-kit	Cultural and chronological attribution
		Direct	Prepared	Working surface(s), localization	Direction of debitage	Platform(s)	Technique			
Korolevo I, layer Ia	*Idem*		Lateral (one-sloped) crest on the ridge between dorsal and ventral flake surfaces, rare bifacial shaping	Wide with slight extension to sides, partially turned, rare turned, Rocourt type (partially turned core with two opposed platforms)	Unidirectional; bidirectional (mostly exhausted)	Plain, several removals, dihedral, tablets	Soft and hard stone percussion	Flake production based on core on flake (transversal reduction of flake slice; ventral face is used as a platform) + trifacial cores	UP (endscrapers, rare burins, retouched blades) prevail over MP (retouched flakes, sidescrapers etc.)	EUP, >25 kBP, 39–37.5 kBP
Usik (1989), Gladilin and Demidenko (1989), Usik et al. (2006a, b) and Sitlivy and Zieba (2006)										
Sokyrnitsa IA, layer 3	Quartzite pebbles	On initial stage of reduction	Mostly on full debitage stage (lateral crest)	Partially turned and narrow (often on massive flakes)	Unidirectional and bidirectional	Plain, rare by several scars	Hard stone percussion		Numerous burins, retouched blades and flakes, some sidescrapers	EUP, 38.8 kBP
Usik (2003) and Sitlivy and Zięba (2006)										

Table 14.2 Eastern Europe. Late middle and early upper Paleolithic blank production

Site	Raw materials	Blade methods: Direct	Prepared	Working surface(s), localization	Direction of debitage	Platform(s)	Technique	Other methods	Tool-kit	Cultural and chronological attribution
Kourdiumovka	Local flint (voluminous cylindrical nodules)	Predominant reduction of naturally voluminous cylindrical nodules	Rare crested flakes (mostly lateral)	Partially turned (from narrow to wide surface; *idem* + second narrow part)	Unidirectional (opposed removals aimed to re-construct flaking surface)	Plain, tablets	Hard stone percussion	Discoidal, centripetal and uni-/bidirectional	Mousterian points on blades, sidescrapers, denticulates, naturally backed knives, truncated facetted pieces	Blade Mousterian, OIS 4, early Glacial
Kolesnik (2000, 2003) and Sitlivy and Zięba (2006)										
Shlyakh, layer 8c	Local flint of mediocre quality	+	Crests: on narrow side, distal part (narrow wedge-shaped cores)	Narrow surface with extension to wide side; successive exploitation of two surfaces (from narrow to wide)	Unidirectional	Plain, elimination of overhang	Hard stone percussion	Recurrent uni-/bidirectional Levallois and rare centripetal, discoidal	Mousterian points, truncated facetted, sidescrapers, backed knives. Rare atypical burins and endscrapers on flakes and blades	Blade Levallois-Mousterian, ~40–42 kBP
Nekhoroshev (2009), Vishnyatsky and Nekhoroshev (2004) and Sitlivy and Zięba (2006)										
Kabazi II, II/A-II/7	High quality local flint (*plaquettes*)	Rare: maintained by cortically backed blades/flakes	Crests: lateral, central (one and two-sloped), neo-crests	Partially turned (wide surface with extension to sides; narrow and wedge-like)	Bidirectional and unidirectional	Facetted, plain, tablets, elimination of overhang	Hard and soft stone percussion	Levallois (lineal for flakes; recurrent mostly bidirectional for blades/flakes), centripetal non-Levallois	Sidescrapers (mostly simple), Mousterian points, some denticulates and notches. UP tools: rare (endscraper, borer) or absent	Blade Levallois-Mousterian, ~40–30 kBP
Kabazi II, II6-A3A	High quality local flint (*plaquettes*)	Rare	*Idem* + bifacial pre-core on flake	Rartially turned (from narrow to wide surface)	Bidirectional and unidirectional	Facetted, Plain, tablets	Hard and soft stone percussion	Recurrent bidirectional Levallois for blades/flakes	*Idem* + MP tools on blades	

Chabai (2004, 2013), Chabai et al. (2004, 2006) and Sitlivy and Zięba (2006)

Table 14.3 Southeast Europe. Late middle and early upper Paleolithic blank production

Site	Raw materials	Blade and bladelet methods						Other methods	Tool-kit	Cultural and chronological attribution
		Direct	Prepared	Working surface(s), localization	Direction of debitage	Platform(s)	Technique			
Klissoura, lowermost layers XVIII–XX	Mostly small-sized radiolarite and flint (chunks, *plaquettes* and pebbles of different quality)	By cortical and *débordant* blades	Rare two-sloped central and one-sloped lateral crests	Narrow-faced and partly turned (from narrow to wide side?)	Unidirectional, rare bidirectional	Plain, rarely facetted	Hard stone percussion	Flake uni-/multidirect, centripetal, discoidal and Levallois (lineal, recurrent for flakes, rarely points and blades)	Dominance of the Mousterian tools: lateral scrapers, high frequency of multiple scrapers and convergent tools, including points	Mousterian, >60–62 14C kBP
Klissoura, uppermost layers VI–VIII	*Idem*	By cortical and *débordant* blades	Rare crests	Narrow-faced and partly turned	Unidirectional, rare bidirectional	Plain, rarely facetted	Mostly hard stone percussion	*Idem* + bipolar (*pièces esquillées*)	*Idem* + considerable quantity of *pièces esquillées*	Mousterian, ~41–50 kBP
Klissoura, layer V	*Idem*	By cortical and *débordant* blades	Often two-sloped central, lateral, secondary crests	Narrow-faced and partly turned	Unidirectional, bidirectional, multidirectional	Plain, rarely facetted, elimination of overhang by abrasion	Direct soft and hard stone percussion	Flake uni-bidirectional, multidirectional, discoidal and flake/bladelet bipolar (*pièces esquillées*)	Rich in curved backed pieces (including lunates on blade/lets) and splintered pieces; sidescrapers, retouched flakes; rare endscrapers, isolated burin	Uluzzian, 40,010 ± 740 14C BP, overlaying CI tephra
Klissoura, layer IV	*Idem* + more medium-sized chunks	1. By bladelet debitage 2. By blade/let debitage	Rare crests	1. Carinated 2. Wide and narrow-faced, partly turned, rarely turned	Unidirectional, rarely bidirectional	Mostly plain	Mostly direct soft stone percussion	Flake uni-multidirectional, discoidal and flake/bladelet bipolar (*pièces esquillées*)	Domination of endscrapers (especially carinated) over retouched blades/flakes, burins, denticulated/notched pieces, sidescrapers	Aurignacian, ~34–32 14C kBP

Sitlivy et al. (2007, 2008a), Koumouzelis et al. (2001), Kaczanowska et al. (2010), Kuhn et al. (2010) and Lowe et al. (2012)										
Româneşti-Dumbrăviţa I, GH3	Mostly medium-sized opal cobbles, nodules, pebbles variable in quality (generally mediocre)	1. By bladelet debitage 2. By bladelet debitage (burin-like (core-on-flakes) 3. By natural "crests"	Mostly lateral crests, neo-crests	1.Carinated (wide-/narrow faced) 2. Narrow-faced 3. Narrow and wide-faced, partly turned	Commonly unidirectional (but also bi-/ multidirectional)	Plain, rarely crudely prepared, platform edge abrasion and trimming by short removals, tablets	Direct soft stone percussion during production phase	Occasional discoidal, Kombewa (recycling)	Rich in non-geometric microliths (mostly Dufours), less frequent burins and retouched blades, rare simple endscrapers, truncations, sidescrapers, notches etc.	Proto-Aurignacian/ Early Aurignacian, OSL and TL ages of ~40–45 kBP

Sitlivy et al. (2012, 2014a, b) and Schmidt et al. (2013)

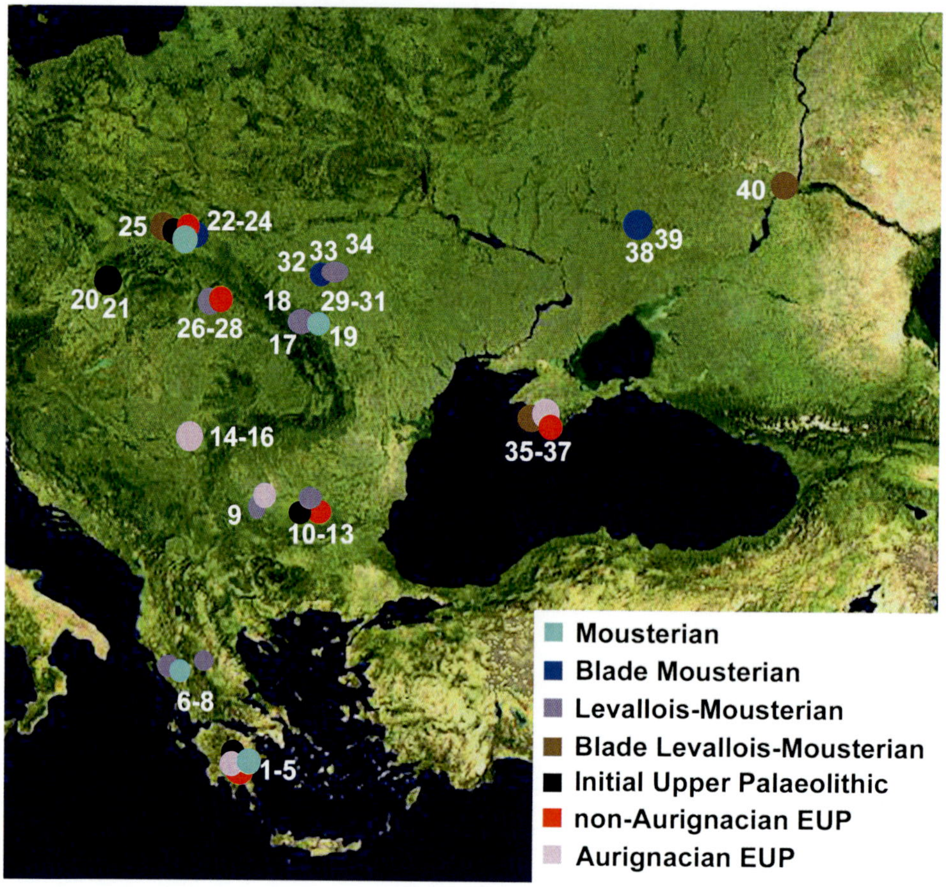

Fig. 14.1 Southeast, Central, and Eastern European cultural entities and sites discussed in the text: Greece—Klissoura, Kephalari, Franchthi, Kalamakia, Lakonis, (1–5), Asprochaliko, Kokkinopilos, Theopetra (6–8); Bulgaria—Kozarnika (9), Bacho Kiro, Temnata, Muselievo, Samuilitsa (10–13); Rumania—Româneşti-Dumbrăviţa, Tincova, Coşava (14–16), Ripiceni-Izvor, Mitoc-Valea Izvorului (17, 18); Moldova—Corpaci (19); Czech Republic—Brno-Bohunice, Stranska skala (20, 21); Poland—Księcia Józefa, Zwierzyniec, Piekary (22–24), Dzierzyslaw (25); Ukraine—Korolevo, Beregovo, Sokirnitsa (26–28), Korman, Molodova, Yezupil (29–31), Bugliv, Proniatin (32–33), Kulichivka (34), Kabazi, Siuren, Buran-Kaya (35–37), Kourdiumovka, Belokuzminovka (38, 39); Russia—Shlyakh (40)

blade/lets with diffused bulbs, and lipped butts) technological *fossiles directeurs*. For example, while trying to place a characteristic crested blade to the "correct side" using solely an analytical approach, it turned out that—after conducting refit analysis—this technical piece may result from not only prismatic techniques, but also discoidal and especially polyhedral core reduction. This phenomenon is observable in the Late Middle Paleolithic industry in layer III at Kraków-Księcia Józefa, Poland (Zięba 2005; Sitlivy et al. 2009a, b, 2014c) and in the Early Middle Paleolithic at Rheindahlen B1, Germany (Bosinski 1966). In some cases involving assemblages reconstructed by refitting (e.g. Księcia Józefa), were able to discuss technologies and (group and individual) human behaviors in terms of concrete reconstructed sequences, which supported many analytical technological conclusions, but which also produced unexpected reduction sequences.

Technological Variability

The MP and the onset of the UP in Southeast Europe and adjacent regions witnessed considerable variability involving the co-occurrence of various "cultural" entities, flaking methods, and techniques. Building on new data collected during previous decades, several inter-regional comparative studies of the LMP and EUP were published recently (e.g. Meignen et al. 2004; Demidenko and Noiret 2012; Tsanova et al. 2012; Kuhn and Zwyns 2014); they serve to emphasise the need for more systematic research. In this chapter, I will demonstrate the technological diversity of lithic production throughout the MP and the changes that paved the way to UP technology through several brief comparisons.

Several primary flaking production systems aimed at producing blanks (flakes and blades) for tool manufacture have been identified across various MP technocomplexes. These debitage methods were used independently, coexisted in a single industry, represented the dominant systems, or were technologically mixed in one reduction sequence. There is no clear spatial and chronological clustering of these technologies and no single evolutionary trend from "archaic" to "developed" methods. Blade manufacture throughout the entire MP is no longer considered exceptional (e.g. Bar-Yosef and Kuhn 1999), and is regarded as common practice, especially in the Near East, Africa, and North-Western Europe. For instance, the UP volumetric concept of systematic blade production is accompanied by mostly Levallois methods, as well as the combination of Middle and Upper Paleolithic tools (e.g. backed blades, burins) in Western Europe since OIS 8–6, and has been documented on several occasions at the beginning and during OIS 3 in other regions of Europe (for a comparative analysis, see Sitlivy and Zięba 2008).

A number of units, expanding during Early (EMP) and Late Middle Paleolithic (LMP) (OIS 7–OIS 3) in Southeast, Central, and Eastern Europe, can be distinguished when considering their technological and typological features: Micoquian, Mousterian, Levallois-Mousterian, Blade Levallois-Mousterian, Blade Mousterian, and Initial/Early Upper Paleolithic, including "transitional industries" (Chabai et al. 2004; Sitlivy and Zięba 2006). The term "transitional industry" will be avoided for several reasons, including the ambiguities associated with the use of this term (Bar-Yosef 2006; Kuhn 2003), and because of the exceedingly long duration of this period (see also Belfer-Cohen and Goring-Morris 2009). It would be more appropriate to speak about "surviving industries"—a term that does not assume directional evolution, especially given the already full UP character of some assemblages reconstructed through refit analysis (e.g. Bohunician or Emiran). The latter pattern also comprises an "archaic" MP component (faceted butts, hard hammer use), which confirms a rather common trend documented throughout the entire Paleolithic, where different Lower/Middle/Upper Paleolithic techno-typological characteristics appear in a single assemblage (Chabai and Sitlivy 1993). The technological analyses of core reduction sequences provide a more detailed subdivision of these entities, which can therefore be used to identify different debitage methods and to trace possible transformations before and during the "Big Transition".

Mousterian (***>OIS 3?–OIS 3***) is characterised by the dominant use of one or several non-Levallois flake methods (Fig. 14.2):

- *dominant discoidal/centripetal* at Bugliv V, layer I (Ukraine), OIS6?/OIS4? (Sytnyk 2000), Asprochaliko, upper layer (Greece; Papaconstantinou 1988), ca. 39 ^{14}C kBP (Bailey et al. 1983);
- *dominant polyhedral, and a considerable variety of flake methods* (e.g. *discoidal, uni-bidirectional*) and *debitage initiated from flakes* (e.g. *Kombewa*), *with scarce Levallois and/or UP blade reconstructed reduction sequences* at Księcia Józefa, layer III (Poland) 44,400 ± 1400 ^{14}C BP (Figs. 14.3, 14.4, and 14.5) (Sitlivy et al. 2009a, b), and less varied at Korman IV, layer 12 (Ukraine; Sytnyk 2011);
- *dominant flake uni-/multidirectional, centripetal/discoidal methods with some flake/blade/point Levallois and UP blade debitage* at Klissoura, layers VI–XX (Greece; Sitlivy et al. 2007, 2008a), throughout the OIS 3 and possibly earlier, as the dates on ABOX pre-treated samples from layers XVIII and XX of 60–62 ^{14}C kBP (Kuhn et al. 2010) should be considered minimum ages (Fig. 14.6);

Blade Mousterian (***OIS 6?–OIS 4***) shows direct and prepared (crest installation) UP/volumetric blade reduction accompanied by non-Levallois flake methods:

- *direct uni-/bidirectional* (several refits) at Bougliv V, trench III, layer II (Ukraine), OIS 6?–OIS 4 (Sytnyk 2000) and Kraków-Zwierzyniec I, layer 2, Poland (Chmielewski et al. 1977);
- *dominant unidirectional reduction of naturally voluminous cylindrical nodules* (*rare crested flakes, mostly lateral cresting—refits*) *and uni-/bidirectional, centripetal non-Levallois with discoidal debitage* at Kourdiumovka (Ukraine), Early Glacial, OIS 4 (Kolesnik 2000, 2003).

Levallois-Mousterian (***OIS 7/6–OIS 3***) is based on recurrent flake or blade/flake Levallois reduction sequences and rarely on exclusive Levallois lineal methods (Fig. 14.7):

- *flake/blade uni-/bidirectional recurrent Levallois and flake lineal or recurrent centripetal reduction sometimes accompanied by bifacial shaping* (*leaf points*) at Korolevo I, layer Va (Ukraine), OIS 7/6 (Gladilin and Sitlivy 1990; Chabai and Sitlivy 1993), Kozarnika Cave, layers XIV–XIII, OIS 7/6 and XII–IX, MIS 5–4 (Guadelli et al. 2005), Muselievo and Samuilitsa II, ^{14}C dated to 42,780 ± 1270 BP (Sirakov 1983; Tsanova 2012) (all Bulgaria), as well as Greek assemblages at Asprochaliko, lower layer (Gowlett and Carter 1997), TL 102 + −14 and 96 + −11 kBP (Huxtable et al. 1992), Theopetra Cave, layer II2 (Panagopoulou 2000), OIS 6/5 BP (Valladas et al. 2007), Lakonis, unit Ib, average ^{14}C date of *ca.* 42 kBP (Panagopoulou et al. 2002–2004; Elefanti et al. 2008);
- *exclusive flake lineal Levallois* at Korolevo I, layer III, OIS 5 (Gladilin 1989; Chabai and Sitlivy 1993; Haesaerts and Koulakovskaya 2006);
- *flake/blade recurrent uni-/bidirectional Levallois* at Yezupil, layer III (Ukraine), OIS 5 (Sytnyk 2000; Boguckyj et al. 2001), and Ripiceni-Izvor, layers 1–3

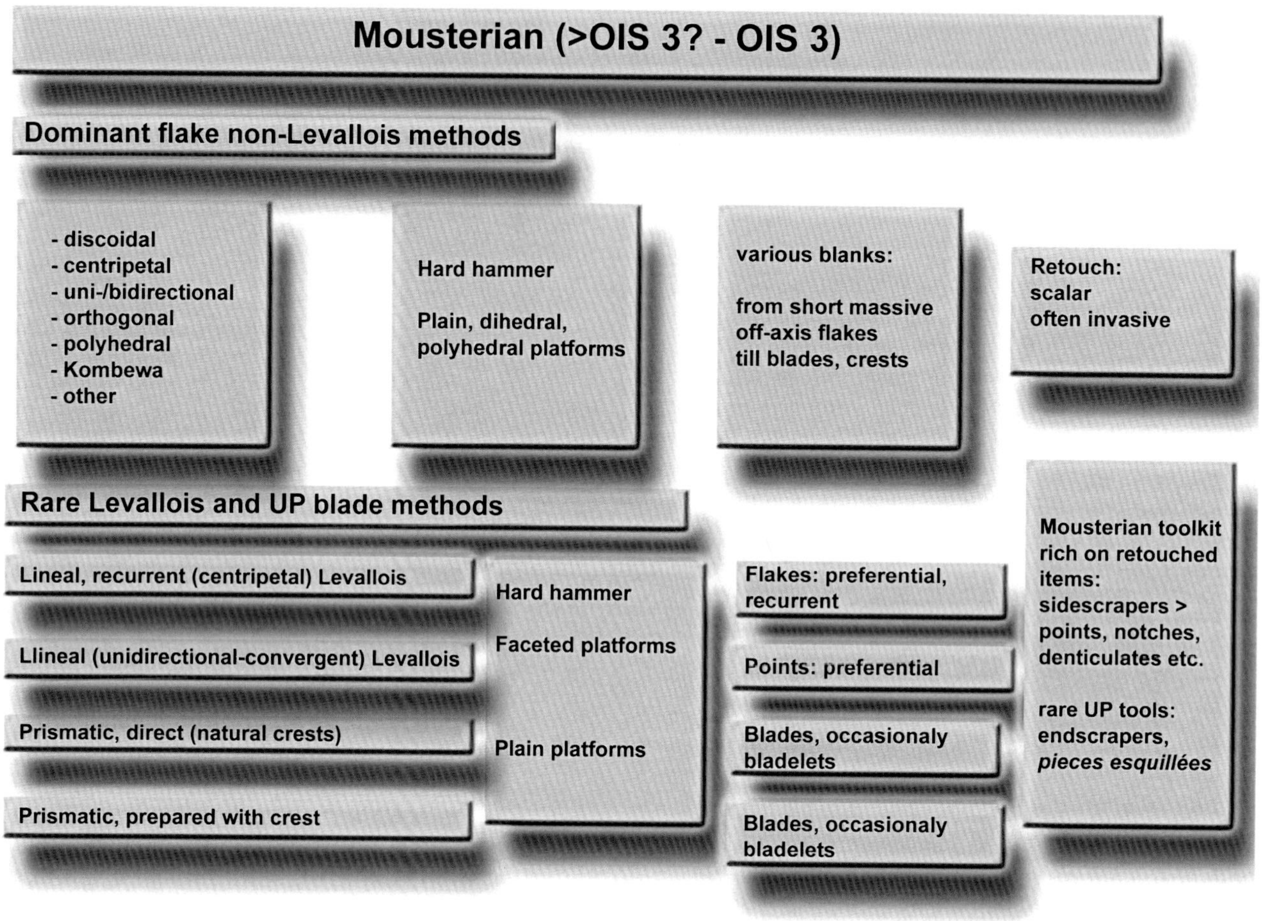

Fig. 14.2 Schematic representation of Mousterian technological structure

(Romania), OIS 3? (Honea 1981; Paunescu 1993), layer 4, >45,500 ^{14}C BP (Doboş and Trinkaus 2012);

- *flake recurrent centripetal Levallois* at Kalamakia, units III and IV (Greece), within the OIS 5e–40 ^{14}C kBP (Darlas and de Lumley 2004; Darlas 2007), and probably at Bacho Kiro, layer 13 (Bulgaria), >47,500 ^{14}C BP (Drobniewicz et al. 1982);
- *lineal Levallois for elongated flakes*—reconstructed sequences at Molodova V, layers 11 and 12 (Ukraine) (Usik 2003), between >45.6 ^{14}C kBP and >35 ^{14}C kBP (Ivanova 1987) or the first part of OIS 3 (Haesaerts et al. 2003) and Proniatin, 87 kBP (Ukraine); TL date on fossil soil (Sytnyk 2000);
- *lineal Levallois convergent, mostly unidirectional for broad-based points*—reconstructed cores at Korolevo I, layer IIb (Usik 1989), early OIS 4 (Gladilin 1989).

Blade Levallois-Mousterian (***OIS 3***) is based on independent use of prepared crested UP/volumetric blade and Levallois flake/blade lineal/recurrent methods, as well as non-Levallois discoidal or centripetal debitage (Fig. 14.8):

- *prepared crested UP blade uni-/bidirectional accompanied by flake lineal Levallois and discoidal* partly reconstructed reduction sequences at Piekary IIa, layer 7c (Poland; Morawski 1992; Zięba et al. 2008), 61–48 kBP, mean 55.0 ± 6.5 kBP (Valladas et al. 2008; Fig. 14.9: 1);
- *lineal convergent unidirectional Levallois for broad-based points and flake recurrent Levallois centripetal with additional discoidal, Kombewa debitage, and rare crested blades* at Piekary IIa, layer 7b (Zięba et al. 2008), 39 kBP (Valladas et al. 2008; Fig. 14.9: 2);
- *prepared crested UP mostly bidirectional reduction together with discoidal and decreased recurrent centripetal Levallois* at Piekary IIa, layer 7a (Zięba et al. 2008), 46–33 kBP (Valladas et al. 2008; Fig. 14.9: 3);
- *prepared crested UP blade and blade/flake mostly bidirectional Levallois reconstructed reduction with/without*

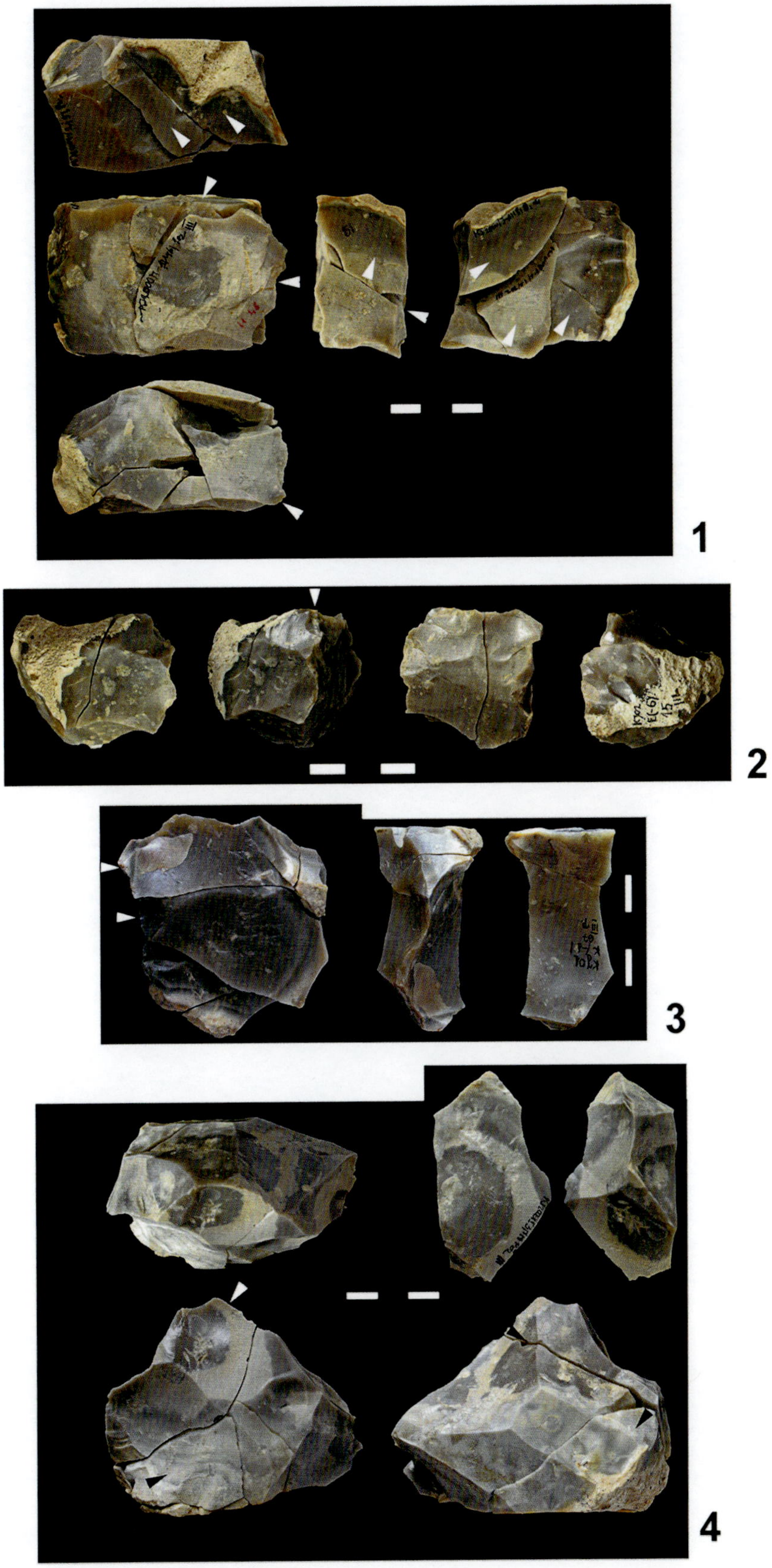

Fig. 14.3 Kraków—Księcia Józefa, Mousterian, layer III. Reconstructed reduction sequences: full debitage episode showing core with parallelepipedal volume formed by six surfaces and right angles between them (*1*), small nodule reduced with change in orientation into cubic core and final invasive removal of a crested flake (*2*), final stage of cubic core reduction and crested plunging blade (*3*), discoidal reduction and removal of crested flake (*4*)

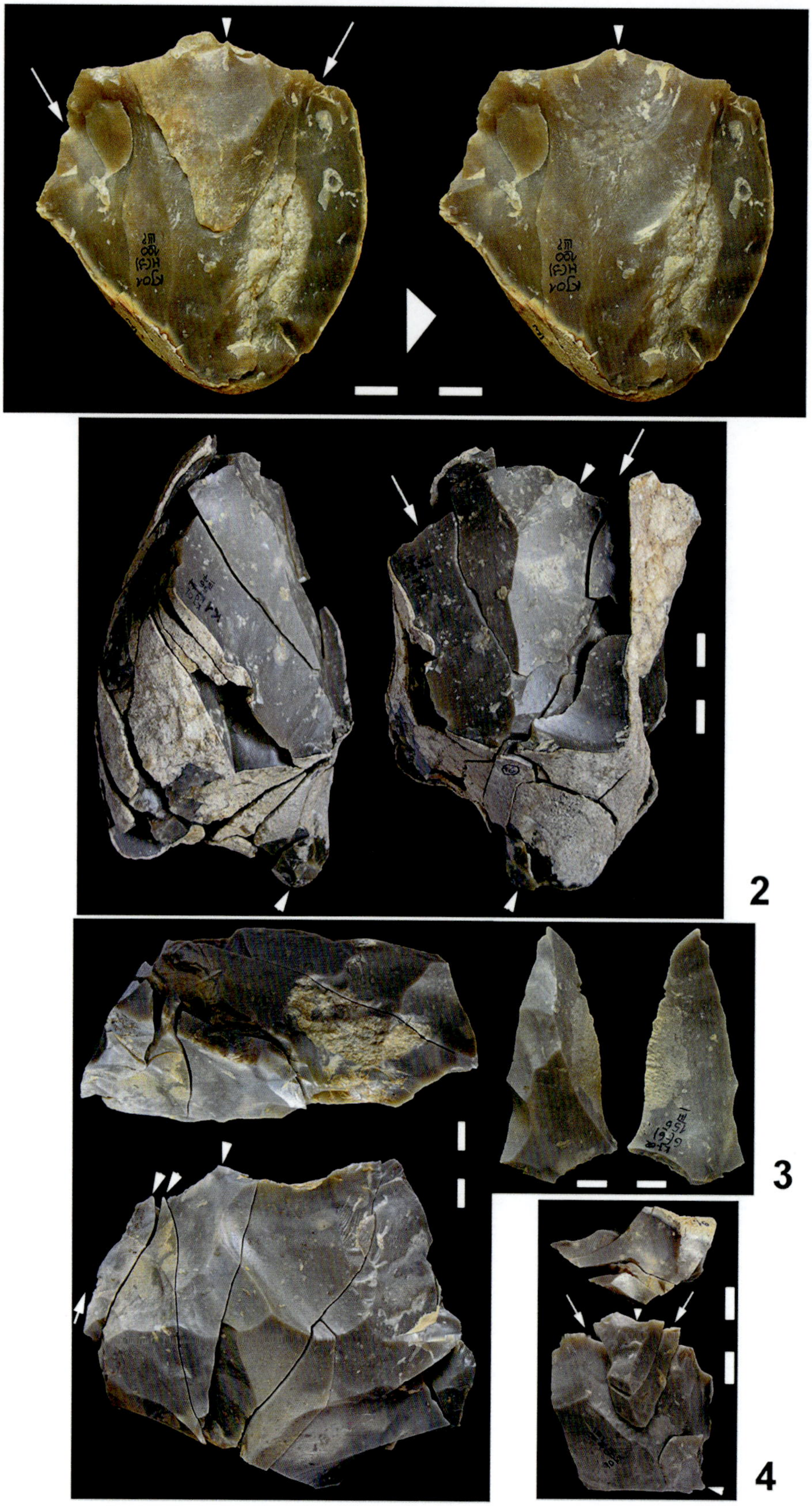

1 2 3 4

Fig. 14.4 Kraków—Księcia Józefa, Mousterian, layer III. Reconstructed Levallois point reduction sequences: convergent unidirectional core and two failed points (*1*), preferential point production followed by bidirectional and secant debitage (*2*), removal of the central crest and next elongated triangular point with faceted *chapeaux de gendarme* butt from wide surface (*3*), core narrowing (lateral crest), and two short points (*4*)

Fig. 14.5 Kraków—Księcia Józefa, Mousterian, layer III. Reconstructed unidirectional blade reduction sequences: unprepared (natural crest) partially turned (*1*), unprepared on narrow-faced core (*2*), prepared (crest) partly turned on core-on-flake (*3*)

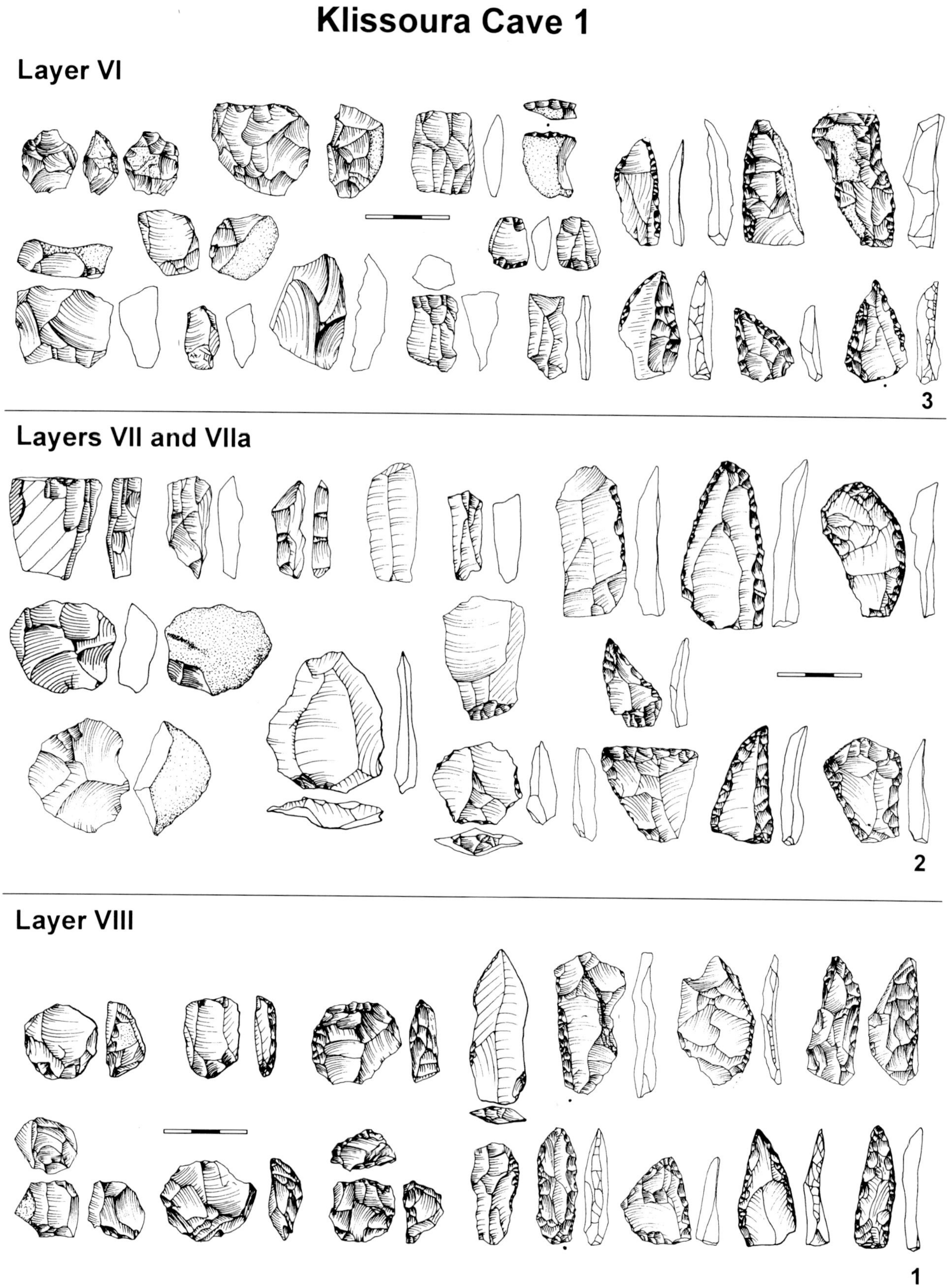

Fig. 14.6 Klissoura Cave 1. Uppermost Mousterian, layers VIII (*1*), VII, VIIa (2) and VI (*3*): cores, debitage and tools (drawings by K. Sobczyk and M. Sudol)

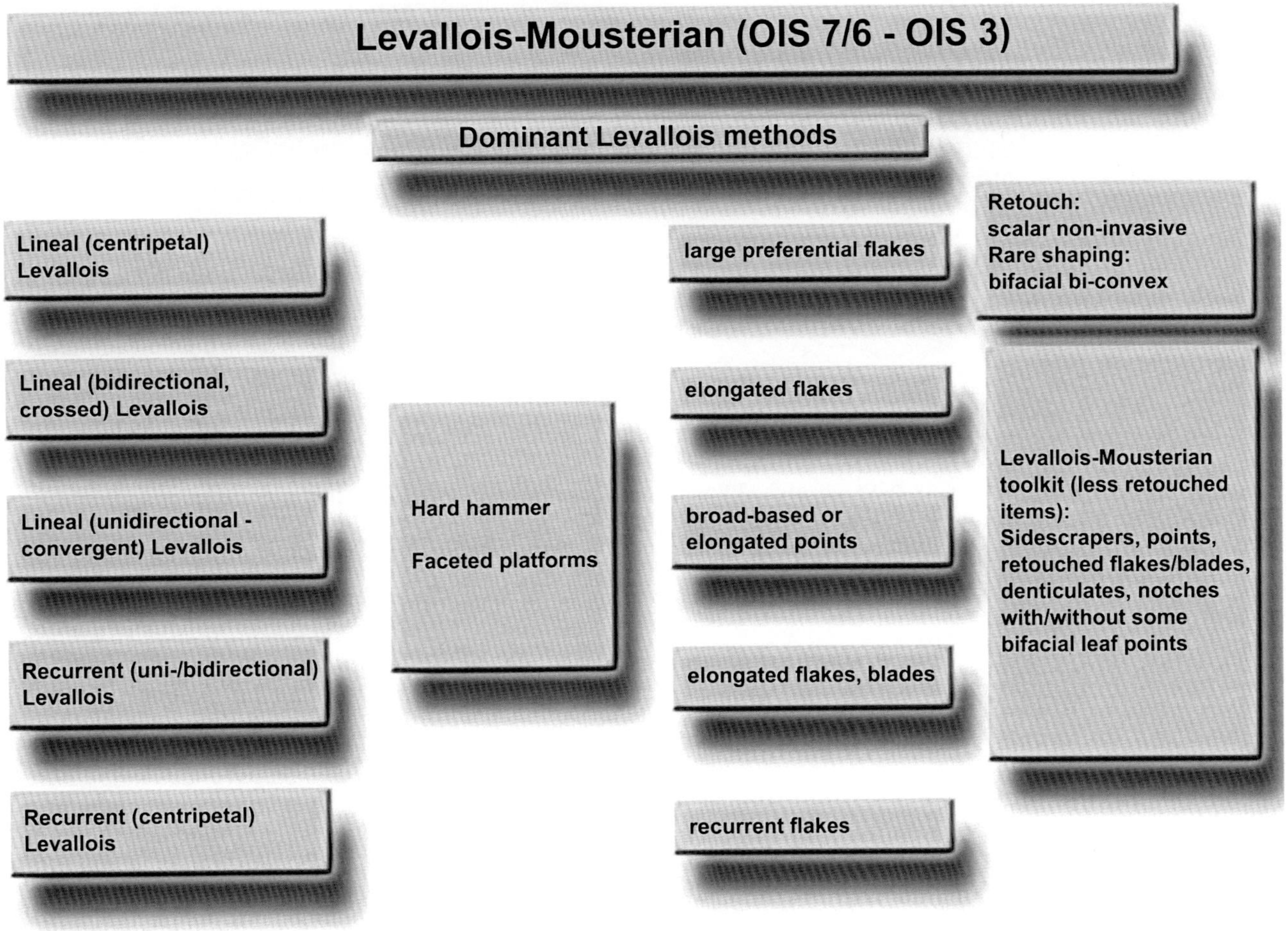

Fig. 14.7 Schematic representation of Levallois-Mousterian technological structure

lineal Levallois for flakes—at Kabazi II, Unit II (Ukraine) ^{14}C dated from *ca.* 40 till 30 kBP (^{14}C/AMS/ESR chronologies and biostratigraphy—Chabai 2013);

- *direct and prepared UP blade unidirectional often narrow-faced/wedge-shaped core reduction and recurrent uni-/bidirectional Levallois* at Shlyakh, layers 8 and 9 (Russia); 40–42 and 42–44 ^{14}C kBP, respectively (Nekhoroshev 2009; Vishnyatsky and Nehoroshev 2004), and Belokuzminovka, layers 2 and 3 (Ukraine), Early Pleniglacial (Kolesnik 2003).

Initial/Early Upper Paleolithic

A large variety of industries, some of which might indicate transition, or at least chrono-stratigraphically precede the "classical" UP, have been recorded. The Initial/Early UP knappers used several models of core reduction: (a) *exclusive/dominant* UP laminar strategies (IUP-blades/unretouched bladelets and EUP-bladelets/blades), (b) a *combination* of the latter with different MP technologies, and (c) a *fusion* in a single reduction sequence of volumetric blade and "flat core" MP methods, especially the Levallois method for producing points (Sitlivy and Zięba 2006). The latter model was originally (Marks and Volkman 1983; Škrdla 1996) verified through refit analysis; it is not widespread and is still poorly documented. When reconstruction is possible, it demonstrates a dynamic shift from one reduction method to another, thus displaying a real *mixing/fusion*. Curiously, this type of reduction began from crest installation (a volumetric approach), followed by bidirectional blade reduction, leaving no possibilities to speculate about an evolution from "archaic" MP Levallois into "revolutionary" UP blade production. This reconstructed trend was documented only during the transitional period in Central Europe (e.g. Brno-Bohunice, Stranska skala, Księcia Józefa, layer III) and the Near East (Boker Tachtit, with a long four-layer sequence). On the contrary, industries with a *double technological structure*, comprising independent UP and MP methods, occurred throughout the entire MP (OIS 8–3). They demonstrate few technological changes and only their

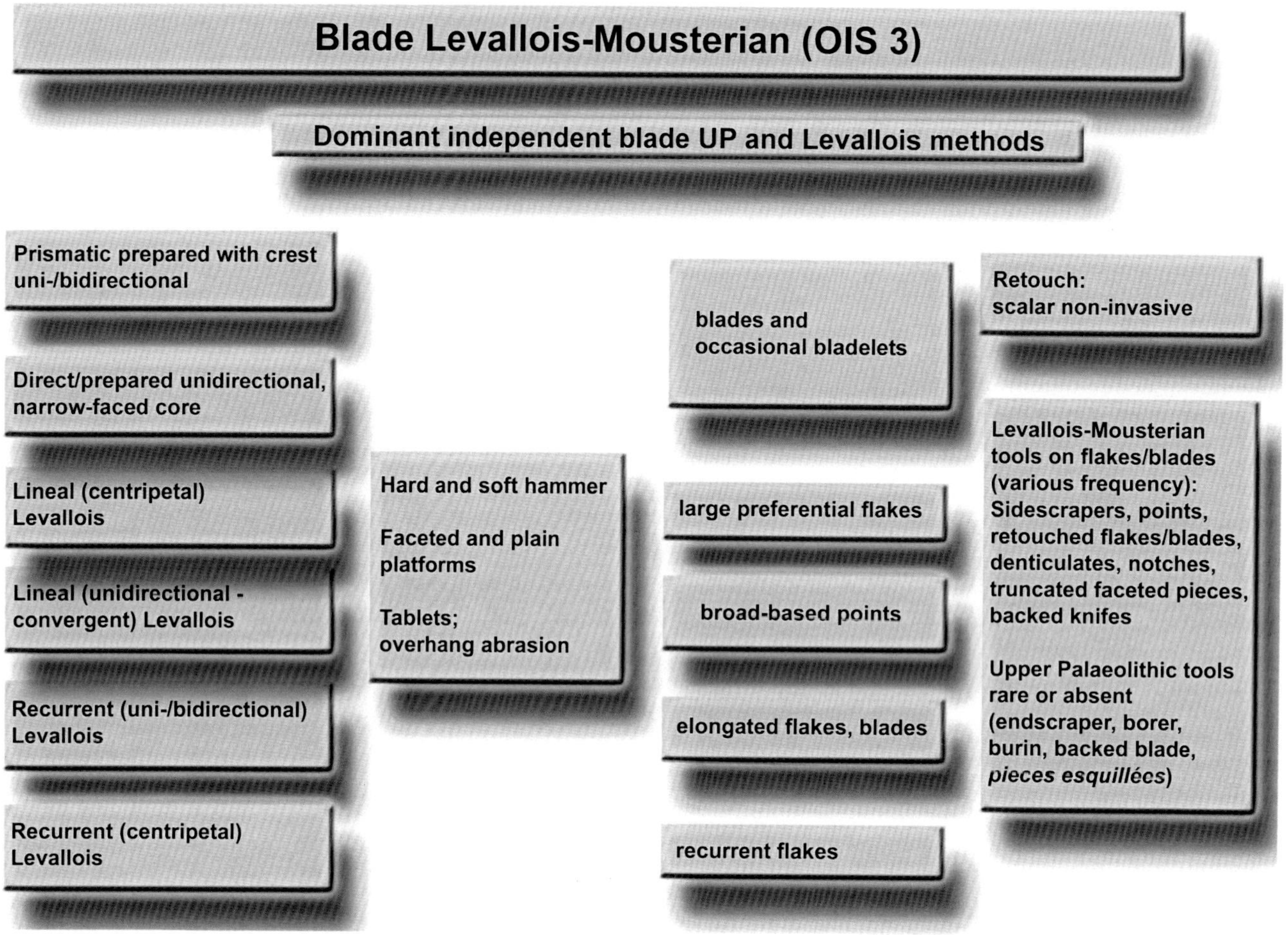

Fig. 14.8 Schematic representation of Blade Levallois-Mousterian technological structure

chrono-stratigraphical position allowed an attribution to the EUP. Tentatively, they even might be assigned to the Bohunician "modern behavior", but once again, only refit analysis could prove this status. The potential risk of contamination/mechanical mixture of MP with UP layers resulting in "transitional" industries is evident (e.g. Anghelinu et al. 2012, for the Romanian record; Valladas et al. 2007, for the revision of the "transitional" industry from Theopetra), this is especially true as "transitional industries" are suspiciously absent in single-layered sites. The case of the open-air site of Korolevo, where EMP, LMP, and two EUP layers were clustered and spatially separated, is unique: as all of these industries were close to the modern surface in a compact geological sequence; otherwise, they would have been found mixed in an excavated trench. Industries with *dominant/exclusive UP laminar debitage* rarely appeared during the MP. They differ, however, from the "classical" UP industries by their "transitional" chronological position (early OIS 3), the persistence in certain cases of some MP typological and/or technological elements (e.g. techniques of blank detachment), and by a generally non-Aurignacian "look". Thus, IUP/EUP industries (early OIS 3) are based on the following technological features (Figs. 14.10 and 14.11):

- *a fusion in one reduced block of prepared crested UP blade with Levallois point/blade/flake reconstructed production sequences* (Fig. 14.10, Model I) at Brno-Bohunice and Stranska skala III, IIIa between 43 and 36 kBP and 41 and 34.5 kBP, respectively (Škrdla 1996; Svoboda 2004); the Bohunician ranges from 41 and at least 33 ^{14}C kBP (Svoboda and Bar-Yosef 2003), but see the mean TL age of 48.2 ± 1.9 kBP (Richter et al. 2008) and new radiocarbon ages between 30 and 40 ^{14}C kBP (Richter et al. 2009);
- *a combination of prepared (crested) and direct UP blade debitage with Levallois bidirectional blade/flake methods* (Fig. 14.10, Model II) could be found in the Dzierzyslaw I, lower layer (Poland) (where the Bohunician attribution by Foltyn and Kozłowski (2003)—needs to be verified by refitting), and probably at Temnata TD-II, layer VI (Bulgaria), *ca.* 50–45 kBP and >38,700 BP (Drobniewicz et al. 2000a), determining, at least in part, if the collection can overcome uncertainties regarding the mechanical

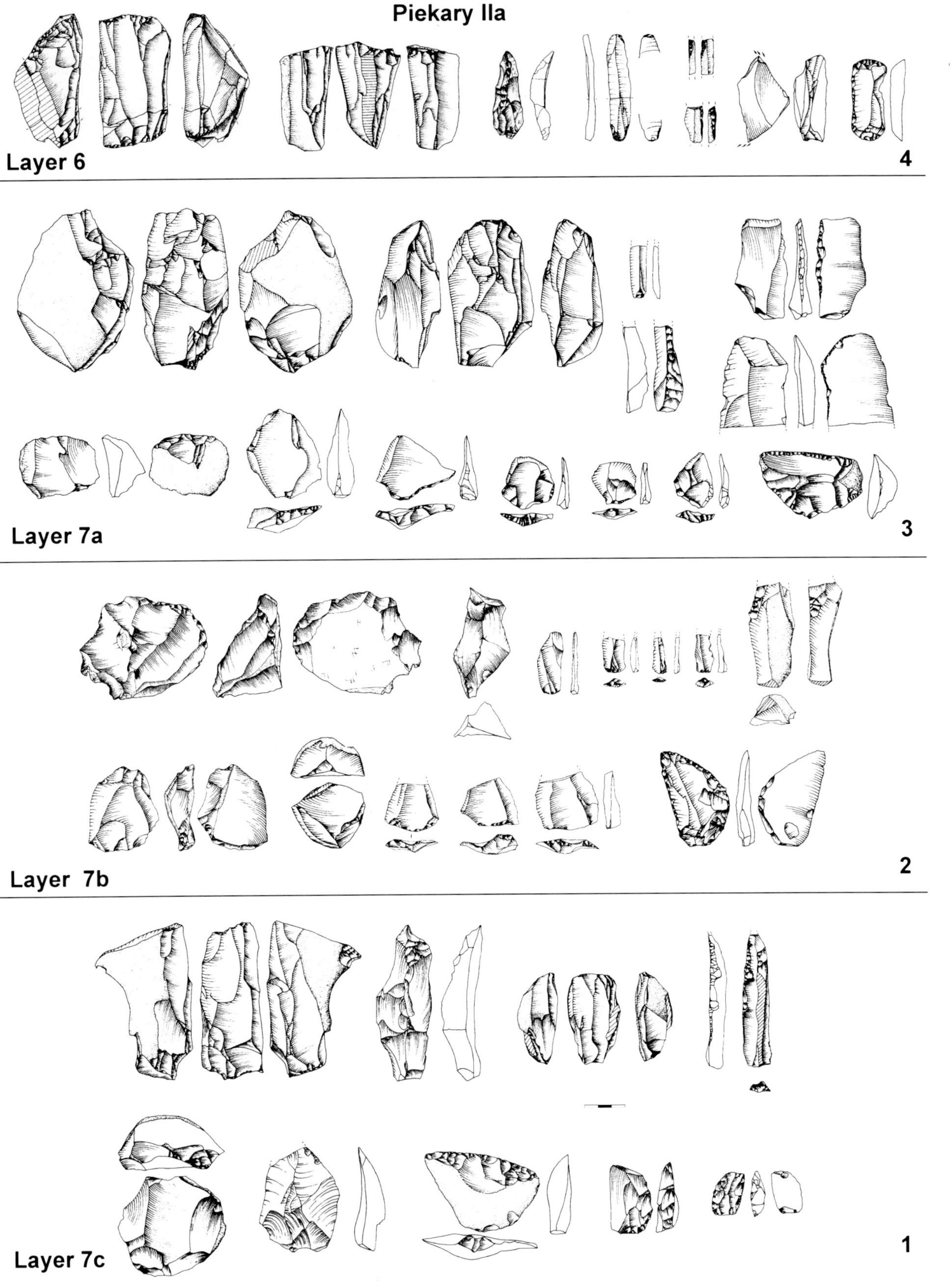

Fig. 14.9 Piekary IIa: layers 7c, 7b, 7a, and 6 showing the technological shift from Blade Levallois-Musterian with independent UP blade and Levallois flake/point production, and rare UP tools (*1*, *2*, *3*) to EUP with exclusive blade production and UP tool-kit (*4*) (drawings by K. Sobczyk and J. Wilczynski)

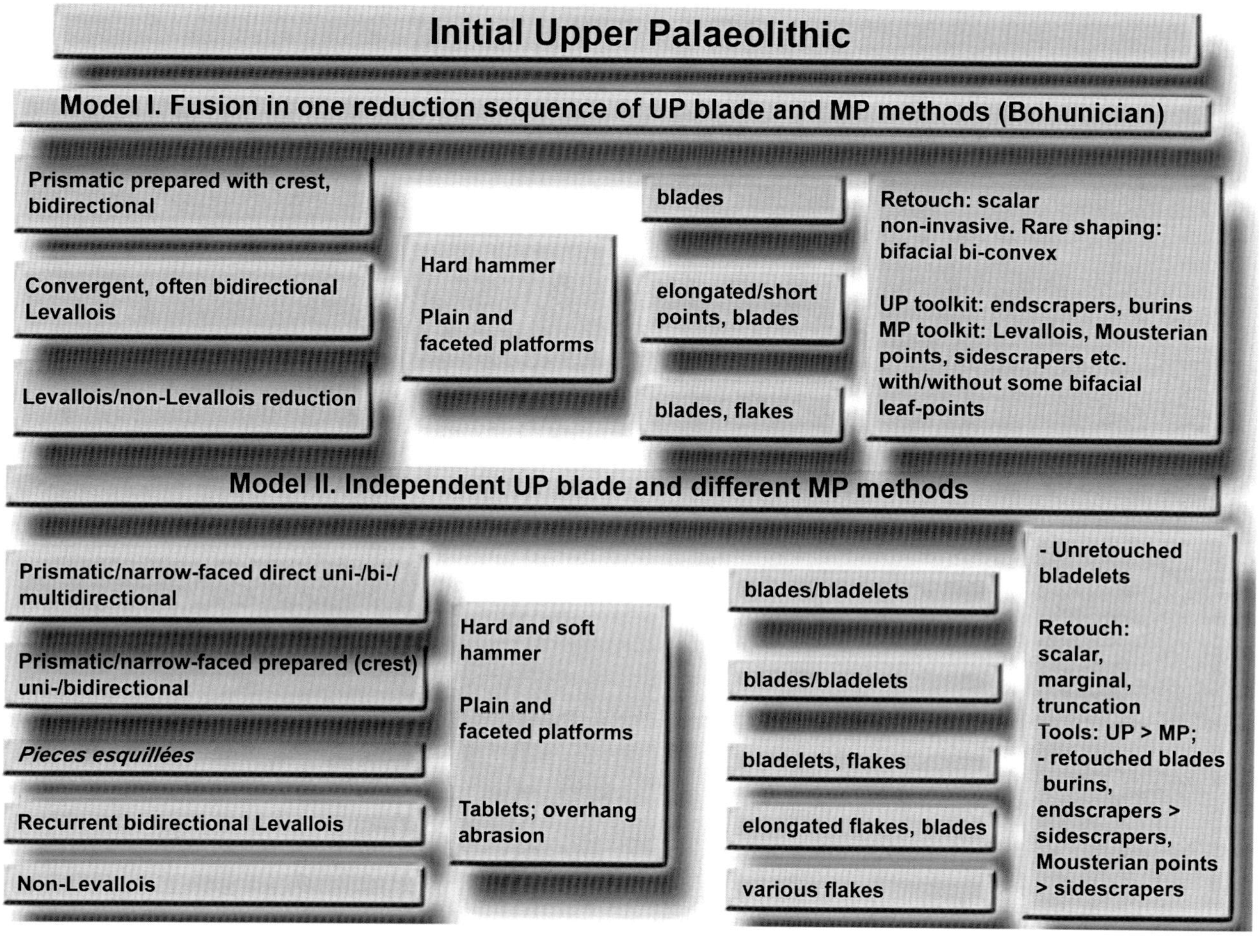

Fig. 14.10 Schematic representation of Initial Upper Paleolithic technological structure

mixing of two different technologies (see the taphonomic analysis—Tsanova 2006). The renowned Temnata collection actually inspired the hypothesis of two opposing technological transformations: (1) from Levallois double-platform cores into UP cores; (2) from UP narrow double-platform cores into flat cores with broad flaking surfaces (Kozłowski 2000). Unfortunately, there are no refits, which would be the only way to positively confirm these transformations (Sitlivy and Zięba 2006; see also Tsanova 2006, for a similar opinion). The existing core transformation data in other industries, as verified by refitting, demonstrate only a single manner of reduction, i.e. from UP crested pre-core *via* flat core, which produced, during or near the end of this reduction sequence, Levallois or non-Levallois blanks (e.g. Meignen 1994; Svoboda and Škrdla 1995; Usik 1989; Sitlivy et al. 2009a, b, 2014c). The refitting data concerning Levallois and UP blade production also suggest an independent status of these debitage methods during the early "blade episodes" by the end of the Middle Paleolithic—e.g. Kabazi II, where double dissociated structure of these reduction strategies show no influence of one over the other throughout a long and detailed sequence (*ca.* 45–30 kBP—Chabai 2004). Another possible candidate for this group of mixed Levallois-based/laminar industries is the revised "ex-pre-Aurignacian" assemblage at Bacho Kiro, layer 11 (Tsanova and Bordes 2003; Kozłowski 2004; Teyssandier 2008). However, this assemblage is definitively an MP-based collection (flat core exploitation, pronounced bulbs, Mousterian points, etc.), displaying only weak signs of the Levallois concept, probably due to heavy reduction (often recycling/*re-debitage*, abundant *pièces esquillées*—Tsanova 2006);

– *a combination of direct and prepared (crested) blade/bladelet uni-/bi-/multidirectional reduction with flake non-Levallois methods* at Klissoura, layer V, with a tool-kit rich in curved backed pieces (Uluzzian) (Fig. 14.11, Model I; Figs. 14.12 and 14.13), radiocarbon dated to 40,010±740 ^{14}C BP (Koumouzelis et al. 2001) and confirmed by the overlaying CI tephra (Lowe et al. 2012). Diagnostic Uluzzian segments were recorded in undated,

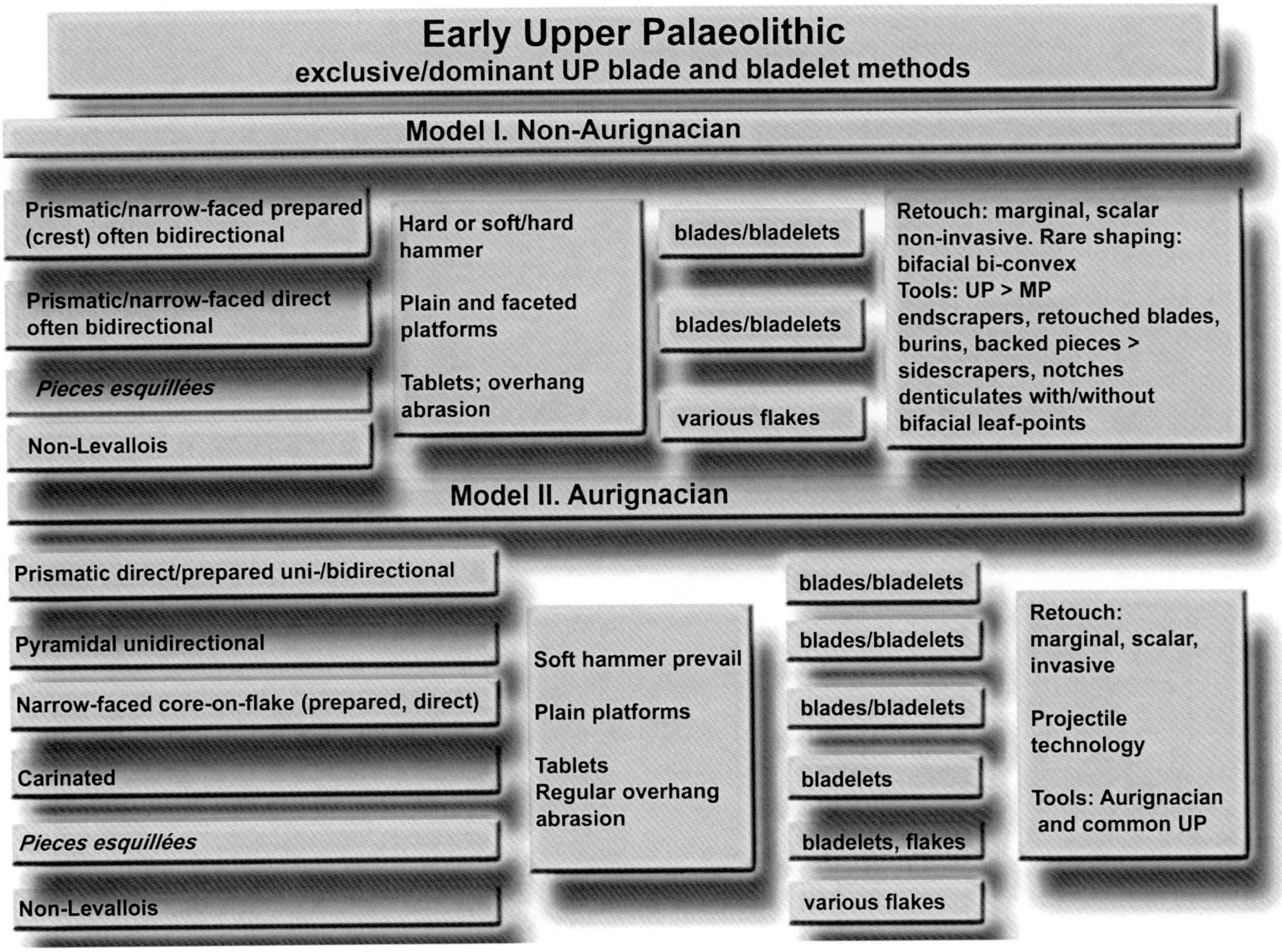

Fig. 14.11 Schematic representation of Early Upper Paleolithic technological structure

more recent and/or variously mixed collections: Zwierzyniecian (Sachse-Kozłowska and Kozłowski 1975), Ripiceni-Izvor IIb (Paunescu 1993), and Corpaci, level 4 (Moldova), 25.250 ± 300 ^{14}C BP (Borziak et al. 1981);

- *exclusively prepared (various crests), often bidirectional UP blade and bladelet (on flake burin-like) reconstructed sequences* at Księcia Józefa, layer II, 40,380 ± 940 ^{14}C BP (Sitlivy et al. 2004, 2009a, b, 2014c; Fig. 14.11, Model I; Fig. 14.14);
- *exclusive UP blade/bladelet continuous direct and prepared (central bifacial or lateral crest) often bidirectional reconstructed core reduction sequences* (Fig. 14.11, Model I) at Korolevo II, level II, *ca.* 38.5 ^{14}C kBP (Usik 1989; Gladilin and Demidenko 1989); Temnata TD-I, layer 4 (phases C, B, A), *ca.* 45–31 ^{14}C kBP (Drobniewicz et al. 2000b); Piekary IIa, layer 6; 32–26 kBP (Zięba et al. 2008; Valladas et al. 2008; Fig. 14.9: 4);
- *a prepared UP mostly unidirectional blade (usually lateral one-sloped crest on the ridge between dorsal and ventral flaking surfaces), with/without non-Levallois flake (transversal reduction of a flake edge from ventral face) reconstructed sequences* at Korolevo I , level Ia, >25 ^{14}C kBP (Usik 1989; Gladilin and Demidenko 1989), recently linked by palynological and geological data to the cold and dry period between 39 and 37.5 kBP; Sokyrnitsa IA, layer 3 (Ukraine), *ca.* 38.8 kBP (Usik 2003; Usik et al. 2006a, b).

Proto-Aurignacian/Early Aurignacian (Fig. 14.11, Model II)

The early phase of the Aurignacian, traditionally viewed as a homogeneous technocomplex, appears to be quite variable across the study area:

- *direct and prepared (proximal or posterior crest) bladelet, unidirectional (pyramidal core) together with bidirectional (prismatic core) and uni-/bidirectional prepared*

Fig. 14.12 Klissoura Cave 1, layer V. Uluzzian. Schematic representation of blade/bladelet and tool production (1, 2—prismatic unidirectional cores; 3—splintered piece; 4—crested blade; 5—blade; 6—bladelet; 7—truncated blade, 8—lunate on blade, 9—lunate on bladelet) (photos by N. Thompson)

(partial crest) reduction of the narrow part of a blank; bifacial shaping (leaf points) at Kozarnika, layer VII, 39–36 ^{14}C kBP (Guadelli et al. 2005; Sirakov et al. 2007; Tsanova 2006) assigned to the "Kozarnikian"/Proto-Aurignacian (Teyssandier 2008; Tsanova et al. 2012);

- *prevalent blade/bladelet/micro-blade production based on independent (long continuous or short) commonly unidirectional (but also bi-/multidirectional) reduction of different cores: carinated, prismatic, and narrow-faced (including burin-like cores-on-flakes and recycled core-on-tools)* in Proto-Aurignacian/Early Aurignacian of Krems-Dufour type in Romania at Româneşti-Dumbrăviţa I, GH3 (OSL and TL ages of *ca.* 40–45 kBP—Schmidt et al. 2013; Sitlivy et al. 2012), Tincova, Coşava (Sitlivy et al. 2014a, b) and Ukraine—Siuren, layer G and H (*ca.* <31 ^{14}C kBP—Demidenko and Noiret 2012), Beregovo I (Usik 2008);
- *bladelet carinated debitage (straight, few curved, and twisted bladelets)* at Franchthi, stratum Q (CI tephra layer) in Greece and *carinated cores with similar lamellar debitage* above stratum R, the lower part being attributed to the Early Aurignacian (Douka et al. 2011);
- *domination of bladelet (carinated), flake/bladelet (splintered pieces), flake (flat unidirectional, discoidal, orthogonal, polyhedral), over blade (mostly unprepared unidirectional partly turned) reduction* at Klissoura, layer IV, between *ca.* 34 and 32 ^{14}C kBP (Koumouzelis et al. 2001)—Early Aurignacian or a local variant of classic Aurignacian (Kuhn et al. 2010).

The technological variability might be underestimated when comparing discarded un-refitted artifacts with fully reconstructed sequences. For instance, the refitted record of the Terminal Mousterian in layer III at Księcia Józefa displays a substantial diversity of reduction sequences. The collected data reflect the complicated nature of raw material exploitation, technological behavior, and site function. The industry is dominated by flake production using a range of *chaînes opératoires*. The reconstructed production sequences for 282 blocks reflect different technological strategies for flake, flake/blade, blade, and point production, which is often polyhedral (45 % of all refitted blocks). Flakes and some blades resulted from "archaic" polyhedral, as well as discoidal and

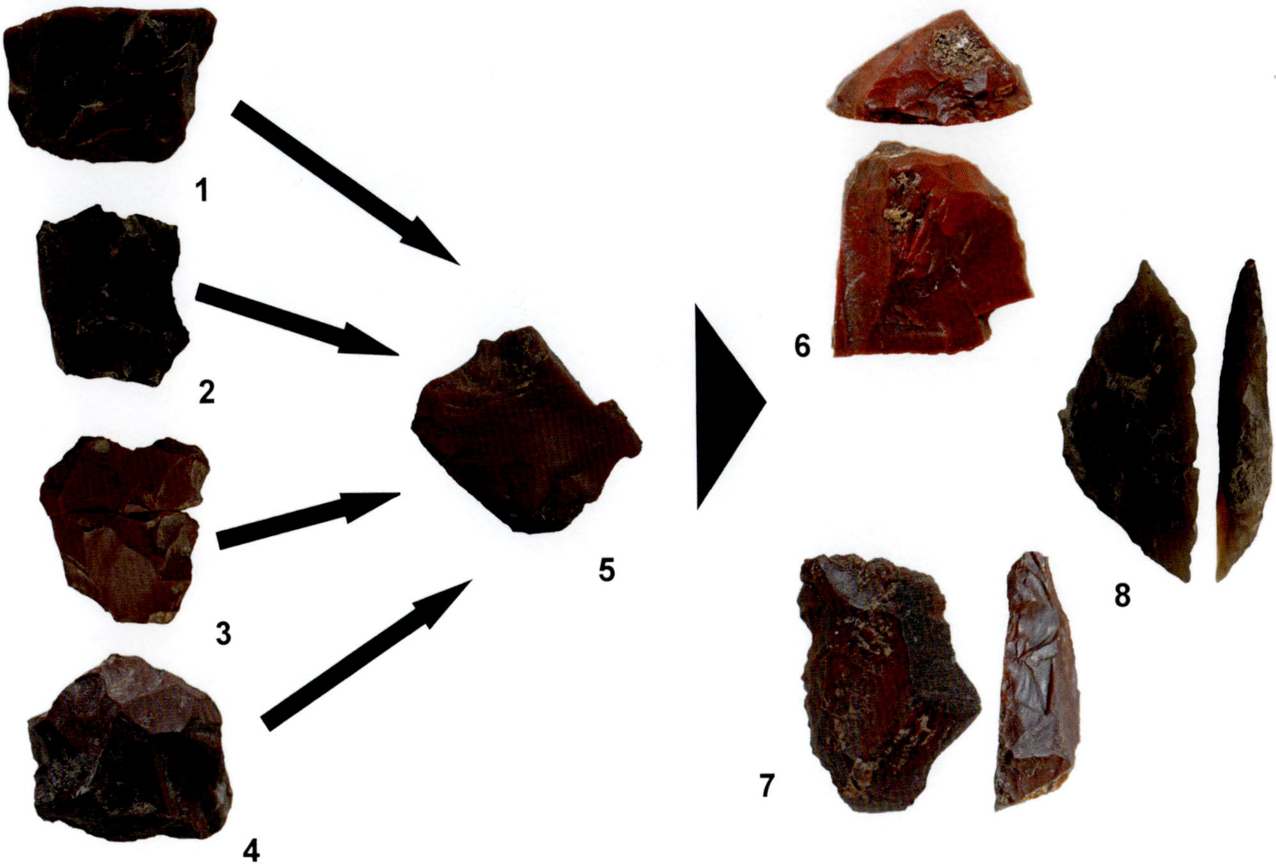

Fig. 14.13 Klissoura Cave 1, layer V. Uluzzian. Schematic representation of flake and tool production (cores: *1*—unidirectional, *2*—bidirectional, *3*—crossed, *4*—polyhedral; *5*—flake; *6*—endscraper; *7*—sidescraper; *8*—lunate on flake) (photos by N. Thompson)

other methods (Fig. 14.3), while some points were obtained in the frame of the "classical" Levallois and Bohunician trend (Fig. 14.4: 3). Thus, the refit data reflect the wider range of technological knowledge, although used with different frequencies. The high degree of technological variability in the layer III assemblage, as well as the combination of overlapped knapping methods and the use of both economical and wasteful production strategies (Zięba 2014; Sitlivy et al. 2014c) are not limited to "cultural" or chronological framework alone. It might be reasonable to explain the technological mosaic at this site in terms of variability of individual technical behavior (e.g. Bodu et al. 1990). Unfortunately, such co-occurrence and fusion of production chains during this time span (*ca.* 40/45 ka BP) and across broader territories of Europe is still insufficiently demonstrated by refit analysis.

Technique

Flakes and points, as well as Levallois blades and often UP laminar products, were obtained by use of the hard hammer technique during the Middle-to-Upper Paleolithic transition. Hard hammer percussion is attested in IUP/EUP assemblages (e.g. Klissoura, layer V, Bohunician, Korolevo II, level II). Both hard and soft hammer percussion for blade production has been documented in some MP (Piekary II, layer 7a or of Kabazi II, level II/7A) and EUP (e.g. Piekary II, layer 6, Sokirnitsa, level 3) laminar industries, while a soft hammer was regularly used by Aurignacian, and some EUP non-Aurignacian knappers (e.g. Korolevo I, level Ia). Platform faceting, documented for the MP manner of core platform preparation, occurred in several LMP, IUP, and EUP industries with and without Levallois production, before the onset of UP, e.g. Bohunician, Korolevo II, layer II (before blade removals), Temnata I, layer 4, Piekary II, layers 7a and 7b, Księcia Józefa, layer II or Blade Levallois-Mousterian at Kabazi II, Unit II. The dominant role of core platform preparation by single removal and rejuvenation by tablets (with corresponding plain and linear butts on obtained blanks) can be attested in the EUP industries, including the Proto-Aurignacian, Aurignacian, Kozarnikian, Ahmarian, Baradostian (e.g. Teyssandier 2003; Tsanova et al. 2012; Sitlivy et al. 2012). The regular and careful butt zone trimming by abrasion and/or faceting became a common practice at that time, while such core platform preparation occurred long before the onset of the UP.

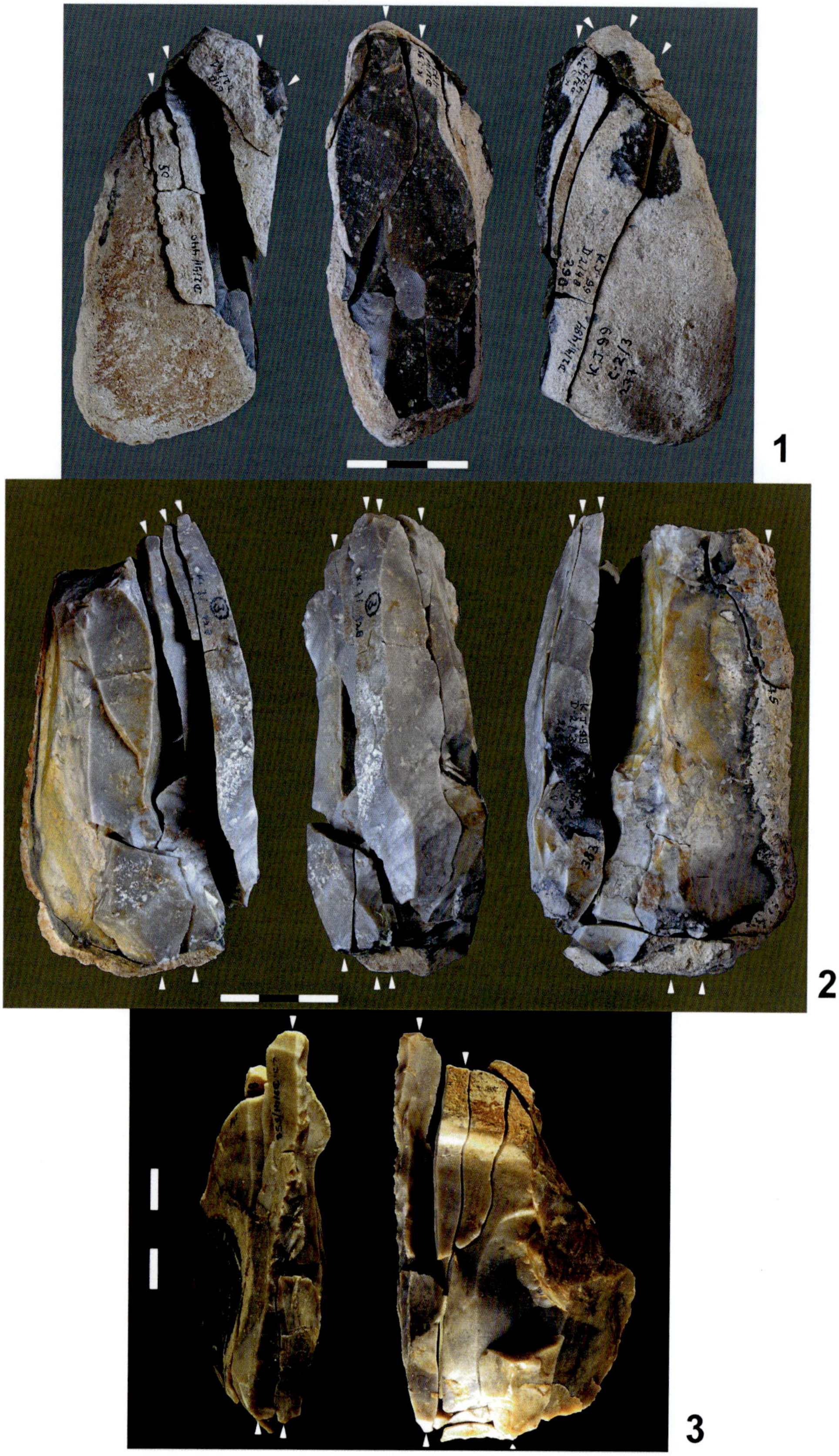

Fig. 14.14 Kraków—Księcia Józefa, EUP, layer II. Reconstructed prepared (crest) blade reduction sequences: partially turned unidirectional, narrow-faced core (*1*) partially turned bidirectional (*2*) and bidirectional, narrow-faced core-on-flake/burin-like (*3*)

The Middle and Early Upper Paleolithic of Greece: a View from Klissoura Cave

Southeast Europe, particularly the Balkans, has long been considered one of the possible routes of human dispersal from the Near East to Europe and *vice versa* (e.g. Kozłowski 1992). While the interior may have represented a geographic crossroads, the coastal Mediterranean regions, separated by mountains, may have acted more as a refugium than as a transit zone. Unlike Spain or Crimea, this zone did not witness a late Neandertal survival, but instead, a hiatus and an extinction of this population, at least on the western coast of Southeast Europe. In contrast, the LMP is followed by the EUP without interruption on the eastern coast (Papagianni 2009). Although the archaeological and anthropological record from this region of Europe and particularly from Greece is constantly increasing (see, e.g. Harvati 2016; Darlas and Psathi 2016), detailed comparative techno-typological studies of the lithic assemblages are rare. Based on the assemblages from Asprochaliko and Kokkinopilos, as well as on several surface collections, the MP variability in Greece has been traditionally represented by Levallois-Mousterian, Micromousterian, and Mousterian with bifacial leaf points (Higgs and Vita-Finzi 1966). To date, five rock shelters/caves with clear unmixed stratigraphic horizons have been analysed: Asprochaliko, Theopetra, Kalamakia, Lakonis, and Klissoura (Darlas 2007). Additional data from Kephalari Cave (Reisch 1980) unfortunately remain unpublished. According to Darlas (2007), these stratified assemblages belong to the "typical" Mousterian (the Quina Mousterian has not been definitively documented). Although dominated by non-Levallois methods, the Levallois (more flake recurrent centripetal than laminar) is more or less frequent, but almost always present (except for the "Micromousterian" at Asprochaliko). It is worth noting that local Levallois industries are less "pronounced" compared to adjacent regions, probably due to the lack of sufficient good quality raw materials.

The available dating record in Greece clusters into two chronological groups: (1) OIS 5e-a, based on luminescence dating and (2) OIS 3, *ca.* 50–35 ^{14}C kBP, according to ^{14}C dates (for dates and references see section "Technological Variability" above). Two Mousterian technological units correspond to each phase: (1) Levallois with laminar blanks and (2) small-scale non-Levallois with flakes. Recently, this two-part chrono-technological subdivision was tested, revealing some inconsistencies (Papagianni 2009). As stressed elsewhere (Sitlivy et al. 2007, 2008a), the Asprochaliko schema applied to the Greek MP does not fit the long multi-layered sequence (up to 6.5 m thick containing 26 archaeological layers) at Klissoura Cave 1. The MP sequence begins and ends with small-sized industries, while the less "microlithic" lowermost layers XVIII–XX with laminar, Levallois, and non-Levallois blanks, currently remain undated. Ages within OIS 3 should be regarded as minimal (Kuhn et al. 2010). Finally, intra-site variability confirms several overlapping technological "oscillations" throughout the Mousterian sequence containing these two broad technological groups. The abundant and representative assemblages from this site (> 100,000 MP artifacts currently studied) demonstrate that Levallois and laminar blanks, butt faceting, and selection of Levallois blanks for retouching persist throughout a long period of the archaeological sequence and become even more significant in some archaeological layers from the middle part of the deposit (Fig. 14.15: 1). This corresponds to the small-sized artifacts and the non-Levallois flake debitage being linked to the reduction of the main group of non-Levallois cores (Fig. 14.15: 2). The presence of Levallois and blade component at Klissoura indicates episodic "ups" and "downs" rather than steady decline from lowermost to uppermost layers, as well as a constant use of different flake and blade technologies. In summary, while the Klissoura analyses are still in progress, based on current data, the above-mentioned generalised technological trends reflect a temporal overlap during early OIS 3, at minimum, and do not represent distinct chronological divisions. At Lakonis I, the Levallois-Mousterian unit Ib, dated to *ca.* 39–43 ^{14}C kBP (Panagopoulou et al. 2002–2004; Elefanti et al. 2008), also contradicts this scheme. Additionally, the published lithics of the basal Mousterian at Asprochaliko (Gowlett and Carter 1997: Fig. 23.9) show mostly "pseudo-Levallois" removals (if the artifacts are oriented according to the debitage axis) rather than elongated final Levallois points. Thus, the technological record in Greece during the MP is much more diversified and raises serious issues regarding the Asprochaliko schema (see also Papagianni 2009). In situ MP industries evidence a comparable parallel presence of different flaking methods, as is also the case across other regions (see below). These MP technologies, as well as tool morphology and quantity (including foliates), with few available stratified sites at hand, fail to confirm any evolutionary trend towards the UP.

Several assemblages containing bifacial leaf points, collected from surface scatters together with Levallois-Mousterian and Aurignacian carinated scrapers, were viewed as a "transitional phase" (Darlas 1994) or used to support an acculturation model (Runnels 1988). They have never been found in stratigraphic contexts and reasonably evoke criticism (Darlas 2007; Papagianni 2009). Moreover, to date, bifacial points have not been documented in the EUP sequences in this area.

Non-Aurignacian EUP industries were recorded at Lakonis I and Klissoura Cave I, while a revised stratigraphy and TL date for the Theopetra layer II4 assemblage (previously interpreted as a "Transitional industry") is now confirmed as a post-depositional

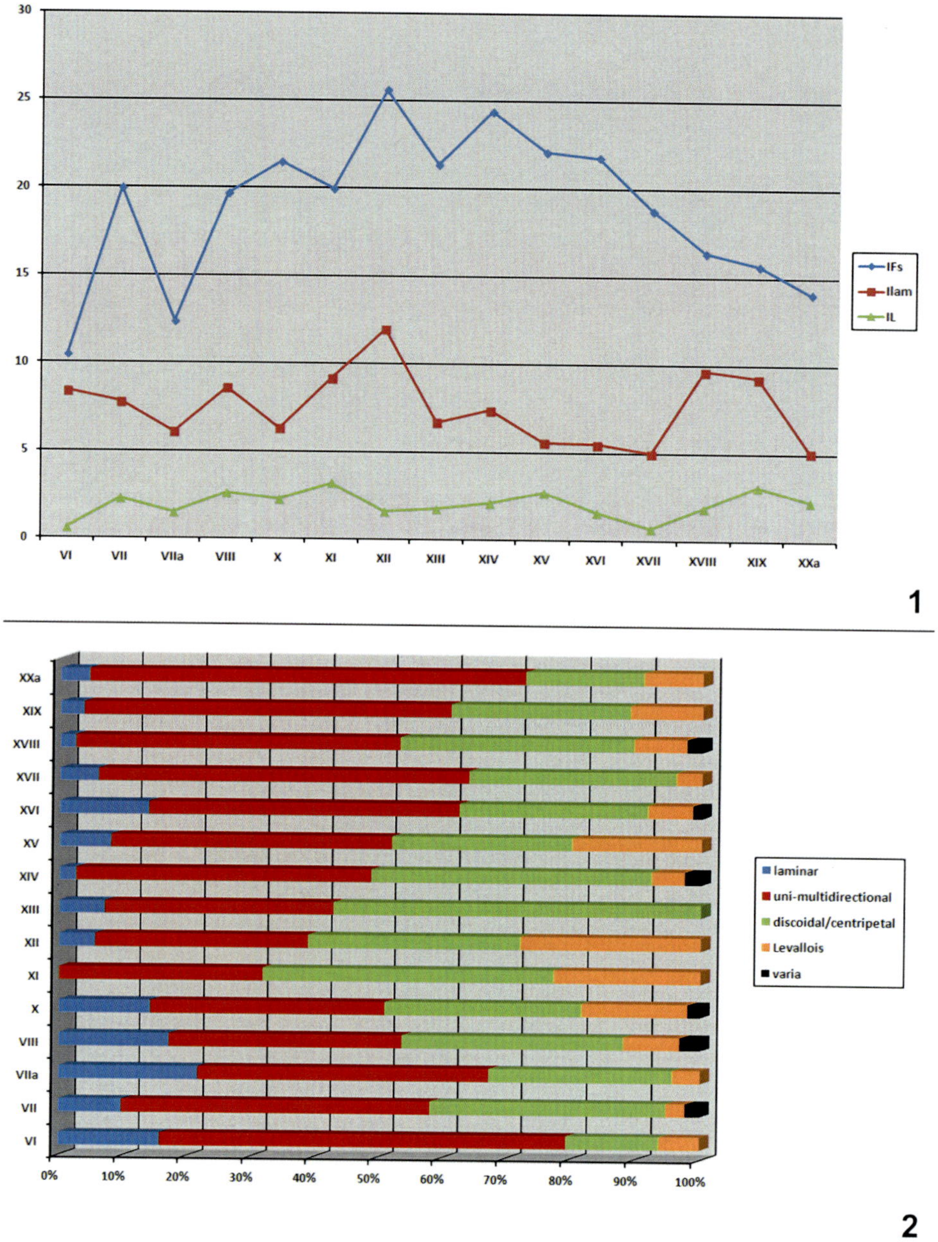

Fig. 14.15 Klissoura Cave 1. Mousterian sequence: platform faceting (IF strict), blade (Ilam), and Levallois (IL) indexes (*1*); core frequencies (*2*)

mixing of cultural material (Valladas et al. 2007). Lakonis I, unit Ia (*ca.* 44–38 ^{14}C kBP) is an important new site with an IUP industry found in association with a Neandertal molar (Panagopoulou et al. 2002–2004; Harvati et al. 2003; 2009; Elefanti et al. 2008). Technological continuity with the lowermost Levallois-Mousterian layer was recorded, as well as an *in situ* development with dominant UP elements in the technology (bladelet uni-/bipolar single-blow cores; core tablets, crested blades) and typology (retouched blades/bladelets, burins, end-scrapers, and truncations). Judging from the preliminary publication, these characteristics do not always show clear UP features (Darlas 2007). Nevertheless, the Levallois background is evident in the techno-typological structure of this industry, which is not the case of the second EUP occurrence found in Klissoura Cave 1.

The Uluzzian-type industry (layer V) at Klissoura was recovered between terminal Mousterian layer VI and Aurignacian layer IV and was radiocarbon dated to 40,010 ± 740 ^{14}C BP (Koumouzelis et al. 2001; see also Kuhn et al. 2010). This early date for the quite "evolved Uluzzian" was recently confirmed by the discovery of CI tephra which caps this industry and occurs at the interface of layers V and IV (Lowe et al. 2012). The flake/blade/bladelet-based Uluzzian assemblage with some discoidal cores (flake cores are "traditionally" uni-/bi-/multidirectional and partially represent final stages of blade core reduction), lacking Levallois elements and carinated bladelet production, does not show technological links with the underlying Mousterian or the overlying Aurignacian (Figs. 14.6, 14.12, and 14.13). Typologically speaking, the abundant curved backed tools (made on blades/bladelets and fewer flakes) distinguishes this assemblage from the Mousterian and the Aurignacian. However, the abundant *pièces esquillées* are a common core/tool class for all three industries, as well as sidescrapers and retouched flakes. Apart from the typological parallels (it should be noted that the tool type frequencies vary from assemblage to assemblage), some technological similarities with the Uluzzian of the middle/"evolved" laminar phase in Italy is evident in the reduction of the common uni-/bidirectional (including bipolar), multidirectional, and polyhedral cores (Palma di Cesnola 1983). As previously mentioned, very little information is available about the Uluzzian technology (Kuhn and Bietti 2000: 60) in spite of recent notable research (e.g. Riel-Salvatore 2007, 2009). Nevertheless, it seems that the difference between the Italian and Greek manifestations of the EUP lies not so much in the quantity of laminar products (flakes dominate the Klissoura debitage products as well), but in the fact that the blade/bladelets were more frequently modified into curved backed pieces/lunates—the *fossiles directeurs* of the Uluzzian (Fig. 14.16).

Several stratified Aurignacian occupations were documented in the uppermost Klissoura sequence, which spans from *ca.* 34–24 ^{14}C kBP (Koumouzelis et al. 2001; Kaczanowska et al. 2010). These assemblages are characterised by use of several independent/dissociated strategies: (a) bladelet (abundant carinated cores/scrapers), (b) blade (commonly based on exploitation of unprepared partly turned single-platform cores), and (c) flake production (unidirectional, discoidal, orthogonal, and polyhedral cores). The high frequency of *pièces esquillées* is notable. When compared to the Uluzzian assemblage from layer V, the technological shift in bladelet production is seen in the dominance of linear butts coupled with the absence of faceting, the significant decrease of straight lateral profiles, and the increase of a twisting pattern. Soft hammer percussion was commonly used: bulbs are mostly diffused or absent (Fig. 14.17).

At Franchthi Cave, the lithic assemblages from the CI tephra units (confirmed by new radiocarbon ages of 35 ^{14}C kBP or 40–39 cal kBP) clearly manifest Early Aurignacian features (Douka et al. 2011), which are techno-typologically similar to the lowermost Aurignacian at Klissoura, and thus blur the distinction between the Proto-Aurignacian and the Aurignacian I. This fact, and the comparable radiocarbon ages of the Early Aurignacian in Eastern and Western Europe (Higham et al. 2012), position the coastal Greek record directly at the centre of debate regarding the routes and the early stages of modern human dispersals in Europe.

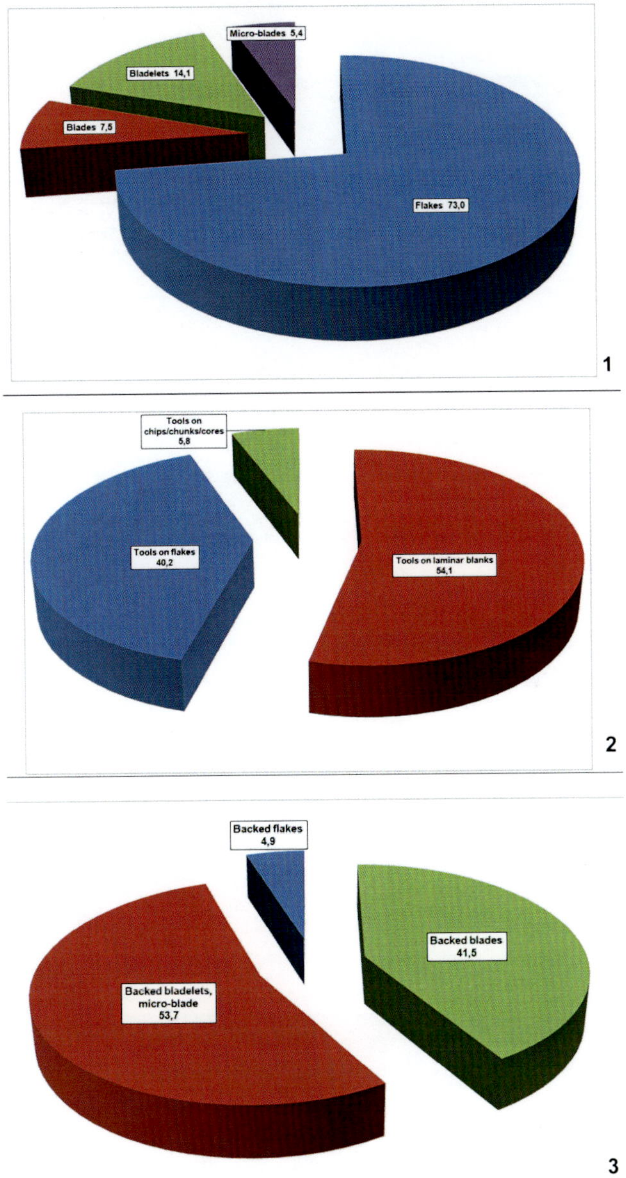

Fig. 14.16 Klissoura Cave 1, layer V. Uluzzian: debitage structure (*1*); tools, blank selection (*2*) and backed pieces, blank selection (*3*)

Discussion

Chronologically, the LMP covered a broad time span from OIS 5–3. During this period, Mousterian, Levallois-Mousterian, Blade Mousterian, Blade Levallois-Mousterian, as well as several IUP and EUP technocomplexes based on prevalent UP laminar production, or with various degrees of involvement (independent or fusion) of Levallois and non-Levallois methods were recorded. These technocomplexes and technologies usually do not show a strict temporal and spatial patterning. The Blade Levallois-Mousterian, which appeared rather late across Southeast, Central, and Eastern Europe during OIS 3, is an exception (Kabazi II, Shlyakh, and Piekary II; Fig. 14.18).

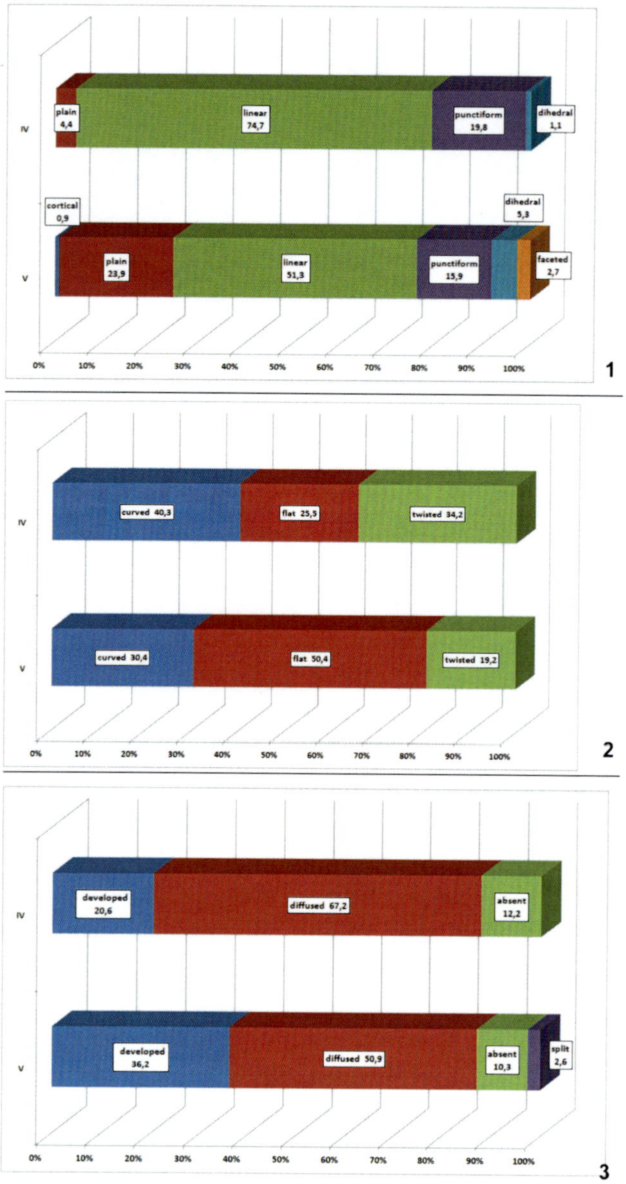

Fig. 14.17 Klissoura Cave 1, Uluzzian (layer V), and Aurignacian (layer IV) bladelet attributes: butts (*1*), lateral profiles (*2*), and bulb patterns (*3*)

The independent appearance and disappearance of prismatic laminar technologies is documented at different moments in time and space over the last 200,000 years. As non-Levallois blade production (Blade Mousterian) was recorded much earlier and before OIS 4 in Western Europe (e.g. Rheindahlen B1), the EMP provides only a scant record across the analysed area (Kraków-Zwierzyniec I, layer 2, Bougliv V/III, layer II, or Kourdiumovka). Research bias is a possible explanation. In contrast to the considerable variability of flake (Levallois and non-Levallois) methods, blade volumetric technology is rather homogeneous and displays no particular differences over time. Blade technologies in the European frame have a clearly chronological pattern, from the earliest in the west to the youngest towards the east. In some cases, these youngest assemblages show a decrease in butt faceting and Levallois component (e.g. in the long sequence of Kabazi II or Piekary II). The use of soft hammers is increasingly more common by the end of the MP. Nevertheless, these and other LMP industries with a developed blade production dating to about 45–30 kBP, remain clearly within the Mousterian frame, both technologically (persistence of various flake methods, e.g. Levallois, discoidal, or polyhedral) and typologically (absence or scarcity of UP tool-kit).

During the transitional period (*ca.* 45–30 kBP) fully MP industries coexisted with the novel Bohunician, as well as fully EUP blade non-Aurignacian and blade/let Aurignacian. Thus, the Bohunician trend documents a technological link between MP and UP technologies. It should be stressed, however, that assemblages containing both Levallois and prismatic blade production do not automatically provide proof of a "transitional" status and/or modernised behavior. Taphonomic, as well as refitting, studies are necessary in order to identify the fusion model of Bohunician/Emiran type. Unreliable excavated/contaminated collections with MP and UP elements cannot be interpreted as "transitional" and should be treated with caution.

Two possible scenarios can be proposed across Southeast, Central, and Eastern Europe during the Middle-to-Upper Paleolithic Transition: (1) continuity, local *in situ* development and (2) discontinuity/replacement or dispersal of technological innovations and different populations. Several examples of these scenarios are briefly discussed below based on selected multi-component sites.

In Central Europe, some archaeological assemblages testify to a local, *in situ* transition from MP to a non-Aurignacian EUP. One such example is located in the Ukrainian Transcarpathia, where two technological innovations led to the origin of the local non-Aurignacian UP and, probably, the Bohunician (Usik et al. 2006a, b: 228). The shift from unidirectional convergent core reduction for the production of short, broad-based Levallois points at Korolevo I, level IIb (OIS 4) to bidirectional blade debitage at Korolevo II, level II (ca. 38.5 kBP), or Bohunician (41–33 ^{14}C kBP and *ca.* 48 kBP TL) seemingly began with the use of short removals from the distal core end (as a starting point at Korolevo I, IIb) in order to prepare/adjust the working convexity of the Levallois point core. Another important technological indication for transition—crest installation—was also recognised at Korolevo II, layer IIb, where, during Levallois point production, *débordant* and naturally crested blades were consistently used. The technological transition apparently went through two separate modes and resulted in two quite different technocomplexes: (a) UP Levallois/Bohunician, where blades were by-products of point technology and (b) non-Levallois Korolevo II, layer II, in which blades became the

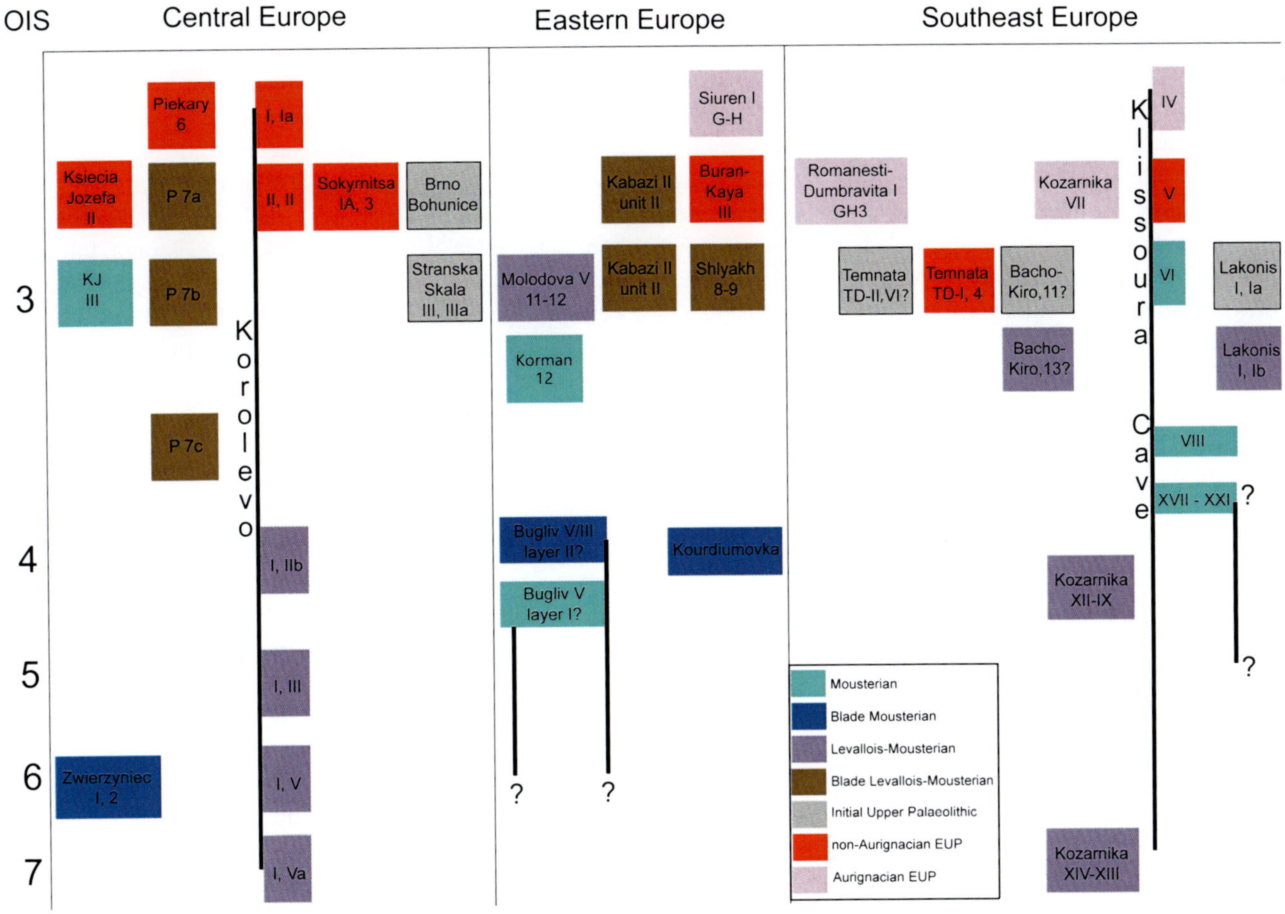

Fig. 14.18 Generalised chronology of analysed Middle and Early Upper Paleolithic assemblages in Central, Eastern, and Southeast Europe

main goal of core reduction (Usik et al. 2006a, b: 229). This local transitional scenario contradicts the migration model and is based on a similar technological trend confirmed by the reconstructed cores in the slightly earlier Boker Tachtit, level 1, in the Levant, and in the later Bohunician (e.g. Škrdla 2003a, b). Keeping in mind this impressive technological similarity, which possibly points to population movements or transmission of innovative ideas, a local development and the onset of a non-Aurignacian EUP in some European areas still remains a credible "parallel" scenario.

Technological continuity was observed during 60/48–32/26 kBP in the Kraków region at Piekary IIa (Fig. 14.9). The blank production throughout three LMP layers (7a, 7b, and 7c) was based on the independent use of UP blade and Levallois concepts (lineal and occasionally recurrent, in order to obtain flakes and points). The use of hard hammer and platform faceting was common. The tool-kit remained that of Mousterian type, with few UP tool types (backed blades, burins) present only in the lowermost assemblage 7c. The terminal Mousterian (level 7a) shows notable blade uni-/bidirectional reduction of prismatic cores with less pronounced Levallois recurrent flake debitage, as well as discoidal and Kombewa methods. Level 6 appeared as fully UP, maintaining the same kind of blade/bladelet core reduction (more bidirectional), and completely lacking the MP technological influence and techniques. The soft hammer percussion and platform rejuvenation by tablets with additional butt zone abrasion (elimination of overhang) was associated with prismatic blade-bladelet production. Thus, the gradual shift at Piekary IIa is even more visible, as the prismatic crested reduction was "invented" before the passage to the UP. The tool-kit in level 6 is a non-Aurignacian UP. Moreover, a few decorated ochre crayons ("lipsticks"—d'Errico and Vanhaeren 2008) were found at this level. It is worth mentioning that the industry of level 6 might be contemporaneous with the neighbouring "classical" Aurignacian cluster with carinated reduction at Piekary I.

The technological transition at Księcia Józefa from the late Mousterian layer III (some volumetric blade and Bohunician refits in a generally MP context) to the EUP layer II (exclusive crested prismatic blade/let production) during a short time interval (*ca.* 44–40 kBP) is less visible, probably due to a different site function (large base camp with knapping activities and restricted debitage zone). To summarise, parallel uni-/bidirectional reconstructed sequences in layer III produced more flakes than blades. However, several refits with unidirectional, partly turned exploitation support the existence of a clear UP model for blade production based on long (volumetric core-on-nodule) or short (narrow core-on-flake) reduction sequences, similar to the chronologically later exclusive laminar production in layer II. Curiously, most crested removals recovered in layer III (expected *fossile directeur* of UP blade production) were conjoined with polyhedral and even with some discoidal cores at different reduction stages. After refitting, the blade production in this assemblage became much less "crested". However, UP laminar "habits" were confirmed by clear reconstructed production sequences. The dominant, wasteful "crude"/"archaic" polyhedral flaking, resulting in volumetric cubic/polyhedral cores, numerous flakes, some blades, and technical removals (UP crests and various *débordants*), could have been related to blade production (e.g. apprenticing). Finally, using all of the available data, and especially the analysis of rich reconstructed material, some of the technological links between these very different assemblages at Księcia Józefa were clearly more visible.

In the Crimea, Levallois-based MP assemblages with advanced volumetric blade production resulted in an exclusively Mousterian tool-kit, without any links to the UP. This small region is unique in Southeast Europe for the temporal coexistence of four MP and EUP technocomplexes. Neandertals are associated with the Micoquian and AMH with the Aurignacian, while the makers of the Levallois-Mousterian and the EUP/Eastern Szeletian Buran-Kaya III, level C type industry, remain unknown (Chabai 2013). This scenario is based on interdisciplinary studies, various radiometric methods and solid stratigraphic contexts, such as overlapping of different entities (e.g. Micoquian above EUP at Buran-Kaya). These data confirm a long survival of the MP and reject the hypothesis of a sharp demise of Neandertals in Eastern Europe due to the "volcanic winter" before 40.0 cal kBP, as proposed recently (Golovanova et al. 2010). Beyond Crimea, the makers of all industries analysed are mainly unknown (except for Lakonis and Kalamakia—Harvati et al. 2003, 2013), due to the scarcity of human fossils, or to the imprecise anthropological definitions (see, e.g. Harvati 2016; Janković et al. 2016). Nevertheless, the Uluzzian in Italy has recently been found to be associated with modern humans (Benazzi et al. 2011), raising the possibility that the Uluzzian-type industry in Klissoura might have been produced by modern humans. However, this assignment as well as the integrity of the Uluzzian assemblage at Grotta del Cavallo (intrusion from overlying Epigravettian layers) has been questioned (Banks et al. 2013; Zilhão et al. 2015; but see Ronchitelli et al. 2014). Thus, as already stated, "*the modern nature of the Uluzzian makers will only be fully demonstrated with the discovery of new palaeontological evidence*" (Hublin 2015).

Taking into consideration that the dating of the long MP sequence at Klissoura is still in progress, the current record of Greece shows both continuity and discontinuity: (a) a rather short (OIS 5) chronological scenario of MP occupations, which are better represented during the latter span (OIS 4–3); (b) an *in situ* UP development from Levallois-based industry at Lakonis; (c) a replacement model with a short, but notable "Uluzzian episode" at Klissoura, followed by the long setting of the Aurignacian and (d) an early (*ca.* 35 ^{14}C kBP) appearance of the Aurignacian at Franchthi. At Lakonis, the shift probably took place between *ca.* 44 and 38 ka BP; at Klissoura, with more precision, *ca.* 40,000 years ago, just prior to the CI eruption. Moreover, evidence from Franchthi cave supports a continuous human occupation, despite the CI eruption. As for the MP, the existing record displays a monotonous presence of generally "typical" Mousterian with a "low-middle" frequency of the Levallois component (more for flakes than for blades and points) and more variable non-Levallois flake rather than blade volumetric technologies. This pattern, often coupled with small-sized industries, most likely stems from the mediocre quality of the available raw materials, which affected the initial sizes and the advances of the reduction processes. The Mousterian industries at Klissoura, with a considerable technological, typological/reduction mosaic and episodic technological "fluctuations", did not evolve into, or were not replaced by, another distinct MP entity. This MP sequence was capped by the late Mousterian (48,990 ± 1770 ^{14}C BP for layer VII and *ca.* 41 ^{14}C kBP for layer VI—Kuhn et al. 2010), showing no credible signs of a techno-typological link to the UP industries. The Mousterian was sharply replaced by the Uluzzian-type industry for a short period[1], and later by Aurignacian migrants, who remained there for quite a long time.

In some parts of the Mediterranean area of Southeast Europe, we can observe discontinuity, hiatus, and an absence of cultural overlap, which lead to hypotheses of depopulation due to migration or local extinction. If Neandertal groups did not return, there were no local Neandertals who survived late to compete with early AMH (Papagianni 2009). This scenario is very different from other regions of the Greater Mediterranean, e.g. Crimea, Italy, or Iberia. Part of

[1]The MSA sites in South Africa and especially late MSA of East Africa (Ambrose 2002; Brandt et al. 2012) are viewed as new "actors" in tracking this issue. Here, a surprising continuity of backed tools/segments together with *pièces esquilles*, similar to Uluzzian typological trend, is visible. Recent archaeological, genetic and demographic data led to the hypothesis of the dispersal of this innovative technology from an East African source into Europe, through the so-called southern Mediterranean route (Mellars 2011; Moroni et al. 2013) pre-dating the "Aurignacian". However these segments should be treated with caution due to younger admixture (piece on Fig. 6, 8 from Monchena Borago in Moroni et al. 2013), irregular backs or attribution of tablets to this tool category (personal communication of Ralf Vogelsang, November 22, 2013).

Greece would then resemble a rather empty cul-de-sac with no local survivors and few occasionally occurring EUP strangers. Does this reflect the state of research or a past reality? As demonstrated above, the so far unique long Middle/Upper Paleolithic sequence at Klissoura, with numerous dense human occupations, coupled with the early presence of a "classical" Aurignacian at Franchthi, as well as the growing regional archaeological record (see other contributions in this volume), suggests that we are dealing with a gap in our empirical data. Finally, tracking the transitions from the Middle to Upper Paleolithic in the Balkans, especially in Greece, requires updated interdisciplinary, radiometric and lithic analyses, including detailed comparative technological analyses, as well as additional intensive field investigations.

Conclusions

Technological analyses of debitage and core reduction sequences supported by numerous refitting data allow us to provide a more detailed subdivision of "cultural" entities during the MP and the Middle-to-Upper Paleolithic transition. Prior to *ca.* 30 kBP, the Mousterian and EUP of Southeast, Central, and Eastern Europe were characterised by their complex technological structure and a diversity of knapping methods and techniques was recorded for selected key assemblages.

Flake and Blade MP Methods

- Levallois lineal method for single preferential flake, or repetitive with "simple" and regular centripetal, orthogonal, rare unidirectional preliminary preparation. This method was seldom used alone and without repetition. It could be linked to the recurrent Levallois methods (represent different stages of core reduction) or it could accompany other technologies. This method occurred throughout most Mousterian industries of the analysed regions. Careful preparation of flaking surfaces and main platform was attested in older industries in Western Ukraine (Korolevo, layer III and Yezupil, layer III–OIS 5) as well as in more recent units in South Ukraine (Kabazi II–OIS 3). Certain variability in shaping, platform preparation, using blanks and sizes of final flakes is recognised: prevalence of classical centripetal tortoise preparation with only one reduction cycle (Korolevo, layer III), combination of unidirectional/opposite distal trimming, multi-convergent/distal trimming or oblique, transversal distal preparation (e.g. Proniatin), or by means of parallel blade or slightly convergent removals with distal preparation in Molodova V, layers 11 and 12, or by lateral and distal trimming from auxiliary platforms in early Unit II of Kabazi II (Usik 2006). Generally, platform faceting is lower in sites from the Kracow region, large desired flakes are rare (e.g. Piekary III), and the reduction displays repetitive character.
- Levallois lineal method for one or two points per prepared surface, or repetitive (several point generations per core) with unidirectional and rare bidirectional preliminary preparation. Levallois points spread generally after the last Intergacial, being scarce in eastern regions. Rare cases of a dominant point method were recorded (Korolevo, layer IIb). Normally, points occurred together with other Levallois debitage or were represented by occasional pieces/refits. Unidirectional method was common; sometimes preparation and maintenance of flaking convexity was achieved from opposed supplementary platform (or without) resulting in distal traces. *Débordant* blades or lateral fragmentation of the core formed rather convex flaking surfaces (the core tends to be partially turned) and were recorded in Korolevo IIb (OIS 4), Piekary 7b (early OIS 3), and occasionally later in Księcia Józefa III. Except in the later Bohunician, they are mostly short, wide based, with flake proportions.
- Levallois recurrent centripetal method for the production of several small short blanks. Centripetal shaping and subsequent production from faceted peripheral platform and re-preparation of flaking surface resulted in the repetitive production of several generations of flakes per core blank. This method was recorded during OIS 5e–OIS 3 (often in Greece, e.g. Kalamakia, units III and IV, Klissoura, layers VI–XX) or OIS 3 (Piekary IIa,7b, 7a) and probably at Bacho Kiro, layer 13.
- Levallois recurrent uni- and bidirectional method for production of elongated flakes and blades. Preliminary centripetal or lateral transversal shaping from auxiliary platforms was used. During reduction from one or opposed faceted platforms, several desired blanks and by-products were obtained giving the flaking surface an exclusively parallel pattern. It can appear together with other Levallois methods, coexist with them, or be technologically mixed in one chain, or dominate (Yezupil, layer III, Kabazi II, Unit II). This method was widespread both in time (OIS 7–OIS 3) and space.

Technique: direct hard hammer was common. However, association of faceted butts with diffuse bulbs and lips suggests soft hammerstone technique in Kabazi II (sandstone hammer).

Non-Levallois flake debitage is mostly recurrent unprepared with flat (centripetal, uni-/bidirectional, crossed, Kombewa), secant (discoidal), and change orientation/multidirectional (polyhedral) exploitation of cores, often resulting in various flakes and some blades. Fine platform faceting is uncommon.

These methods were widespread both in time (OIS 6–OIS 3) and space.

- The centripetal method was based on reduction of a block or flake, showing at the initial stage a Kombewa pattern (common for Greek assemblages, e.g. Klissoura VI–XX, throughout the OIS 3 and possibly earlier; Asprochaliko, upper layer).
- The discoidal method is represented by unifacial and bifacial models with several sequences based on reduction of cores with high section: conical/biconical and even with a tendency to form a pyramidal shape (Bugliv V, layer I, OIS 6?/OIS 4?; Księcia Józefa, layer III, OIS 3).
- The polyhedral method was based on exploitation of entire volume of initial block with continuous turning and moving from one surface to another (adjacent) face and was ensured by the alternation of surface and surface/ridge reduction (Księcia Józefa, layer III, Korman IV, layer 12–OIS 3; possibly earlier in Klissoura).

In many Mousterian industries, these methods were independent from other technologies and display all reduction stages (exceptions are discoidal/polyhedral fusion in Bougliv V, layer I or Księcia Józefa, layer III). These technologies can dominate (polyhedral in Księcia Józefa, layer III), and more often coexist with other methods (Klissoura Mousterian) or play a supplementary role in technological structures of analysed industries (discoidal and centripetal methods accompanied Levallois and/or blade production—Piekary IIa, Kourdiumovka).

Technique: direct percussion with a hard hammerstone was used in all cases.

Blade UP Methods

Blade production (direct and prepared) is typical for all analysed EUP assemblages, but also emerged within MP industries during OIS 6 (?), OIS 4, and early OIS 3 (Tables 14.1, 14.2, and 14.3). Direct secondary blade debitage (re-utilisation of abandoned cores by making an additional flaking surface) occurred mostly in the LMP and IUP (e.g. Samuilitsa II—Tsanova 2006 or Bohunician—Škrdla 1996).

- Direct (with no preliminary shaping of debitage surfaces), uni-, and bidirectional exploitation of voluminous and flat nodules, blocks, flakes, or blades. Preparation of the platform consists mainly in the making of plain platforms (faceted are less frequent) and sometimes in the partial removal of cortex. Exploitation continued due to the natural convexities of initial core blank. Maintenance was insured by retrieval of lateral cortical/natural and full debitage *débordant* removals (blades, flakes), by plunging blades, by bidirectional reduction (with a different order of detachments on one or several alternate surfaces, alternating or twisted), and finally, by changing the debitage orientation. Flaking proceeded both from a large surface with detachment on a narrow side(s) or from narrow side(s) via the large surface. Initial raw material peculiarities often determine the mode of core exploitation. Narrow and wedge-like core reduction was based normally on flat blocks/*plaquettes* and flakes, when turned and partially turned cores with large surfaces were linked to voluminous oblong nodules, pebbles, or chunks. During debitage, platforms were rejuvenated, often by means of tablets; platform/working surface zone was also rearranged by trimming or abrasion. Usually, this model occurred with prepared (by crests) blade manufacturing (but see, e.g. Bougliv V and Kourdiumovka–OIS 4). Also, direct blade reduction sequences were reconstructed by refitting in non-Levallois and non-laminar industries of Księcia Józefa, layer III and Korman IV, layer 12–OIS 3).
- Prepared flaking with (central and lateral) or preliminary bifacial crest installation. Crested blades with two prepared slopes document a UP type of core shaping (narrow-faced and wedge-like cores). Lateral and partial crest preparation was also often used for initialisation of blade debitage and continuation of blade production (neo-crests). Sometimes the beginning of core reduction passed in a direct manner and the crest installation appeared on the subsequent stages. Bifacial shaping occurred rarely in Levallois-Mousterian contexts (large pre-forms in Piekary IIa, layer 7c and 7a), while in the EUP layer II in Korolevo II this model prevailed. Platforms were prepared by one or several removals, often faceted and restored by full or partial tablets. Elimination of overhang by trimming or abrasion is a common practice. Bidirectional mode of core exploitation occurred very often. Localisation of working surfaces and debitage order could be summarised in several main models: a) on the largest side; initiation from lateral edge (rare two edges) via large surface; b) from a large flaking surface on one or two narrow parts; or c) on a narrow edge with or without extension on the side(s). Prepared blade debitage occurred in different parts of analysed areas mostly during the OIS 3 (probably early in the lowermost Mousterian at Klissoura).

Technique: mostly direct hard hammer percussion (MP); direct soft and hard hammerstones (EUP).

Combined Methods: Technological Transition

Technological mixing or fusion of blade UP and Levallois point, blade and flake production in Europe was recorded and documented by numerous refitted reduction sequences at Brno-Bohunice and Stranska skala III, IIIa between 43 and 36 kBP and 41 and 34.5 kBP, respectively (Svoboda and Škrdla 1995; Škrdla 1996; Svoboda 2004). This peculiar technology (confirmed by refitting) emerged at Księcia Józefa, layer III (~44 kBP) in a generally Mousterian technological and typological context. Such fusion of core reduction methods documented a technological shift from MP to UP blank production. Currently, no data exists to support this trend in other "transitional" industries.

Bladelet Methods

Direct and prepared bladelet production usually occurs in EUP non-Aurignacian and especially Proto-Aurignacian/Aurignacian assemblages. Final lamellar blanks were often modified by retouching/truncation into different tools, including innovative projectiles (backed pieces, Dufours). Detachment of bladelets and micro-blades was frequently made by direct soft hammer technique. Several methods based on exploitation of different blade/let or bladelet cores were recorded: prismatic (direct and prepared by crests, uni-/bi-/multidirectional), narrow-faced/burin-like (direct and prepared by crests, uni-/bi-/multidirectional), carinated and splintered pieces (cores/tools). Lamellar production exhibits technological trends which manifested about *ca.* 40 kBP.

- Independent use of non-carinated blade/let (with reduction continuum), bladelet burin-like, and splintered strategies (Uluzzian/Klissourian, layer V; non-Aurignacian of Kraków and Transcarpathia from *ca.* 40 kBP and latter).
- Prevalent prismatic blade/let reduction continuum; rare carination (Kozarnikian; *ca.* 39–36 kBP).
- Independent use of several strategies with blade/let reduction continuum of prismatic, narrow-faced and carinated cores (Banat Proto-/Early Aurignacian; *ca.* 45–40 kBP).
- Carinated bladelet strategy (Early Aurignacian at Franchthi, stratum Q; >35 kBP, within CI tephra).
- Independent use of blade, blade/let (prismatic), and bladelet (carinated, splintered) strategies (e.g. Aurignacian at Klissoura; from 34 till 24 kBP).

The technological analyses of core reduction sequences provide a more detailed subdivision of Mousterian, IUP and EUP units, which can be used to identify different blank production strategies, and to trace possible transformations before and during the transitional period. Several Levallois and non-Levallois methods of flake, blade and point production were documented in Southeast, Central, and Eastern Europe during MP (OIS 7–OIS 3). Debitage methods were used (a) *independently* and coexisted in one industry or represented one dominant system and (b) were *mixed in one reduction sequence*. The Middle Paleolithic technological structure appears to include a series of independent methods of blank production: Levallois and/or non-Levallois, and the Upper Paleolithic laminar production. The Initial Upper Paleolithic and Early Upper Paleolithic technological structures followed several models as well: (a) exclusive/dominant Upper Paleolithic laminar strategies, (b) a combination of these strategies with different Middle Paleolithic technologies and (c) fusion/mixing in one reduction sequence of volumetric blade and "flat" Levallois point concepts. The latter innovative model is central to interpretations regarding the emergence of the UP technological package. There is no specific and clear spatial and chronological clustering of technologies, nor is there a single evolutionary trend from "archaic" to "developed" methods. Industries with UP volumetric concept of systematic prepared (by cresting) blade production accompanied by mostly Levallois methods, as well as the combination of MP and UP tools (e.g. backed blades, burins) occurred at the beginning of OIS 3. Less important typological variability and complex technological structure reflected in lithic production of this area occurred before 30 kBP. The transition from the Middle to Upper Paleolithic in Southeast, Central, and Eastern Europe demonstrates both continuity and discontinuity for specific regions and assemblages within a considerable technological mosaic. This synthesis is intended also to stimulate further inter-regional technological investigations.

Acknowledgements I am grateful to Katerina Harvati for inviting me to the stimulating Symposium, which took place in Tübingen in December 2012 and led to this volume. Many colleagues and institutions contributed to the fieldwork and current analyses in Greece (Institute for Aegean Prehistory) and Romania (Deutsche Forschungsgemeinschaft, CRC 806, Project B1). I express my deep gratitude to Margarita Koumouzelis for her patience and support in Klissoura studies, to Krzysztof Sobczyk for productive collaboration in Poland and Greece, Magdalena Sudol for drawing finalisation, and Nicholas Thompson for photos and English editing, to Mircea Anghelinu, Loredana Niţă, Rebecca Miller, and Victor Chabai for their suggestions, text corrections, and fruitful discussions. Three reviewers, Katerina Harvati and Mirjana Roksandic (editors) helped in consistently improving this chapter. To all people and institutions involved, I express my deep gratitude.

References

Ambrose, S. H. (2002). Small things remembered: Origins of early microlithic industries in sub-Saharan Africa. In Elston, R. & Kuhn, S. (Eds.), *Thinking small: Global perspectives on microlithization. Archaeological Papers of the American Anthropological Association, 12*(1), 9–29.

Anghelinu, M., Niţă, L., Sitlivy, V., Uthmeier, T., & Băltean, I. (2012). Looking around Peştera Cu Oase: The beginnings of Upper Paleolithic in Romania. *Quaternary International, 274*, 136–157.

Bailey, G. N., Carter, P. L., Gamble, C. S., & Higgs, H. P. (1983). Asprochaliko and Kastritsa: Further investigations of Paleolithic settlement and economy in Epirus (North-West Greece). *Proceedings of the Prehistoric Society, 49*, 15–42.

Banks, W. E., d'Errico, F., & Zilhão, J. (2013). Humane climate interaction during the early Upper Paleolithic: Testing the hypothesis of an adaptive shift between the proto-Aurignacian and the early Aurignacian. *Journal of Human Evolution, 64*, 39–55.

Bar-Yosef, O. (2006). Defining the Aurignacian. In O. Bar-Yosef & J. Zilhão (Eds.), *Towards a definition of the Aurignacian* (pp. 11–18). Lisboa: Instituto Português de Arqueologia.

Bar-Yosef, O., & Kuhn, S. L. (1999). The big deal about blades: Laminar technologies and human evolution. *American Anthropologist, 101*(2), 322–328.

Belfer-Cohen, A., & Goring-Morris, A. N. (2009). The shift from the Middle Paleolithic to the Upper Paleolithic: Levantine perspectives. In M. Camps & C. Szmidt (Eds.), *The Mediterranean from 50,000 to 25,000 BP: Turning points and new directions* (pp. 89–100). Oxford: Oxbow Books.

Benazzi, S., Douka, K., Fornai, C., Bauer, C., Kullmer, O., Svoboda, J., Pap. I., Mallegni, F., Bayle, P., Coquerelle M., Condemi, S., Ronchitelli, A., Harvati, K. & Weber, G.W. (2011). Early dispersal of modern humans in Europe and implications for Neanderthal behaviour. *Nature 479*, 526–529.

Bodu, P., Karlin, C., & Ploux, S. (1990). Who's who? The Magdalenian flintknappers of Pincevent (France). In E. Cziesla, S. Eickhoff, N. Arts, & D. Winter (Eds.), *The big puzzle studies in modern archaeology* (Vol. 1, pp. 143–165). Bonn: Holos.

Boguckyj, A., Cyrek, K., Konecka-Betley, K., Lanczont, M., Madeyska, T., Nawrocki, J., et al. (2001). Paleolithic loess-site Yezupil on Dnister (Ukraine)—stratigraphy, environment and cultures. *Studia Quaternaria, 18*, 25–46.

Borziak, I. A., Grigorieva, G. V., & Ketraru, N. A. (1981). *Poselenia Dreviekamennogo Veka na Sevro-Zapade Moldavii*. Kishinev: Shtiintsa (In Russian).

Bosinski, G. (1966). Der paläolitische Fundplatz Rheindahlen, Ziegelei Dreesen-Westwand. *Bonner Jahrbucher, 166*, 319–343.

Brandt, S. A., Fisher, E. C., Hildebrand, E. A., Vogelsang, R., Ambrose, S. H., Lesur, J., et al. (2012). Early MIS 3 occupation of Mochena Borago Rockshelter, Southwest Ethiopian Highlands: Implications for Late Pleistocene archaeology, paleoenvironments and modern human dispersals. *Quaternary International, 274*, 38–54.

Brantingham, P. J., Kuhn, S. L., & Kerry, K. W. (Eds.). (2004). *The Early Upper Paleolithic beyond Western Europe*. Berkley: University of California Press.

Camps, M., & Szmidt, C. (Eds.). (2009). *The Mediterranean from 50,000 to 25,000 BP: Turning points and new directions*. Oxford: Oxbow Books.

Chabai, V. P. (2004). *The Middle Paleolithic of Crimea*. Simferopol: Slyakh.

Chabai, V. P. (2013). The destiny of the last Eastern European Neanderthals. *Ukrainian Archaeology, 2013*, 7–27.

Chabai, V. P., Marks, A. E., & Monigal, K. (2004). Crimea in the context of the Eastern European Middle Paleolithic and Early Upper Paleolithic. In V. Chabai, K. Monigal & A. Marks, *The Middle Paleolithic and Early Upper Paleolithic of Eastern Crimea* (Vol. 3, pp. 419–460). Liège: ERAUL 104, Université de Liège.

Chabai, V. P., Richter, J., & Uthmeier, T. (Eds.). (2006). *Kabazi II: The 70,000 years since the Last Interglacial: Vol. 2. Paleolithic sites of Crimea*. Simferopol: Shlyakh.

Chabai, V. P., & Sitlivy, V. (1993). The periodization of core reduction strategies of the Ancient, Lower, and Middle Paleolithic. *Préhistoire Européenne, 5*, 9–88.

Chmielewski, W., Betley-Konecka, K., & Madeyska, T. (1977). Paleolithic Site Kraków-Zwierzyniec I in the Lights of Investigations 1972–1974. *Biuletyn Instytutu Geologii, 305*, 13–30.

d'Errico, F., & Vanhaeren, M. (2008). Microscopic and technological analysis of decorated ochre crayons from Piekary IIa, Layer 6. Implications for the emergence of symbolism in Europe. In V. Sitlivy, A. Zięba & K. Sobczyk (Eds.), *Middle and Early Upper Paleolithic of Kraków region. Piekary IIa. Vol. 6. Monographie de Préhistoire générale* (pp. 149–158). Bruxelles: Musées Royaux d'Art et d'Histoire.

Darlas, A. (1994). Le Paléolithique inférieur et moyen de Grèce. *L'Anthropologie, 98*, 305–328 (La fin du moyen et le début du Paléolithique supérieur en Grèce).

Darlas, A. (2007). Le Moustérien de Grèce à la lumière des récentes recherches. *L'Anthropologie, 111*, 346–366.

Darlas, A., & de Lumley, H. (2004). La grotte de Kalamakia (Aréopolis, Grèce). Sa contribution à la connaissance du Paléolithique moyen de Grèce. In *XIVe Congrès UISPP, Liège, 2–8 Septembre 2001. Actes de la 5e Section: Le Paléolithique moyen: Vol. 1239. BAR International Series* (pp. 225–233). Oxford: Archaeopress.

Darlas, A., & Psathi, E. (2016). The Middle and Upper Paleolithic on the Western Coast of the Mani Peninsula (Southern Greece). In K. Harvati & M. Roksandic (Eds.), *Paleoanthropology of the Balkans and Anatolia: Human evolution and its context*. Dordrecht: Springer.

Demidenko, Y. E., & Noiret, P. (2012). The Siuren-I Aurignacian of Krems-Dufour type industries in the context of the European Aurignacian. In Y. E. Demidenko, M. Otte & P. Noiret (Eds.), *Siuren I Rock-shelter. From Late Middle and Early Upper Paleolithic to Epi-Paleolithic in Crimea: Vol. 4. The Paleolithic of Crimea* (pp. 343–357). Liège: ERAUL 129, Université de Liège.

Demidenko, Y. E., & Usik, V. I. (1993). The problem of changes in Levallois technique during the technological transition from the Middle to the Upper Paleolithic. *Paléorient, 19*(2), 5–15.

Derevianko, A. P. (Ed.). (2005). The transition from the Middle to Upper Paleolithic in Eurasia: Hypothesis and facts. In *Archaeology, Ethnology & Antropology of Eurasia*, special issue. Novosibirsk.

Derevianko, A. P. (2010). Three scenarios of the Middle to Upper Paleolithic transition. *Archaeology Ethnology & Anthropology of Eurasia, 38*(4), 2–38.

Doboş, A., & Trinkaus, E. (2012). A new AMS radiocarbon date for Middle Paleolithic Layer 4 of Ripiceni-Izvor, Romania. *Materiale şi Cercetări Arheologice (Serie Nouă), 8*, 7–10.

Douka, K., Perlès, C., Valladas, H., Vanhaeren, M., & Hedges, R. E. M. (2011). Franchthi Cave revisited: The age of the Aurignacian in south-eastern Europe. *Antiquity, 85*, 1131–1150.

Drobniewicz, B., Ginter, B., Ivanova, S., & Sirakov, N. (1982). Middle Paleolithic finds. In J. K. Kozłowski (Ed.), *Excavation in the Bacho Kiro Cave (Bulgaria). Final Report* (pp. 81–117). Warszawa: Panstwowe Wydawnictwo Naukowe.

Drobniewicz, B., Ginter, B., Kazior, B., & Kozłowski, J. K. (2000a). "Transitional" industry from layer VI, trench TD-II. In B. Ginter, J. K. Kozłowski, J.-L. Guadelli & H. Laville, *Temnata Cave. Excavations in Karlukovo Karst Area, Bulgaria* (Vol. 2.1, pp. 243–316). Warszawa: Panstwowe Wydawnictwo Naukowe.

Drobniewicz, B., Ginter, B., Kazior, B., & Kozłowski, J. K. (2000b). Early Upper Paleolithic assemblages from layer 4, trench TD-I. In B. Ginter, J. K. Kozłowski, J.-L. Guadelli & H. Laville, *Temnata Cave. Excavations in Karlukovo Karst Area, Bulgaria* (Vol. 2.1, pp. 333–418). Warszawa: Panstwowe Wydawnictwo Naukowe.

Elefanti, P., Panagopoulou, E., & Karkanas, P. (2008). The transition from the Middle to the Upper Paleolithic in the southern Balkans: The evidence from the Lakonis I Cave, Greece. *Eurasian Prehistory, 5*, 85–95.

Foltyn, E., & Kozłowski, J. K. (2003). The Lower Level of the Site of Dierżysław I, Opole voivodship (Silesia, Poland) and the problem of Bohunician. *Eurasian Prehistory, 1*(2), 79–116.

Gladilin, V. N. (1989). The Korolevo Paleolithic site: Research methods, stratigraphy. *Anthropologie (Brno), 27*(2), 93–103.

Gladilin, V. N., & Demidenko, Y. E. (1989). Upper Paleolithic tool complexes from Korolevo. *Anthropologie (Brno), 27*(2–3), 143–178.

Gladilin, V. N., & Sitlivy, V. (1990). *The Acheulian in Central Europe*. Kiev: Naukova Dumka.

Golovanova L.V., Doronichev V.B., Cleghorn N.E., Koulkova M.A., Sapelko T.V., Shackley S. (2010). Significance of Ecological Factors in the Middle to Upper Paleolithic Transition. *Current Anthropology, 51*(5), 655—679.

Gowlett, J., & Carter, P. (1997). The Basal Mousterian of Asprochaliko Rochshelter, Louros Valley. In G. N. Bailey (Ed.), *Klithi: Paleolithic Settlement and Quaternary Landscapes in Northwest Greece* (pp. 441–457). Cambridge: McDonald Institute for Archaeological Research.

Guadelli, J.-L., N. Sirakov, S. Ivanova, S. Sirakova, E. Anastassova, P. Courtaud, I., et al. (2005). Une séquence du Paléolithique inférieur au Paléolithique récent dans les Balkans: la grotte Kozarnika à Orechets (Nord-Ouest de la Bulgarie). In N. Molines, M.-H. Moncel & J.-L.Monnier (Eds.), *Les premiers peuplements en Europe. Colloque international Données récentes sur les modalités de peuplement et sur le cadre chronos-tratigraphique, géologique et paléogéographique des industries du Paléolithique ancien et moyen en Europe, Rennes, 22–25 septembre 2003: Vol. 1364. BAR International Series* (pp. 87–103). Oxford: Archaeopress.

Haesaerts, P., Borziak, I., Chirica, V., Damblon, F., Koulakovska, L., & Van der Plicht, J. (2003). The East Carpathian Loess Record: A reference for the Middle and Late Pleniglacial stratigraphy in Central Europe. *Quarternaire, 14*(3), 163–188.

Haesaerts, P., & Koulakovskaya, L. V. (2006). La séquence paléosedimentaire de Korolevo (Ukraine transcarpathique): Contexte chronostratigraphique et chronologique. In L. V. Koulakovskaya (Ed.), *The European Middle Paleolithic* (pp. 21–38). Kiev: Shlyakh.

Harvati, K. (2016). Paleoanthropology in Greece: Recent finds and implications. In K. Havarti & M. Roksandic (Eds.), *Paleoanthropology of the Balkans and Anatolia: Human evolution and its context* (pp. 3–14). Dordrecht: Springer.

Harvati, K., Darlas, A., Bailey, S. E., Rein, T. R., El Zaatari, S., Fiorenza, L., et al. (2013). New Neanderthal remains from Mani peninsula, Southern Greece: The Kalamakia Middle Paleolithic Cave site. *Journal of Human Evolution, 64*(6), 486–499.

Harvati, K., Panagopoulou, E., & Karkanas, P. (2003). First Neanderthal remains from Greece: The evidence from Lakonis I. *Journal of Human Evolution, 45*, 465–473.

Harvati, K., Panagopoulou, E., & Runnels, C. (2009). The paleoanthropology of Greece. *Evolutionary Anthropology, 18*(4), 131–143.

Higgs, E. S., & Vita-Finzi, C. (1966). The climate, environment and industries of Stone Age Greece: Part II. *Proceedings of the Prehistoric Society, 32*, 1–29.

Higham, T., Bassel, L., Jacobi, R., Wood, R., Bronk Ramsey, C., & Conard, N. J. (2012). Testing models for the beginnings of the Aurignacian and the advent of figurative, art and music: The radiocarbon chronology of Geißenklösterle. *Journal of Human Evolution, 62*, 664–676.

Hoffecker, J. F. (2009). The spread of modern humans in Europe. *Proceedings of the National Academy of Sciences, USA, 106*(38), 16040–16045.

Honea, K. (1981). New Romanian radiocarbon dates: Middle Paleolithic, Mesolithic, Neolithic. *American Journal of Archaeology, 85*(4), 483–486.

Hublin, J.-J. (2015). The modern human colonization of western Eurasia: When and where? *Quaternary Science Reviews, 118*, 194–210.

Huxtable, J., Gowlett, J. A. J., Bailey, G. N., Carter, P. L., & Papaconstantinou, V. (1992). Thermoluminescence dates and new analysis of the early Mousterian from Asprochaliko. *Current Anthropology, 33*, 109–114.

Ivanova, I. K. (1987). Paleogeography and paleoecology of the environment of Stone Age man inhabitation in the Middle Dniester, site of Molodova V. In I. K. Ivanova & S. M. Tseitlin (Eds.), *The multilayered Paleolithic site Molodova V: Stone age man and environment* (pp. 94–124). Moscow: Nauka.

Janković, I., Ahern, J. C. M., Karavanić, I., & Smith, F. H. (2016). The importance of Croatian Pleistocene hominin finds in the study of human evolution. In K. Harvati & M. Roksandic (Eds.), *Paleoanthropology of the Balkans and Anatolia: Human evolution and its context*. Dordrecht: Springer.

Kaczanowska, M., Kozlowski, J. K., & Sobczyk, K. (2010). Upper Paleolithic human occupations and material culture at Klissoura Cave 1. *Eurasian Prehistory, 7*(2), 133–285.

Kolesnik, A. V. (2000). The technological analyzes of Middle Paleolithic flint industry from Kurdumovka (Southeastern Ukraine). *Archaeological Almanac, 9*, 65–80.

Kolesnik, A. V. (2003). The Middle Paleolithic of Donbass. *Archaeological Almanac, 12*, Special issue.

Koumouzelis, M., Kozlowski, J., Escutenaire, C., Sitlivy, V., Sobczyk, K., Valladas, H., et al. (2001). La fin du Paléolithique moyen et le début du Paléolithique supérieur en Grèce: La séquence de la grotte 1 de Klissoura. *L'Anthropologie, 105*, 469–504.

Kozłowski, J. K. (1992). The Balkans in the Middle and Upper Paleolithic: The gate to Europe or a cul-de-sac? *Proceedings of the Prehistoric Society, 58*, 1–20.

Kozłowski, J. K. (2000). The problem of cultural continuity between the Middle and Upper Paleolithic in Central and Eastern Europe. In O. Bar-Yosef & D. Pilbeam (Eds.), *The geography of Neandertals and Modern Humans in Europe and the Greater Mediterranean* (Bulletin 8. pp. 77–105). Boston: Peabody Museum.

Kozłowski, J. K. (2004). Early Upper Paleolithic Backed Blade Industries in Central and Eastern Europe. In P. L. Brantingham, S. L. Kuhn, & K. W. Kerry (Eds.), *The Early Upper Paleolithic beyond Western Europe* (pp. 14–29). Berkley: University of California Press.

Kuhn, S. L. (2003). In what sense is the Levantine Initial Upper Paleolithic a "transitional industry". In J. Zilhão & F. d'Errico (Eds.), *The chronology of the Aurignacian and of the transitional technocomplexes: Dating, stratigraphies, cultural implications. Proceedings of Symposium 6.1 of the XIVth Congress of the UISPP, University of Liège, Belgium, 2–8 September, 2001.* (Trabalhos de Arqueologia, Vol. 33, pp. 61–70). Lisboa: Instituto Português de Arqueologia.

Kuhn, S. L., & Bietti, A. (2000). The late Middle and early Upper Paleolithic in Italy. In O. Bar-Yosef & D. Pilbeam (Eds.), *The geography of Neandertals and modern humans in Europe and the Greater Mediterranean* (Bulletin 8, pp. 49–76). Cambridge: Peabody Museum.

Kuhn, S., Pigati, J., Karkanas, P., Koumouzelis, M., Kozlowski, J., Ntinou, M., et al. (2010). Radiocarbon dating results from the Early Upper Paleolithic of Klissoura 1 Cave. *Eurasian Prehistory, 7*(2), 37–46.

Kuhn, S. L., & Zwyns, N. (2014). Rethinking the initial Upper Paleolithic. *Quaternary International, 347*, 29–38.

Lowe, J., Barton, N., Blockley, S., Ramsey, C. R., Cullen, V. L., Davies, W., et al. (2012). Volcanic ash layers illuminate the resilience of Neanderthals and early modern humans to natural hazards. *Proceedings of the National Academy of Sciences, USA, 109*, 13532–13537.

Marks, A., & Volkman, P. (1983). Changing core reduction strategies: A technological shift from the Middle to the Upper Paleolithic in the Southern Levant. In E. Trinkaus (Ed.), *The Mousterian legacy: Human biocultural change in the Upper Pleistocene: Vol. 164. BAR International Series* (pp. 13–34). Oxford: Archaeopress.

Meignen, L. (1994). Paléolithique moyen au Proche-Orient: Le phénomène laminaire. In S. Révillion & A. Tuffreau (Eds.), *Les industries laminaires au Paléolithique moyen* (pp. 125–159). Paris: CNRS.

Meignen, L., & Bar-Yosef, O. (2002). The Lithic industries of the Middle and Upper Paleolithic of the Levant: Continuity or break? *Archaeology, Ethnology & Antropology of Eurasia, 3*(11), 12–21.

Meignen, L., Geneste, J.-M., Koulakovskaia, L., & Sytnik, A. (2004). Koulichivka and its place in the Middle-Upper Paleolithic transition in Eastern Europe. In P. L. Brantingham, S. L. Kuhn, & K. W. Kerry

(Eds.), *The Early Upper Paleolithic beyond Western Europe* (pp. 50–63). Berkeley: University of California Press.

Mellars, P. (2011). The earliest modern humans in Europe. *Nature, 479*, 483–485.

Morawski, W. (1992). Kompleks stanowisk paleolitycznych w Piekarach. In *Prądnik, Prace i materiały Muzeum im. prof. Władysława Szafera* (Vol. 6, pp. 163–172). Ojców.

Moroni, A., Boscato, P., & Ronchitelli, A. (2013). What roots for the Uluzzian? Modern behaviour in Central-Southern Italy and hypotheses on AMH dispersal routes. *Quaternary International, 316*, 27–44.

Nekhoroshev, P. E. (2009). The end of the Middle Paleolithic on the Russian Plain: the view from Shlyakh. *Archaeological Almanac, 20*, 111–128.

Palma di Cesnola A. (1983). Le leptolithique archaique en Italie. In M. Otte (Ed.), *Aurignacien et Gravettien en Europe* (pp. 119–143). Liège: ERAUL 104, Université de Liège.

Panagopoulou, E., (2000). ,The Middle Paleolithic Assemblages of Theopetra Cave: Technological evolution in the Upper Pleistocene. In N. Kyparissi-Apostolika (Ed.), *Theopetra Cave. twelve years of excavation and research 1987–1998* (pp. 139–161). Athens.

Panagopoulou, E., Karkanas, P., Tsartsidou, G., Kotjabopoulou, E., Harvati, K., & Ntinou, M. (2002–2004). Late Pleistocene archaeological and fossil human evidence from Laconis Cave, Southern Greece. *Journal of Field Archaeology, 29*, 323–349.

Papaconstantinou, V. S. (1988). Micromoustérien: les idées et les pierres. Le Micromoustérien d'Asprochaliko (Grèce) et le problème des industries microlithiques du Moustérien. Ph.D. Dissertation, Université Paris X.

Papagianni, D. (2009). Mediterranean southeastern Europe in the Late Middle and Early Upper Paleolithic: Modern human route to Europe or Neanderthal refugium? In M. Camps & C. Szmidt (Eds.), *The Mediterranean from 50,000 to 25,000 BP: Turning points and new directions* (pp. 115–136). Oxford: Oxbow Books.

Paunescu, A. (1993). *Ripiceni-Izvor. Paleolitic si Mezolitic*. Bucuresti: Editura Academiei Romane.

Reisch, L. (1980). *Pleistozän und Urgeschichte der Peloponnes* (Habilitationschrift, Friedrich-Alexander Universität, Erlangen-Nürnberg).

Richter, D., Tostevin, G., & Škrdla, P. (2008). Bohunician technology and thermoluminescence dating of the type locality of Brno-Bohunice (Czech Republic). *Journal of Human Evolution, 55*, 871–885.

Richter, D., Tostevin, G., Škrdla, P., & Davies, W. (2009). New radiometric ages for the Early Upper Paleolithic type locality of Brno-Bohunice (Czech Republic): Comparison of OSL, IRSL, TL and ^{14}C dating results. *Journal of Archaeological Science, 36*, 708–720.

Riel-Salvatore, J. (2007). The Uluzzian and the Middle–Upper Paleolithic transition in Southern Italy. Ph.D. Dissertation, Arizona State University.

Riel-Salvatore, J. (2009). What Is A 'Transitional' Industry? The Uluzzian of Southern Italy as a Case Study. In M. Camps & P. Chauhan (Eds.), *Sourcebook of Paleolithic transitions: Methods, theories, and interpretation* (pp. 377–396). New York: Springer.

Rigaud, J.-P., & Lucas, G., (2006). The first Aurignacian technocomplexes in Europe: a revision of the Bachokirian. In Bar-Yosef O., Zilhao J. (eds.), *Towards a definition of the Aurignacian. Proceedings of the Symposium held in Lisbon, Portugal, June 25–30, 2002* (Trabalhos de Arqueologia, Vol. 45, pp. 277–284). Lisboa: Instituto Português de Arqueologia.

Ronchitelli, A., Benazzi, S., Boscato, P., Douka, K., & Moroni, A. (2014). Comments on "Human-climate interaction during the Early Upper Paleolithic: Testing the hypothesis of an adaptive shift between the Proto-Aurignacian and the Early Aurignacian" by William E. Banks, Francesco d'Errico, João Zilhão. *Journal of Human Evolution, 73*, 107–111.

Runnels, C. N. (1988). A prehistoric survey of Thessaly: New Light on the Greek Middle Paleolithic. *Journal of Field Archaeology, 15*, 277–290.

Sachse-Kozłowska, E., & Kozłowski, S. K. (1975). Nowa kultura górnopaleolityczna w Europie Środkowej. Ze studiów nad materiałami ze stanowiska Zwierzyniec I. *Archeologia Polski, 20*(2), 275–313.

Savich, V. P. (1975). *The Upper Paleolithic population of Southwestern Volynia*. Kiev: Naukova Dumka.

Schmidt, C., Sitlivy, V., Anghelinu, M., Chabai, V., Kels, H., Uthmeier, T., et al. (2013). First chronometric dates (TL and OSL) for the Aurignacian open-air site of Româneşti-Dumbrăviţa I, Romania. *Journal of Archaeological Science, 40*, 3740–3753.

Sirakov, N. (1983). Reconstruction of the Middle Paleolithic flint assemblages from the Cave Samuilitsa II (northern Bulgaria) and their taxonomical Position seen against the Paleolithic of southeastern Europe. *Folia Quaternaria, 55*, 1–100.

Sirakov, N., Tsanova, T., Sirakova, S., Taneva, S., Krumov, I., Dimitrova, I., et al. (2007). Un nouveau faciès lamellaire du début de Paléolithique supérieur dans les Balkans. *Paléo, 19*, 131–144.

Sitlivy V., Anghelinu M., Chabai V., Niţă L., Uthmeier T., Hauck T., et al. (2014a). Placing the Aurignacian from Banat (Soutwestern Romania) into the European Early Upper Paleolithic Context. In M. Otte & F. Le Brun-Ricalens (Eds.), *Modes de contacts et de déplacements au Paléolithique eurasiatique/Modes of contact and mobility during the Eurasian Paleolithic, ERAUL 140*, 243–277.

Sitlivy, V., Chabai, V., Anghelinu, M., Uthmeier, T., Kels, H., Hilgers, H., et al. (2012). The earliest Aurignacian in Romania: New investigations at the open air site of Româneşti-Dumbrăviţa I (Banat). *Quartär, 59*, 85–130.

Sitlivy, V., Chabai, V., Anghelinu, M., Uthmeier, T., Kels, H., Niţă, L., et al. (2014). Preliminary reassessment of the Aurignacian in Banat (Southwestern Romania). *Quaternary International, 351*, 193–212.

Sitlivy, V., Sobczyk, K., Escutenaire, C., Kalicki, T., Kolenyk, A., Kozłowski, J., et al. (2004). Late middle Paleolithic complexes of Cracow region, Poland. *In Actes du XIVème Congrès UISPP, Section, Le Pateolithique Supérieur.* (BAR International Series, Vol. 1240, pp. 305–317). Oxford: Archaeopress.

Sitlivy, V., Sobczyk, K., Karkanas, P., & Koumouzelis, M. (2007). Middle Paleolithic lithic assemblages of Klissoura Cave, Peloponnesus, Greece: A comparative analysis. *Archaeology, Ethnology & Antropology of Eurasia, 3*(31), 2–15.

Sitlivy, V., Sobczyk, K., Karkanas, P., & Koumouzelis, M. (2008a). Middle Paleolithic Industries of Klissoura Cave, Greece. In A. Darlas & D. Mihailović (Eds.), *The Paleolithic of the Balkans, Proceedings of the XV World Congress of the International Union for Prehistoric and Protohistoric Sciences, Lisbon, 4–9 September 2006.* (BAR International Series, Vol. 1819, pp. 39–49). Oxford: Archaeopress.

Sitlivy, V., & Zięba, A. (2006). Eastern and Central Europe before 30 kyr BP: Mousterian, Levallois & Blade industries. In V. Chabai, J. Richter J. & T. Uthmeier (Eds.), *Kabazi II: The 70,000 years since the Last Interglacial: Vol. 2. The Paleolithic sites of Crimea* (pp. 361–419). Simferopol: Shlyakh.

Sitlivy, V., & Zięba, A. (2008). Comparative analyses. In V. Sitlivy, A. Zięba & K. Sobczyk (Eds.), *Middle and Early Upper Paleolithic of Kraków region. Piekary IIa: Vol. 6. Monographie de Préhistoire générale* (pp. 161–197). Bruxelles: Musées Royaux d'Art et d'Histoire.

Sitlivy, V., Zięba, A., & Sobczyk, K. (Eds.). (2009a). *Middle and Early Upper Paleolithic of Kraków region. Księcia Józefa: Vol. 7. Monographie de Préhistoire générale*. Bruxelles: Musées Royaux d'Art et d'Histoire.

Sitlivy, V., Zięba, A., Sobczyk, K., & Kolesnik, A. (2009b). Lithic Assemblages. In V. Sitlivy, A. Zięba & K. Sobczyk (Eds.), *Middle and Early Upper Paleolithic of Kraków region. Księcia Józefa: Vol. 6. Monographie de Préhistoire générale* (pp. 19–169). Bruxelles: Musées Royaux d'Art et d'Histoire.

Sitlivy, V., Zięba, A., Sobczyk, K., & Kolesnik, A. (2014c). *The Middle-to-Upper Paleolithic Ksiecia Jozefa open-air Site (Krakow, Poland):*

Lithic technology and spatial distribution. Bonn: Universitätsforschungen Zur Prähistorischen Archäologie 250.

Sitlivy, V., Zięba, A., & Sobczyk, K. (Eds.). (2008b). *Middle and Early Upper Paleolithic of Kraków region. Piekary IIa: Vol. 6. Monographie de Préhistoire générale*. Bruxelles: Musées Royaux d'Art et d'Histoire.

Škrdla, P. (1996). The Bogunician reduction strategy. *Quaternaria Nova, 6*, 93–107.

Škrdla, P. (2003a). Bohunician technology. A refitting approach. In J. A. Svoboda & O. Bar-Yosef (Eds.), *Stranska skala. Origins of the Upper Paleolithic in the Brno Basin, Moravia, Czech Republic* (pp. 119–151). Cambridge: Harvard University.

Škrdla, P. (2003b). Comparison of Boker Tachtit and Stranska skala MP/UP transitional industries. *Journal of the Israel Prehistoric Society, 33*, 33–69.

Stepanchuk, V., & Cohen, V. Y. (2000–2001). Kremenician, Middle to Upper Paleolithic transitional industry in the Western Ukraine. *Préhistoire Européenne, 16–17*, 75–110.

Svoboda, J. (2004). Continuities, Discontinuities, and Interactions in Early Upper Paleolithic Technologies. A Viev from the Middle Danube. In P. L. Brantingham, S. L. Kuhn, & K. W. Kerry (Eds.), *The Early Upper Paleolithic beyond Western Europe* (pp. 30–49). Berkeley: University of California Press.

Svoboda, J. (2007). On Modern Human Penetration to Northern Eurasia: The multiple advances hypothesis. In P. Mellars, K. Boyle, O. Bar-Yosef, & C. Stringer (Eds.), *Rethinking the human revolution. New behavioural and biological perspectives on the origin and dispersal of Modern Humans* (pp. 329–340). Cambridge: McDonald Institute Monographs.

Svoboda, J. A., & Bar-Yosef, O. (Eds.). (2003). *Stranska skala. Origins of the Upper Paleolithic in the Brno Basin, Moravia, Czech Republic*. Cambridge: Harvard University.

Svoboda, J., & Škrdla, P. (1995). The Bohunician technology. In H. Dibble & O. Bar-Yosef (Eds.), *The definition and interpretation of Levallois technology* (Monographs in World Archaeology, Vol. 23, pp. 429–438). Wisconsin: Madison.

Sytnyk, O. S. (2000). The Middle Paleolithic of Podolia. Lviv.

Sytnyk, A. S. (2011). Middle Paleolithic assemblages of Korman IV on the Dniester. *Stratum Plus, 1*, 203–236.

Teyssandier, N. (2003). Les débuts de l'Aurignacien en Europe. Discussion à partir des sites de Geissenklösterle, Willendorf II, Krems-Hundssteig et Bacho Kiro. Ph.D. Dissertation, Université de Paris X.

Teyssandier, N. (2008). Revolution or evolution: The emergence of the Upper Paleolithic in Europe. *World Archaeology, 40*(4), 493–519.

Tostevin, G. B., & Škrdla, P. (2006). New excavations at Bohunice and the question of the uniqueness of the type-site for the Bohunician industrial type. *L'Anthropologie, 44*, 31–48.

Tsanova, T. (2006). Les débuts du Paléolithique supérieur dans l'Est des Balkans. Réflexion à partir de l'étude taphonomique et tehcnoéconomique des ensembles lithiques des sites de Bacho Kiro (couche 11), Temnata (couches VI et 4) et Kozarnika (niveau VII). Ph.D. Dissertation, Université Bordeaux 1.

Tsanova, T., & Bordes, J.-G. (2003). Contribution au débat sur l'origine de l'Aurignacien: principaux résultats d'une étude technologique de l'industrie lithique de la couche 11 de Bacho Kiro. In T. Tsonev & E. Montagnari Kokelj (Eds.), *The humanized mineral world: towards social and symbolic evaluation of prehistoric technologies in South-Eastern Europe* (pp. 41–50). Liège: ERAUL 103, Université de Liège.

Tsanova, T., Zwyns, N., Eizenberg, L., Teyssandier, N., Le Brun-Ricalens, F., & Otte, M. (2012). Le plus petit denominateur commun: reflexion sur la variabilite des ensembles lamellaires du Paleolithique superieur ancien d'Eurasie. Un bilan autour des exemples de Kozarnika (Est des Balkans) et Yafteh (Zagros central). *L'Anthropologie, 116*, 469–509.

Tuffreau, A., Chirica, V., Bălescu, S., & Haesaerts, P. (2009). Nouvelles recherches sur le gisement paléolithique de Mitoc-Valea Izvorului (dép. de Botoşani), Fouilles 2003–2004. *Materiale şi Cercetări Arheologice (NS), 5*, 21–31.

Usik, V. I. (1989). Korolevo—transition from lower to Upper Paleolithic according to reconstruction data. *Anthropologie (Brno), 27*(2–3), 179–212.

Usik, V. (2003). Archaeological excavation of Sokirnytsa 1 and Shayan 1 Paleolithic sites in Transcarpathian. *The Stone Age of Ukraine, 2*, 117–125.

Usik, V. I. (2006). The Problem of the Levallois Method in Level II/8 of Kabazi II. Chabai, V. In P. J. Richter & T. Uthmeier, (Eds.), *Kabazi II: The 70000 years since the Last Interglacial* (Vol. 2, pp. 143–168). Simferopol: Shlyakh.

Usik, V. I. (2008). The Upper Paleolithic of Transcarpathia: Chronology and cultural affinity of Beregovo I Aurignacian. *Materials and Investigations of Carpathian and Volyn Archeology, 12*, 49–67.

Usik, V. I., Monigal, K., & Kulakovska, L. V. (2006a). Transkarpathians: New investigations, new data. *The Stone Age of Ukraine, 4*, 247–255.

Usik, V., Monigal, K., & Kulakovskaya, L. (2006b). New perspectives on the Transcarpathian Middle to Upper Paleolithic boundary. In N. J. Conard (Ed.), *When Neanderthals and modern humans met* (pp. 213–232). Tübingen: Kerns Verlag.

Valladas, H., Mercier, N., Froget, L., Joron, J.-L., Reyss, J.-L., Kaltnecker, E., et al. (2008). Radiometric dates for the Middle Paleolithic sequence of Piekary. In V. Sitlivy, A. Zięba & K. Sobczyk (Eds.), *Middle and Early Upper Paleolithic of Kraków region. Piekary IIa: Vol. 6. Préhistoire générale* (pp. 53–61). Bruxelles: Musées Royaux d'Art et d'Histoire.

Valladas, H., Mercier, N., Froget, L., Joron, J. L., Reyss, J. L., Karkanas, P., et al. (2007). TL age-estimates for the Middle Paleolithic layers at the Theopetra cave (Greece). *Quaternary Geochronology, 2*, 303–308.

Vishnyatsky, L. B., & Nehoroshev, P. E. (2004). The Beginning of the Upper Paleolithic on the Russian Plain. In P. L. Brantingham, S. L. Kuhn, & K. W. Kerry (Eds.), *The Early Upper Paleolithic beyond Western Europe* (pp. 80–96). Berkeley: University of California Press.

Zięba, A. (2005). The Middle Paleolithic in Kraków: Piekary IIa, Kraków ul. Księcia Józefa sites in European context. Ph.D. Dissertation, Jagiellonian University of Kraków.

Zięba, A. (2014). Intra-site spatial analysis. In V. Sitlivy, A. Zięba, K. Sobczyk, K. & A. Kolesnik (Eds.), *The Middle-to-Upper Paleolithic Ksiecia Jozefa open-air Site (Krakow, Poland): Lithic technology and spatial distribution* (pp. 529–546). Bonn: Universitätsforschungen Zur Prähistorischen Archäologie 250.

Zięba, A., Sitlivy, V., & Sobczyk, K. (2008). The Lithic Assemblages In V. Sitlivy, A. Zięba & K. Sobczyk (Eds.), *Middle and Early Upper Paleolithic of Kraków region. Piekary IIa: Vol. 6. Préhistoire générale* (pp. 57–148). Bruxelles: Musées Royaux d'Art et d'Histoire.

Zilhão, J., Banks, W. E., d'Errico, F., & Gioia, P. (2015). Analysis of site formation and assemblage integrity does not support attribution of the Uluzzian to modern humans at Grotta del Cavallo. *PLoS One, 10*(7), e0131181. doi:10.1371/journal.pone.0131181.

Part III
Paleoenvironments, Biogeography and Chronology

Chapter 15
The Plio-Pleistocene Large Mammal Record of Greece: Implications for Early Human Dispersals into Europe

George D. Koufos and Dimitris S. Kostopoulos

Abstract Extensive fieldwork and detailed studies during the last three decades have enriched our understanding of the Plio-Pleistocene large mammal record of Greece. While the unearthed material is abundant, it is not evenly distributed throughout the Plio-Pleistocene; therefore, there are time intervals in this period for which the known large mammal fauna is limited and our knowledge is poor. The Greek Plio-Pleistocene large mammal record reveals a paleoenvironmental transition from open woodlands in late Pliocene, to savannah-like landscape during the early Pleistocene, and to open grasslands during the late early Pleistocene. During this environmental shift, several taxa arrived in Greece in their westward expansion, whereas others made their last European appearance. The arrival of *Homo* in Europe is discussed in relation to the Greek faunal record. The available data cannot clearly distinguish between an African or an Asian origin, but the latter is supported by more evidence.

Keywords Villafranchian • Greek mammal faunas • Biochronology • *Homo* dispersal into Europe

Introduction

In comparison with the Miocene, the Plio-Pleistocene fossil mammal record of Greece is relatively poor. Until the late 1980s, the known mammal localities were relatively few and the fossil collections fragmentary, with questionable stratigraphic indications, insufficient to provide taxonomic and/or chronological information suitable for age determination. Moreover, the Middle and Late Pleistocene findings from deposits in-filling caves or fissures cannot provide answers concerning biochronology and/or paleoecology. The published data from this period are therefore rare and sporadic. With the exception of extensive studies of the Megalopolis, Volax, Tourkovounia, and Grevena Basin faunas (Melentis 1961, 1964; Sickenberg 1967, 1968; Symeonidis and de Vos 1976; Steensma 1988), most of the publications report isolated specimens.

During the last three decades, extensive fieldwork in several old and new Plio-Pleistocene fossiliferous sites has provided substantial new collections, which were important for the systematics of the mammal faunas in the region, and for understanding their relationships with the known Eurasian assemblages. Despite the discontinuity of the fossil record, the available faunas provided a good Plio-Pleistocene (Villafranchian) mammal biochronology (Koufos 2001 and ref. cited). These new data carry important paleobiogeographical implications and allow us to put forward paleoecological and paleoenvironmental interpretations.

Here, we summarize the fauna, biochronology, and biostratigraphy of the Plio-Pleistocene large mammal localities of Greece, their paleoenvironmental setting, and their trans-European biogeographic relationships, with special emphasis on their implications for early human dispersals into Europe. Localities with large mammals and good stratigraphic and biochronologic backgrounds are shown in Fig. 15.1. While there are several Pleistocene localities on the islands of the Aegean Sea, they will not be discussed in this article, as they constitute endemic faunas.

G.D. Koufos (✉) • D.S. Kostopoulos
Department of Geology, Laboratory of Geology and Palaeontology, Aristotle University of Thessaloniki, GR-54124 Thessaloniki, Greece
e-mail: koufos@geo.auth.gr; dkostop@geo.auth.gr

Katerina Harvati and Mirjana Roksandic (eds.), *Paleoanthropology of the Balkans and Anatolia*, Vertebrate Paleobiology and Paleoanthropology, DOI 10.1007/978-94-024-0874-4_15

Fig. 15.1 Map of Greece indicating the Plio-Pleistocene large mammal localities; the Pliocene localities marked by an asterisk and the Pleistocene ones by a dot. 1. *SLT* Silata, 2. *MAR* Maramena, 3. *KES* Kessani, 4. *MEV* Megalon Emvolon, 5. *PTL* Ptolemais, 6. *APO* Apolakkia, 7. *MIL* Milia, 8. *DMT* Damatria, 9. *TRV* Tourkovounia 3–5, 10. *SES* Sesklon, 11. *DFN* Dafnero, 12. *VOL* Volax, 13. *VTR* Vatera, 14. *GER* Gerakarou 1, 15. *VSL* Vassiloudi, 16. *KRI* Krimni 1, 17. *LIB* Libakos, 18. *ALK* Alykes, 19. *RVL* Ravin of Voulgarakis, 20. *APL* Apollonia, 21. *KLT* Kalamoto

Large Mammal Faunas and Biochronology

Latest Turolian/Early Ruscinian Faunas (5.3–4.2 Ma)

Large mammal Pliocene localities are rare in Greece and, as a result, the available faunal information is limited. The earliest large mammal Pliocene elements have been recognized in the faunas of Maramena (MAR), Silata (SLT), and Kessani (KES) of Northern Greece (Fig. 15.1). All localities are rich in micromammals, which suggest an age at the Turolian/Ruscinian boundary (Schmidt-Kittler 1995; Vassiliadou et al. 2003; Vasileiadou et al. 2012; Koufos 2006a; Koufos and Vasileiadou 2015; Fig. 15.2).

Fauna: *Mesopithecus* sp., *Martes lefkonensis, Promeles macedonicus, Lutra affinis, Promephitis* sp., *Chasmaporthetes* sp., *Paramachaerodus orientalis, Choerolophodon pentelici,*

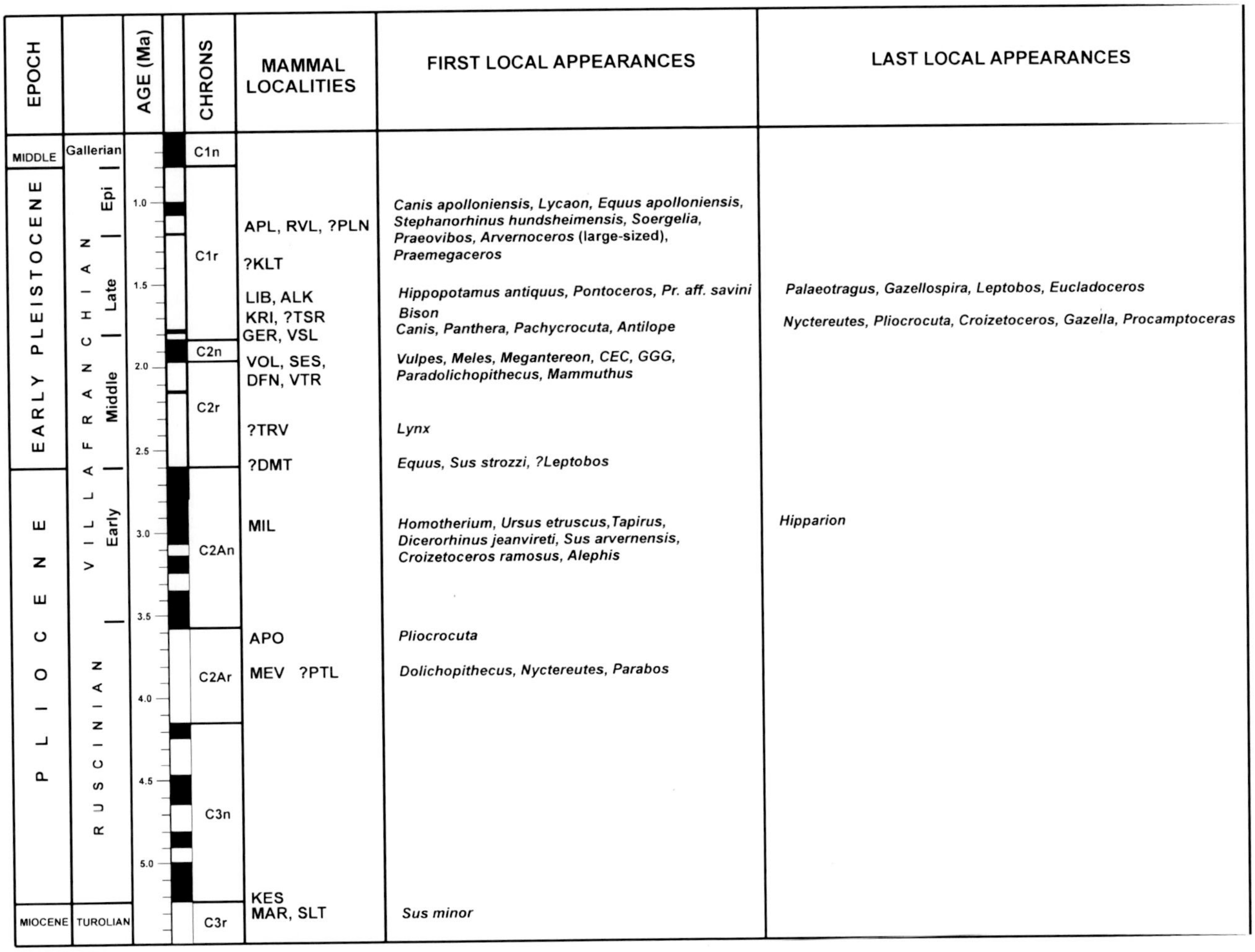

Fig. 15.2 Biostratigraphic table of the Plio-Pleistocene large mammal localities of Greece with the first and last local appearance of the various taxa

Hipparion cf. *mediterraneum, Hipparion* sp. (large), *Hipparion* sp. (small), *"Korynochoerus" palaeochoerus, Microstonyx major, Pliocervus graecus, Norbertia hellenica, Gazella* sp. (Schmidt-Kittler 1995; Koufos 2006a; Vasileiadou et al. 2012).

Some ruminants (*?Helladotherium* sp., *Samotherium* cf. *boissieri, Tragoportax gaudryi, Tragoportax* cf. *amalthea, Ouzocerus* aff. *gracilis*) described as part of the Maramena fauna (Schmidt-Kittler 1995) are poorly documented and their identification is questionable. The fossiliferous site KES in Thrace (Fig. 15.1) provided a rich micromammalian fauna that indicates a Turolian/Ruscinian age; however, large mammals are very few and fragmentary. Among the collected isolated teeth, Vasileiadou et al. (2012) identified the suid *Sus minor* and the equid *Hipparion* sp. together with some indeterminable cervids and bovids.

Late Ruscinian (4.2–3.5 Ma)

Our knowledge of the large mammal fossil record from the Late Rucinian in Greece is also limited, as localities with large mammals from this time period are scarce. The best known such site is Megalon Emvolon (MEV) near Thessaloniki (Fig. 15.1), which has yielded a diverse faunal assemblage (Koufos 2006b). Some large mammal bones are also known from Allatini (AL) near Thessaloniki, and a few isolated specimens have been collected from the lignitic deposits of the Ptolemais Basin (PTL) in Western Macedonia, Greece (Fig. 15.1), as well as from Apolakkia (APO) on the island of Rhodes (Fig. 15.1).

Fauna: *Dolichopithecus ruscinensis* (Spassov and Geraads (2007) refer this taxon to *D. balcanicus*), *Nyctereutes tingi, Eucyon odessanus, Pliocrocuta perrieri, Anancus arvernensis, Hipparion longipes, Hipparion crassum, Rhinocerotidae* indet., *Sus minor, Metacervoceros* cf. *rhenanus, Cervidae* indet., *Parabos macedoniae, Gazella borbonica, Koufotragus bailloudi* (Meulen and Kolfoschoten 1986; Theodorou et al. 2000; Doukas and de Bruijn 2002; Koufos 2006b).

The fauna of MEV (Koufos 2006b), with both small and large mammals, is one of the richest Pliocene assemblages in the Eastern Mediterranean dating to the Late Ruscinian (MN15). The locality of the PTL is very poor in large mammals; however, it has yielded *Dolichopithecus ruscinensis* and *Hipparion crassum* specimens. The exact position of the level from which these fossils originate is unknown. The lignitic-pits of PTL are active, and the fossils possibly originate from the younger levels of the Pliocene deposits, corresponding to Late Ruscinian (Koufos 2001; Doukas and de Bruijn 2002). The locality of APO on the island of Rhodes includes some hipparion remains and is situated below the chronostratigraphically younger locality of Damatria (DMT), which has yielded *Equus* specimens. Therefore, it is possible to associate APO with a Late Ruscinian age (Meulen and Kolfoschoten 1986; Fig. 15.2).

Early Villafranchian (3.5–2.6 Ma)

This time-span covers the Late Pliocene and corresponds to the European Large Mammal zone (MN 16). The end of the Early Villafranchian has been recently defined as the Plio/Pleistocene boundary, dated at 2.58 Ma (Gibbard et al. 2010). Early Villafranchian localities in Greece are scarce and the known fauna is quite poor. Only one locality, Milia (MIL) in the Grevena Basin, is ascertained as Early Villafranchian, while Damatria (DMT) in the island of Rhodes (Fig. 15.1) might also be of this age.

Fauna: *Anancus arvernensis*, *Mammut borsoni*, *Homotherium crenatidens*, *Ursus etruscus*, *Agriotherium* sp., *Hipparion crassum*, *Equus* sp., *Dicerorhinus jeanvireti*, *Tapirus arvernensis arvernensis*, *Sus arvernensis arvernensis*, *Sus strozzii*, *Croizetoceros ramosus*, *Procapreolus cusanus*, Cervidae indet. (medium-sized), *Alephis* sp., *Gazella borbonica*, *Gazella* sp., cf. *Procamptoceras*, Boselaphini indet. (large-sized), Bovini indet., (Meulen and Kolfoschoten 1986; Loghem et al. 2010; Guerin and Tsoukala 2013; Cregut-Bonnoure and Tsoukala 2014; Lazaridis and Tsoukala 2014; Tsoukala et al. 2014).

The age of the MIL locality is considered as Early Villafranchian, dating to *ca.* 3.0–2.5 Ma (Loghem et al. 2010; Guerin and Tsoukala 2013). The locality DMT, which has yielded *Equus*, *Sus strozzii* and probably *Leptobos*, is stratigraphically situated above the *Hipparion*-bearing site APO (Meulen and Kolfoschoten 1986), and may correspond to the first occurrence of modern horses in Greece. The genus *Equus* arrived in Western Europe at *ca.* 2.6 Ma—as recorded in Montopoli fauna (Azzaroli et al. 1988; Rook and Martínez-Navarro 2010)—signaling the beginning of the Middle Villafranchian. Thus, the maximum age of the DMT fauna must be at the Early-Middle Villafranchian boundary (Koufos 2001; Fig. 15.2). This age is consistent with the arrival of *Equus* in Romania (Radulescu and Samson 2001), as well as with the last evidence of *Mammut borsoni*, which is present in Europe until the end of the Early Villafranchian (Spassov 2003).

Middle Villafranchian (2.6–1.8 Ma)

Although the known Greek Middle Villafranchian localities are numerous (Koufos 2001), most of them are old discoveries, identified based on isolated remains that lack stratigraphic information. Several new sites discovered over the last two decades, however, have substantially enriched our understanding of this period in Greece. The collections from Dafnero-1 (DFN), Sesklon (SES) and Vatera (VTR), as well as those from Volax (VOL) and Tourkovounia-3, 5 (TRV) (Fig. 15.1) have yielded a rich and well-studied fauna.

Fauna: *Paradolichopithecus arvernensis*, *Anancus arvernensis*, *Mammuthus meridionalis*, *Chasmaporthetes lunensis*, *Pliohyaena perrieri*, *Baranogale* cf. *helbingi*, *Meles thorali*, *Nyctereutes megamastoides*, *Vulpes alopecoides*, *Vulpes praecorsac*, *Homotherium crenatidens*, *Lynx issiodorensis*, *Megantereon cultridens*, *Ursus* cf. *etruscus*, *Equus stenonis* cf. *vireti*, *Stephanorhinus* cf. *etruscus*, *Palaeotragus inexspectatus*, *Metacervoceros* ex gr. *rhenanus*, *Croizetoceros ramosus*, *Eucladoceros tegulensis*, *Euthyceros thessalicus*, *Gallogoral meneghinii*, *Gazella bouvrainae*, *Gazella aegea*, *Gazella borbonica*, *Gazellospira torticornis*, *?Procamptoceras* sp., *?Caprini* indet. (Koufos 2001, 2006b; de Vos et al. 2002; Athanassiou 2014).

The set of localities VOL, SES, DFN, and VTR all include a similar fauna, dated to the Late-Middle Villafranchian at *ca.* 2.5-2.0 Ma (Koufos and Kostopoulos 1997; Koufos 2001; de Vos et al. 2002; Fig. 15.2), which is referred to as the Middle Villafranchian large mammal assemblage (hereafter MVLMA) of Greece.

Late Villafranchian (1.8–1.2 Ma)

While several Late Villafranchian localities with large mammals are known from Greece, rich assemblages with sufficient stratigraphic information are far fewer. The localities of Gerakarou-1 (GER), Krimni (KRI), Vassiloudi (VSL), Tsiotra Vrissi (TSR) and Platanochori (PLN) situated in the Mygdonia Basin; Alykes (ALK) in Thessaly; and Libakos (LIB) in the Grevena Basin (Fig. 15.1), studied in recent years, have recently provided us with abundant information on a very rich fauna.

Fauna: *Mammuthus meridionalis*, *Canis arnensis*, *Canis etruscus*, *?Homotherium* sp., *Panthera onca toscana*, *Pachycrocuta brevirostris*, *Pliohyaena perrieri*, *Pannonictis* sp., *Meles dimitrius*, *Ursus etruscus*, *Equus stenonis mygdoniensis*, *Stephanorhinus etruscus*, *Sus strozzii*, *Hippopotamus antiquus*, *Palaeotragus martini*, *Croizetoceros ramosus*, *Eucladoceros tegulensis*, *Metacervoceros* aff. *rhenanus*, *Praedama* aff. *savini*, *Pseudodama* cf. *nestii*, *Antilope koufosi*, *Gazella bouvrainae*, *Gazollospira torticornis*,

Leptobos cf. *etruscus, Pontoceros ambiguus, Bison,* sp., *Procamptoceras* sp. (Koufos 2001, 2006b; Van der Made and Tong 2008; Konidaris et al. 2015).

The localities of GER and VSL are from the same stratigraphic horizons of the Mygdonia Basin and mark the earliest occurrence of the genus *Canis* in Greece. The first appearance of *Canis* s.s. in Western Europe is dated at the end of Middle Villafranchian, recognized in the localities of Senèze (France), C. St Giakomo (Italy) and Slivnitsa (Bulgaria), which are all dated to *ca.* 1.95 Ma (Sotnikova and Rook 2010; Kahlke et al. 2011). The GER and VSL faunal assemblages, which have yielded abundant canid remains are slightly younger, and can be correlated to the middle/late Villafranchian transition at *ca.* 1.8 Ma (Koufos 2001).

The large-sized *Pachycrocuta* is a common and widespread taxon, covering all of Eurasia from Spain to China; for this reason the "*Pachycrocuta*-event" was proposed to have replaced the "wolf-event" (Rook and Martínez-Navarro 2010). The coexistence of *Pliocrocuta perrieri* and *Pachycrocuta brevirostris* in GER is indicative of this fauna's transitional character from the Middle to the Late Villafranchian. The faunas from LIB and ALK are similar to the GER fauna, but also include new faunal elements, like *Hippopotamus* and *Pontoceros,* indicating a younger age than GER (Fig. 15.2). The combined fauna from these sites is referred to as the Late Villafranchian large mammal assemblage (hereafter LVLMA) of Greece.

Epi-Villafranchian (1.2–0.8 Ma)

The Epi-Villafranchian localities of Greece are scarce. Only three sites, Apollonia 1 (APL), Platanochori (PLN) Ravin of Voulgarakis (RVL), and probably Kalamoto (KAL) (Fig. 15.1), are well known localities associated with this time period (Fig. 15.2). These collections are new, and all come from localities situated in the Mygdonia Basin (Macedonia, Greece). APL has yielded a rich fauna of large mammals; the RVL fauna includes mainly small mammals of Early Biharian age (Koliadimou and Koufos 1998). The fauna of Kalamoto has not yet been thoroughly studied, but the available preliminary determinations (Tsoukala and Chatzopoulou 2005) suggest more Epi-Villafranchian than Late Villafranchian age. The combined fauna of these localities will be referred to as Epi-Villafranchian large mammal assemblage (hereafter EVLMA).

Fauna: *Mammuthus meridionalis, Canis etruscus, Canis arnensis, Canis apolloniensis, Lycaon (=Xenocyon)* sp., *Vulpes alopecoides, Ursus etruscus, Meles dimitrius, Mustela* sp., *Pachycrocuta brevirostris, Megantereon cultridens* (small variety), *Lynx issiodorensis, Equus apolloniensis, Stephanorhinus hundsheimensis, Hippopotamus antiquus, Praemegaceros pliotarantoides, Arvernoceros* cf. *veretschagini Cervus* sp., *Dama* sp., *Pontoceros ambiguous mediterraneus, Soergelia brigittae, Praeovibos mediterraneus, Bison (Eobison)* sp., *Ovis* sp., *Hemitragus/Capra* sp., cf. *Leptobos etruscus* (Athanassiou and Kostopoulos 2001; Koufos 2001; Tsoukala and Chatzopoulou 2005; Konidaris et al. 2015).

Local Large Mammal Events and Environmental Context

The known Plio-Pleistocene large mammal localities of Greece represent discontinuous evidence, and significant gaps remain between geographic areas, especially for the Pliocene (Fig. 15.3). The oldest known localities are MAR, SLT and KES, with small mammal assemblages that suggest the Miocene/Pliocene boundary, even though the large mammal association retains a more "Miocene character" (Koufos and Vasileiadou, 2015 and references therein). The large mammal faunas of these sites mark the last local occurrence of several taxa such as *Mesopithecus, Choerolophodon, Promeles, Paramachaerodus,* large giraffids, and boselaphine-like bovids. Additionally, the replacement of the large Miocene suid *Microstonyx*—well distributed across Eurasia during the Late Miocene—by a small suine *Sus minor* with possible Asian origin is initially recorded in the KES.

Despite limited data for the Ruscinian of Greece, this rather dramatic faunal turnover represents the terminal phase of a progressive environmental restructuring following the desiccation of the Mediterranean Sea during the Messinian Crisis, between 7.0 and 5.0 Ma (Koufos 2006c; Kostopoulos 2009; Eronen et al. 2009). The faunal renewal concluded with the early Pliocene climatic optimum, which allowed for a warmer and more humid climate (Agustí and Antón 2002). This is well documented by the extensive lignite deposits of this time interval in Northern Greece and surrounding regions. "Wet" conditions seem to continue into the second half of the Ruscinian, which is marked in Greece and southwestern Europe by the arrival of several new taxa, such as the canid *Eucyon odessanus*, the hyaenid *Pliocrocuta perrieri*, the bovine-like bovid *Parabos* and the primate *Dolichopithecus*, the latter of possible African origin (Sotnikova and Rook 2010; Eronen and Rook 2004). The proboscidean taxa *Anancus* and *Mammut* known from the Miocene continue into the Pliocene.

As mentioned above, the Early Villafrancian is poorly represented in Greece. This gap in the large mammal record does not allow us to closely follow any faunal reorganizations that might have ensued after the first glaciation in the Northern Hemisphere and the subsequent increase in thermal seasonality (Agustí and Antón 2002; Kahlke et al. 2011 and ref. cited). Nevertheless, the effect of these major environmental events

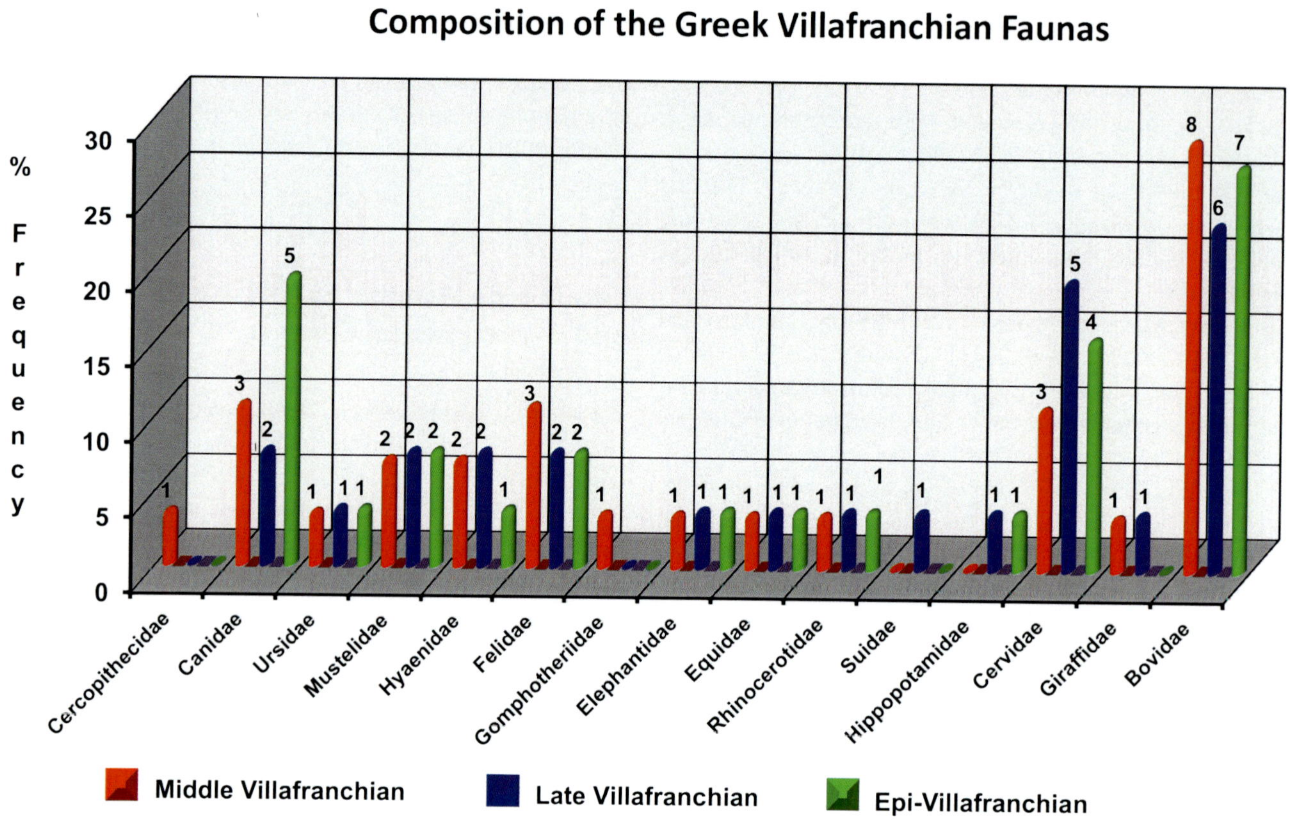

Fig. 15.3 Composition of the Villafranchian faunas of Greece; the number in each bar indicates the number of species found

is evident in the Greek record as a series of extinctions and replacements. Several typical elements of the Late Rucsinian large mammal fauna of Greece (*Dolichopithecus, Parabos, Alephis, Eucyon*) made their last appearance or disappeared. The modern horses (*Equus*) replaced hipparionines, while *Sus minor* was replaced by more advanced suines. *Homotherium, Ursus etruscus, Croizetoceros*, and the Asian *Leptobos* arrived at the same time, enriching the Early Villafranchian mammal community.

The Middle Villafranchian (2.6–1.8 Ma) is characterized by a gradual decrease in global temperature and a progressive aridification of Southern Europe (Kahlke et al. 2011). The effect of these environmental changes has not been fully traced in the Balkan record yet. However, they might be related with the appearance of archaic forms of *Mammuthus* in Romania, Bulgaria, and Greece (Lister and van Essen 2003; Kostopoulos and Koulidou 2015).

The mammal diversity increased significantly in Greece during this period. The MVLMA includes eight bovid taxa (representing 28.6% of the fauna), three cervids (10.7% of the fauna), and various carnivores (11 taxa, ~40% of the fauna; Fig. 15.3). The giraffids, rhinos, equids, elephantids, and gomphotheriids were also present, represented by a single taxon respectively (Fig. 15.3). *Equus* remains predominate in the collected material. The MVLMA is characterized by the first local appearance of the bovid association *Gazella bouvrainae-Gazellospira-Gallogoral* (GGG) and the cervid association *Croizetoceros-Eucladoceros-Metacervoceros* (MEC) together with *Vulpes, Meles, Megantereon, Palaeotragus inexspectatus*, and the large cercopithecine monkey *Paradolichopithecus* with terrestrial lifestyle, which appears for the first time in VTR (Fig. 15.2).

The paleoecological conditions in Greece during the Villafranchian were first studied by Kostopoulos and Koufos (1998, 2000 and ref. therein). The authors analyzed the Villafranchian large mammal faunas and compared them with modern faunas from known environments. The rich mammal fauna of the DFN, VOL, and SES is characterized by the predominance of mixed-feeders and grazers, the presence of a large-sized stenonoid horse, and an equilibrium in the cervid/bovid composition. Dental micro- and meso-wear analyses of the SES herbivores (Rivals and Athanassiou 2008) indicate that the bulk of the taxa are classified among extant mixed feeders, confirming previous results suggesting an open environment similar to the modern woodland savannah (Kostopoulos and Koufos 2000).

The LVLMA has the same taxonomic structure as the MVMLA. Carnivores (nine taxa) constituted ~35% of the assemblage, followed by bovids with six taxa (23.1%) and cervids with five taxa (19.2%). The rest of the families were represented by a single taxon each (Fig. 15.3). The mammal fauna of the GER and VSL date to the lower part of the Late

Villafranchian. They were characterized by an equilibrium of browsers and grazers, a dominance of the intermediate feeders, and more than 50 % of open to open/mixed dwellers, suggesting a savannah-like woodland environment (Kostopoulos and Koufos 2000). During this time interval, however, *Canis* replaced *Nyctereutes*, while *Panthera* and *Pachycrocuta* made their first appearance, the latter coexisting with *Pliohyaena* for a short period. This faunal re-organization, correlated to the strong glacial pulses that begun at *ca.* 1.8 Ma (Agustí and Antón 2002 and ref. therein), continued for some time (as evidenced by the KRI, LIB ALK). It was marked by the last occurrences of *Palaeotragus inexspectatus, Gazella, Gazellospira* and *Leptobos,* and the appearance of new taxa such as the African *Hippopotamus,* as well as the Asian *Bison* and *Pontoceros* (Fig. 15.2). From approximately 2.0 to 1.0 Ma, gradual deforestation and an environmental shift toward more open and grassy landscapes were recorded across Southern Europe, affecting the woodland taxa, which were drastically eliminated (Agustí and Antón 2002; Kahlke et al. 2011).

The last phase of this faunal renewal took place during the Epi-Villafranchian (1.2–0.8 Ma), when the Mid-Late Villafranchian large mammal association collapsed for good and a new faunal association emerged. During the Epi-Villafranchian, the last newcomers (bisons, hippos, etc.) were already firmly established in Greece. Several new immigrants—including the small *Megantereon,* members of the *Praemegaceros* and *Megaloceros* lineages, *Soergelia,* and *Praeovibos*—arrived at this time.

The EVLMA is dominated by the carnivores (11 taxa; 32.2 %), among which the canids prevail (five taxa; 19.2 %). The canids were enriched by the arrivals of *Lycaon* (also known as *Xenocyon*), and *C. apolloniensis*, a form intermediate between *C. etruscus* and *C. mosbachensis* (Fig. 15.2; Koufos and Kostopoulos 1997). At the same time, there was a remarkable increase in the relative body size of herbivores, especially perissodactyls and artiodactyls. The equids were represented by a large-sized species in some respects similar to *E.* ex gr. *süssenbornensis* (Koufos et al. 1997). The cervids were known by giant forms (*Praemegaceros, Arvernoceros*) and the bovids by heavy ovibovines and bovines, like *Bison* (*Eobison*) (Fig. 15.2). Furthermore, the browsers declined significantly and the grazers and intermediate feeders became predominant (Kostopoulos and Koufos 2000). All of these observations indicate open-grassy landscapes subject to mild climatic conditions.

Early Humans and the Greek Mammal Record

According to the most parsimonious biogeographic models, Greece has a crucial position along the most likely route for human dispersals between Africa and Eurasia (i.e., the Levantine Corridor) (O'Regan et al. 2006; Muttoni and Scardia 2010; Tourloukis and Karkanas 2012; Harvati 2016; Spassov 2016; Strait et al. 2016; and ref. cited in all). Several studies emphasized, directly or indirectly, the high paleoanthropological potential of the Greek Early Pleistocene fossil record (Harvati et al. 2009; Dennell 2010; Muttoni et al. 2010; Tourloukis and Karkanas 2012; Tourloukis 2016, and ref. cited in all). However, despite some isolated lithic artifacts of doubtful stratigraphic provenience, Early Paleolithic human remains are currently scarce in Greece (Harvati et al. 2009; Tourloukis and Karkanas 2012). This absence contrasts with both the rich Early-early Middle Pleistocene Greek fossil mammal record, which has yielded several thousands of specimens ascribed to a large number of fossil mammal taxa (Koufos 2001), and the significant and rather continuous Late Miocene to Pliocene and late Middle Pleistocene to Holocene primate fossil record of Greece (Koufos 2006b; Galanidou 2004; Harvati et al. 2009). In our opinion, the apparent Early Pleistocene gap in the human fossil record of Mainland Greece can be ascribed to the interplay of taphonomic factors (see Tourloukis 2016) and inadequate fieldwork, especially given the evidence from the Kozarnika cave in Bulgaria (Guadelli et al. 2012), and the new early Paleolithic sites at Rodafnidia on Lesvos (Galanidou et al. 2013; 2016) and Marathousa 1, Megalopolis on the Peloponnese (Panagopoulou et al. 2015).

At the boundary between the biogeographic zones of Western Asia, Africa/Middle East, and Europe, the large mammal record of Greece and the Southern Balkans remains crucial for the understanding of early human dispersal patterns toward Western Europe, despite the current absence of recorded human activity or fossil remains. Additionally, as it borders land in all directions, it is the most important entrance area for mammals dispersing into Europe from east to west and south to north.

Recent paleoanthropological discoveries place the first, but so far geographically isolated, occurrence of *Homo* in Eurasia at 1.77 Ma at Dmanisi, Georgia (Gabunia et al. 2000; Agustí and Lordkipanidze 2011 and references therein). New data, discussed by Moncel (2010) and Bar-Yosef and Belfer-Cohen (2013) seem to indicate wider early hominin dispersal into Asia at *ca.* 1.8–1.7 Ma ago. Dmanisi would presumably represent the best evidence of it so far. A second important dispersal of humans is evidenced in southern Europe at *ca.* 1.4 Ma. This dispersal is supported by the lithic artifacts of Fuente Nueva 3 and Baranco León 5, Spain, between 1.2 and 0.8 Ma, recorded by solid fossil human evidence in Baranco León 5, Sima del Elefante and Gran Dolina, Spain, Denizli, Turkey, and by lithic artifacts in several other sites around the Northern and Eastern Mediterranean (Martínez-Navarro et al. 1997; Muttoni and Scardia 2010; Muttoni et al. 2011; Dennell 2010; Carbonell et al. 2008; Van der Made 2011; Moncel 2010; Rodriguez et al. 2011; Arzarello et al. 2012; Bar-Yosef and Belfer-Cohen 2013;

Torro Moyano et al. 2013; Lebatard et al. 2014). The Dmanisi "momentum" in the East and the Atapuerca record in the West suggest a delay of about 0.5 Ma in the dispersal of early humans into Europe. This raises questions regarding the cause(s) of this lag, and the evolutionary and geographic path followed by the first "Europeans".

In various scenarios proposed to explain the delayed arrival of *Homo* in Europe, some researchers have focused on the possible dependencies of early humans on other animals, especially mammals (Martínez-Navarro 2010; Finlayson et al. 2010; O'Regan et al. 2011; Van der Made 2011; Van der Made and Mateos 2011; Agustí and Lordkipanidze 2011 and references therein). Early humans are regarded either as a part of a defined paleocommunity, whose chrono-spatial expansion depended upon the climatically controlled adaptation of specific habitats; or, alternatively, as followers of the dispersal paths of other mammals, particularly of (certain or assumed) African origin, as a result of an increasingly carnivorous trophic behavior. Within these conceptual frameworks, a comparison of Dmanisi, Atapuerca, and similar southern European hominin-bearing faunas with the Greek fossil record provides important insights.

The carnivore assemblage of Dmanisi shows a mixture of archaic and modern elements. Among them *Meles, Homotherium crenatidens, Lynx issiodorensis, Ursus etruscus, Vulpes alopecoides*, and *Pliohyaena perrieri* are all present in the MVLMA of Greece, dated before the Olduvai sub-chron. Their coexistence with *Canis* ex gr. *etruscus* (see comment in Sotnikova and Rook 2010), *Panthera onca* and *Pachycrocuta,* is very similar to the early Late Villafranchian carnivore assemblage of Greece, represented mainly by the fauna of Gerakarou. It confirms the biochronologic and geochronologic data, dating both sites to about 1.8 Ma. Therefore, the carnivore assemblage of Dmanisi is comparable to the contemporary southeastern and southwestern European assemblages, demonstrating maximum biozoogeographic similarity. The only exception is the presence at Dmanisi of a small saber-toothed cat of the genus *Megantereon,* which was variously interpreted as representing the African species *M. whitei*, a chronocline of the previous Eurasian *Megantereon cultridens*, or even a new taxon (Hemmer 2000; Palmqvist et al. 2007; Lewis and Werdelin 2010; Martínez-Navarro 2010). This species is absent from the LVLMA of Greece, as well as from the rest of contemporary mammal faunas of Europe, where typical *M. cultridens* may occur.

A similarly advanced small *Megantereon* appears for the first time in the Greek EVLMA of Apollonia-1, as well as in Bugiulesti, Romania, in Urkút, Hungary, in Pirro Nord (Pirro 10) and Monte Argentario, Italy, in Untermassfeld, Germany and in Venta Micena, Spain (Palmqvist et al. 2007; Lewis and Werdelin 2010). This pan-European dispersal is chronologically framed within the 1.4–1.0 Ma time interval. Regardless of the geographic origin of this species, it shows the same delay in its westward expansion that we can observe in early humans compared to Dmanisi. However, the westward dispersion of the small *Meganthereon* is closely followed by multiple Asian lineages of advanced canids—such as the hypercarnivorous *Lycaon (=Xenocyon)*—or members of the *C. mosbachensis* group (Sotnikova and Rook 2010) that penetrated Europe approximately at the same time, suggesting a clear renewal of the carnivore assemblage at 1.4–1.0 Ma.

Like the carnivores, the herbivore assemblage of Dmanisi incorporates both archaic and modern elements. The presence of *Mammuthus meridionalis, Equus stenonis, Stephanorhinus etruscus, Eucladoceros tegulensis, Galogorall meneghinii, Gazellospira torticornis*, and paleotragine giraffids strongly resembles the Greek Middle and Late Villafranchian assemblages and reveals overall biozoogeographic relationships with Europe. *Gallogoral meneghinii* from Dmanisi closely resembles that of Volax (MVLMA) in Greece (Bukshianidze 2005), whereas the last occurrence of the paleotragine *Palaeotragus inexspectatus* in Greece is at the fauna of Libakos (LVLMA).

Despite these similarities, however, the Dmanisi fauna is enriched by a significant number of "fresh" taxa, mostly ungulates of Asian origin that seem to belong to two distinct dispersal events toward the west. The first of these ungulate events included exclusively members of Caprini (*sensu* Hassanin and Douzery 1999), like *Ovis*, *Hemitragus/Capra*, and *Praeovibos*. Their first European appearance at Slivnitsa (Bulgaria), Casa Frata (Italy), Fonelas P1 (Spain), and possibly Senèze (France) (Spassov 2000; Cregut-Bonnoure 2007; Arribas-Herrera 2008) seems sporadic and coincides with the timing of the arrival of *Canis etruscus–Canis arnensis* at about 1.9–1.7 Ma. The second ungulate event between 1.4 and 1.0 Ma included *Equus altidens*, megacerine cervids, early *Bison*, *Soergelia,* and possibly *Pontoceros*, most of which occurred for the first time in Greece in the faunas of Libakos (LVLM), Kalamoto, Platanochori, and Apollonia (EVLMA).

Early bisons emerged before 2.0 Ma in the Indo-Pakistani region and were already present at Dmanisi at *ca.* 1.8 Ma (Kahn et al. 2010; Bukshianidze 2005). Their pan-European expansion took place between 1.6 and 1.0 Ma. Early bisons are recognized in Apollonia and Kalamoto (Greece), Pirro Nord (Italy), Vallonet (France), Venta Micena and Atapuerca (Spain), and Untermassfeld (Germany). The timing and pattern of this dispersal replicates that of *Megantereon* and in most cases coincided with the record of the first Western European human settlements (Van der Made 2011, 2013).

The earliest occurrences of *Equus altidens* in Western Europe are recorded in Venta Micena, Fuente Nueva 3, Baranco León 5 and Huescar 1 (Spain), and Pirro Nord in Italy (Alberdi and Palombo 2013). Equids may have followed the bison's westward expansion from Asia; although, alternative options are equally possible (Van der Made 2013).

The geographic provenance and origin of *Soergelia* are poorly understood. The oldest, and somewhat problematic, record of the genus comes from the possibly Middle Villafranchian site Villany 3 in Hungary. Its next appearance is in Dmanisi, Georgia (Cregut-Bonnoure 2007; Bukshianidze 2005). In any case, *Soergelia* shows a marked south-southwestward expansion between 1.4 and 1.0 Ma, recorded in Apollonia (Greece), Kozarnika (Bulgaria), Trlica (Montenegro), Monte Agrentario (Italy), Vallonet (France), and Venta Micena (Spain) (Cregut-Bonnoure 2007; Martínez-Navarro et al. 2013).

The Caucasus may have been the native area of the spiral-horned "antelope" *Pontoceros* as its earliest known occurrence is at Dmanisi (Bukshianidze 2005). The genus later occupied the North Black Sea shores and dispersed into the Balkans, recorded at Libakos, Platanochori, and Apollonia in Greece. A similar dispersal pattern may have been followed by *Arvernoceros* and other members of the Asian *Megaloceros* and *Praemegaceros* lineages (Vislobokova 2012). During the same time span (1.4–1.0 Ma), *Praedama* aff. *savini* appeared in Europe at Libakos (Van der Made and Tong 2008; Vislobokova 2012), whereas the primitive *Praemegaceros* was originally recorded in Greece (APL, KLT) and further west (Croitor and Kostopoulos 2004; Vislobokova 2012).

Conclusions

The Early Pleistocene mammal record of Greece does not support the direct Afro-European scenario of early human dispersal. As already pointed out the model is oversimplified (Hemmer 2000; Agustí and Lordkipanidze 2011; O'Regan et al. 2006; Lewis and Werdelin 2010), and weakly supported by current evidence, based mainly on a single taxon of unambiguously African origin, the *Hippopotamus*. Instead, both at Dmanisi and in the Western European record the first evidence of humans can be associated with a particular large mammal assemblage, whose elements are either purely Asian in origin—such as bisons, caprines-ovibovines, megacerines, wolf-like canids (Sotnikova and Rook 2010; Van der Made and Mateos 2011; Vislobokova 2012)—or, those for which an Asian period of evolution cannot be excluded, like the small *Megantereon*, *Panthera* and *Pachycrocuta* (Hemmer et al. 2010; Lewis and Werdelin 2010).

The bulk of the European Early Pleistocene carnivore renewal predates the human record of the sub-continent and coincides with the "Dmanisi momentum" and the arrival of several new herbivore taxa, mainly of Asian origin, at the gates of Europe. Notable among them were caprines, which dispersed throughout Southern Europe as far west as Iberian Peninsula. The rather simultaneous appearance in south-western Europe of humans and heavy herbivores—mostly open dwellers living in vast herds that move along plains and plateaus (like bisons, ovibovines, and megacerine deer)—is intriguing, especially when taking into account that this particular human-herbivore pattern was already present in Dmanisi.

As this herbivore fauna is directly related to open habitats, we may assume that the Dmanisi record represents the western edge of the initial or preliminary expansion of this type of environment and related mammal faunas in Eurasia, controlled by major climatic trends. As suggested by Van der Made (2011), the invasion of this fauna into Western Europe might have been prevented by the retention of more closed habitats in the western part of the continent. Given the current stage of research, it is difficult to ascertain whether humans took part in this westward expansion as members of a defined community, or if they appeared at Dmanisi accidentally/opportunistically. Two arguments may speak in favor of the first hypothesis: (1) recent discoveries put the appearance of *Homo* in Asia at the same time as Dmanisi, if not slightly earlier (Bar-Yosef and Belfer-Cohen 2013), and (2) early "expatriate" humans repeatedly show preferential association with a particular Asian herbivore assemblage of a specific habitat (i.e., semi-open savannah type ecotonal and mosaic landscapes; Finlayson et al. 2011). Extended research and continued fieldwork in several of these localities will yield a more complete picture that will help us to better answer these questions.

Acknowledgements Many thanks to the Organizing Committee of the International Symposium: Human Evolution in the Southern Balkans, 6–8 December, 2012, Tübingen, Germany for inviting us to participate. We also thank three anonymous reviewers for their useful comments on the manuscript. We wish to thank M. Roksandic, K. Harvati, and L. McCarty for the linguistic improvement of the manuscript.

References

Agustí, J., & Antón, M. (2002). *Mammoths, sabertooths and hominids*. New York: Columbia University Press.

Agustí, J., & Lordkipanidze, D. (2011). How "African" was the early human dispersal out of Africa? *Quaternary Science Reviews, 30*, 1338–1342.

Alberdi, M.-T., & Palombo, M.-R. (2013). The late Early to early Middle stenonoid horses from Italy. *Quaternary International, 288*, 25–44.

Arribas-Herrera, A. (Ed.). (2008). *Vertebrados den Plioceno superior terminal en el suroeste de Europa: Fonelas P-1 y el proyecto Fonelas: Vol. 10. Cuadernos del Museo Geominero*. Madrid: Publicaciones del Instituto Geologico y Minero de España.

Arzarello, M., Pavia, G., Peretto, C., Petronio, C., & Sardella, R. (2012). Evidence of an Early Pleistocene hominin presence at Pirro Nord (Apricena, Foggia, southern Italy): P13 site. *Quaternary International, 267*, 56–61.

Athanassiou, A. (2014). New giraffid (Artiodactyla) material from the Lower Pleistocene locality of Sésklo (SE Thessaly, Greece): Evidence for an extension of the genus *Palaeotragus* into the Pleistocene. *Zitteliana, B32*, 1–19.

Athanassiou, A., & Kostopoulos, D. S. (2001). Proboscidea of the Greek Pliocene-Early Pleistocene faunas: biochronological and palaeoecological implications. In G. Cavarretta, P. Gioia, M. Mussi, & M.-R. Palombo (Eds.), *The world of Elephants*. Rome: Consiglio Nationale delle Ricerche.

Azzaroli, A., de Giuli, C., Ficcarelli, G., & Torre, D. (1988). Late Pliocene to Early Mid-Pleistocene mammals in Eurasia: Faunal succession and dispersal events. *Palaeogegraphy, Palaeoclimatology, Palaeoecology, 66*, 77–100.

Bar-Yosef, O., & Belfer-Cohen, A. (2013). Following Pleistocene road signs of human dispersals across Eurasia. *Quaternary International, 285*, 30–43.

Bukshianidze, M. (2005). The fossil Bovidae of Dmanisi. Ph.D. Dissertation, University of Ferrara.

Carbonell, E., Bermudes de Castro, J. M., Pares, J. M., Perez-Gonzales, A., Cuenca-Bescos, G., Olle, A., et al. (2008). The first hominin of Europe. *Nature, 452*, 465–470.

Cregut-Bonnoure, E. (2007). Apport des Caprinae et Antilopinae (Mammalia, Bovidae) à la biostratigraphie du Pliocène terminal et du Pléistocène d'Europe. *Quaternaire, 18*, 73–97.

Cregut-Bonnoure, E., & Tsoukala, E. (2014). Bovidae and Cervidae of the Early Villafranchian site of Milia (Grevena, Macedonia, Greece). Abstract Book of the VIth International Conference on Mammoths and their Relatives. *Scientific Annals of the School of Geology, Special Volume 102*, 3.

Croitor, R., & Kostopoulos, D. S. (2004). On the systematic position of the large-sized deer from Apollonia, Early Pleistocene, Greece. *Paläontologische Zeitschrift, 78*(1), 137–159.

Dennell, R. (2010). Out of Africa. In J. G. Fleagle, J. J. Shea, F. E. Grine, A. L. Baden, & R. E. Leakey (Eds.), *Out of Africa I: Who? When? and Where?* Dordrecht: Springer.

de Vos, J., Van der Made, J., Athanassiou, A., Lyras, G., Sondaar, P. Y., & Dermitzakis, M. (2002). Preliminary note on the Late Pliocene fauna from Vatera (Lesvos, Greece). *Annales Géologiques des Pays Helléniques, 39*, 37–70.

Doukas, C. S., & de Bruijn, H. (2002). A new occurrence of *Dolichopithecus* (Mammalia, Primates) in N. Greece. *Annales Géologiques des Pays Helléniques, 39*, 295–297.

Eronen, J. T., Mirzaie Ataabadi, M., Micheels, A., Karme, A., Bernor, R.-L., & Fortelius, M. (2009). Distribution history and climatic controls of the Late Miocene Pikermian chronofauna. *Proceedings of the National Academy of Sciences USA, 106*(29), 11867–11871.

Eronen, J. T., & Rook, L. (2004). The Mio-Pliocene European Primate fossil record: Dynamics and habitat tracking. *Journal of Human Evolution, 47*, 323–341.

Finlayson, C., Carrión, J., Brown, K., Finlayson, G., Sánchez-Marco, A., Fa, D., et al. (2010). The *Homo* habitat niche: Using the avian fossil record to depict ecological characteristics of Palaeolithic Eurasian hominins. *Quaternary Science Reviews, 30*, 1525–1532.

Gabunia, L., Vekua, A., Lordkipanidze, D., Ferring, R., Justus, A., Maisuradze, G., et al. (2000). Current research on the hominid site of Dmanisi. *ERAUL, 92*, 13–27.

Galanidou, N. (2004). Early hominids in the Balkans. In H. I. Griffiths, B. Kristufek, & J. M. Reed (Eds.), *Balkan biodiversity. Pattern and process in the European hotspot*. London: Kluwer Academic Press.

Galanidou, N., Athanassas, C., Cole, J., Iliopoulos, G., Katerinopoulos, A., Magganas, A., & McNabb, J. (2016). The Acheulean site at Rodafnidia, Lisvori on Lesbos, Greece: Part I. In K. Harvati & M. Roksandic (Eds.), *Paleoanthropology of the Balkans and Anatolia: Human evolution and its context* (pp. 119–138). Dordrecht: Springer.

Galanidou, N., Cole, J., Iliopoulos, G., & McNabb, J. (2013). East meets west: The Middle Pleistocene site of Rodafnidia on Lesvos, Greece. Antiquity Project Gallery 87. http://antiquity.ac.uk/projgall/galanidou336/.

Gibbard, P. H., Head, M. J., Michael, J. C., & Walker, M. J. C. (2010). Formal ratification of the quaternary system/period and the pleistocene series/epoch with a base at 2.58 Ma. *Journal of Quaternary Science, 25*(2), 96–102.

Guadelli, J.-L., Sirakov, N., Sirakova, S., et al. (2012). Earliest dispersals and migrations to Europe via Balkans in Lower to Upper Palaeolithic: Evidence from Northern Bulgaria. PAGE Congress, December 2012, Tübingen, Abstracts, p. 9.

Guerin, C., & Tsoukala, E. (2013). The Tapiridae, Rhinocerotidae and Suidae (Mammalia) of the Early Villafranchian site of Milia (Grevena, Macedonia, Greece). *Geodiversitas, 35*(2), 447–489.

Harvati, K. (2016). Paleoanthropology in Greece: Recent finds and implications. In K. Havarti & M. Roksandic (Eds.), *Paleoanthropology of the Balkans and Anatolia: Human evolution and its context* (pp. 3–14). Dordrecht: Springer.

Harvati, K., Panagopoulou, E., & Runnels, C. (2009). The palaeoanthropology of Greece. *Evolutionary Anthropology, 18*, 131–143.

Hassanin, A., & Douzery, E. (1999). The tribal radiation of the family Bovidae (Artiodactyla) and the evolution of the mitochondrial cytochrome b gene. *Molecular Phylogenetics and Evolution, 13*, 227–243.

Hemmer, H. (2000). Out of Asia, a palaeoecological scenario of man and his carnivorous competitors in the European Lower Pleistocene. *ERAUL, 92*, 99–106.

Hemmer, H., Kahlke, R.-D., & Vekua, A. (2010). *Panthera onca georgica* ssp. nov. from the Early Pleistocene of Dmanisi (Republic of Georgia) and the phylogeography of jaguars (Mammalia, Carnivora, Felidae). *Neues Jahrbuch für Geologie und Paläontologie, Abhandlungen, 257*, 115–127.

Kahlke, R. D., García, N., Kostopoulos, D. S., Lacombat, F., Lister, A. M., Mazza, P. P. A., et al. (2011). Western Palaearctic palaeoenvironmental conditions during the Early and early Middle Pleistocene inferred from large mammal communities, and implications for hominin dispersal in Europe. *Quaternary Science Reviews, 30*, 1368–1395.

Kahn, M. A., Kostopoulos, D. S., Akhtar, M., & Nazir, M. (2010). *Bison* remains from the Upper Siwaliks of Pakistan. *Neues Jahrbuch für Geologie und Paläontologie, Abhandlungen, 258*, 121–128.

Koliadimou, K., & Koufos, G. D. (1998). Preliminary report about the Neogene/Quaternary micromammals of the Mygdonia Basin. *Romanian Journal of Stratigraphy, 78*, 75–82.

Konidaris, G. E., Tourloukis, V., Kostopoulos, D. S., Thompson, N., Giusti, D., Michailidis, D., et al. (2015). Two new vertebrate localities from the Early Pleistocene of Mygdonia Basin (Macedonia, Greece): Preliminary results. *Comptes Rendus Palevol, 14*, 353–362.

Kostopoulos, D. S. (2009). The Pikermi event: Temporal and spatial resolution of the Turolian large mammal fauna in SE Europe. *Palaeogeography, Palaeoclimatology, Palaeoecology, 274*, 82–95.

Kostopoulos, D. S., & Koufos, G. D. (1998). Palaeoecological remarks from the Villafranchian mammalian faunas of Macedonia, Greece. *Romanian Journal of Stratigraphy, 78*, 83–90.

Kostopoulos, D. S., & Koufos, G. D. (2000). Palaeoecological remarks on Plio-Pleistocene mammalian faunas. Comparative analysis of several Greek and European assemblages. In G. D. Koufos & Ioakim, C. (Eds.), *Mediterranean Neogene cyclostratigraphy in marine-continental deposits*. Bulletin Geological Society of Greece, *Special Publication, 9*, 139–150.

Kostopoulos, D. S., & Koulidou, I. (2015). An early mammoth maxilla from north-western Greece. *Quaternary International, 379*, 155–163.

Koufos, G. D. (2001). The Villafranchian mammalian faunas and biochronology of Greece. *Bolletino della Societa Paleontologica Italica, 40*(2), 217–223.

Koufos, G. D. (2006a). The large mammals from the Miocene/Pliocene locality of Silata, Macedonia, Greece with implications about the

Latest Miocene palaeoecology. *Beiträge zur Paläontologie, 30*, 293–313.

Koufos, G. D. (2006b). The Neogene mammal localities of Greece: Faunas, chronology and biostratigraphy. *Hellenic Journal of Geosciences, 41*, 183–214.

Koufos, G. D. (2006c). Palaeoecology and chronology of the Vallesian (late Miocene) in the Eastern Mediterranean region. *Palaeogeography, Palaeoclimatology, Palaeoecology, 234*, 127–145.

Koufos, G. D., & Kostopoulos, D. S. (1997). New carnivore material from the Plio-Pleistocene of Macedonia, Greece with the description of a new canid. *Münchener Geowissenschaften Abhandlungen, 34*, 33–63.

Koufos, G. D., Kostopoulos, D. S., & Sylvestrou, I. (1997). *Equus apolloniensis* n. sp. (Mammalia, Equidae) from the latest Villafranchian locality of Apollonia, Macedonia, Greece. *Palaeontologia i Evolucio, 30–31*, 49–76.

Koufos, G. D., Vasileiadou, K. (2015). Miocene/Pliocene mammal faunas of southern Balkans: implications for biostratigraphy and palaeoecology. Palaeobiodiversity and Palaeoenvironments, 95, 285–303.

Lazaridis, G., & Tsoukala, E. (2014). Short note on the Equidae of the Early Villafranchian site of Milia (Grevena, Macedonia, Greece). Abstract Book of the VIth International Conference on Mammoths and their Relatives. *Scientific Annals of the School of Geology, Special Volume 102*, 8.

Lebatard, A. E., Alciçek, M. C., Rochette, P., Khatib, S., Vialet, A., Boulbes, N., et al. (2014). Dating the *Homo erectus* bearing travertine from Kocabaş (Denizli, Turkey) at least 1.1 Ma. *Earth and Planetary Science Letters, 390*, 8–18.

Lewis, M., & Werdelin, L. (2010). Carnivoran dispersal out of Africa during the early Pleistocene: Relevance for hominids? In J. G. Fleagle, J. J. Shea, F. E. Grine, A. L. Baden, & R. E. Leakey (Eds.), *Out of Africa I: Who? When? and Where?* Dordrecht: Springer.

Lister, A. M., & van Essen, H. (2003). *Mammuthus rumanus* (Ştefănescu), the earliest mammoth in Europe. In A. Petrulescu & E. Ştuică (Eds.), *Advances in Vertebrate Palaeontology "Hen to Panda"*. Bucarest.

Loghem, W., Tsoukala, E., & Mol, D. (2010). Milia (Grevena, Macedonia, Greece): A dangerous area for *Mammut borsoni*? *Quaternaire. Hors-Série, 3*, 192–194.

Martínez-Navarro, B. (2010). Early Pleistocene faunas of Eurasia and hominid dispersals. In J. G. Fleagle, J. J. Shea, F. E. Grine, A. L. Baden, & R. E. Leakey (Eds.), *Out of Africa I: Who? When? and Where?* Dordrecht: Springer.

Martínez-Navarro, B., Sardella, R., Rook, L., Bellucci, L., & Ros-Montoya, S. (2013). First occurrence of *Soergelia* (Ovibovini, Bovidae, Mammalia) in the Early Pleistocene of Italy. *Quaternary International, 267*, 98–102.

Martínez-Navarro, B., Turq, A., Agustí, J., & Oms, O. (1997). Fuente Nueva-3 (Orce, Granada, Spain) and the first human occupation of Europe. *Journal of Human Evolution, 33*, 611–620.

Melentis, J. K. (1961). Studien über fossile Vertebraten Griechenlands. 2. Die Dentition der Pleistozänen Proboscidier des Beckens von Megalopolis im Peloponnes (Griechenland). *Annales Géologiques des Pays Helléniques, 12*, 153–262.

Melentis, J. K. (1964). Studien über fossile Vertebraten Griechenlands. 2. Die Dentition der Pleistozänen Cerviden des Beckens von Megalopolis im Peloponnes (Griechenland). *Annales Géologiques des Pays Helléniques, 16*, 1–92.

Meulen, A., & Kolfoschoten, T. (1986). Review of the late Turolian to early biharian mammal faunas from Greece and Turkey. *Memoires Societa Geologica Italiana, 31*, 201–211.

Moncel, M.-H. (2010). Oldest human expansions in Eurasia: Favouring and limiting factors. *Quaternary International, 223–224*, 1–9.

Muttoni, G., & Scardia, G. (2010). Human migration into Europe during the late Early Pleistocene climate transition. *Palaeogeography, Palaeoclimatology, Palaeoecology, 296*, 79–93.

Muttoni, G., Scardia, G., Kent, D. V., Morsiani, E., Tremolada, F., Cremaschi, M., et al. (2011). First dated human occupation of Italy at ~0.85 Ma during the late Early Pleistocene climate transition. *Earth and Planetary Science Letters, 307*, 241–252.

O'Regan, H., Bishop, L., Elton, S., Lamb, A., & Turner, A. (2006). Afro-Eurasian mammalian dispersal routes of the late Pliocene and early Pleistocene, and the bearing on earliest hominin movements. *Courier Forschungsinstitut Senckenberg, 256*, 305–331.

O'Regan, H., Turner, A., Bishop, L., Elton, S., & Lamb, A. (2011). Hominins without fellow travellers? First appearances and inferred dispersals of Afro-Eurasian large-mammals in the Plio-Pleistocene. *Quaternary Science Reviews, 30*, 1343–1352.

Palmqvist, P., Torregrossa, V., Perez-Claros, J. A., Martínez-Navarro, B., & Turner, A. (2007). A re-evaluation of the diversity of *Megantereon* (Mammalia, Carnivora, Machairodontinae) and the problem of species identification in extinct carnivores. *Journal of Vertebrate Paleontology, 27*, 160–175.

Panagopoulou, E., Tourloukis, E., Thompson, N., Athanassiou, A., Tsartsidou, G., Konidaris, G. E., Giusti, D., Karkanas, P., Harvati-Papatheodorou, K. (2015). Marathousa 1 : a new Middle Pleistocene archaeological site from Greece. . Antiquity, 89(343),1–4.

Radulescu, C., & Samson, P. (2001). Biochronology and evolution of the Early Pliocene to the Early Pleistocene mammalian faunas of Romania. *Bolletino della Societa Paleontologica Italica, 40*, 285–291.

Rivals, F., & Athanassiou, A. (2008). Dietary adaptations in an ungulate community from the late Pliocene of Greece. *Palaeogeography, Palaeoclimatology, Palaeoecology, 265*, 134–139.

Rodriguez, J., Burjachs, F., Cuenca-Bescós, G., García, N., Van der Made, J., Pérez González, A., et al. (2011). One million years of cultural evolution in a stable environment at Atapuerca (Burgos, Spain). *Quaternary Science Reviews, 30*, 1396–1412.

Rook, L., & Martínez-Navarro, B. (2010). Villafranchian: The long story of a Plio-Pleistocene European large mammal biochronologic unit. *Quaternary International, 219*, 134–144.

Schmidt-Kittler, N. (Ed.). (1995). The Vertebrate locality Maramena (Macedonia, Greece) at the Turolian/Ruscinian boundary (Neogene). *Münchener Geowissenschaften Abhandlungen, 28*, 1–180.

Sickenberg, O. (1967). Die Unterpleistozäne fauna von Wolaks (Griech.-Mazedonien) I Eine neue Giraffe *Macedonotherium martini* nov. gen. nov. sp. aus dem unteren Pleistozän von Griechenland. *Annales Géologiques des Pays Helléniques, 18*, 314–330.

Sickenberg, O. (1968). Die Unterpleistozäne Fauna von Wolaks (Griech.-Mazedonien) II. Carnivoren. *Annales Géologiques des Pays Helléniques, 19*, 621–646.

Sotnikova, M., & Rook, L. (2010). Dispersal of the Canini (Mammalia, Canidae: Caninae) across Eurasia during the Late Miocene to Early Pleistocene. *Quaternary International, 212*, 86–97.

Spassov, N. (2000). Biochronology and zoogeographic affinities of the Villafranchian faunas of Bulgaria and South Europe. *Historia Naturalis Bulgarica, 12*, 89–128.

Spassov, N. (2003). The Plio-Pleistocene vertebrate fauna in South-Eastern Europe and the megafaunal migratory waves from the east to Europe. *Revue de Paléobiologie, 22*, 197–229.

Spassov, N. (2016). Southeastern Europe as a route for the earliest dispersal of *Homo* towards Europe: Ecological conditions and the timing of the first human occupation of Europe. In K. Harvati & M. Roksandic (Eds.), *Paleoanthropology of the Balkans and Anatolia: Human evolution and its context* (pp. 281–290). Dordrecht: Springer.

Spassov, N., & Geraads, D. (2007). *Dolichopithecus balcanicus* sp. nov., a new Colobinae (Primates, Cercopithecidae) from the Early Pliocene of South-eastern Europe, with a discussion on the taxonomy of the genus. *Journal of Human Evolution, 52*, 434–442.

Steensma, K. J. (1988). Plio-Pleistozäne Grosssaugetiere (Mammalia) aus dem Becken von Kastoria/Grevena südlich von Neapolis-NW Griechenland. Ph.D. Dissertation, Technical University of Clausthal.

Strait, D. S., Orr, C. M., Hodgkins, J., Spassov, N., Gurova, M., Miller, C., & Tzankov, T. (2016). The human fossil record of Bulgaria, and the formulation of biogeographic hypotheses. In K. Harvati & M. Roksandic (Eds.), *Paleoanthropology of the Balkans and Anatolia: Human evolution and its context*. Dordrecht: Springer.

Symeonidis, N., & de Vos, J. (1976). Grosssäuger-Funde aus den Pleistozänen Spatenfülungen von Turkovunia in Athen. *Annales Géologiques des Pays Helléniques, 28*, 35–144.

Theodorou, G., Spjeldnaes, N., Hanken, N. M., Lauritzen, S. E., Velitzelos, E., Athanassiou, A., et al. (2000). Description and taphonomic investigations of Neogene Proboscidea from Rhodos, Greece. *Annales Géologiques des Pays Helléniques, 38*, 133–156.

Torro Moyano, I., Martínez-Navarro, B., Agustí, J., Souday, C., Bermúdez de Castro, J. M., Martinón-Torres, M., et al. (2013). The oldest human fossil in Europe, from Orce (Spain). *Journal of Human Evolution, 65*, 1–9.

Tourloukis, V. (2016). On the spatio-temporal distribution of Mediterranean Lower Paleolithic sites: A geoarchaeological perspective. In K. Harvati & M. Roksandic (Eds.), *Paleoanthropology of the Balkans and Anatolia: Human evolution and its context* (pp. 303–323). Dordrecht: Springer.

Tourloukis, V., & Karkanas, P. (2012). The Middle Pleistocene archaeological record of Greece and the role of the Aegean in hominin dispersals: New data and interpretations. *Quaternary Science Reviews, 43*, 1–15.

Tsoukala, E., & Chatzopoulou, K. (2005). A new Early Pleistocene (Latest Villafranchian) site of large mammals in Kalamotó of the Mygdonia basin (Macedonia, Greece). Preliminary report. *Mitteilungen der Kommission für Quartärforschung Österreichische Akademie Wissenschaften, 14*, 213–233.

Tsoukala, E., van Logchem, W., Lazaridis, G., & Mol, D. (2014). Carnivores of the early Villafranchian site of Milia (Grevena, Macedonia, Greece). Abstract Book of the VIth International Conference on Mammoths and their relatives. *Scientific Annals of the School of Geology, Special Volume, 102*, 6–7.

Van der Made, J. (2011). Biogeography and climatic change as a context to human dispersal out of Africa and within Eurasia. *Quaternary Science Reviews, 30*, 1353–1367.

Van der Made, J. (2013). First description of the large mammals from the locality of Penal, and updated faunal lists for the Atapuerca ungulates—*Equus altidens*, *Bison* and human dispersal into Western Europe. *Quaternary International, 295*, 36–47.

Van der Made, J., & Mateos, A. (2011). Longstanding biogeographic patterns and the dispersal of early *Homo* out of Africa and into Europe. *Quaternary International, 223–224*, 195–200.

Van der Made, J., & Tong, H. (2008). Phylogeny of the giant deer with palmate browtines *Megaloceros* from west and *Sinomegaceros* from Easte Eurosia. *Quaternary International, 179*, 135–162.

Vasileiadou, K., Konidaris, G., & Koufos, G. D. (2012). New data on the micromammalian locality of Kessani (Thrace, Greece) at the Mio-Pliocene boundary. *Palaeobiodiversity and Palaeoenvironments, 92*, 211–237.

Vassiliadou, K., Koufos, G. D., & Syrides, G. E. (2003). Silata, a new locality with micromammals from the Miocene/Pliocene boundary of the Chalkidiki peninsula, Macedonia, Greece. *Deinsea, 10*, 549–562.

Vislobokova, I. A. (2012). Giant Deer: Origin, evolution, role in the biosphere. *Paleontological Journal, 46*, 643–775.

Chapter 16
Southeastern Europe as a Route for the Earliest Dispersal of *Homo* Toward Europe: Ecological Conditions and the Timing of the First Human Occupation of Europe

Nikolai Spassov

Abstract In recent years, an increasing number of discoveries have supported the idea that human occupation of Europe took place earlier than expected, during the Villafranchian and significantly predating 1 Ma. Two hypotheses of dispersal toward Europe seem possible: (1) A direct dispersal from Africa with the earliest possible time frame being *ca.* 2.0–1.95 Ma; (2) A more recent dispersal, possibly from secondary nuclei of speciation in Asia Minor-Caucasus. The earliest "well documented" wave of *Homo* dispersal is probably related to the late Villafranchian/Epivillafranchian boundary, at *ca.* 1.3–1.2 Ma. Two routes of dispersal were possible: via the Bosphorus, or by a circuitous route around the Black Sea basin along the northern peri-Pontic coast. The time of the earliest human appearance in Europe could be related to conditions of increasing aridification and to a domination of open/mosaic landscapes, which roughly correspond to the ecological conditions experienced by African early *Homo*. The early *Homo* populations in Europe were likely not adapted to harsh climates and may have occupied only the southern-most areas of the continent.

Keywords Early Pleistocene *Homo* • First Europeans

N. Spassov (✉)
National Museum of Natural History, Bulgarian Academy of Sciences, 1 Tzar Osvoboditel Boulevard, 1000 Sofia, Bulgaria
e-mail: nspassov@nmnhs.com

The African Origin of the Genus *Homo* and the Timing and Routes of its Earliest Dispersal into Eurasia

The timing and routes of the first human dispersal into Europe are of special interest, as they are related to the origin of the genus *Homo* and its earliest dispersal out of its original area of habitation. Fossil evidence indicates that *Homo* originated in East Africa. The earliest appearance of the genus is documented in Ethiopia at 2.80–2.75 Ma (Villmoare et al. 2015). The species *H. rudolfensis*, *H. habilis*, and *H. ergaster* are generally accepted as the earliest members of the genus (but see also e.g. Collard and Wood 2007) and appear in this area at about 2.5–1.9 Ma (e.g. Schrenk et al. 2007; Tattersall 2007). It is difficult to say which of these hominins undertook the first movement out of Africa. As recently as the middle of the 1990s, it was thought that the earliest appearance of *Homo* in Asia dated to *ca.* 1.25 Ma (Dean and Delson 1995), and most authors at the time agreed that the earliest finds in Europe were probably younger than one million years (e.g. Dennell and Roebroeks 1996; Villa 1996). However, more recent evidence on the presence of *Homo* in Asia suggest considerably earlier ages than previously supposed. Remains confirming the presence of *Homo* at Sangiran, Java, suggest an age of 1.8 Ma (Swisher et al. 1994; Dennell 2004); the proposed age for the artifacts from Riwat, Pakistan is at least 1.9 Ma (Dennell et al. 1988; Dennell 2004; Malassé et al. 2016); a possible hominin presence in China as early as 1.9 Ma has been put forth on the basis of finds from Longgupo, (Huang et al. 1995), although these have been disputed (Etler et al. 2001; Ciochon 2009); finally, the most recent investigations in Dmanisi, Georgia, indicate an age of about 1.85–1.78 Ma for the *Homo* remains from the site (Ferring et al. 2011; Lordkipanidze et al. 2013).

Katerina Harvati and Mirjana Roksandic (eds.), *Paleoanthropology of the Balkans and Anatolia*, Vertebrate Paleobiology and Paleoanthropology, DOI 10.1007/978-94-024-0874-4_16

The Earliest Possible Dispersal Wave into Europe

The Time and Routes of the Earliest Theoretically Possible Dispersal: Argumentation

Until recently, evidence for a human presence in Europe predating 1 Ma was considered unreliable (see above). On the other hand, the discovery of *Homo* remains in Dmanisi as early as *ca.* 1.8 Ma suggests that an early human presence may also be logically expected in Europe. It would be quite strange if during the late Villafranchian early humans dispersed into Asia but not into Europe (and indeed, more recent discoveries have pushed back the timing of human presence in Europe to 1.2 Ma and possibly up to 1.6 Ma; see below)—unless there were insurmountable obstacles to reaching the European continent. The Gibraltar, Sicily, and Asia Minor corridors for dispersal from Africa toward Europe have been discussed several times (e.g. Abbate and Sagri 2012). The hypothesis that early *Homo* could disperse into Europe from Africa via the Iberian Peninsula and Sicily does not seem probable. There are no reliable data for a landbridge via the Gibraltar Straits during the Villafranchian (Turner and O'Regan 2007); furthermore, the fauna of the Iberian Peninsula represents a "cul de sac" compared to the other European fauna, rather than a migratory cross-road. The hypothesis for a dispersal via Sicily is not supported by strong arguments either (Villa 2001; Palombo 2013). The geological data lend strong support only to the hypothesis for an early *Homo* dispersal toward Eurasia via the Arabian Peninsula coastal area, the Near East, and Asia Minor. From there, the dispersal would be possible not only toward the East but also to the West into Europe (Dennell and Roebroeks 1996; Spassov 2001).

Recently, an increasing number of arguments seem to support the hypothesis of a long chronology for the human occupation of Europe with dates much older than 1.0 Ma and related to the Epivillafranchian and even the Late Villafranchian. Two important questions about the earliest human migration waves from the Near East are: (1) Was Europe accessible for dispersal movements at the time of the first migration out of Africa and the first appearance of *Homo* in Asia, i.e. at the beginning of the Late Villafranchian? And, (2) was *Homo* able to enter Europe at this time through the Bosphorus—the shortest path?

Two source centers are possible in relation to this dispersal (Fig. 16.1):

- A direct dispersal from Africa.
- A dispersal from a secondary population source in the Asia Minor-Caucasus area, such as Dmanisi. Several authors have proposed the second hypothesis (e.g. Made 2011; Ferring et al. 2011), and some have even suggested that *H. erectus* may originate from a Western Asian nucleus, from where they could have subsequently dispersed into Europe (Dennell et al. 2010). Yet, the morphological analyses focusing on cranial variation of the Dmanisi specimens, as well as the analyses of the oldest Eurasian artifacts, seem to support an African origin of the first Eurasian *Homo* (Moncel 2010; Lordkipanidze et al. 2013). Early *Homo* likely used the natural dispersal pathways of mammalian megafauna. As a species possibly already adapted to the consumption of animal meat and bone marrow (as suggested by large herbivore bones with cut marks and percussion marks dated to 2.5 Ma from Bouri, Ethiopia (de Heinzelin et al. 1999)), they might have dispersed by following the herbivore herds, similarly to large carnivores. However, the dispersal of African mammals (and especially of African carnivores) to Europe should not be equated with the human dispersal (see Palombo 2013). The patterns of range expansion of early *Homo* may have occurred in shorter episodes, and were most likely determined by parameters unique to our genus' own evolution, demography, and ecology.

Two different routes of dispersal were possible, corresponding to the routes known or hypothesized for the dispersals of fossil and recent mammals. Once in Asia Minor, early *Homo* could have moved toward Europe either directly via the Bosphorus; or by using a circuitous route around the Black Sea Basin, along the northern peri-Pontic coast (Spassov 2002; see also Strait et al. 2016). It should, however, be noted that this was likely not a single event (Dennell 2003; Spassov 2003).

Paleontological and paleoclimatic data can be used to suggest the timing and routes of the first possible migration of *Homo* to Europe. Biochronological analyses of the megafaunal dispersals from the beginning of the Late Villafranchian, based on the recent faunal data from Bulgaria and the Balkans, provide new arguments in favor of the suggestion that the earliest migrations of *Homo* could have an earliest Late Villafranchian age. The first appearance of some new steppe artiodactyls and related carnivores in Europe, to the west of the Black Sea, was documented in the pre-Olduvaian site of Slivnitsa (Bulgaria; see below) and, shortly thereafter, in Gerakarou (Greece) (Spassov 2000, 2003; Kahlke et al. 2011).

In the 1990s, an especially important Bulgarian locality of Villafranchian vertebrate fauna—Slivnitsa—was studied extensively (Spassov 1998, 2000). The micro- and macromammalian species recovered from the site allowed the establishment of rather narrow chronological boundaries for the fauna of the locality, corresponding to the beginning of the Late Villafranchian, between the localities from the MNQ17 zone of the mammalian biochronology (St.-Vallier, Chilhac) and those from the MNQ18 (Olivola Unit). Thus,

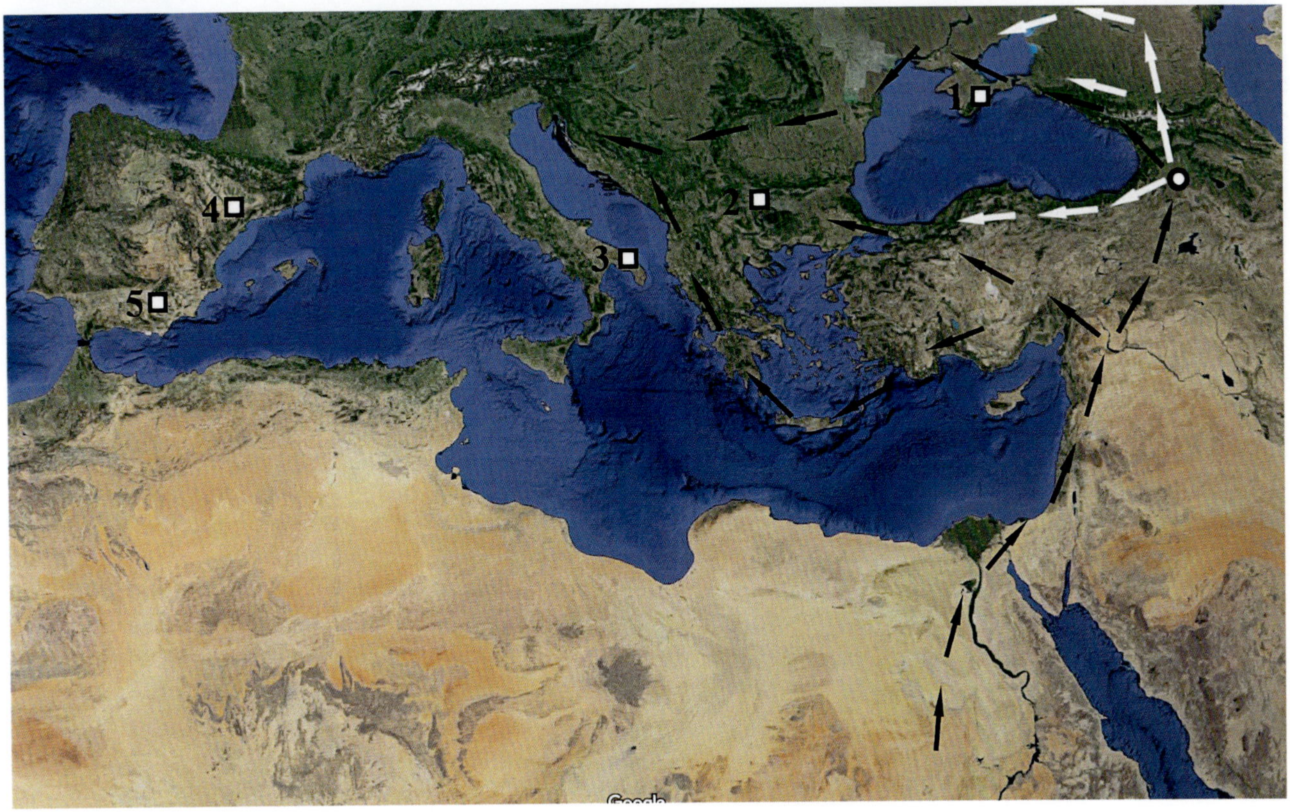

Fig. 16.1 Possible dispersal routes into Europe (map from Google Maps). *Black arrows*: a direct dispersal from Africa with dating of the first possible wave of dispersal at about *ca.* 2.0–1.95 Ma; *White arrows*: a more recent dispersal from a secondary nucleus of speciation in Asia Minor-Caucasus; Main localities (not directly associated on this map with the different type of arrows) of the earliest "well-documented" wave of *Homo* dispersal: (*1*) Sinyaya Balka & Rodniki; (*2*) Kozarnika; (*3*) Pirro Nord; (*4*) Sima del Elefante; (*5*) Barranco León

the locality of Slivnitsa should be referred to the Italian Costa St. Giacomo Unit and the beginning of the mammalian zone MNQ18 (i.e. MNQ18a). The fauna of Slivnitsa has many similarities with the one of Senèze, France, also placed at the beginning of MNQ18 (Spassov 2003). Very recent absolute $^{40}Ar/^{39}Ar$ dates of several important Villafranchian localities in France such as Chilhac and Senèze place the MNQ17/MNQ18 transition to an earlier geological time than was previously thought, indicating an age between 2.21 and 2.09 Ma, near the Reunion subchron (Nomade et al. 2014). Keeping in mind that the fauna from this site indicates an open environment (Nomade et al. 2014), it is more reasonable to accept the later date, which falls into the time of the known pre-Olduvai cooling period (see below).

Following these new age determinations, we could suppose that the age of the Slivnitsa faunal event is also close to *ca.* 2 Ma. The Slivnitsa faunal event must therefore correlate with the climatochronologic zone SCT10 of Zubakov and Borzenkova (1990). This zone, documented in Georgia and the Azov region, has an estimated age immediately predating the Olduvai warming (between the Reunion subchron and the beginning of the Olduvai event, i.e. between *ca.* 2.14 and 1.95 Ma BP). The faunas of several European localities dating to this time span, and especially Slivnitsa, evidence a transition to open environments. During this period, which corresponds to the Meria cooling documented in the Black Sea region, there are indications that some Aegean islands close to the Anatolian coast were connected with Asia Minor and that the Black Sea was a fresh water sea (Zubakov and Borzenkova 1990; Dermitzakis 1990). This suggests a temporary closing of the Bosphorus. The faunal interchange between Asia Minor and Europe that followed explains the presence of some earliest eastern immigrant species in Slivnitsa between *ca.* 2.1 and 1.95 Ma. Through the Balkan migratory route they spread to Western Europe (mainly in the Mediterranean area, where the climate and landscape were similar and where they are found at sites like Olivola, Senèze). The results from the study of the Slivnitsa fauna permitted new analyses and more precise dating of the Late Villafranchian mammalian dispersal events. The fauna of Slivnitsa provides a strong signal for landscape aridification. It records the first dispersal from the east of the wolf-like canids, of *Panthera onca toscana*, as well as a mass Caprinae migration, including the first *Ovis* into Europe (Spassov 2002, 2003). The early "jaguar" *P. onca toscana* probably dispersed from Africa into Europe some 2 Ma through the

Levantine Corridor (Made 2013). Thus, the timing and environmental conditions of the Slivnitsa faunal event allow us to hypothesize that the earliest possible wave of *Homo* dispersal could have occurred at *ca.* 2.0–1.95 Ma, following the routes of megafaunal dispersal from the East toward Europe. Such a dispersal might be evidenced by the earliest *Homo* appearance in Asia (Spassov 2003).

Some new discoveries in the Azov Sea region (e.g., at the site Kermek) are consistent with the hypothesis that the earliest dispersal of humans into Europe was very old (roughly 2 Ma based on the micromammal fauna (Shchelinsky et al. 2010b)). These findings support the second hypothesized route via the Northern peri-Pontic area under the conditions of the pre-Olduvaian cooling. It seems likely that the earliest dispersing human populations could not survive for long, as early *Homo* was probably not well adapted to the climate of the northern latitudes.

Possible Traces of the Ante-Epivillafranchian Hominins in Europe

Recent data on the possible Late Villafranchian presence of *Homo* in Europe have provoked vigorous discussions. Deposits with hominin remains or artifacts are often poorly dated and are usually associated with broad chronological boundaries (Parés et al. 2013). Even if the estimated ages of many of these cases seem questionable, the number of reported finds of Late Villafranchian age has steadily increased, lending some support to the hypothesis of a very early colonization of the continent.

The site of Chilhac 3 (Southern France) is dated on the basis of its fauna to the end of the Middle Villafranchian (*ca.* 1.9 Ma, revised to 2.38 Ma after recently published data (Nomade et al. 2014)). It continues to be cited as a possible site with a very old *Homo* presence (Bonifay 2002; Crochet et al. 2009), although current consensus is that the choppers from Chilhac cannot be considered to be associated with the fauna which defines that locality (Villa 1996).

The site of Lézignan-le-Cèbe (Southern France) is a newly described site of vertebrate fauna associated with lithic artifacts from the Early Pleistocene of the Hérault Valley (Crochet et al. 2009). A basalt layer securely dated at 1.57 Ma overlies the fossiliferous level, but the pebbles presented as artifacts and the faunal remains presented as "intentionally broken" have been met with skepticism and need further examination. Furthermore, the archaeological level and the dated basalt are several meters apart, and it is unclear whether the $^{40}Ar/^{39}Ar$ age represents a *post quem* or *ante quem* date for the archaeological layer (Parés et al. 2013).

The Valdarno region (Northern Italy) is well known for its rich Villafranchian fauna. It has been recently cited as a site of lithic industry with an age of 1.95 Ma, dated through magnetostratigraphic data. However, this information is based on personal communication only, without any other support (Abbate and Sagri 2012).

The site of Dealul Mijlociu (Oltet Valley, Romania) has yielded two lithic "artifacts with a very primitive early Paleolithic technique" (Rădulescu and Samson 1991, p. 232; Rădulescu et al. 1998; see also Doboş and Iovita 2016). It is worth noting that L. Leakey thought that these artifacts were similar to some specimens from the Oldowan lithic industry (L. Leakey pers. comm. in: Rădulescu and Samson 1990, p. 227). At the same time some experts express reservations in relation to their authenticity (E. Delson, pers. comm.). Another problem is the dating of the site, which preserves no fauna (see Doboş and Iovita 2016). Its age, proposed to be close to the fossil locality Valea Graunceanului and related to the Olduvaian warming at approx. 1.8–1.9 Ma (Rădulescu and Samson 1990; Spassov 2003) has been determined only by geological correlation with the sediments of the above-mentioned paleontological locality. On the other hand, it should be noted that some faunal elements show certain affinities with the fauna of Senèze, which must be older than the Olduvai event.

The site of Kermek (Azov sea region, S. Russia) recently became an important area of interest for Pleistocene studies in Eastern Europe. Associations of Early Paleolithic stone artifacts and mammalian fauna were reported a few years ago in the Bogatyri/Sinyaya Balka and Rodniki sites of the region (see below). Recently the 40 m thick sandy member of the deposits with *Allophaiomys deucalion*, paleomagnetically and biostratigraphically dated at about 2.0 Ma, yielded, according to the published data, 279 associated Early Paleolithic stone tools (Shchelinsky et al. 2010b; Shchelinsky 2013). Recent additional investigations have discovered traces of use wear on the lithic material from this site (M. Gurova, pers. comm.; Shchelinsky et al. 2016), lending support to the thesis that the assemblage represents lithic artifacts. The site of Kermek, therefore, may represent evidence supporting the hypothesis that the earliest dispersal of humans into Europe took place before the pre-Olduvaian cooling episode.

The Earliest "Well Documented" Wave of *Homo* Dispersal: Localities and Age

Some Southern European localities have recently received attention due to the discoveries of the oldest, relatively securely dated record of human presence in Europe. Among them are: Barranco León, Orce, Sima del Elefante, Spain; Kozarnika,

Bulgaria; Pirro Nord, Italy and the sites in the region of the Taman Peninsula (Azov sea region) Bogatyri and Rodniki.

Sima del Elefante (Atapuerca, Spain) is the site with the most secure dates of the very early (Late Villafranchian) *Homo* dispersal in Western Europe. The level TE9 at the site of Sima del Elefante, Atapuerca, Spain, has yielded a fragment of a human mandible associated with lithic tools dated to the Early Pleistocene (approximately 1.2–1.1 Ma) on the basis of a combination of paleomagnetism, cosmogenic nuclides, and biostratigraphy (Carbonell et al. 2008). The recent investigations of the micromammals (Cuenca-Bescós et al. 2013) place the lower faunal assemblage (including the human remains) of Sima del Elefante in the *Allophaiomys lavocati* biozone and give to this zone an age between 1.5 and 1.2 Ma: the time span from the late Matuyama to pre-Jaramillo paleomagnetic events. We could note here that, in fact, several localities with *A. lavocati* have an Epivillafranchian age and that the lower boundary of that the Epivillafranchian is placed at *ca.* 1.2 Ma (Kahlke et al. 2011). The micromammals as well as the megafauna of this Spanish site show some affinities with the Italian locality of Pirro Nord: the *Allophaiomys lavocati* biozone in the Iberian Peninsula, present in the lower Unit of Sima del Elefante, may correlate to the *Allophaiomys ruffoi* biozone in Italy. The latter species is known in Pirro Nord (see below).

The very old dates (1.4–1.2 Ma) of some Spanish sites in Orce, such as Barranco León, where a human lower deciduous molar was found recently (Toro-Moyano et al. 2013), and Fuente Nueva 3 (Palmqvist et al. 2005; Crochet et al. 2009) have been questioned (see e.g. Muttoni et al. 2013). According to some authors, the more reliable biochronological and paleomagnetic data provide evidence for an earliest human occupation at Orce at about 1.1–0.9 Ma (e.g. Oms et al. 2000; Muttoni et al. 2013). The Orce sites were argued to be slightly older than Sima del Elefante on the basis of the lack of pigs in Orce, as well as because the *Allophayomys* from the Orce sites seems to be less evolved than the one from Sima del Elefante (Toro-Moyano et al. 2013). However, similarly to Sima del Elefante, these sites were recently dated to the *Allophayomys lavocati* biozone (Cuenca-Bescós et al. 2013). Therefore, together with the older levels in Atapuerca, they could evidence (more or less) the same wave of early human dispersal.

Some lithic artifacts were discovered in the well-known Pirro Nord paleontological site (Apulia, Southern Italy; Arzarello et al. 2007). The fauna of this locality, as mentioned above, also documents mammal migration from the East toward Southern Europe. Together with the lithics, this evidence could be an indication that the Pirro Unit localities preserve traces of a very early human migration (Spassov 2001, 2003). Initially, the time span of the Pirro Unit, which follows the older Farneta Unit (~1.6–1.4 Ma) in the Italian biochronology of the Villafranchian Mammal age, was defined in the approximate frames of 1.2/1.1–1.4 Ma (Gliozzi et al. 1997). However, after the discovery of evidence for human presence at Pirro Nord, an earlier age, ranging from 1.3 to 1.7 Ma was proposed for Pirro and its lithic industry (Arzarello et al. 2007).

Kozarnika is a cave in Northwestern Bulgaria, where the oldest levels (archaeological complexes 11c-14 = biozone B2-2) have yielded stone tools associated with pre-Middle Pleistocene fauna (see also Ivanova 2016). The age is estimated mainly on the basis of biochronological data, but identification of the megafauna is largely based on limited material and the suggested ages are controversial: from 1.9 to 1.8 Ma (MNQ18) on the basis of bovids (Fernandez and Crégut-Bonnoure 2007); to 1.2–1.0 Ma on the basis of small mammals (Popov and Marinska 2007). More recently, Sirakov et al. (2010) estimated the age of the mammal assemblage from Kozarnika's lower level (B2-2) at approximately 1.6–1.4 Ma. It should be noted that a number of taxa are identified with some uncertainty, resulting in rather wide time frames for the Kozarnika faunal assemblage (see also: Kahlke et al. 2011). Most of the securely identified megafaunal elements of the noted level indicate that its maximum age is similar to that of the Italian Pirro Nord Unit and close to the late Villafranchian/Epivillafranchian boundary (see N. Spassov and D. Kostopoulos comments, Kahlke et al. 2011), i.e. about 1.3–1.2 Ma. This estimation is closer to the more reliable age obtained by the micromammals (1.2–1.0 Ma: Popov and Marinska 2007). It is mentioned that the *Mimomys savini/M. pusillus* biozone, corresponding to the oldest levels with artifact presence in Kozarnika, has an age younger than the one of the biozone with *Allophayomys lavocati/A. ruffoi* known in Sima del Elefante and Pirro Nord. On the other hand, the *Mimomys savini/M. pusillus* biozone is also characterized, in some Italian localities, by the presence of *A. ruffoi* (see discussion in: Cuenca-Bescós et al. 2013), so the age of all three mentioned localities from Bulgaria, Italy, and Spain could be rather similar.

The early Paleolithic sites Bogatyri/Sinyaya Balka and Rodniki preserving lithic artifacts (Bogatyri preserves a total of 340 artifacts), were also recently reported from the Taman Peninsula, Azov region, in southern Russia (Shchelinsky et al. 2010a; Shchelinsky 2013). The mammalian fauna and paleomagnetic data indicate, according to the authors, an Early Pleistocene, Early Biharian age of the deposits, in the rather broad frames of 1.6/1.5–1.2 Ma.

In summary, all the localities mentioned in this chapter show general faunal and age similarities and could be related to the time of the Italian Pirro Faunal Unit (see below—"Conclusions").

Ecological Conditions in Southeastern Europe and the Earliest Human Dispersal Toward Europe

Ecological Conditions and the First Possible Wave of Dispersal into Europe

The paleontological data also provide a good idea about the natural conditions in Southeastern Europe, the route of the earliest human dispersals. Data for the time span 2.5–2.3 Ma based on the fauna of Varshets, Bulgaria, and Dafnero and Volakas, Greece (St. Vallier biochronological unit), indicate a relatively humid climate, where the physiognomy of the landscapes was chiefly characterized by forest-steppe (forest-savannah-like) mosaic biotopes or open forests (Spassov 2000; Kostopoulos and Koufos 2000; Kahlke et al. 2011). The faunal turnover, indicating climatic and landscape changes, was registered, as mentioned above, after the fauna of Slivnitsa, Bulgaria. About 750 bones identifiable at the species level were recovered from that locality, representing 11 ungulate species: "*Cervus*" *rhenanus*—"*Dama*" *nestii*, *Eucladoceros ctenoides* cf. *senesensis*, *Gazellospira torticornis*, *Procamptoceras brivatense*, *Gallogoral meneghinii*, *Pliotragus* cf. *ardeus*, *Megalovis* sp., *Hemitragus orientalis*, *Ovis* sp., Bovidae gen et sp. indet (*Soergelia*?), *Equus* cf. *stenonis* (Spassov 1999, 2005). The distribution of these species in different biotopes according to their ecological requirements indicates that only about 13% of the species could be characterized as typical forest animals (Fig. 16.2). The distribution of the bone remains also indicates that the number of remains from the supposed typical forest animals makes up only about 6% of all bones (Fig. 16.3). Remains of "steppe" animals prevail and a relatively large amount of bones likely represents species adapted to forest-steppe environments (the open woodlands). Therefore, the landscape must have been dominated by open areas and possibly by forest-steppe mosaics.

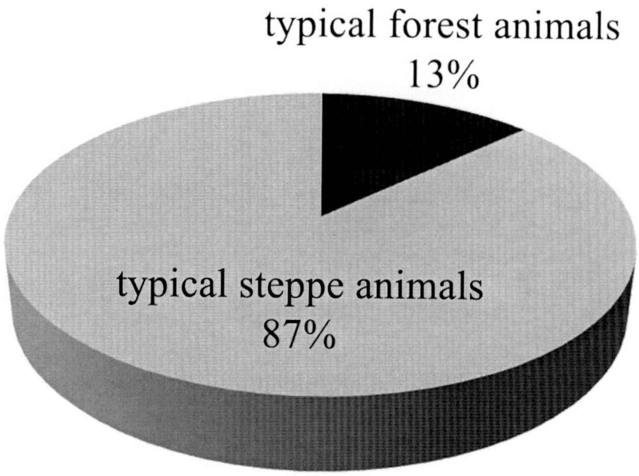

Fig. 16.2 Distribution of the ungulate species of the late Villafranchian locality of Slivnitsa (Bulgaria) after their ecological requirements, determined after morpho-functional characteristics and taxonomic appurtenance in comparison with recent analogs (after Spassov 2002 with additions and modifications—this paper)

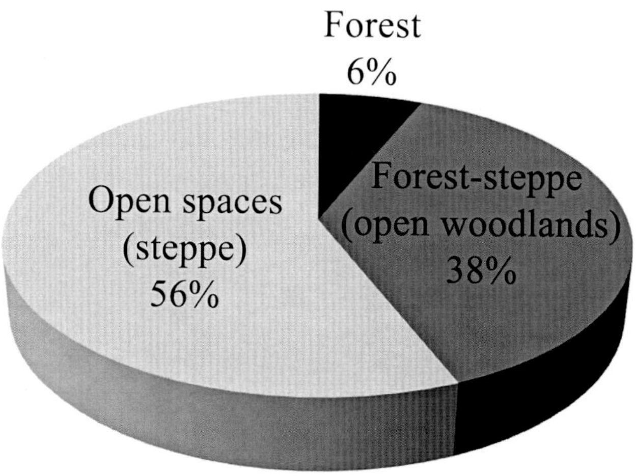

Fig. 16.3 Distribution of the fossil bone remains of the mammals of the Slivnitsa locality. Proportion of the number of bones of ecologically different animals: relative amount of bones of steppe animals, forest animals, and animals from the open woodland

Ecological Conditions of the Earliest "Well Documented" Wave of Homo Dispersal from the East

The paleoenvironmental conditions in the Balkans up to the Final Villafranchian are recorded in some Greek localities, such as Ravin of Voulgarakis, and especially in the well-documented paleontological locality of Apollonia. The age of Apollonia is similar to that of the Italian Pirro Unit (Spassov 2003). Its mammal assemblage represents the most complete fossil assemblage of the Southeastern European Epivillafranchian. The Apollonia/Ravine of Voulgarakis faunas possesses various novel features for the Balkans such as the presence of early bison. The identified ungulates of Apollonia (updated checklist after Kostopoulos 1997; Kostopoulos et al. 2002; Kahlke et al. 2011; see also Koufos and Kostopoulos 2016) include: *Equus apolloniensis*, *Pontoceros ambiguous mediterraneus*, *Soergelia brigittae*, *Hemitragus orientalis*, *Praeovibos mediterraneus*, *Ovis* sp., *Bison* sp., *Praemegaceros pliotarandoides*, and *Arvernoceros sp.* Kostopoulos and Koufos (2000) note that the paleoecological parameters suggest an open grassy landscape with a dominance of intermediate feeders and grazers, as well as an absence of browsers. The prevalence of the open landscapes is also clearly demonstrated in Fig. 16.4 by

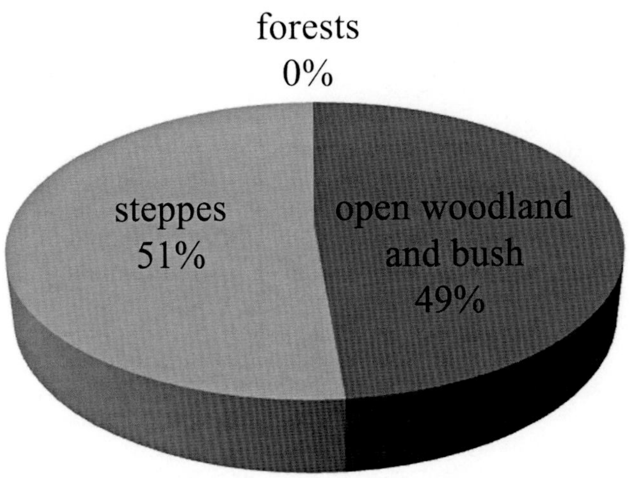

Fig. 16.4 Distribution of the ungulate species of the Final Villafranchian locality of Apollonia (Greece), after their ecological requirements (see text of Fig. 16.2)

the distribution of the ungulate species according to their ecological requirements, which shows the absence of strictly forest-adapted species. The presence of semi-aquatic forms (e.g. *Hippopotamus*), alongside elements with affinities for drier environments, are also possible indicators of moderate climatic conditions (Kahlke et al. 2011).

The landscapes in the region of the Taman Peninsula, Azov sea region, based on the data from the Rodniki and Sinyaya Balka sites, were rather different from those of the Balkans. They were more Nordic and influenced by the wide plain conditions. Nevertheless, they are similar in showing a strong presence of open spaces. The dominant large mammals are specialized grazers such as *Mammuthus* (about 64 % of the finds) and *Elasmotheriun caucasicum* (32 % of the bones). Ninety percent of the analyzed pollen derives from conifers, although pollen from herbaceous taxa (Asteraceae, Chenopodiaceae, Artemisia, Ephedra) was also detected. The pollen spectra and faunal associations indicate widespread forest-steppe and open-steppe landscapes (Shchelinsky et al. 2010a).

Conclusions

The first possible hypothetical migratory wave of *Homo* into Europe could be related to the time interval between the Reunion subchron (2.14 Ma) and the beginning of the Olduvaian event (1.95 Ma), most likely (keeping in mind the age of the site Kermek at the Taman Peninsula) between 2.0 and 1.95 Ma, i.e., at the beginning of the late Villafranchian. This time interval corresponds in the Black sea region with the so-called Meria cooling event, which was accompanied by freshening of the waters of the Black Sea, likely resulting from a short-term closure of the Bosphorus. The establishment of a temporary land bridge between Asia Minor and Europe must have allowed for faunal dispersal (as documented at the Bulgarian Slivnitsa site) and might have also resulted in a human dispersal into the Balkans. The Slivnitsa faunal event, correlated with the MNQ18a mammal zone (*sensu* Spassov 2003) of the Villafranchian mammal age, thus likely documents the first signs of favorable conditions for such a dispersal at the time of the earliest *Homo* movements out of Africa.

Our biochronological comparison of the faunas from Kozarnika (The Balkans), Bogatyri and Rodniki (the Taman Peninsula, Southern Russia), Pirro Nord (Italy), and the Spanish sites in Atapuerca and Orce leads to the suggestion that these sites could document slightly different phases of the same early *Homo* migration toward Europe. These sites could indicate the routes of dispersal and the occupation of the southern part of the continent from East to West. This wave must be related to the Final Villafranchian time (*sensu* Spassov 2003) and probably more precisely to the time of the Italian Pirro Nord Faunal Unit. The most likely time span for this dispersal process is *ca.* 1.3–1.2 Ma, i.e., close to the lower Epivillafranchian (*sensu* Kahlke et al. 2011) boundary, and could be a result of the beginning of considerable landscape aridification.

The most parsimonious interpretation of the earliest human migrations is that Europe was settled in an East to West fashion by two possible routes, also recognized as mammal dispersal routes: via the Balkans and via the Northern peri-Pontic area. The early waves of human dispersal into Europe could be related (as shown in the above examples) to moderate climatic conditions with increasing aridification, leading in turn to the domination of biotopes in which the open/mosaic landscapes played an important role. The reasons explaining the arrivals of hominins in the Levant, and then in all of Eurasia, might therefore be linked to climate drying starting at around 2.5–2 Ma in Africa (Moncel 2010 and references therein).

It is also possible to hypothesize that the earliest dispersing human populations may not have survived for long, given that early *Homo* was probably not well adapted to the climate of the northern latitudes. The area of Europe occupied in early times should therefore have roughly corresponded in its environmental conditions to the regions initially occupied by African early *Homo* (see also Dennell 2003). This line of reasoning would suggest that the early human presence in Europe was limited to the southern latitudes of the continent (Rodríguez et al. 2013).

Judging from the Kozarnika site (Bulgaria) where the lithic industries seem to be present practically without interruption from the older to youngest levels (Guadelli et al. 2005), a more or less constant human presence in Europe (at least in Southern Europe), starting in the Epivillafranchian time, can be hypothesized. It is also possible that the

Epivillafranchian dispersals did not originate directly from Africa, but instead from the Trans-Caucasus—Asia Minor secondary nuclei of humans better adapted to the northern latitudes (see also Ferring et al. 2011).

Acknowledgement The author thanks the three anonymous reviewers, Katerina Harvati and Jana Makedonska, whose valuable comments considerably improved the manuscript.

References

Abbate, E., & Sagri, M. (2012). Early to Middle Pleistocene *Homo* dispersals from Africa to Eurasia: Geological, climatic and environmental constraints. *Quaternary International, 267*, 3–19.

Arzarello, M., Marcolini, F., Pavia, G., Pavia, M., Petronio, C., Petrucci, M., et al. (2007). Evidence of earliest human occurrence in Europe: The site of Pirro Nord (Southern Italy). *Naturwissenschaften, 94*, 107–112.

Bonifay, E. (2002). *Les premiers peuplements de l'Europe*. Paris: La Maison des Roches.

Carbonell, E., Bermúdez de Castro, J. M., Parés, J. M., Pérez-González, A., Cuenca-Bescós, G., Ollé, A., et al. (2008). The first hominin of Europe. *Nature, 452*, 465–470.

Ciochon, R. (2009). The mystery ape of Pleistocene Asia. *Nature, 459*(18), 910–911.

Collard, M., & Wood, B. (2007). Defining the genus *Homo*. In W. Henke & I. Tattersall (Eds.), *Handbook of paleoanthropology* (Vol. 3, pp. 157–1610). Berlin: Springer.

Crochet, J. Y., Welcomme, J.-L., Ivorra, J., Ruffet, G., Boulbes, N., Capdevila, R., et al. (2009). Une nouvelle faune de vertébrés continentaux, associée à des artefacts dans le Pléistocène inférieur de l'Hérault (Sud de la France), vers 1, 57 Ma. *Comptes Rendus Palevol, 8*, 725–736.

Cuenca-Bescós, G., Rofes, J., López-García, J., Blain, H.-A., Rabal-Garcés, R., Sauqué, V., et al. (2013). The small mammals of Sima del Elefante (Atapuerca, Spain) and the first entrance of Homo in Western Europe. *Quaternary International, 295*, 28–35.

de Heinzelin, J., Desmond Clark, J., White, T., Hart, W., Renne, P., Wolde Gabriel, G., et al. (1999). Environment and behavior of 2.5-million-year-old Bouri Hominids. *Science, 284*, 625–629.

Dean, D., & Delson, E. (1995). *Homo* at the gates of Europe. *Nature, 373*, 472–473.

Dennell, R. (2003). Dispersal and colonisation, long and short chronologies, how continuous is the Early Pleistocene record for hominids outside East Africa? *Journal of Human Evolution, 45*, 421–440.

Dennell, R. (2004). Hominid dispersals and Asian biogeography during the Lower and Early Middle Pleistocene, c. 2.0–0.5 Mya. *Asian Perspectives, 43*(2), 205–226.

Dennell, R., Martinón-Torres, M., & Bermudez de Castro, J. M. (2010). Out of Asia: The initial colonisation of Europe in the Early and Middle Pleistocene. *Quaternary International, 439*, 223–224.

Dennell, R., Rendell, H., & Hailwood, E. (1988). Early tool-making in Asia, two-million-year-old artefacts in Pakistan. *Antiquity, 62*, 98–106.

Dennell, R., & Roebroeks, W. (1996). The earliest colonization of Europe, the short chronology revised. *Antiquity, 70*(269), 535–542.

Dermitzakis, M. (1990). The evolution of the Aegeis during the Late Cenozoic. *Geologica Balcanica, 20*(3), 3–16.

Doboş, A., & Iovita, R. (2016). The Lower Paleolithic of Romania revisited: New evidence from the site of Dealul Guran. In K. Harvati & M. Roksandic (Eds.), *Paleoanthropology of the Balkans and Anatolia: Human evolution and its context*. Dordrecht: Springer.

Etler, D., Crummett, T., & Wolpoff, M. (2001). Longuppo, Early *Homo* Colonizer or Late Pliocene *Lufengpithecus* Survivor in South China? *Human Evolution, 16*(1), 1–12.

Fernandez, P., & Crégut-Bonnoure, E. (2007). Les Caprinae (Rupicaprini, Ovibovini, Ovini et Caprini) de la séquence pléistocène de Kozarnika (Bulgarie du Nord): Morphométrie, biochronologie et implications phylogéniques. *Revue de Paléobiologie, Genève, 26*(2), 425–503.

Ferring, R., Oms, O., Agustí, J., Berna, F., Nioradze, M., Shelia, T., et al. (2011). Earliest human occupations at Dmanisi (Georgian Caucasus) dated to 1.85–1.78 Ma. *Proceedings of the National Academy of Sciences USA, 108*(26), 10432–10436.

Gliozzi, E., Abbazzi, L., Argenti, P., Azzaroli, A., Caloi, L., Capasso Barbato, L., et al. (1997). Biochronology of selected Mammals, Molluscs and Ostracods from the Middle Pliocene to the Late Pleistocene in Italy: The state of the art. *Rivista Italiana di Paleontologia e Stratigrafia, 103*(3), 369–388.

Guadelli, J.-L., Sirakov, N., Ivanova, S., Sirakova, Sv., Anastassova, E., Courtaud, P., et al. (2005). Une séquence du Paléolithique inférieur au Paléolithique récent dans les Balkans: La grotte Kozarnika à Oreshets (Nord-Ouest de la Bulgarie). In N. Molines, M.-H. Moncel & J.-L. Monnier. (Eds.), *Les premiers peuplements en Europe. Actes du colloque international "Données récentes sur les modalités de peuplement et sur le cadre chronostratigraphique, géologique et paléogéographique des industries du Paléolithique inférieur et moyen en Europe"*, Rennes, 22–25 septembre 2003 (Vol. 1364, pp. 87–103). Oxford: British Archaeological Research.

Huang, W., Ciochon, H. R., Yumin, G., Larick, R., Qiren, F., Schwarcz, H., et al. (1995). Early *Homo* and associated artefacts from Asia. *Nature, 378*, 275.

Ivanova, S. (2016). A route through the Balkans and implications for the earliest settlement of Europe. In K. Harvati & M. Roksandic (Eds.), *Paleoanthropology of the Balkans and Anatolia: Human evolution and its context*. Dordrecht: Springer.

Kahlke, R.-D., García, N., Kostopoulos, D., Lacombat, F., Lister, A., Mazza, P., et al. (2011). Western Palaearctic palaeoenvironmental conditions during the Early and early Middle Pleistocene inferred from large mammal communities, and implications for hominin dispersal in Europe. In J. Carrión, J. Rose & C. Stringer (Eds.), *Ecological scenarios for human evolution during the Early and Early Middle Pleistocene in the Western Palaearctic: Vol. 30. Quaternary science reviews* (pp. 1368–1395).

Kostopoulos, D. (1997). The Plio-Pleistocene artiodactyls (Vertebrata, Mammalia) of Macedonia. 1. The fossiliferous site "Apollonia-1", Mygdonia bassin of Greece. *Geodiversitas, 19*(4), 845–875.

Kostopoulos, D., & Koufos, G. (2000). Palaeoecological remarks on Plio-Pleistocene mammalian faunas. Comparative analysis of several Greek and European assemblages. *Proceedings Interim Colloquium RCMNS*. Patras, Greece, May 1998. *Geological Society of Greece. Special Publications, 9*, 139–150.

Kostopoulos, D. S., Vassiliadou, K., & Koufos, G. D. (2002). The beginning of Pleistocene in the Balkan area according to the mammal record; palaeozoogeographical approach. Proceedings International Workshop, On Late Plio/Pleistocene extinctions and evolution in the Palearctic. The Vatera site, Vatera-Lesvos. *Annales Géologiques des Pays Helléniques, 39*, 253–278.

Koufos, G. D., & Kostopoulos, D. S. (2016). The Plio-Pleistocene large mammal record of Greece: Implications for early human dispersals into Europe. In K. Harvati & M. Roksandic (Eds.), *Paleoanthropology of the Balkans and Anatolia: Human evolution and its context* (pp. 269–280). Dordrecht: Springer.

Lordkipanidze, D., Ponce de León, M., Margvelashvili, A., Rak, Y., Philip Rightmire, G., Vekua, A., et al. (2013). A complete skull from Dmanisi, Georgia, and the Evolutionary Biology of Early Homo. *Science, 342*, 326–330.

Made, J. van der (2011). Biogeography and climatic change as a context to human dispersal out of Africa and within Eurasia. *Quaternary Science Reviews, 30*, 1353–1367.

Made, J. van der (2013). Faunal exchanges through the Levantine Corridor and human dispersal: The paradox of the late dispersal of the Acheulean industry. *Proceedings of the International Symposium Africa, Cradle of Humanity: Recent discoveries*. Sétif, 26–28 Octobre, 2009. *Travaux du Centre National de Recherches Préhistoriques, Anthropologiques et Historiques, N. Sér. No. 18*, 255–296.

Malassé, A., Moigne, A.-M., Singh, M., Calligaro, T., Karir, B., Gaillard, C., Kaur, A., Bhardwaj, V., Pal, S., Abdessadok, S., Chapon Sao, C., Gargani, J., Tudryn, A., Sanz M. (2016). Intentional cut marks on bovid from the Quranwala zone, 2.6 Ma, Siwalik Frontal Range, northwestern India. *Comptes Rendus Palevol, 15*(3-4),317–339.

Moncel, M.-H. (2010). Oldest human expansions in Eurasia: Favouring and limiting factors. *Quaternary International, 223–224*, 1–9.

Muttoni, G., Scardia, G., & Kent, D. V. (2013). A critique of evidence for human occupation of Europe older than the Jaramillo subchron (~1 Ma): Comment on 'The oldest human fossil in Europe from Orce (Spain)' by Toyo-Moyano et al. (2013). *Journal of Human Evolution, 65*, 746–749.

Nomade, S., Pastre, J., Guillou, H., Faure, M., Guérin, C., Delson, E., et al. (2014). 40Ar/39Ar constraints on some French landmark Late Pliocene to Early Pleistocene large mammalian paleofaunas: Paleoenvironmental and paleoecological implications. *Quaternary Geochronology, 21*, 2–15.

Oms, O., Pare, J. M., Martinez-Navarro, B., Agustı, J., Toro, I., Martinez-Fernandez, G., et al. (2000). Early human occupation of Western Europe: Paleomagnetic dates for two paleolithic sites in Spain. *Proceedings of the National Academy of Sciences USA, 97*(19), 10666–10670.

Palmqvist, P., Martínez-Navarro, B., Toro, I., Patrocinio-Espigares, M., Ros-Montoya, S., Torregrosa, V., et al. (2005). Réévaluation de la présence humaine au Pléistocène inférieur dans le Sud de l'Espagne. *L'Anthropologie, 109*, 411–450.

Palombo, M. R. (2013). What about causal mechanisms promoting early hominin dispersal in Eurasia? A research agenda for answering a hotly debated question. *Quaternary International, 295*, 13–27.

Parés, J. M., Duval, M., & Arnold, L. (2013). New views on an old move: Hominin migration into Eurasia. *Quaternary International, 295*, 5–12.

Popov, V., & Marinska, M. (2007). An almost one million year long (Early to Late Pleistocene) small mammal succession from the archaeological layers of Kozarnika Cave in Northern Bulgaria. *Courier Forschungsinstitut Senckenberg, 259*, 79–92.

Rădulescu, C., & Samson, P. (1990). The Plio-Pleistocene mammalian succession of the Oltet Valley—Dacic Basin, Romania. *Quartärpaläontologie, 8*, 225–232.

Rădulescu, C., & Samson, P. (1991). Traces d'activité humaine à la limite Pliocèene/Pleistocène dans le bassin Dacique (Roumanie). In E. Bonifay & B. Vandermeersch (Eds.), *Les Premiers Européens. Actes du 114 Congrès National des Sociétés Savantes (Paris, 3–9 avril, 1989)* (pp. 203–207). Paris: Éditions du C.T.H.S.

Rădulescu, C., Samson, P., & Stiuca, E. (1998). Cadre biostratigraphique du Paléolithique Inférieur en Roumanie. *Quaternaire, 9*(4), 283–290.

Rodríguez, J., Martín-González, J., Goikoetxea, I., Rodríguez-Gómez, G., & Mateos, A. (2013). Mammalian paleobiogeography and the distribution of *Homo* in early Pleistocene Europe. *Quaternary International, 295*, 48–58.

Schrenk, F., Kullmer, O., & Bromage, T. (2007). The earliest putative *Homo* fossils. In W. Henke & I. Tattersall (Eds.), *Handbook of paleoanthropology* (Vol. 3, pp. 1611–1632). Berlin: Springer.

Shchelinsky, V. (2013). Functional features of Oldovanian sites on the Taman Peninsula in the Southern Sea of Azov Region (geological and archaeological evidence). VIII General Russian Conference: Fundamental problems of the quaternary. Russian Academy of Sciences. Rostov-on-Don. June 10–15, 2015, 713–716.

Shchelinsky, V., Dodonov, A., Baigusheva, V., Kulakov, S., Simakova, A., Tesakov, A., et al. (2010a). Early Palaeolithic sites on the Taman Peninsula (Southern Azov Sea region, Russia): Bogatyri/Sinyaya Balka and Rodniki. *Quaternary International, 223–224*, 28–35.

Shchelinsky, V., Gurova M., Tesakov, A., Titov, V., Frolov P., Simakova, A., 2016. The early Pleistocene site of Kermek in western Ciscaucasia (Southern Russia): Stratigraphy, biotic record and lithic industry (preliminary results), *Quaternary International, 393*, 51–69.

Shchelinsky, V., Tesakov, A., & Titov, V. (2010b). Early Paleolithic sites in the Azov sea region, stratigraphic position, stone associations and new discoveries. Quaternary stratigraphy and paleontology of the southern Russia, connections between Europe, Africa and Asia. Abstract volume. 2010 Annual Meeting INQUA-SEQS. Rostov-on-Don, Russia. June 21–26, 2010, 148–149.

Sirakov, N., Guadelli, J.-L., Ivanova, S., Sirakova, S., Boudadi-Maligne, M., Dimitrova, I., et al. (2010). An ancient continuous human presence in the Balkans and the beginnings of human settlement in Western Eurasia: A Lower Pleistocene example of the Lower Palaeolithic levels in Kozarnika cave (North-Western Bulgaria). *Quaternary International, 223–224*, 94–106.

Spassov, N. (1998). A New Late Villafranchian locality of vertebrate Fauna—Slivnitsa (Bulgaria) and the carnivore dispersal events in Europe on the Pliocene/Pleistocene boundary. *Historia Naturalis Bulgarica, 9*, 101–113.

Spassov, N. (1999). The Mammalian Megafauna from the Late Villafranchian Localities Varshets and Sivnitsa, Bulgaria and the Biochronology of the Villafrancian in S.-E. Europe. Unpublished Thesis. N.M.N.H., Bulg. Acad. Sci. Sofia, 284 pp. (in Bulgarian).

Spassov, N. (2000). Biochronology and Zoogeographic affinities of the Villafranchian faunas of South Europe. *Historia Naturalis Bulgarica, 1*, 89–128.

Spassov, N. (2001). The possible time of the first penetrations of Homo in Europe in the aspect of the paleoclimate analyses and the mammalian Plio-Pleistocene dispersals from the East. *Annual of the Institute and Museum of Archaeology, Sofia, 1*, 16–21.

Spassov, N. (2002). The Late Villafranchian and biochronology of South-East Europe, faunal migrations, environmental changes and possible first appearance of the genus Homo on the continent. In *On Late Plio/Pleistocene Extinctions and Evolution in the Palearctic. The Vatera Site*. Ist International Workshop, Vatera, Lesvos, September 29–October 2, 1999. *Annales Géologiques des Pays Helléniques*, 1 série, *39, A*, 221–243.

Spassov, N. (2003). The Plio-Pleistocene vertebrate fauna in South-Eastern Europe and the megafaunal migratory waves from the east to Europe. *Revue de Paléobiologie, Genève, 22*(1), 197–229.

Spassov, N. (2005). Brief Review of the Pliocene Ungulate Fauna of Bulgaria. In É. Crégut-Bonnoure (Ed.). *Les Ongulés holarctiques du Pliocène et du Pleistocène. Quaternaire*, Hors-Série, 2, 201–212.

Strait, D. S., Orr, C. M., Hodgkins, J., Spassov, N., Gurova, M., Miller, C., & Tzankov, T. (2016). The human fossil record of Bulgaria, and the formulation of biogeographic hypotheses. In K. Harvati & M. Roksandic (Eds.), *Paleoanthropology of the Balkans and Anatolia: Human evolution and its context*. Dordrecht: Springer.

Swisher, C., III, Curtis, G., Jacob, T., Getty, A., Suprijo, A., & Widiasmoro. (1994). Age of the earliest known Hominids in Java, Indonesia. *Science, 263*, 1118–1121.

Tattersall, I. (2007). *Homo ergaster* and its contemporaries. In W. Henke & I. Tattersall (Eds.), *Handbook of paleoanthropology* (Vol. 3, pp. 1633–1654). Berlin: Springer.

Toro-Moyano, I., Martínez-Navarro, B., Augustí, J., Souday, C., Bermúdez de Castro, J. M., Matinón-Torres, M., et al. (2013).

The oldest human fossil in Europe, from Orce (Spain). *Journal of Human Evolution, 65*, 1–9.

Turner, A., & O'Regan, H. (2007). Zoogeography: Primate and early hominin distribution and migration patterns. In W. Henke & I. Tattersal (Eds.), *Handbook of paleoanthropology* (Vol. 1, pp. 271–290). Berlin: Springer.

Villa, P. (1996). Book review, the First Italians. *Lithic Technology, 21*, 71–79.

Villa, P. (2001). Early Italy and the colonization of Western Europe. *Quaternary International, 75*, 113–116.

Villmoare, B., Kimbel, W., Seyoum, C., Campisano, C., DiMaggio, E., Rowan, J., et al. (2015). Early *Homo* at 2.8 Ma from Ledi-Geraru, Afar, Ethiopia. *Science, 347*, 1352–1355.

Zubakov, V., & Borzenkova, I. (1990). *Global Palaeoclimate of the Late Cenozoic. Developments in palaeontology and stratigraphy* (Vol. 12). Amsterdam: Elsevier.

Chapter 17
Small Mammals in the Plio/Pleistocene Sediments of Greece

Constantin S. Doukas and Katerina Papayianni

Abstract Small mammals represent an important source of biostratigraphic and ecological information at Plio/Pleistocene localities. In this paper, we provide an overview of small mammals from Pliocene and Pleistocene sites in Greece, present fossil rodent and insectivore faunal assemblages and their chronology for select loci, and discuss the applicability of Paleomagnetism and Mammal Neogene (MN) zonation for Greek assemblages. Comparisons and contrasts between the faunas of various sites are attempted in order to trace paleoenvironmental and ecotone changes during each period. Small mammal faunas have been studied at only a few archaeological/anthropogenic Pleistocene sites in Greece, and therefore most of the sites mentioned here bear only paleontological information. Our overview, however, demonstrates that small mammals can be an excellent source of supplementary information for the interpretation of archaeological sites.

Keywords Rodents • Insectivores • Dating methods • Paleoecology • Climatic indications

Introduction

The study of fossil small mammal faunas was a late starter in Greece, with the first investigations undertaken as recently as the 1960s. These early studies initially focused on faunas deposited during Oligocene and Miocene up to the Miocene/Pliocene boundary. Later on, Pliocene sediments (e.g. Ptolemais) were also incorporated in the broad faunal sequence of the Neogene. Pleistocene faunas were not studied until the 1970s, beginning with the localities of Tourkobounia near Athens and Megalopolis in Peloponnese. Fossil faunal collections from these sites originated either from karstic fissure fillings in Mesozoic limestone (Tourkobounia) or from lignite open pits (Megalopolis), providing in both cases a sufficient number of well preserved specimens, and thus an adequate source of information toward biostratigraphy and paleoecology. Here, we present the most important Pliocene and Pleistocene small mammal assemblages from the region and discuss their dating and context. We begin with a review of the main dating methods applied to the sites under discussion.

Dating Methods and Small Mammals

Two major groups of dating methods are used in paleontology: those which produce absolute dating and those which provide a relative chronology. In the first group, methods based on the radioactive decay of isotopes, like Ar^{40}/Ar^{40} and Ka^{39}/Ar^{40}, provide absolute dating results. Although they are not absolute dating methods per se, paleomagnetism and cyclostratigraphy can also indicate absolute ages. In terms of relative dating, the most easily applicable method is the use of faunal correlation to distinguish older from younger assemblages, and, therefore, strata. Small mammals are the most frequently used taxa, as they always outnumber large mammals in a faunal assemblage. This is particularly true for the identifiable skeletal elements. Small mammal paleontology is almost uniquely based on teeth, which provide the most accurate identifications. Due to the strong structure of dental enamel, teeth also preserve better than any other part of the skeleton in the fossil record. Furthermore, the distinctive dental morphology of families, genera, and species allows for their straightforward identification.

C.S. Doukas (✉)
Department of Historical Geology and Paleontology, University of Athens, Panepistimiopoli, Ilissia, GR 15784 Athens, Greece
e-mail: cdoukas@geol.uoa.gr

K. Papayianni
Wiener Laboratory, ASCSA, Muséum National d'Histoire Naturelle, UMR 7209 du CNRS, Paris, France
e-mail: katerinapapayiannis@gmail.com

Katerina Harvati and Mirjana Roksandic (eds.), *Paleoanthropology of the Balkans and Anatolia*, Vertebrate Paleobiology and Paleoanthropology, DOI 10.1007/978-94-024-0874-4_17

Table 17.1 Faunal zones and their relationship to geological epochs, mammal ages, magnetostratigraphy, and absolute ages (after Gibbard and Head 2009; Palombo et al. 2008; Steininger et al. 1996)

Faunal zones (Mammal Neogene)	Geochronology		Mammal ages	Absolute age (Ma)	Magneto-stratigraphy
	Holocene				
MNQ 26	Pleistocene/Quaternary	Late	Aurelian	0.1	Brunhes
MNQ 25		Middle		0.2	
MNQ 24				0.3	
MNQ 23				0.4–0.5	
MNQ 22			Galerian	0.6–0.7	
MNQ 21		Early		0.8–0.9	Matuyama
MNQ 20				1.0–1.4	
MNQ 19			Villanyan	1.5–1.9	
MNQ 18	Pliocene	Late		1.9–2.2	Gauss
MN 17		Middle		2.3–2.5	
MN 16				2.5–3.3	
MN 15		Early	Ruscinian	3.4–5.3	
MN 14					
MN 8–MN 13	Miocene	Late	Turolian Vallesian	5.3–11.2	
MN 5–MN 7		Middle	Aragonian Astaracian Orleanian	11.2–17	
MN 1–MN 5		Early	Ramblian Agelian	17–23.8	

The Mammal Neogene (MN) zonation system (Table 17.1) depends on the above variations in mammalian dentition. In order to obtain a relative chronology, small mammal assemblages are referred to an already dated reference faunal list, created mainly for West and Central Europe (De Bruijn et al. 1992). The MN zonation becomes less accurate when the sites are widely separated geographically from those of the reference fauna, because of the observed regional differences between faunas (e.g. Doukas 2003). Therefore, it is critical for researchers to combine all of the above-mentioned techniques. The site of Ptolemais represents the only case in Greece where all techniques have been applied.

Pliocene Faunas

The Ptolemais lignite mines (NE Greece), Vevi (NE Greece), and Tourkobounia 1 (Athens) yielded the most important Pliocene faunas with an age ranging from MN14 to MN16. Other sites from different areas of Greece, including localities in the Dodecanese in south eastern Greece (Karpathos, Archipolis, and Damatria) and localities in northern Greece (Kastoria, Limni, and Spilia) (see van der Meulen and van Kolfschoten 1986; Koufos 2001 for maps; Doukas 2005; Makris 2009) are not reviewed here due to the inadequate number of identifiable specimens. All the names and corresponding abbreviations, as well as the dates obtained (or postulated) for these sites are given in Table 17.2. The faunas of all Pliocene sites are listed in Table 17.3.

Table 17.2 Pliocene localities mentioned in the text, their abbreviations and dating

Locality name	Abbreviation	Date
Kardia	KRD	MN 14
Spilia 1	SPL 1	MN 14
Karpathos?	KRP	MN 14
Komanos 1 low	KOM 1 L	MN 14
Komanos 1 high	KOM 1H	MN 14
Vorio 1	VOR 1	MN 14
Tomea Eksi 3	TOM6-3	MN 15
Vorio 3/3a	VOR 3/3a	MN 15
Notio 1	NOT 1	MN 15
Vevi	VE	MN 15
Spilia 3+4	SPL 3+4	MN 15
Apolakkia 2	APO	MN 15
Archipolis	ARCH	MN 16a
Kastoria 1	KST 1	MN 16
Tourkovounia 1	TB 1	MN 16
Limni 6	LI	MN 16
Damatria	DAM	MN 16

In Ptolemais, the material comes from several lignite mines. Miocene (MN12, 13) and Pliocene (MN14, 15) fossil

Table 17.3 The small mammals of the Pliocene localities

Pliocene species	KRD	SPL1	KRP?	KOM 1 L	KOM 1H	VOR1	TOM6-3	VOR 3/3a	NOT 1	VE	SPL 3+4	APO	ARCH	KST 1	TK 1	LMN 6	DAM
Talpidae										x							
Archeodesmana sp.										x							
Asoriculus sp.										x							
Blarinella sp.										x							
Tamias sp.				x													
Sciurus anomalus									x						cf.		
Pliopetaurista dehneli					cf.			x									
Castoridae indet.					x			x	x								
Microtodon sp.				x													
M. komanensis				x	x			aff.									
Allocricetusehiki				x	x												
Pliospalax tourkobouniensis															x		
Prospalax priscus									x								
Apodemus sp.														x			
A. dominans							?	?	x					x	x	x	
A. cf. *dominans*																	x
A. cf. *mystacinus*																	
A. atavus				x	x	x	x	x	x								
Rhagapodemus frequens									x					x			
R. athenensis/frequens															x		
R. athenensis																x	
R. primaevus				x	x	x	x	x									
Micromys cf. *praeminutus*																x	
M. bendai						x	x	x									
M. steffensi				x	x												
Mus sp.										x							
Occitanomys adroveri				x	x												
O. brailloni						x		x									
O. magnus									x								
Orientalomys similis															x	x	
?Thallomys sp.																	x
Promimomyscor	x	x	x	x	x	x											

(continued)

Table 17.3 (continued)

Pliocene species	KRD	SPL1	KRP?	KOM 1 L	KOM 1H	VOR1	TOM6-3	VOR 3/3a	NOT 1	VE	SPL 3+4	APO	ARCH	KST 1	TK 1	LMN 6	DAM
Mimomys sp.																	
M. occitanus											x	x		x			
M. cf. *hassiacus/septimaaus*															x		
M. gracilis													cf.			cf.	
M. hajnackensis																cf.	cf.
M. davakosi							aff.	x									
Dolomys nehringi									x			x					
Propliomys hungaricus									x								
Pliomys hungaricus														x			
P. graecus															x	x	
Keramidomys sp.														x			
Glissackdillingensis															x		
G. minor				x					x								
Myomimus sp.												x					x
M. roachi																cf.	
M. maritsensis									x								
Dryomimus eliomyoides														cf.	cf.		
Dryomys sp.															x	x	
Eliomys sp.																x	
Hystrix sp.																	x
H. major																	

rodent assemblages have been studied in detail (Doukas 2005; Hordijk and De Bruijn 2009). As can be seen in Table 17.3, MN 14 is represented by different species at different sites; however, the genus *Promimomys* predominates. As the Pliocene advances, it brings about the extinction of *Promimomys* from sites that date to MN 15, 16, and 16a; while the genera *Mimomys*, *Apodemus*, *Keramidomys*, *Rhagapodemus*, *Dryomimomys*, *Pliomys*, *Eliomys*, and *Orientalomys*, among others, become more prevalent. Remarkable in the Vevi assemblage is the first appearance of the genus *Mus*. Unfortunately, this record is based only on one first upper molar (M1), and does not allow more inferences. Absolute dating methods have only been applied at the Ptolemais localities through a high-resolution model, into which the fossil rodent record is integrated. The high-resolution model is based on cyclostratigraphy, paleomagnetism, pollen, and radioactive isotopes ($^{39}Ar/^{40}Ar$; Steenbrink et al. 1999, 2000; Van Vugt et al. 2001). This model facilitates the dating of other sites that have yielded (or could yield in the future) similar faunal remains, but do not necessarily meet all the requirements necessary for the application of a high-resolution dating model like the one used in Ptolemais.

Pleistocene Faunas

Several localities from mainland Greece, Crete, and the Dodecanese dating to the Pleistocene have yielded small mammal assemblages; their names, the corresponding abbreviations and dates are given in Table 17.4. The related small mammal faunas are listed in Tables 17.5, 17.6, 17.7, and 17.8. Some of these localities are ascribed to a MN, but most of them are dated just as "Pleistocene", or sometimes as Middle or Upper Pleistocene (De Bruijn and van der Meulen 1975; Mayhew 1977a; Koufos 2001). We should also mention two cave sites, Kitsos and Franchthi in the mainland, which were excavated by archaeologists; parts of their stratigraphy were dated to the Pleistocene and yielded small mammal material, also included in Tables 17.4 and 17.7 (Jullien 1973; Chaline 1981; Payne 1973, 1982). These are the only Greek Pleistocene sites published to date that document both archaeological and small mammal assemblages (but see discussion of Megalopolis below).

A comparison between the faunas of mainland Greece and Crete indicates a strong island character of the Cretan small mammal assemblages, which belong to a limited number of endemic genera and species (Table 17.8): *Crocidura zimmermanni*, *Kritimys kiridus*, *K. catreus*, *Mus bateae*, and *M. minotaurus*. Among these, the endemic *C. zimmermanni* is the only Pleistocene relic among the modern Greek fauna. Sporadic appearances are made by *Apodemus*, *Glis*, and *Oryctolagus* among the Cretan assemblages, whereas the

Table 17.4 Pleistocene localities mentioned in the text, their abbreviations and dating

Locality name	Abbreviation	Date
Kalavarda 2	KLV 2	MN 16b
RemaAslan	ASL	MN 17
Kardamena	KRM	MN 17
Gerakarou 1	GER	MNQ 18
Kastoria 2	KST	MN 18
Megalopolis TH 1	THO	MNQ 18
Choremi 1	CHO 1	Middle Pleist
Choremi 2	CHO 2	Middle Pleist
Choremi 3	CHO 3	Middle Pleist
Choremi 4	CHO 4	Middle Pleist
Lagada	LGD	MNQ 18
Pyrgos	PRG	?MNQ18
Marathousa	MAR	MNQ 19
Kaiafas	KAF	MNQ 19
Tourkovounia 2	TB 2	MNQ 19
Kalymnos	KLM	MNQ 19
Alikes	ALK	?MNQ19
Ravin Voulgarakis	RVL	MNQ 20
Apollonia 1	APL	MNQ 20
Zeli 2	ZEL	MNQ 20
Zeli 2A + B	ZLI	MNQ 20
Volos	VOL	MNQ 21
Petralona Cave	PTR	Middle Pleist
Arnissa	ARN	Upper Pleist
Kitsos Cave	KTS	40 kBP
Franchthi Cave	FRN	40 kBP
Sphinari	SPH	Upper Pleist
Cave between Canea-Suda	CCS	Upper Pleist
Stavros micro	SID	Upper Pleist
Stavros macro	SA	Upper Pleist
Stavros cave	SG	Upper Pleist
Akrotiri	AKR	Upper Pleist
Cape Maleka 1	MAL 1	Upper Pleist
Cape Maleka 3	MAL 3	Upper Pleist
Liko	LIP	Upper Pleist
Gerani 2	GE2	Upper Pleist
Sourida	SOU	Upper Pleist
Mavromouri	MAV	Upper Pleist
Simonelli cave	SIM	Upper Pleist
Gumbes B	GUB	Upper Pleist
Rethymnon fissure	RES	Upper Pleist
Skaleta	SKA	Upper Pleist
Bali 1	BA1	Upper Pleist
Bali 2	BA2	Upper Pleist
Milatos 1	MI1	Upper Pleist
Milatos 2	MI2	Upper Pleist
Milatos 3	MI3	Upper Pleist
Milatos 4	MI4	Upper Pleist
Sitia 1	SIT 1	Upper Pleist
Kharoumes A	KHA	Upper Pleist
Kharoumes 4	KH4	Upper Pleist
Kharoumes 5	KH5	Upper Pleist
Xeros	XE	Upper Pleist

Table 17.5 The small mammals of the Pleistocene localities MN16–MN18

Species	*KLV 2*	*ASL*	*KRD*	*GER*	*KST 2*	*MGP*	*LGD*	*PRG*
Apodemus dominans			cf.		x			
A. mystacinus				cf.			cf.	
A. sylvaticus/flavicollis							cf.	
Mimomys sp. *polonicus?*				x				
M. reidi			cf.		cf.			
M. pliocaenicus		x			x			
M. pitymyoides					x			
M. newtoni					x			
M. ostramosensis							x	
Jordanomys majori							cf.	x
Myomimus sp.			x					
M. roachi					x		x	
Hystrix major				x				
Pliomys episcopalis					x			
Borsodia sp.				cf.				
Clethrionomys sp.					x			
Kislangia rex						x		

Table 17.6 The small mammals of the Pleistocene localities MN19–MN 21

Species	*MAR*	*KAF*	*TB 2*	*KLM*	*ALK*	*RVL*	*APL*	*ZEL*	*ZLI*	*VOL*
Erinaceus sp.			x							
E. europaeus							x			
E. praeglacialis			x							
Talpa sp.						x				
Desmaninae indet	x									
Crocidura sp.			x							
C. kornfeldi	x		x			x				
Sorex minutus	x					cf.				
S. (Drepanosorex) praearaneus	x					x				
Asoriculus gibberodon	x		x							
A. castellarini						cf.				
Neomys fodiens/anomalus			x							
Beremendia fissidens	x		x			x				
Sciurus sp.		x	cf. anomalus	x						
Spermophilus sp.	x									
S. nogaici						x				
Cricetinus koufosi	x					x				
Cricetulus migratorius		cf.	x							x
Pliospalax senii						x				
S. nehringi				x						
Apodemus mystacinus		x	x			x		x	x	x
A.sylvaticus/flavicollis	x	x		x				x	x	x
A. sylvaticus						x				
A. flavicollis			cf.							
Mimomys sp.	x				x					
M. savini		x				x		x	x	
Micromys minutus									cf.	
Mus aegeaus				x						
Meriones tristrami				x						
Jordanomys majori		x		x						

(continued)

Table 17.6 (continued)

Species	*MAR*	*KAF*	*TB 2*	*KLM*	*ALK*	*RVL*	*APL*	*ZEL*	*ZLI*	*VOL*
Kalymnomys majori			x							
Sicista subtilis	cf.					x				
Myomimus sp.						x				
M. roachi			x	x						
Glis sp.		x	x							
G. sackdillingensis		x								
G. glis						aff.				
Eliomys quercinus			x							
Lagurodon sp.									x	
L. arankae	x	cf.		cf.		x	x	x		
Hystrix refossa				x						
Pliomys episcopalis		x							x	
Kislangia sp.						x				
Microtus sp.				x				x	x	
M. (Allophaiomys) rufoi		cf.								
M. pitymyoides						x				
M. arvalidens										x
M. (Tibericola) eleniae			x							
Lagurus pannonicus										x
L. arankae			x							
Leporidae						x	x			
Oryctolagus lacosti				x						

Table 17.7 The small mammals of the Middle and Upper Pleistocene

Species	*CHO 1*	*CHO 2*	*CHO 3*	*CHO 4*	*PTR*	*ARN*	*KTS*	*FRN*
Erinaceus europaeus/praeglacialis					cf.			
Soricidae	x			x				
Crocidura leucodon						x		
C. russula						x		
Sorex minutus						x		
S. araneus						x		
S. runtonensis					cf.			
Rhinolophus sp.					x			
R. ferrumequinum topali					x			
Myotis sp. I–II					x			
M. blythi oxygnathus					x			
Pipistrellus sp.					x			
Sciurus vulgaris			cf.					
Spermophilus citellus						x	x?	
Castor fiber		x	x					
Cricetulus sp.								x
C. migratorius						x		
Mesocricetus newtoni						x		
Allocricetus bursae simplex					x			
Spalax sp.								x
S. chalkidikae					x			
S. microphthalmus						x		
Apodemus sp.				x		x		

(continued)

Table 17.7 (continued)

Species	*CHO 1*	*CHO 2*	*CHO 3*	*CHO 4*	*PTR*	*ARN*	*KTS*	*FRN*
A. mystacinus					crescendus	x		cf.
A.sylvaticus/flavicollis								x
A. sylvaticus			cf.					
Mimomys sp.		x		x				
M. savini	aff.	aff.	aff.	aff.				
Muridae gen. et. sp. indet.			x					
Mus sp.			x					x
M. spretus				cf.				
Sicista subtilis						x		
Parasminthus brevidens					x			
Dryomys nitedula						x		
Gliridae					x			
Pliomys episcopalis			aff.					
Clethrionomys glareolus						x		
Microtus sp						x		
Microtus (Pitymys) sp.						x		
M. arvalis						x		
M. cf. *arvalis/socialis*								x
M. guentheri						x		
M. (Chionomys) nivalis						x		x
M. (Pallasiinus) praeguentheri					x			
Pitymys sp.								x
Arvicola sp.					x			
A. cantiana/terrestris						x		
Lagurus lagurus						x		
Lepus sp.						x		
L. ?terraerubrae Kretzoi					x			
Oryctolagus sp.					x			
Ochotona pusilla						x		

Rattus sp. should be considered as intrusive, and not of Pleistocene origin (Mayhew 1977b; Reumer 1986).

On mainland Greece, the localities that provide sufficient material for comparisons and statistical analyses are Tourkobounia 2 (TB 2), Megalopolis (MGP), and Arnissa (ARN) (Tables 17.5, 17.6, and 17.7). The remaining sites in Table 17.4, while certainly interesting, did not provide enough specimens for comparative analyses (De Bruijn and van der Meulen 1975; Mayhew 1977a; van der Meulen and Doukas 2001). Tourkobounia 2 dates to the Early Pleistocene, and its assemblage must have been deposited during an interglacial period, as reflected by the dominance of the Muridae family (van der Meulen and Doukas 2001). Arnissa dates to the Late Pleistocene. It contains extant species that still survive in the modern Greek fauna (Mayhew 1977a). These extant species correspond to a steppe/open vegetation, as well as a rocky environment; which, in terms of dating, conforms to glacial conditions in Northern Europe (Mayhew 1977a). Relics of glacial fauna are found within the Arnissa assemblage (*Lagurus lagurus*, *Ochotona pusilla*, *Mesocricetus newtoni*, and *Sicista subtilis*); these species become extinct in Greece after the Pleistocene. This succession and gradual replacement of species marks the advent of the Holocene with different climatic and vegetation conditions.

Among these sites, Megalopolis stands out because its complex fossil record comprises both large and small mammals, as well as evidence of human presence. Mammal faunas that were collected from four levels of the entire Megalopolis basin indicate a Middle Pleistocene age (Sickenberg 1975; Fejfar and Heinrich 1979). Faunal remains from the whole Megalopolis basin were described by Sickenberg (1975) who identified 11 species of large mammals: carnivores, proboscideans, hippopotami, four different cervids, a water buffalo, a horse, and a rhinoceros. A remarkable specimen in the fossil assemblage from the Marathousa Member is an upper third molar (M3) of a hominin (Sickenberg 1975; Marinos 1975; Xirotiris et al. 1979; Harvati 2016). More recently, human presence in the Megalopolis basin has been further attested to by the discovery

Table 17.8 The small mammals of the Cretan Upper Pleistocene localities

	SPH	*CCS*	*SID*	*SA*	*SG*	*AKR*	*MAL 1*	*MAL 3*	*LIP*	*GE2*	*SOU*	*MAV*	*SIM*	*GUB*	*RES*	*SKA*	*BA1*	*BA2*	*MI1*	*MI2*	*MI3*	*MI4*	*KHA*	*KH4*	*KH5*	*XE*	*SIT1*
Soricidae							x		x																		
Crocidura zimmermanni			x	x											x			x		x		x	x		x		
Apodemus sp.																											x
Mus bateae			x			x																					
M. minotaurus		x							x	x	x	x	x	x	x	x				x	x	x					
Kritimys sp.																											x
K. kiridus							x	x									x										aff.
K. catreus	x		x	x	x												x	x	x		x		x	x	x	x	
Rattus sp.										x																	
Chiroptera									x																x		
Gliridae			x																								
Myoxus sp.			cf.																								
Oryctolagus sp.										x																	

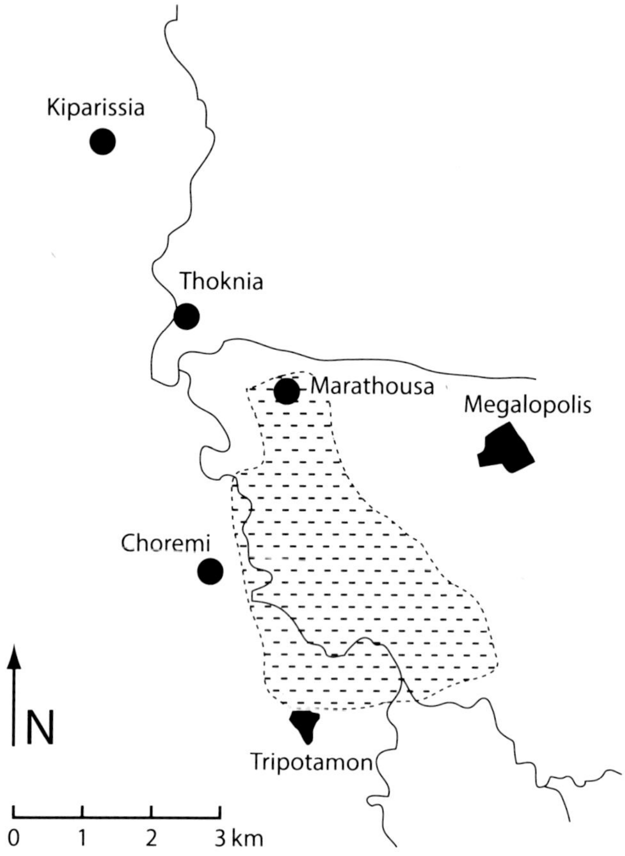

Fig. 17.1 The Megalopolis lignite mine map with localities mentioned in the text

of a new Lower Paleolithic archaeological site, Marathousa 1, in the Megalopolis lignite mine (Fig. 17.1) (Panagopoulou et al. 2015; see also Harvati 2016; Tourloukis 2016).

Sickenberg's (1975) assessment of the stratigraphic position of the Marathousa Member was later modified by Benda et al. (1987), who, using small mammals as their baseline, proposed a late Villanyian age for the lower lignite bed exposed in the Thoknia open cast lignite mine (Table 17.7). There have been multiple successive campaigns in the mines of Thoknia (TH 1–4) and Choremiou (CHO 1–4), which were conducted in the basin of Megalopolis between 1980 and 1995. Their goal was to provide a clearer stratigraphic correlation between these localities. According to Van Vugt (2000) and Van Vugt et al. (2001), the Megalopolis section is dated to the Middle Pleistocene on the basis of small mammal paleontology, and to the lower Brunhes based on magnetostratigraphy. The astronomical tuning of the lignite gives ages of ~900 ka for the base of the section and ~350 ka for the top (Van Vugt 2000). Since we know of no other available magnetostratigraphic data from Pleistocene sediments in Greece, the Megalopolis sequence can offer only faunal grounds for comparisons with other sites. However, based on such comparisons, the Megalopolis magnetostratigraphic dates can be used to date sites with similar faunal remains.

Comparison Between Pliocene and Pleistocene Faunas

There are important differences in small mammal faunal structure between the Pliocene and the Pleistocene in Greece. The Pliocene, especially when compared with the Miocene, is more uniform in composition (Hordijk and De Bruijn 2009). The advent of the Pliocene (MN14) is characterized by the immigration of arvicoline voles (Arvicolinae), which represent a new biostratigraphic tool in addition to the already existing ones (e.g. Murids); the distinction between MN 14 and MN 15 is based only on arvicolines. In the Pleistocene, the dominance of arvicolines varies. There are instances (e.g. Kalymnos, Varkiza) where *Apodemus mystacinus* is dominant. An exception to this is the Arnissa fauna, where *Microtus* predominates (van der Meulen and Doukas 2001).

Evaluation of the Small Mammal Studies

Fossil small mammals provide valuable stratigraphic and ecological information. Various groups of rodents and insectivores have proven to be important stratigraphic and ecological markers, with insectivores being especially important as paleoclimatic indicators (Reumer and Doukas 1985). Arvicolines can be extremely useful in providing dating information for Pleistocene sites (Koenigswald et al. 1992). Murids, represented here by the extant genera *Apodemus* and *Mus*, are used as both stratigraphic and ecologic indicators (Weerd 1976).

When compared with Central and Western Europe, the Eastern Mediterranean faunas show low biodiversity during the Pleistocene, which could have been caused by either low temperatures or aridity. The phenomenon of the Asian Summer Monsoon may be one of the reasons for aridity in the Eastern Mediterranean (Reumer et al. 2002). During the Asian Summer Monsoon, hot humid air rises over South Asia; this heat moves westward through a wave pattern, causing dry air to descend on the Eastern Mediterranean (Reumer et al. 2002). As a result, extremely dry weather conditions occur at a seasonal level (July–August), limiting the availability of food and water resources on which the small mammals rely for their survival.

Final Remarks

There are evident differences in faunal structure between the Pliocene and Pleistocene. The Pleistocene locality of Megalopolis, given the presence of a hominine M3 and Lower Paleolithic archaeological remains, should be further investigated.

Acknowledgments We thank the organizers of the PaGE Symposium in general and Katerina Harvati in particular for putting together Archaeologists, Anthropologists, and Paleontologists in a fruitful discussion mode. George Koufos and Ioanna Sylvestrou provided useful information. Constantine Kostantopoulos was instrumental in writing this manuscript. Mirjana Roksandic made critical suggestions that greatly improved this manuscript. We would also like to thank the reviewers and editors for their comments and corrections.

References

Benda, L., van der Meulen, A. J., Meyer, K. J., & van der Weerd, A. (1987). Biostratigraphic correlations in the Eastern Mediterranean Neogene: 8. Calibration of sporomorph- and rodent-associations from the Megalopolis basin (Peloponnesus, Greece). *Newsletters on Stratigraphy, 17*, 129–141.

Chaline, J. (1981). Les rongeurs de la grotte de Kitsos. In N. Lambert (Ed.), *La grotte préhistorique de Kitsos (Attique)* (Vol. 2, pp. 591–594). Paris: A.D.P.F. École d'Athènes.

De Bruijn, H., Daams, R., Daxner-Hock, G., Fahlbusch, V., Ginsburg, L., Mein, P., et al. (1992). Report of the RCMNS working group on fossil mammals, Reisensburg 1990. *Newsletters on Stratigraphy, 26*(2/3), 65–118.

De Bruijn, H., & van der Meulen, A. J. (1975). The Early Pleistocene rodents from Tourkobounia-1 (Athens, Greece). *Proceedings of the Koninklijke Nederlandse Akademie van Wetenschappen B, 78*(4), 314–338.

Doukas, C. S. (2003). The MN4 faunas of Aliveri and Karydia (Greece). In N. Lopez-Martinez, P. Pelaez Campomanes & M. Hernandez Fernandez (Eds.), *Coloquios de Paleontologia* (Vol. 1, pp. 127–132).

Doukas, C. S., (2005). Greece. In L. W. van den Hoek Ostende, C. S. Doukas & J. W. F. Reumer (Eds.), The fossil record of the Eurasian Neogene insectivores. *Scripta Geologica Special Issue, 5*, 99–112.

Fejfar, O., & Heinrich, W. D. (1979). Biostratigraphical delimitation and division of the continental Quaternary of Europe on the basis of Arvicolids (Rodentia, Mammalia). In V. Šibrava & F. D. Shotton (Eds.), *International Geological Correlation Programme, Project 73/1/24—Quaternary glaciations in the Northern Hemisphere, Report* (Vol. 6, pp. 83–86). Praha: Ústředníústavgeologický.

Gibbard, P., & Head, M. J. (2009). The definition of the Quaternary system/era and the Pleistocene series/epoch. *Quaternaire, 20*(2), 125–133.

Harvati, K. (2016). Paleoanthropology in Greece: Recent finds and implications. In K. Havarti & M. Roksandic (Eds.), *Paleoanthropology of the Balkans and Anatolia: Human evolution and its context* (pp. 3–14). Dordrecht: Springer.

Hordijk, K., & de Bruijn, H. (2009). The succession of rodent faunas from the Mio/Pliocene lacustrine deposits of the Florina-Ptolemais-Servia Basin (Greece). In C. S. Doukas & H. Drinia (Eds.), The Middle Miocene environments and ecosystem dynamics of the Eurasian Neogene (EEDEN) "A bridge to Cross." *Hellenic Journal of Geosciences, 44*, 21–75.

Jullien, R. (1973). Grotte de Kitsos. La faune: mammifères, reptiles et batraciens. *Bulletin de Correspondence Hellénique, 97*, 434–443.

Koenigswald, W. D., Fejfar, O., & Tchernov, E. (1992). Revision einiger alt- und mittelpleistozäner Arvikoliden (Rodentia, Mammalia) aus dem östlichen Mittelmeergebiet ('Ubeidiya, Jerusalem und Kalymnos-Xi). *Neues Jahrbuch für Geologie und Paläontologie, Abhandlungen, 184*, 1–23.

Koufos, G. (2001). The Villafranchian mammalian faunas and biochronology of Greece. *Bolletino della Società Paleontologica Italiana, 40*(2), 217–223.

Makris, S. (2009). Smaller mammals (Insectivora, Rodentia) from the MN 15 locality of Vevi, Northern Greece. Unpublished diploma thesis, University of Athens, Department of Geology (in Greek).

Marinos, G. (1975). Über einen menschlichen Zahn unter den Säugetier-Resten biharischen Alters von Megalopolis. *Annales Géologiques des Pays Helléniques, 27*, 64–65.

Mayhew, D. F. (1977a). Late Pleistocene small mammals from Arnissa (Macedonia, Greece). *Proceedings of the Koninklijke Nederlandse Akademie van Wetenschappen B, 81*(3), 302–321.

Mayhew, D. F. (1977b). The endemic Pleistocene murids of Crete I & II. *Proceedings of the Koninklijke Nederlandse Akademie van Wetenschappen B, 80*(3), 182–214.

Palombo, M. R., Sardella, R., & Novelli, M. (2008). Carnivora dispersal in Western Mediterranean during the last 2.6 Ma. *Quaternary International, 179*, 176–189.

Panagopoulou, E., Tourloukis, V., Thompson, N., Athanassiou, A., Tsartsidou, G., Konidaris, G.E., et al. (2015). Marathousa 1: A new Middle Pleistocene archaeological site from Greece. *Antiquity, 343*, Project Gallery.

Payne, S. (1973). Animal bones, Appendix in Jacobsen, T.W., Excavation in the Franchthi Cave, 1969–1971, I. *Hesperia, 42*(1), 59–66.

Payne, S. (1982). Faunal evidence for environmental/climatic change at Franchthi cave (Southern Argolid, Greece), 25.000 BP to 5.000 BP. Preliminary results. In J. L. Bintliff & V. W. van Zeist (Eds.), Palaeoclimates, palaeoenvironments and human communities in the eastern Mediterranean region in later prehistory: *Vol. 133. BAR International Series* (pp. 133–136). Oxford: Archaeopress.

Reumer, J. W. F. (1986). Notes on the Soricidae (Insectivora, Mammalia) from Crete. I. The Pleistocene species *Crocidura zimmermanni*. *Bonner Zoologische Beiträge, 37*(3), 161–171.

Reumer, J. W. F., & Doukas, C. S. (1985). Early Pleistocene Insectivora (Mammalia) from Tourkobounia (Athens, Greece). *Proceedings of the Koninklijke Nederlandse Akademie van Wetenschappen B, 88*(1), 111–121.

Reumer, J. W. F., Doukas, C. S., & Vassiliadou, K. (2002). Plio-Pleistocene Insectivore diversity in the E. Mediterranean and the Asian summer monsoon. *Annales Géologiques des Pays Helléniques, 39*(a), 245–252.

Sickenberg, O. (1975). Eine Säugetierfauna des tieferen Biharium aus dem Becken von Megalopolis (Peloponnes, Griechenland). *Annales Géologiques des Pays Helléniques, 27*, 25–73.

Steenbrink, J., Van Vugt, N., Hilgen, F. J., Wijbrans, J. R., & Meulenkamp, J. E. (1999). Sedimentary cycles and volcanic ash beds in the Lower Pliocene lacustrine succession of Ptolemais (NW Greece): Discrepancy between ^{39}Arg/^{40}Arg and astronomical ages. *Palaeography, Palaeoclimatology and Palaeoecology, 152*, 283–303.

Steenbrink, J., Van Vugt, N., Kloosterboer-Van, H., & Hilgen, F. J. (2000). Refinement of the Messinian APTS from sedimentary cycle patterns in the lacustrine Lava section. *Earth and Planetary Science Letters, 181*, 161–173.

Steininger, F. F., Berggren, W. A., Kent, D. V., Bernor, R. L., Sen, S., & Agusti, J. (1996). Circum-Mediterranean Neogene (Miocene and Pliocene) Marine-Continental Chronologic correlations of European mammal units. In R. L. Bernor, V. Fahlbusch, & H. W. Mittmann (Eds.), *The evolution of Western Eurasian Neogene mammal faunas* (pp. 7–46). New York: Columbia University Press.

Tourloukis, V. (2016). On the spatio-temporal distribution of Mediterranean Lower Paleolithic sites: A geoarchaeological perspective. In K. Harvati & M. Roksandic (Eds.), *Paleoanthropology of the Balkans and Anatolia: Human evolution and its context* (pp. 303–323). Dordrecht: Springer.

van der Meulen, A. J., & Doukas, C. S. (2001). The Early Biharian rodent fauna from Tourkobounia 2 (Athens, Greece). *Lynx, New Series (Praha), 32*, 255–277.

van der Meulen, A. J., & van Kolfschoten, T. (1986). Review of the late Turolian to early Biharian mammal faunas from Greece and Turkey. *Memorie della Societa Geologica Italiana, 31*, 201–211.

Van Vugt, N. (2000). *Orbital forcing in the late Neogene lacustrine basins from the Mediterranean*. Ph.D. Dissertation, Faculteit Aardweteschappen Universiteit Utrecht 168.

Van Vugt, N., Langereis, C. G., & Hilgen, F. J. (2001). Orbital forcing in Pliocene/Pleistocene Mediterranean lacustrine deposits: Dominant expression of eccentricity versus procession. *Palaeography, Palaeoclimatology and Palaeoecology, 172*, 193–205.

Weerd, van de A. (1976). *Rodent faunas of the Mio-Pliocene continental sediments of the Teruel-Alfambra region, Spain*. Ph.D. Dissertation, University of Utrecht, The Netherlands.

Xirotiris, N., Henke, W., & Symeonidis, N. (1979). Der M3 von Megalopolis—ein Beitrag zu seiner morphologischen Kennzeichnung. *Zeitschrift für Morphologie und Anthropologie, 70*(1), 117–122.

Chapter 18
On the Spatio-Temporal Distribution of Mediterranean Lower Paleolithic Sites: A Geoarchaeological Perspective

Vangelis Tourloukis

Abstract Lower Paleolithic evidence from the Mediterranean region holds a prominent position in discussions about the earliest peopling of Europe. Most studies examining patterns of human occupation focus on purported behavioral capacity, habitat preference, and environmental tolerance of different hominins. This chapter employs a geoarchaeological perspective through the examination of landscape dynamics as a complementary approach. In this context, Lower Paleolithic records of the Mediterranean and the Balkans are reviewed with an emphasis on the geomorphological settings of the best-studied sites. Since most of the oldest, well dated and primary-context material occurs in open-air sites situated in basins, the last part of the chapter explores how basin dynamics could have conditioned the preservation and accessibility of artifact-bearing strata. Spain, Italy, and Greece are used as case-studies and a conceptual model is proposed as a means to assess possible patterned relationships of site locations. A "basin model" offers a working hypothesis for evaluating site distributions and outlines first steps towards a geosciences-based methodology, which can be used to locate new sites.

Keywords Site distributions • Basins • Preservation

Introduction

> *"Such thick and extensive accumulations of sediment may be formed in two ways; either, in profound depths of the sea […] or, sediment may be accumulated to any thickness and extent over a shallow bottom, if it continue slowly to subside […] In fact, this nearly exact balancing between the supply of sediment and the amount of subsidence is probably a rare contingency"*
> C. Darwin 1861: 312, *The Origin of Species*

V. Tourloukis (✉)
Paleoanthropology, Department of Geosciences, Eberhard Karls Universität Tübingen, Tübingen, Germany
e-mail: vangelis.tourloukis@ifu.uni-tuebingen.de

Since Darwin's insightful discussion "On the imperfection of the fossil record," paleontologists have grappled with the intricacies of inferring evolutionary processes from a frustratingly incomplete fossil record. Indeed, the completeness of the geological archive needs to be assessed before biogeographic, phylogenetic, or other patterns are taken at face value (Benton et al. 2000). Paleoanthropologists and Paleolithic archaeologists are also aware of this issue and a number of studies have investigated how the nature of the geological record may be affecting archaeological patterns in space or in time (e.g., Waters and Kuchn 1996; Surovell et al. 2009; Tryon 2010). Ever since the introduction of the concept of "site formation processes" to archaeology, it has been acknowledged that the archaeological record has been shaped by the interaction and cumulative effects of syn- and post-depositional taphonomic processes regardless of the type of environment (Schiffer 1987). Thus, before environmental or behavioral inferences are invoked to explain patterns, archaeologists first need to identify the geomorphic processes that may have potentially influenced the biography of the sedimentary matrix. However, as Burger and colleagues aptly state (2008: 206): "in spite of these realizations at the scale of the individual site, *human action is often given interpretive primacy when studies are conducted at the landscape scale*" (emphasis added).

Observing and understanding patterns at the landscape-scale or at even larger, supra-regional scale is a difficult task, because of significant limitations in the resolution of the relevant chronostratigraphic frameworks. Notwithstanding uncertainties associated with dating, the spatio-temporal distribution of the Early and Middle Pleistocene sites of South Europe offers several emerging patterns. The question is what meaning should be assigned to them. For instance, there is a conspicuous gap in the human fossil record between the times of Dmanisi at *ca.* 1.8 Ma (Ferring et al. 2011) and the next oldest sites with hominin remains, namely Barranco León and Sima del Elefante, which date to *ca.* 1.4 and 1.2 Ma, respectively (Toro-Moyano et al. 2013; Carbonell et al. 2008; see also Dennell 2003). Parés and co-workers (2013) addressed

Katerina Harvati and Mirjana Roksandic (eds.), *Paleoanthropology of the Balkans and Anatolia*, Vertebrate Paleobiology and Paleoanthropology, DOI 10.1007/978-94-024-0874-4_18

the question of whether this gap represents a real hominin absence or, rather, whether it is an artifact of insufficient chronological data, pointing out that it is "somewhat paradoxical [...] that these two oldest Paleolithic sites within Western Eurasia [...] are found at such geographically disparate locations" (2013: 6; Fig. 18.1). Estimates of the number of sites in a particular time frame are highly dependent on what evidence (generally lithic or chronological) one accepts as reliable. From the perspective of geochronologists (Muttoni et al. 2010; Parés et al. 2013) there is indeed only a handful of Early Pleistocene sites with reliable (or, reliably dated) archaeological material between the sites of Ubeidiya in the east (at *ca.* 1.4 Ma; Martínez-Navarro et al. 2009) and Orce in the west (at *ca.* 1.4 Ma; Oms et al. 2011). Even for records as well-studied and rich as that of the Sierra de Atapuerca, and despite Atapuerca's stable environmental conditions throughout a period of almost a million years (Rodríguez et al. 2011), discontinuity in human occupation could be invoked, at least when considering the temporal gaps between Sima del Elefante (at *ca.* 1.2 Ma), Atapuerca TD6 (at *ca.* 0.8 Ma) and Sima de los Huesos (at *ca.* 0.4 Ma; Arnold et al. 2014; Bischoff et al. 2007; cf. MacDonald et al. 2012). In the Northern Caucasus, the earliest human presence dates to around 0.6 Ma, i.e., it is more than a million years younger than the evidence attested in the Southern Caucasus (Dmanisi). The earliest reliably dated Lower Paleolithic sites on the western parts of Eastern Europe date to no earlier than 0.4 Ma (Korolevo I, Layer VI at Ukraine and, probably, Pogreby and Dubossary in Moldova; Doronichev 2008). In continental northern and central Europe, the oldest sites in cool conditions and/or to the east of the river Rhine post-date the Cromerian Complex (i.e., they are younger than MIS 15 or 13) and until MIS 11 the Rhine valley appears to define the eastern limit of the distribution of sites north of the Alps (Cohen et al. 2012: 71).

Arguably, three main factors dictate any apparent spatio-temporal distribution of sites and the 'gaps' associated with them: (1) *research-related parameters*, which touch upon dating constraints, as well as shifts in research interests and fluctuations in the intensity of investigations carried out in different regions; (2) *hominin habitat preference* and *tolerance*, which relate to specific attributes of terrestrial ecosystems, hominin biology and behavioral repertoires that altogether influence whether human populations prefer, or can at least tolerate, the ecological circumstances prevailing at a given area and/or during a particular time-span (see, e.g., Dincer 2016); (3) *taphonomic circumstances*, both syn- and post-depositional eco-geomorphic processes, which act as agents of bias against or in favor of preservation and/or accessibility of the archaeological record.

In regions such as the Balkans, the Lower Paleolithic (hereafter LP) is understudied compared to the rest of Europe, and thus research bias is certainly at play here. Even so, it is difficult to ascertain a particular distribution of sites (or their lack thereof) should be attributed to the paucity of research. On the other end of the spectrum, the discovery of Pakefield in southern England (Parfitt et al. 2005) exemplifies difficulties in deducing hominin absence even when we are dealing with the best-researched places: despite two centuries of investigations of the Cromer Forest-Bed Formation, no artifacts had been found there until 2005.

Fig. 18.1 Main Lower Paleolithic sites of the circum-Mediterranean region discussed in text. (*1*) Dmanisi (*2*) Yarımburgaz (*3*) Dursunlu (*4*) Kaletepe Deresi (*5*) Karain (*6*) Kozarnika (*7*) Monte Poggiolo (*8*) Isernia (*9*) Notarchirico (*10*) Atapuerca sites (Trinchera Dolina, Sima del Elefante, Sima de los Huesos) (*11*) Torralba and Ambrona (*12*) Orce sites (Fuente Nueva 3, Barranco León) (*13*) Sites in the Casablanca area (Thomas Quarry I, Grotte des Rhinocéros) (*14*) Ain Hanech and El-Kherba (*15*) 'Ubeidiya and Gesher Benot Ya'aqov

The examination of types of habitats or ecozones that early hominins would prefer or could tolerate is an equally dynamic field of research, in which new discoveries contradict previous theories. To cite another example from Britain, the recent evidence from Happisburgh (Parfitt et al. 2010) demonstrated that early hominins were able to expand farther than 45° north much earlier than previously thought, and not only during warm intervals, thereby challenging the hypothesis of habitat tracking (Roebroeks 2005). Nonetheless, it is reasonable to assume that site distribution patterns reflect the ecological limits of hominins, to a degree that will always be debated. Biogeographical, environmental (e.g., climatic) and ecological scenarios, usually combined with behavioral explanations, are most commonly advanced to interpret the arrival(s) of humans in Europe and the distribution of early sites in space and time (e.g., Carrion et al. 2011). These models may consider the role of vegetation in the hominin ecological niche (e.g., Hughes et al. 2007), the effects of seasonality and landscape on ecological structure (Hopkinson 2007), the influence of climatic conditions (Agustí et al. 2009), the role of faunal movements and turnovers (van der Made and Mateos 2010), or the carnivorous trophic behavior of hominins (Martínez-Navarro 2010). Other, more "culture-specific," aspects, such as the possible associations of dispersal events with different technological (lithic) systems, have also been put forth to explain distribution of archaeological sites (e.g., Bar-Yosef and Belfer-Cohen 2013).

In contrast to a substantial body of research investigating the importance of environmental, behavioral, and biogeographical parameters, "the role of topography, especially when linked to the study of geomorphological processes, has seldom been studied" (Carrion et al. 2011: 1290). Some researchers (most notably Bailey et al. 2011) have examined the spatial association of archaeological and fossil localities with areas of complex topography and active tectonics, but not from the perspective of *landscape taphonomy*, defined as the processes by which elements of the landscape become selectively removed or buried due to the action of natural or cultural agents, leaving behind a biased record of past landscapes—hence also biased archaeological patterns (Wilkinson 2003: 8; Niknami 2007). As an important additional factor potentially controlling site distributions and the nature of any apparent "gaps" in them, geomorphic agents have significantly influenced the quantity and quality of the uncovered archaeological material. This realization has been accounted for in interpretations of the Mediterranean archaeological record, but less so in appraisals of *Pleistocene* records (with, of course, notable exceptions, such as Barton et al. 2002).

In a recent study that examined landscape dynamics of the early Pleistocene record of Greece, Tourloukis and Karkanas (2012) argued that this fragmented and scant record needs to be understood as an outcome of destructive effects of Quaternary geomorphic processes, and not as an indication of an actual absence of hominins. The article emphasized the potentially central role of the Aegean region in hominin dispersals, since half of the Aegean Sea would have been subaerially exposed during most of the early Pleistocene, and further demonstrated how the study and interpretation of landscape dynamics can alter our understanding of the potential of a region to yield new evidence (Tourloukis and Karkanas 2012). In this chapter, rather than examining the Greek record per se, I will use it as a case study for juxtaposition with the LP records of South Europe and for elaborating further on the use of "landscape taphonomy" as an analytical tool in assessing large-scale archaeological patterns.

The three peninsulas of Southern Europe are well suited for exploring distributional patterns of early hominin sites: (1) based on current evidence, this region witnessed hominin occupation earlier than the rest of the continent, and (2) these southern refugia most likely hosted early Pleistocene core populations which served as "sources" for recolonizations (Dennell et al. 2011). However, the LP of the entirety of Eurasia is characterized by frequent geographic discontinuities and, as Dennell and colleagues have pointed out (2011: 1514), we still need to test the assumption that hominins continuously occupied southern Europe after they first entered it. In order to do that, we need not only to uncover more sites, but also to examine why it is much harder to find evidence of occupation in certain areas than in others.

Mediterranean Lower Paleolithic Records

Italy

The earliest traces of human presence in Italy (Fig. 18.2) are evidenced at the quarry site of Pirro Nord 13, where a Mode 1 lithic assemblage has been recovered from a karst infilling (Arzarello et al. 2012). An age ranging between 1.7/1.6 and 1.2 Ma has been proposed on the basis of biochronological data and paleomagnetic measurements. However, this estimation has been questioned and, due to uncertainties surrounding the age of the Pirro Nord Faunal Unit, it is suggested that the Pirro Nord artifacts may be as young as *ca.* 1.0 Ma (Muttoni et al. 2010); furthermore, a recent paleogeographic reconstruction appears to be in contradiction with the proposed age of the fauna and associated artifacts (Santangelo et al. 2012).

The site of Visogliano is located in a karstic depression on the side of a small doline in the Trieste Karst; it has yielded human remains from the filling of a rockshelter and a breccia outside the rockshelter, and it is dated to between 500 and 300 ka (Abbazzi et al. 2000; Falguères et al. 2008).

Besides the karstic settings of Pirro Nord and Visogliano, the other most important LP sites in Italy are all open-air

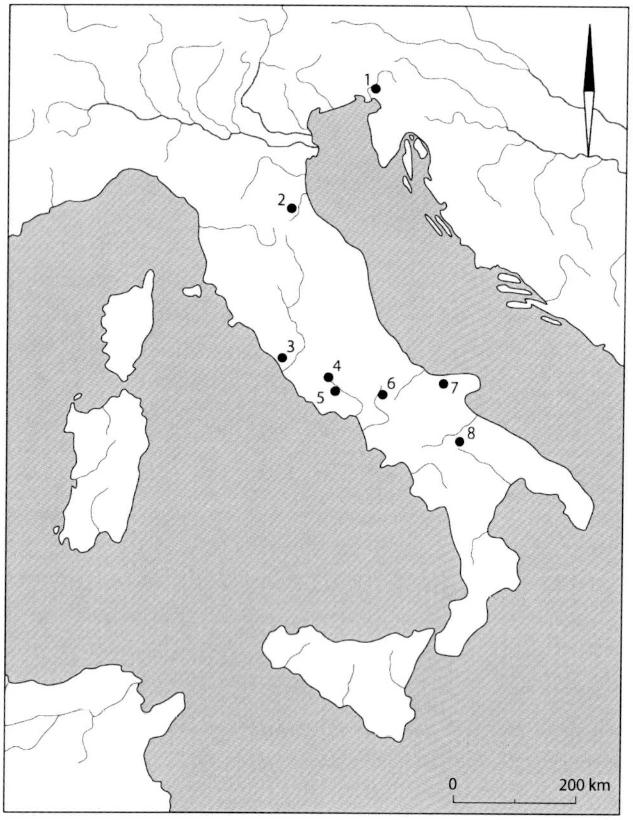

Fig. 18.2 Main Lower Paleolithic sites and drainage networks of Italy. (*1*) Visogliano (*2*) Monte Poggiolo (*3*) Torre in Pietra, Castel di Guido, La Polledrara (*4*) Fontana Ranuccio, Colle Marino (*5*) Ceprano (*6*) Isernia La Pineta (*7*) Pirro Nord (*8*) Venosa Loreto, Notarchirico

sites associated with fluvial, lacustrine, or fluvio-lacustrine depositional settings, within or at the margins of Apenninic basins and/or along former coastlines, such as Monte Poggiolo and the sites of Via Aurelia. Apart from Monte Poggiolo, Isernia and Notarchirico (*ca.* 850, 660 and 640 ka, respectively; Muttoni et al. 2011; Lefevre et al. 2010; but see Villa 2001), the rest of the main LP sites date to the middle and late Middle Pleistocene, with ages generally clustering between *ca.* 500 and 300 ka: Loreto in Venosa basin (Lefevre et al. 2010); Fontana Rannucio and Colle Marino in the Anagni basin (Biddittu et al. 1979; Segre and Ascenzi 1984; but see Villa 2001); Ceprano in the eponymous basin (Ascenzi et al. 1996; Muttoni et al. 2009); and Torre in Pietra, La Polledrara, and Castel di Guido in the valleys of "Via Aurelia" (Anzidei and Arnoldus Huyzendveld 1992; Mussi 1995; Constantini et al. 2001).

All of the Italian LP sites are located below *ca.* 500 m asl. As indicated by the study of tectonic activity and associated geomorphological processes, none of the sites were situated in mountainous areas at the time of occupation; instead, the reconstructed topographic settings suggest flat or gently undulating terrains (Mussi 2001). The emerging pattern of distribution suggests that all of the sites were located in lowland settings and related to water bodies (lakes, rivers, coasts). It remains difficult to assess whether this reflects hominin preferences, a preservation bias, or both. The inner mountainous areas with a rugged relief would have been prone to erosion—especially during glacial periods—and frequently disturbed due to tectonism. In contrast, depressed terrains trapped sediments and protected them from erosion, while sites close to river mouths would have been quickly buried by alluvial deposits.

On the other hand, wherever paleoenvironmental reconstructions are available, they seem to suggest that drainage systems provided habitats rich in resources, hence probably favorable to hominins (Mussi 2001). The Aurelian sites further indicate that not all water-bodies may have been equally attractive: the densely forested, "closed" environment of the Riano Lake appears to have been avoided, while an open landscape of the lacustrine areas of La Polledrara and Castel di Guido seems to have been preferred (Anzidei and Arnoldus Huyzendveld 1992). It is important to note here that hominins continued to exploit those tectonically controlled basins and lakes in the folds of the Apennines during the Middle Paleolithic as well, in environments not very different from those of the earlier periods (Mussi 2001). The observed LP altitudinal threshold of *ca.* 500 m. was exceeded only after *ca.* 300 ka, when Middle Paleolithic (MP) sites are found on hilly, mountainous landscapes as well.

Iberia

In Iberia (Fig. 18.3) only three sites are dated to the Early Pleistocene: Fuente Nueva, Barranco León, and Sima del Elefante; the late Early-early Middle Pleistocene is represented by Atapuerca TD6, Vallparadis, and Sima de los Huesos. The remainder of the Iberian Lower Paleolithic record comprises sites loosely and/or tentatively dated to the late Middle Pleistocene (Santonja and Villa 1990, 2006; Raposo and Santonja 1995; Santonja and Pérez-Gonzáles 2010; Oosterbeek et al. 2010; Parés et al. 2013). The vast majority of those sites (1) occur in the continental interior and mostly on the Meseta; (2) are described as "Acheulean"; (3) are associated with fluvial settings, and (4) usually lack faunal remains.

Intensive surveys carried out on the Iberian river basins have demonstrated an overall scarcity of stratified occurrences in the middle-high and high river terraces; instead, almost all known sites—with or without stratified finds—appear in the levels of the middle terraces (e.g., at +30 m, as is the case with the sites in the Tajo, Duero, and Mino river basins), often associated with high-energy deposits—e.g., Pinedo, La Maya, Torralba. Other sites involve low energy, primary contexts (e.g., Aridos I and certain levels of

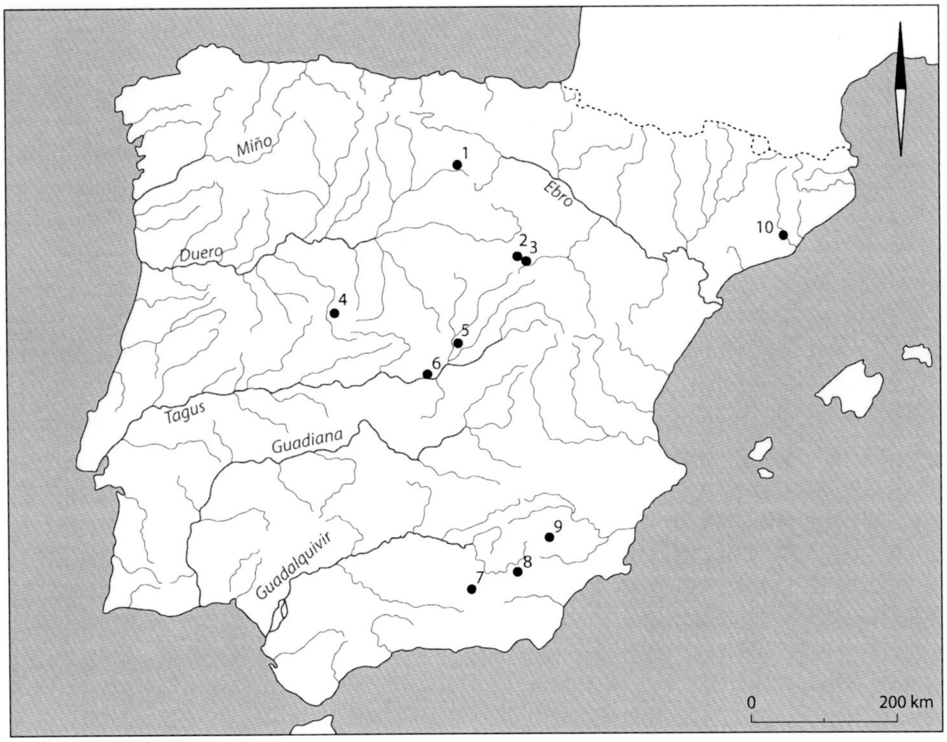

Fig. 18.3 Main Lower Paleolithic sites and drainage systems of Iberia. (*1*) Atapuerca sites: Trinchera Dolina, Sima del Elefante, Sima de los Huesos (*2*) Ambrona (*3*) Torralba (*4*) La Maya (5) Aridos (*6*) Pinedo (*7*) Solana del Zamborino (*8*) Orce sites: Fuente Nueva 3, Barranco León (*9*) Estrecho del Quípar (*10*) Vallparadís

Ambrona; Santonja and Villa 2006). Strong association of sites with fluvial settings is explained as a result of alluvial geomorphic processes that conserve archaeological remains (Santonja and Pérez-Gonzáles 2010), while their common association with the second and third-order confluences of fluvial systems and/or the vestibular areas of secondary valleys, is thought to reflect hominin preferences (Raposo and Santonja 1995: 9). The paucity of the record from, for instance, the Mediterranean and Cantabrian coasts or Galicia, is explained as evidence that Middle and Early Pleistocene river deposits either have not been preserved (Santonja and Pérez-Gonzáles 2010), or that the irregular discharge regime of rivers and the frequent floods did not favor the preservation of archaeological material (Santonja and Villa 2006). In contrast to the latter explanation, syn-sedimentary subsidence could account for the high density of finds in fine-grained floodplain sediments in the terraces of the Manzanares River. However, while the basin of the river Ebro includes well-developed Middle Pleistocene deposits, the area is virtually devoid of early Paleolithic remains (Raposo and Santonja 1995: 15).

The record from Western Iberia is equally puzzling: it is mainly composed of surface assemblages from river terraces with the earliest sites no older than MIS 8–9—creating a large chronological gap between this record and the evidence from Orce in the south, or Atapuerca in the north of the peninsula (Oosterbeek et al. 2010). The earliest-dated and best-preserved sites are associated with lacustrine or karstic settings, while the Iberian LP record is dominated by fluvial depositional environments. Terrace formation is thought to have been controlled more by tectonic processes and the nature of the geological substratum, and less by climatic fluctuations (Raposo and Santonja 1995). Similarly, differences in the surface or stratigraphic positions of artifacts (e.g., in the Duero and Tagus basins, respectively) are seen as reflecting temporal differences in aggradation and incision cycles between the hydrographic systems (Santonja and Villa 2006).

North Africa

To the east of the Atlas Mountains, the sites of Ain Hanech and El-Kherba are located at about 1200 m asl in the Ain Boucherit valley, within the Beni Fouda basin, which is one of several basins of the Eastern Algerian high plateau (Sahnouni 1998). Oldowan-like artifacts have been found in these two localities, which, according to biostratigraphic and paleomagnetic results, date to *ca.* 1.8 Ma (Sahnouni et al. 2013). The sites occur in an outcrop cut by a deep ravine and are surrounded by a series of highlands. Stratigraphically,

El-Kherba and Ain Hanech are laterally equivalent and were formed in the fluvio-lacustrine depositional environment of the Beni Fuda Plio-Pleistocene basin.

The rest of the best-studied North African sites, namely Thomas-I Quarry, Grotte des Rhinocéros, and the Sidi-Abderrahman site-complex, are clustered in the vicinity of Casablanca. The region preserves an exceptional succession of littoral formations exposed in quarries including rich Miocene-Pliocene paleontological sites such as Lissasfa and Ahl-Al-Oughlam (Raynal et al. 2001). The first traces of human presence come from deposits which are substantially later: the late Early Pleistocene layers in the unit L of Thomas Quarry I yielded the oldest lithic assemblages of "Lower Acheulean" artifacts, whereas the first human remains come from the same quarry and were found in Middle Pleistocene deposits, associated with "Middle Acheulean" lithic tools. The terraces, which provide this exceptional record, stretch from 180 m asl to the present sea level, and are associated with intertidal depositional units, dune formations, alteration facies (karsts, paleosols), and reworked deposits (Raynal et al. 2001). Overall, the littoral sequences record transgressions and regressions, which presumably reflect global and local fluctuations in sea level. Besides the aforementioned Algerian sites, as well as Ternifine (in Morocco) (*ca.* 1.0–0.6 Ma), the rest of the North African LP sites chronologically fall into the middle to late Middle Pleistocene (*ca.* 0.6–0.4 Ma).

Levant

In the Levant, the best-dated LP evidence comes from the sites of 'Ubeidiya (*ca.* 1.5/1.4–1.2 Ma; Goren-Inbar 1995; Martínez-Navarro et al. 2009) and Gesher Benot Ya'aqov (GBY) dated to *ca.* 0.8 Ma (Goren-Inbar et al. 2000). Both are located in the Dead Sea Rift with archaeological material recovered from fluvio-lacustrine depositional environments (Goren-Inbar 1995). Overall, the Levantine record appears to be as fragmentary as most of the other circum-Mediterranean ones: between the aforementioned Early Pleistocene and late Middle Pleistocene sites (e.g., from Tabun E, Yabrud I, and Qesem caves), there seems to be a substantial gap (Bar-Yosef 1994, 1998; Goren-Inbar 1995).

Depositional settings in the Levant include lacustrine, fluvial, coastal, and karstic environments. Of particular interest is the area of the Rift Valley, where many lakes were formed throughout the Quaternary. Lacustrine environments associated with archaeological material are primarily limited to this part of the Levant, particularly in the Jordan Valley and its northern segment, the Hula valley, where the sites of 'Ubeidiya and GBY are situated, respectively. Other examples would include the Acheulean artifacts recovered from the fluvial deposits of the Nahariyim Fm, which post-dates 'Ubeidiya, and the assemblages found in the gravels of the Orontes River at Latamne in Syria (Golberg 1995). However, fluvial settings are overall patchy in their spatial and temporal distribution. As Golberg (1995) demonstrated, the temporal distribution of Quaternary landforms and deposits is marked by considerable gaps in all geomorphological settings. For instance, most of the extant cave deposits represent less than 10 % of the Quaternary time-scale. Similarly, lakes were in existence for less than half of the Quaternary, while many of them such as those of the Negev area, appear only in the late Pleistocene (but see Ginat et al. 2003). The geological signature of fluvial and alluvial activity is equally discontinuous, especially with regard to the Middle Pleistocene, for which alluvial occurrences are extremely patchy.

Balkans and Anatolia

Two decades after Darlas' (1995) review of the Balkan Lower Paleolithic, the evidence for an Early and Middle Pleistocene human presence is still sparse and inconclusive in this area (Darlas and Mihailović 2008; Mihailović 2014). Isolated finds and lithic assemblages that were collected at the beginning of the twentieth-century and up until the 1970s suffer from poor documentation, are commonly restricted to a typological description of the specimens, a few drawings, and are often attributed to now-obsolete "cultural periods" (e.g., "Abbevillian," "Clactonian"). Classification according to morphological criteria has dominated the descriptions, outweighing the recording of stratigraphic data (Doboş 2008). A general lack of publications by Balkan scholars in non-local languages, coupled with the above, makes an overview of the "Balkan Lower Paleolithic" a very short one.

A recent review of the LP of Romania illustrates these problems (Doboş 2008; Doboş and Iovita 2016). In some cases, the old terminology—such as "Osteodontokeratic industries" (alleged bone-tools, supposedly preceding the use of stone-tool technology), or the "Tres Ancien Palèolithique" (supposedly preceding the Acheulean) and the "Premousterian"—has not been completely abandoned (Doboş 2008). The artifacts that were found *in situ*, usually limited to 2–3 specimens, have either not been documented adequately (if at all), or their artifactual character is now considered uncertain. Furthermore, with the exception of the "*in situ* finds," most of the locations reported as LP involved disturbed contexts, mainly related to river terraces. "Choppers" and "chopping tools" in these cases are essentially fortuitous finds, many of dubious anthropogenic origin, while the remaining pieces "should not be used as chrono-cultural markers" (Doboş 2008: 230). The recently reported site of Dealul Guran yielded a "Mode I" lithic industry that dates to MIS 11, according to radiometric dates acquired by post IR-IRSL, which makes it the oldest Paleolithic site in Romania and one of the few secure LP contexts in eastern Europe (Iovita et al. 2012). The site is located in the Loess Plateau of the Dobrogea region in south-eastern Romania and is interpreted as a collapsed rockshelter on the flank of a hill (see also Doboş and Iovita 2016).

None of the few Paleolithic sites recorded so far from the (Former Yugoslav) Republic of Macedonia has been securely dated (Kuzman 1993). Possible LP artifacts without a secure provenance have been found at the Kremenac site near Niš (Serbia). Surface material from several sites in the Western Morava valley could belong to the late Lower Paleolithic, but further work is needed to confirm their attribution (D. Mihailović pers. comm. 2013; Mihailović 2014). The redating of the hominin mandible from the cave of Mala Balanica (Serbia) by combined application of ESR/U-series and infrared/post-infrared luminescence dating provided a minimum age between 397 and 525 ka (Rink et al. 2013), which makes the BH-1 mandible (Roksandic et al. 2011; Roksandic 2016) the first human fossil in the Central Balkans recovered from controlled excavations and the easternmost hominin specimen securely dated to the Middle Pleistocene.

The picture is somewhat better in Albania, partly as a result of a recent survey that investigated intensively the hinterland of the Fier Province in central Albania (Runnels et al. 2009). There, 13 artifacts (including three bifaces) were assigned to LP. These are surface finds discovered at four sites, all of which are situated on or between anticlinal ridges that run down to a valley. At one of the sites (Rusinja), an eroded paleosol that is exposed on the summit of an anticlinal ridge is estimated to be older than *ca.* 100 ka (Runnels et al. 2009); the deposition of the artifacts on the surface is thought to predate the formation of the paleosol, which in that case provides a minimum age for the artifacts (Runnels et al. 2009). Apart from this relative dating, the attribution of the artifacts to the LP is based on the typological characteristics of the specimens and the occurrence of certain morphotypes, such as bifaces or core-tools. Choppers and chopping tools found on the surface of river terraces at the site of Baran (Fistani 1993) lack stratigraphic context and have been attributed to the LP on the basis of their "archaic" morphology. Finally, in the cave of Gajtan, a lithic industry with choppers and chopping tools co-occurring with "atypical" or "proto-handaxes" is reportedly associated with a Middle Pleistocene, "Holsteinian" age fauna, but nothing more can be said with certainty (Fistani 1993).

The material from the cave of Kozarnika is currently the oldest-dated evidence of human presence in Bulgaria (Guadelli et al. 2005). The LP artifacts belong to a core-and-flake industry, which essentially comprises simple, unworked flakes, flake fragments, and debitage (see also Ivanova 2016). On the basis of biostratigraphic indications, the geological context of the LP component was initially bracketed between 1.4–0.8 and 0.6–0.4 Ma (Guadelli et al. 2005), but lately the age of the site was pushed back to 1.6 Ma (Sirakov et al. 2010). The latter chronological estimation is based on the attribution of the associated faunal material to a biozone between MNQ17 and MNQ 19. Other researchers argued that the large mammal assemblage from Kozarnika's LP levels is more likely closer to the beginning of the Epivillafranchian (Kahlke et al. 2011; Spassov 2016). Besides Kozarnika, a few isolated finds and sites from the Western Rhodopes and from the middle part of the Arda River in the Eastern Rhodopes have been attributed to the LP on the grounds of their typological characteristics: for instance, this is the case with the cores, core-tools and scrapers from the open-air site of Benkovski (Eastern Rhodopes) and the site of Shiroka Poljana in the Western Rhodopes (Gurova and Ivanova 2008; Ivanova 2008, 2016).

The evidence from the western and central Balkans is equally sparse. In Sandalja I, a chopper was found in a bone breccia with reported Early Pleistocene (Villafranchian) faunal remains (Malez 1980), but its attribution to the LP cannot be confirmed without further study of the contextual and taphonomic conditions of the breccia. The lithic artifacts (including few handaxes) from the open-air sites of Punikve, Donje Pazarište, and Golubovec (Croatia) lack a stratigraphic provenance and their LP status is uncertain, based on typological criteria (Karavanić and Janković 2006).

Even though there is a relatively large number of handaxes and other potential LP artifacts that have been reported as surface finds, the Turkish/Anatolian record essentially includes five LP sites where the material has been retrieved from a secure geological context (see Dincer 2016). Close to Istanbul, the lower chamber in the cave of Yarımburgaz has yielded a lithic industry of mainly flake- and core-tools with little morphological standardization, with many tested pebbles and centripetally worked or discoid cores (Kuhn et al. 1996). The dating of the LP deposits is based on a problematic stratigraphic correlation of lower chamber deposits with those of the upper chamber, and on ESR dates on cave-bear teeth, which range from MIS 6 to 9 (Arsebük and Özbaşaran 1999). Dating uncertainties characterize the cave of Karain, the best-studied Paleolithic site in Turkey. The LP assemblage from Karain E (unit A)—defined as "Clactonian" by the excavators—consists of a few artifacts made on radiolarite; the layers of this unit were estimated to date around 400–370 ka "on the basis of correlation with oxygen isotope stages" (Otte et al. 1998: Table 1), which was in turn based on ESR dates on teeth (120 and 110 ka). In western Turkey, a hominin calvaria attributed to *Homo erectus* was found in a travertine block that was mined from a quarry; the travertine containing the fossil has been dated to *ca.* 1.1–1.3 Ma (Lebatard et al. 2014; see also Aytek and Harvati 2016). In another quarry, close to the village of Dursunlu, located on the Lycaonian plateau in south-central Anatolia, artifacts were recovered from lacustrine sediments of lignite beds (Güleç et al. 2009). A preliminary interpretation of paleomagnetic measurements suggested that the artifact-bearing layers predate the Brunhes-Matuyama boundary (which was, however, not recorded) and post-date the Jaramillo subchron. The age-range of the micro- and macrofauna is seen as supporting this chronological estimation (Güleç et al. 2009). The Dursunlu material is described as a "core-and-flake" assemblage . With the exception of five flint artifacts, the lithic pieces are made on quartz and the researchers stress the difficulty in discriminating artifacts from geofacts (Güleç et al. 2009). In contrast to Dursunlu, the earliest archaeological levels of Kaletepe Deresi 3 (KD 3) yielded Acheulean assemblages with obsidian handaxes, a few cleavers, but also choppers/chopping tools and polyhedrons, made on volcanic raw materials (Slimak et al. 2008; Dincer 2016). KD 3 is located on the bank of a seasonal stream, in a volcanic region of central Anatolia and close to an obsidian source. Here, the artifacts are embedded in a series of alluvial and colluvial layers of volcanic origin. Faunal material is poorly preserved and the age of the Acheulean levels remains uncertain: only the rhyolitic bedrock has been dated to >1.0 Ma providing a maximum age of the finds. Since the LP layers immediately overlie the bedrock, they may provide evidence for an early Acheulean presence in the region. However, their antiquity depends on the amount of time that elapsed between formation of the dated bedrock and deposition of the lowest layers (Slimak et al. 2008).

Greece

A critical re-evaluation of all published reports on alleged LP sites (Fig. 18.4) clearly illustrates that there is no unequivocal evidence for an Early or early Middle Pleistocene human presence in Greece (Tourloukis 2010). The fossil remains from the caves of Petralona (*ca.* 150–350 ka; Grün 1996) and Apidima (*ca.* 105–400 ka; Harvati et al. 2011), as well as an isolated hominin tooth from Megalopolis (likely >300 ka; Sickenberg 1975; van Vugt et al. 2000) demonstrate the presence of humans in Greece during the Middle Pleistocene. The archaeological sites with the most secure (stratified) evidence of Middle Pleistocene human presence are those of Kokkinopilos and Rodia (both of which are older than *ca.* 200 ka; Runnels and van Andel 1993a, 1993b, 2003; Tourloukis 2010; Tourloukis et al. 2015), Rodafnidia (Galanidou et al. 2016), Marathousa 1 (Megalopolis; Panagopoulou et al. 2015) and Crete (Strasser et al. 2010, 2011).

Apidima and Petralona are caves, Kokkinopilos is situated in a tectonic depression (a polje), Rodia is located at the margins of a fluvial basin, Marathousa 1 involves lacustrine deposits of the intramontane basin of Megalopolis, while at Rodafnidia the lithic material is associated with fluvial deposits. The sedimentary sequence of Kokkinopilos accumulated in the environment of an ephemeral lake, whereas the fluvial gravels of Rodia most likely represent point-bar deposits.

Other sites with possible LP material (Tourloukis 2010) are associated with fluvial/alluvial settings (Aliakmon localities, Higgs' handaxe from Palaeokastro and Doumbia in Macedonia, and the sites on the terraces of the Peiros in Peloponnesus); in solution basins with fills of redeposited *terra rossa* or in coastal plains (Alonaki and other sites of Epirus such as Ayios Thomas, Ormos Odysseos); and coastal settings (marine terraces of Nea Skala and the Triadon Bay of Milos). Consequently, in terms of both geomorphological settings and depositional environments, the Greek evidence matches the pattern observed in the rest of the Mediterranean record: the vast majority involves open-air sites in topographic depressions and at low elevations, such as drainage catchments, former lakes, and coastal areas where the archaeological material is commonly associated with fluvial, lacustrine, or fluvio-lacustrine contexts. The location of Rodia at the Rodia Gorge and close to the point where the Titarissios River meets the Pineios is reminiscent of the patterned association of Iberian sites with river confluences and valley entrances. Within the karstic, rugged landscape of Epirus, Kokkinopilos documents repeated visits of hominins at an ephemeral lake close to the river Louros and is reminiscent of the mosaic environments in which the Italian sites are located in the Apennine basins. The Early and Middle Pleistocene basin setting of Megalopolis would be comparable to that of Isernia and Notarchirico (Italy), the sites of the Guadix-Basa basin (Orce, Spain), or the Levantine sites of 'Ubeidiya and Gesher Benot Ya'aqov. Importantly, Middle (and occasionally Upper) Paleolithic evidence from the poljes of Epirus such as Kokkinopilos, Karvounari, Morphi (Tourloukis 2010), and from other depressions, such as the Thessalian basin or that of Mygdonia in North Greece, indicate that hominins continued to exploit the rich resources of those basins in the Late Pleistocene—a pattern already observed for the Italian record.

The scarcity of LP cave/rockshelter sites is as conspicuous in Greece as it is in other areas of the Mediterranean basin and the Balkans, with exceptions such as Petralona (Greece), Yarımburgaz (Turkey), Mala Balanica (Serbia), or Visogliano (Italy). The age estimate for Petralona follows the general

Fig. 18.4 Map of Greece showing key sites that have been reported as (potentially) Lower Paleolithic (after Tourloukis 2010). (*1*) Petrota (*2*) Doumbia (*3*) Siatista (*4*) Palaeokastro (*5*) Rodia (*6*) Korissia (*7*) Alonaki, Ormos Odysseos (*8*) Kokkinopilos (*9*) Nea Skala (*10*) Triadon Bay (*11*) Plakias (*12*) Gavdos (*13*) Rodafnidia. Sites with human remains: *P* Petralona Cave, *A* Apidima Cave, *M* Megalopolis (also: location of open-air site 'Marathousa 1'). Sites with pollen records: *TP* Tenaghi Philippon, *I* Ioannina, *K* Kopais

trend of cave use being a rather late phenomenon—or, as in the case of Mala Balanica, human remains might not have been evidence of human, but rather animal activity in the cave. The cave of Apidima and other cave sites with younger material in southern Peloponnesus such as Lakonis and Kalamakia (Harvati et al. 2009) indicate that, if coastal caves were as important in the Early and Middle Pleistocene as they appear to have been in the Late Pleistocene, then we have certainly lost a lot due to marine inundations of the present or earlier interglacials (see also Darlas and Psathi 2016). The submergence of coastal caves is the most dramatic demonstration of preservation biases. One could argue that the lack of Early Pleistocene evidence of cave occupation could be an "artifact" of preservation, i.e., that cave sediments of such old age do not preserve to the present. However, recent studies show that cave sediments (both clastic and chemogenic) can be up to several millions of years old (Hanja et al. 2010; cf. Pickering et al. 2013).

Large-Scale Spatio-Temporal, Geoarchaeological, and Geomorphological Patterns

As a general rule, most of the Mediterranean and Balkan LP record is dominated by open-air sites that are associated with fluvial, lacustrine, and fluvio-lacustrine depositional environments, the latter occurring in basins, usually of tectonic origin. Most of the sites are located at relatively low elevations, usually below *ca.* 500 m asl. The exceptions to this altitudinal pattern are sites that are situated on upland plateaus, and nearly all of them are at altitudes ranging between 1000 and 1200 m asl: Torralba and Ambrona on the Iberian Meseta, Ain Hanech, and El-Kherba on the eastern Algerian high Plateau, and Dursunlu on the Central Anatolian Plateau. In examining the geography of the European record in the Lower and Middle Paleolithic, Hopkinson (2007) concludes that before around 200 ka, hominins seem to have avoided upland regions. Rather than restricted to the Mediterranean, this altitudinal boundary appears to reflect a wider reality in the LP of Europe. According to Hopkinson (2007), the explanation for this is strongly linked to the ecological dynamics of mosaic landscapes in upland regions. The distributions of plants and animals and the configuration of patches available in these environments were probably eco-environmental barriers that early hominins could not (always or habitually) overcome, thus, they may have been confined to lowland habitats, where resources were distributed closely in space and time. The same author states that the LP occupation of the Italian Apennines is the exception that proves this rule because those localities were associated with fine-grained mosaic landscapes, perhaps benefiting by the positive effects of volcanism. According to Hopkinson (2007), the Apennine record shows that erosional processes did not bias this picture, proving that the above-mentioned pattern is real. Yet, at least in Mediterranean landscapes, the 500/600 m asl contour defines the upland–lowland boundary (cf. Macklin et al. 1995), as well as the boundary between "areas of erosion" and "areas of sedimentation" in river basins (Pinet and Souriau 1988). It could be argued that this boundary holds well also with regard to the ecological structuring of those landscapes. If we accept this to be the lowland–upland boundary, then there is *no* strong evidence of human presence in the Italian mountainous landscapes before *ca.* 300 ka (Mussi 2001) Therefore, the aforementioned *sites on the plateaus* represent the only significant exception to the pattern identified by Hopkinson (2007). The breaking of this altitudinal limitation with the onset of the Middle Paleolithic may have been real; however, the Italian record is probably exemplifying the biasing effects of erosional geomorphic agents active in upland landscapes. Thus, the Lower Paleolithic sites of the Iberian, Maghrebian, and Anatolian upland plateaus can be seen as supporting a preservation-related counterargument, as they are situated on flat or gently undulating terrains, which typically belong to basins.

Wherever equally explored areas with comparable geomorphological and depositional settings can be juxtaposed, it seems that open woodlands close to water bodies were preferred (e.g., Mussi 2001). While the Iberian record stresses the significance of subsidiary fluvial systems, confluences of rivers and entrances of valleys, the Italian evidence points to the importance of mosaic landscapes. Overall, in the Iberian and Italian peninsulas the distribution of sites essentially matches the spatial patterns of fluvial and lacustrine drainage systems. This highlights the importance of drainage catchments in dictating natural routes for inter- and intra-regional human and animal movements, as rivers that dissect bedrock and cut through mountain ranges potentially facilitate dispersal events.

Lakes, swamps, marshes, riverine, and riparian zones are all considered as ecologically highly productive environments; these are commonly hosted within larger topographical depressions, which serve as biogeographic corridors. Alternatively, the strong association of sites with fluvio-lacustrine depositional regimes could be explained by inferring their preservation potential as repositories of early human activity (cf. Mishra et al. 2007). This could imply a positive bias by which sites would be found in those depositional contexts and geomorphological settings (i.e., topographic depressions: river basins, former lakes, coastal areas) because of their specific properties that favor preservation. As discussed with regard to the Levantine record, a negative bias was at play at the same time: such landscape features were discontinuous in space and ephemeral in time for most of the Quaternary. Gaps in the archaeological record may be related to the gaps in geomorphological archive, when the aforementioned landforms were not in existence or were not active (e.g., dry valleys and lakes). Another negative bias, exemplified by the Iberian record, where most of the evidence is associated with the middle terraces of rivers, suggests that, before the time-periods represented by the middle terraces, river behavior was overall too dynamic to allow for the preservation of archaeological material. The latter observation is valid for the vast majority of the Iberian river systems, although terraces in the interior are better preserved than those of the coastal lowlands (Santisteban and Schulte 2007).

The use of caves/rockshelters appears to have been a marginal and chronologically late phenomenon. With few exceptions (most notably Atapuerca, Kozarnika, the caves in the Casablanca area, Arago, perhaps Treugol'naya and possibly Lunel Viel and Payre), the use of caves appears to have gained momentum in the latter half of the Middle Pleistocene. Again, it is difficult to explain whether this is an artifact of preservation or a consequence of hominin preferences for open-air locales in earlier periods. The generally scarce evidence for the use of caves in the Italian LP may reflect a preference for open-air environments; however, neither behavioral constraints nor preservation biases could be excluded. In contrast, the MP of Italy is dominated by cave-sites, with most of them (at least 70) dated to the last glacial (which in itself could be an artifact of preservation; Mussi 1999). Notably, for the time-span between 0.9 and 0.4 Ma the bulk of large mammal assemblages from South-East Europe has been found in caves or karst settings (Kahlke et al. 2011).

Currently, the chronological pattern can be summarized in the following three statements: (1) Early Pleistocene sites are few and their dating is only rarely uncontested. (2) The number of sites increases in the early-middle Middle Pleistocene. (3) It is only from the middle and chiefly the latest part of the Middle Pleistocene that the archaeological signal becomes more substantial.

While this picture could be the result of the fragmentary nature of terrestrial sedimentary archives of the earliest parts of the Pleistocene, it is also a by-product of the methodological limitations of the available dating techniques and datable materials. For example, the gaps in the Turkish record have been attributed to the degree of research intensity and coverage, but also to geological factors: in large parts of the Central Anatolian Plateau, Pleistocene deposits are absent and Miocene strata are exposed on the surface, whereas in other parts the early Pleistocene is buried by thick accumulations of younger sediments. The few known sites from central Anatolia are associated with margins of Pleistocene lakes or with outcrops of limestone or volcanic rocks (Kuhn 2002). Similarly, the scarcity of sites in Albania was attributed to the tectonic activity in the region: "uplift, local faulting, erosion and alluviation in the broader region have together subjected older deposits to destruction" (Runnels et al. 2009: 157).

Effects of Basin Evolution on Archaeological Distributions

Sedimentary basins are topographic depressions of tectonic origin in which sediments accumulate. The bulk of Mediterranean LP sites occur in sedimentary basins, within depositional environments that are fluvial, alluvial, lacustrine, or coastal. This is especially true for the Iberian and Italian peninsulas, which have the richest records. Additionally, sites with primary contexts and/or the oldest-dated material are associated with fluvial/lacustrine settings (e.g., the Orce sites, Isernia, Monte Poggiolo, 'Ubeidiya). This could be related to preservation as well as to hominin preferences for locales adjacent to water resources (cf. Mishra et al. 2007; Dennell 2010). Dennell (2007) suggested that, due to their small foraging ranges, hominins would have faced difficulties in exploiting the vast floodplains of large river systems, such as those of northern India and Pakistan, the Tigris-Euphrates, or the Nile; in contrast, small lake and river basins such as those occurring in Spain or in the Levant (e.g., Orce, 'Ubeidiya) would have been preferred. Korisettar (2007) considered raw material distributions in conjunction with associated ground water resources to propose a basin model for the early Paleolithic settlement of India and notes that "Paleolithic sites [...] have been studied independent of an understanding of basin geometry, geology and geomorphic history" (Korisettar 2007: 72).

Along those lines, I use examples from the Iberian and Italian peninsulas to juxtapose with the picture from Greece, in order to explore: (1) how specific differences in the topographic configuration and tectonic history of basins can shed light to our understanding of processes affecting the *preservation* and *accessibility* of LP sites; and (2) how these differences should be accounted for when examining differences in LP spatio-temporal distributions within and among regional records.

Forty-two percent of Iberia comprises low relief with low-to-medium gradient slopes. The plateaus and plains of the interior (essentially: the Iberian Meseta) add another 23 % of surface with very low topographic roughness and gentle slopes (mean: 2.7°). Accordingly, the low-relief areas cover 65 % of the peninsula (Benito-Calvo et al. 2009), while in Greece, almost the same percentage is represented by high relief areas. Most of the Iberian LP sites are located in these low-relief/low-gradient terrains, with the majority situated on the elevated flat surfaces of the Iberian Meseta. The causes and timing of the uplift of the Meseta is debated, but it probably had two main components: (1) Alpine compressional tectonics, and (2) a recent, Plio-Pleistocene stage of uplift (Casas-Sainz and de Vicente 2009). The major transition affecting the Meseta is most likely related to uplift: the plains and basins (e.g., the Duero, Ebro, and Tagus basins), which were previously endorheic (internally drained), were captured by the fluvial systems and changed to exorheic. The transition from endorheism to exorheism marks the onset of drainage reversal, river incision, and hence dissection and erosion of the basins and plains. The precise timing of this transition is not resolved and it probably differed regionally. Nonetheless, strong incision observed in some basins (e.g., parts of the Duero) is described as occurring in "recent times" (Casas-Sainz and de Vicente 2009). It is quite plausible that well-preserved LP sites in the Iberian basins and plains have remained buried and protected from erosion for most of the Early and Middle Pleistocene, and were only recently (Late Pleistocene to present) exhumed by alluvial/fluvial incision, the latter providing the necessary degree of archaeological accessibility. An example of such a case is given below with respect to the sites near Orce.

The intramontane Guadix-Baza (G-B) basin is situated on a plateau at *ca.* 1000 m asl, now intensely dissected by the river network. A >600 m-thick sequence of fluvio-lacustrine sediments (>2500 m-thick in the center near Baza) accumulated in the enclosed, endorheic depression of G-B (Scott et al. 2007). Activity along the Baza normal fault since *ca.* 8 Ma provided accommodation space for continuous sedimentation in the Baza sub-basin, which was formed in the hanging-wall of this fault (Alfaro et al. 2008). A large lake occupied the depocentre of the latter area, and the archaeological sites (Barranco León, Fuente Nueva) are located at the margins of this paleo-lake (Barsky et al. 2010). Alluvial fans on the borders of the basin were gradually connected with the central lake (Pérez-Peña et al. 2009). Besides the gently sloping fans, the fluvio-lacustrine sediments of Baza lie horizontally and the entire depression is described as an "essentially flat, elevated region" (Pérez-Peña et al. 2009: 206). The central Betic Cordillera, where the G-B basin is located, is currently subjected to regional uplift (Alfaro et al. 2008), but the Pliocene-Pleistocene evolution of the basin was dominated by sedimentary processes largely undisturbed by significant tectonic events (Pérez-Peña et al. 2009). At a certain point, the former endorheic drainage was captured by the Guadalquivir River due to uplift, basin inversion took place and the drainage changed from endorheic to exorheic. From that point on, lacustrine and fluvial sedimentation ended, and erosion predominated in the area (Soria et al. 1998; Díaz-Hernández and Juliá 2006; Alfaro et al. 2008; Pérez-Peña et al. 2009). While the exact age of this change is debated, it probably occurred at *ca.* 200 ka (Oms et al. 2011) or even later, during the Late Pleistocene (Soria et al. 1998), and perhaps later than *ca.* 43 ka (Pérez-Peña et al. 2009). Since the basin was captured by the Guadalquivir, the level of the sea (*ca.* 1000 m lower than the river's level) became the base level of erosion; for the river to adjust its profile to the new conditions, it had to erode the poorly consolidated Neogene-Quaternary sedimentary fill. The incision wave propagated headward very rapidly, but its intensity decreased over time. Most of the erosion has since been concentrated in the Guadix sub-basin because it is close to the capture point (Pérez-Peña et al. 2009). The Baza fault delayed the propagation of erosion into the Baza sub-basin, which explains the large differences in erosion rates between the two sub-basins.

Two points here are significant in geoarchaeological terms: (1) During the Early and Middle Pleistocene, sedimentation was continuous and characterized by high rates (~10–15 cm/ka; Garcia Aguilar and Martin 2000; Scott et al. 2007). It largely consisted of fine-grained material, and it essentially formed a flat-lying terrain. Hence, the most important prerequisites for a good preservation potential were in place: *fine-grained material accumulating fast and continuously in a low-gradient setting*. (2) Erosion started only late in the Late Pleistocene. It was probably vigorous in the beginning (i.e., upon capture of the drainage by the Guadalquivir) but it gradually slowed down. The incision/erosion wave affected mainly the Guadix area, while its propagation to Baza was buffered by the Baza fault. Encroachment of the drainage in Baza only served to expose the Early and Middle Pleistocene sediments, instead of severely eroding them, as is the case with the badlands directly adjacent to the S/SW of Orce (cf. Díaz-Hernández and Juliá 2006). Therefore, for the Orce sites, the most important requirement for today's good archaeological visibility was also present: *erosion starting only late in the Pleistocene, stripping off the uppermost sediments, and exposing the lower artifact-bearing layers*.

Italy provides another instructive example because its topography is much more similar to that of Greece, with alluvial plains and flatlands covering about one quarter of the peninsula (Mussi 2001). The majority of the LP sites are associated with the fluvial and/or lacustrine depositional settings of the Apennine basins. The Italian Late Pliocene and early-middle Pleistocene are characterized by lacustrine environments of low relief in intramontane depressions, which hosted swamps and floodplains of mainly fine-grained sediments. These closed and semi-closed drainage systems were chiefly internally drained because the low relief prevented streams from eroding divides and capturing the drainage (Bartolini 2003). After the Middle Pleistocene, lacustrine sedimentation was significantly reduced, continuing only in a few basins that maintained internal drainage. It is during the Middle and Late Pleistocene that a major rearrangement occurred in the depositional settings of the Apennine depressions: the fluvio-lacustrine environments changed to fluvial–alluvial sequences "in a regionally correlated phase of basin fill incision and drainage integration" (Bartolini et al. 2003: 214). The change from endorheic lacustrine systems to through-going fluvial networks is related to the uplift of the Apennine chain and the creation of the necessary relief that provided the streams with the required energy to capture the drainages. As a result, the older (Early and early Middle Pleistocene) fluvio-lacustrine units were incised and eroded, and they are now overlain by units transitional from low-gradient lacustrine and fluvial environments to coarser deposits of alluvial fans. The drainage-change did not affect all basins, but, as a general pattern, it involved most of them. The basins that were least or not affected are those most distant from the sea (i.e., away from marine base level), where continued normal faulting and associated subsidence prevented their capture from the regional fluvial network. The depositional sequences of these basins (up to 1000 m thick) are today weekly incised and poorly exposed, hence offering inadequate accessibility for archaeological discoveries. For those basins that were captured in the late Pleistocene, we can envisage the low-gradient paleosurfaces being covered and thus protected throughout the Early and Middle Pleistocene. As in the Iberian example, this would have offered better chances for artifact-bearing strata to attain a high degree of preservation and relatively good visibility after dissection and exhumation due to uplift.

For instance, in the area of Isernia La Pineta, the disruption of the Early Pleistocene lacustrine basin occurred due to the Middle Pleistocene extensional tectonics (Coltorti et al. 2005). The fluvio-lacustrine sediments, in which human activity was recorded, were subsequently covered by high-energy stream deposits, generated by a considerable increase in gradient due to these Middle Pleistocene tectonic movements. As a result, "the archaeological levels are sandwiched between the latest episodes of lacustrine sedimentation and the earliest fluvial deposits" (Mussi 2001: 24). Isernia's artifact-bearing deposits remained buried and protected throughout the Middle and most of the Late Pleistocene, until the local faults were re-activated in the latest Pleistocene and most probably in the Holocene, when these sediments were uplifted as part of the up-thrown blocks of the local fault system (see Di Bucci et al. 2002: especially geological cross-section IS2 in Fig. 7 and discussion in the text). Because of this latest morphological inversion, the fluvio-lacustrine deposits are covered by either a relatively thin mantle of Holocene sediments or by an even thinner present-day soil.

The basin of Thessaly, where the LP site of Rodia is situated, is the largest river catchment and lowland area in Greece. Similar to the endorheism–exorheism transition discussed for the Iberian and Italian basins, the Thessalian drainage changed from an internally drained Pliocene lake to a through-going fluvial network (the Pineios river drainage). Yet, in contrast to the basins of Orce or those of the Apennines, this transition in Thessaly occurred in the Late Pliocene-Early Pleistocene, due to uplift related to the first major tectonic phase affecting the region (Fig. 18.5; Caputo et al. 1994). As a fresult of this uplift, as well as due to the subsidence of large parts of the basin during the second phase of tectonic activity (Middle-Late Pleistocene to present), Early-Middle Pleistocene sediments in Thessaly outcrop today at

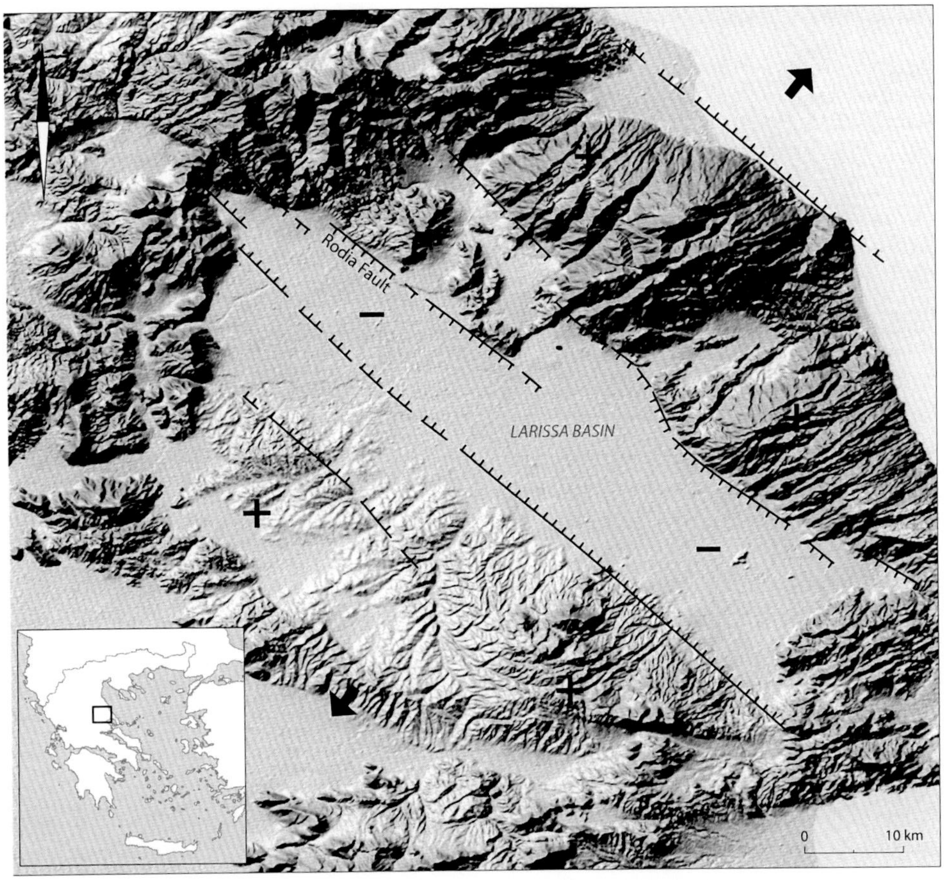

Fig. 18.5 Relief map of Thessaly, showing the Late Pliocene-Early Pleistocene extensional regime (first tectonic phase). *Arrows* indicate the direction of crustal extension, plus and minus signs indicate uplift and subsidence, respectively. The site of Rodia is located to the north-east of the Rodia Fault (on the up-thrown block), which cuts across the entrance of the Rodia Narrows. Exposures of Early and Middle Pleistocene sediments are restricted to limited areas that were uplifted and/or did not experience any subsidence from the Middle Pleistocene to the present (second tectonic phase); the rest of the Early and Middle Pleistocene sediments are now buried at depths of up to 500 m. or more. The relief map is based on SRTM data; tectonic regime after Caputo et al. (1994)

Fig. 18.6 The red-bed badlands of the fault-bounded, intermountainous basin of Kokkinopilos, Epirus. Sometime in the Late Pleistocene, fault re-activation and uplift resulted in the raising and tilting of the basin, the formerly endorheic drainage was captured by the Louros River and headward stream incision initiated. Today, a network of deep gullies exposes artifact-bearing Middle Pleistocene paleosols and sediments associated with the environment of an ephemeral paleo-lake

only 0.8 % of the entire basin. In contrast to Thessaly, the small basin of Kokkinopilos was inverted relatively recently and the intensity of erosion has also accelerated only in very recent times. In the Middle Pleistocene, Kokkinopilos remained a closed depression (a polje), in which terra rossa was being redeposited in the low-gradient, low-energy depositional environment of an ephemeral wetland (Runnels and van Andel 2003; Tourloukis 2009, 2010; Tourloukis et al. 2015). In the Late Pleistocene to Holocene, fault activity and uplift changed the drainage from endorheic to exorheic, initiating dissection and exposing the artifact-bearing sediments (Fig. 18.6).

Such an "advantageous" timing of uplift (Late Pleistocene/Holocene) was rather exceptional for the lowlands of Greece: on the contrary, most basins were affected by uplift already in the Early and Middle Pleistocene. While some (or most) Iberian and Italian basins were experiencing a period of relative tectonic quiescence during the Early Pleistocene, (parts of) the Greek basins changed from "sediment-receiving" to "sediment-producing" areas, in which erosion predominated over deposition. This can be explained by the fact that the last tectonic paroxysm in Greece seems to have begun in the Early and Middle Pleistocene. During the Early and early-Middle Pleistocene, a compressional regime invaded the broader Aegean region, separating the extensional regimes that prevailed before and after that time-span. During this intense compressional phase, the entire Hellenic arc was uplifted and convergence rates increased at its outer circumference from 1 to 3 cm/year (Schattner 2010). In the early Middle Pleistocene, a reorganization of stress trajectories occurred in the southern and northern Aegean (Angelier et al. 1982), and in the north (Florina-Vegoritis-Ptolemais graben) and central mainland Greece (Thessaly; Caputo et al. 1994). A third phase of opening affected the Gulf of Corinth (Rohais et al. 2008), while some basins and coastal areas in Peloponnesus were being uplifted (e.g., Mariolakos et al. 1994). As a whole, these developments are probably related to a major tectonic event that occurred across the entire eastern Mediterranean during the early-to-middle Pleistocene, manifested by a series of synchronous structural deformations that accentuated the topography (Schattner and Lazar 2009; Schattner 2010).

The examples from the Iberian and Italian peninsulas, as well as that of Kokkinopilos (Greece) can be viewed as "windows of opportunity" for a (sub)optimal matching between adequate preservation and accessibility circumstances for today's investigations of basin settings. Avoiding over-generalizations, they also exemplify *first-order differences* between the three Mediterranean peninsulas in terms of basin topography and tectonic history, which are relevant for our understanding of the disparities between the LP records of the latter areas. In that sense, the above-mentioned examples conceptualize initial steps towards a basin model, in which the key factors are the timing of uplift, the inversion of basins (or parts of them) and the duration and intensity of erosion that exposes artifact-bearing deposits. In an ideal situation, Early and Middle Pleistocene low-gradient sediments (e.g., lacustrine sediments) were being protected by burial until the Late Pleistocene. Then, uplift signals the onset of topographic inversion, drainage diversion, fluvial dissection, exposure and, hence, archaeological accessibility. Importantly, the exposed sediments were subjected to the erosional effects of "only" one full glacial-interglacial cycle, and have therefore had a better chance to be preserved until the present. In other words, when uplift occurred in the Early Pleistocene, there was a limited sedimentation in the uplifted area during the Middle Pleistocene, whereas already deposited (Early Pleistocene) sediments were subjected to incision and erosion throughout multiple glacial-interglacial cycles from the Early Pleistocene to the present. Likewise, when uplift started in the Middle Pleistocene, (Middle and) Late Pleistocene sedimentation in the uplifting terrain was reduced and any Middle Pleistocene sediments had low chances of being covered and protected. Instead, soon after their deposition they would have been subjected to the erosional effects of stream dissection throughout more-than-one climatic cycles.

Discussion

Most theoretical frameworks proposed to interpret Lower Paleolithic spatial/temporal site distribution patterns are commonly based on eco-environmental or behavioral/cultural parameters. While not excluding such interpretative scenarios, the basin model sketched here is meant to complement our efforts to understand the Early and Middle Pleistocene human geography of Southern Europe, particularly that of the three Mediterranean peninsulas and their tectonically active areas, to assist in *evaluating distributional patterns from a geomorphological perspective*. However, it is not meant to be applicable to any archaeological scatters in any kind of settings. The model assesses sites in small basins with fluvio-lacustrine deposits and predominantly targets sites in primary contexts *sensu lato* (as e.g., the Orce sites or Isernia). For landscapes such as those of Iberia, Italy, Greece or Asia Minor, tectonic activity and uplift, as well as surface uplift of isostatic origin, should be integral to the model. Factors such as uplift/subsidence, basin inversion, and the endorheism/exorheism transition are significant more as systemic elements within an exemplary scheme of geomorphic processes, and less as all-pervading mechanisms. Hence, it is acknowledged that the above-mentioned factors and conditions (1) may or *may not* all be present in most, let alone *all* basins of a certain region, and (2) they may have had different effects in different parts of the *same* basin, or across different basins. For instance, some basins need not have experienced the endorheism–exorheism transition; similarly, according to the prevailing lithology, aggradation, and/or dissection may be localized in parts of the same basin system. Therefore, there can be significant spatial/temporal inter- or even intra-basin non-uniformity and contrasts in geomorphic or tectonic regimes. The relative importance of uplift/subsidence as mechanisms to-be-modeled is negotiated by time-contingent and regional- or local-specific landscape biographies, which are in turn conditioned by climate, sea-level changes, inherited topography, and other geographical and geological parameters.

The "Mid-Pleistocene Transition" (MPT), i.e., the interval between *ca.* 940 and 640 ka, during which the 41 kyr climatic cycle progressively gave way to a 100 kyr rhythm of periodicity (Head and Gibbard 2005), is thought to have influenced human evolution (e.g., Manzi et al. 2011). Yet, even though the effects of the MPT on the biotic realm (flora, fauna, and hominins) are being investigated (e.g., Carrion et al. 2011 and references therein), concomitant effects of the MPT on the *a-biotic*, geomorphological realm has not been adequately addressed, and less so with regard to spatio-temporal patterns of site distributions. Global-scale studies of climate forcing on surface processes have shown that the most influential change is not so much that of climatic *state*, e.g., from warm and moist to cold and dry, but that of climatic *mode*, i.e., from low-frequency/low amplitude to higher frequency/higher amplitude oscillations (cf. Peizhen et al. 2001). Such a drastic change in climate mode occurred at around the beginning of the Early Pleistocene, with the start of global glaciations, and then again during the MPT. Increased rates of sedimentation—implying increased erosion rates—since 2.0–4.0 Ma, have been attributed to the onset of Milankovitch-scale climate change (Peizhen et al. 2001). Increased rates of uplift are also identified in the Late Pliocene, followed by a relative stability in the Early Pleistocene and then enhanced uplift again during the MPT (since *ca.* 0.9 Ma). This worldwide pattern in the timing of accelerated uplift, coinciding with the MPT, has been explained as a result of coupling between climatic fluctuations and lower crustal flow induced by increased erosion, in turn triggered by the greater severity of the 100 kyr cycles (Bridgland et al. 2007). Causally related to this pattern is a difference observed in fluvial basins: basins formed before *ca.* 900 ka are characterized by larger valley-floor widths, extensive alluviation and wide floodplains; in contrast to basins formed after *ca.* 900 ka, which are associated with greater vertical incision and the development of narrower valleys (Bridgland and Westaway 2008). These MPT-related geomorphological patterns become relevant in the framework of the basin model and its main parameters, as outlined above: the timing of uplift (localized or regional), basin inversion, drainage change (from endorheic to exorheic), erosion, and exposure of older deposits. Some of the most important geomorphic alterations brought about by the MPT included changes in vegetation composition, landscape fragmentation, increases in erosion and sedimentation rates, increases in sediment caliber, reorganization of drainage patterns, changes in the style of fluvial erosion and accumulation, and changes in basin accommodation spaces. Such large-scale changes in landscape would have affected large-scale patterns in the number and distribution of preserved sites, in space (e.g., close or away from major glaciers) and/or in time (pre-, intra-, and post-MPT). Recently, Mosquera and colleagues (2013) suggested that the gap between 900 and 500 ka at Atapuerca may be reflecting a continent-wide situation, rather than a local phenomenon, considering the paucity of European sites from this time-span. In view of the above, it could be argued that this "gap," which broadly encompasses the MPT, is partly (or largely?) mirroring geological circumstances which made it unlikely for sites of this time-span to be preserved, *and* to be accessible at present. In that sense, the MPT could be seen as a geomorphological "window" that remained closed until the switch in climate mode reached its final phase (cf. Maslin and Ridgwell 2005). Following this line of reasoning, the fundamental change in the qualitative and quantitative characteristics of the European archaeological record after *ca.* 650–500 ka (e.g., Roebroeks 2001; McNabb 2005), may be reflecting the opening of this "window", and not only a demographic momentum.

Population dynamics, extinction events, and demographic discontinuities are key concepts in an ongoing debate about evolutionary processes in the European continent and especially its southern, Mediterranean part, which hosts the earliest and richest—yet discontinuous—LP records (e.g., Dennell et al. 2011; MacDonald et al. 2012). Bearing in mind the review of the records presented above, as well as the argumentation unfolded with regard to geomorphic processes controlling the preservation/accessibility interplay, it is worthwhile to coin the concept of *geomorphological bottlenecking*, as a geosciences-derived heuristic tool in assessing spatio-temporal patterns and associated "gaps." The geomorphic processes occurring during the particularly harsh climatic conditions of MIS 22 and MIS 16 may have imposed a geomorphological bottlenecking in terms of site preservation and archaeological visibility. If not MIS 22 and/or 16, which represent the most extreme glacials of the MPT, then MIS 12 and MIS 6 would be the next formidable thresholds that early sites would have had to cross, in order to be protected from erosion until a late Pleistocene or a Holocene event, when they would become exposed again and hence accessible to archaeological investigations. This is a working hypothesis that could be tested by applying a geomorphologically informed research approach, such as the basin model proposed above. The low resolution of the regional chronostratigraphic frameworks and the fragmentary status of the terrestrial records make it extremely difficult to confidently support or reject such hypotheses. However, the effects of landscape dynamics on site distributions need to be explicitly addressed and systematically investigated, instead of being merely acknowledged and circumvented in our theory-building efforts.

Conclusions

There is little doubt that the spatio-temporal distribution of Lower Paleolithic sites in the Mediterranean region reflects a combination of hominin ecological preferences/tolerances and their behavioral/cultural gamut. However, the resultant picture can be misleading if the effects of landscape dynamics on site distribution are not taken into account. Since caves and rockshelters constitute a comparatively small component of the Mediterranean LP record, the focus of this chapter was on open-air sites which constitute the majority of recovered early sites, and which are more vulnerable to the effects of landscape change than the closed settings of caves.

Most of the Mediterranean open-air LP sites are situated in sedimentary basins. These basin settings are structurally related to mostly extensional tectonic movements and are commonly located at relatively low elevations. An early Pleistocene "low elevation site-pattern" may indeed be reflecting a behavioral preference for lowlands, or the inability of hominins to habitually exploit upland regions. However, it certainly also is a pattern biased by geomorphological and tectonic processes. This bias becomes evident when assessing not only the pattern-consistent sites but also the exceptions to this distribution. Namely, some of the Lower Paleolithic sites in uplifted areas would have been at lower elevations at the time of occupation, whereas most of the high-altitude sites occur in basins and plateaus with low relief, which enhances their preservation potential. Therefore, the crossing of this "altitudinal threshold" in the Middle Paleolithic was to a large extent related to geological opportunities for preferential site preservation, and should not be explained by behavioral factors alone. Evidenced throughout the Lower Paleolithic period, the close spatial association of sites with water-bodies is largely the result of geomorphic processes—a fact that tends to be ignored in our reading of site distributional or biogeographical patterns.

The last part of the chapter investigates geological, tectonic, and geographic conditions for the most advantageous matching between archaeological preservation and present-day accessibility of LP sites in basin settings. Syn-sedimentary subsidence and burial ensure preservation, ideally "directly" or "soon enough" (in geological terms) after a site is first abandoned, as this is the time when it is mostly in danger of being destroyed or reworked. Surface uplift exposes artifact-bearing sediments, yet for the site to be adequately accessible and exposed but not severely eroded and disturbed, this should occur in times as recent as possible. As a working hypothesis, it was suggested that this association between good preservation and adequate accessibility can partly explain the distribution of important sites in some Mediterranean settings (most notably in Spain and Italy) and, therefore, it can be modeled to guide future investigations in similar basin settings (such as Greece and the Balkans in general). The critical factors that need to be evaluated in such a model are: (1) the timing and duration of subsidence or equilibrium conditions of the prevailing tectonic regime; (2) the timing of uplift and basin inversion; (3) drainage diversion and/or the (potential) transition from endorheism to exorheism; (4) the timing, duration, and intensity of fluvial/alluvial incision, landscape dissection, and resultant exposure of deposits. If these factors showed a patterned relationship in the settings of already-known LP sites, then basin modeling could be used not only in interpretative scenarios but also as a heuristic tool for future fieldwork, within predictive methodologies that take us beyond chance discoveries. Moreover, basin sequences characterized by repetitive and similar geomorphological conditions may provide a reasonably accurate indication of when hominins first appeared in that area, or the types of conditions they preferred: if some of the aforementioned geomorphological and tectonic factors could be held constant over time, we may be able to identify cases where the absence of evidence reflects a real absence of hominins and is not an artifact of preservation/accessibility.

While this study focuses on Mediterranean settings, a similar model could be pursued for other areas, modified to fit the local/regional circumstances and research questions. Considering the ever-growing geoarchaeological dataset from the Balkans, small inter-mountainous basins in the Balkans or carefully selected parts of larger basins, such as parts of the Danube or the margins of the Panonnian basin, could be some obvious targets. From Central and Eastern Europe and up to Central Asia, extensive areas are covered by loess-paleosol sequences and it is highly likely that the scarcity of LP evidence from these regions is due to the loess that buried sites and made them inaccessible (Iovita et al. 2012; Dennell 2010). Large parts, however, are covered by a relatively thin loess mantle (<2 m.) and sites might be archaeologically accessible, wherever tectonics caused localized and/or regional topographic inversions.

Whether we discuss the scanty LP records of regions such as Central Europe, the Balkans and Greece, or the rich records of areas such as the Iberian and Italian peninsulas, site distributions do not necessarily represent true reflections of human occupation patterns or habitat preferences. Especially in tectonically active parts of the globe, the effects of landscape dynamics should be carefully and systematically assessed, in order to understand large-scale patterns. Inspired by concepts and applications from evolutionary geomorphology, this chapter aspired to sketch out possible pathways towards a methodology in which "the fundamental qualitative behavior of geomorphic systems is more important than the quantitative details" (Phillips 2006: 737). Integrating this kind of methodological tool in a geoarchaeological discourse could improve our understanding of archaeological distributions and, at the same time, enhance the predictability and robustness of model-bound approaches.

Acknowledgements I would like to thank the editors for inviting me to contribute to this volume and for their help in the reviewing process. Moreover, I warmly thank R. Dennell and two anonymous reviewers for helping me to improve the manuscript. Thanks are also due to P. Karkanas and M. Skourtsos for productive discussions on the basin model. Finally, I wish to acknowledge the financial support of the ERC Starting Grant Project PaGE (No. 283503).

References

Abbazzi, L., Fanfani, F., Ferretti, M. P., & Rook, L. (2000). New human remains of archaic *Homo sapiens* and Lower Palaeolithic industries from Visogliano (Duino Aurisina, Trieste, Italy). *Journal of Archaeological Science, 27*, 1173–1186.

Agustí, J., Blain, H.-A., Cuenca-Bescós, G., & Bailon, S. (2009). Climate forcing of the first hominid dispersal in Western Europe. *Journal of Human Evolution, 57*, 815–821.

Alfaro, P., Delgado, J., Sanz de Galdeano, C., Galino-Zaldívar, J., García-Tortosa, F. J., López-Garrido, A. C., et al. (2008). The Baza Fault: A major active extensional fault in the central Betic Cordillera (south Spain). *International Journal of Earth Sciences (Geologische Rundschau), 97*, 1353–1365.

Angelier, J., Lyberis, N., Le Pichon, X., Berrier, E., & Huchon, P. (1982). The tectonic development of the Hellenic arc and the Sea of Crete: A synthesis. *Tectonophysics, 86*, 159–196.

Anzidei, A. P., & Arnoldus Huyzendveld, A. (1992). The Lower Palaeolithic site of La Polledrara di Cecanibbio (Rome, Italy). In E. Herring, R. Whitehouse, & J. Wilkins (Eds.), *New developments in Italian Archaeology, Part I, Papers of the Fourth Conference of Italian Archaeology* (pp. 141–153). London: Accordia Research Institute.

Arnold, L. J., Demuro, M., Pares, J. M., Arsuaga, J. L., Aranburu, A., Bermudez de Castro, J. M., et al. (2014). Luminescence dating and palaeomagnetic age constraint on hominins from Sima de los Huesos, Atapuerca, Spain. *Journal of Human Evolution, 67*, 85–107.

Arsebük, G., & Özbaşaran, M. (1999). Pleistocene archaeology at the cave of Yarımburgaz in eastern Thrace/Turkey: Preliminary results. In G. N. Bailey, E. Adam, E. Panagopoulou, C. Perlès & K. Zachos (Eds.), *The Palaeolithic archaeology of Greece and adjacent areas: Proceedings of the ICOPAG Conference, Ioannina, September 1994* (pp. 59–72). London: British School at Athens Studies 3.

Arzarello, M., Pavia, G., Peretto, C., Petronio, C., & Sardella, R. (2012). Evidence of an Early Pleistocene hominin presence at Pirro Nord (Apricena, Foggia, southern Italy): P13 site. *Quaternary International, 267*, 56–61.

Ascenzi, A., Biddittu, I., Cassoli, P. F., Segre, A. G., & Segre-Naldini, E. (1996). A calvarium of late *Homo erectus* from Ceprano, Italy. *Journal of Human Evolution, 31*, 409–423.

Aytek, I. A., & Harvati, K. (2016). The human fossil record from Turkey. In K. Harvati & M. Roksandic (Eds.), *Paleoanthropology of the Balkans and Anatolia: Human evolution and its context* (pp. 79–91). Dordrecht: Springer.

Bailey, G. N., Reynolds, S. C., & King, C. P. (2011). Landscapes of human evolution: Models and methods of tectonic geomorphology and the reconstruction of hominin landscapes. *Journal of Human Evolution, 60*, 257–280.

Barsky, D., Celiberti, V., Cauche, D., Grégoire, S., Lebègue, F., de Lumley, H., et al. (2010). Raw material discernment and technological aspects of the Barranco León and Fuente Nueva 3 stone assemblages (Orce southern Spain). *Quaternary International, 223–224*, 201–219.

Bartolini, C. (2003). When did the Northern Apennine become a mountain chain? *Quaternary International, 101–102*, 75–80.

Bartolini, C., D'Agostino, N., & Dramis, F. (2003). Topography, exhumation, and drainage network evolution of the Apennines. *Episodes, 26*(3), 212–216.

Barton, M. C., Bernabeu, J., Aura, E., Oreto, G., & La Roca, N. (2002). Dynamic landscapes, artifact taphonomy, and landuse modeling in the western Mediterranean. *Geoarchaeology, 17*, 155–190.

Bar-Yosef, O. (1994). The Lower Palaeolithic of the Near East. *Journal of World Prehistory, 8*, 211–265.

Bar-Yosef, O. (1998). Early colonizations and cultural continuities in the Lower Palaeolithic of western Asia. In M. D. Petraglia & R. Korisettar (Eds.), *Early human behavior in global context: The rise and diversity of the Lower Palaeolithic record* (pp. 215–272). London: Routledge.

Bar-Yosef, O., & Belfer-Cohen, A. (2013). Following Pleistocene road signs of human dispersals across Eurasia. *Quaternary International, 285*, 30–43.

Benito-Calvo, A., Pérez-Gonzáles, A., Magri, O., & Meza, P. (2009). Assessing regional geodiversity: The Iberian Peninsula. *Earth Surface Processes and Landforms, 34*, 1433–1445.

Benton, M. J., Wills, M. A., & Hitchin, R. (2000). Quality of the fossil record through time. *Nature, 403*, 543–547.

Biddittu, I., Cassoli, P. F., Radicati di Brozolo, F., Segre, A. G., Segre-Naldini, E., & Villa, I. (1979). *Anagni, a K/Ar dated Lower and Middle Pleistocene site, Central Italy: Preliminary report* (pp. 53–71). XXI: Quaternaria.

Bischoff, J. L., Williams, R. W., Rosenbauer, R. J., Aramburu, A., Arsuaga, J. L., García, N., & Cuenca- Bescós, G. (2007). High-resolution U-series dates from the Sima de los Huesos hominids yields 600 kyrs: Implications for the evolution of the early Neanderthal lineage. *Journal of Archaeological Science, 34*, 763–770.

Bridgland, D. R., Keen, D., & Westaway, R. (2007). Global correlation of Late Cenozoic fluvial deposits: A synthesis of data from ICCP 449. *Quaternary Science Reviews, 26*, 2694–2700.

Bridgland, D., & Westaway, R. (2008). Climatically controlled river terrace staircases: A worldwide Quaternary phenomenon. *Geomorphology, 98*, 285–315.

Burger, O., Todd, L., & Burnett, P. (2008). The behavior of surface artifacts: Building a landscape taphonomy on the High Plains. In L. L. Scheiber & B. J. Clark (Eds.), *Archaeological landscapes on the high plains* (pp. 203–236). Boulder: University Press of Colorado.

Caputo, R., Bravard, J.-P., & Helly, B. (1994). The Pliocene-Quaternary tecto-sedimentary evolution of the Larissa Plain (Eastern Thessaly, Greece). *Geodinamica Acta, 7*, 219–231.

Carbonell, E., Bermúdez de Castro, J. M., Parés, J. M., Pérez-Gonzales, A., Cuenca-Bescós, G., Ollé, A., et al. (2008). The first hominin in Europe. *Nature, 452*, 465–470.

Carrion, J. S., Rose, J., & Stringer, C. (2011). Early human evolution in the western Palaearctic: Ecological scenarios. *Quaternary Science Reviews, 30*, 1281–1295.

Casas-Sainz, A. M., & de Vicente, G. (2009). On the tectonic origin of Iberian topography. *Tectonophysics, 474*, 214–235.

Cohen, K., MacDonald, K., Joordens, J., Roebroeks, W., & Gibbard, P. L. (2012). The earliest occupation of north-west Europe: A coastal perspective. *Quaternary International, 271*, 70–83.

Coltorti, M., Feraud, G., Marzoli, A., Peretto, C., Ton-That, T., Voinchet, P., et al. (2005). New $^{40}Ar/^{39}Ar$, stratigraphic and palaeoclimatic data on the Isernia La Pineta Lower Palaeolithic site, Molise, Italy. *Quaternary International, 131*, 11–22.

Constantini, R., Ottini, L., Caramiello, S., Palmirotta, R., Mallegni, F., Rossi, A., et al. (2001). Taphonomy of the fossil hominid bones from the Acheulean site of Castel di Guido near Rome, Italy. *Journal of Human Evolution, 41*, 211–225.

Darlas, A. (1995). The earliest occupation of Europe: The Balkans. In W. Roebroeks & T. van Kolfschoten (Eds.), *The earliest occupation of Europe: Proceedings of the European Science Foundation Workshop at Tautavel (France), November 1993* (pp. 51–59). Leiden: Leiden University Press.

Darlas, A., & Mihailović, D. (Eds.). (2008). *The Palaeolithic of the Balkans: Proceedings of the XV World Congress of the IUPPS, Lisbon, 4–9 September 2006*. Oxford: British Archaeological Report 1819.

Darlas, A., & Psathi, E. (2016). The Middle and Upper Paleolithic on the Western Coast of the Mani Peninsula (Southern Greece). In K. Harvati & M. Roksandic (Eds.), *Paleoanthropology of the Balkans and Anatolia: Human evolution and its context*. Dordrecht: Springer.

Darwin, C. (1861). *On the origin of species by means of natural selection, or the preservation of favored races in the struggle for life* (3rd ed.). London: J. Murray.

Dennell, R. (2003). Dispersal and colonization, long and short chronologies: How continuous is the Early Pleistocene record for hominids outside East Africa? *Journal of Human Evolution, 45*, 421–440.

Dennell, R. (2007). "Resource-rich, stone-poor": Early hominin land use in large river systems of northern India and Pakistan. In M. D. Petraglia & B. Allchin (Eds.), *The evolution and history of human populations in South Asia. Vertebrate Paleobiology and Paleoanthropology Series* (pp. 41–68). Dordrecht: Springer.

Dennell, R. (2010). 'Out of Africa I': Current problems and future prospects. In J. G. Fleagle, J. J. Shea, F. E. Grine, A. L. Baden, & R. E. Leaky (Eds.), *Out of Africa I: The first hominin colonization of Eurasia* (pp. 247–273). Dordrecht: Springer.

Dennell, R. W., Martinón-Torres, M., & Bermúdez de Castro, J. M. (2011). Hominin variability, climatic instability and population demography in Middle Pleistocene Europe. *Quaternary Science Reviews, 30*, 1511–1524.

Díaz-Hernández, J. L., & Juliá, R. (2006). Geochronological position of badlands and geomorphological patterns in the Guadix-Baza basin (SE Spain). *Quaternary Research, 65*, 467–477.

Di Bucci, D., Corrado, S., & Naso, G. (2002). Active faults at the boundary between Central and Southern Apennines (Isernia, Italy). *Tectonophysics, 359*, 47–63.

Dincer, B. (2016). The Lower Paleolithic in Turkey: Anatolia and hominin dispersals. In K. Harvati & M. Roksandic (Eds.), *Paleoanthropology of the Balkans and Anatolia: Human evolution and its context* (pp. 213–228). Dordrecht: Springer.

Doboş, A. (2008). The Lower Palaeolithic of Romania: A critical review. *PaleoAnthropology, 2008*, 218–233.

Doboş, A., & Iovita, R. (2016). The Lower Paleolithic of Romania revisited: New evidence from the site of Dealul Guran. In K. Harvati & M. Roksandic (Eds.), *Paleoanthropology of the Balkans and Anatolia: Human evolution and its context* (pp. 171–186). Dordrecht: Springer.

Doronichev, V. (2008). The Lower Palaeolithic in Eastern Europe and the Caucasus: A reappraisal of the data and new approaches. *PaleoAnthropology, 2008*, 107–157.

Falguères, C., Bahain, J.-J., Tozzi, C., Boschian, G., Dolo, J.-M., Mercier, N., et al. (2008). ESR/U-series chronology of the Lower Palaeolithic palaeoanthropological site of Visogliano, Trieste, Italy. *Quaternary Geochronology, 3*, 390–398.

Ferring, R., Oms, O., Agusti, J., Berna, F., Nioradze, M., Shelia, T., et al. (2011). Earliest human occupations at Dmanisi (Georgian Caucasus) dated to 1.85–1.78 Ma. *Proceedings of the National Academy of Sciences USA, 108*, 10432–10436.

Fistani, A. (1993). Human evolution in Albania for the Quaternary period. In B. Sigmon (Ed.), *Before the wall fell: The science of man in socialist Europe* (pp. 141–178). Toronto: Canadian Scholars Press.

Galanidou, N., Athanassas, C., Cole, J., Iliopoulos, G., Katerinopoulos, A., Magganas, A., & McNabb, J. (2016). The Acheulean site at Rodafnidia, Lisvori on Lesbos, Greece: Part I. In K. Harvati & M. Roksandic (Eds.), *Paleoanthropology of the Balkans and Anatolia: Human evolution and its context* (pp. 119–138). Dordrecht: Springer.

Garcia Aguilar, J. M., & Martin, J. M. (2000). Late Neogene to recent continental history and evolution of the Guadix-Baza basin (SE Spain). *Revista de la Sociedad Geológica de España, 13*, 65–77.

Ginat, H., Zilberman, E., & Saragusti, I. (2003). Early Pleistocene lake deposits and Lower Palaeolithic finds in Nahal (wadi) Zihor, Southern Negev Desert, Israel. *Quaternary Research, 59*, 445–458.

Golberg, P. (1995). The changing landscape. In T. E. Levy (Ed.), *The archaeology of society in the Holy Land* (pp. 40–57). London: Leicester University Press.

Goren-Inbar, N. (1995). The Lower Palaeolithic of Israel. In T. E. Levy (Ed.), *The archaeology of society in the Holy Land* (pp. 93–109). London: Leicester University Press.

Goren-Inbar, N., Feibel, C. S., Verosub, K. L., Melamed, Y., Kislef, M. E., Tcernov, E., et al. (2000). Pleistocene milestones on the Out-of-Africa corridor at Gesher Benot Ya'aqov, Israel. *Science, 289*, 944–947.

Grün, R. (1996). A re-analysis of electron spin resonance dating results associated with the Petralona hominid. *Journal of Human Evolution, 30*, 227–241.

Guadelli, J.-L., Sirakov, N., Ivanova, St., Sirakova, Sv., Anastassova, E., Courtaud, P., et al. (2005). Une séquence du Paléolithique inférieur au Paléolithique récent dans les Balkans: La Grotte Kozarnika a Orechets (Nord-Ouest de la Bulgarie). In N. Molines, M.-H. Moncel & J.-L. Monnier (Eds.), *Les Premiers peuplements en Europe: Vol. 1364. British Archaeological Reports International Series* (pp. 87–107). Oxford: Archaeopress.

Güleç, E., White, T., Kuhn, S., Özer, I., Sağir, M., Yilmaz, H., et al. (2009). The Lower Pleistocene lithic assemblage from Dursunlu (Konya), Central Anatolia, Turkey. *Antiquity, 83*, 11–22.

Gurova, M., & Ivanova, S. (2008). The rock shelter of Arkata, Eastern Rhodopes: Preliminary archaeological report. In R. I. Kostov, B. Gaydarska & M. Gurova (Eds.), *Geoarchaeology and Archaeomineralogy. Proceedings of the International Conference, 29–30 October 2008, Sofia* (pp. 22–28). Sofia: St. Ivan Rilski.

Hanja, N. Z., Mihevc, A., Pruner, P., & Bosák, P. (2010). Palaeomagnetic research on karst sediments in Slovakia. *International Journal of Speleology, 39*, 47–60.

Harvati, K., Panagopoulou, E., & Runnels, C. (2009). The paleoanthropology of Greece. *Evolutionary Anthropology, 18*, 131–143.

Harvati, K., Stringer, C., & Karkanas, P. (2011). Multivariate analysis and classification of the Apidima 2 cranium from Mani, Southern Greece. *Journal of Human Evolution, 60*, 246–250.

Head, M. J., & Gibbard, P. L. (2005). Early-Middle Pleistocene transitions: An overview and recommendation for defining the boundary. In M. Head & P. L. Gibbard (Eds.), *Early-Middle Pleistocene transitions: The land-ocean evidence* (pp. 1–18). London: Geological Society Special Publications 247.

Hopkinson, T. (2007). The transition from the Lower to the Middle Palaeolithic in Europe and the incorporation of difference. *Antiquity, 81*, 294–307.

Hughes, J. K., Haywood, A., Mithen, S. J., Sellwood, B. W., & Valdes, P. J. (2007). Investigating early hominin dispersal patterns: Developing a framework for climate data integration. *Journal of Human Evolution, 53*, 465–474.

Iovita, R., Fitzsimmons, K. E., Dobos, A., Hambach, U., Hilgers, A., & Zander, A. (2012). Dealul Guran: Evidence for Lower Palaeolithic (MIS 11) occupation of the Lower Danube loess steppe. *Antiquity, 86*, 973–989.

Ivanova, S. (2008). Early human presence and rock-cut structures in the Eastern Rhodopes. In R. I. Kostov, B. Gaydarska & M. Gurova (Eds.), *Geoarchaeology and Archaeomineralogy. Proceedings of the International Conference, 29–30 October 2008, Sofia* (pp. 185–192). Sofia: St. Ivan Rilski.

Ivanova, S. (2016). A route through the Balkans and implications for the earliest settlement of Europe. In K. Harvati & M. Roksandic (Eds.), *Paleoanthropology of the Balkans and Anatolia: Human evolution and its context* (pp. 187–211). Dordrecht: Springer.

Kahlke, R. D., Garcia, N., Kostopoulos, D. S., Lacombat, F., Lister, A. M., Mazza, P., et al. (2011). Western Palaearctic palaeoenvironmental conditions during the Early and early Middle Pleistocene inferred from large mammal communities, and implications for hominin dispersal in Europe. *Quaternary Science Reviews, 30*, 1368–1395.

Karavanić, I., & Janković, I. (2006). The Middle and Early Upper Palaeolithic in Croatia. *Opuscula Archaeologica, 30*, 21–54.

Korisettar, R. (2007). Towards developing a basin model for Paleolithic settlement of the Indian subcontinent: Geodynamics, monsoon dynamics, habitat diversity and dispersal routes. In M. D. Petraglia & B. Allchin (Eds.), *The evolution and history of human populations in South Asia. Vertebrate Paleobiology and Paleoanthropology Series* (pp. 69–96). Dordrecht: Springer.

Kuhn, S. (2002). Palaeolithic Archaeology in Turkey. *Evolutionary Anthropology, 11*, 198–210.

Kuhn, S., Arsebük, G., & Howell, F. C. (1996). The Middle Pleistocene lithic assemblage from Yarımburgaz Cave, Turkey. *Paléorient, 22*, 31–49.

Kuzman, P. (1993). The Palaeolithic in Macedonia. *Macedonia Acta Archaeologica, 13*, 9–16.

Lebatard, A.-E., Cihat Alçiçek, M., Rochette, P., Khatib, S., Vialet, A., Boulbes, N., et al. (2014). Dating the Homo erectus bearing travertine from Kocabaş (Denizli, Turkey) at at least 1.1 Ma. *Earth and Planetary Science Letters, 390*, 8–18.

Lefevre, D., Raynal, J.-P., Vernet, G., Kiefer, G., & Piperno, M. (2010). Tephro-stratigraphy and the age of ancient Southern Italian Acheulean settlements: The sites of Loreto and Notarchririco (Venosa, Basilicata, Italy). *Quaternary International, 223–224*, 360–368.

MacDonald, K., Martinón-Torres, M., Dennell, R. W., & Bermúdez de Castro, J. M. (2012). Discontinuity in the record for hominin occupation in south-western Europe: Implications for occupation of the middle latitudes of Europe. *Quaternary International, 271*, 84–97.

Macklin, M. G., Lewin, J., & Woodward, J. C. (1995). Quaternary fluvial systems in the Mediterranean basin. In J. Lewin, M. G. Macklin, & J. C. Woodward (Eds.), *Mediterranean Quaternary river environments* (pp. 1–25). Rotterdam: Balkema.

Malez, M. (1980). Sites of fossil men in Croatia. *Collegium Antropologicum, 4*(suppl), 13–29.

Manzi, G., Magri, D., & Palombo, M. R. (2011). Early-Middle Pleistocene environmental changes and human evolution in the Italian peninsula. *Quaternary Science Reviews, 30*, 1420–1438.

Mariolakos, I., Badekas, I., Fountoulis, I., & Theocharis, D. (1994). Reconstruction of the Early Pleistocene paleoshore and paleorelief of SW Peloponnesus area. *Seventh Congress of the Geological Society of Greece, May 1994, Bulletin of the Geological Society of Greece, XXX*, 279–304.

Martínez-Navarro, B. (2010). Early Pleistocene faunas of Eurasia and hominin dispersals. In J. G. Fleagle, J. J. Shea, F. E. Grine, A. L.

Baden, & R. E. Leaky (Eds.), *Out of Africa I: The first hominin colonization of Eurasia* (pp. 207–224). Dordrecht: Springer.

Martínez-Navarro, B., Belmaker, M., & Bar-Yosef, O. (2009). The large carnivores from 'Ubeidiya (early Pleistocene, Israel): Biochronological and biogeographical implications. *Journal of Human Evolution, 56*, 514–524.

Maslin, M. A., & Ridgwell, A. J. (2005). Mid-Pleistocene revolution and the "eccentricity myth". In M. J. Head & P. L Gibbard (Eds.), *Early-Middle Pleistocene transitions: The land-ocean evidence* (pp. 19–34). London: Geological Society Special Publications 247.

McNabb, J. (2005). Hominins and the Early-Middle Pleistocene transition: Evolution, culture and climate in Africa and Europe. In M. Head & P. L. Gibbard (Eds.), *Early-Middle Pleistocene transitions: The land-ocean evidence* (pp. 287–304). London: Geological Society Special Publications 247.

Mihailović, D. (Ed.). (2014). *Palaeolithic and Mesolithic research in the Central Balkans*. Belgrade: Serbian Archaeological Society.

Mishra, S., White, M. J., Beaumont, P., Antoine, P., Bridgeland, D. R., Limondin-Lozouet, N., et al. (2007). Fluvial deposits as an archive of early human activity. *Quaternary Science Reviews, 26*, 2996–3016.

Mosquera, M., Olle, A., & Rodriguez, X. P. (2013). From Atapuerca to Europe: Tracing the earliest peopling of Europe. *Quaternary International, 295*, 130–137.

Mussi, M. (1995). The earliest occupation of Europe: Italy. In W. Roebroeks & T. van Kolfschoten (Eds.), *The earliest occupation of Europe: Proceedings of the European Science Foundation Workshop at Tautavel (France), 1993* (pp. 27–49). Leiden: University of Leiden.

Mussi, M. (1999). The Neanderthals in Italy: A tale of many caves. In W. Roebroeks & C. Gamble (Eds.), *The Middle Palaeolithic occupation of Europe* (pp. 49–80). Leiden: University of Leiden.

Mussi, M. (2001). *Earliest Italy: An overview of the Italian Palaeolithic and Mesolithic*. New York: Kluwer.

Muttoni, G., Scardia, G., Kent, D. V., Morsiani, E., Tremolada, F., Cremaschi, M., et al. (2011). First dated human occupation of Italy at ~ 0.85 Ma during the late Early Pleistocene climate transition. *Earth and Planetary Science Letters, 307*, 241–252.

Muttoni, G., Scardia, G., Kent, D. V., Swisher, C. C., & Manzi, G. (2009). Pleistocene magnetochronology of early hominin sites at Ceprano and Fontana Ranuccio, Italy. *Earth and Planetary Science Letters, 286*, 255–268.

Muttoni, G., Scardia, G., & Kent, D. V. (2010). Human migration into Europe during the late Early Pleistocene climate transition. *Palaeogeography, Palaeoclimatology, Palaeoecology, 296*, 79–93.

Niknami, K. A. (2007). A stochastic model to simulate and predict archaeological landscape taphonomy: Monitoring cultural landscape values based on an Iranian survey project. *Archeologia e Calcolatori, 18*, 101–120.

Oms, O., Anadon, P., Agusti, J., & Julia, R. (2011). Geology and chronology of the continental Pleistocene archaeological and paleontological sites of the Orce area (Baza basin, Spain). *Quaternary International, 243*, 33–43.

Oosterbeek, L., Grimaldi, S., Rosina, P., Cura, S., Pedro, P. C., & Martins, A. (2010). The earliest Pleistocene archaeological sites in western Iberia: Present evidence and research prospects. *Quaternary International, 223–224*, 399–407.

Otte, M., Yalçinkaya, I., Kozlowski, J., Bar-Yosef, O., López Bayón, I., & Taskiran, H. (1998). Long-term technical evolution and human remains in the Anatolian Palaeolithic. *Journal of Human Evolution, 34*, 413–431.

Panagopoulou, E., Tourloukis, V., Thompson, N., Athanassiou, A., Tsartsidou, G., Konidaris, G. E., et al. (2015). Marathousa 1: A new Middle Pleistocene archaeological site from Greece. *Antiquity, 343*, Project Gallery.

Parés, J. M., Duval, M., & Arnold, L. J. (2013). New views on an old move: Hominin migration into Eurasia. *Quaternary International, 295*, 5–12.

Parfitt, S. A., Ashton, N., Lewis, S. G., Abel, R. L., Coope, G. R., Field, M. H., et al. (2010). Early Pleistocene human occupation at the edge of the boreal zone in northwest Europe. *Nature, 466*, 229–233.

Parfitt, S. M., Barendregt, R. W., Breda, M., Candy, I., Collins, M., Coope, G. R., et al. (2005). The earliest record of human activity in northern Europe. *Nature, 438*, 1008–1012.

Peizhen, Z., Molnar, P., & Downs, W. R. (2001). Increased sedimentation rates and grain sizes 2–4 Myr ago due to the influence of climate change on erosion rates. *Nature, 410*, 891–897.

Pérez-Peña, J. V., Azanón, J. M., Azor, A., Tuccimei, P., Della Seta, M., & Soligo, M. (2009). Quaternary landscape evolution and erosion rates for an intramontane Neogene basin (Guadix-Baza basin, SE Spain). *Geomorphology, 106*, 206–218.

Phillips, J. D. (2006). Evolutionary geomorphology: Thresholds and nonlinearity in landform response to environmental change. *Hydrology and Earth System Sciences, 10*, 731–742.

Pickering, R., Jacobs, Z., Herries, A. I. R., Karkanas, P., Bar-Matthews, M., Woodhead, J., et al. (2013). Paleoanthropologically significant South African sea caves dated to 1.1–1.0 million years using a combination of U-Pb, TT-OSL and palaeomagnetism. *Quaternary Science Reviews, 65*, 39–52.

Pinet, P., & Souriau, M. (1988). Continental erosion and large-scale relief. *Tectonics, 7*, 563–582.

Raposo, L., & Santonja, M. (1995). The earliest occupation of Europe: The Iberian peninsula. In W. Roebroeks & T. van Kolfschoten (Eds.), *The earliest occupation of Europe: Proceedings of the European Science Foundation Workshop at Tautavel (France), November 1993* (pp. 7–25). Leiden: Leiden University Press.

Raynal, J.-P., Sbihi-Alaoui, F. Z., Geraads, D., Magoga, L., & Mohi, A. (2001). The earliest occupation of North-Africa: The Moroccan perspective. *Quaternary International, 75*, 65–75.

Rink, W., Mercier, N., Mihailovic, D., Morley, M. W., Thompson, J. W., & Roksandic, M. (2013). New radiometric ages for the BH-1 hominin from Balanica (Serbia): Implications for understanding the role of the Balkans in Middle Pleistocene human evolution. *PLoS One, 8*, 1–7.

Rodríguez, J., Burjachs, F., Cuenca-Bescós, G., García, N., Van der Made, J., Pérez González, A., et al. (2011). One million years of cultural evolution in a stable environment at Atapuerca (Burgos, Spain). *Quaternary Science Reviews, 30*, 1396–1412.

Roebroeks, W. (2001). Hominid behaviour and the earliest occupation of Europe: An exploration. *Journal of Human Evolution, 41*, 437–461.

Roebroeks, W. (2005). Life on the Costa del Cromer. *Nature, 438*, 921–922.

Rohais, S., Eschard, R., & Guillocheau, F. (2008). Depositional model and stratigraphic architecture of rift climax Gilbert-type fan deltas (Gulf of Corinth, Greece). *Sedimentary Geology, 210*, 132–145.

Roksandic, M. (2016). The role of the Central Balkans in the peopling of Europe: Paleoanthropological evidence. In K. Harvati & M. Roksandic (Eds.), *Paleoanthropology of the Balkans and Anatolia: Human evolution and its context* (pp. 15–33) Dordrecht: Springer.

Roksandic, M., Mihailović, D., Mercier, N., Dimitrijević, V., Morley, M. W., Rakočević, Z., et al. (2011). A human mandible (BH-1) from the Pleistocene deposits of Mala Balanica cave (Sićevo Gorge, Niš, Serbia). *Journal of Human Evolution, 61*, 186–196.

Runnels, C., & van Andel, T. H. (1993a). A handaxe from Kokkinopilos, Epirus, and its implications for the Palaeolithic of Greece. *Journal of Field Archaeology, 20*, 191–203.

Runnels, C., & van Andel, T. H. (1993b). The Lower and Middle Palaeolithic of Thessaly, Greece. *Journal of Field Archaeology, 20*, 299–317.

Runnels, C., & van Andel, T. H. (2003). The early Stone age of the Nomos of Preveza: Landscape and settlement. In J. Wiseman & K. Zachos (Eds.), *Landscape Archaeology in Southern Epirus,*

Greece I: Vol. 32. Hesperia Supplement (pp. 47–133). Athens: American School of Classical Studies at Athens.

Runnels, C., Korkuti, M., Galary, M. L., Timpson, M. E., Stocker, S. R., Davis, J. L., et al. (2009). Early prehistoric landscape and land-use in the Fier region of Albania. *Journal of Mediterranean Archaeology, 22*, 151–182.

Sahnouni, M. (1998). *The Lower Palaeolithic of the Maghreb: Excavations and analyses at Ain Hanech, Algeria*. Cambridge Monographs in African Archaeology 42, Oxford: British Archaeological Reports International Series 689.

Sahnouni, M., Rosell, J., van der Made, J., Verges, J. M., Olle, A., Kandi, N., et al. (2013). The first evidence of cut marks and usewear traces from the Plio-Pleistocene locality of El-Kherba (Ain Hanech), Algeria: Implications for early hominin subsistence activities circa 1.8 Ma. *Journal of Human Evolution, 64*, 137–150.

Santangelo, N., Di Donato, V., Lebreton, V., Romano, P., & Ermolli, E. R. (2012). Palaeolandscapes of Southern Apennines during the late Early and the Middle Pleistocene. *Quaternary International, 267*, 20–29.

Santisteban, J. I., & Schulte, L. (2007). Fluvial records of the Iberian Peninsula: A chronological framework. *Quaternary Science Reviews, 26*, 2738–2757.

Santonja, M., & Pérez-Gonzáles, A. (2010). Mid-Pleistocene Acheulean industrial complex in the Iberian Peninsula. *Quaternary International, 223–224*, 154–161.

Santonja, M., & Villa, P. (1990). The Lower Palaeolithic of Spain and Portugal. *Journal of World Prehistory, 4*, 45–94.

Santonja, M., & Villa, P. (2006). The Acheulian of Western Europe. In N. Goren-Inbar & G. Sharon (Eds.), *Axe age: Acheulian tool-making from quarry to discard* (pp. 429–478). London: Equinox.

Schattner, U. (2010). What triggered the early-to-mid Pleistocene tectonic transition across the entire eastern Mediterranean? *Earth and Planetary Science Letters, 289*, 539–548.

Schattner, U., & Lazar, M. (2009). Subduction, collision and initiation of hominin dispersal. *Quaternary Science Reviews, 28*, 1820–1824.

Schiffer, M. B. (1987). *Formation processes of the archaeological record*. Albuquerque: University of New Mexico Press.

Scott, G. R., Gibert, L., & Gibert, J. (2007). Magnetostratigraphy of the Orce region (Baza Basin), SE Spain: New chronologies for Early Pleistocene faunas and hominid occupation sites. *Quaternary Science Reviews, 26*, 415–435.

Segre, A., & Ascenzi, A. (1984). Fontana Ranuccio: Italy's earliest Middle Pleistocene hominid site. *Current Anthropology, 25*, 230–233.

Sickenberg, O. (1975). Eine Säugertierfauna des tieferen Bihariums aus dem Becken von Megalopolis (Peloponnes, Griechenland). *Annales Géologiques des Pays Helléniques, 27*, 25–73.

Sirakov, N., Guadelli, J.-L., Ivanova, S., Sirakova, S., Boudadi-Maligne, M., Dimitrova, I., et al. (2010). An ancient continuous human presence in the Balkans and the beginnings of human settlement in western Eurasia: A Lower Pleistocene example of the Lower Palaeolithic levels in Kozarnika cave (North-western Bulgaria). *Quaternary International, 223–224*, 94–106.

Slimak, L., Kuhn, S., Roche, H., Mouralis, D., Buitenhuis, H., Balkan-Ath, N., et al. (2008). Kaletepe Deresi 3 (Turkey): Archaeological evidence for early human settlement in Central Anatolia. *Journal of Human Evolution, 54*, 99–111.

Soria, J. M., Viseras, C., & Fernandez, J. (1998). Late Miocene-Pleistocene tectono-sedimentary evolution and subsidence history of the central Betic Cordillera (Spain): A case study in the Guadix intramontane basin. *Geological Magazine, 135*, 565–574.

Spassov, N. (2016). Southeastern Europe as a route for the earliest dispersal of *Homo* Towards Europe: Ecological conditions and the timing of the first human occupation of Europe. In K. Harvati & M. Roksandic (Eds.), *Paleoanthropology of the Balkans and Anatolia: Human evolution and its context* (pp. 281–290) Dordrecht: Springer.

Strasser, T. F., Panagopoulou, E., Runnels, C., Murray, P., Thompson, N., Karkanas, P., et al. (2010). Stone Age seafaring in the Mediterranean: Evidence from the Plakias region for Lower Palaeolithic and Mesolithic habitation of Crete. *Hesperia, 79*, 145–190.

Strasser, T. F., Runnels, C., Wegmann, K., Panagopoulou, E., McCoy, F., DiGregorio, C., et al. (2011). Dating Lower Palaeolithic sites in southwestern Crete, Greece. *Journal of Quaternary Science, 26*, 553–560.

Surovell, T. A., Finley, J. B., Smith, G. M., Brantingham, P. J., & Kelly, R. (2009). Correcting temporal frequency distributions for taphonomic bias. *Journal of Archaeological Science, 36*, 1715–1724.

Toro-Moyano, I., Martinez-Navarro, B., Agusti, J., Souday, C., Bermudez de Castro, J. M., Martinon-Torres, M., et al. (2013). The oldest human fossil in Europe dated to ca. 1.4 Ma at Orce (Spain). *Journal of Human Evolution, 65*, 1–9.

Tourloukis, V. (2009). New bifaces from the Palaeolithic site of Kokkinopilos, Greece and their stratigraphic significance. *Antiquity, 83*(320), Project Gallery.

Tourloukis, V. (2010). *The Early and Middle Pleistocene archaeological record of Greece: Current status and future prospects: Vol. 23. Archaeological Studies Leiden University*. Leiden: Leiden University Press.

Tourloukis, V., & Karkanas, P. (2012). The Middle Pleistocene archaeological record of Greece and the role of the Aegean in hominin dispersals: New data and interpretations. *Quaternary Science Reviews, 43*, 1–15.

Tourloukis, V., Karkanas, P., & Wallinga, J. (2015). Revisiting Kokkinopilos: Middle Pleistocene radiometric dates for stratified archaeological material in Greece. *Journal of Archaeological Science, 57*, 355–369.

Tryon, C. A. (2010). How the geological record affects our reconstructions of Middle Stone Age settlement patterns: The case of alluvial fans in Baringo, Kenya. In N. Conard & A. Delagnes (Eds.), *Settlement dynamics of the Middle Paleolithic and Middle Stone Age* (Vol. 3, pp. 39–66). Tübingen: Kerns Verlag.

Van der Made, J., & Mateos, A. (2010). Longstanding biogeographic patterns and the dispersal of early *Homo* out of Africa and into Europe. *Quaternary International, 223–224*, 195–200.

Van Vugt, N., de Bruijn, H., van Kolfschoten, T., & Langereis, C. G. (2000). Magneto- and cyclostratigraphy and mammal fauna's of the Pleistocene lacustrine Megalopolis basin, Peloponnesos, Greece. *Geological Ultrajectina, 189*, 69–92.

Villa, P. (2001). Early Italy and the colonization of Western Europe. *Quaternary International, 75*, 113–130.

Waters, M. R., & Kuehn, D. D. (1996). The Geoarchaeology of place: The effect of geological processes on the preservation and interpretation of the archaeological record. *American Antiquity, 61*, 483–497.

Wilkinson, T. (2003). *Archaeological landscapes of the Near East*. Arizona: The University of Arizona Press.

Index

Katerina Harvati and Mirjana Roksandic (eds.), *Paleoanthropology of the Balkans and Anatolia*, Vertebrate Paleobiology and Paleoanthropology, DOI 10.1007/978-94-024-0874-4